The IMA Volumes in Mathematics and its Applications
Volume 153

For other titles published in this series, go to
www.springer.com/series/811

Institute for Mathematics and its Applications (IMA)

The **Institute for Mathematics and its Applications** was established by a grant from the National Science Foundation to the University of Minnesota in 1982. The primary mission of the IMA is to foster research of a truly interdisciplinary nature, establishing links between mathematics of the highest caliber and important scientific and technological problems from other disciplines and industries. To this end, the IMA organizes a wide variety of programs, ranging from short intense workshops in areas of exceptional interest and opportunity to extensive thematic programs lasting a year. IMA Volumes are used to communicate results of these programs that we believe are of particular value to the broader scientific community.

The full list of IMA books can be found at the Web site of the Institute for Mathematics and its Applications:

http://www.ima.umn.edu/springer/volumes.html.

Presentation materials from the IMA talks are available at

http://www.ima.umn.edu/talks/.

Video library is at http://www.ima.umn.edu/videos/.

Fadil Santosa, Director of the IMA

* * * * * * * * * *

IMA ANNUAL PROGRAMS

1982–1983	Statistical and Continuum Approaches to Phase Transition
1983–1984	Mathematical Models for the Economics of Decentralized Resource Allocation
1984–1985	Continuum Physics and Partial Differential Equations
1985–1986	Stochastic Differential Equations and Their Applications
1986–1987	Scientific Computation
1987–1988	Applied Combinatorics
1988–1989	Nonlinear Waves
1989–1990	Dynamical Systems and Their Applications
1990–1991	Phase Transitions and Free Boundaries
1991–1992	Applied Linear Algebra
1992–1993	Control Theory and its Applications
1993–1994	Emerging Applications of Probability
1994–1995	Waves and Scattering
1995–1996	Mathematical Methods in Material Science
1996–1997	Mathematics of High Performance Computing
1997–1998	Emerging Applications of Dynamical Systems

Continued at the back

Alberto Bressan • Gui-Qiang G. Chen
Marta Lewicka • Dehua Wang
Editors

Nonlinear Conservation Laws and Applications

Springer

Editors
Alberto Bressan
Department of Mathematics
Penn State University
USA
www.math.psu.edu/bressan/

Gui-Qiang G. Chen
Oxford Centre for Nonlinear PDE
Mathematical Institute
University of Oxford
UK
www.maths.ox.ac.uk/node/9869

Marta Lewicka
School of Mathematics
University of Minnesota
USA
www.math.umn.edu/~lewicka/

Dehua Wang
Department of Mathematics
University of Pittsburgh
USA
www.math.pitt.edu/~dwang/

ISSN 0940-6573
ISBN 978-1-4419-9553-7 e-ISBN 978-1-4419-9554-4
DOI 10.1007/978-1-4419-9554-4
Springer New York Dordrecht Heidelberg London

Library of Congress Control Number: 2011926226

Mathematics Subject Classification 2010: 35L65, 35L60, 76N10, 74B20, 35M10, 35-06, 35-02

Printed on acid-free paper

Springer is part of Springer Science+Business Media (www.springer.com)

FOREWORD

This IMA Volume in Mathematics and its Applications

NONLINEAR CONSERVATION LAWS AND APPLICATIONS

contains expository and research papers based on a highly successful three-week IMA Summer Program with the same title. The event was held on July 13-31, 2009. We are grateful to all the participants for making this occasion a very productive and stimulating one.

We would like to thank Alberto Bressan (Department of Mathematics, Penn State University), Gui-Qiang Chen (Oxford Centre for Nonlinear PDE, University of Oxford), Marta Lewicka (School of Mathematics, University of Minnesota), and Dehua Wang (Department of Mathematics, University of Pittsburgh) for their superb role as summer program organizers and editors of this volume.

We take this opportunity to thank the National Science Foundation for its continued support of the IMA.

Series Editors

Fadil Santosa, Director of the IMA

Markus Keel, Deputy Director of the IMA

PREFACE

This volume contains the proceedings of the Summer Program on Nonlinear Conservation Laws and Applications held at the Institute for Mathematics and its Applications on July 13–31, 2009. The program brought together several of the world's leading experts in the field, discussing the most significant theoretical advances and a wide range of applications.

Hyperbolic conservation laws is a classical subject, which has experienced vigorous growth in recent years. For one-dimensional systems with small data, the global existence and uniqueness of entropy weak solutions is well known. In addition, the method of local decomposition of solutions along traveling wave profiles has recently provided an understanding of the convergence of various types of approximations: vanishing viscosity, relaxations, and semi-discrete schemes. On the other hand, the global existence or the finite time blow-up of solution with large data is still a major open problem.

In several space dimensions, the theoretical analysis of hyperbolic conservation laws remains a grand challenge. During the past few years, new techniques have been introduced, resulting in specific advances. Refined measure-theoretical tools have been developed for the study of scalar conservation laws, and for transport equations with rough coefficients. Intriguing counterexamples have been constructed by means of a newly developed Baire category technique. Moreover, major breakthroughs have been achieved in the understanding of shock reflection-diffraction by a wedge in the equation of gas dynamics, and in the existence theory for global weak solutions with large data for the compressible Navier-Stokes equations, in both the isentropic and non-isentropic cases.

Progress in theoretical understanding has been paralleled by an expansion in the applications of hyperbolic conservation laws. Traditional areas of applications in mathematical physics, such as fluid dynamics, magnetohydrodynamics, nonlinear elasticity, combustion models, oil recovery, etc., have experienced sustained growth. In addition, new directions are emerging: continuum models based on conservation laws are increasingly used in the analysis of blood flow and of cell motion, and in the modeling of traffic flow and of large scale supply-chains in economic and industrial applications. A novel aspect in some of these models is that the flow is not only studied on a single road, or pipeline, but on an entire network.

The papers in the present volume provide a comprehensive account of these recent developments, and an outlook on open problems. The initial sections contain contributions from the main lecturers, who gave introductory courses during the first week of the Summer Program. The topics covered include: open questions in the theory of one-dimensional hyperbolic systems (A. Bressan), multidimensional conservation laws in several space variables (G.-Q. Chen), mathematical analysis of fluid motion (E. Feireisl), topics in the approximation of solutions to nonlinear conservation laws (E. Tadmor), and stability and dynamics of viscous shock waves (K. Zumbrun).

The remainder of the volume contains refereed research papers by the invited speakers. Contributions of more theoretical nature cover the following topics: singular limits for viscous systems of conservation laws, basic principles in the modeling of turbulent mixing, hyperbolic problems in two space dimensions related to gas dynamics, transonic flows past an obstacle and a fluid dynamic approach for isometric embedding in geometry, models of nonlinear elasticity, the Monge problem, and transport equations with rough coefficients. In addition, there are a number of papers devoted to applications. These include: models of blood flow, self-gravitating compressible fluids, granular flow, charge transport in fluids, and the modeling and control of traffic flow on networks.

We believe that this volume will provide a timely survey of the state of the art in the exciting field of conservation laws, and a stimulus for its further progress.

As organizers, it is our pleasure to acknowledge the IMA leadership and staff's tremendous support of this summer program. Our thanks go especially to the former and current directors Professors D. Arnold and F. Santosa, and to Ms. P. Brick for her efforts in assembling the proceedings papers. We also gratefully acknowledge the NSF's support. Finally, we thank the authors and the invited speakers for their valuable contribution, and all the participants and attendees for making the program successful.

Alberto Bressan
Department of Mathematics
Penn State University
http://www.math.psu.edu/bressan/

Gui-Qiang G. Chen
Oxford Centre for Nonlinear PDE, Mathematical Institute
University of Oxford
http://www.maths.ox.ac.uk/node/9869

Marta Lewicka
School of Mathematics
University of Minnesota
http://www.math.umn.edu/~lewicka/

Dehua Wang
Department of Mathematics
University of Pittsburgh
http://www.math.pitt.edu/~dwang/

CONTENTS

Part I: General Survey Lectures

OPEN QUESTIONS IN THE THEORY OF ONE DIMENSIONAL HYPERBOLIC CONSERVATION LAWS

ALBERTO BRESSAN*

1. Introduction. These notes are intended to provide an overview of the basic theory of one-dimensional hyperbolic systems of conservation laws, focusing on some major remaining open problems.

After a quick survey of known results on the well-posedness of the Cauchy problem and on vanishing viscosity approximations, in Section 3 we discuss the global existence versus finite time blow up for solutions with large total variation. In Section 4 we review several approximation methods, and their (known or conjectured) convergence properties.

We remark that hyperbolic conservation laws are a class of nonlinear evolution equations. As such, it would be natural to study them also from the point of view of dynamical systems. In particular, this would mean looking at periodic orbits, bifurcations, attractors, chaotic dynamics, etc... At present, however, very little of this is seen, within hyperbolic theory. Apparently, the main reason is that the known existence-uniqueness results are mainly restricted to solutions with small total variation. For such solutions, the asymptotic behavior is, in a sense, trivial. Indeed, as proved by T.P. Liu [52, 53], as $t \to +\infty$ every solution with small total variation approaches the solution to a corresponding Riemann problem.

One conjectures that, for hyperbolic conservation laws with source terms, solutions with large total variation will exhibit a rich dynamic behavior. However, at the time being, this remains largely "off limits" for the present theory.

2. Review of basic theory. A *system of conservation laws* in one space dimension takes the form

$$u_t + f(u)_x = 0\,. \tag{2.1}$$

The components of the vector $u = (u_1, \ldots, u_n) \in \mathbb{R}^n$ are the *conserved quantities*, while the components of the function $f = (f_1, \ldots, f_n) : \mathbb{R}^n \mapsto \mathbb{R}^n$ are the corresponding *fluxes*. For smooth solutions, (1.1) is equivalent to the quasilinear system

$$u_t + A(u)u_x = 0\,, \tag{2.2}$$

where $A(u) \doteq Df(u)$ is the $n \times n$ Jacobian matrix of the flux function f. We say that the system is *strictly hyperbolic* if this Jacobian matrix has

*Department of Mathematics, Penn State University, University Park, PA 16802, USA (bressan@math.psu.edu).

n real distinct eigenvalues, $\lambda_1(u) < \cdots < \lambda_n(u)$ for every $u \in \mathbb{R}^n$. In mathematical physics, the primary example is provided by the equations of non-viscous gases, accounting for the conservation of mass, momentum and energy, see [32].

Toward a rigorous mathematical analysis, the major difficulties stem from the lack of regularity of solutions. Due to the strong nonlinearity of the equations and the absence of diffusion terms with smoothing effect, solutions which are initially regular may become discontinuous within finite time. Global in time solutions can thus be constructed only within a space of discontinuous functions, interpreting the equation (2.1) in distributional sense. More precisely, a vector valued function $u = u(t,x)$ is a *weak solution* of (2.1) if

$$\iint \big[u\,\phi_t + f(u)\,\phi_x\big]\,dxdt = 0$$

for every test function $\phi \in \mathcal{C}^1_c$, continuously differentiable with compact support. In particular, the piecewise constant function

$$u(t,x) \doteq \begin{cases} u^- & \text{if} \quad x < \lambda t\,, \\ u^+ & \text{if} \quad x > \lambda t\,, \end{cases}$$

is a weak solution of (2.1) if and only if the left and right states $u^-, u^+ \in \mathbb{R}^n$ and the speed λ satisfy the Rankine-Hugoniot equations

$$f(u^+) - f(u^-) = \lambda\,(u^+ - u^-)\,.$$

When discontinuities are present, the weak solution of a Cauchy problem may not be unique. To single out a unique "good" solution, additional *entropy conditions* are usually imposed along shocks [43, 50]. These conditions often have a physical motivation, characterizing those solutions which can be recovered from higher order models, letting the diffusion or dispersion coefficients approach zero, see for example [32].

Toward the construction of more general solutions of (2.1), the basic building block is the *Riemann problem*, i.e. the initial value problem where the data are piecewise constant, with a single jump at the origin:

$$u(0,x) = \begin{cases} u^- & \text{if} \quad x < 0\,, \\ u^+ & \text{if} \quad x > 0\,. \end{cases}$$

This problem was first introduced by Riemann (1860) in the context of isentropic gas dynamics.

We observe that the system (2.1) admits a symmetry group. Namely if $u(t,x)$ is a weak solution, then for every $\theta > 0$ the function $u^\theta(t,x) \doteq u(\theta t, \theta x)$ is still another solution. The solutions of a Riemann problem are those which are invariant w.r.t. the above symmetry group.

For a wide class of $n \times n$ hyperbolic systems, assuming that the amplitude $|u^+ - u^-|$ of the jump is small, the solution was constructed by Lax [43], under the additional hypothesis

(H) For each $i = 1, \ldots, n$, the i-th field is either *genuinely nonlinear*, with $D\lambda_i(u) \cdot r_i(u) > 0$ for all u, or else it is *linearly degenerate*, with $D\lambda_i(u) \cdot r_i(u) = 0$ for all u.

Here $D\lambda_i \cdot r_i$ denotes the directional derivative of the i-th eigenvalue λ_i, in the direction of the corresponding eigenvector r_i. The solution of the Riemann problem is self-similar: $u(t,x) = U(x/t)$. It consists of $n+1$ constant states $\omega_0 = u^-, \omega_1, \ldots, \omega_n = u^+$. Each couple of adiacent states ω_{i-1}, ω_i is separated either by a *shock* satisfying the Rankine Hugoniot equations, or else by a *centered rarefaction.* In this second case, the solution u varies continuously between ω_{i-1} and ω_i in a sector of the t-x-plane where the gradient u_x coincides with an i-eigenvector of the matrix $A(u)$. For solutions of the Riemann problem under more general hypotheses, we refer to [49, 8].

Approximate solutions to the Cauchy problem with general initial data

$$u(0,x) = \bar{u}(x) \tag{2.3}$$

can be constructed by patching together several solutions of Riemann problems.

THEOREM 1 (global existence of weak solutions). *Assume that the system (2.1) is strictly hyperbolic, and that each characteristic field is either linearly degenerate or genuinely nonlinear.*

Then there exists a constant $\delta_0 > 0$ such that, for every initial condition $\bar{u} \in \mathbf{L}^1(I\!R; I\!R^n)$ with

$$Tot.Var.\{\bar{u}\} \le \delta_0 \,, \tag{2.4}$$

the Cauchy problem (2.1)–(2.3) has a weak solution $u = u(t,x)$ defined for all $t \ge 0$.

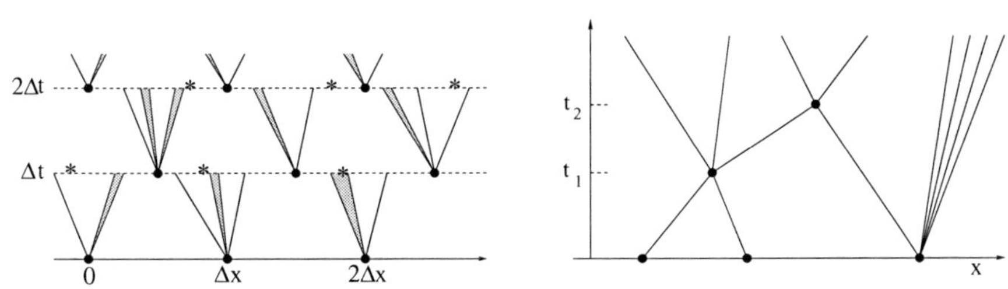

FIG. 1. *Left: an approximate solution constructed by the Glimm scheme. Right: a wave-front tracking approximation.*

In the Glimm scheme [37, 51, 52], one works with a fixed grid in the t-x plane, with mesh sizes Δt, Δx. At time $t = 0$ the initial data is approximated by a piecewise constant function, with jumps at grid points. Solving the corresponding Riemann problems, a solution is constructed up to a time Δt sufficiently small so that waves generated by different Riemann problems do not interact. By a random sampling procedure, the solution $u(\Delta t, \cdot)$ is then approximated by a piecewise constant function having jumps only at grid points. Solving the new Riemann problems at every one of these points, one can prolong the solution to the next time interval $[\Delta t,\ 2\Delta t]$, etc...

An alternative technique for contructing approximate solutions is by wave-front tracking. This method was introduced by Dafermos [31] in the scalar case and later developed by various authors [34, 14, 3, 39]. It now provides an efficient tool in the study of general $n \times n$ systems of conservation laws, both for theoretical and numerical purposes. The initial data is here approximated with a piecewise constant function, and each Riemann problem is solved approximately, within the class of piecewise constant functions. In particular, if the exact solution contains a centered rarefaction, this must be approximated by a *rarefaction fan*, containing several small jumps. At the first time t_1 where two fronts interact, the new Riemann problem is again approximately solved by a piecewise constant function. The solution is then prolonged up to the second interaction time t_2, where the new Riemann problem is solved, etc... The main difference is that in the Glimm scheme one specifies a priori the nodal points where the the Riemann problems are to be solved. On the other hand, in a solution constructed by wave-front tracking the locations of the jumps and of the interaction points depend on the solution itself, and no restarting procedure is needed.

In the end, both algorithms produce a sequence of approximate solutions, whose convergence is proved by a compactness argument based on uniform bounds on the total variation. Assuming that the total variation of the initial data $\bar{u}$ is sufficiently small, one thus obtains the global existence of an entropy weak solution to the Cauchy problem (2.1), (2.3).

The continuous dependence on the initial data of the weak solutions obtained as limits of front-tracking approximations was later established in [19, 20, 25].

THEOREM 2. *There exists a domain $\mathcal{D} \subset \mathbf{L}^1(\mathbb{R};\mathbb{R}^n)$ containing all functions $u : \mathbb{R} \mapsto \mathbb{R}^n$ with sufficiently small total variation, and constants L, L' such that the following holds.*

For every initial data $\bar{u} \in \mathcal{D}$, approximate solutions constructed by front tracking method converge to a unique entropy weak solution $u(t, \cdot) = S_t\bar{u}$ to the Cauchy problem (2.1), (2.3). The map $S : \mathcal{D} \mapsto [0, \infty[\mapsto \mathcal{D}$ thus defined satisfies the semigroup properties

$$S_0 u = u, \qquad S_s S_t u = S_{s+t} u\,, \tag{2.5}$$

together with the uniform Lipschitz estimate

$$\|S_s\bar{u} - S_t\bar{v}\|_{\mathbf{L}^1} \leq L'|t-s| + L\|\bar{u}-\bar{v}\|_{\mathbf{L}^1} \qquad \bar{u},\bar{v}\in\mathcal{D}, \quad s,t\geq 0. \tag{2.6}$$

This well-posedness result has been extended in [2] to hyperbolic systems of balance laws in the presence of source terms, and in [1, 36] to the case of initial-boundary value problems.

Given a Lipschitz semigroup S satisfying (2.5)–(2.6), one can prove a useful estimate on the distance between an arbitrary Lipschitz continuous map $w : [0,T] \mapsto \mathcal{D}$, and the trajectory of the semigroup starting at $w(0)$. Namely, for every $\tau \in [0,T]$ one has

$$\big\|w(\tau) - S_\tau w(0)\big\|_{\mathbf{L}^1} \leq L \cdot \int_0^\tau \left\{ \liminf_{h\to 0+} \frac{\big\|w(t+h) - S_h w(t)\big\|_{\mathbf{L}^1}}{h} \right\} dt. \tag{2.7}$$

In the error estimate (2.7) the integrand can be regarded as the instantaneous error rate for w at time t. Since the flow is uniformly Lipschitz continuous w.r.t. the initial point, during the time interval $[t,\tau]$ this error is amplified at most by a factor L. See [16] for a detailed proof.

Our next concern is to identify a set of conditions that imply uniqueness of the solution to the Cauchy problem

$$u_t + f(u)_x = 0, \qquad\qquad u(0,x) = \bar{u}(x). \tag{2.8}$$

Relying on the fact that a semigroup of solutions has already been constructed, uniqueness will follow if we can show that every weak solution $t \mapsto u(t)$ of (2.8) (satisfying suitable admissibility conditions) coincides with the semigroup trajectory $t \mapsto S_t\bar{u}$. Because of (2.7), this will be the case if

$$\liminf_{h\to 0+} \frac{\big\|u(t+h) - S_h u(t)\big\|_{\mathbf{L}^1}}{h} = 0 \qquad\qquad \text{for a.e. } t > 0. \tag{2.9}$$

A first set of conditions, introduced in [15], is obtained by locally comparing a given solution with two types of approximations.

1. **Comparison with solutions to a Riemann problem.** Let $u = u(t,x)$ be a weak solution. Fix a point (τ,ξ). Define $U^\sharp = U^\sharp_{(\tau,\xi)}$ as the solution of the Riemann problem corresponding to the jump at (τ,ξ):

$$w_t + f(w)_x = 0, \qquad\qquad w(\tau,x) = \begin{cases} u^+ \doteq u(\tau,\xi+) & \text{if} \quad x > \xi \\ u^- \doteq u(\tau,\xi-) & \text{if} \quad x < \xi. \end{cases}$$

We expect that, if u is entropy-admissible, then u will be asymptotically equal to $U^\sharp$ in a forward neighborhood of the point (τ,ξ). More precisely, for every $\hat{\lambda} > 0$, one should have

$$\lim_{h\to 0+} \frac{1}{h}\int_{\xi-h\hat\lambda}^{\xi+h\hat\lambda} \left|u(\tau+h,\ x) - U^\sharp_{(\tau,\xi)}(\tau+h,\ x)\right|\,dx \;=\; 0. \tag{E1}$$

2. **Comparison with solutions to a linear problem with constant coefficients.** Fix again a point (τ,ξ), and choose $\hat\lambda>0$ larger than all wave speeds. Define $U^\flat = U^\flat_{(\tau,\xi)}$ as the solution of the linear Cauchy problem

$$w_t + \widetilde{A} w_x = 0 \qquad\qquad w(\tau,x) = u(\tau,x)$$

with "frozen" coefficients: $\widetilde{A} \doteq A\big(u(\tau,\xi)\big)$. Then, for $a<\xi<b$ and $h>0$, the difference between these two solutions should be estimated by

$$\begin{aligned}\frac{1}{h}\int_{a+\hat\lambda h}^{b-\hat\lambda h} &\left|u(\tau+h,\ x) - U^\flat(\tau+h,\ x)\right|\,dx \\ &\leq\; C\cdot\Big(\text{Tot.Var.}\,\{u(\tau,\cdot);\ \,]a,\,b[\ \}\Big)^2,\end{aligned} \tag{E2}$$

for a constant C depending only on the hyperbolic system (2.1) and the domain $\mathcal{D}$, not on a,b,h, or the solution u.

In [15] it was shown that the two above conditions (E1)–(E2) completely characterize semigroup trajectories.

THEOREM 3 (characterization of semigroup trajectories). *Let $u : [0,T] \mapsto \mathcal{D}$ be Lipschitz continuous w.r.t. the $\mathbf{L}^1$ distance. Then u is a weak solution to the system of conservation laws*

$$u_t + f(u)_x \;=\; 0$$

obtained as limit of front tracking approximations if and only if the estimates (E1)–(E2) are satisfied for a.e. $\tau\in[0,T]$, at every $\xi\in I\!R$.

The conditions (E1)–(E2) have a clear meaning, but are not always easy to check, in the case of solutions obtained by different constructive methods. In alternative, the following regularity assumptions can be used.

(A3) (Tame Oscillation Condition). For some constants C, $\hat\lambda$ the following holds. For every point $x\in I\!R$ and every $t,h>0$ one has

$$\big|u(t+h,x) - u(t,x)\big| \;\leq\; C\cdot \text{Tot.Var.}\Big\{u(t,\cdot);\ \ [x-\hat\lambda h,\ x+\hat\lambda h]\Big\}. \tag{2.10}$$

(A4) (Bounded Variation Condition). There exists $\delta>0$ such that, for every space-like curve $\big\{t = \tau(x)\big\}$ with $|d\tau/dx|\leq\delta$ a.e., the function $x\mapsto u\big(\tau(x),x\big)$ has locally bounded variation.

REMARK 1. The condition (A3) restricts the oscillation of the solution. An equivalent, more intuitive formulation is the following. For some constant $\hat{\lambda}$ larger than all characteristic speeds, given any interval $[a,b]$ and $t \geq 0$, the oscillation of u on the triangle $\Delta \doteq \big\{(s,y):\ \ s \geq t,\ a+\hat{\lambda}(s-t) < y < b - \hat{\lambda}(s-t)\big\}$, defined as

$$\mathrm{Osc}\big\{u;\ \Delta\big\} \doteq \sup_{(s,y),(s',y')\in\Delta} \big|u(s,y) - u(s',y')\big|,$$

is bounded by a constant multiple of the total variation of $u(t,\cdot)$ on $[a,b]$.

The assumption (A4) simply requires that, for some fixed $\delta > 0$, the function u has bounded variation along every space-like curve γ which is "almost horizontal". Indeed, the condition is imposed only along curves of the form $\big\{t = \tau(x);\ x \in [a,b]\big\}$ with

$$\big|\tau(x) - \tau(x')\big| \ \leq\ \delta|x - x'| \qquad\qquad \text{for all } \ x, x' \in [a,b].$$

One can prove that all of the above assumptions are satisfied by every weak solution obtained as limit of Glimm or wave-front tracking approximations [16]. The following result shows that the entropy weak solution of the Cauchy problem (2.8) is unique within the class of functions that satisfy either the additional regularity condition (A3), or (A4).

THEOREM 4 (uniqueness). *Assume that the function* $u : [0,T] \mapsto \mathcal{D}$ *is continuous w.r.t. the* $\mathbf{L}^1$ *distance, and provides an entropy admissible weak solution of the Cauchy problem (2.8), taking values in the domain of the semigroup* S *described in Theorem 2. If either* (A3) *or* (A4) *holds, then*

$$u(t,\cdot) \ = \ S_t\bar{u} \qquad\qquad \textit{for all} \ \ t \in [0,T]. \tag{2.11}$$

In particular, the entropy weak solution that satisfies either one of these conditions is unique.

The above uniqueness result was proved in [21] with the assumption (A3) and in [24] with the assumption (A4). Both of these papers extend the earlier result in [23].

In [15], functions that satisfy the conditions (E1)–(E2) were called *viscosity solutions* of the hyperbolic system (2.1). This name was fully justified a few years later, when it was proved in [11] that these conditions indeed characterize the unique limits of vanishing viscosity approximations:

$$u^\varepsilon_t + f(u^\varepsilon)_x = \varepsilon\, u^\varepsilon_{xx} \qquad\qquad u^\varepsilon(0,x) = \bar{u}(x)\,. \tag{2.12}$$

THEOREM 5 (BV estimates and convergence of vanishing viscosity approximations). *Consider the Cauchy problem for the strictly hyperbolic system with viscosity (2.12). There exist constants* C, L, L' *and* $\delta > 0$ *such that the following holds. If* $\bar{u} \in \mathbf{L}^1$ *with*

$$\textit{Tot.Var.}\{\bar{u}\} < \delta\,, \tag{2.13}$$

then for each $\varepsilon > 0$ the Cauchy problem (2.12) has a unique solution u^ε, defined for all $t \geq 0$. Adopting a semigroup notation, this will be written as $t \mapsto u^\varepsilon(t,\cdot) \doteq S_t^\varepsilon \bar{u}$. In addition, one has:

$$\textbf{BVbounds}: \quad Tot.Var.\{S_t^\varepsilon \bar{u}\} \leq C\, Tot.Var.\{\bar{u}\}. \tag{2.14}$$

$$\mathbf{L}^1\textbf{stability}: \quad \left\|S_t^\varepsilon \bar{u} - S_t^\varepsilon \bar{v}\right\|_{\mathbf{L}^1} \leq L \left\|\bar{u} - \bar{v}\right\|_{\mathbf{L}^1}, \tag{2.15}$$

$$\left\|S_t^\varepsilon \bar{u} - S_s^\varepsilon \bar{u}\right\|_{\mathbf{L}^1} \leq L' \left(|t-s| + \left|\sqrt{\varepsilon t} - \sqrt{\varepsilon s}\,\right|\right). \tag{2.16}$$

Convergence: *As $\varepsilon \to 0+$, the viscous approximations converge in $\mathbf{L}^1$ (uniformly for t in bounded sets) to a unique limit: $u^\varepsilon(t) \to u(t) \doteq S_t \bar{u}$, providing an entropy weak solution to the hyperbolic Cauchy problem (2.8). Moreover, the limit semigroup thus obtained is uniformly Lipschitz continuous:*

$$\left\|S_t \bar{u} - S_s \bar{v}\right\|_{\mathbf{L}^1} \leq L \left\|\bar{u} - \bar{v}\right\|_{\mathbf{L}^1} + L' |t-s|. \tag{2.17}$$

In the genuinely nonlinear case, an estimate on the rate of convergence of these viscous approximations was provided in [27]. Here the Landau symbol $\mathcal{O}(1)$ denotes a uniformly bounded quantity.

THEOREM 6 (convergence rate). *For the strictly hyperbolic system of conservation laws (2.1), assume that every characteristic field is genuinely nonlinear. The difference between the corresponding solutions of (2.12) and (2.8) can be estimated as*

$$\left\|u^\varepsilon(t,\cdot) - u(t,\cdot)\right\|_{\mathbf{L}^1} = \mathcal{O}(1)\cdot(1+t)\sqrt{\varepsilon}|\ln \varepsilon| \text{ Tot.Var.}\{\bar{u}\}.$$

3. Solutions with large total variation. The global existence theorem proved by Glimm [37], as well as the more recent stability results in [19, 20, 25], apply to the case of initial data with small total variation. This assumption plays a key role in the proof of a priori bounds on the total variation of approximate solutions.

Indeed, consider a piecewise constant approximate solution $u = u(t,x)$. As long as fronts do not interact, the total variation of the map $x \mapsto u(t,\cdot)$ remains constant. However, at a time τ where two front interact, Tot.Var.$u(t,\cdot)$ may well increase. Indeed, consider a time τ when two fronts interact. To fix the ideas, let σ_i^-, σ_j^- be the strengths of the incoming fronts (of the i-th and of the j-th family, respectively), before the interaction time. After time τ, the solution is prolonged by solving the Riemann problem generated by the interaction. This solution will contain outgoing waves of strengths $\sigma_1^+, \sigma_2^+, \ldots, \sigma_n^+$. In general, one may well have

$$|\sigma_1^+| + \cdots + |\sigma_n^+| > |\sigma_j^-| + |\sigma_j^-|,$$

and the total variation of the solution will increase. A key estimate proved in [37] is

$$|\sigma_i^+ - \sigma_i^-| + |\sigma_j^+ - \sigma_j^-| + \sum_{k\neq i,j} |\sigma_k^+| \leq C_0 |\sigma_i \sigma_j|. \tag{3.1}$$

for some constant C_0 depending only on the flux function f.

In order to achieve a uniform bound on the total variation, two functional are used. At a fixed time t, let x_α, $\alpha = 1, \ldots, N$, be the locations of the fronts in $u(t, \cdot)$. Moreover, let $|\sigma_\alpha|$ be the strength of the wave-front at x_α. Consider the two functionals

$$V(t) \doteq V\big(u(t)\big) \doteq \sum_\alpha |\sigma_\alpha| \,, \tag{3.2}$$

measuring the *total strength of waves* in $u(t, \cdot)$, and

$$Q(t) \doteq Q\big(u(t)\big) \doteq \sum_{(\alpha,\beta)\in\mathcal{A}} |\sigma_\alpha \sigma_\beta| \,, \tag{3.3}$$

measuring the *wave interaction potential.* Here the summation extends over every couple of *approaching* wave fronts.

Now consider the approximate solution $u = u(t, x)$ constructed by the front tracking algorithm. It is clear that the quantities $V\big(u(t)\big)$, $Q\big(u(t)\big)$ remain constant except at times where an interaction occurs. At a time τ where two fronts of strength $|\sigma_i^-|, |\sigma_j^-|$ collide, the interaction estimate (3.1) yields

$$\Delta V(\tau) \ \doteq\ V(\tau+) - V(\tau-) \ \leq\ C_0\, |\sigma_i^- \sigma_j^-|, \tag{3.4}$$

$$\Delta Q(\tau) \ \doteq\ Q(\tau+) - Q(\tau-) \ \leq -|\sigma_i^- \sigma_j^-| + C_0\, |\sigma_i^- \sigma_j^-| \cdot V(\tau-). \tag{3.5}$$

Indeed (Fig. 2), after time τ the two approaching fronts σ_i^-, σ_j^- are replaced by outgoing fronts $\sigma_1^+, \ldots, \sigma_n^+$ which no longer approach each other. Hence the product $|\sigma_i^- \sigma_j^-|$ is no longer counted within the summation (3.3). On the other hand, the new waves emerging from the interaction (having strength $\leq C_0\, |\sigma_i^- \sigma_j^-|$) can approach all the other fronts not involved in the interaction (which have total strength $\leq V(\tau-)$).

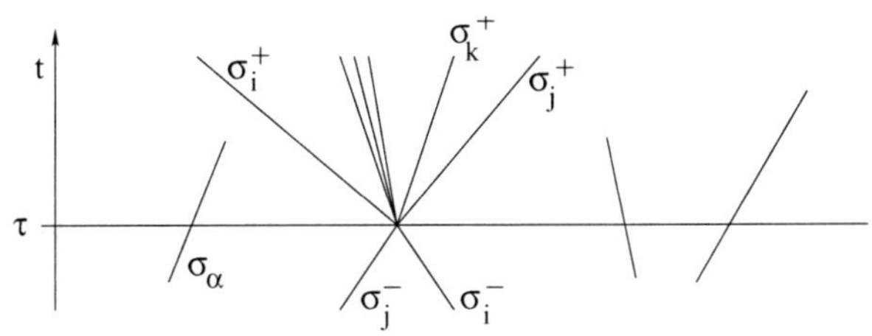

FIG. 2. *Estimating the change in the total variation at a time where two fronts interact.*

If V remains sufficiently small, so that $V(\tau-) \leq (2C_0)^{-1}$, then (3.5) yields

$$Q(\tau+) - Q(\tau-) \ \leq\ -\frac{|\sigma_i^- \sigma_i^-|}{2} \,. \tag{3.6}$$

In particular, if the quantity

$$\Upsilon(t) \doteq V(t) + 2C_0 Q(t)$$

satisfies $\Upsilon(0) < (2C_0)^{-1}$, then Υ will decrease at every interaction, hence

$$V(t) \leq \Upsilon(t) \leq \Upsilon(0) < \frac{1}{2C_0} \qquad \text{for all } t \geq 0.$$

This provides a bound on the total strength of waves, i.e. on the total variation of $u(t,\cdot)$, uniformly valid for all times $t \geq 0$.

On the other hand, if the total variation is large, then $C_0 \cdot V(t) > 1$ and both functionals $V(\cdot)$ and $Q(\cdot)$ in (3.4)–(3.5) can increase at an interaction time.

An example where the total variation actually blows up in finite time was constructed in [40]. It consists of a 3×3 system of the form

$$U_t + F(U)_x = 0, \tag{3.7}$$

where $U = (u, v, w)$ and

$$F(U) = F(u,v,w) = \begin{pmatrix} uv + w \\ g(v) \\ u(1 - v^2) - vw \end{pmatrix}.$$

The scalar function g is strictly convex, with

$$g(0) = 0, \qquad g(v) = g(-v), \qquad -1 < g'(v) < 1 \tag{3.8}$$

for all v. The system is thus strictly hyperbolic. Indeed

$$A(U) \doteq DF(u,v,w) = \begin{pmatrix} v & u & 1 \\ 0 & g'(v) & 0 \\ 1 - v^2 & 2uv - w & -v \end{pmatrix}$$

is a matrix with real distinct eigenvalues

$$\lambda_1 = -1, \qquad \lambda_2 = g'(v), \qquad \lambda_3 = 1.$$

The corresponding right eigenvectors are

$$r_1 = \begin{pmatrix} 1 \\ 0 \\ -1 - v \end{pmatrix}, \qquad r_2 = \begin{pmatrix} 0 \\ 1 \\ 0 \end{pmatrix}, \qquad r_3 = \begin{pmatrix} 1 \\ 0 \\ 1 - v \end{pmatrix}.$$

The first and third characteristic fields are linearly degenerate, the second is genuinely nonlinear.

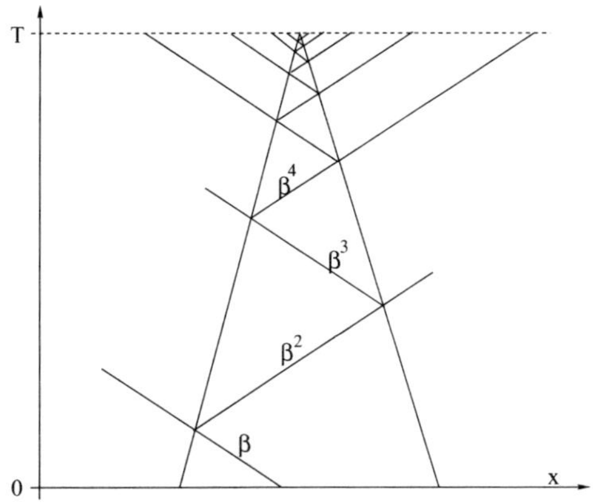

FIG. 3. *An example of finite time blow up: for $\beta > 1$, the strength of reflected wave-fronts becomes arbitrarily large.*

By a suitable choice of the function g, the analysis in [40] shows that one can construct a solution $U = U(t, x)$ of the following form (Fig. 3). The initial data contains two approaching 2-shocks, and a 1-wave between them, of unit strength. Subsequent interactions produce alternatively a 3-wave and a 1-wave, whose strengths increase in a geometric progression. The $\mathbf{L}^\infty$ norm of the solution, and hence also its total variation, approach infinity as $t \to T-$, when the two 2-shocks hit each other.

One might expect that interaction patterns such as the one in Fig. 3 would yield plenty of examples where solutions blow up in finite time, for a wide class of systems. However, this is not the case. Indeed, the analysis in [40, 4] indicates that the interaction pattern in Fig. 3 cannot lead to blow up if the system of conservation laws admits a strictly convex entropy. On the other hand, it is not clear how to rule out a blow up generated by a different, more complicated interaction pattern.

For physical systems, endowed with a strictly convex entropy, it thus remains a largely open problem to determine whether solutions with BV initial data remain with bounded variation for all positive times. The answer is not known, in particular, for the Euler equations of inviscid gas dynamics.

Considerable attention has been devoted to the so-called "p-system", describing isentropic gas dynamics in Lagrangian variables:

$$v_t + u_x = 0\,, \qquad\qquad u_t + p(v)_x = 0\,. \tag{3.9}$$

Here u is the velocity of the gas, v is the specific volume (i.e., the inverse of the density), and $p(v)$ is a function which determines the pressure in terms of of the specific volume. A typical choice is $p(v) = kv^{-\gamma}$, with $\gamma > 1$. See [63, 56] for more details. In this case, the blow up of the v-variable corresponds to appearance of vacuum. It is an outstanding open problem to determine whether this can happen within finite time. For partial results in this direction, see [47, 48]. A detailed analysis of wave interactions, close to the vacuum state, can be found in [64, 65].

We mention that the global existence of large BV solutions is well known in the special case of *Temple class* systems. These are strictly hyperbolic systems of conservation laws such that: (i) for each characteristic family, shock and rarefaction curves coincide, and (ii) there exists a complete system of Riemann invariants $(w_1, \ldots, w_n)$. Using these new dependent variables $w = w(u)$, the quasilinear equations (2.2) thus assume the diagonal form

$$w_{i,t} + \lambda_i(w) w_{i,x} = 0 \qquad\qquad i = 1, \ldots, n.$$

For solutions of Temple class systems, the total variation measured in the w-variables:
$\sum_{i=1}^{n}$ Tot.Var.$\{w_i(t)\}$ is always a non-increasing function of time. See [60] for details.

In general, the issue of global BV bounds versus finite time blow-up can only be resolved by a careful analysis of the production of new waves (due to nonlinear interactions), and of the cancellation of positive and negative waves (determined by to genuine nonlinearity). For genuinely nonlinear 2×2 systems, Glimm and Lax [38] proved that, if the initial data has small $\mathbf{L}^\infty$ norm (but possibly large total variation), then cancellation effects dominate. Hence the Cauchy problem admits a weak solution with bounded variation for all times $t > 0$. An extension of these ideas to $n \times n$ systems can be found in [29].

Remark 2. Due to the finite propagation speed, the uniqueness of large BV solutions is essentially a local problem. We observe that, in a suitably small neighborhood of a given point $\bar{x}$, any BV function $u : \mathbb{R} \mapsto \mathbb{R}^n$ contains at most one big jump (i.e., the one at $\bar{x}$). Indeed, for any $\varepsilon > 0$ one can find $\delta > 0$ such that the total variation of u restricted separately to $]\bar{x} - \delta, \bar{x}[$ and to $]\bar{x}, \bar{x} + \delta[$ is $< \varepsilon$. It thus suffices to study the problem of uniqueness and continuous dependence within a domain of functions obtained by small BV perturbations of a (possibly large) Riemann solution. Results in this direction have been obtained by M. Lewicka in [46] and related papers.

Remark 3. For entropy-weak solutions in $\mathbf{L}^\infty$, with unbounded total variation, examples show that uniqueness and continuous dependence on initial data fail, in general. This is due to the fact that the ODEs determining the location of characteristics in the t-x plane have a discontinuous right hand side, with unbounded directional variation. The example constructed in [26] also exhibits a blow-up of the $\mathbf{L}^\infty$ norm, with a positive amount of mass concentrating at a single point. The Cauchy problem can be globally solved only within a class of measure-valued solutions.

REMARK 4. Toward a general theory of hyperbolic conservations laws in several space dimensions, a key intermediate step is to understand solutions with radial symmetry. These are described by one-dimensional systems of conservation laws defined on the half line $\{r > 0\}$, with a geometric source term that becomes singular as the radius $r \to 0$. In general, these systems do not admit global BV solutions, and the question of existence and continuous dependence of solutions is largely open. See [28, 44] for results in this direction.

4. Stability and convergence of approximate solutions. For the hyperbolic system of conservation laws

$$u_t + f(u)_x = 0, \tag{4.1}$$

the Glimm scheme, as well as front tracking approximations, rely on Riemann solutions as building blocks. They provide a very effective tool for the construction and the analysis of solutions to systems of conservation laws.

There are several other methods to construct approximate solutions to (2.1), or equivalently to (2.2). Some are naturally derived from physical considerations, others lead to more efficient numerical algorithms. For each method, the same natural questions arise:

(i) Does the total variation of the approximate solutions remain uniformly bounded for all times $t > 0$?

(ii) Do the approximate solutions depend continuously on the initial data, in the $\mathbf{L}^1$ distance ?

(iii) As the approximation parameters tend to zero, do the approximate solutions converge to the unique entropy weak solution of the hyperbolic Cauchy problem ?

A number of approximation techniques is examined below.

1. General viscous approximations. We consider here more general viscous approximations, having the form

$$u_t^\varepsilon + A(u^\varepsilon)u_x^\varepsilon = \varepsilon\left(B(u^\varepsilon)u_x^\varepsilon\right)_x. \tag{4.2}$$

If the diffusion matrix $B = B(u)$ is strictly positive definite, then (4.2) becomes a parabolic quasilinear system. In physically relevant models, however, $B(u)$ is only semi-definite matrix, possibly with some null eigenvalues. In any case, letting $\varepsilon \to 0+$, one still expects to recover entropy weak solutions of (2.1) in the limit.

At present, global BV bounds, uniform stability w.r.t. perturbations of the initial data, and convergence to a unique limit solution have been proven only in the case of "artificial viscosity", where $B(u) \equiv I$ is the identity matrix.

The difficulties in extending the result in [11] to more general viscosity matrices lies in the properties of the center manifold of travelling wave profiles. Following [11], one should try to decompose a general solution $u = u(x)$ locally as a superposition of travelling waves, i.e. of solutions to the system of ODEs

$$\big(A(U) - \sigma I\big)U' = \big(B(U)U'\big)', \tag{4.3}$$

chosen within suitable center manifolds. However, when B depends on U, the manifold of travelling viscous shock profiles has weaker regularity properties the in the case $B(u) \equiv I$. This affects the estimates on the size of source terms. As a result, a direct extension of the arguments in [11] is not possible. Progress in this direction will likely come from a careful analysis of travelling profiles for (4.3). See [12, 13] for recent work in this direction.

REMARK 5. The analysis in [11] remains valid for arbitrary quasilinear hyperbolic systems, not necessarily in conservation form. However, in the case of a variable viscosity matrix, the assumption that the system be in conservation form becomes essential. Otherwise, different viscosity matrices can lead to different limit solutions as $\varepsilon \to 0$.

2. Relaxation approximations. Generalizing [55], in [42], Jin and Xin introduced the semilinear system with source

$$\begin{cases} u_t + v_x &= 0\,, \\ v_t + \Lambda\, u_x &= \varepsilon^{-1}\big[f(u) - v\big]\,. \end{cases} \tag{4.4}$$

Here $u, v \in I\!R^n$. For a suitable choice of the constant matrix Λ, as $\varepsilon \to 0+$ one expects that the second equation will force $v = f(u)$ in the limit. In turn, the solution to the first equation should thus approach the one in (2.1). Based on physical considerations, one can also consider fully nonlinear nonlinear relaxation systems of the form

$$\begin{cases} u_t + v_x &= 0\,, \\ v_t + g(u)_x &= \varepsilon^{-1}\big[f(u) - v\big]\,. \end{cases} \tag{4.5}$$

The stability and convergence of the semilinear approximations (4.4) were recently established by S. Bianchini in [10]. The extension of these results to the nonlinear case (4.5) remains an open problem.

3. Semidiscrete numerical schemes. Here we keep the time t as a continuous variable, but we discretize space, choosing a step Δx and setting

$x_k \doteq k\,\Delta x$. The system (2.1) can then be approximated by an infinite set of O.D.E's, describing the evolution of the functions $t \mapsto u_k(t) \doteq u(t, x_k)$. To fix the ideas, assume that all eigenvalues of the Jacobian matrix $Df(u)$ are positive, for every $u \in I\!R^n$. This means that all waves propagate with positive speed. We can then replace (2.1) by a system of countably many O.D.E's:

$$\frac{d}{dt}u_k(t) + \frac{f\big(u_k(t)\big) - f\big(u_{k-1}(t)\big)}{\Delta x} = 0 \qquad k \in Z\!\!Z\,. \tag{4.6}$$

Letting $\Delta x \to 0+$, one expects to recover entropy-admissible weak solutions to the original system (2.1), in the limit. The stability and convergence of these approximations was indeed proved by S. Bianchini in [9].

4. Backward Euler approximations. We now keep x as a continuous variable, but discretize time, choosing a time step $\Delta t > 0$ and defining $t_k \doteq k\,\Delta t$. An approximate solution $U = U_k(x)$ can then be constructed by induction on k. Indeed, given $U_{k-1}(\cdot)$, the function $U_k(\cdot)$ is then implicitly defined by the equation

$$U_k - U_{k-1} + \Delta t \cdot \big[f(U_k)\big]_x = 0\,. \tag{4.7}$$

Letting $\Delta t \to 0+$, we expect the $\mathbf{L}^1$-convergence $U_k(\cdot) \to u(k\,\Delta t, \cdot)$, where u is the corresponding entropy weak solution of (2.1). In the case of a scalar conservation law, (4.7) corresponds to the well known Yosida approximations. Their stability and convergence follows from standard techniques in the theory of contractive semigroups and accretive operators [30]. It is not known whether the approximations (4.7) still converge, in the case of $n \times n$ strictly hyperbolic systems. See [7] for a proof in the special case of systems where rarefaction curves are straight lines.

5. Fully discrete numerical schemes. Here we discretize both space and time, and construct a grid in the t-x plane with mesh $\Delta t, \Delta x$. An approximate solution $U_{k,j} \approx u(k\,\Delta t\,,\ j\,\Delta x)$ is obtained by replacing partial derivatives in (1.1) with finite differences. For example, if for all u the eigenvalues of the Jacobian matrix $Df(u)$ satisfy

$$\big|\lambda_i(u)\big| < \Delta x/\Delta t \qquad i = 1, \ldots, n\,,$$

then an appropriate difference scheme is

$$U_{k+1,j} = U_{k,j} + \frac{\Delta t}{2\Delta x}\big[f(U_{k,j-1}) - f(U_{k,j+1})\big]\,. \tag{4.8}$$

In the case where

$$0 < \lambda_i(u) < \Delta x/\Delta t \qquad i = 1, \ldots, n\,,$$

one can also use the upwind Godunov scheme:

$$U_{k+1,j} = U_{k,j} + \frac{\Delta t}{\Delta x}\big[f(U_{k,j-1}) - f(U_{k,j+})\big]\,. \tag{4.9}$$

It has been a long standing open problem to understand the stability and convergence of these schemes, for solutions to $n \times n$ hyperbolic systems in the presence of shocks. In the light of recent results valid for other approximations schemes, one may try to decompose a general solution in terms of travelling waves. We recall here the main definitions. A *discrete travelling wave profile* for the difference scheme (4.8) is a continuous function $U : \mathbb{R} \mapsto \mathbb{R}^n$ such that, by sampling the function $u(t,x) = U(x - \lambda t)$ along point of the grid $\Delta t\,\mathbb{N} \times \Delta x\,\mathbb{Z}$ one obtains a solution of (4.8). In other words, setting

$$U_{k,j} \doteq U\big(j\Delta x - \lambda k \Delta t\big) \qquad\qquad k \in \mathbb{N}\,, \quad j \in \mathbb{Z}\,,$$

we obtain a solution of (4.8). In the case where

$$\lim_{s\to-\infty} U(s) \;=\; u^-, \qquad\qquad \lim_{s\to+\infty} U(s) \;=\; u^+,$$

we say that U is a *discrete shock profile* connecting the states $u^-, u^+ \in \mathbb{R}^n$. Here λ is the speed of the shock, which satisfies the Rankine-Hugoniot equations $\lambda(u^+ - u^-) = f(u^+) - f(u^-)$. Similar definitions apply to other difference schemes, such as (4.9).

At this stage, for each $i = 1, \ldots, n$, one should construct a "center manifold" of discrete travelling wave profiles, of dimension $n+2$, containing all discrete shock profiles together with other (unbounded) discrete profiles. In order to prove suitable interaction estimates, one now needs to analyze the regularity of this manifold of discrete travelling profiles. It is here that a key obstruction is encountered. Indeed, as proved by D. Serre [60, 62], for general strictly hyperbolic systems the family of discrete shock profiles cannot depend continuously on the wave speed, and hence on the states u^-, u^+, w.r.t. the BV norm

$$\|U\|_{BV} \;\doteq\; \|U\|_{\mathbf{L}^1} + \text{Tot.Var.}\{U\}\,.$$

This accounts for the substantially different ways in which discrete shock profiles are constructed, in the case of case of rational or irrational speed [58, 57]. A specific example, illustrating how discrete shock profiles fail to depend continuously on the speed λ, was constructed in [5]. It consists of a 2×2 system in triangular form

$$\left\{\begin{array}{rcl} u_t + f(u)_x &=& 0, \\ v_t + g(u)_x &=& 0. \end{array}\right. \tag{4.10}$$

The characteristic speeds are 0 and $f'(u)$ and the system is strictly hyperbolic provided $f'(u) > 0$. In the specific example, the flux function f satisfies $1/4 < f'(u) < 1$, while g is constant outside a bounded interval.

The Lax-Friedrichs scheme for (4.10) with $\Delta x = \Delta t$ takes the form

$$\begin{cases} u_{n+1,j} &= \frac{1}{2}\big(u_{n,j+1} + u_{n,j-1}\big) - \frac{1}{2}\left[f(u_{n,j+1}) - f(u_{n,j-1})\right], \\ v_{n+1,j} &= \frac{1}{2}\big(v_{n,j+1} + v_{n,j-1}\big) - \frac{1}{2}\left[g(u_{n,j+1}) - g(u_{n,j-1})\right]. \end{cases} \tag{4.11}$$

A *discrete shock profile* (DSP) with speed λ for (4.11) is a pair of functions

$$\big(U(x), V(x)\big) = \big(U^{(\lambda)}(x), V^{(\lambda)}(x)\big)$$

satisfying

$$\begin{cases} U(x-\lambda) &= \frac{U(x+1)+U(x-1)}{2} - \frac{f(U(x+1))-f(U(x-1))}{2}, \\ V(x-\lambda) &= \frac{V(x+1)+V(x-1)}{2} - \frac{g(U(x+1))-g(U(x-1))}{2}. \end{cases}$$

In [5], a discrete shock profile (U, V) is constructed, with a rational speed $\lambda \in]\frac{1}{4},\ 1[$, connecting the left and right states (u^-, v^-) and (u^+, v^+). In addition, a sequence of a discrete shock profiles $\big(U_k, V_k\big)$ is given, with rational speeds $\lambda + \epsilon_k$, connecting the left and right states (u^-, v^-) and (u_k^+, v_k^+). As $k \to \infty$, one has the convergence $\epsilon_k \to 0$ and $(u_k^+, v_k^+) \to (u^+, v^+)$. However, a detailed analysis show that all these discrete shock profiles have a uniformly positive amount of total variation, on intervals $I_k \doteq [-\epsilon_k^{-2},\ -\epsilon_k^{-2}/2]$. More precisely, there exists $\delta > 0$ such that

$$\text{Tot.Var.}\left\{V_k\,;\quad [-\epsilon_n^{-2},\ -\epsilon_n^{-2}/2]\right\} \;\geq\; \delta.$$

Notice that, as $k \to \infty$, the intervals I_k shift to the left, toward $-\infty$. This excludes the possibility that the sequence V_k has any limit in the BV norm. In particular, the sequence of discrete shock profiles (U_k, V_k) does not converge to the discrete shock profile (U, V).

Roughly speaking, these downstream oscillations are caused by resonances between the speed of the shock and the ratio $\Delta x/\Delta t = 1$ of the grid. They are noticeable when the speed is close (but not exactly equal) to a given rational number λ. It is expected that the same behavior will be observed for any other discrete scheme.

By further developing these ideas, in [6] an example is constructed, where the total variation of the solution computed by the Godunov scheme becomes arbitrarily large, for large times. The hyperbolic system has again the triangular form

$$\begin{cases} u_t + \big(\ln(1+e^u)\big)_x &= 0, \\ v_t + \frac{1}{2}v_x + g(u)_x &= 0. \end{cases} \tag{4.12}$$

The special choice of the flux function for the u-component is motivated by an observation of P. Lax [43]. If $z_{n,j} > 0$ provide a solution to the linear difference equation

$$z_{n+1,j} = \frac{z_{n,j} + z_{n,j-1}}{2}, \tag{4.13}$$

then the nonlinear transformation

$$u_{n,j} = \ln\left(\frac{z_{n,j-1}}{z_{n,j}}\right),$$

provides a solution to

$$u_{n+1,j} = u_{n,j} + \Big(\ln\big(1+e^{u_{n,j-1}}\big) - \ln\big(1+e^{u_{n,j}}\big)\Big).$$

This provides an analogue of the Hopf-Cole transformation for the viscous Burgers' equation, now valid for the discrete Godunov scheme. Using this formula, one can explicitly construct and analyze in detail the discrete approximations generated by the Godunov scheme. The main result proved in [6] can be stated as follows.

PROPOSITION 1. *Let $0 < u^+ < u^-$ be the right and left states of a shock for the scalar conservation law*

$$u_t + \big[\ln(1+e^u)\big]_x = 0\,,$$

having a rational speed λ, with $1/2 < \lambda < 1$. Then there exists a smooth flux function g and a sequence of smooth perturbations $\phi_\nu : I\!R \mapsto I\!R$ such that

$$Tot.Var.\{\phi_\nu\} \to 0\,, \qquad \|\phi_\nu\|_{\mathcal{C}^k} \to 0 \qquad as \quad \nu\to\infty$$

for all $k \geq 1$, and such that the following holds. The Godunov approximations $(u_\nu\,,\ v_\nu)$ to the Cauchy problem for (4.12) with initial data

$$u_\nu(0,x) = \begin{cases} u^- + \phi_\nu(x) & if \quad x<0\,,\\ \quad u^+ & if \quad x>0\,,\end{cases} \qquad v_\nu(0,x) = 0\,, \tag{4.14}$$

satisfy

$$Tot.Var.\big\{v_\nu(T_\nu,\cdot)\big\} \to \infty \qquad as \quad \nu\to\infty\,,$$

for some sequence of times $T_\nu \to \infty$.

We observe that, for the exact solutions having (4.14) as initial data, the total variation remains uniformly bounded for all times. However, the small perturbations ϕ_ν slightly change the speed of the shock in the u-component of the solution, thus producing a resonance with the grid. Over a large interval of time, this determines an arbitrarily large amount of downstream oscillations in the numerically computed solution.

Counterexamples of this type point to a fundamental limitation of rigorous theoretical analysis. For exact solutions of the general $n\times n$ Cauchy problem, the main theorems on existence, uniqueness and continuous dependence are based on a priori bounds on the total variation. However, any attempt to analyze finite difference schemes cannot rely on a priori BV bounds, simply because these are not valid, in general.

On the other hand, it seems unlikely that these resonances in the numerically computed solutions will be observed in practice. Indeed, these oscillations are mild, in the sense that they are spread out over a large number of grid points and should not prevent the convergence to the exact solution. Moreover, they apparently occur only for a small set of "prepared" initial data.

At present, positive results on the stability and convergence of numerical schemes for systems of conservation laws are known in two main cases:

1. For the 2×2 system modelling isentropic gas dynamics, convergence of a subsequence of finite difference approximations has been proved in [33], by the method of compensated compactness.

2. For Temple class systems, and more generally for $n \times n$ systems where all shock curves are straight lines, uniform BV bounds, stability and convergence of numerical approximations were proved in [45] and in [22], respectively.

Establishing the convergence of any finite difference scheme, for solutions with shocks of general $n \times n$ hyperbolic systems, remains an outstanding open problem.

REFERENCES

[1] D. Amadori and R.M. Colombo, *Continuous dependence for 2×2 conservation laws with boundary*, J. Differential Equations **138** (1997), 229–266.

[2] D. Amadori, L. Gosse, and G. Guerra, *Global BV entropy solutions and uniqueness for hyperbolic systems of balance laws*, Arch. Ration. Mech. Anal. **162** (2002), 327–366.

[3] P. Baiti and H.K. Jenssen, *On the front tracking algorithm*, J. Math. Anal. Appl. **217** (1998), 395–404.

[4] P. Baiti and H.K. Jenssen, *Blow-up in $\mathbf{L}^\infty$ for a class of genuinely nonlinear systems of conservation laws*, Discr. Cont. Dynam. Syst. **7** (2001), 837–853.

[5] P. Baiti, A. Bressan, and H.K. Jenssen, *Instability of travelling wave profiles for the Lax-Friedrichs scheme*, Discr. Cont. Dynam. Syst. **13** (2005), 877–899.

[6] P. Baiti, A. Bressan, and H.K. Jenssen, *BV instability of the Godunov scheme*, Comm. Pure Appl. Math. **59** (2006), 1604–1638.

[7] S. Bianchini, *A note on singular limits to hyperbolic systems of conservation laws*, Comm. Pure Appl. Analysis **2** (2003), 51–64.

[8] S. Bianchini, *A note on the Riemann problem for non-conservative hyperbolic systems*, Arch. Rational Mech. Anal. **166** (2003), 1–26.

[9] S. Bianchini, *BV solutions of the semidiscrete upwind scheme*, Arch. Rational Mech. Anal. **167** (2003), 1–81.

[10] S. Bianchini, *Hyperbolic limit of the Jin-Xin relaxation model*, Comm. Pure Appl. Math. **59** (2006), 688–753.

[11] S. Bianchini and A. Bressan, *Vanishing viscosity solutions of nonlinear hyperbolic systems*, Annals of Mathematics **161** (2005), 223–342.

[12] S. Bianchini and L. Spinolo, *The boundary Riemann solver coming from the real vanishing viscosity approximation*, *Arch. Ration. Mech. Anal.* **191** (2009), 1–96.

[13] S. Bianchini and L. Spinolo, *Invariant manifolds for a singular ordinary differential equation*, to appear in *J. Differential Equations.*

[14] A. Bressan, *Global solutions to systems of conservation laws by wave-front tracking*, *J. Math. Anal. Appl.* **170** (1992), 414–432.

[15] A. Bressan, *The unique limit of the Glimm scheme*, *Arch. Rational Mech. Anal.* **130** (1995), 205–230.

[16] A. Bressan, *Hyperbolic Systems of Conservation Laws, The One Dimensional Cauchy Problem*, Oxford University Press, 2000.

[17] A. Bressan, *BV solutions to systems of conservation laws by vanishing viscosity*, C.I.M.E. course in Cetraro, 2003, P.Marcati ed., *Springer Lecture Notes in Math.* **1911** (2007), 1–78.

[18] A. Bressan, *A tutorial on the Center Manifold Theorem, C.I.M.E. course in Cetraro*, 2003, P.Marcati ed., *Springer Lecture Notes in Math.* **1911** (2007), 327–344.

[19] A. Bressan and R.M. Colombo, *The semigroup generated by 2×2 conservation laws*, *Arch. Rational Mech. Anal.* **133** (1995), 1–75.

[20] A. Bressan, G. Crasta, and B. Piccoli, *Well posedness of the Cauchy problem for $n \times n$ conservation laws*, *Amer. Math. Soc. Memoir* **694** (2000).

[21] A. Bressan and P. Goatin, *Oleinik type estimates and uniqueness for $n \times n$ conservation laws*, *J. Diff. Equat.* **156** (1999), 26–49.

[22] A. Bressan and H.K. Jenssen, *On the convergence of Godunov scheme for nonlinear hyperbolic systems*, *Chinese Ann. Math.* **B-21** (2000), 1–16.

[23] A. Bressan and P. LeFloch, *Uniqueness of weak solutions to systems of conservation laws*, *Arch. Rat. Mech. Anal.* **140** (1997), 301–317.

[24] A. Bressan and M. Lewicka, *A uniqueness condition for hyperbolic systems of conservation laws*, *Discr. Cont. Dynam. Syst.* **6** (2000), 673–682.

[25] A. Bressan, T.P. Liu, and T. Yang, *L^1 stability estimates for $n \times n$ conservation laws*, *Arch. Rational Mech. Anal.* **149** (1999), 1–22.

[26] A. Bressan and W. Shen, *Uniqueness for discontinuous O.D.E. and conservation laws*, *Nonlinear Analysis, T.M.A.* **34** (1998), 637–652.

[27] A. Bressan and T. Yang, *On the convergence rate of vanishing viscosity approximations*, *Comm. Pure Appl. Math* **57** (2004), 1075–1109.

[28] G.Q. Chen and J. Glimm, *Global solutions to the compressible Euler equations with geometrical structure*, *Comm. Math. Phys.* **180** (1996), 153–193.

[29] C. Cheverry, *Systèmes de lois de conservation et stabilité BV*, *Mémoire Soc. Math. France* **75** (1998).

[30] M.G. Crandall, *The semigroup approach to first-order quasilinear equations in several space variables*, *Israel J. Math.* **12** (1972), 108–132.

[31] C. Dafermos, *Polygonal approximations of solutions of the initial value problem for a conservation law*, *J. Math. Anal. Appl.* **38** (1972), 33–41.

[32] C. Dafermos, *Hyperbolic Conservation Laws in Continuum Physics*, Dafermos, Constantine M. Hyperbolic conservation laws in continuum physics. Second edition. Springer-Verlag, Berlin, 2005.

[33] X. Ding, G.Q. Chen, and P. Luo, *Convergence of the fractional step Lax-Friedrichs scheme and Godunov scheme for the isentropic system of gas dynamics*, *Comm. Math. Phys.* **121** (1989), 63–84.

[34] R. DiPerna, *Global existence of solutions to nonlinear hyperbolic systems of conservation laws*, *J. Diff. Equat.* **20** (1976), 187–212.

[35] R. DiPerna, *Convergence of approximate solutions to conservation laws*, *Arch. Rational Mech. Anal.* **82** (1983), 27–70.

[36] C. Donadello and A. Marson, *Stability of front tracking solutions to the initial and boundary value problem for systems of conservation laws*, *Nonlin. Diff. Equat. Appl.* **14** (2007), 569–592.

[37] J. GLIMM, *Solutions in the large for nonlinear hyperbolic systems of equations*, *Comm. Pure Appl. Math.* **18** (1965), 697–715.

[38] J. GLIMM AND P. LAX, *Decay of solutions of systems of nonlinear hyperbolic conservation laws*, *Amer. Math. Soc. Memoir* **101** (1970).

[39] H. HOLDEN AND N. H. RISEBRO, *Front Tracking for Hyperbolic Conservation Laws*, Springer Verlag, New York 2002.

[40] H.K. JENSSEN, *Blowup for systems of conservation laws*, *SIAM J. Math. Anal.* **31** (2000), 894–908.

[41] H.K. JENSSEN AND R. YOUNG, *Gradient driven and singular flux blowup of smooth solutions to hyperbolic systems of conservation laws*, *J. Hyp. Diff. Eq.* **1** (2004), 627–641.

[42] S. JIN AND Z. XIN, *The relaxation schemes for systems of conservation laws in arbitrary space dimensions*, *Comm. Pure Appl. Math.* **48** (1955), 235–277.

[43] P. LAX, *Hyperbolic systems of conservation laws II*, *Comm. Pure Appl. Math.* **10** (1957), 537–566.

[44] P.G. LEFLOCH AND M. WESTDICKENBERG, *Finite energy solutions to the isentropic Euler equations with geometric effects*, *J. Math. Pures Appl.* **88** (2007), 389–429.

[45] R.J. LEVEQUE AND B. TEMPLE, *Stability of Godunov's method for a class of 2×2 systems of conservation laws*, *Trans. Amer. Math. Soc.* **288** (1985), 115–123.

[46] M. LEWICKA, *Well-posedness for hyperbolic systems of conservation laws with large BV data*, *Arch. Rational Mech. Anal.* **173** (2004), 415–445.

[47] L. LIN, *On the vacuum state for the equations of isentropic gas dynamics*, *J. Math. Anal. Appl.* **121** (1987), 406–425.

[48] L. LIN, *Survey of the vacuum theory and conjecture*, *AMS/IP Stud. Adv. Math.* **3** (1997), 293–305.

[49] T.P. LIU, *The Riemann problem for general systems of conservation laws*, *J. Differential Equations* **18** (1975), 218–234.

[50] T.P. LIU, *The entropy condition and the admissibility of shocks*, *J. Math. Anal. Appl.* **53** (1976), 78–88.

[51] T.P. LIU, *The deterministic version of the Glimm scheme*, *Comm. Math. Phys.* **57** (1977), 135–148.

[52] T. P. LIU, *Linear and nonlinear large-time behavior of solutions of general systems of hyperbolic conservation laws*, *Comm. Pure Appl. Math.* **30** (1977), 767–796.

[53] T.P. LIU, *Asymptotic behavior of solutions of general system of nonlinear hyperbolic conservation laws*, *Indiana Univ. Math. J.* **27** (1978), 211–253.

[54] T.P. LIU, *Admissible solutions of hyperbolic conservation laws*, *Amer. Math. Soc. Memoir* **240** (1981).

[55] T.P. LIU, *Hyperbolic conservation laws with relaxation*, *Comm. Math. Phys.* **108** (1987), 153–175.

[56] T.P. LIU AND J. SMOLLER, *On the vacuum state for the isentropic gas dynamics equations*, *Advances Pure Appl. Math.* **1** (1980), 345–359.

[57] T.P. LIU AND S.H. YU, *Continuum shock profiles for discrete conservation laws I. Construction*, *Comm. Pure Appl. Math.* **52** (1999), 85–127.

[58] A. MAJDA AND J. RALSTON, *Discrete shock profiles for systems of conservation laws*, *Comm. Pure Appl. Math.* **32** (1979), 445–482.

[59] S. SCHOCHET, *Sufficient conditions for local existence via Glimm's scheme for large BV data*, *J. Differential Equations* **89** (1991), 317–354.

[60] D. SERRE, *Remarks about the discrete profiles of shock waves*, *Mat. Contemp.* **11** (1996), 153–170.

[61] D. SERRE, *Systems of Conservation Laws I, II*, Cambridge University Press, 2000.

[62] D. SERRE, *Discrete shock profiles: existence and stability*, C.I.M.E. course in Cetraro, 2003, P. Marcati ed., *Springer Lecture Notes in Math.* **1911** (2007), 79–158.

[63] J. SMOLLER, *Shock Waves and Reaction-Diffusion Equations*, Springer-Verlag, New York, 1983.

[64] R. YOUNG, *Global wave interactions in isentropic gas dynamics, Preprint 2008.*
[65] R. YOUNG, *Composite wave interactions and the collapse of vacuums in gas dynamics*, Preprint 2009.

MULTIDIMENSIONAL CONSERVATION LAWS: OVERVIEW, PROBLEMS, AND PERSPECTIVE

GUI-QIANG G. CHEN*

Abstract. Some of recent important developments are overviewed, several longstanding open problems are discussed, and a perspective is presented for the mathematical theory of multidimensional conservation laws. Some basic features and phenomena of multidimensional hyperbolic conservation laws are revealed, and some samples of multidimensional systems/models and related important problems are presented and analyzed with emphasis on the prototypes that have been solved or may be expected to be solved rigorously at least for some cases. In particular, multidimensional steady supersonic problems and transonic problems, shock reflection-diffraction problems, and related effective nonlinear approaches are analyzed. A theory of divergence-measure vector fields and related analytical frameworks for the analysis of entropy solutions are discussed.

Key words. Multidimension (M-D), conservation laws, hyperbolicity, Euler equations, shock, rarefaction wave, vortex sheet, vorticity wave, entropy solution, singularity, uniqueness, reflection, diffraction, divergence-measure fields, systems, models, steady, supersonic, subsonic, self-similar, mixed hyperbolic-elliptic type, free boundary, iteration, partial hodograph, implicit function, weak convergence, numerical scheme, compensated compactness.

AMS(MOS) subject classifications. Primary: 35-02, 35L65, 35L67, 35L10, 35F50, 35M10, 35M30, 76H05, 76J20, 76L05, 76G25, 76N15, 76N10, 57R40, 53C42, 74B20, 26B12; Secondary: 35L80, 35L60, 57R42.

1. Introduction. We overview some of recent important developments, discuss several longstanding open problems, and present a perspective for the mathematical theory of multidimensional (M-D, for short) conservation laws.

Hyperbolic Conservation Laws, quasilinear hyperbolic systems in divergence form, are one of the most important classes of nonlinear partial differential equations (PDEs, for short). They typically take the form:

$$\partial_t \mathbf{u} + \nabla_{\mathbf{x}} \cdot \mathbf{f}(\mathbf{u}) = 0, \qquad \mathbf{u} \in \mathbb{R}^m, \tag{1.1}$$

for $(t, \mathbf{x}) \in \mathbb{R}^{d+1}_+ := \mathbb{R}_+ \times \mathbb{R}^d := [0, \infty) \times \mathbb{R}^d$, where $\nabla_{\mathbf{x}} = (\partial_{x_1}, \ldots, \partial_{x_d})$, and $\mathbf{f} = (\mathbf{f}_1, \ldots, \mathbf{f}_d) : \mathbb{R}^m \to (\mathbb{R}^m)^d$ is a nonlinear mapping with $\mathbf{f}_i : \mathbb{R}^m \to \mathbb{R}^m$ for $i = 1, \ldots, d$. Another prototypical form is

$$\partial_t A_0(u_t, \nabla_{\mathbf{x}} u) + \nabla_{\mathbf{x}} \cdot \mathbf{A}(u_t, \nabla_{\mathbf{x}} u) = 0, \tag{1.2}$$

*Oxford Centre for Nonlinear PDE, Mathematical Institute, University of Oxford, Oxford, OX1, 3LB, UK; and Department of Mathematics, Northwestern University, Evanston, IL 60208, USA. The work of Gui-Qiang G. Chen was supported in part by NSF grants DMS-0935967, DMS-0807551, the Royal Society–Wolfson Research Merit Award (UK), and the EPSRC Science and Innovation award to the Oxford Centre for Nonlinear PDE (EP/E035027/1).

where $A_0(p_0, \mathbf{p}) : \mathbb{R} \times \mathbb{R}^d \to \mathbb{R}$ and $\mathbf{A}(p_0, \mathbf{p}) : \mathbb{R} \times \mathbb{R}^d \to \mathbb{R}^d$ with $\mathbf{p} = (p_1, ..., p_d) \in \mathbb{R}^d$ are C^1 nonlinear mappings.

The hyperbolicity for (1.1) requires that, for any $\mathbf{n} \in \mathcal{S}^{d-1}$,

$$(\nabla_{\mathbf{w}}\mathbf{f}(\mathbf{w}) \cdot \mathbf{n})_{m\times m} \text{have } m \text{ real eigenvalues } \lambda_i(\mathbf{w}; \mathbf{n}),\ 1 \le i \le m. \quad (1.3)$$

We say that system (1.1) is hyperbolic in a state domain $\mathcal{D}$ if condition (1.3) holds for any $\mathbf{w} \in \mathcal{D}$ and $\mathbf{n} \in \mathcal{S}^{d-1}$.

The hyperbolicity for (1.2) requires that, rewriting (1.2) into the second-order nondivergence form such that the terms $u_{x_i x_j}$ have corresponding coefficients $a_{ij}, i, j = 0, 1, ..., d$, all the eigenvalues of the matrix $(a_{ij})_{d\times d}$ are real, and one eigenvalue has a different sign from the other d eigenvalues.

An archetype of nonlinear hyperbolic systems of conservation laws is the Euler equations for compressible fluids in $\mathbb{R}^d$, which are a system of $d + 2$ conservation laws:

$$\begin{cases} \partial_t \rho + \nabla_{\mathbf{x}} \cdot \mathbf{m} = 0, \\ \partial_t \mathbf{m} + \nabla_{\mathbf{x}} \cdot (\dfrac{\mathbf{m} \otimes \mathbf{m}}{\rho}) + \nabla_{\mathbf{x}} p = 0, \\ \partial_t(\rho E) + \nabla_{\mathbf{x}} \cdot \big(\mathbf{m}(E + p/\rho)\big) = 0. \end{cases} \quad (1.4)$$

System (1.4) is closed by the constitutive relations:

$$p = p(\rho, e), \qquad E = \frac{1}{2}\frac{|\mathbf{m}|^2}{\rho^2} + e. \quad (1.5)$$

In (1.4)–(1.5), $\tau = 1/\rho$ is the deformation gradient (the specific volume for fluids, the strain for solids), $\mathbf{m} = (m_1, \dots, m_d)^\top$ is the fluid momentum vector with $\mathbf{v} = \mathbf{m}/\rho$ the fluid velocity, p is the scalar pressure, and E is the total energy with e the internal energy, which is a given function of (τ, p) or (ρ, p) defined through thermodynamical relations. The notation $\mathbf{a} \otimes \mathbf{b}$ denotes the tensor product of the vectors $\mathbf{a}$ and $\mathbf{b}$. The other two thermodynamic variables are the temperature θ and the entropy S. If (ρ, S) are chosen as independent variables, then the constitutive relations can be written as $(e, p, \theta) = (e(\rho, S), p(\rho, S), \theta(\rho, S))$ governed by the Gibbs relation:

$$\theta dS = de + p d\tau = de - \frac{p}{\rho^2} d\rho. \quad (1.6)$$

For a polytropic gas,

$$p = R\rho\theta = \kappa\rho^\gamma e^{\frac{S}{c_v}}, \quad e = c_v\theta = \frac{\kappa}{\gamma - 1}\rho^{\gamma-1}e^{\frac{S}{c_v}}, \qquad \gamma = 1 + \frac{R}{c_v}, \quad (1.7)$$

where $R > 0$ may be taken as the universal gas constant divided by the effective molecular weight of the particular gas, $c_v > 0$ is the specific heat at constant volume, $\gamma > 1$ is the adiabatic exponent, and $\kappa > 0$ can be any positive constant by scaling. The sonic speed is $c := \sqrt{p_\rho(\rho, S)}$. For polytropic gas, $c = \sqrt{\gamma p/\rho}$.

As indicated in §2.4 below, no matter how smooth the initial function is for the Cauchy problem, the solution of (1.4) generically develops singularities in a finite time. Then system (1.4) is complemented by the Clausius-Duhem inequality:

$$\partial_t(\rho S) + \nabla_{\mathbf{x}} \cdot (\mathbf{m} S) \geq 0 \tag{1.8}$$

in the sense of distributions in order to single out physical discontinuous solutions, so-called *entropy solutions.*

When a flow is isentropic (i.e., the entropy S is a uniform constant S_0 in the flow), the Euler equations take the following simpler form:

$$\begin{cases} \partial_t \rho + \nabla_{\mathbf{x}} \cdot \mathbf{m} = 0, \\ \partial_t \mathbf{m} + \nabla_{\mathbf{x}} \cdot \big(\dfrac{\mathbf{m} \otimes \mathbf{m}}{\rho}\big) + \nabla_{\mathbf{x}} p = 0, \end{cases} \tag{1.9}$$

where the pressure is regarded as a function of density, $p = p(\rho, S_0)$, with constant S_0. For a polytropic gas,

$$p(\rho) = \kappa \rho^\gamma, \qquad \gamma > 1, \tag{1.10}$$

where $\kappa > 0$ can be any positive constant under scaling. When the entropy is initially a uniform constant, a smooth solution of (1.4) satisfies the equations in (1.9). Furthermore, it should be observed that the solutions of system (1.9) are also close to the solutions of system (1.4) even after shocks form, since the entropy increases across a shock to third-order in wave strength for solutions of (1.4), while in (1.9) the entropy S is constant. Moreover, system (1.9) is an excellent model for the isothermal fluid flow with $\gamma = 1$ and for the shallow water flow with $\gamma = 2$.

An important example for (1.2) is the potential flow equation:

$$\partial_t \rho(\Phi_t, \nabla_{\mathbf{x}}\Phi) + \nabla_{\mathbf{x}} \cdot \big(\rho(\Phi_t, \nabla_{\mathbf{x}}\Phi)\nabla_{\mathbf{x}}\Phi\big) = 0, \tag{1.11}$$

where $\rho(\Phi_t, \nabla_{\mathbf{x}}\Phi) = h^{-1}\big(K - (\partial_t\Phi + \frac{1}{2}|\nabla_{\mathbf{x}}\Phi|^2)\big)$, $h'(\rho) = p'(\rho)/\rho$, for the pressure function $p(\rho)$, which typically takes form (1.10) with $\gamma \geq 1$.

The importance of the potential flow equation (1.11) in the time-dependent Euler flows (1.4) with weak discontinuities was observed by Hadamard [123]. Also see Bers [16], Cole-Cook [81], Courant-Friedrichs [84], Majda-Thomas [159], and Morwawetz [165].

2. Basic features and phenomena. We first reveal some basic features and phenomena of M-D hyperbolic conservation laws with state domain $\mathcal{D}$, i.e., $\mathbf{u} \in \mathcal{D}$.

2.1. Convex entropy and symmetrization. A function $\eta : \mathcal{D} \to \mathbb{R}$ is called an entropy of system (1.1) if there exists a vector function $\mathbf{q} : \mathcal{D} \to \mathbb{R}^d, \mathbf{q} = (\mathbf{q}_1, \dots, \mathbf{q}_d)$, satisfying

$$\nabla \mathbf{q}_i(\mathbf{u}) = \nabla \eta(\mathbf{u}) \nabla \mathbf{f}_i(\mathbf{u}), \qquad i = 1, \dots, d. \tag{2.1}$$

Then $\mathbf{q}$ is called the corresponding entropy flux, and $(\eta, \mathbf{q})$ is simply called an entropy pair. An entropy $\eta(\mathbf{u})$ is called a convex entropy in $\mathcal{D}$ if $\nabla^2 \eta(\mathbf{u}) \geq 0$ for any $\mathbf{u} \in \mathcal{D}$ and a strictly convex entropy in $\mathcal{D}$ if $\nabla^2 \eta(\mathbf{u}) \geq c_0 I$ with a constant $c_0 > 0$ uniform for $\mathbf{u} \in \mathcal{D}_1$ for any $\mathcal{D}_1 \subset \bar{\mathcal{D}}_1 \Subset \mathcal{D}$, where I is the $m \times m$ identity matrix. Then the correspondence of the Clausius-Duhem inequality (1.8) in the context of hyperbolic conservation laws is the Lax entropy inequality:

$$\partial_t \eta(\mathbf{u}) + \nabla_{\mathbf{x}} \cdot \mathbf{q}(\mathbf{u}) \leq 0 \tag{2.2}$$

in the sense of distributions for any C^2 convex entropy pair $(\eta, \mathbf{q})$.

The following important observation is due to Friedrich-Lax [109], Godunov [117], and Boillat [21] (also see Ruggeri-Strumia [179]).

THEOREM 2.1. *A system in* (1.1) *endowed with a strictly convex entropy in $\mathcal{D}$ must be symmetrizable and hence hyperbolic in $\mathcal{D}$.*

This theorem is particularly useful for determining whether a large physical system is symmetrizable and hence hyperbolic, since most of physical systems from continuum physics are endowed with a strictly convex entropy. In particular, for system (1.4),

$$(\eta_*, \mathbf{q}_*) = (-\rho S, -\mathbf{m} S) \tag{2.3}$$

is a strictly convex entropy pair when $\rho > 0$ and $p > 0$; while, for system (1.9), the mechanical energy and energy flux pair:

$$(\eta_*, \mathbf{q}_*) = \big(\rho(\frac{1}{2}\frac{|\mathbf{m}|^2}{\rho^2} + e),\ \mathbf{m}(\frac{1}{2}\frac{|\mathbf{m}|^2}{\rho^2} + e(\rho) + \frac{p(\rho)}{\rho})\big) \tag{2.4}$$

is a strictly convex entropy pair when $\rho > 0$ for polytropic gases. For a hyperbolic system of conservation laws without a strictly convex entropy, it is possible to enlarge the system so that the enlarged system is endowed with a globally defined, strictly convex entropy. See Brenier [24], Dafermos [91], Demoulini-Stuart-Tzavaras [94], Qin [176], and Serre [185].

There are several direct, important applications of Theorem 2.1 based on the symmetry property of system (1.1) endowed with a strictly convex entropy. We list three of them below.

Local existence of classical solutions. Consider the Cauchy problem for a general hyperbolic system (1.1) with a strictly convex entropy:

$$\mathbf{u}|_{t=0} = \mathbf{u}_0. \tag{2.5}$$

THEOREM 2.2. *Assume that* $\mathbf{u}_0 : \mathbb{R}^d \to \mathcal{D}$ *is in* $H^s \cap L^\infty$ *with* $s > d/2+1$. *Then, for the Cauchy problem* (1.1) *and* (2.5), *there exists a finite time* $T = T(\|\mathbf{u}_0\|_s, \|\mathbf{u}_0\|_{L^\infty}) \in (0, \infty)$ *such that there exists a unique bounded classical solution* $\mathbf{u} \in C^1([0,T]\times\mathbb{R}^d)$ *with* $\mathbf{u}(t,\mathbf{x}) \in \mathcal{D}$ *for* $(t,\mathbf{x}) \in [0,T]\times\mathbb{R}^d$ *and* $\mathbf{u} \in C([0,T]; H^s) \cap C^1([0,T]; H^{s-1})$.

Kato [131] first formulated and applied a basic idea in the semigroup theory to yield the local existence of smooth solutions to (1.1). Majda in [157] provided a proof which relies solely on the elementary linear existence theory for symmetric hyperbolic systems with smooth coefficients via a classical iteration scheme (cf. [85]) by using the symmetry of system (1.1). Moreover, a sharp continuation principle was also provided: For $\mathbf{u}_0 \in H^s$ with $s > d/2+1$, the interval $[0,T)$ with $T < \infty$ is the maximal interval of the classical H^s existence for (1.1) if and only if either $\mathbf{u}(t,\mathbf{x})$ escapes every compact subset $K \Subset \mathcal{D}$ as $t \to T$, or $\|(\mathbf{u}_t, D\mathbf{u})(t,\cdot)\|_{L^\infty} \to \infty$ as $t \to T$. The first catastrophe is associated with a blow-up phenomenon such as focusing and concentration, and the second in this principle is associated with the formation of shocks and vorticity waves, among others, in the smooth solutions. Also see Makino-Ukai-Kawashima [160] and Chemin [37] for the local existence of classical solutions of the Cauchy problem for the M-D Euler equations.

2.1.2. Stability of Lipschitz solutions, rarefaction waves, and vacuum states in the class of entropy solutions in L^∞. Assume that system (1.1) is endowed with a strictly convex entropy on compact subsets of $\mathcal{D}$.

THEOREM 2.3 (Dafermos [90, 91]). *Suppose that* $\mathbf{w}$ *is a Lipschitz solution of* (1.1) *on* $[0,T)$, *taking values in a convex compact subset* K *of* $\mathcal{D}$, *with initial data* $\mathbf{w}_0$. *Let* $\mathbf{u}$ *be any entropy solution of* (1.1) *on* $[0,T)$, *taking values in* K, *with initial data* $\mathbf{u}_0$. *Then*

$$\int_{|\mathbf{x}|<R} |\mathbf{u}(t,\mathbf{x}) - \mathbf{w}(t,\mathbf{x})|^2 d\mathbf{x} \le C(T) \int_{|\mathbf{x}|<R+Lt} |\mathbf{u}_0(\mathbf{x}) - \mathbf{w}_0(\mathbf{x})|^2 d\mathbf{x}$$

holds for any $R > 0$ *and* $t \in [0,T)$, *with* $L > 0$ *depending solely on* K *and the Lipschitz constant of* $\mathbf{w}$.

The results can be extended to M-D hyperbolic systems of conservation laws with partially convex entropies and involutions; see Dafermos [91] (also see [22]). Some further ideas have been developed [42] to show the stability of planar rarefaction waves and vacuum states in the class of entropy solutions in L^∞ for the M-D Euler equations by using the theory of divergence-measure fields (see §7).

THEOREM 2.4 (Chen-Chen [42]). *Let* $\mathbf{n} \in \mathcal{S}^{d-1}$. *Let* $\mathbf{R}(t,\mathbf{x}) = (\hat\rho, \hat{\mathbf{m}})(\frac{\mathbf{x}\cdot\mathbf{n}}{t})$ *be a planar solution, consisting of planar rarefaction waves and possible vacuum states, of the Riemann problem:*

$$\mathbf{R}|_{t=0} = (\rho_\pm, \mathbf{m}_\pm), \qquad \pm\mathbf{x}\cdot\mathbf{n} \ge 0,$$

with constant states $(\rho_\pm, \mathbf{m}_\pm)$. Suppose that $\mathbf{u}(t,\mathbf{x}) = (\rho, \mathbf{m})(t,\mathbf{x})$ is an entropy solution in L^∞ of (1.9) *that may contain vacuum. Then, for any $R > 0$ and $t \in [0,\infty)$,*

$$\int_{|\mathbf{x}|<R} \alpha(\mathbf{u},\mathbf{R})(t,\mathbf{x})\, d\mathbf{x} \leq \int_{|\mathbf{x}|<R+Lt} \alpha(\mathbf{u},\mathbf{R})(0,\mathbf{x})\, d\mathbf{x},$$

where $L > 0$ depends solely on the bounds of the solutions $\mathbf{u}$ and $\mathbf{R}$, and $\alpha(\mathbf{u},\mathbf{R}) = (\mathbf{u}-\mathbf{R})^\top(\int_0^1 \nabla^2\eta_*(\mathbf{R}+\tau(\mathbf{u}-\mathbf{R}))d\tau)(\mathbf{u}-\mathbf{R})$ with the mechanical energy $\eta_*(\mathbf{u})$ in (2.4).

A similar theorem to Theorem 2.4 was also established for the adiabatic Euler equations (1.4) with appropriate chosen entropy function in Chen-Chen [42]; also cf. Chen [41] and Chen-Frid-Li [55]. Also see Perthame [174] for the time-decay of the internal energy and the density for entropy solutions to (1.4) for polytropic gases (also cf. [37]).

2.1.3. Local existence of shock-front solutions. Shock-front solutions, the simplest type of discontinuous solutions, are the most important discontinuous nonlinear progressing wave solutions of conservation laws (1.1). Shock-front solutions are discontinuous piecewise smooth entropy solutions with the following structure:

(i) There exist a C^2 space-time hypersurface $\mathcal{S}(t)$ defined in $(t,\mathbf{x})$ for $0 \leq t \leq T$ with space-time normal $(\mathbf{n}_t, \mathbf{n}_\mathbf{x}) = (\mathbf{n}_t, \mathbf{n}_1, \ldots, \mathbf{n}_d)$ and two C^1 vector-valued functions: $\mathbf{u}^\pm(t,\mathbf{x})$, defined on respective domains $\mathcal{D}^\pm$ on either side of the hypersurface $\mathcal{S}(t)$, and satisfying

$$\partial_t \mathbf{u}^\pm + \nabla\cdot\mathbf{f}(\mathbf{u}^\pm) = 0 \qquad \text{in } \mathcal{D}^\pm. \tag{2.6}$$

(ii) The jump across $\mathcal{S}(t)$ satisfies the Rankine-Hugoniot condition:

$$\big(\mathbf{n}_t(\mathbf{u}^+ - \mathbf{u}^-) + \mathbf{n}_\mathbf{x}\cdot(\mathbf{f}(\mathbf{u}^+) - \mathbf{f}(\mathbf{u}^-))\big)\big|_{\mathcal{S}} = 0. \tag{2.7}$$

Since (1.1) is nonlinear, the surface $\mathcal{S}$ is not known in advance and must be determined as a part of the solution of the problem; thus the equations in (2.6)–(2.7) describe a M-D free-boundary value problem for (1.1).

The initial functions yielding shock-front solutions are defined as follows. Let $\mathcal{S}_0$ be a smooth hypersurface parameterized by α, and let $\mathbf{n}(\alpha) = (\mathbf{n}_1, \ldots, \mathbf{n}_d)(\alpha)$ be a unit normal to $\mathcal{S}_0$. Define the piecewise smooth initial functions $\mathbf{u}_0^\pm(\mathbf{x})$ for respective domains $\mathcal{D}_0^\pm$ on either side of the hypersurface $\mathcal{S}_0$. It is assumed that the initial jump satisfies the Rankine-Hugoniot condition, i.e., there is a smooth scalar function $\sigma(\alpha)$ so that

$$-\sigma(\alpha)\big(\mathbf{u}_0^+(\alpha) - \mathbf{u}_0^-(\alpha)\big) + \mathbf{n}(\alpha)\cdot\big(\mathbf{f}(\mathbf{u}_0^+(\alpha)) - \mathbf{f}(\mathbf{u}_0^-(\alpha))\big) = 0 \tag{2.8}$$

and that $\sigma(\alpha)$ does not define a characteristic direction, i.e.,

$$\sigma(\alpha) \neq \lambda_i(\mathbf{u}_0^\pm), \qquad \alpha \in \overline{\mathcal{S}_0}, \quad 1 \leq i \leq m, \tag{2.9}$$

where λ_i, $i = 1, \ldots, m$, are the eigenvalues of (1.1). It is natural to require that $\mathcal{S}(0) = \mathcal{S}_0$.

Consider the 3-D full Euler equations in (1.4) for $\mathbf{u} = (\rho, \mathbf{m}, \rho E)$, away from vacuum, with piecewise smooth initial data:

$$\mathbf{u}|_{t=0} = \mathbf{u}_0^{\pm}(\mathbf{x}), \qquad \mathbf{x} \in \mathcal{D}_0^{\pm}. \tag{2.10}$$

THEOREM 2.5 (Majda [156]). *Assume that $\mathcal{S}_0$ is a smooth hypersurface in $\mathbb{R}^3$ and that $\mathbf{u}_0^+(\mathbf{x})$ belongs to the uniform local Sobolev space $H^s_{ul}(\mathcal{D}_0^+)$, while $\mathbf{u}_0^-(\mathbf{x})$ belongs to the Sobolev space $H^s(\mathcal{D}_0^-)$, for some fixed $s \geq 10$. Assume also that there is a function $\sigma(\alpha) \in H^s(\mathcal{S}_0)$ so that (2.8)–(2.9) hold, and the compatibility conditions up to order $s-1$ are satisfied on $\mathcal{S}_0$ by the initial data, together with the entropy condition:*

$$\frac{\mathbf{m}_0^+}{\rho} \cdot \mathbf{n}(\alpha) + c(\rho_0^+, S_0^+) < \sigma(\alpha) < \frac{\mathbf{m}_0^-}{\rho} \cdot \mathbf{n}(\alpha) + c(\rho_0^-, S_0^-), \tag{2.11}$$

and the Majda stability condition:

$$\Big(\frac{p_S(\rho^+, S^+)}{\rho^+\theta^+} + 1\Big)\frac{p(\rho^+, S^+) - p(\rho^-, S^-)}{\rho^+ c^2(\rho^+, S^+)} < 1. \tag{2.12}$$

Then there exists $T > 0$ and a C^2-hypersurface $\mathcal{S}(t)$ together with C^1–functions $\mathbf{u}^{\pm}(t, \mathbf{x})$ defined for $t \in [0, T]$ so that $\mathbf{u}(t, \mathbf{x}) = \mathbf{u}^{\pm}(t, \mathbf{x}), (t, \mathbf{x}) \in \mathcal{D}^{\pm}$, is the discontinuous shock-front solution of the Cauchy problem (1.4) and (2.10) satisfying (2.6)–(2.7).

Condition in (2.12) is always satisfied for shocks of any strength for polytropic gas with $\gamma > 1$ and for sufficiently weak shocks for general equation of state. In Theorem 2.5, $c = c(\rho, S)$ is the sonic speed. The Sobolev space $H^s_{ul}(\mathcal{D}_0^+)$ is defined as follows: A vector function $\mathbf{u}$ is in H^s_{ul}, provided that there exists some $r > 0$ such that

$$\max_{\mathbf{y} \in \mathbb{R}^d} \|\omega_{r,\mathbf{y}} \mathbf{u}\|_{H^s} < \infty \qquad \text{with } \omega_{r,\mathbf{y}}(\mathbf{x}) = \omega(\frac{\mathbf{x} - \mathbf{y}}{r}),$$

where $\omega \in C_0^\infty(\mathbb{R}^d)$ is a function so that $\omega(\mathbf{x}) \geq 0$, $\omega(\mathbf{x}) = 1$ when $|\mathbf{x}| \leq \frac{1}{2}$, and $\omega(\mathbf{x}) = 0$ when $|\mathbf{x}| > 1$.

The compatibility conditions in Theorem 2.5 are defined in [156] and needed in order to avoid the formation of discontinuities in higher derivatives along other characteristic surfaces emanating from $\mathcal{S}_0$: Once the main condition in (2.8) is satisfied, the compatibility conditions are automatically guaranteed for a wide class of initial functions. Further studies on the local existence and stability of shock-front solutions can be found in Majda [156, 157]. The existence of shock-front solutions whose lifespan is uniform with respect to the shock strength was obtained in Métivier [162]. Also see Blokhin-Trokhinin [20] for further discussions.

The idea of the proof is similar to that for Theorem 2.2 by using the existence of a strictly convex entropy and the symmetrization of (1.1), but the technical details are quite different due to the unusual features of the problem in Theorem 2.5. For more details, see [156].

The Navier-Stokes regularization of M-D shocks for the Euler equations has been established in Guès-Métiver-Williams-Zumbrun [120]. The local existence of rarefaction wave front-solutions for M-D hyperbolic systems of conservation laws has also been established in Alinhac [1]. Also see Benzoni-Gavage–Serre [14].

2.2. Hyperbolicity. There are many examples of $m \times m$ hyperbolic systems of conservation laws for $d = 2$ which are strictly hyperbolic; that is, they have simple characteristics. However, for $d = 3$, the situation is different. The following result for the case $m \equiv 2 (mod\, 4)$ is due to Lax [138] and for the more general case $m \equiv \pm 2, \pm 3, \pm 4 (mod\, 8)$ due to Friedland-Robin-Sylvester [107].

THEOREM 2.6. *Let* $\mathbf{A}, \mathbf{B}$, *and* $\mathbf{C}$ *be three* $m \times m$ *matrices such that* $\alpha\mathbf{A} + \beta\mathbf{B} + \gamma\mathbf{C}$ *has real eigenvalues for any real* α, β, *and* γ. *If* $m \equiv \pm 2, \pm 3, \pm 4\, (mod\, 8)$, *then there exist* α_0, β_0, *and* γ_0 *with* $\alpha_0^2 + \beta_0^2 + \gamma_0^2 \neq 0$ *such that* $\alpha_0\mathbf{A} + \beta_0\mathbf{B} + \gamma_0\mathbf{C}$ *is degenerate, that is, there are at least two eigenvalues which coincide.*

This implies that, for $d = 3$, there are no strictly hyperbolic systems if $m \equiv \pm 2, \pm 3, \pm 4\, (mod\, 8)$. Such multiple characteristics influence the propagation of singularities.

Consider the *isentropic Euler equations* (1.9). When $d = 2$ and $m = 3$, the system is strictly hyperbolic with three real eigenvalues $\lambda_- < \lambda_0 < \lambda_+$:

$$\lambda_0 = n_1 v_1 + n_2 v_2, \qquad \lambda_\pm = n_1 v_1 + n_2 v_2 \pm c(\rho), \qquad \rho > 0.$$

Strict hyperbolicity fails at the vacuum states $\rho = 0$. However, when $d = 3$ and $m = 4$, the system is no longer strictly hyperbolic even when $\rho > 0$ since the eigenvalue $\lambda_0 = n_1 v_1 + n_2 v_2 + n_3 v_3$ has double multiplicity, although the other eigenvalues $\lambda_\pm = n_1 v_1 + n_2 v_2 + n_3 v_3 \pm c(\rho)$ are simple when $\rho > 0$.

Consider the *adiabatic Euler equations* (1.4). When $d = 2$ and $m = 4$, the system is nonstrictly hyperbolic, since the eigenvalue $\lambda_0 = n_1 v_1 + n_2 v_2$ has double multiplicity, although $\lambda_\pm = n_1 v_1 + n_2 v_2 \pm c(\rho, S)$ are simple when $\rho > 0$. When $d = 3$ and $m = 5$, the system is again nonstrictly hyperbolic, since the eigenvalue $\lambda_0 = n_1 v_1 + n_2 v_2 + n_3 v_3$ has triple multiplicity, although $\lambda_\pm = n_1 v_1 + n_2 v_2 + n_3 v_3 \pm c(\rho, S)$ are simple when $\rho > 0$.

2.3. Genuine nonlinearity. The jth-characteristic field of system (1.1) in $\mathcal{D}$ is called *genuinely nonlinear* if, for each fixed $\mathbf{n} \in S^{d-1}$, the jth-eigenvalue $\lambda_j(\mathbf{u}; \mathbf{n})$ and the corresponding eigenvector $\mathbf{r}_j(\mathbf{u}; \mathbf{n})$ determined by $(\nabla \mathbf{f}(\mathbf{u}) \cdot \mathbf{n})\mathbf{r}_j(\mathbf{u}; \mathbf{n}) = \lambda_j(\mathbf{u}; \mathbf{n})\mathbf{r}_j(\mathbf{u}; \mathbf{n})$ satisfy

$$\nabla_{\mathbf{u}} \lambda_j(\mathbf{u}; \mathbf{n}) \cdot \mathbf{r}_j(\mathbf{u}; \mathbf{n}) \neq 0 \qquad \text{for any } \mathbf{u} \in \mathcal{D},\ \mathbf{n} \in \mathcal{S}^{d-1}. \tag{2.13}$$

The jth-characteristic field of (1.1) in $\mathcal{D}$ is called linearly degenerate if

$$\nabla_{\mathbf{u}}\lambda_j(\mathbf{u};\mathbf{n})\cdot\mathbf{r}_j(\mathbf{u};\mathbf{n})\equiv 0 \qquad \text{for any } \mathbf{u}\in\mathcal{D}. \tag{2.14}$$

Then any *scalar* quasilinear conservation law in $\mathbb{R}^d, d\geq 2$, is never genuinely nonlinear in all directions. This is because, in this case, $\lambda(u;\mathbf{n}) = \mathbf{f}''(u)\cdot\mathbf{n}$ and $r(u;\mathbf{n})=1$, and $\lambda'(u;\mathbf{n})r(u;\mathbf{n})=\mathbf{f}''(u)\cdot\mathbf{n}$, which is impossible to make this never equal to zero in all directions. A M-D version of genuine nonlinearity for scalar conservation laws is that the set

$\{u\,:\,\tau+\mathbf{f}'(u)\cdot\mathbf{n}=0\}$ *has zero Lebesgue measure for any* $(\tau,\mathbf{n})\in\mathcal{S}^d$,

which is a generalization of (2.13). Under this generalized nonlinearity, the following have been established: (i) Solution operators are compact as $t>0$ in Lions-Perthame-Tadmor [152] (also see [53, 193]); (ii) Decay of periodic solutions (Chen-Frid [52]; also see Engquist-E [102]); (iii) Strong initial and boundary traces of entropy solutions (Chen-Rascle [60], Vasseur [202]; also see Panov [173]); (iv) BV-like structure of L^∞ entropy solutions (De Lellis-Otto-Westdickenberg [92]). Furthermore, we have

THEOREM 2.7 (Lax [138]). *Every real, strictly hyperbolic quasilinear system for $m=2k$, $k\geq 1$ odd, and $d=2$ is linearly degenerate in some direction.*

Quite often, linear degeneracy results from the loss of strict hyperbolicity. For example, even in the 1-D case, if there exists $i\neq j$ such that $\lambda_i(\mathbf{u})=\lambda_j(\mathbf{u})$ for all $\mathbf{u}\in K$, then Boillat [22] proved that the ith- and j^{th}-characteristic fields are linearly degenerate.

For the *isentropic Euler equations* (1.9) with $d=2$ and $m=3$ for polytropic gases with the eigenvalues:

$$\lambda_0 = n_1v_1+n_2v_2, \qquad \lambda_\pm = n_1v_1+n_2v_2\pm c(\rho),$$

and the corresponding eigenvectors $\mathbf{r}_0$ and $\mathbf{r}_\pm$, we have $\nabla\lambda_0\cdot\mathbf{r}_0\equiv 0$, which is linearly degenerate, and $\nabla\lambda_\pm\cdot\mathbf{r}_\pm\neq 0$ for $\rho\in(0,\infty)$, which are genuinely nonlinear. For the *adiabatic Euler equations* (1.4) with $d=2$ and $m=4$ for polytropic gases with the eigenvalues:

$$\lambda_0 = n_1v_1+n_2v_2, \qquad \lambda_\pm = n_1v_1+n_2v_2\pm c(\rho,S),$$

and the corresponding eigenvectors $\mathbf{r}_0$ and $\mathbf{r}_\pm$, we have $\nabla\lambda_0\cdot\mathbf{r}_0\equiv 0$, which is linearly degenerate, and $\nabla\lambda_\pm\cdot\mathbf{r}_\pm\neq 0$ for $\rho, S\in(0,\infty)$, which are genuinely nonlinear.

2.4. Singularities. For the 1-D case, singularities include the formation of shocks, contact discontinuities, and the development of vacuum states, at least for gas dynamics. For the M-D case, the situation is much more complicated: besides shocks, contact discontinuities, and vacuum states, singularities may include vorticity waves, focusing waves, concentration waves, complicated wave interactions, among others. Consider the

Cauchy problem of the Euler equations in (1.4) in $\mathbb{R}^3$ for polytropic gases with smooth initial data:

$$\mathbf{u}|_{t=0} = \mathbf{u}_0(\mathbf{x}) \qquad \text{with } \rho_0(\mathbf{x}) > 0, \tag{2.15}$$

satisfying $(\rho_0, \mathbf{m}_0, S_0)(\mathbf{x}) = (\bar{\rho}, 0, \bar{S})$ for $|\mathbf{x}| \geq L$, where $\bar{\rho} > 0$, $\bar{S}$, and L are constants. The equations in (1.4) possess a unique local C^1-solution $\mathbf{u}(t, \mathbf{x})$ with $\rho(t, \mathbf{x}) > 0$ provided that the initial function (2.15) is sufficiently regular (Theorem 2.2). The support of the smooth disturbance $(\rho_0(\mathbf{x}) - \bar{\rho}, \mathbf{m}_0(\mathbf{x}), S_0(\mathbf{x}) - \bar{S})$ propagates with speed at most $\bar{c} = \sqrt{p_\rho(\bar{\rho}, \bar{S})}$ (the sound speed) that is, $(\rho, \mathbf{m}, S)(t, \mathbf{x}) = (\bar{\rho}, 0, \bar{S})$ if $|\mathbf{x}| \geq L + \bar{c}t$. Take $\bar{p} = p(\bar{\rho}, \bar{S})$. Define

$$P(t) = \int_{\mathbb{R}^3} \big(p(t,\mathbf{x})^{1/\gamma} - \bar{p}^{1/\gamma}\big) d\mathbf{x}, \qquad F(t) = \int_{\mathbb{R}^3} \mathbf{m}(t, \mathbf{x}) \cdot \mathbf{x}\, d\mathbf{x},$$

which measure the entropy and the radial component of momentum.

THEOREM 2.8 (Sideris [190]). *Suppose that $(\rho, \mathbf{m}, S)(t, \mathbf{x})$ is a C^1-solution of (1.4) and (2.15) for $0 < t < T$ and*

$$P(0) \geq 0, \qquad F(0) > \frac{16}{3}\pi \bar{c} L^4 \max_{\mathbf{x}} \big(\rho_0(\mathbf{x})\big). \tag{2.16}$$

Then the lifespan T of the C^1-solution is finite.

In particular, when $\rho_0 = \bar{\rho}$ and $S_0 = \bar{S}$, $P(0) = 0$ and the second condition in (2.16) holds if the initial velocity $\mathbf{v}_0(\mathbf{x}) = \frac{\mathbf{m}_0(\mathbf{x})}{\rho_0(\mathbf{x})}$ satisfies $\int_{|\mathbf{x}|<L} \mathbf{v}_0(x) \cdot \mathbf{x}\, d\mathbf{x} > \bar{c}\sigma L^4$. From this, one finds that the initial velocity must be supersonic in some region relative to $\bar{c}$. The formation of a singularity (presumably a shock) is detected as the disturbance overtakes the wave-front forcing the front to propagate with supersonic speed. The formation of singularities occurs even without condition of largeness such as (2.16). For more details, see [190].

In Christodoulou [79], the relativistic Euler equations for a perfect fluid with an arbitrary equation of state have been analyzed. The initial function is imposed on a given spacelike hyperplane and is constant outside a compact set. Attention is restricted to the evolution of the solution within a region limited by two concentric spheres. Given a smooth solution, the geometry of the boundary of its domain of definition is studied, that is, the locus where shocks may form. Furthermore, under certain smallness assumptions on the size of the initial data, a remarkable and complete picture of the formation of shocks in $\mathbb{R}^3$ has been obtained. In addition, sharp sufficient conditions on the initial data for the formation of shocks in the evolution have been identified, and sharp lower and upper bounds for the time of existence of a smooth solution have been derived. Also see Christodoulou [80] for the formation of black holes in general relativity.

2.5. BV bound. For 1-D strictly hyperbolic systems, Glimm's theorem [113] indicates that, as long as $\|\mathbf{u}_0\|_{BV}$ is sufficiently small, the solution $\mathbf{u}(t,x)$ satisfies the following stability estimate:

$$\|\mathbf{u}(t,\cdot)\|_{BV} \le C\,\|\mathbf{u}_0\|_{BV}. \tag{2.17}$$

Even more strongly, for two solutions $\mathbf{u}(t,x)$ and $\mathbf{w}(t,x)$ obtained by either the Glimm scheme, wave-front tracking method, or vanishing viscosity method with small total variation,

$$\|\mathbf{u}(t,\cdot)-\mathbf{w}(t,\cdot)\|_{L^1(\mathbb{R})} \le C\,\|\mathbf{w}(0,\cdot)-\mathbf{v}(0,\cdot)\|_{L^1(\mathbb{R})}.$$

See Bianchini-Bressan [18] and Bressan [27]; also see Dafermos [91], Holden-Risebro [126], LeFloch [141], Liu-Yang [154], and the references cited therein.

The recent great progress for entropy solutions to 1-D hyperbolic conservation laws based on BV estimates and trace theorems of BV fields naturally arises the expectation that a similar approach may also be effective for M-D hyperbolic systems of conservation laws, that is, whether entropy solutions satisfy the relatively modest stability estimate (2.17). Unfortunately, this is not the case. Rauch [177] showed that the necessary condition for (2.17) to be held is

$$\nabla\mathbf{f}_k(\mathbf{u})\nabla\mathbf{f}_l(\mathbf{u}) = \nabla\mathbf{f}_l(\mathbf{u})\nabla\mathbf{f}_k(\mathbf{u}) \qquad \text{for all } k,l=1,2,\dots,d. \tag{2.18}$$

The analysis above suggests that only systems in which the commutativity relation (2.18) holds offer some hope for treatment in the BV framework. This special case includes the scalar case $m=1$ and the 1-D case $d=1$. Beyond that, it contains very few systems of physical interest. An example is the system with fluxes: $\mathbf{f}_k(\mathbf{u})=\phi_k(|\mathbf{u}|^2)\mathbf{u}$, $k=1,2,\dots,d$, which governs the flow of a fluid in an anisotropic porous medium. However, the recent study in Bressan [28] and Ambrosio-DeLellis [4] shows that, even in this case, the space BV is not a well-posed space (also cf. Jenssen [130]). Moreover, entropy solutions generally do not have even the relatively modest stability: $\|\mathbf{u}(t,\cdot)-\bar{\mathbf{u}}\|_{L^p} \le C_p\|\mathbf{u}_0-\bar{\mathbf{u}}_0\|_{L^p}$, $p\neq 2$, based on the linear theory by Brenner [26].

In this regard, it is important to identify good analytical frameworks for the analysis of entropy solutions of M-D conservation laws (1.1), which are not in BV, or even in L^p. A general framework is the space of divergence-measure fields, formulated recently in [54, 64, 66], which is based on entropy solutions satisfying the Lax entropy inequality (2.2). See §7 for more details.

Another important notion of solutions is the notion of the measure-valued entropy solutions, based on the Young measure representation of a weak convergent sequence (cf. [195, 12]), proposed by DiPerna [97]. An effort has been made to establish the existence of measure-valued solutions to the M-D Euler equations for compressible fluids; see Ganbo-Westdickenberg [111].

2.6. Nonuniqueness. Another important feature is the nonuniqueness of entropy solutions in general. In particular, De Lellis-Székelyhidi [93] recently showed the following remarkable fact.

THEOREM 2.9. *Let $d \geq 2$. Then, for any given function $p = p(\rho)$ with $p'(\rho) > 0$ when $\rho > 0$, there exist bounded initial functions $(\rho_0, \mathbf{v}_0) = (\rho_0, \frac{\mathbf{m}_0}{\rho_0})$ with $\rho_0(\mathbf{x}) \geq \delta_0 > 0$ for which there exist infinitely many bounded solutions $(\rho, \mathbf{m})$ of (1.9) with $\rho \geq \delta$ for some $\delta > 0$, satisfying the energy identity in the sense of distributions:*

$$\partial_t(\rho E) + \nabla_{\mathbf{x}} \cdot \big(\mathbf{m}(E + p/\rho)\big) = 0. \tag{2.19}$$

In fact, the same result also holds for the full Euler system, since the solutions constructed satisfy the energy equality so that there is no energy production at all. The main point for the result is that the solutions constructed contain only vortex sheets and vorticity waves which keep the energy identity (2.19) even for weak solutions in the sense of distributions, while the vortex sheets do not appear in the 1-D case. The construction of the infinitely many solutions is based on a variant of the Baire category method for differential inclusions. Therefore, for the uniqueness issue, we have to narrow down further the class of entropy solutions to single out physical relevant solutions, at least for the Euler equations.

3. Multidimensional systems and models. M-D problems are extremely rich and complicated. Some great developments have been made in the recent decades through strong and close interdisciplinary interactions and diverse approaches, including

(i) Experimental data;
(ii) Large and small scale computing by a search for effective numerical methods;
(iii) Asymptotic and qualitative modeling;
(iv) Rigorous proofs for prototype problems and an understanding of the solutions.

In some sense, the developments made by using approach (iv) are still behind those by using the other approaches (i)–(iii); however, most scientific problems are considered to be solved satisfactorily only after approach (iv) is achieved. In this section, together with Sections 4–7, we present some samples of M-D systems/models and related important problems with emphasis on those prototypes that have been solved or may be expected to be solved rigorously at least for some cases.

Since the M-D problems are so complicated in general, a natural strategy to attack these problems as a first step is to study (i) simpler nonlinear systems with strong physical motivations and (ii) special, concrete nonlinear physical problems. Meanwhile, extend the results and ideas from the first step to (i) study the Euler equations in gas dynamics and elasticity; (ii) study more general problems; (iii) study nonlinear systems that the Euler equations are the main subsystem or describe the dynamics of macroscopic variables such as Navier-Stokes equations, MHD equations, combustion equations, Euler-Poisson equations, kinetic equations especially including the Boltzmann equation, among others.

Now we first focus on some samples of M-D systems and models.

3.1. Unsteady transonic small disturbance equation. A simple model in transonic aerodynamics is the UTSD equation or so-called the 2-D inviscid Burgers equation (see Cole-Cook [81]):

$$\begin{cases} \partial_t u + \partial_x\big(\frac{1}{2}u^2\big) + \partial_y v = 0, \\ \partial_y u - \partial_x v = 0, \end{cases} \tag{3.1}$$

or in the form of Zabolotskaya-Khokhlov equation [208]:

$$\partial_t(\partial_t u + u\partial_x u) + \partial_{yy} u = 0. \tag{3.2}$$

The equations in (3.1) describe the potential flow field near the reflection point in weak shock reflection, which determines the leading order approximation of geometric optical expansions; and it can also be used to formulate asymptotic equations for the transition from regular to Mach reflection for weak shocks. See Morawetz [165], Hunter [128], and the references cited therein. Equation (3.2) also arises in many different situations; see [128, 199, 208].

3.2. Pressure-gradient equations and nonlinear wave equations. The inviscid fluid motions are driven mainly by the pressure gradient and the fluid convection (i.e., transport). As for modeling, it is natural to study first the effect of the two driving factors separately.

Separating the pressure gradient from the Euler equations (1.4), we first have the pressure-gradient system:

$$\partial_t\rho = 0, \quad \partial_t(\rho\mathbf{v}) + \nabla_\mathbf{x} p = 0, \quad \partial_t(\rho e) + p\nabla_\mathbf{x}\cdot\mathbf{v} = 0. \tag{3.3}$$

We may choose $\rho = 1$. Setting $p = (\gamma-1)P$ and $t = \frac{s}{\gamma-1}$, and eliminating the velocity $\mathbf{v}$, we obtain the following nonlinear wave equation for P:

$$\partial_{ss}(\ln P) - \Delta_\mathbf{x} P = 0. \tag{3.4}$$

Although system (3.3) is obtained from the splitting idea, it is a good approximation to the full Euler equations, especially when the velocity $\mathbf{v}$ is small and the adiabatic exponent $\gamma > 1$ is large. See Zheng [215].

Another related model is the following nonlinear wave equation proposed by Canic-Keyfitz-Kim [32]:

$$\partial_{tt}\rho - \Delta_\mathbf{x} p(\rho) = 0, \tag{3.5}$$

where $p = p(\rho)$ is the pressure-density relation for fluids. Equation (3.5) is obtained from (1.9) by neglecting the inertial terms, i.e., the quadratic terms in $\mathbf{m}$. This yields the following system:

$$\partial_t \rho + \nabla_{\mathbf{x}} \cdot \mathbf{m} = 0, \quad \partial_t \mathbf{m} + \nabla_{\mathbf{x}} p(\rho) = 0, \tag{3.6}$$

which leads to (3.5) by eliminating $\mathbf{m}$ in the system.

3.3. Pressureless Euler equations. With the pressure-gradient equations (3.3), the convection (i.e., transport) part of fluid flow forms the pressureless Euler equations:

$$\begin{cases} \partial_t \rho + \nabla_{\mathbf{x}} \cdot (\rho \mathbf{v}) = 0, \\ \partial_t (\rho \mathbf{v}) + \nabla_{\mathbf{x}} \cdot (\rho \mathbf{v} \otimes \mathbf{v}) = 0, \\ \partial_t (\rho E) + \nabla_{\mathbf{x}} \cdot (\rho E \mathbf{v}) = 0. \end{cases} \tag{3.7}$$

This system also models the motion of free particles which stick under collision; see Brenier-Grenier [25], E-Rykov-Sinai [99], and Zeldovich [210]. In general, solutions of (3.7) become measure solutions.

System (3.7) has been analyzed extensively; cf. [23, 25, 59, 99, 118, 127, 142, 188, 204] and the references cited therein. In particular, the existence of measure solutions of the Riemann problem was first presented in Bouchut [23] for the 1-D case. It has been shown that δ-shocks and vacuum states do occur in the Riemann solutions even in the 1-D case. Since the two eigenvalues of the transport equations coincide, the occurrence as $t > 0$ can be regarded as a result of resonance between the two characteristic fields. Such phenomena can also be regarded as the phenomena of concentration and cavitation in solutions as the pressure vanishes. It has been rigorously shown in Chen-Liu [59] for $\gamma > 1$ and Li [142] for $\gamma = 1$ that, as the pressure vanishes, any two-shock Riemann solution to the Euler equations tends to a δ-shock solution to (3.7) and the intermediate densities between the two shocks tend to a weighted δ-measure that forms the δ-shock. By contrast, any two-rarefaction-wave Riemann solution of the Euler equations has been shown in [59] to tend to a two-contact-discontinuity solution to (3.7), whose intermediate state between the two contact discontinuities is a vacuum state, even when the initial density stays away from the vacuum.

3.4. Incompressible Euler equations. The incompressible Euler equations take the form:

$$\begin{cases} \partial_t \mathbf{v} + \nabla \cdot (\mathbf{v} \otimes \mathbf{v}) + \nabla p = 0, \\ \nabla \cdot \mathbf{v} = 0 \end{cases} \tag{3.8}$$

for the density-independent case, and

$$\begin{cases} \partial_t \rho + \nabla \cdot (\rho \mathbf{v}) = 0, \\ \partial_t (\rho \mathbf{v}) + \nabla \cdot (\rho \mathbf{v} \otimes \mathbf{v}) + \nabla p = 0, \\ \nabla \cdot \mathbf{v} = 0 \end{cases} \tag{3.9}$$

for the density-dependent case, where p should be regarded as an unknown. These systems can be obtained by formal asymptotics for low Mach number expansions from the Euler equations (1.4). For more details, see Chorin [78], Constantin [82], Hoff [125], Lions [150], Lions-Masmoudi [151], Majda-Bertozzi [158], and the references cited therein.

System (3.8) or (3.9) is an excellent model to capture M-D vorticity waves by ignoring the shocks in fluid flow, while system (1.11) is an excellent model to capture M-D shocks by ignoring the vorticity waves.

3.5. Euler equations in nonlinear elastodynamics. The equations of nonlinear elastodynamics provide another excellent example of the rich special structure one encounters when dealing with hyperbolic conservation laws. In $\mathbb{R}^3$, the state vector is $(\mathbf{v}, \mathbf{F})$, where $\mathbf{v} \in \mathbb{R}^3$ is the velocity vector and $\mathbf{F}$ is the 3×3 matrix-valued deformation gradient constrained by the requirement that $\det \mathbf{F} > 0$. The system of conservation laws, which express the integrability conditions between $\mathbf{v}$ and $\mathbf{F}$ and the balance of linear momentum, reads

$$\begin{cases} \partial_t F_{i\alpha} - \partial_{x_\alpha} v_i = 0, & i, \alpha = 1, 2, 3, \\ \partial_t v_j - \sum_{\beta=1}^{3} \partial_{x_\beta} S_{j\beta}(F) = 0, & j = 1, 2, 3. \end{cases} \tag{3.10}$$

The symbol $\mathbf{S}$ stands for the *Piola-Kirchhoff stress tensor*, which is determined by the (scalar-valued) *strain energy function* $\sigma(\mathbf{F})$: $S_{j\beta}(\mathbf{F}) = \frac{\partial \sigma(\mathbf{F})}{\partial F_{j\beta}}$. System (3.10) is hyperbolic if and only if, for any vectors $\xi, \mathbf{n} \in \mathbf{S}^3$,

$$\sum_{1 \le i,j,\alpha,\beta \le 3} \frac{\partial^2 \sigma(\mathbf{F})}{\partial F_{i\alpha} \partial F_{j\beta}} \xi_i \xi_j n_\alpha n_\beta > 0. \tag{3.11}$$

System (3.10) is endowed with an entropy pair:

$$\eta = \sigma(\mathbf{F}) + \frac{1}{2}|\mathbf{v}|^2, \qquad q_\alpha = -\sum_{1 \le j \le 3} v_j S_{j\alpha}(\mathbf{F}).$$

However, the laws of physics do not allow $\sigma(\mathbf{F})$, and thereby η, to be convex functions. Indeed, the convexity of σ would violate the principle of *material frame indifference* $\sigma(\mathbf{OF}) = \sigma(\mathbf{F})$ for all $\mathbf{O} \in SO(3)$ and would also be incompatible with the natural requirement that $\sigma(\mathbf{F}) \to \infty$ as $\det \mathbf{F} \downarrow 0$ or $\det \mathbf{F} \uparrow \infty$ (see Dafermos [89]).

The failure of σ to be convex is also the main source of complication in elastostatics, where one is seeking to determine equilibrium configurations of the body by minimizing the total strain energy $\int \sigma(\mathbf{F}) d\mathbf{x}$. The following alternative conditions, weaker than convexity and physically reasonable, are relevant in that context:

(i) *Polyconvexity* in the sense of Ball [11]: $\sigma(\mathbf{F}) = g(\mathbf{F}, \mathbf{F}^*, \det \mathbf{F})$, where $\mathbf{F}^* = (\det \mathbf{F})\mathbf{F}^{-1}$ is the adjugate of $\mathbf{F}$ (the matrix of cofactors of $\mathbf{F}$) and $g(\mathbf{F}, \mathbf{G}, w)$ is a convex function of 19 variables;
(ii) *Quasiconvexity* in the sense of Morrey [167];
(iii) *Rank-one convexity*, expressed by (3.11).

It is known that convexity ⇒ polyconvexity ⇒ quasiconvexity ⇒ rank-one convexity, however, none of the converse statements is generally valid.

It is important to investigate the relevance of the above conditions in elastodynamics. A first start was made in Dafermos [89] where it was shown that rank-one convexity suffices for the local existence of classical solutions, quasiconvexity yields the uniqueness of classical solutions in the class of entropy solutions in L^∞, and polyconvexity renders the system symmetrizable (also see [176]). To achieve this for polyconvexity, one of the main ideas is to enlarge system (3.10) into a large, albeit equivalent, system for the new state vector $(\mathbf{v}, \mathbf{F}, \mathbf{F}^*, w)$ with $w = \det \mathbf{F}$:

$$\partial_t w = \sum_{1\le\alpha,i\le3} \partial_{x_\alpha}(F^*_{\alpha i} v_i), \tag{3.12}$$

$$\partial_t F^*_{\gamma k} = \sum_{1\le\alpha,\beta,i,j\le3} \partial_{x_\alpha}(\epsilon_{\alpha\beta\gamma}\epsilon_{ijk}F_{j\beta}v_i), \quad \gamma, k = 1,2,3, \tag{3.13}$$

where $\epsilon_{\alpha\beta\gamma}$ and ϵ_{ijk} denote the standard permutation symbols. Then the enlarged system (3.10) and (3.12)–(3.13) with 21 equations is endowed a uniformly convex entropy $\eta = \sigma(\mathbf{F}, \mathbf{F}^*, w) + \frac{1}{2}|\mathbf{v}|^2$ so that the local existence of classical solutions and the stability of Lipschitz solutions may be inferred directly from Theorem 2.3. See Dafermos [91], Demoulini-Stuart-Tzavaras [94], and Qin [176] for more details.

3.6. Born-infeld system in electromagnetism. The Born-Infeld system is a nonlinear version of the Maxwell equations:

$$\partial_t B + \mathrm{curl}\big(\frac{\partial W}{\partial D}\big) = 0, \quad \partial_t D - \mathrm{curl}\big(\frac{\partial W}{\partial B}\big) = 0, \tag{3.14}$$

where $W : \mathbb{R}^3 \times \mathbb{R}^3 \to \mathbb{R}$ is the given energy density. The Born–Infeld model corresponds to the special case

$$W_{BI}(B, D) = \sqrt{1 + |B|^2 + |D|^2 + |P|^2}.$$

When W is strongly convex (i.e., $D^2W > 0$), system (3.14) is endowed with a strictly convex entropy. However, $W_{\rm BI}$ is not convex for a large enough field. As in §3.5, the Born-Infeld model is enlarged from 6 to 10 equations in Brenier [24], by an adjunction of the conservation laws satisfied by $P := B \times D$ and W so that the augmented system turns out to be a set of conservation laws in the unknowns $(h, B, D, P) \in \mathbb{R} \times \mathbb{R}^3 \times \mathbb{R}^3 \times \mathbb{R}^3$, which is in the physical region:

$$\{(h, B, D, P) : P = D \times B, h = \sqrt{1 + |B|^2 + |D|^2 + |P|^2} > 0\}.$$

Then the enlarged system is

$$\begin{cases} \partial_t h + \mathrm{div} P = 0, \\ \partial_t B + \mathrm{curl}(\dfrac{P \times B + D}{h}) = 0, \\ \partial_t D + \mathrm{curl}(\dfrac{P \times D - B}{h}) = 0, \\ \partial_t P + \mathrm{div}(\dfrac{P \otimes P - B \otimes B - D \otimes D - I}{h}) = 0, \end{cases} \tag{3.15}$$

which is endowed with a strongly convex entropy, where I is the 3×3 identity matrix. Also see Serre [185] for another enlarged system consisting of 9 scalar evolution equations in 9 unknowns (B, D, P), where P stands for the relaxation of the expression $D \times B$.

3.7. Lax systems. Let $f(\mathbf{u})$ be an analytic function of a single complex variable $\mathbf{u} = u + vi$. We impose on the complex-valued function $\mathbf{u} = \mathbf{u}(t, z)$, $z = x + yi$, and the real variable t the following nonlinear PDE:

$$\partial_t \bar{\mathbf{u}} + \partial_z f(\mathbf{u}) = 0, \tag{3.16}$$

where the bar denotes the complex conjugate and $\partial_z = \frac{1}{2}(\partial_x - i\partial_y)$. We may express this equation in terms of the real and imaginary parts of $\mathbf{u}$ and $\frac{1}{2} f(\mathbf{u}) = a(u, v) + b(u, v)i$. Then (3.16) gives

$$\begin{cases} \partial_t u + \partial_x a(u, v) + \partial_y b(u, v) = 0, \\ \partial_t v - \partial_x b(u, v) + \partial_y a(u, v) = 0. \end{cases} \tag{3.17}$$

In particular, when $f(\mathbf{u}) = \mathbf{u}^2 = u^2 + v^2 + 2uvi$, system (3.16) is called the complex Burger equation, which becomes

$$\begin{cases} \partial_t u + \dfrac{1}{2}\partial_x(u^2 + v^2) + \partial_y(uv) = 0, \\ \partial_t v - \partial_x(uv) + \dfrac{1}{2}\partial_y(u^2 + v^2) = 0. \end{cases} \tag{3.18}$$

System (3.17) is a symmetric hyperbolic system of conservation laws with a strictly convex entropy $\eta(u, v) = u^2 + v^2$; see Lax [139] for more details. For the 1-D case, this system is an archetype of hyperbolic conservation laws with umbilic degeneracy, which has been analyzed in Chen-Kan [56], Schaeffer-Shearer [181], and the references cited therein.

3.8. Gauss-Codazzi system for isometric embedding. A fundamental problem in differential geometry is to characterize intrinsic metrics on a 2-D Riemannian manifold $\mathcal{M}^2$ which can be realized as isometric immersions into $\mathbb{R}^3$ (cf. Yau [207]; also see [124]). For this, it suffices to solve the Gauss-Codazzi system, which can be written as (cf. [62, 124])

$$\begin{cases} \partial_x M - \partial_y L &= \Gamma_{22}^{(2)} L - 2\Gamma_{12}^{(2)} M + \Gamma_{11}^{(2)} N, \\ \partial_x N - \partial_y M &= -\Gamma_{22}^{(1)} L + 2\Gamma_{12}^{(1)} M - \Gamma_{11}^{(1)} N, \end{cases} \tag{3.19}$$

and

$$LN - M^2 = K. \tag{3.20}$$

Here $K(x, y)$ is the given Gauss curvature, and $\Gamma_{ij}^{(k)}$ is the given Christoffel symbol. We now follow Chen-Slemrod-Wang [62] to present the fluid dynamical formulation of system (3.19)–(3.20). Set $L = \rho v^2 + p, M = -\rho uv, N = \rho u^2 + p$, and set $q^2 = u^2 + v^2$ as usual. Then the Codazzi equations (3.19) become the familiar balance laws of momentum:

$$\begin{cases} \partial_x(\rho uv) + \partial_y(\rho v^2 + p) = -\Gamma_{22}^{(2)}(\rho v^2 + p) - 2\Gamma_{12}^{(2)} \rho uv - \Gamma_{11}^{(2)}(\rho u^2 + p), \\ \partial_x(\rho u^2 + p) + \partial_y(\rho uv) = -\Gamma_{22}^{(1)}(\rho v^2 + p) - 2\Gamma_{12}^{(1)} \rho uv - \Gamma_{11}^{(1)}(\rho u^2 + p), \end{cases} \tag{3.21}$$

and the Gauss equation (3.20) becomes $\rho p q^2 + p^2 = K$. We choose p to be the Chaplygin gas-type to allow the Gauss curvature K to change sign: $p = -\frac{1}{\rho}$. Then we obtain the "Bernoulli" relation:

$$\rho = \frac{1}{\sqrt{q^2 + K}}, \qquad \text{or} \qquad p = -\sqrt{q^2 + K}. \tag{3.22}$$

In general, system (3.21)–(3.22) for unknown (u, v) is of mixed hyperbolic-elliptic type determined by the sign of Gauss curvature (hyperbolic when $K < 0$ and elliptic when $K > 0$).

For the Gauss-Codazzi-Ricci equations for isometric embedding of higher dimensional Riemannian manifolds, see Chen-Slemrod-Wang [63].

4. Multidimensional steady supersonic problems. M-D steady problems for the Euler equations are fundamental in fluid mechanics. In particular, understanding of these problems helps us to understand the asymptotic behavior of evolution solutions for large time, especially global attractors. One of the excellent sources of steady problems is Courant-Friedrichs's book [84]. In this section we first discuss some of recent developments in the analysis of M-D steady supersonic problems. The M-D steady Euler flows are governed by

$$\begin{cases} \nabla_{\mathbf{x}} \cdot (\rho \mathbf{v}) = 0, \\ \nabla_{\mathbf{x}}(\rho \mathbf{v} \otimes \mathbf{v}) + \nabla_{\mathbf{x}} p = 0, \\ \nabla_{\mathbf{x}} \cdot \big(\rho(E + \frac{p}{\rho})\mathbf{v}\big) = 0, \end{cases} \tag{4.1}$$

where $\mathbf{v}$ is the velocity, E is the total energy, and the constitutive relations among the thermodynamical variables ρ, p, e, θ, and S are determined by (1.5)–(1.7).

For the barotropic (isentropic or isothermal) case, $p = p(\rho)$ is determined by (1.10) with $\gamma \geq 1$, and then the first $d+1$ equations in (4.1) form a self-contained system, the Euler system for steady barotropic fluids.

System (4.1) governing a supersonic flow (i.e., $|\mathbf{v}|^2 > c^2$) has all real eigenvalues and is hyperbolic, while system (4.1) governing a subsonic flow (i.e., $|\mathbf{v}|^2 < c^2$) has complex eigenvalues and is elliptic-hyperbolic mixed and composite.

4.1. Wedge problems involving supersonic shock-fronts. The analysis of 2-D steady supersonic flows past wedges whose vertex angles are less than the critical angle can date back to the 1940s since the stability of such flows is fundamental in applications (cf. [84, 205]). Local solutions around the wedge vertex were first constructed in Gu [119], Li [146], and Schaeffer [180]. Also see Zhang [212] and the references cited therein for global potential solutions when the wedges are a small perturbation of the straight-sided wedge. For the wedge problem, when the vertex angle is suitably large, the flow contains a large shock-front and, for this case, the full Euler equations (4.1) are required to describe the physical flow. When a wedge is straight and its vertex angle is less than the critical angle ω_c, there exists a supersonic shock-front emanating from the wedge vertex so that the constant states on both sides of the shock are supersonic; the critical angle condition is necessary and sufficient for the existence of the supersonic shock (also see [69, 84]).

Consider 2-D steady supersonic Euler flows past a 2-D Lipschitz curved wedge $|x_2| \leq g(x_1), x_1 > 0$, with $g \in Lip(\mathbb{R}_+)$ and $g(0) = 0$, whose vertex angle $\omega_0 := \arctan(g'(0+))$ is less than the critical angle ω_c, along which $TV\{g'(\cdot); \mathbb{R}_+\} \leq \varepsilon$ for some constant $\varepsilon > 0$. Denote

$$\Omega := \{\mathbf{x} : x_2 > g(x_1),\, x_1 \geq 0\}, \quad \Gamma := \{\mathbf{x} : x_2 = g(x_1),\, x_1 \geq 0\},$$

and $\mathbf{n}(x_1\pm) = \frac{(-g'(x_1\pm),1)}{\sqrt{(g'(x_1\pm))^2+1}}$ are the outer normal vectors to Γ at points $x_1\pm$, respectively. The uniform upstream flow $\mathbf{u}_- = (\rho_-, \rho_- u_-, 0, \rho_- E_-)$ satisfies $u_- > c(\rho_-, S_-)$ so that a strong supersonic shock-front emanates from the wedge vertex. Since the problem is symmetric with respect to the x_2-axis, the wedge problem can be formulated into the following problem of initial-boundary value type for system (4.1) in Ω:

$$\textit{Cauchy Condition:} \qquad \mathbf{u}|_{x_1=0} = \mathbf{u}_-; \tag{4.2}$$

$$\textit{Boundary Condition:} \qquad \mathbf{v}\cdot\mathbf{n} = 0 \quad \text{on } \Gamma. \tag{4.3}$$

In Chen-Zhang-Zhu [69], it has been established that there exist $\varepsilon_0 > 0$ and $C > 0$ such that, when $\varepsilon \leq \varepsilon_0$, there exists a pair of functions

$$\mathbf{u} = (\rho, \mathbf{v}, \rho E) \in BV(\mathbb{R}; \mathbb{R}^2 \times \mathbb{R}_+ \times \mathbb{R}_+), \qquad \sigma \in BV(\mathbb{R}_+; \mathbb{R})$$

with $\chi = \int_0^{x_1} \sigma(s)ds \in Lip(\mathbb{R}_+; \mathbb{R}_+)$ such that $\mathbf{u}$ is a global entropy solution of problem (4.1)–(4.3) in Ω with

$$TV\{\mathbf{u}(x_1, \cdot) : [g(x_1), -\infty)\} \leq C\,TV\{g'(\cdot)\} \quad \text{for every } x_1 \in \mathbb{R}_+, \quad (4.4)$$

and the strong shock-front $x_2 = \chi(x_1)$ emanating from the wedge vertex is nonlinearly stable in structure. Furthermore, the global L^1-stability of entropy solutions with respect to the incoming flow at $x_1 = 0$ in L^1 has also been established in Chen-Li [58]. This asserts that any supersonic shock-front for the wedge problem is nonlinearly L^1–stable with respect to the BV perturbation of the incoming flow and the wedge boundary.

In order to achieve this, we have first developed an adaptation of the Glimm scheme whose mesh grids are designed to follow the slope of the Lipschitz wedge boundary so that the lateral Riemann building blocks contain only one shock or rarefaction wave emanating from the mesh points on the boundary. Such a design makes the BV estimates more convenient for the Glimm approximate solutions. Then careful interaction estimates have been made. One of the essential estimates is the estimate of the strength δ_1 of the reflected 1-waves in the interaction between the 4-strong shock-front and weak waves $(\alpha_1, \beta_2, \beta_3, \beta_4)$, that is,

$$\delta_1 = \alpha_1 + K_{s1}\beta_4 + O(1)|\alpha_1|(|\beta_2| + |\beta_3|) \qquad \text{with } |K_{s1}| < 1.$$

The second essential estimate is the interaction estimate between the wedge boundary and weak waves. Based on the construction of the approximation solutions and interaction estimates, we have successfully identified a Glimm-type functional to incorporate the curved wedge boundary and the strong shock-front naturally and to trace the interactions not only between the wedge boundary and weak waves, but also between the strong shock-front and weak waves. With the aid of the important fact that $|K_{s1}| < 1$, we have showed that the identified Glimm functional monotonically decreases in the flow direction. Another essential estimate is to trace the approximate strong shock-fronts in order to establish the nonlinear stability and asymptotic behavior of the strong shock-front emanating from the wedge vertex under the BV wedge perturbation.

For the 3-D **cone problem**, the nonlinear stability of a self-similar 3-D full gas flow past an infinite cone is another important problem. See Lien-Liu [140] for the cones with small vertex angle. Also see [71, 77] for the construction of piecewise smooth potential flows under smooth perturbation of the straight-sided cone.

In Elling-Liu [101], an evidence has been provided that the steady supersonic weak shock solution is dynamically stable indeed, in the sense that it describes the long-time behavior of an unsteady flow.

4.2. Stability of supersonic vortex sheets. Another natural problem is the stability of supersonic vortex sheets above the Lipschitz wall $x_2 = g(x_1), x_1 \geq 0$, with

$$g \in Lip(\mathbb{R}_+;\mathbb{R}), \ \ g(0) = g'(0+) = 0, \ \ \lim_{x_1\to\infty} \arctan(g'(x_1+)) = 0,$$

and $g' \in BV(\mathbb{R}_+;\mathbb{R})$ such that $TV\{g'(\cdot)\} \le \varepsilon$ for some $\varepsilon > 0$. Denote again $\Omega = \{\mathbf{x} : x_2 > g(x_1), x_1 \ge 0\}$ and $\Gamma = \{\mathbf{x} : x_2 = g(x_1), x_1 \ge 0\}$. The upstream flow consists of one supersonic straight vortex sheet $x_2 = y_0 > 0$ and two constant vectors $\mathbf{u}_0 = (\rho_0, \rho_0 u_0, 0, \rho_0 E_0)$ when $x_2 > y_0 > 0$ and $\mathbf{u}_1 = (\rho_1, \rho_1 u_1, 0, \rho_1 E_1)$ when $0 < x_2 < y_0$ satisfying $u_1 > u_0 > 0$ and $u_i > c(\rho_i, S_i)$ for $i = 0, 1$. Then the vortex sheet problem can be formulated into the following problem of initial-boundary value type for system (4.1):

$$\textit{Cauchy Condition:} \qquad \mathbf{u}|_{x_1=0} = \begin{cases} \mathbf{u}_0, & 0 < x_2 < y_0, \\ \mathbf{u}_1, & x_2 > y_0; \end{cases} \tag{4.5}$$

$$\textit{Slip Boundary Condition}: \qquad \mathbf{v}\cdot\mathbf{n} = 0 \qquad \text{on } \Gamma. \tag{4.6}$$

It has been proved that steady supersonic vortex sheets, as time-asymptotics, are stable in structure globally, even under the *BV* perturbation of the Lipschitz walls in Chen-Zhang-Zhu [70]. The result indicates that the strong supersonic vortex sheets are nonlinearly stable in structure globally under the *BV* perturbation of the Lipschitz wall, although there may be weak shocks and supersonic vortex sheets away from the strong vortex sheet. In order to establish this theorem, as in §4.1, we first developed an adaption of the Glimm scheme whose mesh grids are designed to follow the slope of the Lipschitz boundary. For this case, one of the essential estimates is the estimate of the strength δ_1 of the reflected 1-wave in the interaction between the 4-weak wave α_4 and the strong vortex sheet from below is less than one: $\delta_1 = K_{01}\alpha_4$ and $|K_{01}| < 1$. The second new essential estimate is the estimate of the strength δ_4 of the reflected 4-wave in the interaction between the 1-weak wave β_1 and the strong vortex sheet from above is also less than one: $\delta_4 = K_{11}\beta_1$ and $|K_{11}| < 1$. Another essential estimate is to trace the approximate supersonic vortex sheets under the *BV* boundary perturbation. For more details, see Chen-Zhang-Zhu [70]. The nonlinear L^1-stability of entropy solutions with respect to the incoming flow at $x_1 = 0$ is under current investigation.

5. Multidimensional steady transonic problems. In this section we discuss another important class of M-D steady problems: transonic problems. In the last decade, a program has been initiated on the existence and stability of M-D transonic shock-fronts, and some new analytical approaches including techniques, methods, and ideas have been developed. For clear presentation, we focus mainly on the celebrated steady potential flow equation of aerodynamics for the velocity potential $\varphi : \Omega \subset \mathbb{R}^d \to \mathbb{R}$, which is a second-order nonlinear PDE of mixed elliptic-hyperbolic type:

$$\nabla_{\mathbf{x}} \cdot (\rho(|\nabla_{\mathbf{x}}\varphi|^2)\nabla_{\mathbf{x}}\varphi) = 0, \qquad \mathbf{x} \in \Omega \subset \mathbb{R}^d, \tag{5.1}$$

where the density $\rho(q^2)$ by scaling is

$$\rho(q^2) = \left(1 - \frac{\gamma-1}{2}q^2\right)^{\frac{1}{\gamma-1}} \tag{5.2}$$

with adiabatic exponent $\gamma > 1$. Equation (5.1) is elliptic at $\nabla_{\mathbf{x}}\varphi$ with $|\nabla_{\mathbf{x}}\varphi| = q$ if $\rho(q^2) + 2q^2\rho'(q^2) > 0$, which is equivalent to

$$q < q_* := \sqrt{2/(\gamma+1)},$$

i.e., the flow is subsonic. Equation (5.1) is hyperbolic if $\rho(q^2)+2q^2\rho'(q^2) < 0$, i.e., $q > q_*$, that is, the flow is supersonic.

Let Ω^+ and Ω^- be open subsets of Ω such that $\Omega^+\cap\Omega^- = \emptyset$, $\overline{\Omega^+}\cup\overline{\Omega^-} = \overline{\Omega}$, and $\mathcal{S} = \partial\Omega^+ \cap \Omega$. Let $\varphi \in C^{0,1}(\Omega) \cap C^1(\overline{\Omega^\pm})$ be a weak solution of (5.1), which satisfies $|\nabla_{\mathbf{x}}\varphi| \le q_{cav} := \sqrt{2/(\gamma-1)}$, so that $\nabla_{\mathbf{x}}\varphi$ experiences a jump across $\mathcal{S}$ that is a $(d-1)$-D smooth surface. Set $\varphi^\pm = \varphi|_{\Omega^\pm}$. Then φ satisfies the following Rankine-Hugoniot conditions on $\mathcal{S}$:

$$\varphi^+ = \varphi^-, \tag{5.3}$$

$$\rho(|\nabla_{\mathbf{x}}\varphi^+|^2)\nabla_{\mathbf{x}}\varphi^+ \cdot \mathbf{n} = \rho(|\nabla_{\mathbf{x}}\varphi^-|^2)\nabla_{\mathbf{x}}\varphi^- \cdot \mathbf{n}. \tag{5.4}$$

Suppose that $\varphi \in C^1(\overline{\Omega^\pm})$ is a weak solution satisfying

$$|\nabla_{\mathbf{x}}\varphi| < q_* \text{ in } \Omega^+, \qquad |\nabla_{\mathbf{x}}\varphi| > q_* \text{ in } \Omega^-, \qquad \nabla_{\mathbf{x}}\varphi^\pm \cdot \mathbf{n}|_{\mathcal{S}} > 0. \tag{5.5}$$

Then φ is a *transonic shock-front solution* with *transonic shock-front* $\mathcal{S}$ dividing Ω into the *subsonic region* Ω^+ and the *supersonic region* Ω^- and satisfying the physical entropy condition (see Courant-Friedrichs [84]):

$$\rho(|\nabla_{\mathbf{x}}\varphi^-|^2) < \rho(|\nabla_{\mathbf{x}}\varphi^+|^2) \qquad \text{along } \mathcal{S}. \tag{5.6}$$

Note that equation (5.1) is elliptic in the subsonic region and hyperbolic in the supersonic region.

As an example, let $(x_1, \mathbf{x}')$ be the coordinates in $\mathbb{R}^d$, where $x_1 \in \mathbb{R}$ and $\mathbf{x}' = (x_2, \dots, x_d) \in \mathbb{R}^{d-1}$. Fix $\mathbf{V}_0 \in \mathbb{R}^d$, and let $\varphi_0(\mathbf{x}) := \mathbf{V}_0 \cdot \mathbf{x}$, $\mathbf{x} \in \mathbb{R}^d$. If $|\mathbf{V}_0| \in (0, q_*)$ (resp. $|\mathbf{V}_0| \in (q_*, q_{cav})$), then $\varphi_0(\mathbf{x})$ is a subsonic (resp. supersonic) solution in $\mathbb{R}^d$, and $\mathbf{V}_0 = \nabla_{\mathbf{x}}\varphi_0$ is its velocity.

Let $q_0^- > 0$ and $\mathbf{V}_0' \in \mathbb{R}^{d-1}$ be such that the vector $\mathbf{V}_0^- := (q_0^-, \mathbf{V}_0')$ satisfies $|\mathbf{V}_0^-| > q_*$. Then there exists a unique $q_0^+ > 0$ such that

$$\left(1 - \frac{\gamma-1}{2}(|q_0^+|^2+|\mathbf{V}_0'|^2)\right)^{\frac{1}{\gamma-1}} q_0^+ = \left(1 - \frac{\gamma-1}{2}(|q_0^-|^2+|\mathbf{V}_0'|^2)\right)^{\frac{1}{\gamma-1}} q_0^-. \tag{5.7}$$

The entropy condition (5.6) implies $q_0^+ < q_0^-$. By denoting $\mathbf{V}_0^+ := (q_0^+, \mathbf{V}_0')$ and defining the functions $\varphi_0^\pm(\mathbf{x}) := V_0^\pm \cdot \mathbf{x}$ on $\mathbb{R}^d$, then φ_0^+ (resp. φ_0^-) is a subsonic (resp. supersonic) solution. Furthermore, from (5.4) and (5.7), the function

$$\varphi_0(\mathbf{x}) := \min(\varphi_0^-(\mathbf{x}), \varphi_0^+(\mathbf{x})) = V_0^\pm \cdot \mathbf{x}, \quad \mathbf{x} \in \Omega_0^\pm := \{\pm x_1 > 0\}, \tag{5.8}$$

is a plane transonic shock-front solution in $\mathbb{R}^d$, Ω_0^- and Ω_0^+ are respectively its supersonic and subsonic regions, and $\mathcal{S} = \{x_1 = 0\}$ is a transonic shock-front. Note that, if $\mathbf{V}_0' = 0$, the velocities $V_0^\pm$ are orthogonal to the shock-front $\mathcal{S}$ and, if $\mathbf{V}_0' \neq 0$, the velocities are not orthogonal to $\mathcal{S}$.

5.1. Transonic shock-front problems in $\mathbb{R}^d$. Consider M-D perturbations of the uniform transonic shock-front solution (5.8) in $\mathbb{R}^d$ with $d \geq 3$. Since it suffices to specify the supersonic perturbation φ^- only in a neighborhood of the unperturbed shock-front $\{x_1 = 0\}$, we introduce domains $\Omega := (-1, \infty) \times \mathbb{R}^{d-1}, \Omega_1 := (-1, 1) \times \mathbb{R}^{d-1}$. Note that we expect the subsonic region Ω^+ to be close to the half-space $\Omega_0^+ = \{x_1 > 0\}$.

Problem 5.1. Given a supersonic solution $\varphi^-(\mathbf{x})$ of (5.1) in Ω_1, find a transonic shock-front solution $\varphi(\mathbf{x})$ in Ω such that

$$\Omega^- \subset \Omega_1, \qquad \varphi(\mathbf{x}) = \varphi^-(\mathbf{x}) \quad \text{in } \Omega^-,$$

where $\Omega^- := \Omega \setminus \Omega^+$ and $\Omega^+ := \{\mathbf{x} \in \Omega \, : \, |\nabla_{\mathbf{x}}\varphi(\mathbf{x})| < q_*\}$, and

$$\varphi = \varphi^-, \quad \partial_{x_1}\varphi = \partial_{x_1}\varphi^- \qquad \text{on } \{x_1 = -1\}, \tag{5.9}$$

$$\lim_{R\to\infty} \|\varphi - \varphi_0^+\|_{C^1(\Omega^+\setminus B_R(0))} = 0. \tag{5.10}$$

Condition (5.9) determines that the solution has supersonic upstream, while condition (5.10) especially determines that the uniform velocity state at infinity in the downstream direction is equal to the unperturbed downstream velocity state. The additional requirement in (5.10) that $\varphi \to \varphi_0^+$ at infinity within Ω^+ fixes the position of shock-front at infinity. This allows us to determine the solution of *Problem 5.1* uniquely. In Chen-Feldman [47], we have employed the free boundary approach first developed for the potential flow equation (5.1) to solve the stability problem, *Problem 5.1*. This existence result can be extended to the case that the regularity of the steady perturbation φ^- is only $C^{1,1}$. It would be interesting to establish similar results for the full Euler equations (1.4).

5.2. Nozzle problems involving transonic shock-fronts. We now consider M-D transonic shock-fronts in the following infinite nozzle Ω with arbitrary smooth cross-sections: $\Omega = \Psi(\Lambda \times \mathbb{R}) \cap \{x_1 > -1\}$, where $\Lambda \subset \mathbb{R}^{d-1}$ is an open bounded connected set with a smooth boundary, and $\Psi : \mathbb{R}^d \to \mathbb{R}^d$ is a smooth map, which is close to the identity map. For simplicity, we assume that $\partial\Lambda$ is in $C^{[\frac{d}{2}]+3,\alpha}$ and $\|\Psi - I\|_{[\frac{d}{2}]+3,\alpha,\mathbb{R}^d} \leq \sigma$ for some $\alpha \in (0,1)$ and small $\sigma > 0$, where $[s]$ is the integer part of s, $I : \mathbb{R}^d \to \mathbb{R}^d$ is the identity map, and $\partial_l\Omega := \Psi(\mathbb{R} \times \partial\Lambda) \cap \{x_1 > -1\}$. For concreteness, we also assume that there exists $L > 1$ such that $\Psi(\mathbf{x}) = \mathbf{x}$ for any $\mathbf{x} = (x_1, \mathbf{x}')$ with $x_1 > L$.

In the 2-D case, $\Omega = \{(x_1, x_2) \ : \ x_1 > -1, \ b^-(x_2) < x_2 < b^+(x_2)\}$, where $\|b^\pm - b^\pm_\infty\|_{4,\alpha,\mathbb{R}} \le \sigma$ and $b^\pm \equiv b^\pm_\infty$ on $[L, \infty)$ for some constants $b^\pm_\infty$ satisfying $b^+_\infty > b^-_\infty$. For the M-D case, the geometry of the nozzles is much richer. Note that our setup implies that $\partial\Omega = \overline{\partial_o\Omega} \cup \partial_l\Omega$ with

$$\partial_l\Omega := \Psi[(-\infty, \infty) \times \partial\Lambda] \cap \{(x_1, \mathbf{x}') \ : \ x_1 > -1\},$$
$$\partial_o\Omega := \Psi((-\infty, \infty) \times \Lambda) \cap \{(x_1, \mathbf{x}') \ : \ x_1 = -1\}.$$

Then our transonic nozzle problem can be formulated as

Problem 5.2: Transonic Nozzle Problem. Given the supersonic upstream flow at the entrance $\partial_o\Omega$:

$$\varphi = \varphi_e^-, \quad \varphi_{x_1} = \psi_e^- \qquad \text{on } \partial_o\Omega, \tag{5.11}$$

the slip boundary condition on the nozzle boundary $\partial_l\Omega$:

$$\nabla_{\mathbf{x}}\varphi \cdot \mathbf{n} = 0 \qquad \text{on } \partial_l\Omega, \tag{5.12}$$

and the uniform subsonic flow condition at the infinite exit $x_1 = \infty$:

$$\|\varphi(\cdot) - q_\infty x_1\|_{C^1(\Omega\cap\{x_1>R\})} \to 0 \quad \text{ as } R \to \infty \text{ for some } q_\infty \in (0, q_*), \tag{5.13}$$

find a M-D transonic flow φ of problem (5.1) and (5.11)–(5.13) in Ω.

The standard local existence theory of smooth solutions for the initial-boundary value problem (5.11)–(5.12) for second-order quasilinear hyperbolic equations implies that, as σ is sufficiently small, there exists a supersonic solution φ^- of (5.1) in $\Omega_2 := \{-1 \le x_1 \le 1\}$, which is a C^{l+1} perturbation of $\varphi_0^- = q_0^- x_1$: For any $\alpha \in (0, 1]$,

$$\|\varphi^- - \varphi_0^-\|_{l,\alpha,\Omega_2} \le C_0\sigma, \qquad l = 1, 2, \tag{5.14}$$

for some constant $C_0 > 0$, and satisfies $\nabla\varphi^- \cdot \mathbf{n} = 0$ on $\partial_l\Omega_2$, provided that (φ_e^-, ψ_e^-) on $\partial_o\Omega$ satisfying

$$\|\varphi_e^- - q_0^- x_1\|_{H^{s+l}} + \|\psi_e^- - q_0^-\|_{H^{s+l-1}} \le \sigma, \qquad l = 1, 2, \tag{5.15}$$

for some integer $s > d/2 + 1$ and the compatibility conditions up to order $s + 1$, where the norm $\|\cdot\|_{H^s}$ is the Sobolev norm with $H^s = W^{s,2}$.

In Chen-Feldman [48], *Problem 5.2* has been solved. More precisely, let $q_0^- \in (q_*, q_{cav})$ and $q_0^+ \in (0, q_*)$ satisfy (5.7), and let φ_0 be the transonic shock-front solution (5.8) with $\mathbf{V}' = 0$. Then there exist $\sigma_0 > 0$ and C, depending only on d, α, γ, q_0^-, Λ, and L, such that, for every $\sigma \in (0, \sigma_0)$, any map Ψ introduced above, and any supersonic upstream flow (φ_e^-, ψ_e^-)

on $\partial_o\Omega$ satisfying (5.15) with $l = 1$, there exists a solution $\varphi \in C^{0,1}(\Omega) \cap C^{1,\alpha}(\overline{\Omega^+})$ satisfying

$$\Omega^+(\varphi) = \{x_1 > f(\mathbf{x}')\}, \qquad \Omega^-(\varphi) = \{x_1 < f(\mathbf{x}')\},$$
$$\|\varphi - q^\pm x_1\|_{1,\alpha,\Omega^\pm} \le C\sigma, \qquad \|\varphi - q_\infty x_1\|_{C^1(\Omega\cap\{x_1>R\})} \to 0 \text{ as } R \to \infty,$$

where $q_\infty \in (0, q_*)$ is the unique solution of the equation

$$\rho((q^+)^2)q^+ = Q^+ := \frac{1}{|\Lambda|}\int_{\partial_o\Omega} \rho(|\nabla_{\mathbf{x}'}\varphi_e^-|^2 + (\psi_e^-)^2)\psi_e^- \, d\mathcal{H}^{d-1}. \tag{5.16}$$

In addition, if the supersonic uniform flow (φ_e^-, ψ_e^-) on $\partial_o\Omega$ satisfies (5.15) with $l = 2$, then $\varphi \in C^{2,\alpha}(\overline{\Omega^+})$ with $\|\varphi - q^+x_1\|_{2,\alpha,\Omega_+} \le C\sigma$, and the solution with a transonic shock-front is unique and stable with respect to the nozzle boundary and the smooth supersonic upstream flow at the entrance. The techniques have been extended to solving the nozzle problem for the 2-D full Euler equations first in Chen-Chen-Feldman [43].

In the previous setting, the location of the transonic shock-front in the solution is not unique in general if we prescribe only the pressure at the nozzle exit, since the flat shock-front between uniform states can be translated along the flat nozzle which does not change the flow parameters at the entrance and the pressure at the exit. The analysis of the relation among the unique shock-front location, the flow parameters, and the geometry of the diverging nozzle has been made for various Euler systems in different dimensions; see Chen [73] and Lin-Yuan [153] for the 2-D full Euler equations, Li-Xin-Yin [145] for the 2-D and 3-D axisymmetric Euler systems with narrow divergent nozzles, and Bae-Feldman [9] for perturbed diverging cone-shaped nozzle of arbitrary cross-section for the non-isentropic potential flow system in any dimension.

A major open problem is the *physical de Laval nozzle problem*: Consider a nozzle Ω which is flat (i.e., $\Psi = I$) between $-2 \le x_1 \le -1$ and $1 \le x_1 \le \infty$, and has some special geometry between $-1 \le x_1 \le 1$ to make the part $\{-1 \le x_1 \le 0\}$ convergent and the part $\{0 \le x_1 \le 1\}$ divergent. Given certain incoming subsonic flow at $x_1 = -2$, find a transonic flow containing transonic shock-fronts in the nozzle such that the downstream at $x_1 = \infty$ is subsonic. A related reference is Kuz'min's book [135].

5.3. Wedge/cone problems involving transonic shock-fronts. The existence and stability of transonic flows past infinite wedges or cones are further long-standing open transonic problems. Some progress has been made for the wedge case in 2-D steady flow in Chen-Fang [75], Fang [104], and Chen-Chen-Feldman [44]. In [104], it was proved that the transonic shock-front is conditionally stable under perturbation of the upstream flow and/or the wedge boundary in some weak Sobolev norms. In [44], the existence and stability of transonic flows past the curved wedge have been established in the strong Hölder norms for the full Euler equations.

Conical flow (i.e. cylindrically symmetric flow with respect to an axis, say, the x_1-axis) in $\mathbb{R}^3$ occurs in many physical situations (cf. [84]). Unlike the 2-D case, the governing equations for the 3-D conical case have a singularity at the cone vertex and the flow past the straight-sided cone is self-similar, but is no longer piecewise constant. These have resulted in additional difficulties for the stability problem. In Chen-Fang [46], we have developed techniques to handle the singular terms in the equations and the singularity of the solutions. Our main results indicate that the self-similar transonic shock-front is conditionally stable with respect to the conical perturbation of the cone boundary and the upstream flow in appropriate function spaces. That is, the transonic shock-front and downstream flow in our solutions are close to the unperturbed self-similar transonic shock-front and downstream flow under the conical perturbation, and the slope of the shock-front asymptotically tends to the slope of the unperturbed self-similar shock-front at infinity.

In order to achieve these results, we have first formulated the stability problem as a free boundary problem and have then introduced a coordinate transformation to reduce the free boundary problem into a fixed boundary value problem for a singular nonlinear elliptic system. We have developed an iteration scheme that consists of two iteration mappings: one is for an iteration of approximate transonic shock-fronts; and the other is for an iteration of the corresponding boundary value problems for the singular nonlinear systems for given approximate shock-fronts. To ensure the well-definedness and contraction property of the iteration mappings, it is essential to establish the well-posedness for a corresponding singular linearized elliptic equation, especially the stability with respect to the coefficients of the equation, and to obtain the estimates of its solutions reflecting their singularity at the cone vertex and decay at infinity. The approach is to employ key features of the equation, introduce appropriate solution spaces, and apply a Fredholm-type theorem in Maz'ya-Plamenevskivı [161] to establish the existence of solutions by showing the uniqueness in the solution spaces. Also see Cui-Yin [87] for related results.

5.4. Airfoil/obstacle problems: Subsonic flows past an airfoil or an obstacle. The Euler equations for potential flows (5.1)–(5.2) in $\mathbb{R}^2$ can be rewritten as

$$\begin{cases} \partial_{x_1} v - \partial_{x_2} u = 0, \\ \partial_{x_1}(\rho u) + \partial_{x_2}(\rho v) = 0, \end{cases} \tag{5.17}$$

with $(u, v) = \nabla_{\mathbf{x}}\varphi$ and $q = |\nabla_{\mathbf{x}}\varphi|$, where ρ is again determined by the Bernoulli relation (5.1). An important problem is subsonic flows past an airfoil or an obstacle. Shiffman [189], Bers [15], Finn-Gilbarg [106], and Dong [98] studied subsonic (elliptic) solutions of (5.1) outside an obstacle when the upstream flows are sufficiently subsonic; also see Chen-Dafermos-Slemrod-Wang [45] via compensated compactness argument. Morawetz in

[166] first showed that the flows of (5.17) past an obstacle may contain transonic shocks in general. A further problem is to construct global entropy solutions of the airfoil problem (see [163, 110]).

In Chen-Slemrod-Wang [61], we have introduced the usual flow angle $\theta = \tan^{-1}(\frac{v}{u})$ and written the irrotationality and mass conservation equation as an artificially viscous problem:

$$\begin{cases} \partial_{x_1} v^\varepsilon - \partial_{x_2} u^\varepsilon = \varepsilon \Delta \theta^\varepsilon, \\ \partial_{x_1}(\rho^\varepsilon u^\varepsilon) + \partial_{x_2}(\rho^\varepsilon v^\varepsilon) = \varepsilon \nabla \cdot (\sigma(\rho^\varepsilon) \nabla \rho^\varepsilon), \end{cases} \tag{5.18}$$

where $\sigma(\rho)$ is suitably chosen, and appropriate boundary conditions are imposed for this regularized "viscous" problem. The crucial new discovery is that a uniformly L^∞ bound in q^ε can be obtained when $1 \le \gamma < 3$ which uniformly prevents cavitation. However, in this formulation, a uniform bound in the flow angle θ^ε and a uniform lower bound in q^ε in any fixed region disjoint from the profile must be assumed apriori. By making further careful energy estimates, Morawetz's argument [163] then applies, and the strong convergence in $L^1_{\text{loc}}(\Omega)$ of our approximating sequence is achieved.

An important open problem is whether two remaining conditions on $(q^\varepsilon, \theta^\varepsilon)$ can be removed. See Chen-Slemrod-Wang [61] for more details.

5.5. Nonlinear approaches. We now discuss several nonlinear approaches to deal with steady transonic problems.

5.5.1. Free boundary approaches. We first describe two of the free boundary approaches for *Problems 5.1–5.2*, developed in [47, 48].

Free Boundary Problems. The transonic shock-front problems can be formulated into a one-phase free boundary problem for a nonlinear elliptic PDE: Given $\varphi^- \in C^{1,\alpha}(\overline{\Omega})$, find a function φ that is continuous in Ω and satisfies

$$\varphi \le \varphi^- \qquad \text{in } \bar{\Omega}, \tag{5.19}$$

equation (5.1), the ellipticity condition in the non-coincidence set $\Omega^+ = \{\varphi < \varphi^-\}$, the free boundary condition (5.4) on the boundary $\mathcal{S} = \partial\Omega^+ \cap \Omega$, as well as the prescribed conditions on the fixed boundary $\partial\Omega$ and at infinity. These conditions are different in different problems, for example, conditions (5.9)–(5.10) for *Problem 5.1* and (5.11)–(5.13) for *Problem 5.2.*

The free boundary is the location of the shock-front, and the free boundary conditions (5.3)–(5.4) are the Rankine-Hugoniot conditions. Note that condition (5.19) is motivated by the similar property (5.8) of unperturbed shock-fronts; and (5.19), locally on the shock-front, is equivalent to the entropy condition (5.6). Condition (5.19) transforms the transonic shock-front problem, in which the subsonic region Ω^+ is determined by the gradient condition $|\nabla_{\mathbf{x}}\varphi(\mathbf{x})| < q_*$, into a free boundary problem in which Ω^+ is the non-coincidence set. In order to solve this free boundary problem, equation (5.1) is modified to be uniformly elliptic and then the free boundary condition (5.4) is correspondingly modified. The problem becomes a one-phase free boundary problem for the uniformly elliptic equation which we can solve. Since φ^- is a small $C^{1,\alpha}$–perturbation of φ_0^-, the solution φ of the free boundary problem is shown to be a small $C^{1,\alpha}$–perturbation of the given subsonic shock-front solution φ_0^+ in Ω^+. In particular, the gradient estimate implies that φ in fact satisfies the original free boundary problem, hence the transonic shock-front problem, *Problem 5.1* (*Problem 5.2*, respectively).

The modified free boundary problem does not directly fit into the variational framework of Alt-Caffarelli [2] and Alt-Caffarelli-Friedman [3], and the regularization framework of Berestycki-Caffarelli-Nirenberg [17]. Also, the nonlinearity of the free boundary problem makes it difficult to apply the Harnack inequality approach of Caffarelli [30]. In particular, a boundary comparison principle for positive solutions of nonlinear elliptic equations in Lipschitz domains is not available yet for the equations that are not homogeneous with respect to $(\nabla_{\mathbf{x}}^2 u, \nabla_{\mathbf{x}} u, u)$, which is our case.

Iteration approach. The first approach we developed in Chen-Feldman [47, 48] for the steady potential flow equation is an iteration scheme based on the non-degeneracy of the free boundary condition: the jump of the normal derivative of a solution across the free boundary has a strictly positive lower bound. Our iteration process is as follows. Suppose that the domain Ω_k^+ is given so that $S_k := \partial\Omega_k^+ \setminus \partial\Omega$ is $C^{1,\alpha}$. Consider the oblique derivative problem in Ω_k^+ obtained by rewriting the (modified) equation (5.1) and free boundary condition (5.4) in terms of the function $u := \varphi - \varphi_0^+$. Then the problem has the following form:

$$\begin{aligned} &\nabla_{\mathbf{x}} \cdot \mathbf{A}(\mathbf{x}, \nabla_{\mathbf{x}} u) = F(\mathbf{x}) && \text{in } \Omega_k^+ := \{u > 0\}, \\ &\mathbf{A}(\mathbf{x}, \nabla_{\mathbf{x}} u) \cdot \mathbf{n} = G(\mathbf{x}, \mathbf{n}) && \text{on } \mathcal{S} := \partial\Omega_k^+ \setminus \partial\Omega, \end{aligned} \tag{5.20}$$

plus the fixed boundary conditions on $\partial\Omega_k^+ \cap \partial\Omega$ and the conditions at infinity. The equation is quasilinear, uniformly elliptic, $\mathbf{A}(\mathbf{x}, 0) \equiv 0$, while $G(\mathbf{x}, \mathbf{n})$ has a certain structure. Let $u_k \in C^{1,\alpha}(\overline{\Omega_k^+})$ be the solution of (5.20). Then $\|u_k\|_{1,\alpha,\Omega_k^+}$ is estimated to be small if the perturbation is small, where appropriate weighted Hölder norms are actually needed in the unbounded domains. The function $\varphi_k := \varphi_0^+ + u_k$ from Ω_k^+ is extended to Ω so that the $C^{1,\alpha}$–norm of $\varphi_k - \varphi_0^+$ in Ω is controlled by $\|u_k\|_{1,\alpha,\Omega_k^+}$. For the next step, define $\Omega_{k+1}^+ := \{\mathbf{x} \in \Omega : \varphi_k(\mathbf{x}) < \varphi^-(\mathbf{x})\}$. Note that, since $\|\varphi_k - \varphi_0^+\|_{1,\alpha,\Omega}$ and $\|\varphi^- - \varphi_0^-\|_{1,\alpha,\Omega}$ are small, we have $|\nabla_{\mathbf{x}}\varphi^-| - |\nabla_{\mathbf{x}}\varphi_k| \geq \delta > 0$ in Ω, and this nondegeneracy implies that $S_{k+1} := \partial\Omega_{k+1}^+ \setminus \partial\Omega$ is $C^{1,\alpha}$ and its norm is estimated in terms of the data of the problem.

The fixed point Ω^+ of this process determines a solution of the free boundary problem since the corresponding solution φ satisfies $\Omega^+ = \{\varphi < \varphi^-\}$ and the Rankine-Hugoniot condition (5.4) holds on $\mathcal{S} := \partial\Omega^+ \cap \Omega$. On the other hand, the elliptic estimates alone are not sufficient to get the existence of a fixed point, because the right-hand side of the boundary condition in problem (5.20) depends on the unit normal $\mathbf{n}$ of the free boundary. One way is to require the orthogonality of the flat shock-fronts so that $\rho(|\nabla_{\mathbf{x}}\varphi_0^+|^2)\nabla_{\mathbf{x}}\varphi_0^+ = \rho(|\nabla_{\mathbf{x}}\varphi_0^-|^2)\nabla_{\mathbf{x}}\varphi_0^-$ in Ω to obtain better estimates for the iteration and prove the existence of a fixed point. For more details, see Chen-Feldman [47, 48].

Partial hodograph approach. The second approach we have developed in [47] is a partial hodograph procedure, with which we can handle the existence and stability of M-D transonic shock-fronts that are not nearly orthogonal to the flow direction. One of the main ingredients in this new approach is to employ a partial hodograph transform to reduce the free boundary problem to a conormal boundary value problem for the corresponding nonlinear second-order elliptic equation of divergence form and then to develop techniques to solve the conormal boundary value problem. To achieve this, the strategy is to construct first solutions in the intersection domains between the physical unbounded domain under consideration and a series of half balls with radius R, then make uniform estimates in R, and finally send $R \to \infty$. It requires delicate apriori estimates to achieve this. A uniform bound in a weighted L^∞-norm can be achieved by both employing a comparison principle and identifying a global function with the same decay rate as the fundamental solution of the elliptic equation with constant coefficients which controls the solutions. Then, by scaling arguments, the uniform estimates can be obtained in a weighted Hölder norm for the solutions, which lead to the existence of a solution in the unbounded domain with some decay rate at infinity. For such decaying solutions, a comparison principle holds, which implies the uniqueness for the conormal problem. Finally, by the gradient estimate, the limit function can be shown to be a solution of the M-D transonic shock problem, and then the existence result can be extended to the case that the regularity of the steady perturbation is only $C^{1,1}$. We can further prove that the M-D transonic shock-front solution is stable with respect to the $C^{2,\alpha}$ supersonic perturbation.

When the regularity of the steady perturbation is $C^{3,\alpha}$ or higher: $\|\varphi^- - \varphi_0^-\|_{3,\alpha,\Omega_1}^{(d-1)} \le \sigma$, we have introduced another simpler approach, the implicit function approach, to deal with the existence and stability problem in Chen-Feldman [47].

Other approaches can be found in Canic-Keyfitz-Lieberman [31], Canic-Keyfitz-Kim [33], Chen-Chen-Feldman [43], Chen [72, 73], Zheng [215], and the references cited therein.

5.5.2. Weak convergence approaches. Recent investigations on conservation laws based on weak convergence methods suggested that the method of compensated compactness be amenable to flows which exhibit both elliptic and hyperbolic regimes. In [163] (also see [164]), Morawetz layed out a program for proving the existence of the steady transonic flow problem about a bump profile in the upper half plane (which is equivalent to a symmetric profile in the whole plane). It was shown that, if the key hypotheses of the method of compensated compactness could be satisfied, now known as a "compactness framework" (see Chen [39]), then indeed there would exist a weak solution to the problem of flow over a bump which is exhibited by subsonic and supersonic regimes, i.e., transonic flow.

The compactness framework for system (5.17) can be formulated as follows. Let a sequence of functions $\mathbf{w}^\varepsilon(\mathbf{x}) = (u^\varepsilon, v^\varepsilon)(\mathbf{x})$ defined on an open set $\Omega \subset \mathbb{R}^2$ satisfy the following set of conditions:

(A.1) $q^\varepsilon(\mathbf{x}) = |\mathbf{w}^\varepsilon(\mathbf{x})| \le q_\sharp$ a.e. in Ω for some positive constant $q_\sharp < q_{cav}$;

(A.2) $\nabla_\mathbf{x} \cdot \mathbf{Q}_\pm(\mathbf{w}^\varepsilon)$ are confined in a compact set in $H^{-1}_{\rm loc}(\Omega)$ for entropy pairs $\mathbf{Q}_\pm = (Q_{1\pm}, Q_{2\pm})$, where $\mathbf{Q}_\pm(\mathbf{w}^\varepsilon)$ are confined to a bounded set uniformly in $L^\infty_{\rm loc}(\Omega)$.

When (A.1) and (A.2) hold, the Young measure $\nu_\mathbf{x}(\mathbf{w}), \mathbf{w} = (u, v)$, determined by the uniformly bounded sequence of functions $\mathbf{w}^\varepsilon$ is constrained by the following commutator relation:

$$\begin{aligned}\langle \nu_\mathbf{x},\, Q_{1+}Q_{2-} - Q_{1-}Q_{2+}\rangle = \;& \langle \nu_\mathbf{x},\, Q_{1+}\rangle\langle \nu_\mathbf{x},\, Q_{2-}\rangle \\ & -\langle \nu_\mathbf{x},\, Q_{1-}\rangle\langle \nu_\mathbf{x},\, Q_{2+}\rangle.\end{aligned} \tag{5.21}$$

The main point for the compensated compactness framework is to prove that $\nu_\mathbf{x}$ is a Dirac mass by using entropy pairs, which implies the compactness of the sequence $\mathbf{w}^\varepsilon(\mathbf{x}) = (u^\varepsilon, v^\varepsilon)(\mathbf{x})$ in $L^1_{\rm loc}(\Omega)$. In this context, It is needed Morawetz [163] to presume the existence of an approximating sequence parameterized by ε to their problems satisfying (A.1) and (A.2) so that they could exploit the commutator identity and obtain the strong convergence in $L^1_{\rm loc}(\Omega)$ to a weak solution of their problems.

As it turns out, there is a classical problem where (A.1) and (A.2) hold trivially, i.e., the sonic limit of subsonic flows. In that case, we return to the result by Bers [15] and Shiffman [189], which says that, if the speed at infinity, q_∞, is less than some $\hat{q}$, there is a smooth unique solution of the problem and ask what happens as $q_\infty \nearrow \hat{q}$. In this case, the flow develops sonic points and the governing equations become degenerate elliptic. Thus, if we set $\varepsilon = \hat{q} - q_\infty$ and examine a sequence of exact smooth solutions to our system, we see trivially that (A.1) is satisfied since $|q^\varepsilon| \le q_*$, and (A.2) is also satisfied since $\nabla_\mathbf{x} \cdot \mathbf{Q}_\pm(\mathbf{w}^\varepsilon) = 0$ along our solution sequence. The effort is in finding entropy pairs which guarantee the Young measure $\nu_\mathbf{x}$ reduces to a Dirac mass. Ironically, the original conservation equations of momentum in fact provide two sets of entropy pairs, while the irrotationality and mass conservation equations provide another two sets. This observation has been explored in detail in Chen-Dafermos-Slemrod-Wang [45].

What then about the fully transonic problem of flow past an obstacle or bump where $q_\infty > \hat{q}$? In Chen-Slemrod-Wang [61], we have provided some of the ingredients to satisfying (A.1) and (A.2) as explained in §5.4. On the other hand, (A.2) is easily obtained from the viscous formulation by using a special entropy pair of Osher-Hafez-Whitlow [172]. In fact, this entropy pair is very important: It guarantees that the inviscid limit of the above viscous system satisfies a physically meaningful "entropy" condition (Theorem 2 of [172]). With (A.1) and (A.2) satisfied, the compensated compactness argument then applies, to yield the strong convergence in $L^1_{\mathrm{loc}}(\Omega)$ of our approximate solutions.

Some further compensated compactness frameworks have been developed for solving the weak continuity of solutions to the Gauss-Codazzi-Ricci equations in Chen-Slemrod-Wang [62, 63]. Also see Dacorogna [88], Evans [103], and the references cited therein for various weak convergence methods and techniques.

6. Shock reflection-diffraction and self-similar solutions. One of the most challenging problems is to study solutions with data that give rise to self-similar solutions (such solutions include Riemann solutions). For the Euler equations (1.4) for $\mathbf{x} \in \mathbb{R}^2$, the self-similar solutions

$$(\rho, \mathbf{v}, p) = (\rho, \mathbf{v}, p)(\xi, \eta), \qquad (\xi, \eta) = \mathbf{x}/t, \tag{6.1}$$

are determined by

$$\begin{cases} \partial_\xi(\rho U) + \partial_\eta(\rho V) = -2\rho, \\ \partial_\xi(\rho U^2 + p) + \partial_\eta(\rho UV) = -3\rho U, \\ \partial_\xi(\rho UV) + \partial_\eta(\rho V^2 + p) = -3\rho V, \\ \partial_\xi(\rho U(E + p/\rho)) + \partial_\eta(\rho V(E + p/\rho)) = -2\rho(E + p/\rho), \end{cases} \tag{6.2}$$

where $(U, V) = \mathbf{v} - (\xi, \eta)$ is the pseudo-velocity and $E = e + \frac{1}{2}(U^2 + V^2)$.

The four eigenvalues are

$$\lambda_0 = \frac{V}{U} \text{ (2-multiplicity)}, \qquad \lambda_\pm = \frac{UV \pm c\sqrt{U^2 + V^2 - c^2}}{U^2 - c^2},$$

where $c = \sqrt{p_\rho(\rho, S)}$ is the sonic speed. When $U^2 + V^2 > c^2$, system (6.2) is hyperbolic with four real eigenvalues and the flow is called pseudo-supersonic, simply called supersonic without confusion. When $U^2 + V^2 < c^2$, system (6.2) is hyperbolic-elliptic composite type (two repeated eigenvalues are real and the other two are complex): two equations are hyperbolic and the other two are elliptic. The region $U^2 + V^2 = c^2$ in the (ξ, η)-plane is simply called the sonic region in the flow. In general, system (6.2) is hyperbolic-elliptic mixed and composite type, and the flow is transonic. For a bounded solution $(\rho, \mathbf{v}, p)$, the flow must be supersonic when $\xi^2 + \eta^2 \to \infty$.

An important prototype problem for both practical applications and the theory of M-D complex wave patterns is the problem of diffraction of a shock-front which is incident along an inclined ramp. When a plane shock-front hits a wedge head on, a self-similar reflected shock-front moves outward as the original shock-front moves forward (cf. [13, 34, 84, 116, 165, 183, 201]). Then the problem of shock reflection-diffraction by a wedge can be formulated as follows:

Problem 6.1 (Initial-boundary value problem). *Seek a solution of system* (1.4) *satisfying the initial condition at* $t = 0$*:*

$$(\rho, \mathbf{v}, p) = \begin{cases} (\rho_0, 0, 0, p_0), & |x_2| > x_1 \tan\theta_w, x_1 > 0, \\ (\rho_0, u_1, 0, p_1), & x_1 < 0; \end{cases} \tag{6.3}$$

and the slip boundary condition along the wedge boundary:

$$\mathbf{v} \cdot \mathbf{n} = 0, \tag{6.4}$$

where $\mathbf{n}$ *is the exterior unit normal to the wedge boundary, and state* (0) *and* (1) *satisfy*

$$u_1 = \sqrt{\frac{(p_1 - p_0)(\rho_1 - \rho_0)}{\rho_0 \rho_1}}, \quad \frac{p_1}{p_0} = \frac{(\gamma+1)\rho_1 - (\gamma-1)\rho_0}{(\gamma+1)\rho_0 - (\gamma-1)\rho_1}, \quad \rho_1 > \rho_0. \tag{6.5}$$

Given ρ_0, p_0, ρ_1, and $\gamma > 1$, the other variables u_1 and p_1 are determined by (6.5). In particular, the Mach number $M_1 = \frac{u_1}{c_1}$ for state (1) is determined by $M_1^2 = \frac{2(\rho_1 - \rho_0)^2}{\rho_0((\gamma+1)\rho_1 - (\gamma-1)\rho_0)}$.

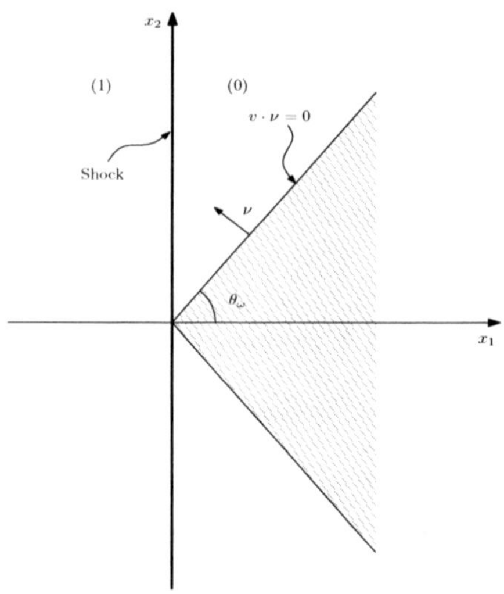

FIG. 1. *Initial-boundary value problem.*

Since the initial-boundary value problem, *Problem 6.1*, is invariant under the self-similar scaling, we seek self-similar solutions (6.1) governed by system (6.2). Since the problem is symmetric with respect to the axis $\eta = 0$, it suffices to consider the problem in the half-plane $\eta > 0$ outside the half-wedge:

$$\Lambda := \{\xi < 0, \eta > 0\} \cup \{\eta > \xi \tan\theta_w,\ \xi > 0\}.$$

Then *Problem 6.1* in the $(t, \mathbf{x})$–coordinates can be formulated as the following boundary value problem in the self-similar coordinates (ξ, η):

Problem 6.2 (Boundary value problem in the unbounded domain). *Seek a solution to system* (6.2) *satisfying the slip boundary condition on the wedge boundary:* $(U, V) \cdot \mathbf{n} = 0$ *on* $\partial\Lambda = \{\xi \le 0, \eta = 0\} \cup \{\xi > 0, \eta \ge \xi \tan\theta_w\}$, *the asymptotic boundary condition as* $\xi^2 + \eta^2 \to \infty$:

$$(\rho, U + \xi, V + \eta, p) \longrightarrow \begin{cases} (\rho_0, 0, 0, p_0), & \xi > \xi_0, \eta > \xi \tan\theta_w, \\ (\rho_1, u_1, 0, p_1), & \xi < \xi_0, \eta > 0. \end{cases}$$

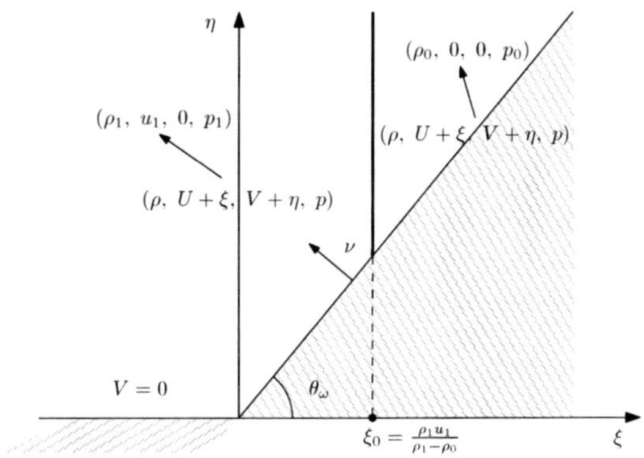

FIG. 2. *Boundary value problem in the unbounded domain* Λ.

For our problem, since φ_1 does not satisfy the slip boundary condition (6.11), the solution must differ from φ_1 in $\{\xi < \xi_0\} \cap \Lambda$, thus a shock diffraction-diffraction by the wedge vertex occurs. The experimental, computational, and asymptotic analysis shows that various patterns of reflected shock-fronts may occur, including the regular and Mach reflections (cf. [13, 84, 116, 121, 128, 129, 132, 133, 148, 155, 165, 191, 192, 197, 198, 199, 201, 209]). It is expected that the solutions of *Problem 6.2* contain all possible patterns of shock reflection-diffraction configurations as observed. For a wedge angle $\theta_w \in (0, \frac{\pi}{2})$, different reflection-diffraction patterns may occur. Various criteria and conjectures have been proposed for the existence of configurations for the patterns (cf. Ben-Dor [13]). One of the most important conjectures made by von Neumann [170, 171] in 1943 is the *detachment conjecture*, which states that the regular reflection-diffraction configuration may exist globally whenever the two-shock configuration (one is the incident shock-front and the other the reflected shock-front) exists locally around the point P_0 (see Fig. 3). The following theorem was rigorously shown in Chang-Chen [34] (also see Sheng-Yin [187], Bleakney-Taub [19], Neumann [170, 171]).

THEOREM 6.1 (Local theory). *There exists $\theta_d = \theta_d(\rho_0, \rho_1, \gamma) \in (0, \frac{\pi}{2})$ such that, when $\theta_w \in (\theta_d, \frac{\pi}{2})$, there are two states (2):$(\rho_2^a, U_2^a, V_2^a, p_2^a)$ and $(\rho_2^b, U_2^b, V_2^b, p_2^b)$ such that $|(U_2^a, V_2^a)| > |(U_2^b, V_2^b)|$ and $|(U_2^b, V_2^b)| < c(\rho_2^b, S_2^b)$.*

The von Neumann Detachment Conjecture ([170, 171]): *There exists a global regular reflection-diffraction configuration whenever the wedge angle θ_w is in $(\theta_d, \frac{\pi}{2})$.*

It is clear that the regular reflection-diffraction configuration is not possible without a local two-shock configuration at the reflection point on the wedge, so this is the weakest possible criterion. In this case, the local theory indicates that there are two possible choices for state (2). There had been a long debate to determine which one is more physical for the local theory; see Courant-Friedrichs [84], Ben-Dor [13], and the references cited therein. Since the reflection-diffraction problem is not a local problem, we take a different point of view that the selection of state (2) should be determined by the global features of the problem, more precisely, by the stability of the configuration with respect to the wedge angle θ_w, rather than the local features of the problem.

Stability Criterion to Select the Correct State (2) (Chen-Feldman [49]): *Since the solution is unique when the wedge angle $\theta_w = \frac{\pi}{2}$, it is required that the global regular reflection-diffraction configuration be stable and converge to the unique normal reflection solution when $\theta_w \to \frac{\pi}{2}$, provided that such a global configuration can be constructed.*

We employ this stability criterion to conclude that our choice for state (2) must be $(\rho_2^a, U_2^a, V_2^a, p_2^a)$. In general, $(\rho_2^a, U_2^a, V_2^a, p_2^a)$ may be supersonic or subsonic. If it is supersonic, the propagation speeds are finite and state (2) is completely determined by the local information: state (1), state (0), and the location of the point P_0. This is, any information from the reflected region, especially the disturbance at the corner P_3, cannot travel towards the reflection point P_0. However, if it is subsonic, the information can reach P_0 and interact with it, potentially altering the reflection-diffraction type. This argument motivated the following conjecture:

The von Neumann Sonic Conjecture [170, 171]: *There exists a regular reflection-diffraction configuration when $\theta_w \in (\theta_s, \frac{\pi}{2})$ for $\theta_s > \theta_d$ such that $|(U_2^a, V_2^a)| > c_2^a$ at P_0.*

If state (2) is sonic when $\theta_w = \theta_s$, then $|(U_2^a, V_2^a)| > c_2^a$ for any $\theta_w \in (\theta_s, \frac{\pi}{2})$. This sonic conjecture is stronger than the detachment one. In fact, the regime between the angles θ_s and θ_d is very narrow and is only fraction of a degree apart; see Sheng-Yin [187].

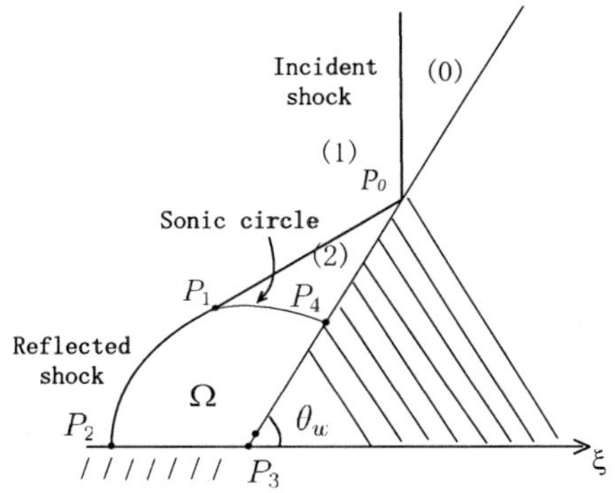

FIG. 3. *Regular reflection-diffraction configuration.*

Following the argument in Chen-Feldman [50], we have

THEOREM 6.2 (Chen-Feldman [50]). *Let (ρ, U, V, p) be a solution of Problem 6.2 such that (ρ, U, V, p) is $C^{0,1}$ in the open region $P_0P_1P_2P_3$ and the gradient of the tangential component of (U, V) is continuous across the sonic arc Γ_{sonic}. Let Ω_1 be the subregion of Ω formed by the fluid trajectories past the sonic arc Γ_{sonic}. Then, in Ω_1, the potential flow equation for self-similar solutions:*

$$\nabla \cdot \big(\rho(\nabla\varphi, \varphi)\nabla\varphi\big) + 2\rho(\nabla\varphi, \varphi) = 0, \tag{6.6}$$

with $\rho(|\nabla\varphi|^2, \varphi) = \big(\rho_0^{\gamma-1} - (\gamma-1)(\varphi + \frac{1}{2}|\nabla\varphi|^2)\big)^{\frac{1}{\gamma-1}}$, coincides with the full Euler equations (6.2), that is, equation (6.6) is exact in the domain Ω_1.

The regions such as Ω_1 also exist in various Mach reflection-diffraction configurations. Theorem 6.2 applies to such regions whenever the solution (ρ, U, V, p) is $C^{0,1}$ and the gradient of the tangential component of (U, V) is continuous. In fact, Theorem 6.2 indicates that, for the solutions φ of (6.6), the $C^{1,1}$–regularity of φ and the continuity of the tangential component of the velocity field $(U, V) = \nabla\varphi$ are optimal across the sonic arc Γ_{sonic}.

Equation (6.6) is a nonlinear equation of mixed elliptic-hyperbolic type. It is elliptic if and only if $|\nabla\varphi| < c(|\nabla\varphi|^2, \varphi, \rho_0^{\gamma-1})$, which is equivalent to

$$|\nabla\varphi| < q_*(\varphi, \rho_0, \gamma) := \sqrt{\frac{2}{\gamma+1}\big(\rho_0^{\gamma-1} - (\gamma-1)\varphi\big)}. \tag{6.7}$$

For the potential equation (6.6), shock-fronts are discontinuities in the pseudo-velocity $\nabla\varphi$. That is, if D^+ and $D^- := D \setminus \overline{D^+}$ are two nonempty open subsets of $D \subset \mathbb{R}^2$, and $\mathcal{S} := \partial D^+ \cap D$ is a C^1–curve where $D\varphi$ has a jump, then $\varphi \in W^{1,1}_{loc}(D) \cap C^1(D^\pm \cup S) \cap C^2(D^\pm)$ is a global weak solution of (6.6) in D if and only if φ is in $W^{1,\infty}_{loc}(D)$ and satisfies equation (6.6) in $D^\pm$ and the Rankine-Hugoniot condition on $\mathcal{S}$:

$$\big[\rho(|\nabla\varphi|^2, \varphi)\nabla\varphi \cdot \mathbf{n}\big]_{\mathcal{S}} = 0. \tag{6.8}$$

Then the plane incident shock-front solution in the $(t, \mathbf{x})$–coordinates with states $(\rho, \nabla_{\mathbf{x}}\Phi) = (\rho_0, 0, 0)$ and $(\rho_1, u_1, 0)$ corresponds to a continuous weak solution φ of (6.6) in the self-similar coordinates (ξ, η) with the following form:

$$\varphi_0(\xi,\eta) = -\frac{1}{2}(\xi^2+\eta^2) \qquad \text{for } \xi > \xi_0, \tag{6.9}$$

$$\varphi_1(\xi,\eta) = -\frac{1}{2}(\xi^2+\eta^2) + u_1(\xi-\xi_0) \qquad \text{for } \xi < \xi_0, \tag{6.10}$$

respectively, where $u_1 = \sqrt{\frac{2(\rho_1-\rho_0)(\rho_1^{\gamma-1}-\rho_0^{\gamma-1})}{(\gamma-1)(\rho_1+\rho_0)}} > 0$ and $\xi_0 = \frac{\rho_1 u_1}{\rho_1-\rho_0} > 0$ are the velocity of state (1) and the location of the incident shock-front, uniquely determined by (ρ_0, ρ_1, γ) through (6.8). Then $P_0 = (\xi_0, \xi_0 \tan\theta_w)$ in Fig. 2, and *Problem 6.2* in the context of the potential flow equation can be formulated as

Problem 6.3 (Boundary value problem) (cf. Fig. 2). *Seek a solution φ of* (6.6) *in the self-similar domain Λ with the boundary condition on $\partial\Lambda$:*

$$\nabla\varphi \cdot \mathbf{n}|_{\partial\Lambda} = 0, \tag{6.11}$$

and the asymptotic boundary condition at infinity:

$$\varphi \to \bar{\varphi} := \begin{cases} \varphi_0 & \text{for } \xi > \xi_0, \eta > \xi\tan\theta_w, \\ \varphi_1 & \text{for } \xi < \xi_0, \eta > 0, \end{cases} \qquad \text{when } \xi^2+\eta^2 \to \infty, \tag{6.12}$$

where (6.12) *holds in the sense that* $\lim_{R\to\infty} \|\varphi - \overline{\varphi}\|_{C(\Lambda\setminus B_R(0))} = 0$.

In Chen-Feldman [49], we first followed the von Neumann criterion and the stability criterion introduced above to establish a local existence theory of regular shock reflection near the reflection point P_0 in the level of potential flow, when the wedge angle is large and close to $\frac{\pi}{2}$. In this case, the vertical line is the incident shock-front that hits the wedge at $P_0 = (\xi_0, \xi_0\tan\theta_w)$, and state (0) and state (1) ahead of and behind S are given by φ_0 and φ_1 defined in (6.9) and (6.10), respectively. The solutions φ and φ_1 differ only in the domain $P_0P_1P_2P_3$ because of shock diffraction by the wedge, where the curve $P_0P_1P_2$ is the reflected shock-front with the straight segment P_0P_1. State (2) behind P_0P_1 can be computed explicitly with the form:

$$\varphi_2(\xi,\eta) = -\frac{1}{2}(\xi^2+\eta^2) + u_2(\xi-\xi_0) + (\eta - \xi_0\tan\theta_w)u_2\tan\theta_w, \tag{6.13}$$

which satisfies $\nabla\varphi\cdot\mathbf{n} = 0$ on $\partial\Lambda \cap \{\xi > 0\}$; the constant velocity u_2 and the angle θ_s between P_0P_1 and the ξ–axis are determined by $(\theta_w, \rho_0, \rho_1, \gamma)$ from the two algebraic equations expressing (6.8) and continuous matching of states (1) and (2) across P_0P_1 as in Theorem 6.1. Moreover, φ_2 is the unique solution in the domain $P_0P_1P_4$, as argued in [34, 183]. Hence $P_1P_4 := \Gamma_{sonic} = \partial\Omega \cap \partial B_{c_2}(u_2, u_2 \tan\theta_w)$ is the sonic arc of state (2) with center $(u_2, u_2 \tan\theta_w)$ and radius c_2.

It should be noted that, in order that the solution φ in the domain Ω is a part of the global solution to *Problem 6.3*, that is, φ satisfies the equation in the sense of distributions in Λ, especially across the sonic arc P_1P_4, it is required that $\nabla(\varphi - \varphi_2) \cdot \mathbf{n}|_{P_1P_4} = 0$. That is, we have to match our solution with state (2), which is the necessary condition for our solution in the domain Ω to be a part of the global solution. To achieve this, we have to show that our solution is at least C^1 with $\nabla(\varphi - \varphi_2) = 0$ across P_1P_4. Then *Problem 6.3* can be reformulated as the following free boundary problem:

Problem 6.4 (Free boundary problem). *Seek a solution φ and a free boundary $P_1P_2 = \{\xi = f(\eta)\}$ such that*

(i) *$f \in C^{1,\alpha}$ and*

$$\Omega := \{\xi > f(\eta)\} \cap \Lambda = \{\varphi < \varphi_1\} \cap \Lambda; \tag{6.14}$$

(ii) *φ satisfies the free boundary condition (6.8) along P_1P_2;*

(iii) *$\varphi \in C^{1,\alpha}(\overline{\Omega}) \cap C^2(\Omega)$ solves (6.6) in Ω, is subsonic in Ω, and satisfies*

$$(\varphi - \varphi_2, \nabla(\varphi - \varphi_2) \cdot \mathbf{n})|_{P_1P_2} = 0, \tag{6.15}$$

$$\nabla\varphi \cdot \mathbf{n}|_{P_3P_4 \cup \Gamma_{symm}} = 0. \tag{6.16}$$

The boundary condition on Γ_{symm} implies that $f'(0) = 0$ and thus ensures the orthogonality of the free boundary with the ξ-axis. Formulation (6.14) implies that the free boundary is determined by the level set $\varphi = \varphi_1$. The free boundary condition (6.8) along P_1P_2 is the conormal boundary condition on P_1P_2. Condition (6.15) ensures that the solution of the free boundary problem in Ω is a part of the global solution. Condition (6.16) is the slip boundary condition. *Problem 6.4* involves two types of transonic flow: one is a continuous transition through the sonic arc P_1P_4 as a fixed boundary from the supersonic region (2) to the subsonic region Ω; the other is a jump transition through the transonic shock-front as a free boundary from the supersonic region (1) to the subsonic region Ω.

In Chen-Feldman [49, 50, 51], we have developed a rigorous mathematical approach to solve the von Neumann sonic conjecture, *Problem 6.4*, and established a global theory for solutions of regular reflection-diffraction up to the sonic angle, which converge to the unique solution of the normal shock reflection when θ_w tends to $\pi/2$. For more details, see Chen-Feldman [49, 50, 51].

The mathematical existence of Mach reflection-diffraction configurations is still open; see [13, 74, 84]. Some progress has been made in the recent years in the study of the 2-D Riemann problem for hyperbolic conservation laws; see [35, 36, 39, 114, 115, 116, 122, 134, 140, 143, 144, 182, 184, 186, 188, 194, 213, 214] and the references cited therein.

Some recent developments in the study of the local nonlinear stability of multidimensional compressible vortex sheets can be found in Coulombel-Secchi [83], Bolkhin-Trakhinin [20], Trakhinin [200], Chen-Wang [68], and the references cited therein. Also see Artola-Majda [7]. For the construction of the non-self-similar global solutions for some multidimensional systems, see Chen-Wang-Yang [67] and the references cited therein.

7. Divergence-measure vector fields and multidimensional conservation laws. Naturally, we want to approach the questions of existence, stability, uniqueness, and long-time behavior of entropy solutions for M-D hyperbolic conservation laws with neither specific reference to any particular method for constructing the solutions nor additional regularity assumptions. Some recent efforts have been in developing a theory of divergence-measure fields to construct a global framework for the analysis of solutions of M-D hyperbolic systems of conservation laws.

Consider system (1.1) in $\mathbb{R}^d$. As discussed in §2.5, the BV bound generically fails for the M-D case. In general, for M-D conservation laws, especially the Euler equations, solutions of (1.1) are expected to be in the following class of entropy solutions:

(i) $\mathbf{u}(t,\mathbf{x}) \in \mathcal{M}(\mathbb{R}^{d+1}_+)$, or $L^p(\mathbb{R}^{d+1}_+)$, $1 \le p \le \infty$;

(ii) $\mathbf{u}(t,\mathbf{x})$ satisfies the Lax entropy inequality:

$$\mu_\eta := \partial_t \eta(\mathbf{u}(t,\mathbf{x})) + \nabla_{\mathbf{x}} \cdot \mathbf{q}(\mathbf{u}(t,\mathbf{x})) \le 0 \tag{7.1}$$

in the sense of distributions for any convex entropy pair $(\eta, \mathbf{q}) : \mathbb{R}^n \to \mathbb{R} \times \mathbb{R}^d$ so that $\eta(\mathbf{u}(t,\mathbf{x}))$ and $\mathbf{q}(\mathbf{u}(t,\mathbf{x}))$ are distributions.

The Schwartz Lemma infers from (7.1) that the distribution μ_η is in fact a Radon measure: $\mathrm{div}_{(t,\mathbf{x})}(\eta(\mathbf{u}(t,\mathbf{x})), \mathbf{q}(\mathbf{u}(t,\mathbf{x}))) \in \mathcal{M}(\mathbb{R}^{d+1}_+)$. Furthermore, when $\mathbf{u} \in L^\infty$, this is also true for any C^2 entropy-entropy flux pair $(\eta, \mathbf{q})$ (η not necessarily convex) if (1.1) has a strictly convex entropy, which was first observed in Chen [38]. More generally, we have

DEFINITION. *Let $\mathcal{D} \subset \mathbb{R}^N$ be open. For $1 \le p \le \infty$, $\mathbf{F}$ is called a $\mathcal{DM}^p(\mathcal{D})$-field if $\mathbf{F} \in L^p(\mathcal{D}; \mathbb{R}^N)$ and*

$$\|\mathbf{F}\|_{\mathcal{DM}^p(\mathcal{D})} := \|\mathbf{F}\|_{L^p(\mathcal{D};\mathbb{R}^N)} + \|div\ \mathbf{F}\|_{\mathcal{M}(\mathcal{D})} < \infty; \tag{7.2}$$

and the field $\mathbf{F}$ is called a $\mathcal{DM}^{ext}(\mathcal{D})$-field if $\mathbf{F} \in \mathcal{M}(\mathcal{D}; \mathbb{R}^N)$ and

$$\|\mathbf{F}\|_{\mathcal{DM}^{ext}(\mathcal{D})} := \|(\mathbf{F}, div\ \mathbf{F})\|_{\mathcal{M}(\mathcal{D})} < \infty. \tag{7.3}$$

Furthermore, for any bounded open set $\mathcal{D} \subset \mathbb{R}^N$, $\mathbf{F}$ is called a $\mathcal{DM}^p_{loc}(\mathbb{R}^N)$-field if $\mathbf{F} \in \mathcal{DM}^p(\mathcal{D})$; and $\mathbf{F}$ is called a $\mathcal{DM}^{ext}_{loc}(\mathbb{R}^N)$-field if $\mathbf{F} \in \mathcal{DM}^{ext}(\mathcal{D})$. A field $\mathbf{F}$ is simply called a $\mathcal{DM}$-field in $\mathcal{D}$ if $\mathbf{F} \in \mathcal{DM}^p(\mathcal{D}), 1 \le p \le \infty$, or $\mathbf{F} \in \mathcal{DM}^{ext}(\mathcal{D})$.

It is easy to check that these spaces, under the respective norms $\|\mathbf{F}\|_{\mathcal{DM}^p(\mathcal{D})}$ and $\|\mathbf{F}\|_{\mathcal{DM}^{ext}(\mathcal{D})}$, are Banach spaces. These spaces are larger than the space of BV-fields. The establishment of the Gauss-Green theorem, traces, and other properties of BV functions in the 1950s (cf. Federer [105]; also Ambrosio-Fusco-Pallara [5], Giusti [112], and Volpert [203]) has significantly advanced our understanding of solutions of nonlinear PDEs and related problems in the calculus of variations, differential geometry, and other areas, especially for the 1-D theory of hyperbolic conservation laws. A natural question is whether the $\mathcal{DM}$–fields have similar properties, especially the normal traces and the Gauss-Green formula to deal with entropy solutions for M-D conservation laws.

On the other hand, motivated by various nonlinear problems from conservation laws, as well as for rigorous derivation of systems of balance laws with measure source terms from the physical principle of balance law and the recovery of Cauchy entropy flux through the Lax entropy inequality for entropy solutions of hyperbolic conservation laws by capturing the entropy dissipation, a suitable notion of normal traces and corresponding Gauss-Green formula for divergence-measure fields are required.

Some earlier efforts were made on generalizing the Gauss-Green theorem for some special situations, and relevant results can be found in Anzellotti [6] for an abstract formulation for $\mathbf{F} \in L^\infty$, Rodrigues [178] for $\mathbf{F} \in L^2$, and Ziemer [216] for a related problem for $\operatorname{div}\mathbf{F} \in L^1$; also see Baiocchi-Capelo [10] and Brezzi-Fortin [29]. In Chen-Frid [54], an explicit way to calculate the suitable normal traces was first observed for $\mathbf{F} \in \mathcal{DM}^\infty$, under which a generalized Gauss-Green theorem was shown to hold, which has motivated the development of a theory of divergence-measure fields in [54, 64, 66].

Some entropy methods based on the theory of divergence-measure fields presented above have been developed and applied for solving various nonlinear problems for hyperbolic conservation laws and related nonlinear PDEs. These problems especially include (i) Stability of Riemann solutions, which may contain rarefaction waves, contact discontinuities, and/or vacuum states, in the class of entropy solutions of the Euler equations in [42, 54, 55]; (ii) Decay of periodic entropy solutions in [52]; (iii) Initial and boundary layer problems in [54, 60, 64, 202]; (iv) Rigorous derivation of systems of balance laws from the physical principle of balance law and the recovery of Cauchy entropy flux through the Lax entropy inequality for entropy solutions of hyperbolic conservation laws by capturing the entropy dissipation in [66];

It would be interesting to develop further the theory of divergence-measure fields and more efficient entropy methods for solving more various problems in PDEs and related areas whose solutions are only measures or L^p functions. For more details, see [54, 64, 66].

We also refer the reader to the related papers on other aspects of multidimensional conservation laws in this volume.

REFERENCES

[1] S. Alinhac, *Existence of rarefaction waves for multidimensional hyperbolic quasilinear systems*, Comm. Partial Diff. Eqs. **14** (1989), pp. 173–230.

[2] H.W. Alt and L.A. Caffarelli, *Existence and regularity for a minimum problem with free boundary*, J. Reine Angew. Math. **325** (1981), pp. 105–144.

[3] H.W. Alt, L.A. Caffarelli, and A. Friedman, *A free-boundary problem for quasilinear elliptic equations*, Ann. Scuola Norm. Sup. Pisa Cl. Sci. (4), **11** (1984), pp. 1–44.

[4] L. Ambrosio and C. De Lellis, *Existence of solutions for a class of hyperbolic systems of conservation laws in several space dimensions*, Int. Math. Res. Not. **2003** (2003), pp. 2205–2220.

[5] L. Ambrosio, N. Fusco, and D. Pallara, *Functions of Bounded Variation and Free Discontinuity Problems*, Oxford Mathematical Monographs, The Clarendon Press, Oxford University Press: New York, 2000.

[6] G. Anzellotti, *Pairings between measures and bounded functions and compensated compactness,* Ann. Mat. Pura Appl. (4) **135** (1983), pp. 293–318.

[7] M. Artola and A. Majda, *Nonlinear development of instabilities in supersonic vortex sheets, I: the basic kink modes*, Phys. D. **28** (1987), pp. 253–281; *II: Resonant interaction among kink modes*, SIAM J. Appl. Math. **49** (1989), pp. 1310–1349; *Nonlinear kink modes for supersonic vortex sheets*, Phys. Fluids A, **1** (1989), pp. 583–596.

[8] M.-J. Bae, G.-Q. Chen, and M. Feldman, *Regularity of solutions to regular shock reflection for potential flow*, Invent. Math. **175** (2009), pp. 505–543.

[9] M.-J. Bae and M. Feldman, *Transonic shocks in multidimensional divergent nozzles*, Preprint, 2010.

[10] C. Baiocchi and A. Capelo, *Variational and Quasi-Variational Inequalities, Applications to Free-Boundary Problems*, Vols. **1, 2**, John Wiley: New York, 1984.

[11] J.M. Ball, *Convexity conditions and existence theorems in nonlinear elasticity*, Arch. Ration. Mech. Anal. **63** (1977), pp. 337–403.

[12] J.M. Ball, *A version of the fundamental theorem for Young measures*, In: PDEs and Continuum Models of Phase Transitions (Nice, 1988), Lecture Notes in Phys. **344** (1989), pp. 207–215, Springer-Verlag: Berlin-New York.

[13] G. Ben-Dor, *Shock Wave Reflection Phenomena*, 2nd Edition, Springer-Verlag: New York, 2007.

[14] S. Benzoni-Gavage and D. Serre, *Multidimensional Hyperbolic Partial Differential Equations. First-Order Systems and Applications*, The Clarendon Press and Oxford University Press: Oxford, 2007.

[15] L. Bers, *Existence and uniqueness of subsonic flows past a given profile*, Comm. Pure Appl. Math. **7** (1954), pp. 441–504.

[16] L. Bers, *Mathematical Aspects of Subsonic and Transonic Gas Dynamics*, John Wiley & Sons, Inc.: New York; Chapman & Hall, Ltd.: London 1958.

[17] H. Berestycki, L. Caffarelli, and L. Nirenberg, *Uniform estimates for regularization of free boundary problems*, In: Analysis and Partial Differential Equations, pp. 567–619, Dekker: New York, 1990.

[18] S. Bianchini and A. Bressan, *Vanshing viscosity solutions of nonlinear hyperbolic systems*, Ann. Math. (2), **161** (2005), 223–342.

[19] W. Bleakney and A.H. Taub, *Interaction of shock waves*, Rev. Modern Phys. **21** (1949), pp. 584–605.

[20] A. Blokhin, and Y. Trakhinin, *Stability of strong discontinuities in fluids and MHD*, In: Handbook of Mathematical Fluid Dynamics, Vol. **1**, pp. 545–652, Elsevier Science B. V.: Amsterdam, The Netherlands, 2002.

[21] G. Boillat, *La Propagation des Ondes*, Gauthier-Villars: Paris, 1965.

[22] G. Boillat, *Chocs caractéristiques*, C. R. Acad. Sci. Paris Sér. A–B, **274A** (1972), pp. 1018–1021; *Involutions des systèmes conservatifs*, C. R. Acad. Sci. Paris, Sèrie I, **307** (1988), pp. 891–894.

[23] F. Bouchut, *On zero pressure gas dynamics*, In: Advances in Kinetic Theory and Computing, pp. 171–190, World Sci. Publishing: River Edge, NJ, 1994.

[24] Y. Brenier, *Hydrodynamic structure of the augmented Born–Infeld equations*, Arch. Ration. Mech. Anal. **172** (2004), pp. 65–91.

[25] Y. Brenier and E. Grenier, *Sticky particles and scalar conservation laws*, SIAM J. Numer. Anal. **35** (1998), pp. 2317–2328.

[26] P. Brenner, *The Cauchy problem for the symmetric hyperbolic systems in L_p*, Math. Scand. **19** (1966), pp. 27–37.

[27] A. Bressan, *Hyperbolic Systems of Conservation Laws: The One Dimensional Cauchy Problem*, Oxford University Press: Oxford, 2000.

[28] A. Bressan, *An ill-posed Cauchy problem for a hyperbolic system in two space dimensions*, Rend. Sem. Mat. Univ. Padova, **110** (2003), pp. 103–117.

[29] F. Brezzi and M. Fortin, *Mixed and Hydrid Finite Element Methods*, Springer-Verlag: New York, 1991.

[30] L.A. Caffarelli, *A Harnack inequality approach to the regularity of free boundaries, I. Lipschitz free boundaries are $C^{1,\alpha}$*, Rev. Mat. Iberoamericana, **3** (1987), pp. 139–162; *II. Flat free boundaries are Lipschitz*, Comm. Pure Appl. Math. **42** (1989), pp. 55–78; *III. Existence theory, compactness, and dependence on X*, Ann. Scuola Norm. Sup. Pisa Cl. Sci. (4), **15** (1989), pp. 583–602.

[31] S. Čanić, B.L. Keyfitz, and G. Lieberman, *A proof of existence of perturbed steady transonic shocks via a free boundary problem*, Comm. Pure Appl. Math. **53** (2000), pp. 484–511.

[32] S. Čanić, B.L. Keyfitz, and E.H. Kim, *Free boundary problems for nonlinear wave systems: Mach stems for interacting shocks*, SIAM J. Math. Anal. **37** (2006), 1947–1977.

[33] S. Čanić, B.L. Keyfitz, and E.H. Kim, *Free boundary problems for a quasilinear degenerate elliptic equation: regular reflection of weak shocks*, Comm. Pure Appl. Math. **55** (2002), pp. 71–92.

[34] T. Chang and G.-Q. Chen, *Diffraction of planar shock along the compressive corner*, Acta Math. Sci. **6** (1986), pp. 241–257.

[35] T. Chang, G.-Q. Chen, and S. Yang, *On the Riemann Problem for two-dimensional Euler equations I: Interaction of shocks and rarefaction waves; II: Interaction of contact discontinuities*, Discrete Contin. Dynam. Systems, **1** (1995), pp. 555–584; **6** (2000), pp. 419–430.

[36] T. Chang and L. Hsiao, *The Riemann Problem and Interaction of Waves in Gas Dynamics,* Longman Scientific & Technical, Harlow; and John Wiley & Sons, Inc.: New York, 1989.

[37] J.-Y. Chemin, *Dynamique des gaz à masse totale finie*, Asymptotic Anal. **3** (1990), pp. 215–220.

[38] G.-Q. Chen, *Hyperbolic systems of conservation laws with a symmetry*, Commun. Partial Diff. Eqs. **16** (1991), pp. 1461–1487.

[39] G.-Q. Chen *Euler Equations and Related Hyperbolic Conservation Laws*, In: Handbook of Differential Equations: Evolutionary Differential Equations, Vol. **2**, pp. 1–104, 2005, Elsevier Science B.V: Amsterdam, The Netherlands.

[40] G.-Q. Chen, *Remarks on global solutions to the compressible Euler equations with spherical symmetry*, Proc. Royal Soc. Edinburgh, **127A** (1997), pp. 243–259.

[41] G.-Q. Chen, *Vacuum states and stability of rarefaction waves for compressible flow*, Meth. Appl. Anal. **7** (2000), pp. 337–362.

[42] G.-Q. Chen and J. Chen, *Vacuum states and global stability of rarefaction waves for the Euler equations for compressible fluids*, J. Hyper. Diff. Eqns. **4** (2007), pp. 105–122.

[43] G.-Q. Chen, J. Chen, and M. Feldman, *Transonic shocks and free boundary problems for the full Euler equations in infinite nozzles*, J. Math. Pures Appl. **88** (2007), pp. 191–218.

[44] G.-Q. Chen, J. Chen, and M. Feldman, *Existence and stability of transonic shocks in the full Euler flows past wedges*, Preprint, November 2010.

[45] G.-Q. Chen, C.M. Dafermos, M. Slemrod, and D. Wang, *On two-dimensional sonic-subsonic flow*, Commun. Math. Phys. **271** (2007), pp. 635–647.

[46] G.-Q. Chen and B. Fang, *Stability of transonic shock-fronts in steady potential flow past a perturbed cone*, Discrete Conti. Dynamical Systems, **23** (2009), pp. 85–114.

[47] G.-Q. Chen and M. Feldman, *Multidimensional transonic shocks and free boundary problems for nonlinear equations of mixed type*, J. Amer. Math. Soc. **16** (2003), pp. 461–494; *Steady transonic shocks and free boundary problems in infinite cylinders for the Euler equations*, Comm. Pure Appl. Math. **57** (2004), pp. 310–356; *Free boundary problems and transonic shocks for the Euler equations in unbounded domains*, Ann. Scuola Norm. Sup. Pisa Cl. Sci. (5), **3** (2004), pp. 827–869.

[48] G.-Q. Chen and M. Feldman, *Existence and stability of multidimensional transonic flows through an infinite nozzle of arbitrary cross-sections*, Arch. Ration. Mech. Anal. **184** (2007), pp. 185–242

[49] G.-Q. Chen and M. Feldman, *Potential theory for shock reflection by a large-angle wedge*, Proceedings of the National Academy of Sciences USA (PNAS), **102** (2005), pp. 15368–15372; *Global solutions to shock reflection by large-angle wedges for potential flow*, Annals of Math. **171** (2010), 1067–1182.

[50] G.-Q. Chen and M. Feldman, *Shock reflection-diffraction and multidimensional conservation laws*, In: Hyperbolic Problems: Theory, Numerics, and Applications, pp. 25–51, Proc. Sympos. Appl. Math. **67**, AMS: Providence, RI, 2010.

[51] G.-Q. Chen and M. Feldman, *Shock reflection-diffraction and nonlinear partial differential equations of mixed type*, Contemporary Mathematics, AMS: Providence (to appear); *Potential Theory for Shock Reflection-Diffraction and von Neumann's Conjectures*, Preprint, July 2010.

[52] G.-Q. Chen and H. Frid, *Decay of entropy solutions of nonlinear conservation laws*, Arch. Ration. Mech. Anal. **146** (1999), pp. 95–127; *Large-time behavior of entropy solutions of conservation laws*, J. Diff. Eqs. **152** (1999), pp. 308–357.

[53] G.-Q. Chen and H. Frid, *Large-time behavior of entropy solutions in L^∞ for multidimensional conservation laws*, In: Advances in Nonlinear PDE and Related Areas, pp. 28–44, World Scientific, Singapore, 1998.

[54] G.-Q. Chen and H. Frid, *Divergence-measure fields and hyperbolic conservation laws*, Arch. Ration. Mech. Anal. **147** (1999), pp. 89–118; *Extended divergence-measure fields and the Euler equations for gas dynamics*, Commun. Math. Phys. **236** (2003), pp. 251–280.

[55] G.-Q. Chen, H. Frid, and Y. Li, *Uniqueness and stability of Riemann solutions with large oscillation in gas dynamics,* Commun. Math. Phys. **228** (2002), pp. 201–217.

[56] G.-Q. Chen and P.-T. Kan, *Hyperbolic conservation laws with umbilic degeneracy (I)–(II)*: Arch. Ration. Mech. Anal. **130** (1995), pp. 231–276; **160** (2001), pp. 325–354.

[57] G.-Q. Chen and P. LeFloch, *Compressible Euler equations with general pressure law*, Arch. Ration. Mech. Anal. **153** (2000), pp. 221–259; *Existence theory for the isentropic Euler equations*, Arch. Ration. Mech. Anal. **166** (2003), pp. 81–98.

[58] G.-Q. Chen and T.-H. Li *Well-posedness for two-dimensional steady supersonic Euler flows past a Lipschitz wedge*, J. Diff. Eqs. **244** (2008), pp. 1521–1550.

[59] G.-Q. Chen and H. Liu, *Formation of delta-shocks and vacuum states in the vanishing pressure limit of solutions to the Euler equations for isentropic fluids*, SIAM J. Math. Anal. **34** (2003), pp. 925–938; *Concentration and cavitation in the vanishing pressure limit of solutions to the Euler equations for nonisentropic fluids*, Phys. D, **189** (2004), pp. 141–165.

[60] G.-Q. Chen and M. Rascle, *Initial layers and uniqueness of weak entropy solutions to hyperbolic conservation laws*, Arch. Ration. Mech. Anal. **153** (2000), pp. 205–220.

[61] G.-Q. Chen, M. Slemrod, and D. Wang, *Vanishing viscosity method for transonic flow*, Arch. Ration. Mech. Anal. **189** (2008), pp. 159–188.

[62] G.-Q. Chen, M. Slemrod, and D. Wang, *Isometric immersion and compensated compactness*, Commun. Math. Phys. **294** (2010), pp. 411–437.

[63] G.-Q. Chen, M. Slemrod, and D. Wang, *Weak continuity of the Gauss-Codazzi-Ricci equations for isometric embedding*, Proc. Amer. Math. Soc. **138** (2010), pp. 1843–1852.

[64] G.-Q. Chen and M. Torres, *Divergence-measure fields, sets of finite perimeter, and conservation laws*, Arch. Ration. Mech. Anal. **175** (2005), pp. 245–267.

[65] G.-Q. Chen and M. Torres, *On entropy solutions without bounded variation for hyperbolic systems of conservation laws*, Preprint, 2010.

[66] G.-Q. Chen, M. Torres, and W. Ziemer, *Gauss-Green theorem for weakly differentiable fields, sets of finite perimeter, and balance laws*, Comm. Pure Appl. Math. **62** (2009), pp. 242–304; *Measure-theoretical analysis and nonlinear conservation laws*, Pure Appl. Math. Quarterly, **3** (2007), pp. 841–879.

[67] G.-Q. Chen, D.-H. Wang, and X.-Z. Yang, *Evolution of discontinuity and formation of triple-shock pattern in solutions to a two-dimensional hyperbolic system of conservation laws*, SIAM J. Math. Anal. **41** (2009), 1–25.

[68] G.-Q. Chen and Y.-G. Wang, *Existence and stability of compressible current-vortex sheets in three-dimensional magnetohydrodynamics*, Arch. Ration. Mech. Anal. **187** (2007), pp. 369–408; *Analysis of compressible current-vortex sheets in three-dimensional magnetohydrodynamics*, Preprint, November 2010.

[69] G.-Q. Chen, Y. Zhang, and D. Zhu, *Existence and stability of supersonic Euler flows past Lipschitz wedges*, Arch. Ration. Mech. Anal. **181** (2006), pp. 261–310.

[70] G.-Q. Chen, Y. Zhang, and D. Zhu, *Existence and stability of supersonic vortex sheets in Euler flows past Lipschitz walls*, SIAM J. Math. Anal. **38** (2007), pp. 1660–1693.

[71] S.-X. Chen, *Existence of stationary supersonic flows past a point body*, Arch. Ration. Mech. Anal. **156** (2001), pp. 141–181.

[72] S.-X. Chen, *Stability of transonic shock fronts in two-dimensional Euler systems*, Trans. Amer. Math. Soc. **357** (2005), pp. 287–308; *Transonic shocks in 3-D compressible flow passing a duct with a general section for Euler systems*, Trans. Amer. Math. Soc. **360** (2008), pp. 5265–5289.

[73] S.-X. Chen, *Compressible flow and transonic shock in a diverging nozzle*, Commun. Math. Phys. **289** (2009), pp. 75–106.

[74] S.-X. Chen *Stability of a Mach configuration*, Comm. Pure Appl. Math. **59** (2006), pp. 1–35; *Mach configuration in pseudo-stationary compressible flow*, J. Amer. Math. Soc. **21** (2008), pp. 63–100.

[75] S.-X. Chen and B. Fang, *Stability of transonic shocks in supersonic flow past a wedge*, J. Diff. Eqs. **233** (2007), pp. 105–135.

[76] S.-X. Chen and H. Yuan, *Transonic shocks in compressible flow passing a duct for three-dimensional Euler systems*, Arch. Ration. Mech. Anal. **187** (2008), pp. 523–556.

[77] S.-X. Chen, Z. Xin, and H. Yin, *Global shock waves for the supersonic flow past a perturbed cone*, Commun. Math. Phys. **228** (2002), pp. 47–84.

[78] A.J. Chorin, *Vorticity and Turbulence*, Springer-Verlag: New York, 1994.

[79] D. Christodoulou, *The Formation of Shocks in 3-Dimensional Fluids*, European Mathematical Society (EMS): Zürich, 2007.

[80] D. Christodoulou, *The Formation of Black Holes in General Relativity*, European Mathematical Society (EMS): Zürich, 2009.

[81] J.D. Cole and L.P. Cook, *Transonic Aerodynamics*, Elsevier: Amsterdam, 1986.

[82] P. Constantin, *Some mathematical problems of fluid mechanics*, Proceedings of the International Congress of Mathematicians, Vols. **1, 2** (Zürich, 1994), pp. 1086–1095, Birkhäuser: Basel, 1995.

[83] J.F. Coulombel and P. Secchi, *The stability of compressible vortex sheets in two space dimensions*, Indiana Univ. Math. J. **53** (2004), 941–1012; *Nonlinear compressible vortex sheets in two space dimensions*, Ann. Sci. Éc. Norm. Supér. (4), **41** (2008), 85–139; *Uniqueness of 2-D compressible vortex sheets*, Commun. Pure Appl. Anal. **8** (2009), 1439–1450.

[84] R. Courant and K.O. Friedrichs, *Supersonic Flow and Shock Waves*, Springer-Verlag: New York, 1962.

[85] R. Courant and D. Hilbert, *Methods of Mathematical Physics*, Interscience Publishers, Inc.: New York, 1953.

[86] M.S. Cramer and A.R. Seebass, *Focusing of a weak shock at an arâte*, J. Fluid Mech. **88** (1978), pp. 209–222.

[87] D. Cui and H. Yin, *Global supersonic conic shock wave for the steady supersonic flow past a cone: polytropic gas*, J. Diff. Eqs. **246** (2009), pp. 641–669.

[88] B. Dacorogna, *Weak Continuity and Weak Lower Semicontinuity of Nonlinear Functionals*, Lecture Notes in Math. **922**, Springer-Verlag: New York, 1982.

[89] C.M. Dafermos, *Quasilinear hyperbolic systems with involutions*, Arch. Ration. Mech. Anal. **94** (1986), 373–389.

[90] C.M. Dafermos, *Entropy and the stability of classical solutions of hyperbolic systems of conservation laws*, Lecture Notes in Math. **1640**, pp. 48–69, Springer-Verlag: Berlin, 1996.

[91] C.M. Dafermos, *Hyperbolic Conservation Laws in Continuum Physics*, Springer-Verlag: Berlin, 2010.

[92] C. De Lellis, F. Otto, and M. Westdickenberg, *Structure of entropy solutions for multidimensional scalar conservation laws*, Arch. Ration. Mech. Anal. **170** (2003), pp. 137–184.

[93] C. De Leliss and L. Székelyhidi Jr. *The Euler equations as a differential inclusion*, Ann. of Math. (2), **170** (2009), 1417–1436; *On admissibility criteria for weak solutions of the Euler equations*, Arch. Ration. Mech. Anal. **195** (2010), 225–260.

[94] S. Demoulini, D.M. Stuart, and A.E. Tzavaras, *A variational approximation scheme for three-dimensional elastodynamics with polyconvex energy*, Arch. Ration. Mech. Anal. **157** (2001), pp. 325–344.

[95] R.J. DiPerna, *Compensated compactness and general systems of conservation laws*, Trans. Amer. Math. Soc. **292** (1985), pp. 383–420.

[96] R.J. DiPerna, *Convergence of approximate solutions to conservation laws*, Arch. Ration. Mech. Anal. **88** (1985), pp. 223–270; *Convergence of viscosity method for isentropic gas dynamics*, Commun. Math. Phys. **91** (1983), pp. 1–30.

[97] R.J. DiPerna, *Measure-valued solutions to conservation laws*, Arch. Ration. Mech. Anal. **88** (1985), 223–270.

[98] G. Dong, *Nonlinear Partial Differential Equations of Second Order*, Transl. Math. Monographs, **95**, AMS: Providence, RI, 1991.

[99] W. E, Y. Rykov, and Y. Sinai, *Generalized variational principles, global weak solutions, and behavior with random initial data for systems of conservation laws arising in adhesion particle dynamics*, Commun. Math. Phys. **177** (1996), pp. 349–380.

[100] V. Elling and T.-P. Liu, *The ellipticity principle for self-similar potential flows*, J. Hyperbolic Differential Equations, **2** (2005), 909–917.

[101] V. Elling and T.-P. Liu, *Supersonic flow onto a solid wedge*, Comm. Pure Appl. Math. **61** (2008), pp. 1347–1448.

[102] B. Engquist and W. E, *Large time behavior and homogenization of solutions of two-dimensional conservation laws*, Comm. Pure Appl. Math. **46** (1993), pp. 1–26.

[103] L.C. Evans, *Weak Convergence Methods for Nonlinear Partial Differential Equations*, CBMS **72**, AMS: Providence, RI, 1990.

[104] B. Fang, *Stability of transonic shocks for the full Euler system in supersonic flow past a wedge*, Math. Methods Appl. Sci. **29** (2006), pp. 1–26.

[105] H. Federer, *Geometric Measure Theory*, Springer-Verlag: Berlin-Heidelberg-New York, 1969.

[106] R. Finn and D. Gilbarg, *Three-dimensional subsonic flows, and asymptotic estimates for elliptic partial differential equations*, Acta Math. **98** (1957), pp. 265–296.

[107] S. Friedland, J.W. Robbin, and J. Sylvester, *On the crossing rule*, Comm. Pure Appl. Math. **37** (1984), pp. 19–38.

[108] K.O. Friedrichs, *Symmetric hyperbolic linear differential equations*, Comm. Pure Appl. Math. **7** (1954), pp. 345–392.

[109] K. Friedrichs and P. Lax, *Systems of conservation laws with a convex extension*, Proc. Nat. Acad. Sci., U.S.A. **68** (1971), pp. 1686–1688.

[110] I.M. Gamba and C.S. Morawetz, *A viscous approximation for a 2-D steady semiconductor or transonic gas dynamic flow: existence theorem for potential flow*, Comm. Pure Appl. Math. **49** (1996), pp. 999–1049.

[111] W. Gangbo and M. Westdickenberg, *Optimal transport for the system of isentropic Euler equations*, Comm. Partial Diff. Eqs. **34** (2009), 1041–1073.

[112] E. Giusti, *Minimal Surfaces and Functions of Bounded Variation*, Birkhäuser Verlag: Basel, 1984.

[113] J. Glimm, *Solutions in the large for nonlinear hyperbolic system of equations*, Comm. Pure Appl. Math. **18** (1965), pp. 95–105.

[114] J. Glimm, X. Ji, J. Li, X. Li, P. Zhang, T. Zhang, and Y. Zheng, *Transonic shock formation in a rarefaction Riemann problem for the 2D compressible Euler equations*, SIAM J. Appl. Math. **69** (2008), pp. 720–742.

[115] J. Glimm, C. Klingenberg, O. McBryan, B. Plohr, D. Sharp, and S. Yaniv, *Front tracking and two-dimensional Riemann problems*, Adv. Appl. Math. **6** (1985), pp. 259–290.

[116] J. Glimm and A. Majda, *Multidimensional Hyperbolic Problems and Computations*, IMA Volumes in Mathematics and its Applications, **29**, Springer-Verlag: New York, 1991.

[117] S. Godunov, *An interesting class of quasilinear systems*, Dokl. Akad. Nauk SSSR. **139** (1961), pp. 521–523, English transl: Soviet Math. **2** (1961), pp. 947–949; *Elements of Continuum Mechanics*, Moscow: Nauka, 1978; *Lois de conservation et integrals d'énergie des équations hyperboliques*, Lecture Notes in Math. **1270** (1987), pp. 135–149, Berlin: Springer.

[118] E. Grenier, *Existence globale pour le système des gaz sans pression*, C. R. Acad. Sci. Paris Sér. I Math. **321** (1995), pp. 171–174.

[119] C.-H. Gu, *A method for solving the supersonic flow past a curved wedge (in Chinese)*, Fudan Univ. J. **7** (1962), pp. 11–14.

[120] O. Guès, G. Métiver, M. Williams, and K. Zumbrun, *Navier-Stokes regularization of multidimensional Euler shocks*, Ann. Sci. École Norm. Sup. (4) **39** (2006), pp. 75–175.

[121] K.G. Guderley, *The Theory of Transonic Flow*, Pergamon Press: Oxford-London-Paris-Frankfurt; Addison-Wesley Publishing Co. Inc.: Reading, Mass. 1962.

[122] L. Guo, W. Sheng, and T. Zhang, *The two-dimensional Riemann problem for isentropic Chaplygin gas dynamic system*, Commun. Pure Appl. Anal. **9** (2010), pp. 431–458.

[123] J. Hadamard, *Leçons sur la Propagation des Ondes et les Équations de l'Hydrodynamique*, Hermann, Paris, 1903 (Reprinted by Chelsea 1949).

[124] Q. Han and J.-X. Hong, *Isometric embedding of Riemannian manifolds in Euclidean spaces*, AMS: Providence, RI, 2006.

[125] D. HOFF, *The zero-Mach limit of compressible flows*, Comm. Math. Phys. **192** (1998), pp. 543–554.

[126] H. HOLDEN AND N.H. RISEBRO, *Front Tracking for Hyperbolic Conservation Laws*, Springer-Verlag: New York, 2002.

[127] F. HUANG AND Z. WANG, *Well posedness for pressureless flow*, Commun. Math. Phys. **222** (2001), pp. 117–146.

[128] J. HUNTER, *Transverse diffraction of nonlinear waves and singular rays,* SIAM J. Appl. Math. **48** (1988), pp. 1–37; *Nonlinear wave diffraction*, In: Geometrical Optics and Related Topics (Cortona, 1996), pp. 221–243, Birkhäuser Boston: Boston, MA, 1997.

[129] J. HUNTER AND J. KELLER, *Weak shock diffraction*, Wave Motion, **6** (1984), pp. 79–89 and pp. 321.

[130] H.K. JENSSEN, *Blowup for systems of conservation laws*, SIAM J. Math. Anal. **31** (2000), pp. 894–908.

[131] T. KATO, *The Cauchy problem for quasi-linear symmetric hyperbolic systems*, Arch. Ration. Mech. Anal. **58** (1975), pp. 181–205; *Quasilinear equations of evolution with applications to partial differential equations*, Lecture Notes in Math. **448**, pp. 25–70, Springer-Verlag: Berlin, 1975.

[132] J.B. KELLER AND A. BLANK, *Diffraction and reflection of pulses by wedges and corners*, Comm. Pure Appl. Math. **4** (1951), pp. 75–94.

[133] J.B. KELLER AND L. TING, *Periodic vibrations of systems governed by nonlinear partial differential equations*, Comm. Pure Appl. Math. **19** (1966), pp. 371–420.

[134] A. KURGANOV AND E. TADMOR, *Solution of two-dimensional Riemann problems for gas dynamics without Riemann problem solvers*, Numer. Methods Partial Diff. Eqs. **18** (2002), pp. 584–608.

[135] A.G. KUZ'MIN, *Boundary Value Problems for Transonic Flow*, John Wiley & Sons, 2002.

[136] P.D. LAX, *Shock waves and entropy*, In: Contributions to Nonlinear Functional Analysis, E. A. Zarantonello (ed.), pp. 603–634, Academic Press: New York, 1971.

[137] P.D. LAX, *Hyperbolic Systems of Conservation Laws and the Mathematical Theory of Shock Waves*, SIAM: Philadelphia, 1973.

[138] P.D. LAX, *The multiplicity of eigenvalues*, Bull. Amer. Math. Soc. (N.S.), **6** (1982), pp. 213–214.

[139] P.D. LAX, *Hyperbolic systems of conservation laws in several space variables*, In: Current Topics in PDEs, pp. 327–341, Kinokuniya: Tokyo, 1986.

[140] P.D. LAX AND X.-D. LIU, *Solutions of two dimensional Riemann problem of gas dynamics by positive schemes*, SIAM J. Sci. Comput. **19** (1998), pp. 319–340.

[141] P.G. LEFLOCH, *Hyperbolic Systems of Conservation Laws : The theory of classical and nonclassical shock waves,* Birkhäuser: Zürich, 2002.

[142] J. LI, *Note on the compressible Euler equations with zero temperature*, Appl. Math. Lett. **14** (2001), pp. 519–523.

[143] J. LI, T. ZHANG, AND S. YANG, *The Two-Dimensional Riemann Problem in Gas Dynamics*, Longman (Pitman Monographs **98**): Essex, 1998.

[144] J. LI AND Y. ZHENG, *Interaction of four rarefaction waves in the bi-symmetric class of the two-dimensional Euler equations*, Comm. Math. Phys. **296** (2010), pp. 303–321.

[145] J. LI, Z. XIN, AND H. YIN, *On transonic shocks in a nozzle with variable end pressure*, Commun. Math. Phys. **291** (2009), pp. 111–150.

[146] T.-T. LI, *On a free boundary problem*, Chinese Ann. Math. **1** (1980), pp. 351–358.

[147] T.-T. LI AND W.-C. YU, *Boundary Value Problems for Quasilinear Hyperbolic Systems*, Duke University Press: Durham, 1985.

[148] M.J. LIGHTHILL, *The diffraction of a blast I*, Proc. Royal Soc. London, **198** (1949), pp. 454–47; *The diffraction of a blast II*, Proc. Roy. Soc. London, **200A** (1950), pp. 554–565.

[149] W.-C. LIEN AND T.-P. LIU, *Nonlinear stability of a self-similar 3-dimensional gas flow*, Commun. Math. Phys. **204** (1999), pp. 525–549.

[150] P.-L. LIONS, *Mathematical Topics in Fluid Mechanics*, Vols. **1–2**, Oxford University Press: Oxford, 1998.

[151] P.-L. LIONS AND N. MASMOUDI, *Incompressible limit for a viscous compressible fluid*, J. Math. Pures Appl. (9), **77** (1998), pp. 585–627.

[152] P.-L. LIONS, B. PERTHAME, AND E. TADMOR, *A kinetic formulation of multidimensional scalar conservation laws and related equations*, J. Amer. Math. Soc. **7** (1994), pp. 169–191.

[153] L. LIU AND H.R. YUAN, *Stability of cylindrical transonic shocks for two-dimensional steadt compressible Euler system*, J. Hyper. Diff. Eqs. **5** (2008), pp. 347–379.

[154] T.-P. LIU AND T. YANG, *L^1 stability of weak solutions for 2×2 systems of hyperbolic conservation laws*, J. Amer. Math. Soc. **12** (1999), pp. 729–774; *Well-posedness theory for hyperbolic conservation laws*, Comm. Pure Appl. Math. **52** (1999), pp. 1553–1586.

[155] E. MACH, *Über den verlauf von funkenwellenin der ebene und im raume*, Sitzungsber. Akad. Wiss. Wien, **78** (1878), pp. 819–838.

[156] A. MAJDA, *The stability of multidimensional shock fronts*, Mem. Amer. Math. Soc. **275**, AMS: Providence, 1983; *The existence of multidimensional shock fronts*, Mem. Amer. Math. Soc. **281**, AMS: Providence, 1983.

[157] A. MAJDA, *Compressible Fluid Flow and Systems of Conservation Laws in Several Space Variables*, Springer-Verlag: New York, 1984.

[158] A. MAJDA AND A. BERTOZZI, *Vorticity and Incompressible Flow*, Cambridge University Press: Cambridge, 2002.

[159] A. MAJDA AND E. THOMANN, *Multidimensional shock fronts for second order wave equations*, Comm. Partial Diff. Eqs. **12** (1987), pp. 777–828.

[160] T. MAKINO, S. UKAI, AND S. KAWASHIMA, *Sur la solution à support compact de l'iquations d'Euler compressible*, Japan J. Appl. Math. **3** (1986), pp. 249–257.

[161] V.G. MAZ'YA AND B.A. PLAMENEVSKIvi, *Estimates in L_p and in Hölder classes and the Miranda-Agman maximun principle for solutions of elliptic boundary value problems in domains with sigular points on the boundary*, Amer. Math. Soc. Transl., **123** (1984), pp. 1–56.

[162] G. MÉTIVIER, *Stability of multi-dimensional weak shocks*, Comm. Partial Diff. Eqs. **15** (1990), pp. 983–1028.

[163] C.S. MORAWETZ, *On a weak solution for a transonic flow problem*, Comm. Pure Appl. Math. **38** (1985), pp. 797–818. *On steady transonic flow by compensated compactness*, Meth. Appl. Anal. **2** (1995), pp. 257–268.

[164] C.S. MORAWETZ, *Mixed equations and transonic flow*, J. Hyper. Diff. Eqns. **1** (2004), pp. 1–26.

[165] C.S. MORAWETZ, *Potential theory for regular and Mach reflection of a shock at a wedge*, Comm. Pure Appl. Math. **47** (1994), pp. 593–624.

[166] C.S. MORAWETZ, *On the non-existence of continuous transonic flows past profiles I,II,III*, Comm. Pure Appl. Math. **9** (1956), pp. 45–68; **10** (1957), pp. 107–131; **11** (1958), pp. 129–144.

[167] C.B. MORREY, *Quasiconvexity and the lower semicontinuity of multiple integrals*, Pacific J. Math. **2** (1952), pp. 25–53.

[168] F. MURAT, *The injection of the positive cone of H^{-1} in $W^{-1,q}$ is completely continuous for all $q < 2$*, J. Math. Pures Appl. (9), **60** (1981), pp. 309–322.

[169] F. MURAT, *Compacité par compensation*, Ann. Suola Norm. Pisa (4), **5** (1978), pp. 489–507; *Compacité par compensation, condition nécessaire et suffisante de continuité faible sous une hypothèse de range constant*, Ann. Scuola Norm. Sup. Pisa (4), **8** (1981), pp. 69–102; *A survey on compensated compactness*, In: Contributions to Modern Calculus of Variations (Bologna, 1985), pp. 145–183, Longman Sci. Tech.: Harlow, 1987.

[170] J. VON NEUMANN, *Oblique reflection of shocks*, Navy Department, Bureau of Ordance, Explosive Research Report, No. **12**, 1943.

[171] J. VON NEUMANN, *Collect Works*, Vol. **5**, Pergamon: New York, 1963.

[172] S. OSHER, M. HAFEZ, AND W. WHITLOW, *Entropy condition satisfying approximations for the full potential equation of transonic flow*, Math. Comp. **44** (1985), pp. 1–29.

[173] E. YU. PANOV, *Existence of strong traces for quasi-solutions of multidimensional conservation laws*, J. Hyper. Diff. Eqs. **4** (2007), pp. 729–770.

[174] B. PERTHAME, *Time decays: an analogy between kinetic transport, Schrödinger and gas dynamics equations*, In: Nonlinear Partial Differential Equations and Their Applications, pp. 281–293, Longman: Harlow, 1998.

[175] B. PERTHAME, *Kinetic Formations of Conservation Laws*, Oxford University Press: Oxford, 2002.

[176] T. QIN, *Symmetrizing the nonlinear elastodynamic system*, J. Elasticity, **50** (1998), pp. 245–252.

[177] J. RAUCH, *BV estimates fail for most quasilinear hyperbolic systems in dimension great than one*, Commun. Math. Phys. **106** (1986), pp. 481–484.

[178] J.-F. RODRIGUES, *Obstacle Problems in Mathematical Physics*, Elsevier Science Publishers B.V. 1987.

[179] T. RUGGERI AND A. STRUMIA, *Main field and convex covariant density for quasi-linear hyperbolic systems*, Ann. Inst. Henri Poincaré, Section A, **34** (1981), pp. 65–84.

[180] D.G. SCHAEFFER, *Supersonic flow past a nearly straight wedge*, Duke Math. J. **43** (1976), pp. 637–670.

[181] D. SCHAEFFER AND M. SHEARER, *The classification of* 2×2 *systems of nonstrictly hyperbolic conservation laws, with application to oil recovery*, Comm. Pure Appl. Math. **40** (1987), pp. 141–178; *Riemann problems for nonstrictly hyperbolic* 2×2 *systems of conservation laws*, Trans. Amer. Math. Soc. **304** (1987), pp. 267–306.

[182] C.W. SCHULZ-RINNE, J.P. COLLINS, AND H.M. GLAZ, *Numerical solution of the Riemann problem for two-dimensional gas dynamics*, SIAM J. Sci. Comput. **14** (1993), pp. 1394–1414.

[183] D. SERRE, *Écoulements de fluides parfaits en deux variables indépendantes de type espace. Rèflexion d'un choc plan par un dièdre compressif (French)*, Arch. Ration. Mech. Anal. **132** (1995), pp. 15–36.

[184] D. SERRE, *Spiral waves for the two-dimensional Riemann problem for a compressible fluid*, Ann. Fac. Sci. Toulouse Math. (6), **5** (1996), pp. 125–135.

[185] D. SERRE, *Hyperbolicity of the nonlinear models of Maxwell's equations*, Arch. Ration. Mech. Anal. **172** (2004), pp. 309–331.

[186] D. SERRE, *Multidimensional shock interaction for a Chaplygin gas*, Arch. Ration. Mech. Anal. **191** (2009), pp. 539–577.

[187] W.-C. SHENG AND G. YIN, *Transonic shock and supersonic shock in the regular reflection of a planar shock*, Z. Angew. Math. Phys. **60** (2009), pp. 438–449.

[188] W.-C. SHENG AND T. ZHANG, *The Riemann Problem for the Transportation Equations in Gas Dynamics*, Mem. Amer. Math. Soc. **654**, AMS: Providence, 1999.

[189] M. SHIFFMAN, *On the existence of subsonic flows of a compressible fluid*, J. Ration. Mech. Anal. **1** (1952), pp. 605–652.

[190] T.C. SIDERIS, *Formation of singularities in three-dimensional compressible fluids*, Commun. Math. Phys. **101** (1985), pp. 475–485.

[191] B. SKEWS AND J. ASHWORTH, *The physical nature of weak shock wave reflection*, J. Fluid Mech. **542** (2005), pp. 105–114.

[192] E. TABAK AND R. ROSALES, *Focusing of weak shock waves and the von Neumann paradox of oblique shock reflection*, Phys. Fluids, **6** (1994), pp. 1874–1892.

[193] E. TADMOR, M. RASCLE, AND P. BAGNERINI, *Compensated compactness for 2D conservation laws*, J. Hype. Diff. Eqs. **2** (2005), 697–712.

[194] D. Tan and T. Zhang, *Two dimensional Riemann problem for a hyperbolic system of nonlinear conservation laws I: four-J cases,* J. Diff. Eqs. **111** (1994), pp. 203–254; *II: Initial data involving some rarefaction waves*, J. Diff. Eqs. **111** (1994), pp. 255–283.

[195] L. Tartar, *Compensated compactness and applications to partial differential equations.* In: Res. Notes in Math. **39**, pp. 136–212, Pitman: Boston-London, 1979.

[196] L. Tartar, *Une nouvelle méthode de resolution d'equations aux derivées partielles nonlinéaires*, Lecture Notes in Math. **665**, pp. 228–241, Springer-Verlag: Berlin, 1977; *The compensated compactness method applied to systems of conservation laws*, In: Systems of Nonlinear P.D.E., J.M. Ball (eds.), pp. 263–285, NATO Series, C. Reidel publishing Co. 1983.

[197] A. Tesdall and J.K. Hunter, *Self-similar solutions for weak shock reflection*, SIAM J. Appl. Math. **63** (2002), pp. 42–61.

[198] A.M. Tesdall, R. Sanders, and B. Keyfitz, *Self-similar solutions for the triple point paradox in gasdynamics*, SIAM J. Appl. Math. **68** (2008), pp. 1360–1377.

[199] R. Timman, *Unsteady motion in transonic flow*, In: Symposium Transsonicum (IUTAM, Aachen, Sept. 1962), pp. 394–401, ed. K. Oswatitsch, Springer-Verlag: Berlin, 1964.

[200] Y. Trakhinin, *Existence of compressible current-vortex sheets: variable coefficients linear analysis*, Arch. Ration. Mech. Anal. **177** (2005), 331–366; *The existence of current-vortex sheets in ideal compressible magnetohydrodynamics*, Arch. Ration. Mech. Anal. **191** (2009), 245–310.

[201] M. Van Dyke, *An Album of Fluid Motion*, The Parabolic Press: Stanford, 1982.

[202] A. Vasseur, *Strong traces for solutions to multidimensional scalar conservation laws*, Arch. Ration. Mech. Anal. **160** (2001), pp. 181–193.

[203] A. Volpert, *The Space BV and quasilinear equations*, Mat. Sb. (N.S.) **73** (1967), pp. 255–302; Math. USSR Sbornik, **2** (1967), pp. 225–267 (in English).

[204] Z. Wang, F. Huang, and X. Ding, *On the Cauchy problem of transportation equations*, Acta Math. Appl. Sinica (English Ser.), **13** (1997), pp. 113–122.

[205] G.B. Whitham, *Linear and Nonlinear Waves*, Wiley-Interscience: New York, 1973.

[206] Z. Xin, W. Yan, and H. Yin, *Transonic shock problem for the Euler system in a nozzle*, Arch. Ration. Mech. Anal. **194** (2009), 1–47.

[207] S.-T. Yau, *Review of geometry and analysis*, In: Mathematics: Frontiers and Perspectives, pp. 353–401, International Mathematics Union, Eds. V. Arnold, M. Atiyah, P. Lax, and B. Mazur, AMS: Providence, 2000.

[208] G.I. Zabolotskaya and R.V. Khokhlov, *Quasi-plane waves in the nonlinear acoustics of confined beams*, Sov. Phys.-Acoustics, **15** (1969), pp. 35–40.

[209] A.R. Zakharian, M. Brio, J.K. Hunter, and G.M. Webb, *The von Neumann paradox in weak shock reflection*, J. Fluid Mech. **422** (2000), pp. 193–205.

[210] Ya.B. Zeldovich, *Gravitational instability: an approximate theory for large density perturbations*, Astronaut. Astrophys. **5** (1970), pp. 84–89.

[211] T. Zhang and Y. Zheng, *Conjecture on the structure of solutions of the Riemann problem for two-dimensional gas dynamics*, SIAM J. Math. Anal. **21** (1990), pp. 593–630.

[212] Y.-Q. Zhang, *Global existence of steady supersonic potential flow past a curved wedge with a piecewise smooth boundary,* SIAM J. Math. Anal. **31** (1999), pp. 166–183; *Steady supersonic flow past an almost straight wedge with large vertex angle*, J. Diff. Eqs. **192** (2003), pp. 1–46.

[213] Y. Zheng, *Systems of Conservation Laws: Two-Dimensional Riemann Problems*, Birkhäuser: Boston, 2001.

[214] Y. Zheng, *A global solution to a two-dimensional Riemann problem involving shocks as free boundaries*, Acta Math. Appl. Sin. Engl. Ser. **19** (2003), pp. 559–572.

[215] Y. ZHENG, *Two-dimensional regular shock reflection for the pressure gradient system of conservation laws*, Acta Math. Appl. Sin. Engl. Ser. **22** (2006), pp. 177–210.

[216] W.P. ZIEMER, *Cauchy flux and sets of finite perimeter*, Arch. Ration. Mech. Anal. **84** (1983), pp. 189–201.

MATHEMATICAL ANALYSIS OF FLUIDS IN MOTION

EDUARD FEIREISL*

Abstract. Continuum fluid mechanics is a phenomenological theory based on macroscopic observable state variables, the time evolution of which is described by means of systems of partial differential equations. The resulting mathematical problems are highly non-linear and rather complex, even in the simplest physically relevant situations. We discuss several recent results and newly developed methods based on the concept of weak solution. The class of weak solutions is happily large enough in order to guarantee the existence of global-in-time solutions without any essential restrictions on the size of the relevant data. On the other hand, the underlying structural hypotheses impose quite severe restrictions on the specific form of constitutive relations. The best known open problems - hypothetical presence of vacuum zones, propagation of density oscillations, sequential stability of the temperature field, among others - are discussed. The final part of the study addresses the simplified problems, in particular, the incompressible Navier-Stokes system.

Contents

*Mathematical Institute of the Academy of Sciences of the Czech Republic, Žitná 25, 115 67 Praha 1, Czech Republic. The work of E. Feireisl was supported by Grant 201/08/0315 of GA ČR as a part of the general research programme of the Academy of Sciences of the Czech Republic, Institutional Research Plan AV0Z10190503.

1. Introduction. Although ignored in some cases, there is no doubt that fluid dynamics plays a significant role in virtually all processes observed in the real world. *Fluid* is a material that continually deforms, or flows, when a force is applied. Typical examples of fluids are gases and liquids. *Continuum fluid mechanics* is a phenomenological theory that describes fluids by means of macroscopic observable quantities - fields. Examples of fields are *density*, *velocity*, *temperature*, among others. The time evolution of fields in continuum fluid mechanics is governed by a system of partial differential equations derived from the basic principles of *classical physics* (see Batchelor [7], Gallavotti [33], Truesdell [65], and Truesdell and Rajagopal [64], among others). The enormous progress in applications of fluid mechanics in recent years is mainly due to the availability of powerful computer resources, allowing numerical simulation of complicated processes in meteorology, geophysics, or astrophysics carried out in real time. In this survey, we focus on purely theoretical aspects of problems arising in the studies of fluids in motion.

1.1. Balance laws in continuum fluid mechanics. Stresses in a fluid are given by *Stokes' law*

$$\mathbb{T} = \mathbb{S} - p\mathbb{I}, \tag{1.1}$$

where $\mathbb{T}$ is the Cauchy stress tensor, $\mathbb{S}$ is the viscous stress tensor, and p denotes the pressure. Constitutive relation (1.1) may be viewed as a mathematical definition of fluid (see Truesdell and Rajagopal [64]).

Basic physical principles in continuum mechanics are expressed in terms of *balance laws*. If $\varrho = \varrho(t,x)$ is the fluid density at the time t and the spatial position $x \in R^3$, and $\mathbf{u} = \mathbf{u}(t,x)$ the velocity field, the physical principle of *conservation of mass* may be stated in the form of an integral identity

$$\int_B \varrho(t_2,x)\,\mathrm{d}x - \int_B \varrho(t_1,x)\,\mathrm{d}x = -\int_{t_1}^{t_2}\int_{\partial B} \varrho(t,x)\mathbf{u}(t,x)\cdot\mathbf{n}(x)\,\mathrm{dS}_x\,\mathrm{d}t \tag{1.2}$$

for any volume element $B \subset R^3$ and any time interval (t_1,t_2), where $\mathbf{n}$ denotes the outer normal vector to ∂B. If all quantities appearing in (1.2) are *smooth*, or at least continuously differentiable, we can perform the limit $t_1, t_2 \to t$, $B = U(x) \to x$ to obtain the standard *equation of continuity*

$$\partial_t \varrho(t,x) + \mathrm{div}_x(\varrho(t,x)\mathbf{u}(t,x)) = 0. \tag{1.3}$$

However, the physical quantities like the density ϱ need not be even continuous therefore we should always keep in mind that the "correct" form of the principle of mass conservation is given by the family of integral identities (1.2) rather than the partial differential equation (1.3).

A general balance law reads

$$\int_B d(t_2,x)\ \mathrm{d}x - \int_B d(t_1,x)\ \mathrm{d}x + \int_{t_1}^{t_2}\int_{\partial B} \mathbf{F}(t,x)\cdot\mathbf{n}(x)\ \mathrm{dS}_x\ \mathrm{d}t \\ = \int_{t_1}^{t_2}\int_B s(t,x)\ \mathrm{d}x\ \mathrm{d}t, \tag{1.4}$$

where d stands for the volumetric density of an observable quantity, the vector field $\mathbf{F}$ is its flux, and s is a source term. The expression on the left-hand side of (1.4) may be viewed as a *normal trace* of the four-component vector field $[d,\mathbf{F}]$ on the time-space cylinder $[t_1,t_2]\times B$.

Similarly to (1.3), we can rewrite (1.4) in the "differential" form

$$\partial_t d(t,x) + \mathrm{div}_x\mathbf{F}(t,x) = s(t,x), \tag{1.5}$$

or, more concisely,

$$\mathrm{DIV}_{t,x}[d,F] = s.$$

The requirement for $[d,F]$ to possess a well-defined normal trace on any, say, open set imposes certain restrictions on the source term, namely, s must be at least a signed measure in the time-space. The corresponding theory of vector fields of divergence measure has been developed in a series of papers by Chen et al. [14], [15], [16]. It turns out that vector fields with divergence measure possess a well-defined normal trace that can be viewed as a bounded linear functional on the space of functions that are Lipschitz continuous on ∂B. Accordingly, relation (1.4) may be replaced by

$$-\lim_{\varepsilon\to 0}\int_{t_1}^{t_2}\int_B \Big(d(t,x)\partial_t\chi_\varepsilon(t,x) + \mathbf{F}(t,x)\cdot\nabla_x\chi_\varepsilon(t,x)\Big)\ \mathrm{d}x\ \mathrm{d}t \\ = \lim_{\varepsilon\to 0}\int_{t_1}^{t_2}\int_B s(t,x)\chi_\varepsilon(t,x)\ \mathrm{d}x\ \mathrm{d}t, \tag{1.6}$$

where $\chi_\varepsilon\in C_c^\infty((t_1,t_2)\times B)$, $\chi_\varepsilon\nearrow 1$ for $\varepsilon\to 0$.

Formula (1.7) can be rewritten in a seemingly stronger form

$$\int_0^T\int_\Omega \Big(d\ \partial_t\varphi + \mathbf{F}\cdot\nabla_x\varphi\Big)\ \mathrm{d}x\ \mathrm{d}t + < s;\varphi >= 0 \tag{1.7}$$

for any *test function* $\varphi\in C_c^\infty((0,T)\times\Omega)$, where $<;>$ denotes the duality between the space of measures $\mathcal{M}$ and C, and Ω is a reference spatial domain. The family of integral identities (1.7) is usually termed a *weak formulation* of equation (1.5) posed in the open set $(0,T)\times\Omega$.

1.2. Navier-Stokes-Fourier system. Newton's second law yields *momentum equation* in the form

$$\partial_t(\varrho\mathbf{u}) + \mathrm{div}_x(\varrho\mathbf{u}\otimes\mathbf{u}) + \nabla_x p = \mathrm{div}_x\mathbb{S} + \varrho\mathbf{f} + \mathbf{g}, \tag{1.8}$$

where we have tacitly assumed that all quantities in question are smooth. Alternatively, the same equation can be considered in a weak form similar to (1.7). The symbols $\mathbf{f}$ and $\mathbf{g}$ stand for external forces acting on the fluid.

Taking the scalar product of (1.8) with $\mathbf{u}$, we obtain the kinetic energy balance

$$\begin{aligned}\partial_t\left(\frac{1}{2}\varrho|\mathbf{u}|^2\right) + \mathrm{div}_x\left(\frac{1}{2}\varrho|\mathbf{u}|^2\mathbf{u} + p\mathbf{u}\right) - \mathrm{div}_x(\mathbb{S}\mathbf{u})&\\ = p\mathrm{div}_x\mathbf{u} - \mathbb{S}:\nabla_x\mathbf{u} + \varrho\mathbf{f}\cdot\mathbf{u} + \mathbf{g}\cdot\mathbf{u}.&\end{aligned} \tag{1.9}$$

Clearly, equation (1.9) contains a source term, namely $p\mathrm{div}_x\mathbf{u} - \mathbb{S}:\nabla_x\mathbf{u}$, even if $\mathbf{f} = \mathbf{g} = 0$, meaning, even in the total absence of any external energy supply.

On the other hand, the *First law of thermodynamics* asserts that the *total energy* of the fluid is a conserved quantity. In other words, the total energy density of the system must consist of at least two components, namely

$$E_{\mathrm{tot}} = \frac{1}{2}\varrho|\mathbf{u}|^2 + \varrho e,$$

where e is the (specific) *internal energy* satisfying

$$\partial_t(\varrho e) + \mathrm{div}_x(\varrho e\mathbf{u}) + \mathrm{div}_x\mathbf{q} = \mathbb{S}:\nabla_x\mathbf{u} - p\mathrm{div}_x\mathbf{u} + r, \tag{1.10}$$

where $\mathbf{q}$ is the internal energy (diffusive) flux, and r is a source term.

For given source terms $\mathbf{f}$, $\mathbf{g}$, and r, we may *close* the system of equations (1.3), (1.8), (1.10) by adding specific *constitutive relations* for $\mathbb{S}$ and $\mathbf{q}$. Here, we assume that $\mathbb{S}$ obeys *Newton's rheological law*

$$\mathbb{S} = \mu\Big(\nabla_x\mathbf{u} + \nabla_x^t\mathbf{u} - \frac{2}{3}\mathrm{div}_x\mathbf{u}\mathbb{I}\Big) + \eta\mathrm{div}_x\mathbf{u}\mathbb{I}, \tag{1.11}$$

where μ is the shear viscosity and η the bulk viscosity coefficient. Moreover, introducing the *absolute temperature* ϑ, we postulate *Fourier's law*

$$\mathbf{q} = -\kappa\nabla_x\vartheta, \tag{1.12}$$

with the heat conductivity coefficient κ.

Equations (1.3), (1.8), (1.10), supplemented with the constitutive relations (1.11), (1.12), are called *Navier-Stokes-Fourier system.*

1.3. Initial and boundary conditions. In the remaining part of this study, the quantities ϱ, $\mathbf{u}$, and ϑ are taken to be the fundamental *state* variables that characterize *completely* the state of the fluid at a given instant t. The initial state of the system is therefore uniquely determined through a family of initial conditions

$$\varrho(0,x) = \varrho_0(x), \ \mathbf{u}(0,x) = \mathbf{u}_0(x), \ \vartheta(0,x) = \vartheta_0(x). \tag{1.13}$$

Moreover, classical physical principles require ϑ_0 to be strictly positive, while ϱ_0 may vanish on some part of the physical space Ω. However, we should always keep in mind that the Navier-Stokes-Fourier system is a model for *non-dilute* fluids therefore the fluid density should be definitely bounded below away from zero. The hypothetical appearance of *vacuum zones* represents one of the major problems of the mathematical theory and its resolution is a great challenge. We come to this interesting issue in the forthcoming section.

If Ω has a boundary, it is necessary to specify the behavior of certain quantities on $\partial\Omega$. Prescribing the velocity field

$$\mathbf{u}|_{\partial\Omega} = \mathbf{u}_b,$$

we have to determine the values of the density ϱ on the part of $\partial\Omega$ where $\mathbf{u}_b \cdot \mathbf{n} < 0$. This becomes irrelevant under the *impermeability condition*

$$\mathbf{u} \cdot \mathbf{n}|_{\partial\Omega} = 0. \tag{1.14}$$

Viscous fluids are believed to adhere completely to a rigid boundary; whence (1.15) is usually enforced by the *no-slip boundary condition*

$$\mathbf{u}|_{\partial\Omega} = 0. \tag{1.15}$$

Alternatively, we may impose *Navier's partial slip* boundary conditions

$$[\mathbb{S}\mathbf{n}]_{\tan} + \beta\mathbf{u}|_{\partial\Omega} = 0, \tag{1.16}$$

where $\beta > 0$ plays a role of a friction coefficient. Note that (1.16) yields the *complete slip* boundary condition

$$[\mathbb{S}\mathbf{n}] \times \mathbf{n}|_{\partial\Omega} = 0 \tag{1.17}$$

for $\beta = 0$ and reduces to (1.15) in the asymptotic limit $\beta \to \infty$.

Finally, we may prescribe the value ϑ_b of the absolute temperature on $\partial\Omega$, or the heat flux

$$\mathbf{q} \cdot \mathbf{n}|_{\partial\Omega} = q_b. \tag{1.18}$$

Of course, other physically relevant boundary conditions may be imposed as the case may be.

2. A priori bounds and related problems. Equations (1.3), (1.8), (1.10) are highly *non-linear*, with "geometrical" nonlinearities represented by the convective terms $\varrho\mathbf{u}$, $\varrho\mathbf{u}\otimes\mathbf{u}$, $\varrho e\mathbf{u}$, "dissipative" nonlinearities like $\mathbb{S}:\nabla_x\mathbf{u}$, and "constitutive" nonlinearities characterizing the physical properties of a specific fluid. Solvability of a non-linear problem is closely related to the available *a priori* estimates. These are formal bounds imposed on the set of all admissible solutions by the data, the equations, and other constraints as the case may be. As is well known, *a priori* bounds determine the function spaces framework, where the solutions are looked for, together with a suitable concept of solution itself - classical, strong, weak, very weak, etc.

2.1. First law of thermodynamics, total energy. Basically all *a priori* bounds for the Navier-Stokes-Fourier system introduced in Section 1 follow from the total dissipation balance. We focus on energetically closed systems, for which the total energy of the fluid is a constant of motion. To this end, we impose the no-slip boundary condition for the velocity

$$\mathbf{u}|_{\partial\Omega}=0, \tag{2.1}$$

together with the no-flux condition for the heat flux

$$\mathbf{q}\cdot\mathbf{n}|_{\partial\Omega}=0. \tag{2.2}$$

In addition, for the sake of simplicity, we set $\mathbf{f}=\nabla_x F$, $\mathbf{g}=0$, $r=0$ in (1.8), (1.10), where $F=F(x)$ is a scalar potential independent of time. Finally, we shall assume that the pressure $p=p(\varrho,\vartheta)$ and the specific internal energy $e=e(\varrho,\vartheta)$ are given functions of the thermostatic state variables ϱ and ϑ.

Integrating (1.9), (1.10) over Ω yields the *total energy balance* in the form

$$\frac{\mathrm{d}}{\mathrm{d}t}\int_\Omega\left(\frac{1}{2}\varrho|\mathbf{u}|^2+\varrho e(\varrho,\vartheta)-\varrho F\right)(t,\cdot)\ \mathrm{d}x=0. \tag{2.3}$$

Similarly, we deduce from (1.3) that the *total mass* of the fluid is a conserved quantity, specifically,

$$\frac{\mathrm{d}}{\mathrm{d}t}\int_\Omega\varrho(t,\cdot)\ \mathrm{d}x=0. \tag{2.4}$$

Relations (2.3), (2.4) imply that

$$\sup_{t\in[0,T]}\|\varrho(t,\cdot)\|_{L^1(\Omega)}=M_0, \tag{2.5}$$

$$\sup_{t\in[0,T]}\|\varrho|\mathbf{u}|^2(t,\cdot)\|_{L^1(\Omega)}\le c(M_0,E_0,F_0), \tag{2.6}$$

$$\sup_{t\in[0,T]}\|\varrho e(\varrho,\vartheta)(t,\cdot)\|_{L^1(\Omega)}\le c(M_0,E_0,F_0), \tag{2.7}$$

where we have denoted

$$M_0 = \int_\Omega \varrho_0 \ \mathrm{d}x, \ E_0 = \int_\Omega \left(\frac{1}{2} \varrho_0 |\mathbf{u}_0|^2 + \varrho_0 e(\varrho_0, \vartheta_0) \right) \ \mathrm{d}x, \ F_0 = \|F\|_{L^1(\Omega)}.$$

Thus any solution emanating from initial data of finite mass and energy admits *a priori* bounds (2.5 - 2.7).

2.2. Second law of thermodynamics, entropy. Evidently, the bounds established in (2.5 - 2.7) are very weak. They imply, in the best case, *integrability* of the state variables but do not provide any information on possible *oscillations* of these quantities. In order to deduce more estimates, we use the Second law of thermodynamics, in particular, we introduce the concept of *entropy*. The specific entropy $s = s(\varrho, \vartheta)$ is a function of the state variable ϱ, ϑ with the following properties (see Callen [12], Rajagopal and Srinivasa [56]):

- The specific entropy s is an increasing function of the internal energy e,

$$\vartheta = \left(\frac{\partial s}{\partial e} \right)^{-1}.$$

- The static (equilibrium) states of the system maximize the total entropy $\int_\Omega \varrho s \ \mathrm{d}x$.
- The entropy tends to zero if ϑ tends to zero.
- The entropy remains constant in the processes when the fluid responds "elastically".

In accordance with the general principles stated above, we suppose that the functions p, e, and s are interrelated through *Gibbs' equation*

$$\vartheta D s(\varrho, \vartheta) = D e(\varrho, \vartheta) + p(\varrho, \vartheta) D \left(\frac{1}{\varrho} \right) \tag{2.8}$$

(see Callen [12], Gallavotti [33]).

Dividing the internal energy balance (1.10) on ϑ and using Gibbs' equation (2.8), we deduce the *entropy balance*

$$\partial_t (\varrho s(\varrho, \vartheta)) + \mathrm{div}_x (\varrho s(\varrho, \vartheta) \mathbf{u}) + \mathrm{div}_x \left(\frac{\mathbf{q}}{\vartheta} \right) = \sigma, \tag{2.9}$$

with the *entropy production rate*

$$\sigma = \frac{1}{\vartheta} \left(\mathbb{S} : \nabla_x \mathbf{u} - \frac{\mathbf{q} \cdot \nabla_x \vartheta}{\vartheta} \right). \tag{2.10}$$

2.3. Equilibrium states, thermodynamics stability. Equilibrium states are solutions that minimize the entropy production σ. In view of the constitutive relations (1.11), (1.12), this amounts to

$$\nabla_x \mathbf{u} + \nabla_x \mathbf{u}^t - \frac{2}{3} \mathrm{div}_x \mathbf{u} \mathbb{I} = 0, \ \nabla_x \vartheta = 0$$

provided $\mu > 0$, $\kappa > 0$. Since the velocity vanishes on $\partial\Omega$, we may use Korn's inequality to conclude that the set of equilibrium states of the Navier-Stokes-Fourier system describing a viscous and heat conducting fluid consists of *static states* $\mathbf{u} = 0$, $\vartheta = \overline{\vartheta}$, and $\varrho = \tilde{\varrho}$,

$$\nabla_x p(\tilde{\varrho}, \overline{\vartheta}) = \tilde{\varrho} \nabla_x F,$$

with the constant equilibrium temperature $\overline{\vartheta} > 0$.

In order to identify the static distribution of the density $\tilde{\varrho}$, we suppose, in addition to Gibbs' relation (2.8), the *hypothesis of thermodynamics stability*

$$\frac{\partial p(\varrho, \vartheta)}{\partial \varrho} > 0, \ \frac{\partial e(\varrho, \vartheta)}{\partial \vartheta} > 0, \tag{2.11}$$

together with

$$\liminf_{\varrho \to 0} \frac{\partial p(\varrho, \vartheta)}{\partial \varrho} > 0 \text{ for any } \vartheta > 0. \tag{2.12}$$

Under the hypotheses (2.11), (2.12), equilibrium solutions are uniquely determined by their mass and energy. More specifically, we report the following result [29, Theorem 4.1].

THEOREM 2.1. *Let $\Omega \subset R^3$ be a bounded domain. Assume that p, e, s are continuously differentiable functions of $(\varrho, \vartheta) \in (0, \infty)^2$ satisfying (2.8), (2.11), and (2.12). Let $F \in W^{1,\infty}(\Omega)$.*

Then, given $M_0 > 0$, E_0, there exists at most one solution $\tilde{\varrho} = \tilde{\varrho}(x) \geq 0$, $\overline{\vartheta} > 0$ of the static problem

$$\nabla_x p(\tilde{\varrho}, \overline{\vartheta}) = \tilde{\varrho} \nabla_x F, \ \overline{\vartheta} > 0 \ \textit{a constant} \tag{2.13}$$

satisfying

$$\int_\Omega \tilde{\varrho} \ \mathrm{d}x = M_0, \ \int_\Omega \Big(\tilde{\varrho} e(\tilde{\varrho}, \overline{\vartheta}) - \tilde{\varrho} F \Big) \ \mathrm{d}x = E_0. \tag{2.14}$$

Moreover, $\tilde{\varrho}$ is strictly positive in $\overline{\Omega}$, and

$$\int_\Omega \tilde{\varrho} s(\tilde{\varrho}, \overline{\vartheta}) \ \mathrm{d}x \geq \int_\Omega \varrho s(\varrho, \vartheta) \ \mathrm{d}x$$

for any couple of $\varrho \geq 0$, $\vartheta > 0$ of measurable functions satisfying (2.14).

2.4. Total dissipation balance. Having collected the necessary preliminary material, we are ready to formulate the total dissipation balance, which is the main and practically also the only source of *a priori* bounds for the Navier-Stokes-Fourier system. Multiplying the entropy balance (2.9) on $\overline{\vartheta}$, integrating over Ω, and adding the resulting expression to (2.3), we obtain

$$\frac{\mathrm{d}}{\mathrm{d}t}\int_\Omega\left(\frac{1}{2}\varrho|\mathbf{u}|^2+\varrho e(\varrho,\vartheta)-\overline{\vartheta}\varrho s(\varrho,\vartheta)-\varrho F\right)(t,\cdot)\ \mathrm{d}x+\overline{\vartheta}\int_\Omega\sigma\ \mathrm{d}x=0. \quad (2.15)$$

In order to exploit (2.15), we introduce *Helmholtz function*

$$H(\varrho,\vartheta)=\varrho e(\varrho,\vartheta)-\overline{\vartheta}\varrho s(\varrho,\vartheta), \tag{2.16}$$

reminiscent of the Helmholtz free energy $\varrho e-\vartheta\varrho s$. A straightforward computation yields

$$\frac{\partial^2 H(\varrho,\overline{\vartheta})}{\partial\varrho^2}=\frac{1}{\varrho}\frac{\partial p(\varrho,\overline{\vartheta})}{\partial\varrho}, \tag{2.17}$$

and

$$\frac{\partial H(\varrho,\vartheta)}{\partial\vartheta}=\frac{\varrho}{\vartheta}(\vartheta-\overline{\vartheta})\frac{\partial e(\varrho,\vartheta)}{\partial\vartheta}.$$

Thus hypothesis of thermodynamic stability (2.11) implies that

- $\varrho\mapsto H(\varrho,\overline{\vartheta})$ is strictly convex,
- $\vartheta\mapsto H(\varrho,\vartheta)$ is decreasing for $\vartheta<\overline{\vartheta}$ and strictly increasing for $\vartheta>\overline{\vartheta}$ for any ϱ.

Let $\tilde{\varrho}$, $\overline{\vartheta}$ be the static solution introduced in the previous section. It follows from (2.17) that

$$\frac{\partial H(\tilde{\varrho},\overline{\vartheta})}{\partial\varrho}=F+\text{const},$$

therefore the total dissipation balance (2.15) can be rewritten in the form

$$\begin{aligned}\frac{\mathrm{d}}{\mathrm{d}t}\int_\Omega\Big(\frac{1}{2}\varrho|\mathbf{u}|^2+H(\varrho,\vartheta)-\frac{\partial H(\tilde{\varrho},\overline{\vartheta})}{\partial\varrho}(\tilde{\varrho},\overline{\vartheta})(\varrho-\tilde{\varrho})-H(\tilde{\varrho},\overline{\vartheta})\\ -(\varrho-\tilde{\varrho})F\Big)(t,\cdot)\ \mathrm{d}x+\overline{\vartheta}\int_\Omega\sigma\ \mathrm{d}x=0,\end{aligned} \tag{2.18}$$

where we have assumed that

$$\int_\Omega(\varrho-\tilde{\varrho})\ \mathrm{d}x=0,\quad \int_\Omega H(\tilde{\varrho},\overline{\vartheta})\ \mathrm{d}x<\infty. \tag{2.19}$$

Note that relation (2.19) obviously holds if $\Omega\subset R^3$ is a *bounded* domain.

It can be shown (see [28, Lemma 4.1]) that

$$\begin{aligned}H(\varrho,\vartheta)-\frac{\partial H(\tilde{\varrho},\overline{\vartheta})}{\partial\varrho}(\tilde{\varrho},\overline{\vartheta})(\varrho-\tilde{\varrho})-H(\tilde{\varrho},\overline{\vartheta})\geq c(K)\Big(|\varrho-\tilde{\varrho}|^2+|\vartheta-\overline{\vartheta}|^2\Big)\\ \text{for any } (\varrho,\vartheta)\in K,\end{aligned} \tag{2.20}$$

$$H(\varrho,\vartheta)-\frac{\partial H(\tilde\varrho,\overline{\vartheta})}{\partial\varrho}(\tilde\varrho,\overline{\vartheta})(\varrho-\tilde\varrho)-H(\tilde\varrho,\overline{\vartheta})\geq c(K)\Big(\varrho e(\varrho,\vartheta)+\varrho|s(\varrho,\vartheta)|+1\Big)$$
$$\text{whenever } (\varrho,\vartheta)\in[0,\infty)^2\setminus K, \tag{2.21}$$

for any *compact set* $K\subset(0,\infty)^2$.

Thus we infer that relation (2.18) yields uniform *a priori* bounds provided the initial data are chosen in such a way that

$$\varrho_0|\mathbf{u}_0|^2\in L^1(\Omega),\ (\varrho_0-\tilde\varrho)\in L^1(\Omega),$$
$$H(\varrho_0,\vartheta_0)-\frac{\partial H(\tilde\varrho,\overline{\vartheta})}{\partial\varrho}(\tilde\varrho,\overline{\vartheta})(\varrho_0-\tilde\varrho)-H(\tilde\varrho,\overline{\vartheta})\in L^1(\Omega).$$

Accordingly, at least for a bounded domain Ω, we conclude that

$$\sup_{t\in[0,T]}\int_\Omega\Big(\varrho e(\varrho,\vartheta)+\varrho|s(\varrho,\vartheta)|\Big)(t,\cdot)\ \mathrm{d}x\leq c. \tag{2.22}$$

If Ω is unbounded, similar estimates may be deduced depending on the behavior of ϱ, ϑ when $|x|\to\infty$.

Another consequence of (2.18) is that the entropy production rate is bounded in the space of integrable functions. In particular,

$$\int_0^T\int_\Omega\left(\frac{\mu}{\vartheta}\left|\nabla_x\mathbf{u}+\nabla_x\mathbf{u}^t-\frac{2}{3}\mathrm{div}_x\mathbf{u}\right|^2+\frac{\kappa|\nabla_x\vartheta|^2}{\vartheta^2}\right)\ \mathrm{d}x\leq c, \tag{2.23}$$

where the constant depends only on the size of initial data. Thus if we assume that $\mu(\vartheta)\approx\vartheta$, $\kappa(\vartheta)\approx\vartheta^2$, relation (2.23), together with (2.1), yield the bounds

$$\nabla_x\mathbf{u}\in L^2(0,T;L^2(\Omega;R^{3\times3})),\ \nabla_x\vartheta\in L^2(0,T;L^2(\Omega;R^3)). \tag{2.24}$$

2.5. State equation of a monoatomic gas. In order to get more estimates from (2.22), certain restrictions must be imposed on the relation between s, e, and ϱ, ϑ. We assume that e and p are interrelated through

$$p(\varrho,\vartheta)=\frac{2}{3}\varrho e(\varrho,\vartheta), \tag{2.25}$$

which characterizes *monoatomic gases* (see Eliezer et al. [25]).

Combining (2.25) with Gibbs' equation (2.8) we deduce that the only admissible form of the pressure reads

$$p(\varrho,\vartheta)=\vartheta^{5/2}P\left(\frac{\varrho}{\vartheta^{3/2}}\right) \tag{2.26}$$

for a certain function P. We assume that $P(0)=0$, and, in accordance with hypothesis of thermodynamics stability (2.11),

$$P'(Z)>0,\ \frac{\frac{5}{3}P(Z)-P'(Z)Z}{Z}>0 \text{ for all } Z\geq0. \tag{2.27}$$

In particular, the function $Z \mapsto P(Z)/Z^{5/3}$ is non-increasing, and we suppose

$$\lim_{Z\to\infty} \frac{P(Z)}{Z^{5/3}} = p_\infty > 0. \tag{2.28}$$

If (2.28) holds, the uniform bounds established in (2.22) imply

$$\sup_{t\in[0,T]} \|\varrho(t,\cdot)\|_{L^{5/3}(\Omega)} \leq c. \tag{2.29}$$

2.6. A priori bounds and the vacuum problem. Summarizing the *a priori* bounds obtained in the previous part we may conclude, very roughly indeed, that ϱ, $\mathbf{u}$, ϑ, together with the gradients $\nabla_x \mathbf{u}$, $\nabla_x \vartheta$ are bounded in certain L^p-spaces, with a suitable $p > 1$ depending on the structural properties of the constitutive functions. This is apparently not enough, not even to deduce a.a. pointwise convergence of these quantities constructed by means of a suitable approximation scheme. Indeed a Aubin-Lions type argument *does not* apply as the time derivatives $\partial_t(\varrho \mathbf{u})$, $\partial_t(\varrho s(\varrho, \vartheta))$ contain the density ϱ that may vanish on a set of positive measure.

Let us examine this so-called *vacuum problem* more closely. Equation of continuity (1.3) can be written in the form

$$\partial_t \varrho + \mathbf{u}\cdot\nabla_x \varrho = -\varrho \mathrm{div}_x \mathbf{u}; \tag{2.30}$$

whence,

$$\begin{aligned} &\inf_{x\in\Omega} \varrho_0(x) \exp\left(-\int_0^\tau \|\mathrm{div}_x \mathbf{u}\|_{L^\infty(\Omega)} \ \mathrm{d}t \right) \\ &\qquad \leq \varrho(\tau, y) \leq \sup_{x\in\Omega} \varrho_0(x) \exp\left(\int_0^\tau \|\mathrm{div}_x \mathbf{u}\|_{L^\infty(\Omega)} \ \mathrm{d}t \right) \end{aligned} \tag{2.31}$$

for any $\tau \geq 0$, $y \in \Omega$. Thus *uniform a priori* bounds on ϱ in terms of the initial data are conditioned by integrability of $\|\mathrm{div}_x \mathbf{u}\|_{L^\infty(\Omega)}$! As a matter of fact, a similar condition guarantees uniqueness of weak solutions to transport equations like (2.30), and of the associated system

$$\mathbf{X}'(t) = \mathbf{u}(t, \mathbf{X}(t)), \ \mathbf{X} = \mathbf{x}_0$$

describing the streamlines (particle paths) in the fluid (see DiPerna and Lions [23]). **Absence of uniform bounds on $\mathrm{div}_x\mathbf{u}$ represents one of the major open problems of the existing mathematical theory of compressible fluids.** The hypothetical possibility of a vacuum occurring in a finite time may hamper the analysis of the system considerably. In particular, we loose control over possible *time oscillations* of the velocity field as well as the temperature. Thus, in contrast with a common belief, the velocity field $\mathbf{u}$ is not (known to be) precompact in the Lebesgue space L^1.

2.7. Radiation pressure. To begin, we point out that the effect of radiation on the fluid motion is a complex problem that requires much more sophisticated treatment than we present here.

In the "zero-th order" approximation, the pressure p is simply augmented by a *radiation* component proportional to $a\vartheta^4/3$, where $a > 0$ is the Stefan-Boltzman constant. In accordance with Gibbs' relation (2.8), the internal energy contains a new term $a\vartheta^4/3\varrho$, while the specific entropy s is augmented by $4\vartheta^3/3\varrho$. This yields, together with (2.26),

$$p(\varrho,\vartheta) = \vartheta^{5/2} P\left(\frac{\varrho}{\vartheta^{3/2}}\right) + \frac{a}{3}\vartheta^4, \tag{2.32}$$

$$\varrho e(\varrho,\vartheta) = \frac{3}{2}\vartheta^{5/2} P\left(\frac{\varrho}{\vartheta^{3/2}}\right) + a\vartheta^4, \tag{2.33}$$

and

$$\begin{aligned} \varrho s(\varrho,\vartheta) &= \varrho S\left(\frac{\varrho}{\vartheta^{3/2}}\right) + \frac{4}{3}\vartheta^3, \\ S'(Z) &= -\frac{3}{2}\frac{\frac{5}{3}P(Z) - P'(Z)Z}{Z^2}. \end{aligned} \tag{2.34}$$

Similarly, the heat conductivity coefficient κ in (1.12) should be supplemented by a "radiation" component $\kappa_R(\vartheta) = d\vartheta^4$.

Including radiation effects into the model is a suitable compromise between the physical background and the mathematical necessity dictated by the available mathematical tools and our present state on knowledge. The presence of the intensive component $a\vartheta^4$ in the internal energy, or $a\vartheta^3/3$ in the entropy balance, yields the necessary bounds on the time derivative of ϑ required by the Aubin-Lions type argument. We shall discuss this point in detail in Section 4.

3. Weak formulation of the Navier-Stokes-Fourier system. A suitable weak formulation of the Navier-Stokes-Fourier system is based on the entropy balance equation (2.9). At the level of classical solutions, we have seen that (2.9) is equivalent to the internal energy balance (1.10), or the *total energy balance*

$$\begin{aligned} \partial_t\Big(\frac{1}{2}\varrho|\mathbf{u}|^2 + \varrho e(\varrho,\vartheta)\Big) + \mathrm{div}_x\Big(\Big(\frac{1}{2}\varrho|\mathbf{u}|^2 + \varrho e(\varrho,\vartheta) + p(\varrho,\vartheta)\Big)\mathbf{u}\Big) \\ + \mathrm{div}_x(\mathbf{q} - \mathbb{S}\mathbf{u}) = \varrho\nabla_x F\cdot\mathbf{u}. \end{aligned} \tag{3.1}$$

However, equation (1.10) contains the term $p\mathrm{div}_x\mathbf{u}$ that is not (known to be) *a priori* bounded, while (3.1) entails similar technical difficulties. Moreover, equation (3.1) contains the term $\mathbb{S}\mathbf{u}$, the oscillations and integrability of which are not controlled on the hypothetical vacuum sets.

3.1. Weak formulation based on the Second law of thermodynamics. We shall say that the functions ϱ, ϑ, $\mathbf{u}$ represent a weak solution to the *Navier-Stokes-Fourier system* if:

$$\int_0^T \int_\Omega \Big(\varrho \partial_t \varphi + \varrho \mathbf{u} \cdot \nabla_x \varphi \Big) \ \mathrm{d}x \ \mathrm{d}t = - \int_\Omega \varrho_0 \varphi(0, \cdot) \ \mathrm{d}x \tag{3.2}$$

for any $\varphi \in C_c^\infty([0,T) \times \overline{\Omega})$;

$$\begin{aligned} &\int_0^T \int_\Omega \Big(\varrho \mathbf{u} \cdot \partial_t \varphi + (\varrho \mathbf{u} \otimes \mathbf{u}) : \nabla_x \varphi + p(\varrho, \vartheta) \mathrm{div}_x \varphi \Big) \ \mathrm{d}x \mathrm{d}t \\ &= \int_0^T \int_\Omega \mathbb{S} : \nabla_x \varphi \ \mathrm{d}x \ \mathrm{d}t - \int_0^T \int_\Omega \varrho \nabla_x F \cdot \varphi \ \mathrm{d}x - \int_\Omega \varrho_0 \mathbf{u}_0 \cdot \varphi(0, \cdot) \ \mathrm{d}x \end{aligned} \tag{3.3}$$

for any $\varphi \in C_c^\infty([0,T) \times \Omega; R^3)$;

$$\begin{aligned} &\int_0^T \int_\Omega \Big(\varrho s(\varrho, \vartheta) \partial_t \varphi + \varrho s(\varrho, \vartheta) \mathbf{u} \cdot \nabla_x \varphi + \frac{\mathbf{q}}{\vartheta} \cdot \nabla_x \varphi \Big) \ \mathrm{d}x \ \mathrm{d}t + < \sigma ; \varphi > \\ &\qquad = - \int_\Omega \varrho_0 s(\varrho_0, \vartheta_0) \varphi(0, \cdot) \ \mathrm{d}x \end{aligned} \tag{3.4}$$

for any $\varphi \in C_c^\infty([0,T) \times \overline{\Omega})$, where σ is a non-negative Borel measure on $[0,T] \times \overline{\Omega}$ satisfying

$$\sigma \geq \frac{1}{\vartheta} \left(\mathbb{S} : \nabla_x \mathbf{u} - \frac{\mathbf{q} \cdot \nabla_x \vartheta}{\vartheta} \right); \tag{3.5}$$

$$\begin{aligned} &\int_0^T \int_\Omega \left(\frac{1}{2} \varrho |\mathbf{u}|^2 + \varrho e(\varrho, \vartheta) - \varrho F \right) \partial_t \varphi \ \mathrm{d}x \ \mathrm{d}t = - \int_\Omega E_0 \varphi(0) \ \mathrm{d}x, \\ &\qquad E_0 = \frac{1}{2} \varrho_0 |\mathbf{u}_0|^2 + \varrho_0 e(\varrho_0, \vartheta_0) - \varrho_0 F \end{aligned} \tag{3.6}$$

for any $\varphi \in C_c^\infty[0,T)$.

The weak formulation (3.2 - 3.6) already includes the satisfaction of the initial conditions and the energetically insulated boundary conditions

$$\mathbf{u}|_{\partial\Omega} = 0, \ \mathbf{q} \cdot \mathbf{n}|_{\partial\Omega} = 0$$

provided we assume that

$$\mathbf{u} \in L^p(0,T; W_0^{1,q}(\Omega; R^3)) \text{ for certain } p, q > 1. \tag{3.7}$$

The most remarkable feature of the weak formulation introduced through (3.2 - 3.7) is replacing (2.10) by *inequality* (3.5). As we have seen in Section 2, the available *a priori* estimates guarantee only *integrability* of the quadratic terms in the entropy production proportional to $|\nabla_x \mathbf{u}|^2$ and $|\nabla_x \vartheta|^2$. Since quadratic functional are only weakly lower semi-continuous, the resulting entropy production rate σ may contain a contribution that is not absolutely continuous with respect to the standard Lebesgue measure, in particular, σ may exceed the value indicated by (2.10). Such uncontrolled dissipation of the kinetic energy is well-known in the theory of *inviscid* fluids described by the Euler equations, however, it represents a rather hypothetical possibility for a *viscous* fluid. Finally, thanks to the total energy balance (3.6) included in the weak formulation, the inequality (3.5) reduces to (2.10) provided the weak solution is smooth (see [28]). **It is a very interesting open problem whether (2.10) holds also in the class of weak solutions to the Navier-Stokes-Fourier system.**

4. Weak sequential stability: Temperature. The question of *weak sequential stability* (compactness) is crucial in the study of any non-linear problem. For a sequence of solutions $\{\varrho_n, \vartheta_n, \mathbf{u}_n\}$ satisfying the *a priori* bounds established in Section 2, the goal is to show that

$$\varrho_n \to \varrho,\ \vartheta_n \to \vartheta,\ \mathbf{u}_n \to \mathbf{u}$$

in a certain sense, where ϱ, ϑ, $\mathbf{u}$ is a (weak) solution of the same problem. To this end, as the constitutive relations are non-linear functions of ϱ, ϑ, we have to show that, at least,

$$\varrho_n \to \varrho \text{ a.a. in } (0,T) \times \Omega,$$
$$\vartheta_n \to \vartheta \text{ a.a. in } (0,T) \times \Omega.$$

In this part, we discuss the pointwise (a.a.) convergence of the temperature field $\{\vartheta_n\}_{n=1}^{\infty}$. The arguments are sketched only, the reader may consult the monograph [28] for details. For $v_n \to v$ weakly in L^1, it is convenient to introduce the notation $\overline{b(v)}$ for a weak limit (in L^1) of the sequence $\{b(v_n)\}_{n=1}^{\infty}$.

4.1. Weak compactness results. The classical *Aubin-Lions lemma* reads as follows (see Simon [60]):

LEMMA 4.1. *Let $X \subset Y \subset Z$ be a trio of Banach spaces such that the embedding $X \hookrightarrow Y$ is compact and $Y \hookrightarrow Z$ continuously. Let*

$$\{v_n\}_{n=1}^{\infty} \text{ be bounded in } L^p(0,T;X),$$

and

$$\{\partial_t v_n\}_{n=1}^{\infty} \text{ be bounded in } L^q(0,T;Z),$$

$p \geq q > 1$.

Then

$$\{v_n\}_{n=1}^{\infty} \text{ is precompact in } L^q(0,T;Y).$$

The strength of Lemma 4.1 lies in the fact that the space Z can be quite large. Still, we need a more delicate argument to establish weak sequential stability of the temperature field in the Navier-Stokes-Fourier system, namely, the celebrated *Div-Curl lemma* of Murat and Tartar [50], [61]:

LEMMA 4.2. *Let Q be a bounded domain in R^N,*

$$\mathbf{U}_n \to \mathbf{U} \text{ weakly in } L^p(Q; R^N),$$

$$\mathbf{V}_n \to \mathbf{V} \text{ weakly in } L^q(Q; R^N),$$

where

$$\{\text{div}\mathbf{U}_n\}_{n=1}^{\infty} \text{ is precompact in } W^{-1,s}(Q),$$

$$\{\mathbf{curl}\ \mathbf{V}_n\}_{n=1}^{\infty} \text{ is precompact in } W^{-1,s}(Q; R^{N\times N}),$$

$$s > 1,\ \frac{1}{p} + \frac{1}{q} = \frac{1}{r} < 1.$$

Then

$$\mathbf{U}_n \cdot \mathbf{V}_n \to \mathbf{U} \cdot \mathbf{V} \text{ weakly in } L^r(Q).$$

4.2. Application of Div-Curl lemma. Keeping in mind the *a priori bounds* established in Section 2, we can say that $\{\vartheta_n\}_{n=1}^{\infty}$, together with the spatial gradients $\{\nabla_x \vartheta_n\}_{n=1}^{\infty}$ are bounded in the Lebesgue space L^p for a certain $p > 1$. If we could establish a bound on $\{\partial_t \vartheta_n\}_{n=1}^{\infty}$, we would apply Aubin-Lions lemma in order to obtain a.a. convergence of $\{\vartheta_n\}_{n=1}^{\infty}$. Instead, we use the entropy balance equation (3.4), together with Div-Curl lemma 4.2, to deduce that

$$\overline{\varrho s(\varrho, \vartheta)\vartheta} = \overline{\varrho s(\varrho, \vartheta)}\vartheta, \tag{4.1}$$

where ϑ denotes a weak limit of $\{\vartheta_n\}_{n=1}^{\infty}$.

On the other hand, as s is strictly increasing in ϑ, we have

$$\int_0^T \int_\Omega \Big(\varrho_n s(\varrho_n, \vartheta_n) - \varrho_n s(\varrho_n, \vartheta)\Big)(\vartheta_n - \vartheta)\ \mathrm{d}x\ \mathrm{d}t \geq 0;$$

and, thanks to the radiation component of the entropy proportional to ϑ^3 (see (2.34)), it is enough to show that

$$\int_0^T \int_\Omega \Big(\varrho_n s(\varrho_n, \vartheta_n) - \varrho_n s(\varrho_n, \vartheta)\Big)(\vartheta_n - \vartheta)\ \mathrm{d}x\ \mathrm{d}t \to 0.$$

Indeed, we have

$$\begin{aligned} &\liminf_{n\to\infty} \int_0^T \int_\Omega \Big(\varrho_n s(\varrho_n, \vartheta_n) - \varrho_n s(\varrho_n, \vartheta)\Big)(\vartheta_n - \vartheta)\ \mathrm{d}x\ \mathrm{d}t \\ &\qquad \geq c\ \liminf_{n\to\infty} \int_0^T \int_\Omega |\vartheta_n - \vartheta|^4\ \mathrm{d}x\ \mathrm{d}t, \text{ with } c > 0. \end{aligned} \tag{4.2}$$

In view of (4.1), (4.2), the proof of pointwise convergence of $\{\vartheta_n\}_{n=1}^{\infty}$ reduces to showing

$$\int_0^T \int_\Omega \varrho_n s(\varrho_n, \vartheta)(\vartheta_n - \vartheta) \ \mathrm{d}x \ \mathrm{d}t \to 0. \tag{4.3}$$

In order to see (4.3), we need the concept of *renormalized* solution to the equation of continuity (1.3) introduced by DiPerna and Lions [23].

4.3. Renormalized solutions of the equation of continuity. Renormalized solutions to equation of continuity (1.3) were introduced by DiPerna and Lions [23] in order to obtain a more precise description of possible oscillations that solutions may develop in a finite time. Assuming ϱ, $\mathbf{u}$ is a classical solution of (1.3), we easily deduce that

$$\partial_t b(\varrho) + \mathrm{div}_x(b(\varrho)\mathbf{u}) + \Big(b'(\varrho)\varrho - b(\varrho)\Big)\mathrm{div}_x \mathbf{u} = 0 \tag{4.4}$$

for any (nonlinear) function b. A weak formulation of (4.4) reads

$$\int_0^T \int_\Omega \Big(b(\varrho)\partial_t\varphi + b(\varrho)\mathbf{u}\cdot\nabla_x\varphi + \Big(b(\varrho) - b'(\varrho)\varrho\Big)\mathrm{div}_x\mathbf{u}\varphi\Big) \ \mathrm{d}x \ \mathrm{d}t = -\int_\Omega b(\varrho_0)\varphi(0,\cdot) \ \mathrm{d}x \tag{4.5}$$

for any test function $\varphi \in C_c^\infty([0,T) \times \overline{\Omega}; R^3)$.

We shall say that ϱ, $\mathbf{u}$ represents a *renormalized solution* of equation (1.3) if the integral identity (4.5) holds. Although (4.5) may be reminiscent of the concept of *entropy solution* introduce for scalar conservation laws by Kruzhkov [39], its meaning is slightly different. In particular, equation (4.5) is related to *concentrations* rather than oscillations in the family of solutions. Indeed any weak solution ϱ, $\mathbf{u}$ of (1.3) is a renormalized solution as soon as $\varrho \in L^\infty(0,T;L^p(\Omega))$, $\mathbf{u} \in L^q(0,T;W^{1,q}(\Omega;R^3))$, where

$$\frac{1}{p} + \frac{1}{q} \le 1$$

(see DiPerna and Lions [23]).

As a matter of fact, for (4.5) to make sense, it is enough to assume that $\mathrm{div}_x\mathbf{u} \in L^1((0,T)\times\Omega)$ provided we restrict ourselves to the class of functions b such that $b' \in C_c^\infty[0,\infty)$.

Going back to relation (4.3), we revoke (2.24) to deduce

$$\vartheta_n \to \vartheta \text{ weakly in } L^2(0,T;W^{1,2}(\Omega)), \tag{4.6}$$

and, supposing the ϱ_n, $\mathbf{u}_n$ are renormalized solutions of (1.3),

$$b(\varrho_n) \to \overline{b(\varrho)} \text{ in } C_{\mathrm{weak}}([0,T];L^1(\Omega)). \tag{4.7}$$

In view of relations (4.6), (4.7), the desired conclusion (4.3) follows from a general result concerning families of parametrized (Young) measures. Let $Q \subset R^N$ be a domain. We say that $\psi : Q \times R^M$ is a Caratheodory function on $Q \times R^M$ if

$$\left\{ \begin{array}{l} \text{for a. a. } x \in Q, \text{ the function } \lambda \mapsto \psi(x, \lambda) \text{ is continuous on } R^M; \\ \text{for all } \lambda \in R^M, \text{ the function } x \mapsto \psi(x, \lambda) \text{ is measurable on } Q. \end{array} \right\} \quad (4.8)$$

We say that $\{\nu_x\}_{x \in Q}$ is a *family of parametrized measures* if ν_x is a probability measure for a.a. $x \in Q$, and if

$$\left\{ \begin{array}{c} \text{the function } x \to \int_{R^M} \phi(\lambda)\, \mathrm{d}\nu_x(\lambda) := <\nu_x, \phi> \text{ is measurable on } Q \\ \text{for all } \phi : R^M \to R,\ \phi \in C(R^M) \cap L^\infty(R^M)). \end{array} \right\} (4.9)$$

We report the following result (see Pedregal [51, Chapter 6, Theorem 6.2]).

THEOREM 4.1. *Let $\{\mathbf{v}_n\}_{n=1}^\infty$, $\mathbf{v}_n : Q \subset R^N \to R^M$ be a sequence of functions bounded in $L^1(Q; R^M)$, where Q is a domain in R^N.*

Then there exist a subsequence (not relabeled) and a parameterized family $\{\nu_y\}_{y \in Q}$ of probability measures on R^M depending measurably on $y \in Q$ with the following property:

For any Caratheodory function $\Phi = \Phi(y, z)$, $y \in Q$, $z \in R^M$ such that

$$\Phi(\cdot, \mathbf{v}_n) \to \overline{\Phi} \text{ weakly in } L^1(Q),$$

we have

$$\overline{\Phi}(y) = \int_{R^M} \psi(y, z)\, \mathrm{d}\nu_y(z) \text{ for a.a. } y \in Q.$$

Theorem 4.1, together with (4.6), (4.7), completes the proof of (4.3), therefore the strong pointwise (a.a.) convergence of the sequence $\{\vartheta_n\}_{n=1}^\infty$. **We should keep in mind that compactness of the temperature field follows from the presence of the radiation pressure component in the state equation (2.34).**

5. Weak sequential stability: Density. Strong pointwise convergence of the density field $\{\varrho_n\}_{n=1}^\infty$ is the central issue in the mathematical theory of compressible fluids. Note that, unlike the velocity field $\mathbf{u}$ and the temperature ϑ, the density satisfies a transport equation (1.3). Accordingly, we have no control over possible time and *space* oscillations of the sequence $\{\varrho_n\}_{n=1}^\infty$. As we will see in this section, however, the density oscillations can be effectively controlled by combining (1.3) with momentum equation (1.8). The basic idea goes back to P.-L.Lions [42], [43], with the so-called *effective viscous pressure* playing a crucial role. We first recall the facts obtained in the previous part of the paper. Let $\{\varrho_n, \vartheta_n, \mathbf{u}_n\}_{n=1}^\infty$ be a sequence of solutions of the Navier-Stokes-Fourier system satisfying the uniform bounds established in Section 2. In accordance with Section 4, we have

$$\varrho_n \to \varrho \text{ in } C_{\text{weak}}([0,T]; L^p(\Omega)), \tag{5.1}$$

$$\vartheta_n \to \vartheta \text{ in } L^4((0,T)\times\Omega),\ \nabla_x\vartheta_n \to \nabla_x\vartheta \text{ weakly in } L^2((0,T)\times\Omega; R^3), \tag{5.2}$$

$$\mathbf{u}_n \to \mathbf{u} \text{ weakly in } L^q((0,T)\times\Omega; R^3) \tag{5.3}$$

for certain $p, q > 1$. For the sake of simplicity, we shall assume that p, q are as large as we need although this is clearly not the case in applications.

5.1. Effective viscous pressure. The basic idea is to "compute" the pressure in the momentum equation (1.8). For the sake of simplicity, we set $\mathbf{f} = \mathbf{g} = 0$. Formally, we have

$$p - \nabla_x\Delta^{-1}\nabla_x : \mathbb{S} = -\nabla_x\Delta^{-1}\nabla_x : (\varrho\mathbf{u}\otimes\mathbf{u}) - \Delta^{-1}\text{div}_x(\partial_t(\varrho\mathbf{u})), \tag{5.4}$$

where we have ignored the boundary conditions. Assume, for a while, that the viscosity coefficients μ, η in Newton's law (1.11) are constant. Accordingly, we have

$$\nabla_x\Delta^{-1}\nabla_x\mathbb{S} = \left(\frac{4\mu}{3} + \eta\right)\text{div}_x\mathbf{u}; \tag{5.5}$$

the quantity

$$p - \nabla_x\Delta^{-1}\nabla_x : \mathbb{S} = p - \left(\frac{4\mu}{3} + \eta\right)\text{div}_x\mathbf{u}$$

is termed *effective viscous flux.*

The quantity on the right-hand side of (5.4) possesses a remarkable property of *weak continuity.* Multiplying the right-hand side of (5.4) on ϱ, integrating the resulting expression by parts, and ignoring the boundary integrals, we arrive at the following expression:

$$\int_0^T\int_\Omega \varrho\mathbf{u}\cdot\nabla_x\Delta^{-1}\text{div}_x(\varrho\mathbf{u})\ dx\ dt - \int_0^T\int_\Omega(\varrho\mathbf{u}\otimes\mathbf{u}) : (\nabla_x\Delta^{-1}\nabla_x)[\varrho]\ dx\ dt.$$

Thus, for any sequence of weak solutions $\{\varrho_n, \vartheta_n, \mathbf{u}_n\}_{n=1}^{\infty}$,

$$\lim_{n\to\infty}\int_0^T\int_\Omega\left(p(\varrho_n,\vartheta_n) - (4\mu/3+\eta)\text{div}_x\mathbf{u}_n\right)\varrho_n\ dx\ dt \tag{5.6}$$

$$= \lim_{n\to\infty}\int_0^T\int_\Omega \varrho_n\mathbf{u}_n\cdot\nabla_x\Delta^{-1}\text{div}_x(\varrho_n\mathbf{u}_n)\ dx\ dt$$

$$- \lim_{n\to\infty}\int_0^T\int_\Omega(\varrho_n\mathbf{u}_n\otimes\mathbf{u}_n) : (\nabla_x\Delta^{-1}\nabla_x)[\varrho_n]\ dx\ dt.$$

At this stage, we need the following result that may be viewed as a direct consequence of Div-Curl lemma.

LEMMA 5.1. *Let*

$$\mathbf{U}_n \to \mathbf{U} \text{ weakly in } L^p(R^N;R^N),$$
$$\mathbf{V}_n \to \mathbf{V} \text{ weakly in } L^q(R^N;R^N),$$

where

$$\frac{1}{p}+\frac{1}{q}=\frac{1}{r}<1.$$

Then

$$\mathbf{U}_n\cdot(\nabla_x\Delta^{-1}\mathrm{div}_x)[\mathbf{V}_n]-(\nabla_x\Delta^{-1}\mathrm{div}_x)[\mathbf{U}_n]\cdot\mathbf{V}_n$$
$$\to$$
$$\mathbf{U}\cdot(\nabla_x\Delta^{-1}\mathrm{div}_x)[\mathbf{V}]-(\nabla_x\Delta^{-1}\mathrm{div}_x)[\mathbf{U}]\cdot\mathbf{V}$$

weakly in $L^r(R^N)$.

Proof. Rewrite

$$\begin{aligned}\mathbf{U}_n\cdot(\nabla_x\Delta^{-1}\mathrm{div}_x)[\mathbf{V}_n]&-(\nabla_x\Delta^{-1}\mathrm{div}_x)[\mathbf{U}_n]\cdot\mathbf{V}_n\\&=\Big(\mathbf{U}_n-(\nabla_x\Delta^{-1}\mathrm{div}_x)[\mathbf{U}_n]\Big)\cdot(\nabla_x\Delta^{-1}\mathrm{div}_x)[\mathbf{V}_n]\\&\quad-(\nabla_x\Delta^{-1}\mathrm{div}_x)[\mathbf{U}_n]\cdot\Big(\mathbf{V}_n-(\nabla_x\Delta^{-1}\mathrm{div}_x)[\mathbf{V}_n]\Big).\end{aligned}$$

Since

$$\mathrm{div}_x(\mathbf{U}_n-(\nabla_x\Delta^{-1}\mathrm{div}_x)[\mathbf{U}_n]\Big)=\mathrm{div}_x(\mathbf{V}_n-(\nabla_x\Delta^{-1}\mathrm{div}_x)[\mathbf{V}_n])=0,$$

the desired conclusion follows from Div-Curl lemma (Lemma 4.2). □

Combining (5.6), together with Lemma 5.1, and the fact that

$$\varrho_n\mathbf{u}_n\to\varrho\mathbf{u} \text{ in } C_{\mathrm{weak}}([0,T];L^q(\Omega))$$

for a certain $q>1$, implies that

$$(4\mu/3+\eta)\Big(\overline{\varrho\mathrm{div}_x\mathbf{u}}-\varrho\mathrm{div}_x\mathbf{u}\Big)=\overline{p(\varrho,\vartheta)\varrho}-\overline{p(\varrho,\vartheta)}\varrho. \tag{5.7}$$

5.2. Strong convergence of the density - renormalized equation. Relation (5.6), together with the renormalized equation of continuity (4.5), give rise to strong (a.a. pointwise) convergence of the densities $\{\varrho_n\}_{n=1}^\infty$. Indeed evoking the renormalized equation for $b(\varrho)=\varrho log(\varrho)$, we get

$$\begin{aligned}\int_\Omega\varrho_n\log(\varrho_n)(\tau,\cdot)\ \mathrm{d}x+\int_0^\tau\int_\Omega\varrho_n\mathrm{div}_x\mathbf{u}_n\ \mathrm{d}x\ \mathrm{d}t&\\=\int_\Omega\varrho_n\log(\varrho_n)(0,\cdot)\ \mathrm{d}x.&\end{aligned}\tag{5.8}$$

Assuming the *a priori bounds* to be strong enough to imply higher integrability of ϱ, $\nabla_x \mathbf{u}$, we may apply the theory developed by DiPerna and Lions [23] to deduce that the limit functions ϱ, $\mathbf{u}$ also satisfy the renormalized equation of continuity (4.5). In other words,

$$\int_\Omega \varrho \log(\varrho)(\tau,\cdot)\ \mathrm{d}x + \int_0^\tau \int_\Omega \varrho \mathrm{div}_x \mathbf{u}\ \mathrm{d}x\ \mathrm{d}t \\ = \int_\Omega \varrho \log(\varrho)(0,\cdot)\ \mathrm{d}x. \tag{5.9}$$

Combining (5.6 - 5.8) we may infer that

$$\limsup_{n\to\infty} \int_\Omega \Big(\varrho_n \log(\varrho_n)(\tau,\cdot) - \varrho \log(\varrho)(\tau,\cdot) \Big)\ \mathrm{d}x \le 0, \tag{5.10}$$

for any $\tau \ge 0$ provided

$$\int_\Omega \varrho_n \log(\varrho_n)(0,\cdot)\ \mathrm{d}x \to \int_\Omega \varrho \log(\varrho)(0,\cdot)\ \mathrm{d}x \text{ as } n \to \infty.$$

As the function $\varrho \mapsto \varrho \log(\varrho)$ is strictly convex, relation (5.9) implies *strong a.a. pointwise* convergence of the sequence $\{\varrho_n\}_{n=1}^\infty$.

5.3. Open problems. As a matter of fact, the arguments of Section 5 are very sketchy. In particular, we did not clarify completely the available integrability bounds on the density and the velocity, with respect to constitutive relations. The first question that arises is whether or not we can go beyond DiPerna and Lions theory to deduce that a couple of functions ϱ, $\mathbf{u}$ solves the renormalized equation (4.5). Following [27] we introduce the oscillations defect measure of a sequence

$$\varrho_n \to \varrho \text{ weakly in } L^1(Q),\ Q \subset R^4,$$
$$\mathbf{osc}_q[\varrho_n \to \varrho](Q) = \lim_{k\to\infty} \Big(\limsup_{n\to\infty} \int_Q |T_k(\varrho_n) - T_k(\varrho)|^q\ dy \Big),$$

where T_k are the cut-off functions

$$T_k(\varrho) = \begin{cases} \varrho \text{ if } |\varrho| \le k, \\ k \text{ if } \varrho > k, \\ -k \text{ if } \varrho < -k. \end{cases}$$

PROPOSITION 5.1. *Let $Q \subset R^4$ be a bounded open set. Let $\{\varrho_n\}_{n=1}^\infty$, $\{\mathbf{u}_n\}_{n=1}^\infty$ be a sequence of renormalized solutions of (4.5) such that*

$$\varrho_n \to \varrho \textit{ weakly in } L^1(Q),$$
$$\mathbf{u}_n \to \mathbf{u} \textit{ weakly in } L^q(Q;R^3),\ \nabla_x \mathbf{u}_n \to \nabla_x \mathbf{u} \textit{ weakly in } L^q(Q;R^3),$$

such that

$$\mathbf{osc}_p[\varrho_n \to \varrho](Q) < \infty$$

for

$$\frac{1}{p} + \frac{1}{q} = \frac{1}{r} < 1.$$

Then ϱ, $\mathbf{u}$ is a renormalized solution of (4.5) in Q.

The proof of Proposition 5.1 can be found in [27, Chapter 5]. Proposition 5.1 shows certain kind of *stability* of the class of renormalized solutions under very mild assumptions imposed on $\{\varrho_n\}_{n=1}^{\infty}$, $\{\mathbf{u}_n\}_{n=1}^{\infty}$. If the pressure p is given by the equation of state (2.32), where P satisfies (2.28), and the velocity fields belong to the class (2.24), the resulting bounds imposed on a sequence of weak solutions $\{\varrho_n\}_{n=1}^{\infty}$, $\{\mathbf{u}_n\}_{n=1}^{\infty}$ are strong enough for Proposition 5.1 to apply. Accordingly, the class of weak solutions introduced in Section 3, supplemented with the renormalized equation of continuity

$$\int_0^T \int_\Omega \Big(b(\varrho)\partial_t \varphi + b(\varrho)\mathbf{u}\cdot\nabla_x\varphi + \Big(b(\varrho) - b'(\varrho)\varrho\Big)\mathrm{div}_x\mathbf{u}\varphi \Big) \ \mathrm{d}x \ \mathrm{d}t = - \int_\Omega b(\varrho_0)\varphi(0,\cdot) \ \mathrm{d}x \tag{5.11}$$

satisfied for any test function $\varphi \in C_c^\infty([0,T)\times\overline{\Omega}; R^3)$, is *weakly sequentially stable*. This means that any family $\{\varrho_n, \vartheta_n, \mathbf{u}_n\}_{n=}^{\infty}$ of weak solutions subject to the uniform bounds following from the total energy balance contains a subsequence weakly converging to another solution of the same problem (see [28]). As (5.11) plays a crucial role in the analysis of the Navier-Stokes-Fourier system, it is customary to replace (3.2) by (5.11) in the definition of weak solutions.

Combining the analysis of Section 5 with Proposition 5.1 we can therefore control the density oscillations for a vast class of constitutive equations. However, the reader will have noticed that the crucial relation (5.6) was obtained under the hypothesis that both the shear viscosity coefficient μ and the bulk viscosity coefficient η are constant. In many applications, the transport coefficients μ, η, and κ depend on the state variables ϱ and ϑ. In particular, they are effective functions of the absolute temperature in the case of gases. Recalling relation (5.5) that holds for constant μ and η, we examine the *commutator*

$$\nabla_x \Delta^{-1} \nabla_x : \mathbb{S} - \left(\frac{4\mu}{3} + \eta\right) \mathrm{div}_x \mathbf{u} \tag{5.12}$$

of the operator $\nabla_x\Delta^{-1}\nabla_x$ with multiplication by μ. A partial result in this direction is provided by the following lemma (see [28, Theorem 10.28]).

LEMMA 5.2. *Let* $w \in W^{1,r}(R^N)$ *and* $\mathbf{V} \in L^p(R^N;R^N)$ *be given, with*

$$1 < r < N,\ 1 < p < \infty,\ \frac{1}{r} + \frac{1}{p} - \frac{1}{N} < 1.$$

Then

$$\left\| \nabla_x \Delta^{-1} \nabla_x [w\mathbf{V}] - w \nabla_x \Delta^{-1} \nabla_x [\mathbf{V}] \right\|_{W^{\beta,s}(R^N;R^N)} \leq c \|w\|_{W^{1,r}(R^N)} \|\mathbf{V}\|_{L^p(R^N;R^N)}$$

for any

$$\frac{1}{r} + \frac{1}{p} - \frac{1}{N} < \frac{1}{s} < 1,$$
$$\frac{\beta}{N} = \frac{1}{s} + \frac{1}{N} - \frac{1}{p} - \frac{1}{r}.$$

In view of Lemma 5.2, we can still recover the same result as (5.5) provided the viscosity coefficient μ belongs to the Sobolev space $W^{1,r}$ for any time $t \in (0,T)$. This is the case (see (2.24)) if μ and η depend only on the temperature ϑ. The relevant existence results for the Navier-Stokes-Fourier system are shown in [28]. **On the other hand, it is an open problem if similar results can be obtained if μ and η depend on both ϱ and ϑ.** Positive results in the case when μ and η depend *only* on the density ϱ were obtained by Bresch et al. [9], [10], Mellet and Vasseur [48], [49].

Another interesting question is to which extent the previous analysis depends on the *isotropic* character of the rheological properties of the fluid. As is well known (see, for instance, Chapman and Cowling [13]), the "anisotropic" viscosity appears when a conductive fluid is subject to an electromagnetic field. The resulting formula for $\mathbb{S}$ is rather complex, containing at least 7 different viscosity coefficients. **Compactness of the density for a non-isotropic viscosity is an open problem, even in the seemingly simplest case of two shear viscosity in different directions.**

6. Reduced models.

6.1. Incompressible Navier-Stokes system. One of the most challenging problems in the theory of partial differential equations is the following "incompressible" Navier-Stokes system:

$$\mathrm{div}_x \mathbf{u} = 0, \tag{6.1}$$
$$\partial_t \mathbf{u} + \mathrm{div}_x(\mathbf{u} \otimes \mathbf{u}) + \nabla_x p = \nu \Delta \mathbf{u} + \mathbf{f} \tag{6.2}$$

for the unknowns $\mathbf{u} = \mathbf{u}(t,x)$ and $p(t,x)$. System (6.1), (6.2) is a mathematical model of an *incompressible viscous fluid*, where $\mathbf{u}$ is the fluid velocity and p the pressure, depending on the time t and the spatial position $x \in \Omega \subset R^3$. The symbol $\mathbf{f} = \mathbf{f}(t,x)$ denotes a driving force imposed on the fluid. Similarly to the more complex Navier-Stokes-Fourier system discussed in the preceding part of this text, the term

$$\nu \Delta \mathbf{u}, \text{ correctly interpreted as } \mathrm{div}_x \nu \Big(\nabla_x \mathbf{u} + \nabla_x^t \mathbf{u} \Big)$$

represents the viscous forces acting on the fluid in motion, where $\nu > 0$ is a positive constant.

We suppose the no-slip boundary condition

$$\mathbf{u}|_{\partial\Omega} = 0, \tag{6.3}$$

together with the initial condition

$$\mathbf{u}(0,\cdot) = \mathbf{u}_0 \text{ in } \Omega, \tag{6.4}$$

where $\mathbf{u}_0$ is a given vector field. Note that no initial value is prescribed for the pressure p.

Problem (6.1), (6.2), supplemented with (6.3), (6.4), is widely used in both real world applications and theoretical studies as a mathematical description of the motion of an *incompressible viscous fluid.* Despite its apparent simplicity, the Navier-Stokes system (6.1), (6.2) admits solutions of great complexity related to chaotic motions and phenomena directly connected to turbulence. The presence of viscosity possessed by any real fluid ensures that the equations are ultimately irreversible in time, in accordance with the second law of thermodynamics. In addition, there are many arguments to believe that the incompressible Navier-Stokes system is the right model to be studied by mathematical fluid mechanics:

- Golse and Saint-Raymond [35], [36] (see also Bardos et al. [4], [5], [6], Golse and Levermore [34], Masmoudi and Saint-Raymond [47], among others), showed that system (6.1), (6.2) can be obtained as an asymptotic limit of the kinetic Boltzmann equation for certain choice of singular parameters.
- System (6.1), (6.2) can be viewed as an incompressible limit of the full Navier-Stokes-Fourier system discussed in the previous part when the speed of sound tends to infinity (low Mach number limit), see Alazard [1], [2], Danchin [21], [28], Klainerman and Majda [38], Lions and Masmoudi [44], Masmoudi [46], Schochet [57], and the references cited therein.
- System (6.1), (6.2) is a "natural" viscosity regularization of the Euler system of a perfect fluid.
- Validity of the mathematical models based on (6.1), (6.2) has been successfully tested in a huge number of practical applications and their numerical analysis, see Feistauer at al. [30], among a vast amount of the relevant literature.

Despite a number of explicitly known solutions representing certain "basic" fluid flows, a universal solution formula that would provide solutions to (6.1 - 6.4) in terms of the data $\mathbf{f}$, $\mathbf{u}_0$ lies definitely beyond the scope of all available analytical methods. **What is more, the mere *existence* of regular solutions for any admissible choice of the data and any time interval (T_1, T_2) represents an outstanding open problem - a variant of Fermat's last theorem of the modern theory of partial differential equations, see Fefferman [26].**

In 1933, Jean Leray [41] introduced what is now known as *weak solution* to problem (6.1), (6.2), considered in his work in the whole space R^3. Later Hopf [37] extended Leray's result to a general bounded domain $\Omega \subset R^3$ for a suitable choice of boundary conditions. Leray's solutions were called "turbulent", however, in light of the more recent regularity results discovered by Prodi [55], Serrin [59], and many others, the phenomena related to turbulence in viscous fluids are nowadays explained on the basis of chaotic behavior of otherwise perfectly regular solutions (see Davidson [22], Eckmann and Ruelle [24]).

It seems that the right object to be studied in the context of well-posedness of the incompressible Navier-Stokes system is the *vorticity* $\mathbf{curl}_x \mathbf{u}$, see Majda and Bertozzi [45]. The complex relation between vorticity and regularity of solutions to (6.1), (6.2) was revealed by the celebrated result of Beale et al. [8] (see also Constantin and Feffermann [17]), namely solutions of the Navier-Stokes system remain regular as long as we are able to control the integral

$$\int_0^T \|\mathbf{curl}_x(t,\cdot)\|_{L^\infty(\Omega)} \, \mathrm{d}t.$$

In particular, as vorticity can be controlled in the $2-$*dimensional flow*, a complete and fully satisfactory existence theory for the Navier-Stokes system can be established for $\Omega \subset R^2$ (see Ladyzhenskaya [40]). As already pointed out, the situation is rather different in the physically relevant $3D-$case, where, despite the concerted effort of generations of excellent mathematicians, Leray's result is still the best available at least at the level of global-in-time solutions emanating from arbitrarily large initial data.

6.2. Compressible barotropic fluids. Ignoring the influence of the temperature on the motion we obtain the *barotropic Navier-Stokes system* in the form

$$\partial_t \varrho + \mathrm{div}_x(\varrho \mathbf{u}) = 0, \tag{6.5}$$

$$\partial_t(\varrho \mathbf{u}) + \mathrm{div}_x(\varrho \mathbf{u} \otimes \mathbf{u}) + \nabla_x p(\varrho) = \mathrm{div}_x \mathbb{S} + \varrho \mathbf{f}. \tag{6.6}$$

The incompressibility constraint (6.1) is replaced by *equation of continuity* (6.5), where ϱ denotes the fluid density, while the pressure p is now an explicitly given function of ϱ. Note that (6.5), (6.6) can be viewed as a particular case of the full Navier-Stokes-Fourier system, where the pressure p depends only on the density ϱ and the viscosity coefficients are constant.

Similarly to the preceding section, suitable boundary conditions must be imposed on the velocity field $\mathbf{u}$, and also on the density ϱ, provided $\partial\Omega$ is non-empty. Since (6.5) represents a first order transport equation, the boundary distribution of the density can be prescribed only on the part of the boundary where $\mathbf{u} \cdot \mathbf{n} < 0$. Recall that $\mathbf{n}$ denotes the *outer* normal vector. We assume that the boundary is impermeable, meaning

$$\mathbf{u} \cdot \mathbf{n}|_{\partial\Omega} = 0; \tag{6.7}$$

whence the boundary values of ϱ become irrelevant. Condition (6.7) may be supplemented, as in the preceding part, with the no-slip boundary constraint

$$[\mathbf{u}]_\tau|_{\partial\Omega} = 0. \tag{6.8}$$

In accordance with (6.7 - 6.8), the boundary of the physical space Ω is energetically insulated, in particular, the total energy balance

$$\frac{\mathrm{d}}{\mathrm{d}t} \int_\Omega \Big(\frac{1}{2} \varrho |\mathbf{u}|^2 + H(\varrho) \Big) \ \mathrm{d}x + \int_\Omega \mathbb{S} : \nabla_x \mathbf{u} \ \mathrm{d}x = \int_\Omega \varrho \mathbf{f} \cdot \mathbf{u} \ \mathrm{d}x, \tag{6.9}$$

or rather its "weak" counterpart

$$\frac{\mathrm{d}}{\mathrm{d}t} \int_\Omega \Big(\frac{1}{2} \varrho |\mathbf{u}|^2 + H(\varrho) \Big) \ \mathrm{d}x + \int_\Omega \mathbb{S} : \nabla_x \mathbf{u} \ \mathrm{d}x \leq \int_\Omega \varrho \mathbf{f} \cdot \mathbf{u} \ \mathrm{d}x \tag{6.10}$$

holds, with

$$H(\varrho) = \varrho \int_1^\varrho \frac{p(z)}{z^2} \ \mathrm{d}z$$

(cf. the entropy inequality (3.5)).

In accordance with (6.10), the available *a priori* bounds and the related mathematical problems concerning weak sequential stability of the densities are very similar to the full Navier-Stokes-Fourier system discussed in Section 5. In particular, for the *isentropic* pressure-density state equation

$$p(\varrho) = a\varrho^\gamma, \ a > 0,$$

the *existence* of global-in-time weak solutions can be established for any $\gamma > 3/2$ (see [27], Lions [43]). The result is slightly better in the stationary case (see Březina and Novotný [11], Freshe et al. [32], Plotnikov and Sokolowski [54], [52], [53]). **The existence of global-in-time weak solutions remains completely open for the physically relevant *isothermal* case $\gamma = 1$.**

The simplified models, introduced in this part, are, strictly speaking, physically relevant only on medium time scales, where the inevitable influence of the *Second law of thermodynamics* can be ignored. Still they provide a suitable alternative to the more complex Navier-Stokes-Fourier system discussed in the first part of this study. Even the Navier-Stokes-Fourier system represents a drastical simplification of the real world as the physical boundaries were assumed to be energetically isolated and the number of state variables was reduced to three. On the other hand, the reduced systems may be viewed as prototypes of more complex ones (see Babin and Vishik [3], Constantin et al. [18], [19], [20], Foias et al. [31], Sell and You [58], Temam [63], [62], among many others).

REFERENCES

[1] T. Alazard. Low Mach number flows and combustion. *SIAM J. Math. Anal.*, **38**(4):1186–1213 (electronic), 2006.

[2] T. Alazard. Low Mach number limit of the full Navier-Stokes equations. *Arch. Rational Mech. Anal.*, **180**:1–73, 2006.

[3] A.V. Babin and M.I. Vishik. *Attractors of evolution equations.* North-Holland, Amsterdam, 1992.

[4] C. Bardos, F. Golse, and C.D. Levermore. Fluid dynamical limits of kinetic equations, I : Formal derivation. *J. Statist. Phys.*, **63**:323–344, 1991.

[5] C. Bardos, F. Golse, and C.D. Levermore. Fluid dynamical limits of kinetic equations, II : Convergence proofs for the Boltzman equation. *Comm. Pure Appl. Math.*, **46**:667–753, 1993.

[6] C. Bardos, C.D. Levermore, S. Ukai, and T. Yang. Kinetic equations: fluid dynamical limits and viscous heating. *Bull. Inst. Math. Acad. Sin. (N.S.)*, **3**(1):1–49, 2008.

[7] G.K. Batchelor. *An introduction to fluid dynamics.* Cambridge University Press, Cambridge, 1967.

[8] J.T. Beale, T. Kato, and A. Majda. Remarks on the breakdown of smooth solutions for the 3-D Euler equations. *Comm. Math. Phys.*, **94**(1):61–66, 1984.

[9] D. Bresch and B. Desjardins. On the existence of global weak solutions to the Navier-Stokes equations for viscous compressible and heat conducting fluids. *J. Math. Pures Appl.*, **87**:57–90, 2007.

[10] D. Bresch, B. Desjardins, and D. Gérard-Varet. On compressible Navier-Stokes equations with density dependent viscosities in bounded domains. *J. Math. Pures Appl.*, **87**:227–235, 2007.

[11] J. Březina and A. Novotný. On weak solutions of steady navier-stokes equations for monatomic gas. *Preprint Nečas Center for Mathematical Modeling*, 2007.

[12] H. Callen. *Thermodynamics and an Introduction to Thermostatistics.* Wiley, New Yoerk, 1985.

[13] S. Chapman and T.G. Cowling. *Mathematical theory of non-uniform gases.* Cambridge Univ. Press, Cambridge, 1990.

[14] G.-Q. Chen and H. Frid. On the theory of divergence-measure fields and its applications. *Bol. Soc. Brasil. Mat. (N.S.)*, **32**(3):401–433, 2001. Dedicated to Constantine Dafermos on his 60th birthday.

[15] G.-Q. Chen and M. Torres. Divergence-measure fields, sets of finite perimeter, and conservation laws. *Arch. Ration. Mech. Anal.*, **175**(2):245–267, 2005.

[16] G.-Q. Chen, M. Torres, and W.P. Ziemer. Gauss-Green theorem for weakly differentiable vector fields, sets of finite perimeter, and balance laws. *Comm. Pure Appl. Math.*, **62**(2):242–304, 2009.

[17] P. Constantin and C. Fefferman. Direction of vorticity and the problem of global regularity for the Navier-Stokes equations. *Indiana Univ. Math. J.*, **42**(3):775–789, 1993.

[18] P. Constantin and C. Foias. *Navier-Stokes equations.* Chicago Lectures in Mathematics. University of Chicago Press, Chicago, IL, 1988.

[19] P. Constantin, C. Foias, and R. an Temam. *Attractors representing turbulent flows.* Mem. Amer. Math. Soc., **53**, Providence, 1985.

[20] P. Constantin, C. Foias, B. Nicolaenko, and R. Temam. *Integral and inertial manifolds for dissipative partial differential equations.* Springer-Verlag, New York, 1988.

[21] R. Danchin. Low Mach number limit for viscous compressible flows. *M2AN Math. Model Numer. Anal.*, **39**:459–475, 2005.

[22] P.A. Davidson. *Turbulence:An introduction for scientists and engineers.* Oxford University Press, Oxford, 2004.

[23] R.J. DiPerna and P.-L. Lions. Ordinary differential equations, transport theory and Sobolev spaces. *Invent. Math.*, **98**:511–547, 1989.

[24] J.-P. Eckmann and D. Ruelle. Ergodic theory of chaos and strange attractors. *Rev. Modern Phys.*, **57**(3, Part 1):617–656, 1985.

[25] S. Eliezer, A. Ghatak, and H. Hora. *An introduction to equations of states, theory and applications.* Cambridge University Press, Cambridge, 1986.

[26] C.L. Fefferman. Existence and smoothness of the Navier-Stokes equation. In *The millennium prize problems*, pp. 57–67. Clay Math. Inst., Cambridge, MA, 2006.

[27] E. Feireisl. *Dynamics of viscous compressible fluids.* Oxford University Press, Oxford, 2004.

[28] E. Feireisl and A. Novotný. *Singular limits in thermodynamics of viscous fluids.* Birkhäuser-Verlag, Basel, 2009.

[29] E. Feireisl and D. Pražák. *Asymptotic behavior of dynamical systems in fluid mechanics.* AIMS on Applied Mathematics, Vol. **4**, Springfield 2010.

[30] M Feistauer, J. Felcman, and I. Straškraba. *Mathematical and compunational methods for compressible flow.* Oxford Science Publications, Oxford, 2003.

[31] C. Foias, O. Manley, R. Rosa, and R. Temam. *Navier-Stokes equations and turbulence*, volume 83 of *Encyclopedia of Mathematics and its Applications.* Cambridge University Press, Cambridge, 2001.

[32] J. Frehse, S. Goj, and M. Steinhauer. L^p - estimates for the Navier-Stokes equations for steady compressible flow. *Manuscripta Math.*, **116**:265–275, 2005.

[33] G. Gallavotti. *Foundations of fluid dynamics.* Springer-Verlag, New York, 2002.

[34] F. Golse and C.D. Levermore. The Stokes-Fourier and acoustic limits for the Boltzmann equation. *Commun. on Pure and Appl. Math.*, **55**, 2002.

[35] F. Golse and L. Saint-Raymond. The Navier-Stokes limit of the Boltzmann equation for bounded collision kernels. *Invent. Math.*, **155**:81–161, 2004.

[36] F. Golse and L. Saint-Raymond. The incompressible Navier-Stokes limit of the Boltzmann equation for hard cutoff potentials. *J. Math. Pures Appl. (9)*, **91**(5):508–552, 2009.

[37] E. Hopf. Über die Anfangswertaufgabe für die hydrodynamischen Grundgleichungen. *Math. Nachr.*, **4**:213–231, 1951.

[38] S. Klainerman and A. Majda. Singular limits of quasilinear hyperbolic systems with large parameters and the incompressible limit of compressible fluids. *Comm. Pure Appl. Math.*, **34**:481–524, 1981.

[39] S.N. Kruzhkov. First order quasilinear equations in several space variables (in Russian). *Math. Sbornik*, **81**:217–243, 1970.

[40] O.A. Ladyzhenskaya. *The mathematical theory of viscous incompressible flow.* Gordon and Breach, New York, 1969.

[41] J. Leray. Sur le mouvement d'un liquide visqueux emplissant l'espace. *Acta Math.*, **63**:193–248, 1934.

[42] P.-L. LIONS. *Mathematical topics in fluid dynamics*, Vol.**1**, *Incompressible models.* Oxford Science Publication, Oxford, 1996.

[43] P.-L. LIONS. *Mathematical topics in fluid dynamics*, Vol. **2**, *Compressible models.* Oxford Science Publication, Oxford, 1998.

[44] P.-L. LIONS AND N. MASMOUDI. Incompressible limit for a viscous compressible fluid. *J. Math. Pures Appl.*, **77**:585–627, 1998.

[45] A.J. MAJDA AND A.L. BERTOZZI. *Vorticity and incompressible flow*, Vol. **27** of *Cambridge Texts in Applied Mathematics.* Cambridge University Press, Cambridge, 2002.

[46] N. MASMOUDI. Examples of singular limits in hydrodynamics. *In Handbook of Differential Equations, III, C. Dafermos, E. Feireisl Eds., Elsevier, Amsterdam,* 2006.

[47] N. MASMOUDI AND L. SAINT-RAYMOND. From the Boltzmann equation to the Stokes-Fourier system in a bounded domain. *Comm. Pure Appl. Math.*, **56**(9):1263–1293, 2003.

[48] A. MELLET AND A. VASSEUR. On the barotropic compressible Navier-Stokes equations. *Comm. Partial Differential Equations*, **32**(1-3):431–452, 2007.

[49] A. MELLET AND A. VASSEUR. Existence and uniqueness of global strong solutions for one-dimensional compressible Navier-Stokes equations. *SIAM J. Math. Anal.*, **39**(4):1344–1365, 2007/08.

[50] F. MURAT. Compacité par compensation. *Ann. Sc. Norm. Sup. Pisa, Cl. Sci. Ser. 5*, **IV**:489–507, 1978.

[51] P. PEDREGAL. *Parametrized measures and variational principles.* Birkhäuser, Basel, 1997.

[52] P.I. PLOTNIKOV AND J. SOKOLOWSKI. Concentrations of stationary solutions to compressible Navier-Stokes equations. *Comm. Math. Phys.*, **258**, 2005.

[53] P.I. PLOTNIKOV AND J. SOKOLOWSKI. Stationary solutions of Navier-Stokes equations for diatomic gases. *Russian Math. Surveys*, **62**:3, 2007.

[54] P.I. PLOTNIKOV AND J. SOKOLOWSKI. On compactness domain dependence and existence of steady state solutions to compressible isothermal Navier-Stokes equations. 2002. Preprint.

[55] G. PRODI. Un teorema di unicità per le equazioni di Navier-Stokes. *Ann. Mat. Pura Appl.*, **48**:173–182, 1959.

[56] K.R. RAJAGOPAL AND A.R. SRINIVASA. On thermodynamical restrictions of continua. *Proc. Royal Soc. London*, **A 460**:631–651, 2004.

[57] S. SCHOCHET. The mathematical theory of low Mach number flows. *M2ANMath. Model Numer. anal.*, **39**:441–458, 2005.

[58] G.R. SELL AND Y. YOU. *Dynamics of evolutionary equations.* Springer-Verlag, Berlin, Heidelberg, 2002.

[59] J. SERRIN. On the interior regularity of weak solutions of the Navier-Stokes equations. *Arch. Rational Mech. Anal.*, **9**:187–195, 1962.

[60] J. SIMON. Compact sets and the space $L^p(0, T; B)$. *Ann. Mat. Pura Appl.*, **146**:65–96, 1987.

[61] L. TARTAR. Compensated compactness and applications to partial differential equations. *Nonlinear Anal. and Mech., Heriot-Watt Sympos., L.J. Knopps editor, Research Notes in Math 39, Pitman, Boston*, pp. 136–211, 1975.

[62] R. TEMAM. *Navier-Stokes equations.* North-Holland, Amsterdam, 1977.

[63] R. TEMAM. *Infinite-dimensional dynamical systems in mechanics and physics.* Springer-Verlag, New York, 1988.

[64] C. TRUESDELL AND K.R. RAJAGOPAL. *An introduction to the mechanics of fluids.* Birkhäuser, Boston, 2000.

[65] L. TRUESDELL. Notes on the history of the general equations of hydrodynamics. *Amer. Math. Monthly*, **60**:445–458, 1953.

SELECTED TOPICS IN APPROXIMATE SOLUTIONS OF NONLINEAR CONSERVATION LAWS. HIGH-RESOLUTION CENTRAL SCHEMES

EITAN TADMOR*

Central schemes offer a simple and versatile approach for computing approximate solutions of non-linear systems of hyperbolic conservation laws and related PDEs. The solution of such problems often involves the spontaneous evolution of steep gradients. The multiscale aspect of these gradients poses a main computational challenge for their numerical solution. Central schemes utilize a minimal amount of information on the propagation speeds associated with the problems, in order to accurately detect these steep gradients. This information is then coupled with high-order, non-oscillatory reconstruction of the approximate solution in 'the direction of smoothness': that is, information of smoothness does not cross regions of steep gradients. The use of central stencils enables us to realize the reconstructed solutions through simple quadratures. In this manner, central schemes avoid the intricate and time-consuming details of the eigen-structure of the underlying PDEs, and in particular, the use of (approximate) Riemann solvers, dimensional splitting, etc. The resulting family of central schemes offers relatively simple, "black-box" solvers for a wide variety of problems governed by multi-dimensional systems of non-linear hyperbolic conservation laws and related convection-diffusion problems.

We highlight several features of this new class of central schemes. `Scalar equations`. Both the second- and third-order schemes were shown to have variation bounds, which in turn yield convergence with precise error estimates, as well as entropy and (multidimensional) L^∞-stability estimates. `Systems of equations`. Extension to systems is carried out by *component-wise* application of the scalar framework. It is in this context that our central schemes offer a remarkable advantage over the corresponding upwind framework. `Multidimensional problems`. Since we bypass the need for (approximate) Riemann solvers, multidimensional problems are solved *without* dimensional splitting. In fact, the class of central schemes is utilized for a variety of nonlinear transport equations. A partial list of more than 120 references can be found in **CentPack**, [4]. CentPack is a collection of freely distributed C++ routines that implement a number of high-order, non-oscillatory central schemes for hyperbolic systems of conservation laws in one- and two-space dimensions, $\mathbf{u}_t+\mathbf{f}(\mathbf{u})_x+\mathbf{g}(\mathbf{u})_y = 0$.

---

*Department of Mathematics, Institute for Physical Science and Technology and Center of Scientific Computation and Mathematical Modeling (CSCAMM), University of Maryland, College Park, MD 20742, USA (`tadmor@cscamm.umd.edu`; `http://www.cscamm.umd.edu/~tadmor`). Research was supported in part by NSF grants FRG07-57227, DMS10-08397, and ONR grant N000140910385.

The numerical algorithm for the implementation of central schemes consists of two main steps: (i) a non-oscillatory piecewise polynomial reconstruction of point values from their cell averages; followed by (ii) time evolution of the reconstructed polynomial, which is governed by the flux functions $\mathbf{f}(\cdot)$ and $\mathbf{g}(\cdot)$.

The efficiency and versatility of central schemes resides, mainly, in their simplicity: they eliminate the need for Riemann solvers and avoid dimensional splitting, yielding a generic formulation valid for any hyperbolic system that can be written in the above form. Only information specific to the model and problem to be solved needs to be provided; namely, a description of the flux functions, $\mathbf{f}(\mathbf{u})$ and $\mathbf{g}(\mathbf{u})$, the maximal propagation speed, and the appropriate initial and boundary conditions.

1. Introduction. In recent years, central schemes for approximating solutions of hyperbolic conservation laws, received a considerable amount of renewed attention. A family of high-resolution, non-oscillatory, *central* schemes, was developed to handle such problems. Compared with the 'classical' *upwind* schemes, these *central* schemes were shown to be both simple and stable for a large variety of problems ranging from one-dimensional scalar problems to multi-dimensional systems of conservation laws. They were successfully implemented for a variety of other related problems, such as, e.g., the incompressible Euler equations [30], [24], [22], [23], the magneto-hydrodynamics equations [5], viscoelastic flows—[22] hyperbolic systems with relaxation source terms [6], [43], [1] non-linear optics, [10], traffic flow [26], and a host of other applications listed on the "central-station" site, `http://www.cscamm.umd.edu/centpack/publications/`.

The family of high-order *central* schemes we deal with, can be viewed as a direct extension to the first-order, Lax-Friedrichs (LxF) scheme [12], which on one hand is robust and stable, but on the other hand suffers from excessive dissipation. To address this problematic property of the LxF scheme, a Godunov-like second-order central scheme was developed by Nessyahu and Tadmor (NT) in [38] (see also [45]). It was extended to higher-order of accuracy as well as for more space dimensions (consult [2], [19], [3] and [23], for the two-dimensional case, and [44], [17], [34] for the third-order schemes).

The NT scheme is based on reconstructing, in each time step, a piecewise-polynomial interpolant from the cell-averages computed in the previous time step. This interpolant is then (exactly) evolved in time, and finally, it is projected on its staggered averages, resulting with the staggered cell-averages at the next time-step. The one- and two-dimensional second-order schemes, are based on a piecewise-linear MUSCL-type reconstruction, whereas the third-order schemes are based on the non-oscillatory piecewise-parabolic reconstruction [33], [34]. Higher orders schemes are treated in [7], [28], [29]. Schemes base on staggered stencils, such as the NT scheme, are necessarily *redundant*. The use of redundant stencils was extended to multi-dimensional *overlapping cells*, and as an example, we mention in this context the recent works on central discontinuous Galerkin methods, [35], [36], [37].

Like *upwind* schemes, the reconstructed piecewise-polynomials used by the central schemes, also make use of non-linear limiters which guarantee the overall non-oscillatory nature of the approximate solution. But unlike the upwind schemes, central schemes avoid the intricate and time consuming Riemann solvers; this advantage is particularly important in the multi-dimensional setup, where no such Riemann solvers are available.

2. A short guide to Godunov-type schemes. We want to solve the hyperbolic system of conservation laws

$$u_t + f(u)_x = 0 \tag{2.1}$$

by Godunov-type schemes. To this end we proceed in two steps. First, we introduce a small spatial scale, Δx, and we consider the corresponding (Steklov) sliding average of $u(\cdot, t)$,

$$\bar{u}(x,t) := \frac{1}{|I_x|}\int_{I_x} u(\xi,t)d\xi, \qquad I_x = \left\{\xi \;\middle|\; |\xi - x| \leq \frac{\Delta x}{2}\right\}.$$

The sliding average of (2.1) then yields

$$\bar{u}_t(x,t) + \frac{1}{\Delta x}\left[f(u(x+\frac{\Delta x}{2},t)) - f(u(x-\frac{\Delta x}{2},t))\right] = 0. \tag{2.2}$$

Next, we introduce a small time-step, Δt, and integrate over the slab $t \leq \tau \leq t + \Delta t$,

$$\begin{aligned}\bar{u}(x,t+\Delta t) &= \bar{u}(x,t)\\ &-\frac{1}{\Delta x}\Big[\int_{\tau=t}^{t+\Delta t} f(u(x+\frac{\Delta x}{2},\tau))d\tau - \int_{\tau=t}^{t+\Delta t} f(u(x-\frac{\Delta x}{2},\tau))d\tau\Big].\end{aligned} \tag{2.3}$$

We end up with an equivalent reformulation of the conservation law (2.1): it expresses the precise relation between the sliding averages, $\bar{u}(\cdot, t)$, and their underlying pointvalues, $u(\cdot, t)$. We shall use this reformulation, (2.3), as the starting point for the construction of Godunov-type schemes.

We construct an approximate solution, $w(\cdot, t^n)$, at the discrete time-levels, $t^n = n\Delta t$. Here, $w(x, t^n)$ is a piecewise polynomial written in the form

$$w(x,t^n) = \sum p_j(x)\chi_j(x), \quad \chi_j(x) := 1_{I_j},$$

where $p_j(x)$ are algebraic polynomials supported at the discrete cells, $I_j = I_{x_j}$, centered around the midpoints, $x_j := j\Delta x$. An *exact* evolution of $w(\cdot, t^n)$ based on (2.3), reads

$$\bar{w}(x,t^{n+1}) = \bar{w}(x,t^n) \\ -\frac{1}{\Delta x}\Big[\int_{t^n}^{t^{n+1}} f(w(x+\frac{\Delta x}{2},\tau))d\tau - \int_{t^n}^{t^{n+1}} f(w(x-\frac{\Delta x}{2},\tau))d\tau\Big]. \tag{2.4}$$

To construct a Godunov-type scheme, we *realize* (2.4) — or at least an accurate approximation of it, at discrete gridpoints. Here, we distinguish between the main methods, according to their way of *sampling* (2.4): these two main sampling methods correspond to upwind schemes and central schemes.

2.1. Upwind schemes. Let $\bar{w}_j^n$ abbreviates the cell averages, $\bar{w}_j^n := \frac{1}{\Delta x}\int_{I_j} w(\xi,t^n)d\xi$. By sampling (2.4) at the *mid-cells*, $x = x_j$, we obtain an evolution scheme for these averages, which reads

$$\bar{w}_j^{n+1} = \bar{w}_j^n - \frac{1}{\Delta x}\Big[\int_{\tau=t^n}^{t^{n+1}} f(w(x_{j+\frac{1}{2}},\tau))d\tau - \int_{\tau=t^n}^{t^{n+1}} f(w(x_{j-\frac{1}{2}},\tau))d\tau\Big]. \tag{2.5}$$

Here, it remains to recover the *pointvalues*, $\{w(x_{j+\frac{1}{2}},\tau)\}_j$, $t^n \le \tau \le t^{n+1}$, in terms of their known cell averages, $\{\bar{w}_j^n\}_j$, and to this end we proceed in two steps:

- First, the *reconstruction* – we recover the pointwise values of $w(\cdot,\tau)$ at $\tau = t^n$, by a reconstruction of a piecewise polynomial approximation

$$w(x,t^n) = \sum_j p_j(x)\chi_j(x), \quad \bar{p}_j(x_j) = \bar{w}_j^n. \tag{2.6}$$

- Second, the *evolution* — $w(x_{j+\frac{1}{2}}, \ \tau \ge t^n)$ are determined as the solutions of the generalized Riemann problems

$$w_t + f(w)_x = 0, \quad t \ge t^n; \quad w(x,t^n) = \begin{cases} p_j(x) & x < x_{j+\frac{1}{2}}, \\ p_{j+1}(x) & x > x_{j+\frac{1}{2}}. \end{cases} \tag{2.7}$$

The solution of (2.7) is composed of a family of nonlinear waves – left-going and right-going waves. An exact Riemann solver, or at least an approximate one is used to distribute these nonlinear waves between the two neighboring cells, I_j and I_{j+1}. It is this distribution of waves according to their direction which is responsible for *upwind differencing*, consult Figure 2.1. We briefly recall few canonical examples for this category of upwind Godunov-type schemes.

The original Godunov scheme is based on piecewise-constant reconstruction, $w(x,t^n) = \Sigma\bar{w}_j^n\chi_j$, followed by an exact Riemann solver. This results in a first-order accurate upwind method [14], which is the forerunner for all other Godunov-type schemes. A second-order extension was introduced by van Leer [21]: his MUSCL scheme reconstructs a piecewise linear approximation, $w(x,t^n) = \Sigma p_j(x)\chi_j(x)$, with linear pieces of

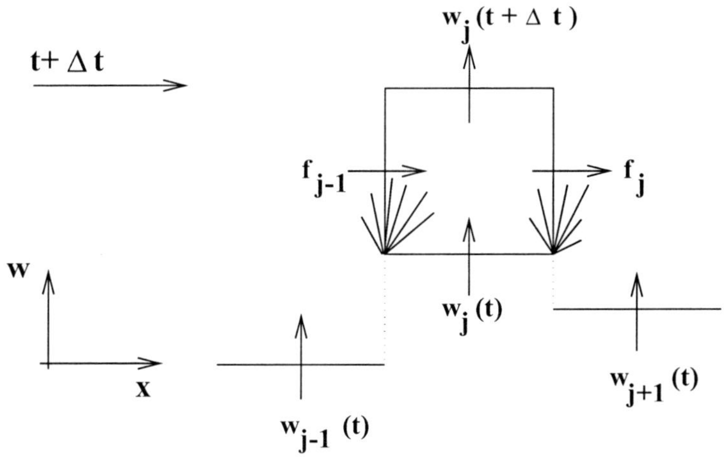

FIG. 1. *Upwind differencing by Godunov-type scheme.*

the form $p_j(x) = \bar{w}_j^n + w_j'\left(\frac{x-x_j}{\Delta x}\right)$ so that $\bar{p}_j(x_j) = \bar{w}_j^n$. Here the w_j'-s are possibly limited slopes which are reconstructed from the known cell-averages, $w_j' = \{(w_j^n)'\} = \{w'(\bar{w}_k^n)_{k=j-1}^{j+1}\}$. (Throughout this lecture we use primes, $w_j', w_j'', \ldots$, to denote *discrete* derivatives, which approximate the corresponding differential ones). A whole library of limiters is available in this context, so that the co-monotonicity of $w(x, t^n)$ with $\Sigma\bar{w}_j\chi_j$ is guaranteed, e.g., [46]. The Piecewise-Parabolic Method (PPM) of Colella-Woodward [9] and respectively, ENO schemes of Harten et.al. [16], offer, respectively, third- and higher-order Godunov-type upwind schemes. (A detailed account of ENO schemes can be found in lectures of C.W. Shu in this volume). Finally, we should not give the impression that limiters are used exclusively in conjunction with Godunov-type schemes. The *positive schemes* of Liu and Lax, [32], offer simple and fast upwind schemes for multidimensional systems, based on an alternative positivity principle.

2.2. Central schemes. As before, we seek a piecewise-polynomial, $w(x, t^n) = \Sigma p_j(x)\chi_j(x)$, which serves as an approximate solution to the *exact* evolution of sliding averages in (2.4),

$$\begin{aligned}\bar{w}(x, t^{n+1}) &= \bar{w}(x, t^n)\\ &-\frac{1}{\Delta x}\Big[\int_{t^n}^{t^{n+1}} f(w(x+\frac{\Delta x}{2}, \tau))d\tau - \int_{t^n}^{t^{n+1}} f(w(x-\frac{\Delta x}{2}, \tau))d\tau\Big].\end{aligned} \tag{2.8}$$

Note that the polynomial pieces of $w(x, t^n)$ are supported in the cells, $I_j = \left\{\xi \;\middle|\; |\xi - x_j| \leq \frac{\Delta x}{2}\right\}$, with interfacing breakpoints at the half-integers gridpoints, $x_{j+\frac{1}{2}} = \left(j + \frac{1}{2}\right)\Delta x$.

We recall that upwind schemes (2.5) were based on sampling (2.4) in the *midcells*, $x = x_j$. In contrast, central schemes are based on sampling (2.8) at the *interfacing breakpoints*, $x = x_{j+\frac{1}{2}}$, which yields

$$\bar{w}^{n+1}_{j+\frac{1}{2}} = \bar{w}^{n}_{j+\frac{1}{2}} - \frac{1}{\Delta x}\left[\int_{\tau=t^n}^{t^{n+1}} f(w(x_{j+1},\tau))d\tau - \int_{\tau=t^n}^{t^{n+1}} f(w(x_j,\tau))d\tau\right]. \quad (2.9)$$

We want to utilize (2.9) in terms of the known cell averages at time level $\tau = t^n$, $\{\bar{w}^n_j\}_j$. The remaining task is therefore to recover the *pointvalues* $\{w(\cdot,\tau)|\ t^n \leq \tau \leq t^{n+1}\}$, and in particular, the *staggered averages*, $\{\bar{w}^n_{j+\frac{1}{2}}\}$. As before, this task is accomplished in two main steps:

- First, we use the given cell averages $\{\bar{w}^n_j\}_j$, to *reconstruct* the pointvalues of $w(\cdot,\tau = t^n)$ as piecewise polynomial approximation

$$w(x,t^n) = \sum_j p_j(x)\chi_j(x), \quad \bar{p}_j(x_j) = \bar{w}^n_j. \quad (2.10)$$

  In particular, the staggered averages on the right of (2.9) are given by

$$\bar{w}^n_{j+\frac{1}{2}} = \frac{1}{\Delta x}\left[\int_{x_j}^{x_{j+\frac{1}{2}}} p_j(x)dx + \int_{x_{j+\frac{1}{2}}}^{x_{j+1}} p_{j+1}(x)dx\right]. \quad (2.11)$$

  The resulting central scheme (2.9) then reads

$$\begin{aligned}\bar{w}^{n+1}_{j+\frac{1}{2}} &= \frac{1}{\Delta x}\left[\int_{x_j}^{x_{j+\frac{1}{2}}} p_j(x)dx + \int_{x_{j+\frac{1}{2}}}^{x_{j+1}} p_{j+1}(x)dx\right] + \quad (2.12)\\ &- \frac{1}{\Delta x}\left[\int_{\tau=t^n}^{t^{n+1}} f(w(x_{j+1},\tau))d\tau - \int_{\tau=t^n}^{t^{n+1}} f(w(x_j,\tau))d\tau\right].\end{aligned}$$

- Second, we follow the *evolution* of the pointvalues along the mid-cells, $x = x_j$, $\{w(x_j,\tau \geq t^n)\}_j$, which are governed by

$$w_t + f(w)_x = 0, \quad \tau \geq t^n; \quad w(x,t^n) = p_j(x) \quad x \in I_j. \quad (2.13)$$

  Let $\{a_k(u)\}_k$ denote the eigenvalues of the Jacobian $A(u) := \frac{\partial f}{\partial u}$. By hyperbolicity, information regarding the interfacing discontinuities at $(x_{j\pm\frac{1}{2}},t^n)$ propagates no faster than $\max_k|a_k(u)|$. Hence, the mid-cells values governed by (2.13), $\{w(x_j,\tau \geq t^n)\}_j$, remain free of discontinuities, at least for sufficiently small time step dictated by the CFL condition $\Delta t \leq \frac{1}{2}\Delta x \cdot \max_k|a_k(u)|$. Consequently, since the numerical fluxes on the right of (2.12), $\int_{\tau=t^n}^{t^{n+1}} f(w(x_j,\tau))d\tau$, involve only smooth integrands, they can be computed within any degree of desired accuracy by an appropriate quadrature rule.

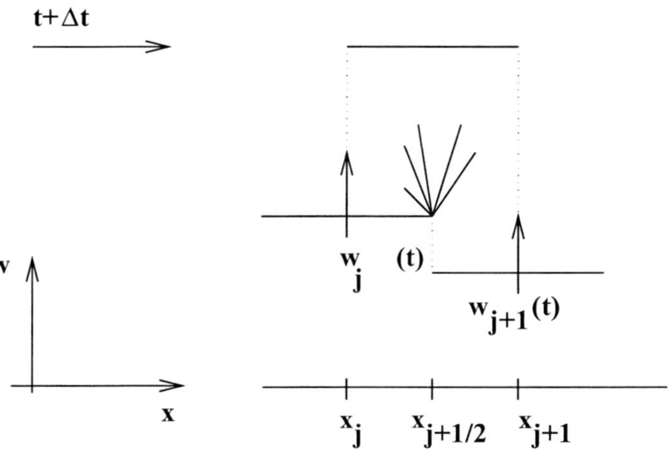

FIG. 2. *Central differencing by Godunov-type scheme.*

It is the *staggered* averaging over the fan of left-going and right-going waves centered at the half-integered interfaces, $(x_{j+\frac{1}{2}}, t^n)$, which characterizes the *central* differencing, consult Figure 2.2. A main feature of these central schemes – in contrast to upwind ones, is the computation of *smooth* numerical fluxes along the mid-cells, $(x = x_j, \tau \geq t^n)$, which avoids the costly (approximate) Riemann solvers. A couple of examples of central Godunov-type schemes is in order.

The first-order Lax-Friedrichs (LxF) approximation is the forerunner for such central schemes — it is based on piecewise constant reconstruction, $w(x, t^n) = \Sigma p_j(x)\chi_j(x)$ with $p_j(x) = \bar{w}_j^n$. The resulting central scheme, (2.12), then reads (with the usual fixed mesh ratio $\lambda := \frac{\Delta t}{\Delta x}$)

$$\bar{w}_{j+\frac{1}{2}}^{n+1} = \frac{1}{2}(\bar{w}_j + \bar{w}_{j+1}) - \lambda\Big[f(\bar{w}_{j+1}) - f(\bar{w}_j)\Big]. \tag{2.14}$$

Our main focus in the rest of this chapter is on non-oscillatory higher-order extensions of the LxF schemes.

3. Central schemes in one-space dimension.

3.1. The second-order Nessyahu-Tadmor scheme. In this section we overview the construction of high-resolution central schemes in one-space dimension. We begin with the reconstruction of the second-order, non-oscillatory Nessyahu and Tadmor (NT) scheme, [38]. To approximate solutions of (2.1), we introduce a piecewise-linear approximate solution at the discrete time levels, $t^n = n\Delta t$, based on linear functions $p_j(x, t^n)$ which are supported at the cells I_j (see Figure 3),

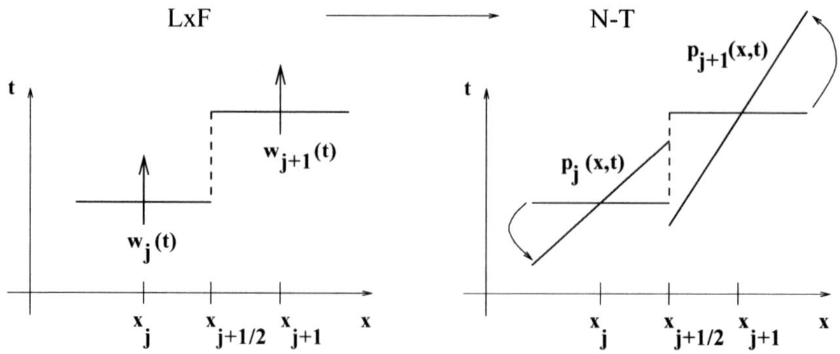

FIG. 3. *The second-order reconstruction.*

$$
\begin{aligned}
w(x,t)|_{t=t^n} &= \sum_j p_j(x,t^n)\chi_j(x) := \sum_j \left[\bar{w}_j^n + w_j'\left(\frac{x-x_j}{\Delta x}\right)\right]\chi_j(x),\\
\chi_j(x) &:= 1_{I_j}.
\end{aligned} \tag{3.1}
$$

Second-order of accuracy is guaranteed if the discrete slopes approximate the corresponding derivatives, $w_j' \sim \Delta x \cdot \partial_x w(x_j, t^n) + O(\Delta x)^2$. At the same time, the second-order reconstruction is sought to be *non-oscillatory* in a manner which is properly quantified in terms of a maximum principle, total variation bound etc. To maintain both – second-order accuracy and the non-oscillatory character of the reconstruction, one may choose from a large class of *n*onlinear limiters, e.g., [21], [15], [46], [33]. We mention here the canonical class of limiters of the form

$$
w_j' = MM\{\theta(\bar{w}_{j+1}^n - \bar{w}_j^n), \frac{1}{2}(\bar{w}_{j+1}^n - \bar{w}_{j-1}^n), \theta(\bar{w}_j^n - \bar{w}_{j-1}^n)\}. \tag{3.2}
$$

Here and below, $\theta \in [1,2]$ is a free parameter which limits the maximal reconstructed slope, and MM denotes the so-called min-mod function

$$
MM\{x_1, x_2, ...\} = \begin{cases} \min_i\{x_i\} & \text{if } x_i > 0, \forall i \\ \max_i\{x_i\} & \text{if } x_i < 0, \forall i \\ 0 & \text{otherwise.} \end{cases}
$$

An *exact* evolution of w, based on integration of the conservation law over the staggered cell, $I_{j+\frac{1}{2}}$, then reads, (2.9)

$$
\bar{w}_{j+\frac{1}{2}}^{n+1} = \frac{1}{\Delta x}\int_{I_{j+\frac{1}{2}}} w(x,t^n)dx - \frac{1}{\Delta x}\int_{\tau=t^n}^{t^{n+1}} \left[f(w(x_{j+1},\tau)) - f(w(x_j,\tau))\right] d\tau.
$$

The first integral is the staggered cell-average at time t^n, $\bar{w}_{j+\frac{1}{2}}^n$, which can be computed directly from the above reconstruction,

$$
\bar{w}_{j+\frac{1}{2}}^n := \frac{1}{\Delta x}\int_{x_j}^{x_{n+1}} w(x,t^n)dx = \frac{1}{2}(\bar{w}_j^n + \bar{w}_{j+1}^n) + \frac{1}{8}(w_j' - w_{j+1}'). \tag{3.3}
$$

The time integrals of the flux are computed by the second-order accurate mid-point quadrature rule

$$\int_{\tau=t^n}^{t^{n+1}} f(w(x_j,\tau))d\tau \sim \Delta t \cdot f(w(x_j,t^{n+\frac{1}{2}})).$$

Here, the Taylor expansion is being used to predict the required mid-values of w

$$\begin{aligned} w(x_j,t^{n+\frac{1}{2}}) &\sim w(x_j,t) + \frac{\Delta t}{2} w_t(x_j,t^n) \\ &= \bar{w}_j^n - \frac{\Delta t}{2} A(\bar{w}_j^n)(p_j(x_j,t^n))_x = \bar{w}_j^n - \frac{\lambda}{2} A_j^n w_j'. \end{aligned}$$

In summary, we end up with the central scheme, [38], which consists of a first-order *predictor step*,

$$w_j^{n+\frac{1}{2}} = \bar{w}_j^n - \frac{\lambda}{2} A_j^n w_j', \qquad A_j^n := A(\bar{w}_j^n), \tag{3.4}$$

followed by the second-order *corrector step*, (2.12),

$$\bar{w}_{j+\frac{1}{2}}^{n+1} = \frac{1}{2}(\bar{w}_j^n + \bar{w}_{j+1}^n) + \frac{1}{8}(w_j' - w_{j+1}') - \lambda \left[f(w_{j+1}^{n+\frac{1}{2}}) - f(w_j^{n+\frac{1}{2}}) \right]. \tag{3.5}$$

The *scalar* non-oscillatory properties of (3.4)–(3.5) were proved in [38], including the TVD property, cell entropy inequality, $L^1_{loc}-$ error estimates, etc. Moreover, the numerical experiments, reported in [38], [3], [5], [43], [1], [7], with one-dimensional *systems* of conservation laws, show that such second-order central schemes enjoy the same high-resolution as the corresponding second-order upwind schemes do. The main difference lies in the resolution of linear contact waves, where upwind differencing in the *characteristic* eigen-directions yields improved resolution; but see [25], [31] for example of enhancing the resolution of contact discontinuities in central schemes. Thus, the excessive smearing typical to the first-order LxF central scheme is compensated here by the second-order accurate MUSCL reconstruction.

In Figure 4 we compare, side by side, the upwind ULT scheme of Harten, [15], with our central scheme (3.4)–(3.5). The comparable high-resolution of this so called Lax's Riemann problem is evident.

At the same time, the central scheme (3.4)–(3.5) has the advantage over the corresponding upwind schemes, in that no (approximate) Riemann solvers, as in (2.7), are required. Hence, these Riemann-free central schemes provide an efficient high-resolution alternative in the one-dimensional case, and a particularly advantageous framework for multidimensional computations, e.g., [3], [19]. This advantage in the multidimensional case will be explored in the next section. Also, *staggered* central differencing, along the lines of the Riemann-free Nessyahu-Tadmor scheme (3.4)–(3.5), admits

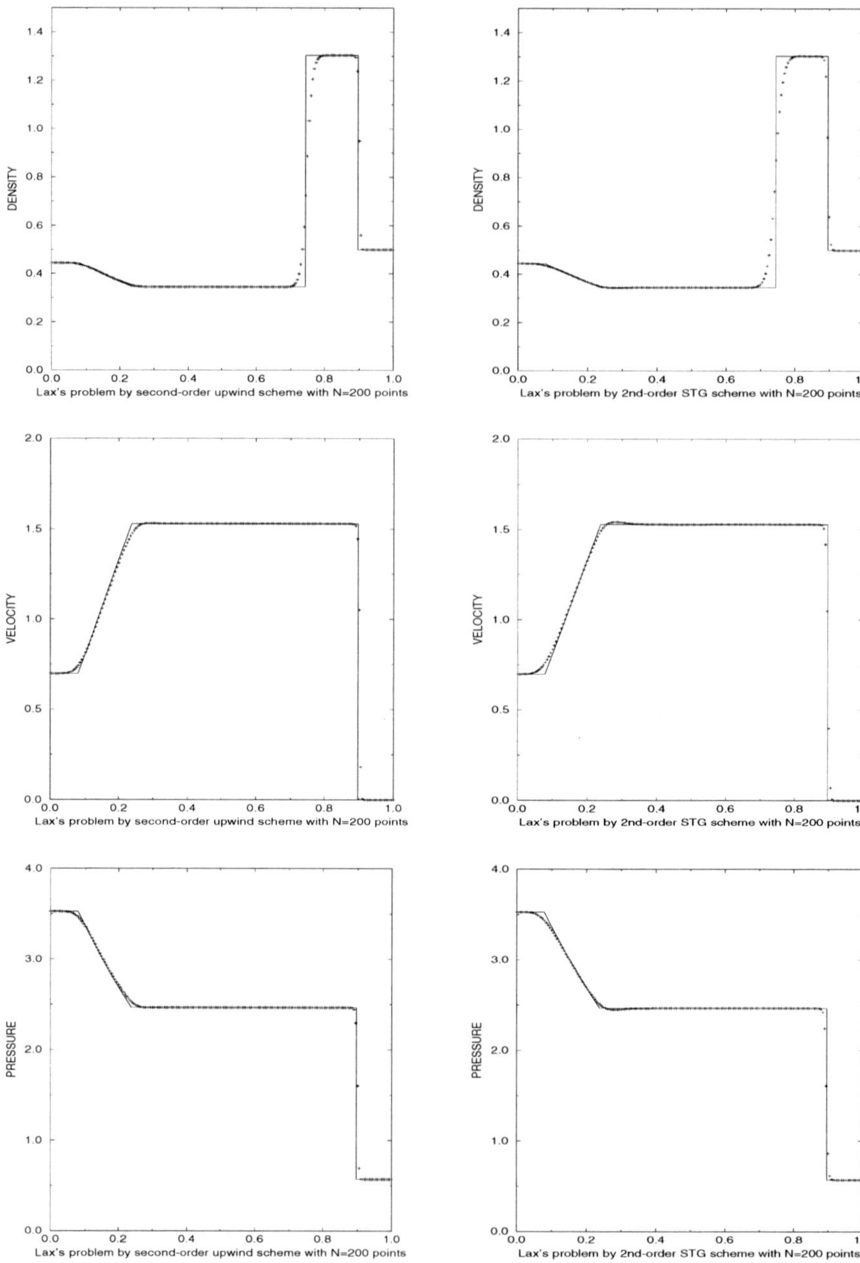

FIG. 4. *2nd order: central (STG) vs. upwind (ULT) — Lax's Riemann problem.*

simple efficient extensions in the presence of general source terms, [11], [40] and in particular, stiff source terms. Indeed, it is a key ingredient behind the relaxation schemes studied in [20].

It should be noted, however, that the component-wise version of these central schemes might result in deterioration of resolution at the computed extrema. The second-order computation presented in Figure 5 below demonstrates this point. (this will be corrected by higher order central methods). Of course, this – so called extrema clipping, is typical to high-resolution upwind schemes as well; but it is more pronounced with our central schemes due to the built-in extrema-switching to the dissipative LxF scheme. Indeed, once an extrema cell, I_j, is detected (by the limiter), it sets a zero slope, $w'_j = 0$, in which case the second-order scheme (3.4)–(3.5) is reduced back to the first-order LxF, (2.14).

3.2. The third-order central scheme. Following the framework outlined in §3.1, the upgrade to third-order central scheme consists of two main ingredients:

(i) A third-order accurate, piecewise-quadratic polynomial reconstruction which enjoys desirable non-oscillatory properties;

(ii) An appropriate quadrature rule to approximate the numerical fluxes along cells' interfaces.

Following [34], we proceed as follows. The piecewise-parabolic reconstruction takes the form

$$p_j(x) = w_j^n + w'_j\left(\frac{x-x_j}{\Delta x}\right) + \frac{1}{2}w''_j\left(\frac{x-x_j}{\Delta x}\right)^2. \tag{3.6}$$

Here, w''_j are the (pointvalues of) the *reconstructed second derivatives*

$$w''_j := \theta_j \Delta_+\Delta_- \bar{w}_j^n; \tag{3.7}$$

w'_j are the (pointvalues of) the *reconstructed slopes*,

$$w'_j := \theta_j \Delta_0 \bar{w}_j^n; \tag{3.8}$$

and w_j^n are the *reconstructed pointvalues*

$$w_j^n := \bar{w}_j^n - \frac{w''_j}{24}. \tag{3.9}$$

Observe that, starting with third- (and higher-) order accurate methods, pointwise values *cannot* be interchanged with cell averages, $w_j^n \neq \bar{w}_j^n$.

Here, θ_j are appropriate nonlinear limiters which guarantee the non-oscillatory behavior of the third-order reconstruction; its precise form can be found in [33], [34]. They guarantee that the reconstruction (3.6) is non-oscillatory in the sense that $N(w(\cdot,t^n))$ — the number of extrema of $w(x,t^n)$, does not exceed that of its piecewise-constant projection, $N(\Sigma\bar{w}_j^n\chi_j(\cdot))$,

$$N(w(\cdot,t^n)) \leq N(\Sigma\bar{w}_j^n\chi_j(\cdot)). \tag{3.10}$$

Next we turn to the evolution of the piecewise-parabolic reconstructed solution. To this end we need to evaluate the staggered averages, $\{\bar{w}_{j+\frac{1}{2}}^n\}$, and to approximate the interface fluxes, $\left\{\int_{\tau=t^n}^{t^{n+1}} f(w(x_j,\tau))d\tau\right\}$.

With $p_j(x) = w_j^n + w_j'\left(\frac{x-x_j}{\Delta x}\right) + \frac{1}{2}w_j''\left(\frac{x-x_j}{\Delta x}\right)^2$ specified in (3.6)–(3.9), one evaluates the staggered averages of the third order reconstruction $w(x,t^n) = \Sigma p_j(x)\ \chi_j(x)$

$$\bar{w}_{j+\frac{1}{2}}^n = \frac{1}{\Delta x}\int_{x_j}^{x_{j+1}} w(x,t^n)dx = \frac{1}{2}(\bar{w}_j + \bar{w}_{j+1}) + \frac{1}{8}(w_j' - w_{j+1}'). \tag{3.11}$$

Remarkably, we obtain here the same formula for the staggered averages as in the second-order cases, consult (3.3); the only difference is the use of the new limited slopes in (3.8), $w_j' = \theta_j\Delta_0\bar{w}_j^n$.

Next, we approximate the (exact) numerical fluxes by Simpson's quadrature rule, which is (more than) sufficient for retaining the overall third-order accuracy,

$$\frac{1}{\Delta x}\int_{\tau=t^n}^{t^{n+1}} f(w(x_j,\tau))d\tau \sim \frac{\lambda}{6}\left[f(w_j^n) + 4f(w_j^{n+\frac{1}{2}}) + f(w_j^{n+1})\right]. \tag{3.12}$$

This in turn, requires the three approximate *pointvalues* on the right, $w_j^{n+\beta} \sim w(x_j,t^{n+\beta})$ for $\beta = 0, \frac{1}{2}, 1$. Following our approach in the second-order case, [38], we use Taylor expansion to *predict*

$$w_j^n = \bar{w}_j^n - \frac{w_j''}{24}; \tag{3.13}$$

$$\begin{aligned}\dot{w}_j^n \equiv (\Delta x\cdot\partial_t)w(x_j,t^n) &= -\Delta x\cdot\partial_x f(w(x_j,t^n))\\ &= -a(w_j^n)\cdot w_j',;\end{aligned} \tag{3.14}$$

$$\begin{aligned}\ddot{w}_j^n \equiv (\Delta x\cdot\partial_t)^2 w(x_j,t^n) &= \Delta x\cdot\partial_x\left[a(w_j^n)\Delta x\cdot\partial_x f(w(x_j,t^n))\right]\\ &= a^2(w_j^n)w_j'' + 2a(w_j^n)a'(w_j^n)(w_j')^2.\end{aligned} \tag{3.15}$$

In summary of the scalar setup, we end up with a two step scheme where, starting with the reconstructed pointvalues

$$w_j^n = \bar{w}_j^n - \frac{w_j''}{24}, \tag{3.16}$$

we *predict* the pointvalues $w_j^{n+\beta}$ by, e.g. Taylor expansions,

$$w_j^{n+\beta} = w_j^n + \lambda\beta \dot{w}_j^n + \frac{(\lambda\beta)^2}{2}\ddot{w}_j^n, \qquad \beta = \frac{1}{2}, 1; \tag{3.17}$$

this is followed by the *corrector* step

$$\begin{aligned} \bar{w}_{j+\frac{1}{2}}^n &= \frac{1}{2}(\bar{w}_j^n + \bar{w}_{j+1}^n) + \frac{1}{8}(w_j' - w_{j+1}') \\ &-\frac{\lambda}{6}\left\{\left[f(w_{j+1}^n) + 4f(w_{j+1}^{n+\frac{1}{2}}) + f(w_{j+1}^{n+1})\right]\right. \\ &\left.\qquad -\left[f(w_j^n) + 4f(w_j^{n+\frac{1}{2}}) + f(w_j^{n+1})\right]\right\}. \end{aligned} \tag{3.18}$$

In Figure 5 we revisit the so called Woodward-Colella problem, [49], where we compare the second vs. the third-order results. The improvement in resolving the density field is evident.

We conclude this section with several remarks.

REMARK.

1. Stability.
   We briefly mention the stability results for the scalar central schemes. In the second order case, the NT scheme was shown to be both TVD and entropy stable in the sense of satisfying a cell entropy inequality – consult [38]. The third-order scalar central scheme is stable in the sense of satisfying the NED property, (3.10), namely
   THEOREM 3.1 ([34]). *Consider the central scheme (3.16), (3.17), (3.18), based on the third-order accurate quadratic reconstruction, (3.6)–(3.9). Then it satisfies the so-called Number of Extrema Diminishing (NED) property, in the sense that*

$$N\left(\sum_\nu \bar{w}_{v+\frac{1}{2}}^{n+1}\chi_{\nu+\frac{1}{2}}(x)\right) \le N\left(\sum_\nu \bar{w}_\nu^n \chi_\nu(x)\right). \tag{3.19}$$

2. Source terms, radial coordinates, ...
   Extensions of the central framework which deal with both, stiff and non-stiff source terms can be found in [43], [1], [11], [6]. In particular, Kupferman in [22], [23] developed the central framework within the radial coordinates which require to handle both – variable coefficients + source terms.
3. Higher order central schemes.
   We refer to [7], where a high-order ENO reconstruction is realized by a staggered cell averaging. Here, intricate Riemann solvers are replaced by high order quadrature rules. and for this purpose, one can effectively use the RK method (rather than the Taylor expansion outlined above):

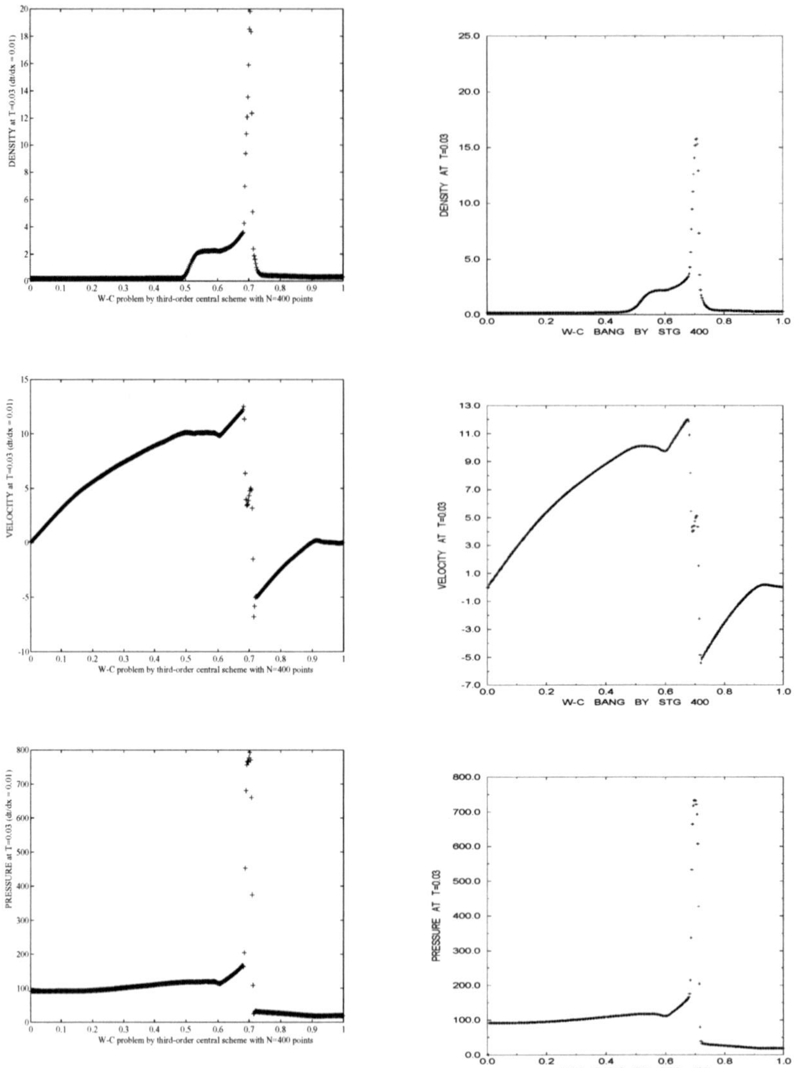

FIG. 5. 3rd *vs.* 2nd *order central schemes — Woodward-Colella problem at* $t = 0.03$.

4. Taylor vs. Runge-Kutta.
   The evaluations of Taylor expansions could be substituted by the more economical Runge-Kutta integrations; the simplicity becomes more pronounced with *systems*. A particular useful approach in this context was proposed in [7], [28], [29] using the natural continuous extensions of RK schemes.

5. Systems.
   One of the main advantages of our central-staggered framework over that of the upwind schemes, is that expensive and time-consuming characteristic decompositions can be avoided. Specifically, all the non-oscillatory computations can be carried out with diagonal limiters, based on a *component-wise* extension of the scalar limiters outlined above.

4. Central schemes in two space dimensions. Following the one dimensional setup, one can derive a non-oscillatory, two-dimensional central scheme. Here we sketch the construction of the second-order two-dimensional scheme following [19] (see also [2]). For the two-dimensional third-order accurate scheme, we refer to [28].

We consider the two-dimensional hyperbolic system of conservation laws

$$u_t + f(u)_x + g(u)_y = 0. \tag{4.1}$$

To approximate a solution to (4.1), we start with a two-dimensional linear reconstruction

$$\begin{aligned} w(x,y,t^n) &= \sum_{j,k} p_{j,k}(x,y)\chi_{j,k}(x,y), \\ p_{j,k}(x,y) &= \bar{w}^n_{j,k} + w'_{j,k}\left(\frac{x-x_j}{\Delta x}\right) + w^{\backprime}_{j,k}\left(\frac{y-y_k}{\Delta y}\right). \end{aligned} \tag{4.2}$$

Here, the discrete slopes in the x and in the y direction approximate the corresponding derivatives, $w'_{j,k} \sim \Delta x \cdot w_x(x_j, y_k, t^n) + O(\Delta x)^2$, $w^{\backprime}_{j,k} \sim \Delta y \cdot w_y(x_j, y_k, t^n) + O(\Delta y)^2$, and $\chi_{j,k}(x,y)$ is the characteristic function of the cell $C_{j,k} := \left\{(\xi,\eta)\big||\xi - x_j| \le \frac{\Delta x}{2}, |\eta - y_k| \le \frac{\Delta y}{2}\right\} = I_j \otimes J_k$. Of course, it is essential to reconstruct the discrete slopes, w' and $w^{\backprime}$, with *limiters*, which guarantee the non-oscillatory character of the reconstruction; the family of min-mod limiters is a prototype example

$$\begin{aligned} w'_{jk} = MM\{\theta(\bar{w}^n_{j+1,k} - \bar{w}^n_{j,k}), \frac{1}{2}(\bar{w}^n_{j+1,k} - \bar{w}^n_{j-1,k}), \theta(\bar{w}^n_{j,k} - \bar{w}^n_{j-1,k})\}, \\ 1 \le \theta \le 2, \end{aligned} \tag{4.3$'$}$$

$$\begin{aligned} w^{\backprime}_{jk} = MM\{\theta(\bar{w}^n_{j,k+1} - \bar{w}^n_{j,k}), \frac{1}{2}(\bar{w}^n_{j,k+1} - \bar{w}^n_{j,k-1}), \theta(\bar{w}^n_{j,k} - \bar{w}^n_{j,k-1})\}, \\ 1 \le \theta \le 2. \end{aligned} \tag{4.3$^{\backprime}$}$$

An exact evolution of this reconstruction, which is based on integration of the conservation law over the staggered volume yields

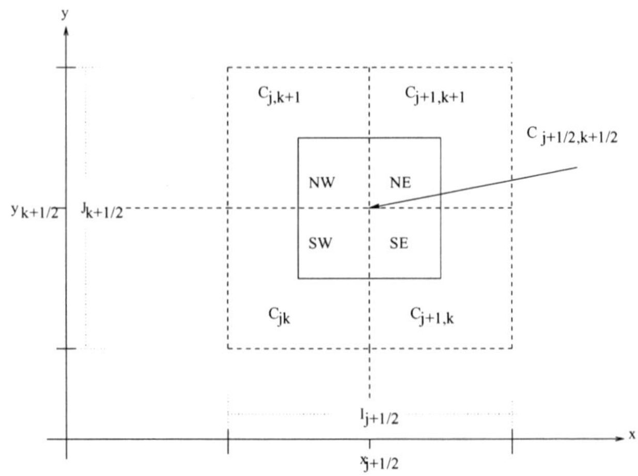

FIG. 6. *Floor plan of the staggered grid.*

$$\begin{aligned}\bar{w}^{n+1}_{j+\frac{1}{2},k+\frac{1}{2}} &= \fint_{C_{j+\frac{1}{2},k+\frac{1}{2}}} w(x,y,t^n)dxdy \qquad (4.4)\\ &- \lambda\left\{\fint_{\tau=t^n}^{t^{n+1}} \fint_{y\in J_{k+\frac{1}{2}}} \left[f(w(x_{j+1},y,\tau)) - f(w(x_j,y,\tau))\right] dyd\tau\right\}\\ &- \mu\left\{\fint_{\tau=t^n}^{t^{n+1}} \fint_{x\in I_{j+\frac{1}{2}}} \left[g(w(x,y_{k+1},\tau)) - g(w(x,y_k,\tau))\right] dxd\tau\right\}.\end{aligned}$$

Here and below, $\fint$ denotes the normalized integral, $\displaystyle\fint_\Omega := \frac{1}{|\Omega|}\int_\Omega$.

The exact averages at t^n – consult the floor plan in Figure 6 yields

$$\begin{aligned}\bar{w}^{n}_{j+\frac{1}{2},k+\frac{1}{2}} &:= \fint_{C_{j+\frac{1}{2},k+\frac{1}{2}}} w(x,y,t^n)dxdy \qquad (4.5)\\ &= \frac{1}{4}(\bar{w}^n_{jk} + \bar{w}^n_{j+1,k} + \bar{w}^n_{j,k+1} + \bar{w}^n_{j+1,k+1})\\ &+\frac{1}{16}\Big\{(w'_{jk} - w'_{j+1,k}) + (w'_{j,k+1} - w'_{j+1,k+1})\\ &+(w^{\grave{}}_{jk} - w^{\grave{}}_{j,k+1}) + (w^{\grave{}}_{j+1,k} - w^{\grave{}}_{j+1,k+1})\Big\}.\end{aligned}$$

So far everything is *exact*. We now turn to *approximate* the four fluxes on the right of (4.4), starting with the one along the East face, consult Figure 7, $\displaystyle\fint_{t^n}^{t^{n+1}} \fint_{J_{k+\frac{1}{2}}} f(w(x_{j+1},y,\tau))dyd\tau$. We use the midpoint

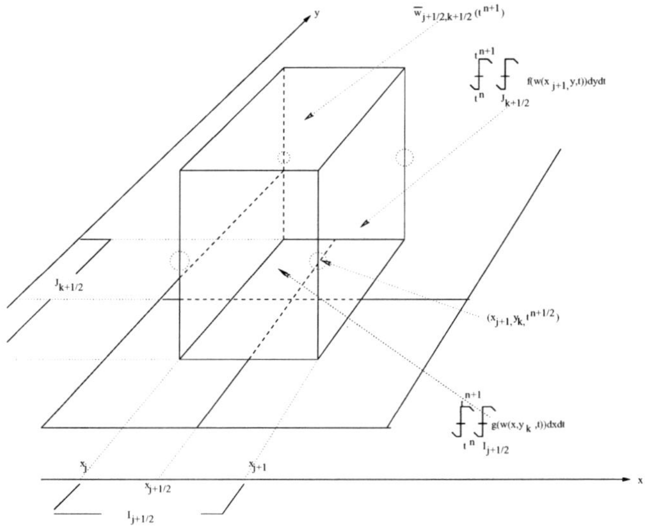

FIG. 7. *The staggered stencil in two dimensions.*

quadrature rule for second-order approximation of the temporal integral, $\fint_{y\in J_{k+\frac{1}{2}}} f(w(x_{j+1},y,t^{n+\frac{1}{2}}))dy$; and, for reasons to be clarified below, we use the second-order rectangular quadrature rule for the spatial integration across the y-axis, yielding

$$\fint_{t^n}^{t^{n+1}} \fint_{y\in J_{k+\frac{1}{2}}} f(w(x_{j+1},y,\tau))dyd\tau \sim \frac{1}{2}\left[f(w_{j+1,k}^{n+\frac{1}{2}}) + f(w_{j+1,k+1}^{n+\frac{1}{2}})\right]. \quad (4.6)$$

In a similar manner we approximate the remaining fluxes.

These approximate fluxes make use of the midpoint values, $w_{jk}^{n+\frac{1}{2}} \equiv w(x_j,y_k,t^{n+\frac{1}{2}})$, and it is here that we take advantage of utilizing these midvalues for the spatial integration by the rectangular rule. Namely, since these midvalues are secured at the smooth center of their cells, C_{jk}, bounded away from the jump discontinuities along the edges, we may use Taylor expansion, $w(x_j,y_k,t^{n+\frac{1}{2}}) = \bar{w}_{jk}^n + \frac{\Delta t}{2}w_t(x_j,y_k,t^n) + \mathcal{O}(\Delta t)^2$. Finally, we use the conservation law (4.1) to express the time derivative, w_t, in terms of the spatial derivatives, $f(w)'$ and $g(w)^{\backprime}$,

$$w_{jk}^{n+\frac{1}{2}} = \bar{w}_{jk}^n - \frac{\lambda}{2}f(w)'_{jk} - \frac{\mu}{2}g(w)^{\backprime}_{jk}. \quad (4.7)$$

Here, $f(w)'_{jk} \sim \Delta x \cdot f(w(x_j,y_k,t^n))_x$ and $g(w)^{\backprime}_{jk} \sim \Delta y \cdot g(w(x_j,y_k,t^n))_y$, are one-dimensional discrete slopes in the x- and y-directions, of the type reconstructed in (4.3′)–(4.3`); for example, multiplication by the corresponding Jacobians A and B yields

$$f(w)'_{jk} = A(\bar{w}^n_{jk})w'_{jk}, \qquad g(w)^{\grave{}}_{jk} = B(\bar{w}^n_{jk})w^{\grave{}}_{jk}.$$

Equipped with the midvalues (4.7), we can now evaluate the approximate fluxes, e,g., (4.6). Inserting these values, together with the staggered average computed in (4.6), into (4.4), we conclude with new staggered averages at $t = t^{n+1}$, given by

$$\begin{aligned}
\bar{w}^{n+1}_{j+\frac{1}{2},k+\frac{1}{2}} &= \frac{1}{4}(\bar{w}^n_{jk} + \bar{w}^n_{j+1,k} + \bar{w}^n_{j,k+1} + \bar{w}^n_{j+1,k+1}) \\
&+\frac{1}{16}(w'_{jk} - w'_{j+1,k}) - \frac{\lambda}{2}\left[f(w^{n+\frac{1}{2}}_{j+1,k}) - f(w^{n+\frac{1}{2}}_{j,k})\right] \\
&+\frac{1}{16}(w'_{j,k+1} - w'_{j+1,k+1}) - \frac{\lambda}{2}\left[f(w^{n+\frac{1}{2}}_{j+1,k+1}) - f(w^{n+\frac{1}{2}}_{j,k+1})\right] \\
&+\frac{1}{16}(w^{\grave{}}_{jk} - w^{\grave{}}_{j,k+1}) - \frac{\mu}{2}\left[g(w^{n+\frac{1}{2}}_{j,k+1}) - g(w^{n+\frac{1}{2}}_{j,k})\right] \\
&+\frac{1}{16}(w^{\grave{}}_{j+1,k} - w^{\grave{}}_{j+1,k+1}) - \frac{\mu}{2}\left[g(w^{n+\frac{1}{2}}_{j+1,k+1}) - g(w^{n+\frac{1}{2}}_{j+1,k})\right].
\end{aligned} \tag{4.8}$$

In summary, we end up with a simple two-step predictor-corrector scheme which could be conveniently expressed in terms on the one-dimensional staggered averaging notations

$$\langle w_{j,\cdot}\rangle_{k+\frac{1}{2}} := \frac{1}{2}(w_{j,k} + w_{j,k+1}), \quad \langle w_{\cdot,k}\rangle_{j+\frac{1}{2}} := \frac{1}{2}(w_{j,k} + w_{j+1,k}).$$

Our scheme consists of a *predictor step*

$$w^{n+\frac{1}{2}}_{j,k} = w^n_{j,k} - \frac{\lambda}{2}f'_{j,k} - \frac{\mu}{2}g^{\grave{}}_{j,k}, \tag{4.9}$$

followed by the *corrector step*

$$\begin{aligned}
\bar{w}^{n+1}_{j+\frac{1}{2},k+\frac{1}{2}} &= \langle\frac{1}{4}(\bar{w}^n_{j,\cdot} + \bar{w}^n_{j+1,\cdot}) + \frac{1}{8}(w'_{j,\cdot} - w'_{j+1,\cdot}) - \lambda(f^{n+\frac{1}{2}}_{j+1,\cdot} - f^{n+\frac{1}{2}}_{j,\cdot})\rangle_{k+\frac{1}{2}} \\
&+ \langle\frac{1}{4}(\bar{w}^n_{\cdot,k} + \bar{w}^n_{\cdot,k+1}) + \frac{1}{8}(w^{\grave{}}_{\cdot,k} - w^{\grave{}}_{\cdot,k+1}) - \mu(g^{n+\frac{1}{2}}_{\cdot,k+1} - g^{n+\frac{1}{2}}_{\cdot,k})\rangle_{j+\frac{1}{2}}.
\end{aligned}$$

In Figures 8 taken from [19], we present the two-dimensional computation of a double-Mach reflection problem; in Figure 9 we quote from [5] the two-dimensional computation of MHD solution of Kelvin-Helmholtz instability due to shear flow. The computations are based on our second-order central scheme. It is remarkable that such a simple 'two-lines' algorithm, with no characteristic decompositions and no dimensional splitting, approximates the rather complicated double Mach reflection problem with such high resolution. Couple of remarks are in order.

- Two-dimensional computations using central schemes are sensitive to the choice of limiter being used. In the context of the double Mach reflection problem, for exmple, the MM_2 (consult (3.2) with $\theta = 2$) seems to yield the sharper results. the one-dimensional central scheme can be as sensitive as the 2D schemes.

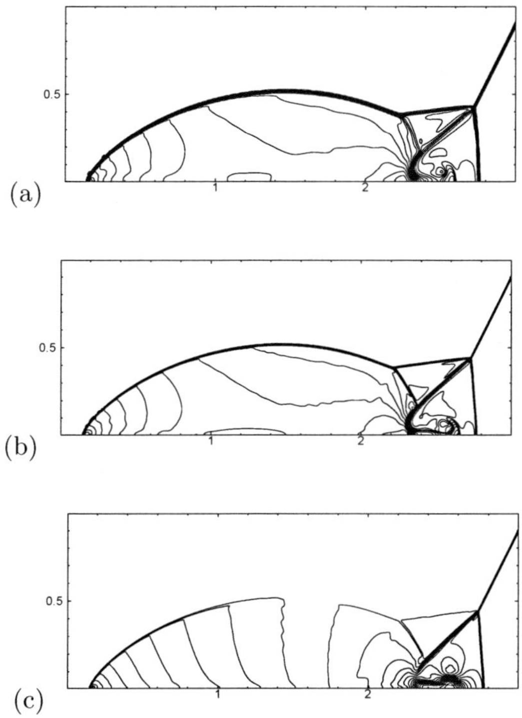

FIG. 8. *Double Mach reflection problem computed with the central scheme using MM_2 limiter with CFL=0.475 at* $t = 0.2$ *(a) density computed with* 480×120 *cells (b) density computed with* 960×240 *cells (c) x-velocity computed with* 960×240 *cells.*

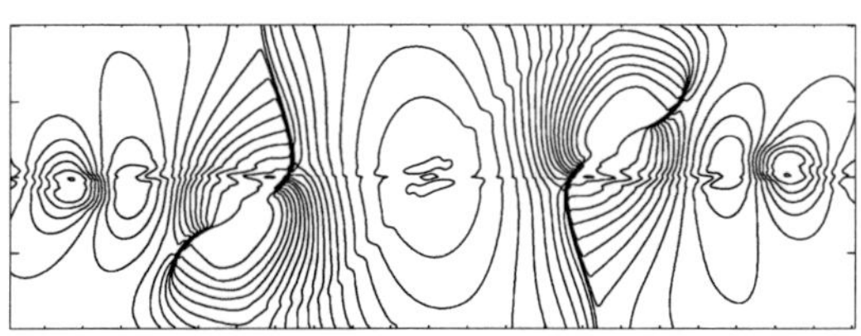

FIG. 9. *Kelvin-Helmholtz instability due to shear flow. Transverse configuration (B perpendicular to v). Pressure contours at* $t = 140$.

- No effort was made to optimize the boundary treatment. The staggered stencils require a different treatment for even-odd cells intersecting with the boundaries. The lack of boundary resolution could be observed at the bottom of the two Mach stems.

We conclude this section with brief remarks on further results related to central schemes.

REMARK.

1. Simplicity.
   Again, we would like to highlight the simplicity of the central schemes, which is particularly evident in the multidimensional setup: no characteristic information is required – in fact, even the exact Jacobians of the fluxes are not required; also, since no (approximate) Riemann solvers are involved, the central schemes require no dimensional splitting; as an example we refer to the approximation of the incompressible equations by central schemes, [24, 30]; the results in [10] provide another example of a *weakly* hyperbolic multidimensional system which could be efficiently solved in term of central schemes, by avoiding dimensional splitting.
2. Non-staggering. We refer to [18] for a non-staggered version of the central schemes.
3. Stability.
   The following maximum principle holds for the nonoscillatory scalar central schemes:
   THEOREM 4.1 ([19]). *Consider the two-dimensional scalar scheme (4.7–4.8), with minmod slopes, w' and $w^{\grave{}}$, in (4.3′–4.3`)). Then for any $\theta < 2$ there exists a sufficiently small CFL number, C_θ (– e.g. $C_1 = (\sqrt{7}-2)/6 \sim 0.1$), such that if the CFL condition is fulfilled,*

$$\max(\lambda \cdot \max_u |f_u(u)|, \mu \cdot \max_u |g_u(u)|) \leq C_\theta,$$

then the following local maximum principle holds

$$\min_{\substack{|p-(j+\frac{1}{2})|=\frac{1}{2} \\ |q-(k+\frac{1}{2})|=\frac{1}{2}}} \{\bar{w}^n_{p,q}\} \leq \bar{w}^{n+1}_{j+\frac{1}{2},k+\frac{1}{2}} \leq \max_{\substack{|p-(j+\frac{1}{2})|=\frac{1}{2} \\ |q-(k+\frac{1}{2})|=\frac{1}{2}}} \{\bar{w}^n_{p,q}\}. \tag{4.10}$$

REFERENCES

[1] A.M. ANILE, V. ROMANO, AND G. RUSSO, *Extended hydrodymnamical model of carrier transport in semiconductors*, SIAM Journal Applied Mathematics, **61**, 2000, 74–101.

[2] P. ARMINJON, D. STANESCU, AND M.-C. VIALLON, *A Two-Dimensional Finite Volume Extension of the Lax-Friedrichs and Nessyahu-Tadmor Schemes for Compressible Flow*, Proc. of the 6th Internat. Symp. on Comp. Fluid Dynamics, (M. Hafez, K. Oshima eds.), Lake Tahoe, Nevada, September 1995, Vol. **IV**, pp. 7–14.

[3] P. ARMINJON AND M.-C. VIALLON, *Généralisation du Schéma de Nessyahu-Tadmor pour Une Équation Hyperbolique à Deux Dimensions D'espace*, C.R. Acad. Sci. Paris, **t. 320** , série I. (1995), pp. 85–88.

[4] J. BALBAS AND E. TADMOR, CentPack: A package of high-resolution central schemes for nonlinear conservation laws and related problems, `http://www.cscamm.umd.edu/centpack/`.

[5] J. BALBAS, E. TADMOR, AND C.-C. WU Non-oscillatory central schemes for one- and two-dimensional MHD equations Journal of Computational Physics, **201**, 2004, 261–285.

[6] F. BEREUX AND L. SAINSAULIEU, *A Roe-type Riemann Solver for Hyperbolic Systems with Relaxation Based on Time-Dependent Wave Decomposition*, Numer. Math., **77**, 1997, 143–185.

[7] F. BIANCO, G. PUPPO, AND G. RUSSO, *High order central schemes for hyperbolic systems of conservation laws*, SIAM J. Sci. Comp., **21**, 1999, 294–322.

[8] D. L. BROWN, AND M. L. MINION *Performance of under-resolved two-dimensional incompressible flow simulations*, J. Comp. Phys., **122**, 1985, 165–183.

[9] P. COLELLA AND P. WOODWARD, *The piecewise parabolic method (PPM) for gas-dynamical simulations*, JCP, **54**, 1984, 174–201.

[10] B. ENGQUIST AND O. RUNBORG, *Multi-phase computations in geometrical optics*, Journal of Computational and Applied Mathematics, **74**, 1996, 175–192.

[11] ERBES, *A high-resolution Lax-Friedrichs scheme for Hyperbolic conservation laws with source term. Application to the Shallow Water equations.* Preprint.

[12] K.O. FRIEDRICHS AND P.D. LAX, *Systems of Conservation Equations with a Convex Extension*, Proc. Nat. Acad. Sci., **68**, 1971, 1686–1688.

[13] E. GODLEWSKI, AND P.-A. RAVIART, *Hyperbolic Systems of Conservation Laws*, Mathematics & Applications, Ellipses, Paris, 1991.

[14] S.K. GODUNOV, *A finite difference method for the numerical computation of discontinuous solutions of the equations of fluid dynamics*, Mat. Sb., **47**, 1959, 271–290.

[15] A. HARTEN, *High Resolution Schemes for Hyperbolic Conservation Laws*, JCP, **49**, 1983, 357–393.

[16] A. HARTEN, B. ENGQUIST, S. OSHER, AND S.R. CHAKRAVARTHY, *Uniformly high order accurate essentially non-oscillatory schemes. III*, JCP, **71**, 1982, 231–303.

[17] HUYNH, *A piecewise-parabolic dual-mesh method for the Euler equations*, AIAA-95-1739-CP, The 12th AIAA CFD Conf., 1995.

[18] G.-S. JIANG, D. LEVY, C.-T. LIN, S. OSHER, AND E. TADMOR, *High-resolution Non-Oscillatory Central Schemes with Non-Staggered Grids for Hyperbolic Conservation Laws*, SIAM Journal on Num. Anal., to appear.

[19] G.-S JIANG AND E. TADMOR, *Nonoscillatory Central Schemes for Multidimensional Hyperbolic Conservation Laws*, SIAM Journal on Scientific Computing, **19**, 1998, 1892–1917.

[20] S. JIN AND Z. XIN, *The relaxing schemes for systems of conservation laws in arbitrary space dimensions*, Comm. Pure Appl. Math., **48**, 1995, 235–277.

[21] B. VAN LEER, *Towards the Ultimate Conservative Difference Scheme, V. A Second-Order Sequel to Godunov's Method*, JCP, **32**, 1979, 101–136.

[22] R. KUPFERMAN, *Simulation of viscoelastic fluids: Couette-Taylor flow*, J. Comp. Phys., **147**, 1998, 22–59

[23] R. KUPFERMAN, *A numerical study of the axisymmetric Couette-Taylor problem using a fast high-resolution second-order central scheme*, SIAM. J. Sci. Comp., **20**, 1998, 858–877.

[24] R. KUPFERMAN AND E. TADMOR, *A Fast High-Resolution Second-Order Central Scheme for Incompressible Flow s*, Proc. Nat. Acad. Sci., **94**, 1997, 4848–4852.

[25] A. KURGANOV AND G. PETROVA, *Central schemes and contact discontinuities* Mathematical Modelling and Numerical Analysis, **34**, 2000, 1259–1275.

[26] A. KURGANOV AND A. POLIZZI, *Non-oscillatory central schemes for traffic flow models with Arrhenius look-ahead dynamics*, Netowrks and Heterogeneous Media, **4**(3), 2009, 431–452.

[27] R.J. LEVEQUE, *Numerical Methods for Conservation Laws*, Lectures in Mathematics, Birkhauser Verlag, Basel, 1992.

[28] D. LEVY, G. PUPPO, AND G. RUSSO, *A third order central WENO scheme for 2D conservation laws* Applied Numerical Mathematics, **33**, 2000, 415–421.

[29] D. Levy, G. Puppo, and G. Russo, *A fourth-order central WENO scheme for multi-dimensional hyperbolic systems of conservation laws*, SIAM Journal on Scientific Computing, **24**, 2002, 480–506.

[30] D. Levy and E. Tadmor, *Non-oscillatory Central Schemes for the Incompressible 2-D Euler Equations*, Math. Res. Let., **4**, 1997, 321–340.

[31] K.-A. Lie and S. Noelle, *On the artificial compression method for second-order nonoscillatory central difference schemes for systems of conservation laws*, SIAM Journal on Scientific Computation, **24**, 2003, 1157–1174.

[32] X.-D. Liu and P.D. Lax, *Positive Schemes for Solving Multi-dimensional Hyperbolic Systems of Conservation Laws*, J. Comp. Fluid Dynamics, **5**(2), 1996, 133–156.

[33] X.-D. Liu and S. Osher, *Nonoscillatory High Order Accurate Self-Similar Maximum Principle Satisfying Shock Capturing Schemes I*, SINUM, **33**(2), 1996, 760–779.

[34] X.-D. Liu and E. Tadmor, *Third Order Nonoscillatory Central Scheme for Hyperbolic Conservation Laws*, Numerische Mathematik, **79**, 1998, 397–425.

[35] Y.-J. Liu, C.-W. Shu, E. Tadmor, and M. Zhang, *Central discontinuous Galerkin methods on overlapping cells with a non-oscillatory hierarchical reconstruction*, SIAM Journal on Numerical Analysis, **45**(6), 2007, 2442–2467.

[36] Y.-J. Liu, C.-W. Shu, E. Tadmor, and M. Zhang, *Non-oscillatory hierarchical reconstruction for central and finite volume schemes*, Communications in Mathematical Physics, **2**(5), 2007, 933–963.

[37] Y.-J. Liu, C.-W. Shu, E. Tadmor, and M. Zhang, L2-stability analysis of the central discontinuous Galerkin method, Mathematical Modelling and Numerical Analysis, **42**, 2008, 593–607.

[38] H. Nessyahu and E. Tadmor, *Non-oscillatory Central Differencing for Hyperbolic Conservation Laws*, JCP, **87**(2), 1990, 408–463.

[39] S. Osher and E. Tadmor, *On the Convergence of Difference Approximations to Scalar Conservation Laws*, Math. Comp., **50**(181), 1988, 19–51.

[40] L. Pareschi and G. Russo, *Implicit Explicit Runge Kutta Schemes and Applications to Hyperbolic Systems with Relaxation*, J. Sci. Computing, **25**(1), 2005, 129–155.

[41] P.L. Roe, *Approximate Riemann Solvers, Parameter Vectors, and Difference Schemes*, JCP, **43**, 1981, 357–372.

[42] A. Rogerson and E. Meiburg, *A numerical study of the convergence properties of ENO schemes*, J. Sci. Comput., **5**, 1990, 127–149.

[43] V. Romano and G. Russo, *Numerical solution for hydrodynamical models of semiconductors*, Mathematical Models and Applications in Applied Sciences, **10**(7), 2000, 1099–1120.

[44] R. Sanders, *A Third-order Accurate Variation Nonexpansive Difference Scheme for Single Conservation Laws*, Math. Comp., **41**, 1988, 535–558.

[45] R. Sanders and A. Weiser, *A High Resolution Staggered Mesh Approach for Nonlinear Hyperbolic Systems of Conservation Laws*, JCP, **1010**, 1992, 314–329.

[46] P.K. Sweby, *High Resolution Schemes Using Flux Limiters for Hyperbolic Conservation Laws*, SINUM, **21**(5), 1984, 995–1011.

[47] C.-W. Shu, *Numerical experiments on the accuracy of ENO and modified ENO schemes*, JCP, **5**, 1990, 127–149.

[48] G. Sod, *A survey of several finite difference methods for systems of nonlinear hyperbolic conservation laws*, JCP, **22**, 1978, 1–31.

[49] P. Woodward and P. Colella, *The numerical simulation of two-dimensional fluid flow with strong shocks*, JCP, **54**, 1988, 115–173.

STABILITY AND DYNAMICS OF VISCOUS SHOCK WAVES

KEVIN ZUMBRUN*

Abstract. We examine from a classical dynamical systems point of view stability, dynamics, and bifurcation of viscous shock waves and related solutions of nonlinear pde. The central object of our investigations is the Evans function: its meaning, numerical approximation, and behavior in various asymptotic limits.

1. Introduction and motivating examples. The object of this course is to study, using techniques related to the Evans function, the stability and dynamics of traveling-wave solutions $u(x,t) = \bar{u}(x-st)$ of various systems of viscous conservation laws arising in continuum mechanics:

$$u_t + f(u)_x = (B(u)u_x)_x. \quad u \in \mathbb{R}^n. \tag{1.1}$$

Our emphasis is on concrete applications: in particular, numerical verification of stability/bifurcation conditions for viscous shock waves. The techniques we shall describe should have application also to other problems involving stability of traveling waves, particularly for large systems, or systems with no spectral gap between decaying and neutral modes of the linearized operator about the wave. We begin by introducing some equations from Biology and Physics possessing traveling wave solutions.

Reaction–diffusion equation. Consider a non-conservative problem:

$$u_t + g(u) = u_{xx}, \qquad u = u(x,t) \in \mathbb{R},\ x \in \mathbb{R},\ t \in \mathbb{R}^+, \tag{1.2}$$

where:

$$g(u) = dG(u) = (1/2)(u^2-1)u \quad \text{and} \quad G(u) = (1/8)(u^2-1)^2. \tag{1.3}$$

This equation is instructive to study, as it combines the effects of: (i) spatial diffusion, exemplified by the heat equation $u_t = u_{xx}$, with (ii) reaction, exemplified by the gradient flow $u_t = -dG(u)$ for which G represents a potential (chemical, for example) of the state associated with the value u (for example, densities of various chemical constituents).

Equation (1.2) models propagation of electro-chemical signals along a nerve axon, pattern-formation in chemical systems, and population dynamics. It also serves as a model for phase-transitional phenomena in material science, where u is an "order parameter" and G a free energy potential.

Burgers equation. The following scalar viscous conservation law:

$$u_t + f(u)_x = u_{xx}, \qquad u = u(x,t) \in \mathbb{R},\ x \in \mathbb{R},\ t \in \mathbb{R}^+, \tag{1.4}$$

*Indiana University, Bloomington, IN 47405 (kzumbrun@indiana.edu). Research of K. Zumbrun was partially supported under NSF grants no. DMS-0300487 and DMS-0801745.

where:

$$f(u) = u^2/2, \tag{1.5}$$

serves as a simple model for gas dynamics, traffic flow, or shallow-water waves, with u representing the density of some conserved quantity and f its flux through a fixed point x.

Isentropic gas dynamics. The equations of an isentropic compressible gas, take the form in Lagrangian coordinates (that is, coordinates in which x represents initial particle location, as opposed to Eulerian coordinates, in which x represents fixed location):

$$v_t - u_x = 0, \qquad u_t + p(u)_x = \left(\frac{u_x}{v}\right)_x, \tag{1.6}$$

where u represents specific volume (one over density), v velocity, and p pressure. System (1.6) models ideal gas dynamics when $p' < 0$. In the case $p' \gtrless 0$, it models phase-transitional gas dynamics or elasticity; thus generalizing both of the previous two examples.

Here, we consider mainly a γ-law pressure, modeling an ideal isentropic polytropic gas:

$$p(v) = av^{-\gamma}, \tag{1.7}$$

where $a > 0$ and $\gamma > 1$ are constants that characterize the gas.

Isentropic MHD. In Lagrangian coordinates, the equations for planar compressible isentropic magnetohydrodynamics take the form:

$$\begin{cases} v_t - u_{1x} = 0, \\ u_{1t} + (p + (1/2\mu_0)(B_2^2 + B_3^2))_x = ((1/v)u_{1x})_x, \\ u_{2t} - ((1/\mu_0)B_1^* B_2)_x = ((\tilde{\mu}/v)u_{2x})_x, \\ u_{3t} - ((1/\mu_0)B_1^* B_3)_x = ((\tilde{\mu}/v)u_{3x})_x, \\ (vB_2)_t - (B_1^* u_2)_x = ((1/\sigma\mu_0 v)B_{2x})_x, \\ (vB_3)_t - (B_1^* u_3)_x = ((1/\sigma\mu_0 v)B_{3x})_x, \end{cases} \tag{1.8}$$

where v denotes specific volume, $u = (u_1, u_2, u_3)$ velocity, $p = p(v)$ pressure, $B = (B_1^*, B_2, B_3)$ magnetic induction, B_1^* constant (by the divergence-free condition $\nabla_x \cdot B \equiv 0$), $\tilde{\mu} = \frac{\mu}{2\mu+\eta}$ where $\mu > 0$ and $\eta > 0$ are the respective coefficients of Newtonian and dynamic viscosity, $\mu_0 > 0$ the magnetic permeability, and $\sigma > 0$ the electrical resistivity [A, C, J].

We take again a γ-law pressure (1.7). Though we do not specify η, we have in mind mainly the value $\eta = -2\mu/3$ typically prescribed for (nonmagnetic) gas dynamics [Ba],[1] or $\tilde{\mu} = 3/4$. An interesting, somewhat simpler, subcase is that of infinite electrical resistivity $\sigma = \infty$, in which the last two equations of (1.8) are replaced by:

$$(vB_2)_t - (B_1^* u_2)_x = 0, \qquad (vB_3)_t - (B_1^* u_3)_x = 0, \tag{1.9}$$

[1] Predicted by Chapman-Enskog expansion of Boltzmann's equation.

and only the velocity variables (u_1, u_2, u_3) experience parabolic smoothing, through viscosity. Here, we are assuming a planar multidimensional flow, i.e., one that depends only on a single space-direction x. In the simplest, *parallel case* when $u_2 = u_3 = B_2 = B_3 \equiv 0$, the equations reduce to those of isentropic gas dynamics (1.6).

2. Traveling and standing waves. We shall study *traveling waves* to (1.1), which are solutions of the form:

$$u(x,t) = \bar{u}(x - st), \tag{2.1}$$

where s is a constant *speed* of the *profile* $\bar{u}(\cdot)$ connecting *endstates*:

$$\lim_{z\to\pm\infty} \bar{u}(z) = u_\pm. \tag{2.2}$$

Other types of profiles, such as waves that are oscillatory at infinity, are also interesting but beyond our scope. If $s = 0$ the wave (2.1) is called a *standing wave* of the associated evolution equation. Note that a traveling wave may always be converted to a standing wave by the change of coordinates $x \mapsto x - st$, corresponding to frame moving with the speed s.

Traveling waves have practical importance, as the simplest solutions that maintain their form while still reflecting the spatial dynamics of the problem; in particular, except in the trivial case $u \equiv$ const, they depend nontrivially on the variable x. We derive below nontrivial traveling-wave solutions for each of the models just described, choosing parameters carefully in order to have explicit solutions.

Generally, one must approximate profiles numerically, since for arbitrary choices of f, g, p, the profile equations are not explicitly soluble. However, the properties of the solutions remain similar, in particular, *profiles decay exponentially as $x \to \pm\infty$*, which is the principle property we will need in what follows. Numerical approximation of profiles is another interesting problem, but one that has already been well studied [Be].

Reaction–diffusion equation. Observing that

$$\partial_t \bar{u}(x - st) = -s\bar{u}', \quad \partial_x \bar{u}(x - st) = \bar{u}', \quad \partial_x^2 \bar{u}(x - st) = \bar{u}'', \tag{2.3}$$

we obtain for a solution (2.1) the *profile equation*: $-s\bar{u}' + dG(\bar{u}) = \bar{u}''$. Taking for simplicity $s = 0$, we find that it becomes a *nonlinear oscillator*, which is a Hamiltonian equation:

$$dG(\bar{u}) = \bar{u}'', \tag{2.4}$$

that may be reduced to first order. Specifically, multiplying by $\bar{u}'$ on both sides, we obtain:

$$dG(\bar{u})\bar{u}' = \bar{u}'\bar{u}'' \quad \text{where} \quad G(\bar{u})' = [(\bar{u}')^2/2]'.$$

Integrating from $-\infty$ to x then yields: $(\bar{u}')^2 = 2[G(\bar{u}) - G(u_-)]$. Thanks to (2.2), we have $\bar{u}'(\pm\infty) = 0$, hence, recalling (2.4), we have $dG(u_\pm) = 0$, where $u_\pm = \pm 1, 0$. As $u = 0$ is a nonlinear center admitting no orbital connections, we have: $u_\pm = \pm 1, \quad G(u_\pm) = 0$ and

$$\bar{u}' = \pm\sqrt{2G(\bar{u})} = \pm\sqrt{(1/4)(\bar{u}^2 - 1)^2} = \pm(1/2)(\bar{u}^2 - 1).$$

Choosing the sign $+$, we obtain finally, the profile solution (unique up to translation in x):

$$\bar{u}(x) = -\tanh(x/2). \tag{2.5}$$

Burgers equation. Using again (2.3), we obtain as the equation for a Burgers profile:

$$-s\bar{u}' + f(\bar{u})' = \bar{u}''. \tag{2.6}$$

Integrating from $-\infty$ to x reduces (2.6) to a first-order equation:

$$\bar{u}' = (f(\bar{u}) - s\bar{u}) - (f(u_-) - su_-).$$

For definiteness taking again $s = 0$, $u_- = 1$, we obtain $\bar{u}' = (1/2)(1 - \bar{u}^2)$ as in the reaction–diffusion case, and thus the same tanh solution (2.5), again unique up to translation.

Gas dynamics. Similarly, we find for (1.6) the profile equations:

$$-s\bar{v}' - \bar{u}' = 0 \qquad -s\bar{u}' + p(\bar{v})' = (\bar{v}^{-1}\bar{u}')'. \tag{2.7}$$

Substituting the first equation into the second, and integrating from $-\infty$ to x, we obtain:

$$\bar{v}' = (-s)^{-1}\bar{v}[(p(\bar{v}) + s^2\bar{v}) - (p(v_-) + s^2 v_-)]. \tag{2.8}$$

Choosing $s = -1$, $v_- = 1$, $a = v_+$, taking a isothermal, or constant-temperature, pressure $p(v) = av^{-1}$, and setting $v_m := \frac{v_- + v_+}{2}$, $\delta := v_- - v_+$, and $\bar{v} := v_m + (\delta/2)\bar{\theta}$, we obtain, finally, $\bar{\theta}' = (\delta/2)(\bar{\theta}^2 - 1)$, hence $\bar{\theta}(x) = -\tanh((\delta/2)x)$, giving a solution [Tay]:

$$\bar{v} := v_m - (\delta/2)\tanh((\delta/2)x) \tag{2.9}$$

similar to the Burgers profile (2.5), with $\bar{u} = \bar{v} + C$, C an arbitrary constant.

MHD. In the "parallel case" the equations of isentropic MHD reduce to those of isentropic gas dynamics, and so we have the same family of traveling waves already studied. However, from the standpoint of behavior, there is a major difference that we wish to point out. Namely, taking $v_- = 1$,

$u_{1,-} = 0$, $s = -1$, and for simplicity taking $\sigma = \infty$, but *without making the parallel assumptions*, we obtain the more general profile equation:

$$\begin{aligned} v' &= v(v-1) + v(p - p_-) + \frac{1}{2\mu_0 v}((B_- + B_1^* w)^2 - v^2 B_-^2), \\ \tilde{\mu} w' &= vw - \frac{B_1^*}{\mu_0}(B_-(1-v) + B_1^* w), \end{aligned} \tag{2.10}$$

where $w = (u_2, u_3)$ denotes transverse velocity components, and $B \equiv v^{-1}(B_- + B_1^* w)$.

The point we wish to emphasize is that even in the case of a parallel traveling-wave profile, for which $B_{2,\pm} = B_{3,\pm} = u_{2,\pm} = u_{3,\pm} = 0$, there may exist other traveling-wave profiles connecting the same endstates, but which are not of parallel type. Indeed, this depends precisely on the type of the rest points at $\pm\infty$, i.e., the number of stable (negative real part) and unstable (positive real part) eigenvalues of the coefficient of the linearized equations about these equilibria, which are readily determined by the observation that v and w equations decouple in the parallel case [BHZ].

3. Spectral stability and the eigenvalue equations. A fundamental issue in physical applications is *nonlinear* (time-evolutionary) *stability* of traveling-wave solutions $\bar{u}$ in (2.1). The question is whether or not a perturbed solution $\tilde{u} = \bar{u} + u$, with perturbation u sufficiently small at $t = 0$, approaches as $t \to +\infty$ to some translate $\bar{u}(x - st - x_0)$ of the original wave $\bar{u}(\cdot)$.

Observe that a perturbation $u(x,t) = \bar{u}(x - x_0 - st) - \bar{u}(x - st)$ is a traveling wave itself, hence does not decay. Thus, approach to a translate "orbital stability" is all that we can expect. Unstable solutions are impermanent, since they disappear under small perturbations, and typically play a secondary role if any in behavior: for example, as transition states lying on separatrices between basins of attraction of neighboring stable solutions, or as background organizing centers from which other solutions bifurcate.

Let us briefly recall the study of asymptotic stability of an equilibrium $\bar{U} \in \mathbb{R}^n$ of an autonomous ODE in $\mathbb{R}^n$:

$$U_t = \mathcal{F}(U), \qquad \mathcal{F}(\bar{U}) = 0. \tag{3.1}$$

We first linearize about $\bar{U}$, to obtain:

$$U_t = LU := d\mathcal{F}(\bar{U})U,$$

where the operator L has constant coefficients, since $\bar{U}$ is constant and $\mathcal{F}$ is autonomous. The formal justification for this step is that if perturbation $U = \tilde{U} - \bar{U}$ is small, then terms of second order are much smaller than terms of first order in the perturbation, and so can be ignored. If the solutions of (3.1) decay as $t \to \infty$, we say that $\bar{U}$ is *linearly* (asymptotically) *stable*.

Linear stability is nearly but not completely necessary for *nonlinear stability*, defined as decay as $t \to \infty$ of perturbations $U = \tilde{U} - \bar{U}$ that are sufficiently small at $t = 0$, for a solution $\tilde{U}$ of (3.1). A necessary condition for linear stability is *spectral stability*, or nonexistence of eigenvalues λ of L with $\Re\lambda \geq 0$. For, if there are $W \in \mathbb{R}^n$, $\lambda \in \mathbb{C}$ such that $LW = \lambda W$ with $\Re\lambda \geq 0$, then $U(t) = e^{\lambda t}W$ is a *growing mode* of (3.1), or a solution that does not decay.

These three notions of stability are linked in $\mathbb{R}^n$ by *Lyapunov's Theorem*, asserting that linearized and spectral stability are equivalent, and imply nonlinear exponential stability. One generally studies the simplest condition of spectral stability which corresponds to nonexistence of solutions W, λ with $\Re\lambda \geq 0$ of the eigenvalue equation:

$$LW = \lambda W. \tag{3.2}$$

This may be studied by Nyquist diagram, Routh–Hurwitz stability criterion [Hol], or other methods based on the characteristic polynomial $p(\lambda) = \det(\lambda \mathrm{Id} - L)$ associated with L.

We will pursue a similar program in the case of interest for PDEs, linearizing and studying the associated eigenvalue equations with the aid of the *Evans function*, which is an infinite-dimensional generalization of the characteristic polynomial. As a first step, we derive here the linearized and eigenvalue equations for the 4 described models, and then put them in a common framework as (complex) systems of first-order ODE:

$$W' = A(\lambda, x)W, \quad W \in \mathbb{C}^n,\ A \in \mathbb{C}^{n \times n}. \tag{3.3}$$

Reaction–diffusion equation. Let $\tilde{u}$ and $\bar{u}$ be two solutions of (1.2). The variable $u = \tilde{u} - \bar{u}$ satisfies the *nonlinear perturbation equation*:

$$u_t + (g(\bar{u} + u) - g(\bar{u})) = u_{xx}. \tag{3.4}$$

Taylor expanding $(g(\bar{u}+u)-g(\bar{u})) = dg(\bar{u})u+O(u^2)$ and dropping nonlinear terms of higher order, we obtain the linearized equation:

$$u_t + dg(\bar{u})u = u_{xx} \tag{3.5}$$

where $dg(\bar{u}) = d^2G(\bar{u}) = (1/2)(3\bar{u}^2 - 1)$, $dg(\bar{u}) = d^2G(\bar{u}) = (1/2)(3\bar{u}^2 - 1)$, and $\bar{u}(x) = -\tanh(x/2)$. Seeking *normal modes* $u(x,t) = e^{\lambda t}w(x)$, we obtain the *eigenvalue equation*:

$$\lambda w + dg(\bar{u})w = w''. \tag{3.6}$$

Observe now a complication that (3.4), by translational-invariance of the original equation (1.2), admits stationary solutions $u(x,t) \equiv \bar{u}(x+c) - \bar{u}(x)$ which do not decay. Likewise, the linearized equation (3.5) admits the stationary solutions $u(x,t) = c\bar{u}'(x)$ and $\bar{u}(x+c) - \bar{u}(x) = c\bar{u}'(x) + O(c^2)$, hence the two (nonlinear vs. linear) solutions approach each other as $c \to 0$. Also, as remarked above, we cannot expect in terms of stability more than approach to a translation for (3.4) and a multiple $c\bar{u}'$ of $\bar{u}'$ for (3.5). As spectral stability, therefore, we require that *there exist no solution* $w \in L^2$, $\lambda \in \mathbb{C}$ *of the eigenvalue equation* (3.6) *with* $\Re\lambda \geq 0$ *except for the "translational solution"* $w = c\bar{u}'$, $\lambda = 0$.

Adjoining first derivatives, (3.6) may be written in the form (3.3) as:

$$\begin{pmatrix} w \\ w' \end{pmatrix}' = \begin{pmatrix} 0 & 1 \\ \lambda + dg(\bar{u}) & 0 \end{pmatrix} \begin{pmatrix} w \\ w' \end{pmatrix}, \tag{3.7}$$

where $dg(\bar{u}) = (1/2)(3\bar{u}^2 - 1)$ and $\bar{u} = -\tanh(x/2)$.

Burgers equation. We obtain the perturbation equation:

$$u_t + (f(\bar{u} + u) - f(\bar{u}))_x = u_{xx}.$$

Expanding $f(\bar{u} + u) - f(\bar{u}) = df(\bar{u})u + O(u^2)$ and discarding the nonlinear part, we get the linearized equation $u_t + (df(\bar{u})u)_x = u_{xx}$, and:

$$\lambda w + (df(\bar{u})w)' = w'', \qquad \text{where } df(\bar{u}) = \bar{u}, \quad \bar{u}(x) = -\tanh(x/2), \tag{3.8}$$

which is the eigenvalue equation. A standard trick is to integrate (3.8), substitute $\breve{w}(x) = \int_{-\infty}^{x} w(z)\,dz$, and recover the *integrated eigenvalue equation*:

$$\lambda \breve{w} + df(\bar{u})\breve{w}' = \breve{w}''. \tag{3.9}$$

LEMMA 3.1 ([ZH]). *The eigenvalues of* (3.8) *and* (3.9) *for* $\Re\lambda \geq 0$ *coincide, except at* $\lambda = 0$.

The advantage of (3.9) vs. (3.8) is that there is no bounded solution $\breve{w}$ for $\lambda = 0$, that is, we have removed the eigenvalue $\lambda = 0$ by integration. As spectral stability for Burgers equation, we hence require that *there is no solution* $\breve{w} \in L^2$, $\lambda \in \mathbb{C}$ *of* (3.9) *with* $\Re\lambda \geq 0$.

The equation (3.9) has the equivalent form (3.3) as:

$$\begin{pmatrix} \breve{w} \\ \breve{w}' \end{pmatrix}' = \begin{pmatrix} 0 & 1 \\ \lambda & df(\bar{u}) \end{pmatrix} \begin{pmatrix} \breve{w} \\ \breve{w}' \end{pmatrix}, \qquad df(\bar{u}) = \bar{u} = -\tanh(x/2). \tag{3.10}$$

Gas dynamics. Reasoning as before, we obtain in the frame $\tilde{x} = x - st$:

$$\begin{aligned} v_t - s v_{\tilde{x}} - u_{\tilde{x}} &= 0 \\ u_t - s u_{\tilde{x}} + \big((dp(\bar{v}) + \bar{v}^{-2}\bar{u}_{\tilde{x}})v\big)_{\tilde{x}} &= ((1/\bar{v})u_{\tilde{x}})_{\tilde{x}}, \end{aligned} \tag{3.11}$$

and the eigenvalue system:

$$\begin{aligned} \lambda v - s v' - u' &= 0 \\ \lambda u - s u' + \big((dp(\bar{v}) + \bar{v}^{-2}\bar{u}')v\big)' &= ((1/\bar{v})u')'. \end{aligned} \tag{3.12}$$

Integrating, we obtain finally the integrated eigenvalue equations

$$\begin{aligned} \lambda\check{v} + \check{v}' - \check{u}' &= 0 \\ \lambda\check{u} + \check{u}' + (dp(\bar{v}) + \bar{v}^{-2}\bar{u}')\tilde{v}' &= (1/\bar{v})\tilde{u}''. \end{aligned} \tag{3.13}$$

As for the Burgers equation, we require as spectral stability *that there exist no solution $\check{w} \in L^2$, $\lambda \in \mathbb{C}$ of the system* (3.13) *with* $\Re\lambda \geq 0$.

Following [HLZ], we get that (3.13) can be put in the form (3.3) as:

$$\begin{pmatrix} \check{u} \\ \check{v} \\ \check{v}' \end{pmatrix}' = \begin{pmatrix} 0 & \lambda & 1 \\ 0 & 0 & 1 \\ \lambda\bar{v} & \lambda\bar{v} & f(\bar{v}) - \lambda \end{pmatrix} \begin{pmatrix} \check{u} \\ \check{v} \\ \check{v}' \end{pmatrix}, \tag{3.14}$$

where:

$$f(\bar{v}) = 2\bar{v} - (\gamma - 1)\Big(\frac{1 - v_+}{1 - v_+^\gamma}\Big)\Big(\frac{v_+}{\bar{v}}\Big)^\gamma - \Big(\frac{1 - v_+}{1 - v_+^\gamma}\Big)v_+^\gamma - 1. \tag{3.15}$$

MHD. Linearizing (1.8) about a parallel shock $(\bar{v}, \bar{u}_1, 0, 0, B_1^*, 0, 0)$, we get:

$$\begin{cases} v_t + v_x - u_{1x} = 0 \\ u_{1t} + u_{1x} - a\gamma\left(\bar{v}^{-\gamma-1}v\right)_x = \left(\dfrac{u_{1x}}{\bar{v}} + \dfrac{\bar{u}_{1x}}{\bar{v}^2}v\right)_x \\ u_{2t} + u_{2x} - \dfrac{1}{\mu_0}(B_1^*B_2)_x = \tilde{\mu}\left(\dfrac{u_{2x}}{\bar{v}}\right)_x \\ u_{3t} + u_{3x} - \dfrac{1}{\mu_0}(B_1^*B_3)_x = \tilde{\mu}\left(\dfrac{u_{3x}}{\bar{v}}\right)_x \\ (\bar{v}B_2)_t + (\bar{v}B_2)_x - (B_1^*u_2)_x = \left(\left(\dfrac{1}{\sigma\mu_0}\right)\dfrac{B_{2x}}{\bar{v}}\right)_x \\ (\bar{v}B_3)_t + (\bar{v}B_3)_x - (B_1^*u_3)_x = \left(\left(\dfrac{1}{\sigma\mu_0}\right)\dfrac{B_{3x}}{\bar{v}}\right)_x, \end{cases} \tag{3.16}$$

which is a decoupled system, consisting of the linearized isentropic gas dynamic equations in (v, u_1) about profile $(\bar{v}, \bar{u}_1)$, and two copies of an equation in variables $(u_j, \bar{v}B_j)$, $j = 2, 3$.

Introducing integrated variables $\tilde{v} := \int v$, $\tilde{u} := \int u_1$ and $w_j := \int u_j$, $\alpha_j := \int \bar{v}B_j$, $j = 2, 3$, we find that the integrated linearized eigenvalue equations decouple into the integrated linearized eigenvalue equations for gas dynamics (3.13) and two copies of:

$$\begin{cases} \lambda w + w' - \dfrac{B_1^*\alpha'}{\mu_0\bar{v}} = \mu\dfrac{w''}{\bar{v}} \\ \lambda\alpha + \alpha' - B_1^*w' = \dfrac{1}{\sigma\mu_0\bar{v}}\left(\dfrac{\alpha'}{\bar{v}}\right)' \end{cases} \tag{3.17}$$

in variables (w_j, α_j), $j = 2, 3$. As spectral stability, therefore, we require the pair of conditions that *there exist no solution* $\check{w} \in L^2$, $\lambda \in \mathbb{C}$ *with* $\Re\lambda \geq 0$ *of either* (3.13) *or the integrated transverse system* (3.17).

The transverse eigenvalue equations (3.17) may be written as a first-order system (3.3), indexed by the three parameters B_1^*, σ, and $\tilde{\mu}_0 := \sigma\mu_0$:

$$\begin{pmatrix} w \\ \tilde{\mu} w' \\ \alpha \\ \frac{\alpha'}{\sigma\mu_0\hat{v}} \end{pmatrix}' = \begin{pmatrix} 0 & 1/\tilde{\mu} & 0 & 0 \\ \lambda\hat{v} & \hat{v}/\tilde{\mu} & 0 & -\sigma B_1^*\hat{v} \\ 0 & 0 & 0 & \sigma\mu_0\hat{v} \\ 0 & -B_1^*\hat{v}/\tilde{\mu} & \lambda\hat{v} & \sigma\mu_0\hat{v}^2 \end{pmatrix} \begin{pmatrix} w \\ \mu w' \\ \alpha \\ \frac{\alpha'}{\sigma\mu_0\hat{v}} \end{pmatrix}. \tag{3.18}$$

For the four models of physical interest, we have found traveling-wave solutions and characterized spectral stability in terms of existence of solutions decaying as $x \to \pm\infty$ of an associated eigenvalue equation of linear ordinary differential type. However, such solution in general cannot be carried out explicitly. Following, we will study the question of existence numerically by an efficient "shooting" method based on the Evans function, and analytically by the examination of various asymptotic limits.

4. The Evans function: construction and an example. We now present the Evans function introduced in [E1]–[E4] and elaborated and generalized in [AGJ, PW, GZ, HuZ2]. Let L be a linear differential operator with asymptotically constant coefficients along a preferred spatial direction x. Assume that its eigenvalue equation $(L - \lambda\mathrm{Id})w = 0$ may be expressed as a 1st order ODE in an appropriate phase space:

$$W' = A(x,\lambda)W, \qquad \lim_{x\to\pm\infty} A(x,\lambda) = A_\pm(\lambda), \tag{4.1}$$

with A analytic in λ as a function from $\mathbb{C}$ to $C^1(\mathbb{R}, \mathbb{C}^{n\times n})$. Further, assume:

(h0) For $\Re\lambda > 0$, the dimension k of the stable subspace S_+ of $A_+(\lambda)$ and dimension $n-k$ of the unstable subspace U_- of $A_-(\lambda)$ sums to the dimension n of the entire phase space ("consistent splitting" [AGJ]) and that associated eigenprojections $\Pi_\pm$ (which must be automatically analytic on $\Re\lambda > 0$) extend analytically to $\Re\lambda = 0$.

(h1) In a neighborhood of any $\lambda_0 \in \{\Re\lambda \geq 0\}$, for fixed C, $\theta > 0$:

$$|A - A_\pm|(x,\lambda) \leq Ce^{-\theta|x|} \quad \text{for } x \gtrless 0.$$

Condition (h1) holds in the traveling-wave context whenever the underlying wave is a connection between hyperbolic rest points, defined as equilibria whose linearized equations have no center subspace.

Introduce now the complex ODE:

$$R' = \Pi' R, \quad R(\lambda_*) = R_*, \tag{4.2}$$

where $'$ denotes $d/d\lambda$, $\lambda_* \in \{\Re\lambda \geq 0\}$ is fixed, $\Pi = \Pi_\pm$, and $R = R_\pm$ with R_+ and R_- being $n \times k$ and $n \times (n-k)$ complex matrices. Let R_* be full rank and satisfy $\Pi(\lambda_*)R_* = R_*$, its columns constituting a basis for the stable (resp. unstable) subspace of A_+ (resp. A_-).

LEMMA 4.1 ([K, Z10, Z11]). *For $\Lambda \subset \{\Re\lambda \geq 0\}$ simply connected, there exists a unique global analytic solution R of* (4.2) *on Λ such that (i)* $\operatorname{rank} R \equiv \operatorname{rank} R^*$, *(ii)* $\Pi R \equiv R$, *(iii)* $\Pi R' \equiv 0$.

Property (iii) indicates that the Kato basis R is an optimal choice in the sense that it involves its minimal variation. From (ii) and (iii), we obtain the theoretically useful alternative formulation: $R' = [\Pi', \Pi]R$, where $[P, Q]$ denotes the commutator $PQ - QP$.

The following fundamental result is known as the *conjugation lemma*:

LEMMA 4.2 ([MeZ1, PZ]). *Assuming (h1), there exist coordinate changes $W = P^{\pm}Z_{\pm}$, $P_{\pm} = Id + \Theta^{\pm}$, defined and uniformly invertible on $x \gtrless 0$, with:*

$$|(\partial/\partial\lambda)^j \Theta^p_{\pm}| \leq C(j)e^{-\bar{\theta}|x|} \quad \text{for } x \gtrless 0, \quad 0 \leq j, \tag{4.3}$$

for any $0 < \bar{\theta} < \theta$, converting (4.1) *to the constant-coefficient limiting systems $Z' = A_{\pm}Z$.*

Consequently, we see that on $\Re\lambda > 0$, the manifolds of solutions of (4.1) decaying as $x \to \pm\infty$ are spanned by $\{W_j^{\pm} := (PR_j)^{\pm}\}$, where $R_j^{\pm}$ are the columns of $R^{\pm}$ as in Lemma 4.1.

We now introduce the main:

DEFINITION 4.1 ([AGJ, PW, GZ, Z1, HuZ2, HSZ]). *Let $W_1^+, \ldots, W_k^+$ and $W_{k+1}^-, \ldots, W_n^-$ be analytically-chosen (in λ) bases of the manifolds of solutions decaying as $x \to +\infty$ and $-\infty$. The Evans function on Λ is:*

$$\begin{aligned} D(\lambda) &= \det(P^+R_1^+, \cdots, P^+R_k^+, P^-R_{k+1}^-, P^-R_n^-)|_{x=0}, \\ &= \det\begin{pmatrix} W_1^+ & \cdots & W_k^+ & W_{k+1}^- & \cdots & W_n^- \end{pmatrix}_{|x=0}. \end{aligned} \tag{4.4}$$

PROPOSITION 4.1 ([GJ1, GJ2, MaZ3]). *The Evans function is analytic in λ on $\Re\lambda \geq 0$. On $\Re\lambda > 0$, its zeros agree in location and multiplicity with eigenvalues of L.*

Note that the above is indeed "the" Evans function, and not "an" Evans function, since we have uniquely specified a choice of asymptotic bases $R_j^{\pm}$. In general, the Evans function, similarly to the background profile, *is only numerically evaluable.*

4.1. A fundamental example. Consider Burgers' equation, $u_t + (u^2)_x = u_{xx}$, and the family of stationary viscous shocks:

$$\hat{u}^\epsilon(x) := -\epsilon \tanh(\epsilon x/2), \qquad \lim_{z \to \pm\infty} \hat{u}^\epsilon(z) = \mp\epsilon \tag{4.5}$$

of amplitude $|u_+ - u_-| = 2\epsilon$. The integrated eigenvalue equation (3.9) appears as $w'' = \hat{u}^\epsilon w' + \lambda w$. It reduces by the linearized Hopf–Cole transformation $w = \operatorname{sech}(\epsilon x/2)z$ to the constant-coefficient linear oscillator $z'' = (\lambda + \epsilon^2/4)z$, yielding exact solutions:

$$w^{\pm}(x,\lambda) = \operatorname{sech}(\epsilon x/2)e^{\mp\sqrt{\epsilon^2/4+\lambda}x}$$

which decay as $x \to \pm\infty$, with asymptotic behavior:

$$\mathcal{W}^{\pm}(x,\lambda) \sim e^{\mu_{\pm}(\lambda)x}\mathcal{V}_{\pm}(\lambda),$$

where $\mu_{\pm}(\lambda) := \mp(\epsilon/2 + \sqrt{\epsilon^2/4+\lambda}$ and $\mathcal{V}_{\pm} := (1, \mu_{\pm}(\lambda))^T$ are the eigenvalues and eigenvectors of the limiting constant-coefficient equations at $x = \pm\infty$ for $\mathcal{W}_{\pm} := (w, w')^T_{\pm}$.

Defining an Evans function $\mathcal{D}(\lambda) = \det(\mathcal{W}^-, \mathcal{W}^+)_{|x=0}$, we may compute explicitly:

$$\mathcal{D}(\lambda) = -2\sqrt{\epsilon^2/4+\lambda}.$$

However, this is not "the" Evans function $D(\lambda) = \det(W^-, W^+)_{|x=0}$ specified in the previous sections, which is constructed, rather, from a special basis $W_{\pm} = c_{\pm}(\lambda)\mathcal{W}^{\pm} \sim e^{\mu_{\pm}x}V^{\pm}$, where $V_{\pm} = c_{\pm}(\lambda)\mathcal{V}^{\pm}$ are "Kato" eigenvectors determined uniquely (up to a constant factor independent of λ) by the property [K, GZ, HSZ] that there exist left eigenvectors $\tilde{V}^{\pm}$ with:

$$(\tilde{V}\cdot V)^{\pm} \equiv \text{constant}, \qquad (\tilde{V}\cdot V')^{\pm} \equiv 0. \tag{4.6}$$

Computing dual eigenvectors $\tilde{\mathcal{V}}^{\pm} = (\lambda + \mu^2)^{-1}(\lambda, \mu_{\pm})$ satisfying $(\tilde{\mathcal{V}}\cdot\mathcal{V})^{\pm} \equiv 1$, and setting $V^{\pm} = c_{\pm}\mathcal{V}^{\pm}$, $\tilde{V}^{\pm} = \mathcal{V}^{\pm}/c_{\pm}$, we find that (4.6) is equivalent to the complex ODE:

$$\dot{c}_{\pm} = -\Big(\frac{\tilde{V}\cdot\dot{V}}{\tilde{V}\cdot V}\Big)^{\pm} c_{\pm} = -\Big(\frac{\dot{\mu}}{2\mu-\epsilon}\Big)_{\pm} c_{\pm}, \tag{4.7}$$

which may be solved by exponentiation, yielding the general solution:

$$c_{\pm}(\lambda) = C(\epsilon^2/4+\lambda)^{-1/4}.$$

Initializing at a fixed nonzero point, say $c_{\pm}(1) = 1$, and noting that $D_{\epsilon}(\lambda) = c_- c_+ \mathcal{D}_{\epsilon}(\lambda)$, we thus obtain the remarkable formula:

$$D_{\epsilon}(\lambda) \equiv -2\sqrt{\epsilon^2/4+1}. \tag{4.8}$$

That is, with the "Kato" normalization, the Evans function associated with a Burgers shock is not only stable (nonvanishing), but *identically constant.*

5. Abstract stability theorem: the Evans condition. Consider a viscous shock solution:

$$U(x,t) = \bar{U}(x-st), \qquad \lim_{z\to\pm\infty}\bar{U}(z) = U_{\pm}, \tag{5.1}$$

of a hyperbolic–parabolic system of conservation laws:

$$U_t + F(U)_x = (B(U)U_x)_x, \qquad x\in\mathbb{R},\ U,\ F\in\mathbb{R}^n,\ B\in\mathbb{R}^{n\times n}. \tag{5.2}$$

Profile $\bar{U}$ satisfies the traveling-wave ODE:

$$B(U)U' = F(U) - F(U_-) - s(U - U_-). \tag{5.3}$$

In particular, the condition that $U_\pm$ be rest points implies the *Rankine–Hugoniot condtions*:

$$F(U_+) - F(U_-) = s(U_+ - U_-) = 0 \tag{5.4}$$

of inviscid shock theory. Following this analogy, we define the *degree of compressivity* d as the total number of incoming hyperbolic characteristics toward the shock profile minus the dimension of the system, that is $d = \dim U(A_-) + \dim S(A_+) - n$, where $d = 1$ corresponds to a standard Lax shock, $d \le 0$ to an undercompressive shock, and $d \ge 2$ to an overcompressive shock. Denote $A(U) = DF(U)$ and let:

$$B_\pm := \lim_{z \to \pm\infty} B(z) = B(U_\pm), \qquad A_\pm := \lim_{z \to \pm\infty} A(z) = DF(U_\pm). \tag{5.5}$$

Following [TZ3, Z2, Z3], we introduce the structural assumptions:

(A1) $U = \begin{pmatrix} U_1 \\ U_2 \end{pmatrix}, \quad F = \begin{pmatrix} F_1 \\ F_2 \end{pmatrix}, \quad B = \begin{pmatrix} 0 & 0 \\ 0 & b \end{pmatrix}, \quad b$ nonsingular, where $U \in \mathbb{R}^n$, $U_1 \in \mathbb{R}^{n-r}$, $U_2 \in \mathbb{R}^r$, and $b \in \mathbb{R}^{r \times r}$. Moreover, $F_1(U)$ is *linear in* U (strong block structure).

(A2) There exists a smooth, positive definite matrix field $A^0(U)$, without loss of generality block-diagonal, such that $A^0_{11}A_{11}$ is symmetric, $A^0_{22}b$ is positive definite (not necessarily symmetric), and $(A^0A)_\pm$ is symmetric.

(A3) No eigenvector of $A_\pm$ lies in $\mathrm{Ker} B_\pm$ (genuine coupling [Kaw, KSh]).

In all of the examples, $A_{11} = \alpha \mathrm{Id}$, corresponding with simple transport along fluid particle paths, with genuine coupling equivalent to A_{12} full rank.

To (A1)–(A3), we add the following hypotheses. Here and elsewhere, $\sigma(M)$ denotes the spectrum of a matrix or linear operator M.

(H0) $F, B \in C^k$, $k \ge 4$.

(H1) $\sigma(A_{11})$ is real, with constant multiplicity, $\sigma(A_{11}) < s$ or $\sigma(A_{11}) > s$.

(H2) $\sigma(A_\pm)$ is real, semisimple, and does not include s.

(H3) Considered as connecting orbits of (5.3), $\bar{U}$ lies in an ℓ-dimensional manifold, $\ell \ge 1$, of solutions (5.1) connecting fixed pair of endstates $U_\pm$.

Conditions (A1)–(A3) and (H0)–(H3) are a slightly strengthened version of the corresponding hypotheses of [MaZ4, Z2, Z3] for general systems with "real", or partially parabolic viscosity, the difference lying in the assumed in (A1) linearity of the U_1 equation. The class of equations satisfying our assumptions, though not complete, is sufficiently broad to include many models of physical interest, in particular the compressible

Navier–Stokes equations and the equations of compressible magnetohydrodynamics (MHD), expressed in Lagrangian coordinates, with either ideal or "real" van der Waals-type equation of state.

We remark that much more general assumptions (in particular, allowing Eulerian coordinates) are sufficient for the stability results reported here [MaZ4, Z1]. Lagrangian coordinates are important for later material on conditional stability and bifurcation, and so we make the stronger assumptions here for the sake of a common presentation.

Also, we notice that by the change of coordinates $x \mapsto x - st$, we may assume without loss of generality $s = 0$ as we shall do from now on. Linearizing (5.2) about $\bar U$ yields:

$$U_t = LU := -(AU)_x + (BU_x)_x, \tag{5.6}$$
$$B(x) := B(\bar U(x)), \quad A(x)V := dF(\bar U(x))V - (dB(\bar U(x))V)\bar U_x, \tag{5.7}$$

for which the generator L possesses [He, Sa, ZH] both a translational zero-eigenvalue and essential spectrum tangent at zero to the imaginary axis.

Conditions (H1)–(H2) imply that $U_\pm$ are nonhyperbolic rest points of ODE (5.3) expressed in terms of the U_2-coordinate, whence:

$$|\partial_x^r(\bar U - U_\pm)(x)| \le Ce^{-\eta|x|}, \qquad 0 \le r \le k+1, \tag{5.8}$$

for $x \gtrless 0$, some η, $C > 0$. Let d_2 be the dimension of the unstable manifold at $-\infty$ of the traveling-wave ODE expressed in the U_2 coordinate plus the dimension of the stable one at $+\infty$, minus the dimension r of the ODE.

PROPOSITION 5.1 ([MaZ3]). *Under (A1)–(A3), (H0)–(H2), d_2 is equal to the degree of compressivity d, hence, if the traveling-wave connection is transversal, the set of all traveling-wave connections between $U_\pm$ form an ℓ-dimensional manifold $\{\bar U^\alpha\}$, $\alpha \in \mathbb{R}^\ell$, in the vicinity of $\bar U$, with $\ell = d$. In any case, for Lax or overcompressive shocks ($d \ge 1$), the linearized equations have a d-dimensional set of stationary solutions decaying as $x \to \pm\infty$, given in the transversal case by* $\mathrm{Span}\{\partial_{\alpha_j}\bar U^\alpha\}$. *Undercompressive connections ($d \le 0$) are never transversal.*

The following is another fundamental fact:

LEMMA 5.1. *Under (A1)–(A3), (H0)–(H2):*

$$\Re\sigma(-i\xi A - \xi^2 B)_\pm \le -\frac{\theta|\xi|^2}{1+|\xi|^2} \qquad \theta > 0, \quad \forall \xi \in \mathbb{R}. \tag{5.9}$$

Consequently: (i) the essential spectrum $\sigma_{ess}(L)$ lies to the left of a curve $\lambda(\xi) = -\frac{\theta|\xi|^2}{1+|\xi|^2}$, $\theta > 0$, (ii) there exist essential spectra $\lambda_*(\xi)$ lying to the right of $\lambda(\xi) = -\frac{\theta-2|\xi|^2}{1+|\xi|^2}$, $\theta_2 > \theta > 0$, (iii) the eigenvalue equations satisfy (h0)–(h1).

COROLLARY 5.1. *The Evans function $D(\lambda)$ is well defined and is analytic on $\{\Re\lambda \ge 0\}$, with at least ℓ zeros at $\lambda = 0$.*

DEFINITION 5.1. *We define the Evans stability condition as:*

(D) $D(\cdot)$ *has precisely* ℓ *zeros on* $\{\Re\lambda \geq 0\}$, *necessarily at* $\lambda = 0$.

We see that it is more difficult to verify the stability condition than disprove it, since stability is verified on an uncountable set of λ whereas, to show instability it suffices to find one eigenvalue with $\Re\lambda > 0$.

5.1. Relation to inviscid stability. As $\lambda = 0$ is an embedded eigenvalue in the essential spectrum of L, the meaning of the Evans function at $\lambda = 0$ is not immediately clear. The answer is given by the following fundamental lemma describing the low-frequency behavior of the Evans function. Recall in the Lax case that inviscid shock stability is determined [Er3, M1, M2, M3, Me] by the *Lopatinski determinant*:

$$\Delta(\lambda) := \lambda \det(r_1^-, \ldots, r_{p-1}^-, r_{p+1}^+, \ldots, r_n^+, (U_+ - U_-)), \tag{5.10}$$

where $r_1^-, \ldots r_{p-1}^-$ denote eigenvectors of $dF(U_-)$ associated with negative eigenvalues and $r_{p+1}^+, \ldots r_n^+$ denote eigenvectors of $dF(U_+)$ associated with positive eigenvalues. The shock is stable if $\Delta/\lambda \neq 0$.

LEMMA 5.2 ([GZ, ZS]). *Under (A1)–(A3), (H0)–(H3),*

$$D(\lambda) = \beta\Delta(\lambda) + o(|\lambda|^\ell), \tag{5.11}$$

where β *is a constant transversality coefficient which in the Lax or overcompressive case is a Wronskian of the linearized traveling-wave ODE measuring transverality of the connecting orbit* $\bar U(\cdot)$, *with* $\beta \neq 0$ *equivalent to transversality, and* $\Delta(\cdot)$ *is a positive homogeneous degree* ℓ *function which in the Lax or undercompressive case is exactly the Lopatinski determinant determing inviscid stability. In the undercompressive case,* β *measures "maximal transversality", or transversality of connections within the subspace spanned by the tangent spaces of stable and unstable manifolds.*

It follows that a shock wave satisfies the viscous (Evans) stabilty condition *only if* it satisfies the inviscid (Lopatinski) condition along with transversality of the viscous traveling-wave connection [GZ, ZS].

5.2. Basic nonlinear stability result. We now introduce a basic viscous stability result applying to the Lax or overcompressive case. For the undercompressive case, more complicated pointwise analyses [HZ, RZ] again confirm that Evans implies nonlinear stability. For a proof, see [MaZ4], or the discussion in Section 8.1 of the more general Theorem 8.2.

THEOREM 5.1 ([MaZ4, R]). *For Lax or overcompressive shocks, under (A1)–(A3), (H0)–(H3), and (D),* $\bar U$ *is nonlinearly orbitally stable in* $L^1 \cap H^3$ *with respect to perturbations with* $L^1 \cap H^3$ *norm and* L^1*-first moment* $E_0 := \|\tilde U(\cdot, 0) - \hat U\|_{L^1\cap H^3}$ *and* $E_1 := \||x|\,|\tilde U(\cdot, 0) - \hat U|\|_{L^1}$ *sufficiently small, in the sense that, for some* $\alpha(\cdot) \in \mathbb{R}^\ell$, *any* $\varepsilon > 0$, *and all* $p \geq 1$,

$$\begin{aligned}
|U(x,t)-\bar{U}^{\alpha(t)}(x)|_{L^p} &\le C(1+t)^{-\frac{1}{2}(1-\frac{1}{p})}(E_0+E_1),\\
|U(x,t)-\bar{U}^{\alpha(t)}(x)|_{H^3} &\le C(1+t)^{-\frac{1}{4}}(E_0+E_1),\\
\dot{\alpha}(t) &\le C(1+t)^{-1+\varepsilon}(E_0+E_1),\\
\alpha(t) &\le C(1+t)^{-1/2+\varepsilon}(E_0+E_1).
\end{aligned} \tag{5.12}$$

Note that the rates of convergence given in (5.12) are time-algebraic rather than -exponential, in accordance with the absence of a spectral gap between neutral and decaying modes. This is a substantial technical difference from the finite-dimensional ODE theory described in Section 3.

For shock waves with $|U_+ - U_-|$ sufficiently small, and $U_\pm$ approaching a base state U_0 for which $dF(U_0)$ has a single "genuinely nonlinear" zero eigenvalue, the existence of traveling waves may be studied as a bifurcation from the constant solution, both for inviscid [L, Sm] and viscous [MP, Pe] shock waves, reducing by the Implicit Function Theorem to the scalar, Burgers case – in particular, a classical Lax-type wave. In this case, the conditions in Theorem 5.1 may be verified via the same reduction [PZ, FS, HuZ1]. Small-amplitude waves bifurcating from multiple zero eigenvalues are typically of nonclassical over- or undercompressive type, and may be unstable [AMPZ, GZ, Z6]. However, their stability should be determinable in principle by similar reduction/singular perturbation techniques to those used in the Lax case, an interesting open problem.

Except in isolated special cases [Z6], stability of large-amplitude shock waves for the moment must be verified either numerically or else in certain asymptotic limits. Whether or not there exist structural conditions for large-amplitude stability, connected with existence of a convex entropy perhaps, remains a fundamental open question, as does even the profile existence problem for large-amplitude shocks [Z1, Z2]. In what follows, we describe a workable general approach based on numerical approximation and asymptotic ODE theory.

6. Numerical approximation of the Evans function. We now discuss the general question of numerical stability analysis by approximation of the Evans function. An efficient method introduced in [EF, Er1, Er2] is the "Nyquist diagram" approach familiar from control theory, based on the Principle of the Argument of Complex Analysis. Specifically, taking the winding number around a contour $\Gamma = \partial\Lambda \subset \{\Re\lambda \ge 0\}$, where Λ is a set outside which eigenvalues may be excluded by other methods (e.g. energy estimates or asymptotic ODE theory), counts the number of unstable eigenvalues in Λ of the linearized operator about the wave, with zero winding number corresponding to stability [Br1, Br2, BrZ, BDG, HSZ, HuZ2, BHRZ, HLZ, CHNZ, HLyZ1, HLyZ2, BHZ]. Alternatively, one may use Mueller's method or any number of root-finding methods for analytic functions to locate individual roots [OZ1, LS].

A first important detail is the choice of approximate spatial infinities $L_\pm$ determining the computational domain $x \in [-L_-, L_+]$. These must be sufficiently small that computational costs remain reasonable, but sufficiently large that the solution of the variable-coefficient Evans system remains within a desired relative error of its value at $\pm\infty$ (which is a constant solution of the limiting constant-coefficient system). Indeed, the same fixed-point argument by which the exact, variable coefficient is defined (the conjugation lemma) shows that a good rule of thumb for $|A - A_\pm| \sim C_\pm e^{-\theta_\pm |x|}$, is:

$$L_\pm \sim (\log C_\pm + |\log TOL|)/\theta_\pm \tag{6.1}$$

where TOL is desired relative error tolerance [HLyZ1].

Further, the numerical approximation of Evans function breaks into 2 steps below. In both, it is important to preserve analyticity in λ, which concerns *numerical propagation of subspaces*, thus tying into large bodies of theory in numerical linear algebra [ACR, DDF, DE1, DE2, DF] and hydrodynamic stability theory [Dr, Da, NR1, NR2, NR3, NR4, Ba].

Step (i). Computation of analytic bases for stable (resp. unstable) subspaces of A_+ (resp. A_-) [BrZ, HSZ, HuZ2, Z10, Z11]. Choose a set of mesh points λ_j, $j = 0, \ldots, J$ along a path $\Gamma \subset \Lambda$ and denote by $\Pi_j := \Pi(\lambda_j)$ and R_j the approximation of $R(\lambda_j)$. Typically, $\lambda_0 = \lambda_J$, i.e., Γ is a closed contour.

Given a matrix A, one may efficiently ($\sim 32n^3$ operations [GvL, SB]) compute by "ordered" Schur decomposition ($A = QUQ^{-1}$, Q orthogonal and U upper triangular, for which also the diagonal entries of U are ordered in increasing real part) an orthonormal basis $\check{R}_u := (Q_{k+1}, \ldots, Q_n)$, of its unstable subspace, where $Q_{k+1}, \ldots, Q_n$ are the last $n-k$ columns of Q, and $n-k$ is the dimension of the unstable subspace. Performing the same procedure for $-A$, A^*, and $-A^*$ we obtain orthonormal bases $\check{R}_s$, $\check{L}_u$, $\check{L}_s$ also for the stable subspace of A and the unstable and stable subspaces of A^*, from which we compute the stable and unstable eigenprojections in numerically well-conditioned manner via:

$$\Pi_s := \check{R}_s(\check{L}_s^*\check{R}_s)^{-1}\check{L}_s^*, \quad \Pi_u := \check{R}_u(\check{L}_u^*\check{R}_u)^{-1}\check{L}_u^*. \tag{6.2}$$

Applying this to matrices $A_j^\pm := A_\pm(\lambda_j)$, we obtain the projectors $\Pi_j^\pm := \Pi_\pm(\lambda_j)$. Hereafter, we consider $\Pi_j^\pm$ as known quantities.

Approximating $\Pi'(\lambda_j)$ to first order by the finite difference $(\Pi_{j+1} - \Pi_j)/(\lambda_{j+1} - \lambda_j)$ and substituting this into a first-order Euler scheme gives:

$$R_{j+1} = R_j + (\lambda_{j+1} - \lambda_j)\frac{\Pi_{j+1} - \Pi_j}{\lambda_{j+1} - \lambda_j}R_j,$$

or $R_{j+1} = R_j + \Pi_{j+1}R_j - \Pi_j R_j$, yielding by the property $\Pi_j R_j = R_j$ (preserved exactly by the scheme) the simple greedy algorithm:

$$R_{j+1} = \Pi_{j+1}R_j. \tag{6.3}$$

It is a remarkable fact [Z10, Z11] (a consequence of Lemma 4.1) that, up to numerical error, evolution of (6.3) about a closed loop $\lambda_0 = \lambda_J$ yields the original value $R_J = R_0$.

To obtain a second-order discretization of (4.2), we approximate

$$R_{j+1} - R_j \approx \Delta\lambda_j \Pi'_{j+1/2} R_{j+1/2},$$

where $\Delta\lambda_j := \lambda_{j+1} - \lambda_j$. Noting that $R_{j+1/2} \approx \Pi_{j+1/2} R_j$ to second order by (6.3), and approximating $\Pi_{j+1/2} \approx \frac{1}{2}(\Pi_{j+1} + \Pi_j)$, and $\Pi'_{j+1/2} \approx (\Pi_{j+1} - \Pi_j)/\Delta\lambda_j$, we obtain, combining and rearranging: $R_{j+1} = R_j + \frac{1}{2}(\Pi_{j+1} - \Pi_j)(\Pi_{j+1} + \Pi_j)R_j$. Stabilizing by following with a projection Π_{j+1}, we obtain the reduced second-order explicit scheme:

$$R_{j+1} = \Pi_{j+1}[I + \frac{1}{2}\Pi_j(I - \Pi_{j+1})]R_j. \tag{6.4}$$

This is the version we recommend for serious computations. For individual numerical experiments the simpler greedy algorithm (6.3) suffices.

Step (ii). Propagation of the bases by ODE (4.1) on a sufficiently large interval $x \in [M, 0]$ (resp. $x \in [-M, 0]$). Our basic principles for efficient numerical integration of (4.1) are readily motivated by consideration of the simpler constant-coefficient case $W_x = AW$ with $A \equiv \text{const}$. We must avoid two main potential pitfalls, which are: the wrong direction of integration, and existence of "parasitic modes" [Z10].

For general systems of equations, the dimension of the stable subspace of A typically involves two or more eigenmodes, with distinct decay rates. In practice, parasitic faster-decaying modes will tend to take over slower-decaying modes, preventing their resolution. Degradation of results from parasitic modes is of the same rough order as that resulting from integrating in the wrong spatial direction.

Resolution one: the centered exterior product scheme. Denote $R^+ = (R_1^+, \dots, R_k^+)$ and $R^- = (R_1^-, \dots, R_{n-k}^-)$. Define the initializing wedge products:

$$\mathcal{R}_{S_+} := R_1^+ \wedge \cdots \wedge R_k^+ \quad \text{and} \quad \mathcal{R}_{S_-} := R_1^- \wedge \cdots \wedge R_{n-k}^-. \tag{6.5}$$

The Evans determinant is then recovered through the isomorphism:

$$\begin{aligned} &\det(W_1^+, \cdots, W_k^+, W_{k+1}^-, \cdots, W_n^-) \\ &\qquad\qquad \sim (W_1^+ \wedge \cdots \wedge W_k^+) \wedge (W_{k+1}^- \wedge \cdots \wedge W_n^-); \end{aligned} \tag{6.6}$$

see [AS, Br1, Br2, BrZ, BDG, AlB] and ancestors [GB, NR1, NR2, NR3, NR4]. This reduces the problem to the case $k = 1$. We may further optimize by factoring out the expected asymptotic decay rate $e^{\mu x}$ of the single decaying mode and solving the "centered" equation:

$$Z' = (A - \mu \text{Id})Z, \quad Z(+\infty) = r : A_+ r = \mu r \tag{6.7}$$

for $Z = e^{-\mu x}W$, which is now asymptotically an equilibrium as $x \to +\infty$. With these preparations, one obtains excellent results [BDG, HuZ2]; however, omitting any one of them leads to a loss of efficiency of as much as an order of magnitude [HuZ3, BZ].

Resolution two: the polar coordinate method. Unfortunately, the dimension $\binom{n}{k}$ of the phase space for the exterior product in center exterior product scheme grows exponentially with n, since k is $\sim n/2$ in typical applications. This limits its usefulness to $n \le 10$ or so, whereas the Evans system arising in compressible MHD is size $n = 15$, $k = 7$ [BHZ], giving a phase space of size $\binom{n}{k} = 6,435$: clearly impractical. A more compact, but nonlinear, alternative is the polar coordinate method of [HuZ2], in which the exterior products of the columns of $W_\pm$ are represented in "polar coordinates" $(\Omega, \gamma)_\pm$, where the columns of $\Omega_+ \in \mathbb{C}^{n\times k}$ and $\Omega_- \in \mathbb{C}^{(n-k)\times k}$ are orthonormal bases of the subspaces spanned by the columns of $W_+ := \begin{pmatrix} W_1^+ & \cdots & W_k^+ \end{pmatrix}$ and $W_- := \begin{pmatrix} W_{k+1}^- & \cdots & W_n^- \end{pmatrix}$.

For each $\lambda \in \Lambda$, compute matrices $\Omega_\pm(\lambda)$ whose columns form orthonormal bases for $S_\pm$, by the same ordered Schur decomposition used in the computation of $\Pi_\pm$. Equating: $\Omega_+\tilde\alpha_+(\lambda) = R_+(\lambda)$, $\Omega_-\tilde\alpha_-(\lambda) = R_-(\lambda)$, for some $\tilde\alpha_\pm$, we obtain:

$$\tilde\alpha_+(\lambda) = \Omega_+^* R_+(\lambda), \quad \tilde\alpha_-(\lambda) = \Omega_-^* R_-(\lambda),$$

and therefore the exterior product of the columns of $R_\pm$ is equal to the exterior product of the columns of $\Omega_\pm$ times $\tilde\gamma_\pm(\lambda) := \det(\Omega^* R)_\pm(\lambda)$.

Imposing now the choice $\Omega^*\Omega'$ to fix a specific realization (similar to $PP' = 0$ determining Kato's ODE), we obtain after a brief calculation [HuZ2] the block-triangular system:

$$\begin{aligned} \Omega' &= (I - \Omega\Omega^*)A\Omega, \\ (\log\tilde\gamma)' &= \text{trace }(\Omega^* A\Omega) - \text{trace }(\Omega^* A\Omega)(\pm\infty), \end{aligned} \tag{6.8}$$

$\tilde\gamma := \tilde\gamma e^{-\text{trace }(\Omega^* A\Omega)(\pm\infty)x}$, for which the "angular" Ω-equation is exactly the continuous orthogonalization method of Drury [Dr, Da], and the "radial" $\tilde\gamma$-equation, given Ω, may be solved by simple quadrature. Note that for constant A, the invariant subspaces Ω of A are equilibria of the flow, and solutions of the $\tilde\gamma$-equation are constant, so that any first-order or higher numerical scheme resolves $\log\tilde\gamma$ exactly; thus, the $\tilde\gamma$-equation may in practice be solved together with the Ω-equation with good results.

The Evans function is recovered, finally, through the relation:

$$\begin{aligned} D(\lambda) &= \det(W_1^+, \cdots, W_k^+, W_{k+1}^-, \cdots, W_n^-)|_{x=0} \\ &= \tilde\gamma_+\tilde\gamma_- \det(\Omega^+, \Omega^-)|_{x=0}. \end{aligned} \tag{6.9}$$

This method performs comparably to the exterior product method for small systems [HuZ2], but does not suffer from exponential growth in complexity with respect to the system's size.

As the integration of Kato's ODE (6.8) is carried out on a bounded closed curve, standard error estimates apply and convergence is essentially automatic. In contrast to integration in x of the Evans system, integration in λ of the Kato system is a one-time cost, so not a rate-determining factor in the performance of the overall code. However, the computation time *is* sensitive to the number of mesh points in λ at which the Evans function is computed, so this should be held down as much as possible.

7. Global stability analysis: a case study. To show the full power of the Evans function approach, we now carry out a global stability analysis for the equations of isentropic gas dynamics (1.6) by a combination of energy estimates, numerical Evans function evaluation, and asymptotic ODE. The result is that for an isentropic γ-law gas, $\gamma \in [1,3]$, viscous shock profiles are (numerically) unconditionally stable.

Step (i). Preliminary estimates. Taking $(x,t,v,u,a_0) \mapsto (-\varepsilon s(x-st), \varepsilon s^2 t, v/\varepsilon, -u/(\varepsilon s), a_0\varepsilon^{-\gamma-1}s^{-2})$, with ε so that $0 < v_+ < v_- = 1$, we consider stationary solutions $(\bar v, \bar u)(x)$ of:

$$v_t + v_x - u_x = 0, \qquad u_t + u_x + (av^{-\gamma})_x = \left(\frac{u_x}{v}\right)_x \tag{7.1}$$

where $a = a_0\varepsilon^{-\gamma-1}s^{-2}$. Steady shock profiles of (7.1) satisfy:

$$v' = H(v, v_+) := v(v - 1 + a(v^{-\gamma} - 1)), \tag{7.2}$$

where a is found by $H(v_+, v_+) = 0$, yielding Rankine-Hugoniot condition:

$$a = -\frac{v_+ - 1}{v_+^{-\gamma} - 1} = v_+^\gamma \frac{1 - v_+}{1 - v_+^\gamma}. \tag{7.3}$$

Evidently, $a \to \gamma^{-1}$ in the weak shock limit $v_+ \to 1$, while $a \sim v_+^\gamma$ in the strong shock limit $v_+ \to 0$. In this scaling, large-amplitude limit corresponds to $v_+ \to 0$, or $\rho_+ = 1/v_+ \to \infty$. Linearizing (7.1) about the profile $(\bar v, \bar u)$ and integrating with respect to x, we obtain:

$$\lambda v + v' - u' = 0, \qquad \lambda u + u' - \frac{h(\bar v)}{\bar v^{\gamma+1}} v' = \frac{u''}{\bar v}, \tag{7.4}$$

where $h(\bar v) = -\bar v^{\gamma+1} + a(\gamma - 1) + (a+1)\bar v^\gamma$. Spectral stability of $\hat v$ corresponds to nonexistence of solutions of (7.4) decaying at $x = \pm\infty$ for $\Re\lambda \geq 0$ [HuZ2, BHRZ, HLZ].

PROPOSITION 7.1 ([BHRZ]). *For each $\gamma \geq 1$, $0 < v_+ \leq 1$, (7.2) has a unique (up to translation) monotone decreasing solution $\hat v$ decaying to its endstates with a uniform exponential rate. For $0 < v_+ \leq \frac{1}{12}$ and $\hat v(0) = v_+ + \frac{1}{12}$, there holds:*

$$|\bar v(x) - v_+| \leq \left(\frac{1}{12}\right) e^{-\frac{3x}{4}} \;\; \forall x \geq 0, \quad |\bar v(x) - v_-| \leq \left(\frac{1}{4}\right) e^{\frac{x+12}{2}} \;\; \forall x \leq 0. \tag{7.5}$$

Above, existence and monotonicity follow trivially by the fact that (7.2) is a scalar first-order ODE with convex righthand side. Exponential convergence as $x \to +\infty$ follows by:

$$H(v, v_+) = (v - v_+) \left(v - \left(\frac{1 - v_+}{1 - v_+^\gamma} \right) \left(\frac{1 - (\frac{v_+}{v})^\gamma}{1 - (\frac{v_+}{v})} \right) \right),$$

whence $v - \gamma \leq \frac{H(v,v_+)}{v - v_+} \leq v - (1 - v_+)$ by $1 \leq \frac{1 - x^\gamma}{1 - x} \leq \gamma$ for $0 \leq x \leq 1$.

Writing now (7.4) as $U_t + A(x)U_x = B(x)U_{xx}$, with $B = \begin{pmatrix} 0 & 0 \\ 0 & \frac{1}{\bar{v}} \end{pmatrix}$ and $A = \begin{pmatrix} 1 & -1 \\ -\frac{h(\bar{v})}{\bar{v}^{\gamma+1}} & 1 \end{pmatrix}$, we see that $S = \begin{pmatrix} 1 & 0 \\ 0 & \frac{\bar{v}^{\gamma+1}}{h(\bar{v})} \end{pmatrix}$ symmetrizes A, B. Taking the L^2 complex inner product of SU against the equations yields $\Re\lambda\langle U, SU\rangle + \langle U', SBU'\rangle = -\langle u, g(\hat{v})u\rangle$, where $g > 0$ under condition (7.6) below; see [MN, BHRZ]. Consequently:

PROPOSITION 7.2 ([MN]). *Viscous shocks of* (1.6) *are spectrally stable whenever:*

$$\left(\frac{v_+^{\gamma+1}}{a\gamma}\right)^2 + 2(\gamma - 1)\left(\frac{v_+^{\gamma+1}}{a\gamma}\right) - (\gamma - 1) \geq 0. \tag{7.6}$$

In particular, stability condition (7.6) holds for $|v_+ - 1| << 1$. Further energy estimates related to those of Proposition 7.2 yield:

PROPOSITION 7.3 ([BHRZ]). *Nonstable eigenvalues λ of* (7.4), *i.e., eigenvalues with nonnegative real part, are confined for any $\gamma \geq 1$, $0 < v_+ \leq 1$ to the region Λ defined by*

$$\Re(\lambda) + |\Im(\lambda)| \leq \left(\sqrt{\gamma} + \frac{1}{2}\right)^2. \tag{7.7}$$

Step (ii). The Evans function formulation. We express (7.4) as a first-order system: $W' = A(x, \lambda)W$, where:

$$A(x, \lambda) = \begin{pmatrix} 0 & \lambda & 1 \\ 0 & 0 & 1 \\ \lambda\bar{v} & \lambda\bar{v} & f(\bar{v}) - \lambda \end{pmatrix}, \quad W = \begin{pmatrix} u \\ v \\ v' \end{pmatrix}, \quad {}' = \frac{d}{dx}, \tag{7.8}$$

$$f(\bar{v}) = 2\bar{v} - (\gamma - 1)\left(\frac{1 - v_+}{1 - v_+^\gamma}\right)\left(\frac{v_+}{\bar{v}}\right)^\gamma - \left(\frac{1 - v_+}{1 - v_+^\gamma}\right)v_+^\gamma - 1. \tag{7.9}$$

Eigenvalues of (7.4) correspond to nontrivial solutions W for which the boundary conditions $W(\pm\infty) = 0$ are satisfied. Because $A(x, \lambda)$ as a function of $\bar{v}$ is asymptotically constant in x, the behavior of (7.8) near $x = \pm\infty$ is governed by the limiting systems: $W' = A_\pm(\lambda)W$, where $A_\pm(\lambda) = A(\pm\infty, \lambda)$. From this we readily find on the (nonstable) domain $\Re\lambda \geq 0$, $\lambda \neq 0$ that there is a one-dimensional unstable manifold $W_1^-(x)$ of solutions decaying at $x = -\infty$ and a two-dimensional stable manifold $W_2^+(x) \wedge W_3^+(x)$ of solutions decaying at $x = +\infty$, analytic in λ, with asymptotic behavior $W_j^\pm(x,\lambda) \sim e^{\mu_\pm(\lambda)x}V_j^\pm(\lambda)$ as $x \to \pm\infty$, where $\mu_\pm(\lambda)$ and $V_j^\pm(\lambda)$ are eigenvalues and analytically chosen eigenvectors of $A_\pm(\lambda)$.

A standard choice of eigenvectors $V_j^\pm$ [BrZ], uniquely specifying D (up to constant factor) is obtained by Kato's ODE, as in section 4.1, whose solution can be alternatively characterized by the property that there exist corresponding left eigenvectors $\tilde{V}_j^\pm$ such that:

$$(\tilde{V} \cdot V)^\pm \equiv const, \quad (\tilde{V} \cdot V')^\pm \equiv 0.$$

Defining the *Evans function* D associated with operator L as:

$$D(\lambda) = \det(W_1^- W_2^+ W_3^+)_{|x=0}, \tag{7.10}$$

we find [Z2] that D is analytic on $\Re\lambda \geq 0$, with eigenvalues of L corresponding in location and multiplicity to zeroes of D.

7.1. Main results. Taking a formal limit as $v_+ \to 0$ of (7.1) and recalling that $a \sim v_+^\gamma$, we get:

$$v_t + v_x - u_x = 0, \qquad u_t + u_x = \left(\frac{u_x}{v}\right)_x \tag{7.11}$$

corresponding to a *pressure-less gas*, or $\gamma = 1$. The limiting profile equation $v' = v(v-1)$ has solution $\hat{v}_0(x) = (1 - \tanh(x/2))/2$, and the limiting eigenvalue system is $W' = A^0(x,\lambda)W$:

$$\begin{aligned} A^0(x,\lambda) &= \begin{pmatrix} 0 & \lambda & 1 \\ 0 & 0 & 1 \\ \lambda\bar{v}_0 & \lambda\bar{v}_0 & f_0(\bar{v}_0) - \lambda \end{pmatrix}, \\ f_0(\bar{v}_0) &= 2\bar{v}_0 - 1 = -\tanh(x/2). \end{aligned} \tag{7.12}$$

Observe that $A_+^0(\lambda) := A^0(+\infty, \lambda)$ is nonhyperbolic for all λ, having eigenvalues $0, 0, -1-\lambda$. In particular, the stable manifold drops to dimension one in the limit $v_+ \to 0$, and so the prescription of an associated Evans function is *underdetermined.*

This difficulty is resolved by a careful boundary-layer analysis in [HLZ], determining a special "slow stable" mode $V_2^+ \pm (1,0,0)^T$ – the common limiting direction of slow stable and unstable modes as $v_+ \to 0$, collapsing to a Jordan block– augmenting the "fast stable" mode $V_3 := (a^{-1}(\lambda/a + 1), a^{-1}, 1)^T$ associated with the single stable eigenvalue $a = -1-\lambda$ of A_+^0. This determines a *limiting Evans function* $D^0(\lambda)$ by the prescription (7.10).

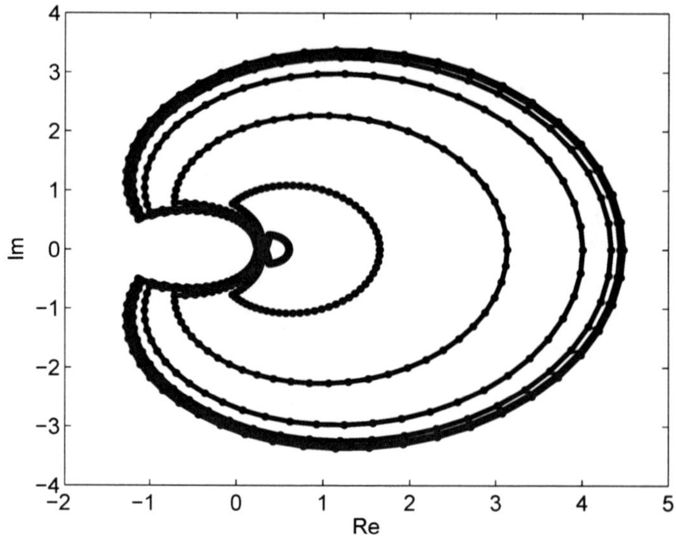

FIG. 1. *Convergence to the limiting Evans function as $v_+ \to 0$ for a monatomic gas (reproduced from [HLZ] with permission of the authors).*

THEOREM 7.1 ([HLZ]). *For λ in any compact subset of $\Re\lambda \geq 0$, $D(\lambda)$ converges uniformly to $D^0(\lambda)$ as $v_+ \to 0$. Moreover, the limiting function D^0 is nonzero on $\Re\lambda \geq 0$.*

Consequently, for any $\gamma \geq 1$, isentropic Navier–Stokes shocks are stable in the strong shock limit, i.e., for v_+ sufficiently small.

Stability in the weak shock limit is known [MN].[2] For the intermediate strong shocks, one needs to consider a bounded parameter range on which the Evans function may be effiently computed numerically [BrZ]. Specifically, we may map a semicircle $\partial\{\Re\lambda \geq 0\} \cap \{|\lambda| \leq 10\}$ enclosing Λ for $\gamma \in [1,3]$ by D^0 and compute the winding number of its image about the origin to determine the number of zeroes of D^0 within the semicircle, and thus within Λ. Such a study was carried out systematically in [BHRZ] on the parameter range $\gamma \in [1,3]$, for shocks with Mach number $M \in [1, 3,000]$, which corresponds on $\gamma \in [1, 2.5]$ to $v_+ \geq 10^{-3}$, with the result of stability for all values tested.

In Figure 1, we superimpose on the numerically computed image of a semicircle by D^0, its image by D, for a monatomic gas $\gamma \approx 1.66$ at successively higher Mach numbers $v+$ =`1e-1,1e-2,1e-3,1e-4,1e-5,1e-6`, showing convergence of D to D^0 in the strong shock limit as $v_+ \to 0$ and its convergence to a constant in the weak shock limit $v_+ \to 1$.

[2] For an extension to general hyperbolic-parabolic systems, see [HuZ1].

The displayed contours are, to the scale visible by eye, "monotone" in v_+, or nested, one within the other, with lower-Mach number contours essentially "trapped" within higher-Mach number contours, and all contours interpolating smoothly between this and the inner, constant limit. Behavior for other $\gamma \in [0, 3]$ is entirely similar [HLZ].

8. Conditional stability of viscous shock waves [Z4, Z7, Z8]. We next consider the situation of a viscous shock *with one or more strictly unstable but no neutrally unstable eigenvalues*, and seek to describe the nearby phase portrait in terms of invariant manifolds and behavior therein. Specifically, for shock waves of systems of conservation laws with artificial viscosity, we construct a center stable manifold and show that the shock is conditionally (nonlinearly) stable with respect to this codimension p set of initial data, where p is the number of unstable eigenvalues. Note that this includes a rigorous *nonlinear instability* result in the case of unstable eigenvalues. Such conditionally stable shock waves play an important role in asymptotic behavior as metastable states [AMPZ, GZ].

For simplicity of exposition, we treat the easier case of a Lax shock of a semilinear, identity viscosity system [Z7]. For the general case, including in particular the class of equations described in Section 5, see [Z8, Z9].

Consider a viscous shock $u(x,t) = \bar{u}(x)$, $\lim_{z\to\pm\infty} \bar{u}(z) = u_\pm$ of:

$$u_t + f(u)_x = u_{xx}, \qquad u,\, f \in \mathbb{R}^n, \quad x\,, t \in \mathbb{R}, \tag{8.1}$$

under the basic assumptions:

(H0) $f \in C^{k+2}$, $k \geq 2$.

(H1) $A_\pm := df(u_\pm)$ have simple, real, nonzero eigenvalues.

Linearizing (8.1) about $\bar{u}$ yields equations with generator L [He, Sa, ZH]:

$$u_t = Lu := -(df(\bar{u})u)_x + u_{xx}, \tag{8.2}$$

possessing both a translational zero-eigenvalue and essential spectrum tangent at zero to the imaginary axis. The absence of a spectral gap between neutral (zero real part) and stable (negative real part) spectra of L prevents the usual ODE-type decomposition of the flow near $\bar{u}$ into invariant stable, center, and unstable manifolds. Our first result asserts that we can still determine a center stable manifold, and that this may be chosen to respect the underlying translation-invariance of (8.1). We omit the proof of this standard observation; for a particularly short treatment, see [Z7].

THEOREM 8.1 ([Z7]). *Under assumptions (H0)–(H1), there exists, in an H^2 neighborhood of the set of translates of $\bar{u}$, a codimension-p translation invariant C^k (with respect to H^2) center stable manifold $\mathcal{M}_{cs}$, tangent at $\bar{u}$ to the center stable subspace Σ_{cs} of L. It is (locally) invariant under the forward time-evolution of* (8.1) *and contains all solutions that remain bounded and sufficiently close to a translate of $\bar{u}$ in forward time, where p is the (necessarily finite) number of unstable eigenvalues of L.*

We add to (H0)–(H1) the hypothesis that $\bar{u}$ be a *Lax-type shock*:

(H2) The dimensions of the unstable subspace of $df(u_-)$ and the stable subspace of $df(u_+)$ sum to $n+1$.

We assume further the following *spectral genericity* conditions.

(D1) L has no nonzero imaginary eigenvalues.

(D2) The orbit $\bar{u}(\cdot)$ is a transversal connection of standing wave equation $\bar{u}_x = f(\bar{u}) - f(u_-)$.

(D3) The associated inviscid shock (u_-, u_+) is hyperbolically stable, i.e.,

$$\det(r_1^-, \dots, r_{p-1}^-, r_{p+1}^+, \dots, r_n^+, (u_+ - u_-)) \neq 0, \tag{8.3}$$

where $r_1^-, \dots r_{p-1}^-$ denote eigenvectors of $df(u_-)$ associated with negative eigenvalues and $r_{p+1}^+, \dots r_n^+$ denote eigenvectors of $df(u_+)$ associated with positive eigenvalues.

As discussed in [ZH, MaZ1], (D2)–(D3) correspond in the absence of a spectral gap to a generalized notion of simplicity of the embedded eigenvalue $\lambda = 0$ of L. Thus, (D1)–(D3) together yield that there are no additional (usual or generalized) eigenvalues on the imaginary axis other than the translational eigenvalue at $\lambda = 0$. Hence the shock is not in transition between different degrees of stability, but has stability properties that are insensitive to small variations in parameters.

With these assumptions, we obtain our second main result characterizing stability properties of $\bar{u}$. In the case $p = 0$, this reduces to the nonlinear orbital stability result established in [ZH, MaZ1, MaZ2, MaZ3, Z2, Z4].

THEOREM 8.2 ([Z7]). *Under (H0)–(H2) and (D1)–(D3), $\bar{u}$ is nonlinearly orbitally stable for all its sufficiently small perturbations in $L^1 \cap H^2$ lying on $\mathcal{M}_{cs}$ and its translates. That is:*

$$\begin{aligned}
|u(x,t) - \bar{u}(x - \alpha(t))|_{L^p} &\leq C(1+t)^{-\frac{1}{2}(1-\frac{1}{p})}|u(x,0) - \bar{u}(x)|_{L^1 \cap H^2},\\
|u(x,t) - \bar{u}(x - \alpha(t))|_{H^2} &\leq C(1+t)^{-\frac{1}{4}}|u(x,0) - \bar{u}(x)|_{L^1 \cap H^2},\\
\dot{\alpha}(t) &\leq C(1+t)^{-\frac{1}{2}}|u(x,0) - \bar{u}(x)|_{L^1 \cap H^2},\\
\alpha(t) &\leq C|u(x,0) - \bar{u}(x)|_{L^1 \cap H^2},
\end{aligned} \tag{8.4}$$

for some $\alpha(\cdot)$, and all L^p. Moreover, $\bar{u}$ is orbitally unstable with respect to small H^2 perturbations not lying in $\mathcal{M}_{cs}$, in the sense that the corresponding solution leaves a fixed-radius neighborhood of the set of translates of $\bar{u}$ in finite time.

8.1. Conditional stability analysis. Define the perturbation variable $v(x,t) := u(x + \alpha(t), t) - \bar{u}(x)$ for u a solution of (8.1), where α is to be specified later in a way appropriate for the task at hand. Subtracting the equations for $u(x + \alpha(t), t)$ and $\bar{u}(x)$, we obtain:

$$v_t - Lv = N(v)_x + \partial_t \alpha(\phi + \partial_x v), \tag{8.5}$$

where $L := -\partial_x df(\bar{u}) + \partial_x^2$ as in (8.2) denotes the linearized operator about $\bar{u}$, $\phi = \bar{u}_x$, and:

$$N(v) := -(f(\bar{u}+v) - f(\bar{u}) - df(\bar{u})v). \tag{8.6}$$

So long as $|v|_{H^1}$ (hence $|v|_{L^\infty}$ and $|u|_{L^\infty}$) remains bounded, there holds:

$$N(v)=O(|v|^2),\ \ \partial_x N(v)=O(|v||\partial_x v|),\ \ \partial_x^2 N(v)=O(|\partial_x|^2+|v||\partial_x^2 v|). \tag{8.7}$$

LEMMA 8.1 ([Z7]). *Let Π_u denote the eigenprojection of L onto its unstable subspace Σ_u, and $\Pi_{cs} = \mathrm{Id} - \Pi_u$ the eigenprojection onto its center stable subspace Σ_{cs}. Assuming (H0)–(H1), there exists $\tilde{\Pi}_j$ such that $\Pi_j \partial_x = \partial_x \tilde{\Pi}_j$ for $j = u$, cs and, for all $1 \le p \le \infty$, $0 \le r \le 4$:*

$$|\Pi_u|_{L^p \to W^{r,p}}, |\tilde{\Pi}_u|_{L^p \to W^{r,p}} \le C, \quad |\tilde{\Pi}_{cs}|_{W^{r,p} \to W^{r,p}}, |\tilde{\Pi}_{cs}|_{W^{r,p} \to W^{r,p}} \le C.$$

Let now $G_{cs}(x,t;y) := \Pi_{cs} e^{Lt} \delta_y(x)$ denote the Green kernel of the linearized solution operator on the center stable subspace Σ_{cs}. The following is a consequence of the detailed pointwise bounds established in [TZ2, MaZ1].

THEOREM 8.3 ([TZ2, MaZ1]). *Assuming (H0)–(H2), (D1)–D(3), the center stable Green function may be decomposed as $G_{cs} = E + \tilde{G}$, where*

$$E(x,t;y) = \partial_x \bar{u}(x) e(y,t), \tag{8.8}$$

$$e(y,t) = \sum_{a_k^- > 0} \left(errfn \left(\frac{y + a_k^- t}{\sqrt{4(t+1)}} \right) - errfn \left(\frac{y - a_k^- t}{\sqrt{4(t+1)}} \right) \right) l_k^-(y) \tag{8.9}$$

for $y \le 0$ and symmetrically for $y \ge 0$, $l_k^- \in \mathbb{R}^n$ constant. Above, $a_j^\pm$ are the eigenvalues of $df(u_\pm)$, $x^\pm$ denotes the positive/negative part of x. Moreover:

$$\left| \int_{-\infty}^{+\infty} \partial_x^s \tilde{G}(\cdot,t;y) f(y) dy \right|_{L^p} \le C(1 + t^{-\frac{s}{2}}) t^{-\frac{1}{2}(\frac{1}{q} - \frac{1}{p})} |f|_{L^q}, \tag{8.10}$$

$$\left| \int_{-\infty}^{+\infty} \partial_x^s \tilde{G}_y(\cdot,t;y) f(y) dy \right|_{L^p} \le C(1 + t^{-\frac{s}{2}}) t^{-\frac{1}{2}(\frac{1}{q} - \frac{1}{p}) - \frac{1}{2}} |f|_{L^q}, \tag{8.11}$$

for all $t \ge 0$, $0 \le s \le 2$, some $C > 0$, for any $1 \le q \le p$ (equivalently, $1 \le r \le p$) and $f \in L^q$, where $1/r + 1/q = 1 + 1/p$. The kernel e satisfies:

$$|e_y(\cdot,t)|_{L^p}, |e_t(\cdot,t)|_{L^p} \le Ct^{-\frac{1}{2}(1-1/p)}, \qquad |e_{ty}(\cdot,t)|_{L^p} \le Ct^{-\frac{1}{2}(1-1/p)-1/2},$$

for all $t > 0$. Moreover, for $y \le 0$ we have the pointwise bounds:

$$|e_y(y,t)|, |e_t(y,t)| \le C(t+1)^{-\frac{1}{2}} \sum_{a_k^- > 0} \left(e^{-\frac{(y+a_k^- t)^2}{M(t+1)}} + e^{-\frac{(y-a_k^- t)^2}{M(t+1)}} \right),$$

$$|e_{ty}(y,t)| \le C(t+1)^{-1} \sum_{a_k^->0} \left(e^{-\frac{(y+a_k^- t)^2}{M(t+1)}} + e^{-\frac{(y-a_k^- t)^2}{(t+1)t}}\right),$$

for $M>0$ sufficiently large, and symmetrically for $y \ge 0$.

8.2. Reduced equations and shock location. Recalling that $\partial_x \bar u$ is a stationary solution of the equations $u_t = Lu$, so that $L\partial_x \bar u = 0$, or:

$$\int_{-\infty}^{\infty} G(x,t;y)\bar u_x(y)dy = e^{Lt}\bar u_x(x) = \partial_x \bar u(x),$$

we have, applying Duhamel's principle to (8.5):

$$\begin{aligned} v(x,t) = &\int_{-\infty}^{\infty} G(x,t;y)v_0(y)\,dy \\ &- \int_0^t \int_{-\infty}^{\infty} G_y(x,t-s;y)(N(v)+\dot\alpha v)(y,s)\,dy\,ds + \alpha(t)\partial_x \bar u(x). \end{aligned}$$

Following [ZH, Z4, MaZ2, MaZ3], define (with e as in (8.9)):

$$\begin{aligned} \alpha(t) = \quad & -\int_{-\infty}^{\infty} e(y,t)v_0(y)\,dy \\ & + \int_0^t \int_{-\infty}^{+\infty} e_y(y,t-s)(N(v)+\dot\alpha\, v)(y,s)dyds. \end{aligned} \tag{8.12}$$

Recalling that $G = E + G_u + \tilde G$, we obtain the *reduced equations*:

$$\begin{aligned} v(x,t) = &\int_{-\infty}^{\infty} (G_u+\tilde G)(x,t;y)v_0(y)\,dy \\ &- \int_0^t \int_{-\infty}^{\infty} (G_u+\tilde G)_y(x,t-s;y)(N(v)+\dot\alpha v)(y,s)dy\,ds. \end{aligned} \tag{8.13}$$

Differentiating (8.12) in t, and noting that $e_y(y,s)\to 0$ as $s\to 0$, we get:

$$\begin{aligned} \dot\alpha(t) = \quad & -\int_{-\infty}^{\infty} e_t(y,t)v_0(y)\,dy \\ & + \int_0^t \int_{-\infty}^{+\infty} e_{yt}(y,t-s)(N(v)+\dot\alpha v)(y,s)\,dy\,ds. \end{aligned} \tag{8.14}$$

As discussed in [Go, Z4, MaZ2, MaZ3, GMWZ1, BeSZ], α may be considered as defining a notion of approximate shock location.

We further obtain the following nonlinear damping estimate:

PROPOSITION 8.1 ([MaZ3]). *Assume (H0)-(H3), let $v_0 \in H^2$, and suppose that for $0 \le t \le T$, the H^2 norm of a perturbation v remains bounded by a sufficiently small constant, for u a solution of (8.1). Then, for some constants $\theta_{1,2}>0$, for all $0\le t\le T$:*

$$\|v(t)\|_{H^2}^2 \le Ce^{-\theta_1 t}\|v(0)\|_{H^2}^2 + C\int_0^t e^{-\theta_2(t-s)}(|v|_{L^2}^2 + |\dot\alpha|^2)(s)\,ds. \tag{8.15}$$

Proof. ([Z7]) Subtracting the equations for $u(x+\alpha(t),t)$ and $\bar u(x)$, we may write the perturbation equation for v alternatively as:

$$v_t + \left(\int_0^1 df\left(\bar u(x) + \tau v(x,t)\right) d\tau\, v \right)_x - v_{xx} = \dot\alpha(t)\partial_x \bar u(x) + \dot\alpha(t)\partial_x v. \quad (8.16)$$

Observing that $\partial_x^j(\partial_x \bar u)(x) = O(e^{-\eta|x|})$ is bounded in L^1 norm for $j \le 2$, we take the L^2 product in x of $\sum_{j=0}^2 \partial_x^{2j} v$ against (8.16), integrate by parts and arrive at:

$$\partial_t \|v\|_{H^2}^2(t) \le -\theta \|\partial_x^3 v\|_{L^2}^2 + C\left(\|v\|_{H^2}^2 + |\dot\alpha(t)|^2\right),$$

for $\theta > 0$, for $C > 0$ sufficiently large, and so long as $\|v\|_{H^2}$ remains bounded. Using the Sobolev interpolation inequality

$$\|v\|_{H^2}^2 \le \tilde C^{-1} \|\partial_x^3 v\|_{L^2}^2 + \tilde C \|v\|_{L^2}^2$$

for $\tilde C > 0$ sufficiently large, we obtain:

$$\partial_t \|v\|_{H^2}^2(t) \le -\tilde\theta \|v\|_{H^2}^2 + C\left(\|v\|_{L^2}^2 + |\dot\alpha(t)|^2\right),$$

from which (8.15) follows by Gronwall's inequality. □

8.3. Proof of Theorem 8.2 ([Z7]]. Decompose now the nonlinear perturbation v as: $v(x,t) = w(x,t) + z(x,t)$, where $w = \Pi_{cs} v$, $z = \Pi_u v$. Applying Π_{cs} to (8.13) and by Lemma 8.1:

$$\begin{aligned} w(x,t) &= \int_{-\infty}^{\infty} \tilde G(x,t;y) w_0(y)\, dy \\ &\quad - \int_0^t \int_{-\infty}^{\infty} \tilde G_y(x,t-s;y) \tilde\Pi_{cs}(N(v) + \dot\alpha v)(y,s) dy\, ds \end{aligned} \quad (8.17)$$

for the flow along the center stable manifold, parametrized by $w \in \Sigma_{cs}$. On the other hand:

LEMMA 8.2 ([Z7]). *Assuming (H0)–(H1), for v lying initially on the center stable manifold $\mathcal{M}_{cs}$:*

$$|z|_{W^{r,p}} \le C|w|_{H^2}^2 \quad \textit{for some } C > 0 \quad \forall 1 \le p \le \infty,\ 0 \le r \le 4, \quad (8.18)$$

so long as $|w|_{H^2}$ remains sufficiently small.

Indeed, by tangency of the center stable manifold to Σ_{cs}, we have immediately $|z|_{H^2} \le C|w|_{H^2}^2$, whreas (8.18) follows by equivalence of norms for finite-dimensional vector spaces, applied to the p-dimensional space Σ_u.

Recalling by Theorem 8.1 that solutions remaining for all time in a sufficiently small radius neighborhood $\mathcal{N}$ of the set of translates of $\bar u$ lie in the center stable manifold $\mathcal{M}_{cs}$, we obtain that solutions not originating in $\mathcal{M}_{cs}$ must exit $\mathcal{N}$ in finite time, verifying the final assertion of orbital instability with respect to perturbations not in $\mathcal{M}_{cs}$.

Consider now a solution $v \in \mathcal{M}_{cs}$, or, equivalently, a solution $w \in \Sigma_{cs}$ of (8.17) with $z = \Phi_{cs}(w) \in \Sigma_u$. Define:

$$\zeta(t) := \sup_{0\le s\le t} \Big(|w|_{H^2}(1+s)^{\frac14} + (|w|_{L^\infty} + |\dot\alpha(s)|)(1+s)^{\frac12} \Big). \tag{8.19}$$

We shall establish that for all $t \ge 0$ for which a solution exists with ζ uniformly bounded by some fixed, sufficiently small constant, there holds:

$$\zeta(t) \le C_2(E_0 + \zeta(t)^2) \quad \text{for} \quad E_0 := |v_0|_{L^1\cap H^2}. \tag{8.20}$$

From this result, provided $E_0 < 1/4C_2^2$, we infer that $\zeta(t) \le 2C_2E_0$ implies $\zeta(t) < 2C_2E_0$, and so we may conclude by continuous induction that $\zeta(t) < 2C_2E_0$for all $t \ge 0$, from which we readily obtain the stated bounds by continuous induction.

It remains to prove the claim (8.20). By Lemma 8.1, $|w_0|_{L^1\cap H^2} = |\Pi_{cs}v_0|_{L^1\cap H^2} \le CE_0$. Likewise, by Lemma 8.2, (8.19), (8.7), and Lemma 8.1, for $0 \le s \le t$:

$$|\tilde\Pi_{cs}(N(v) + \dot\alpha v)(y,s)|_{L^2} \le C\zeta(t)^2(1+s)^{-\frac34}. \tag{8.21}$$

Combining the latter bounds with (8.17) and (8.14) and applying Theorem 8.3, we obtain:

$$\begin{aligned} |w(x,t)|_{L^p} &\le \Big| \int_{-\infty}^{\infty} \tilde G(x,t;y) w_0(y)\, dy \Big|_{L^p} \\ &\quad + \Big| \int_0^t \int_{-\infty}^{\infty} \tilde G_y(x,t-s;y) \tilde\Pi_{cs}(N(v) + \dot\alpha v)(y,s) dy\, ds \Big|_{L^p} \\ &\le E_0(1+t)^{-\frac12(1-\frac1p)} + C\zeta(t)^2 \int_0^t (t-s)^{-\frac34+\frac{1}{2p}}(1+s)^{-\frac34} dy\, ds \\ &\le C(E_0 + \zeta(t)^2)(1+t)^{-\frac12(1-\frac1p)}, \end{aligned} \tag{8.22}$$

and, similarly, using Hölder's inequality and applying further Theorem 8.3:

$$\begin{aligned} |\dot\alpha(t)| &\le \int_{-\infty}^{\infty} |e_t(y,t)||v_0(y)|\, dy \\ &\quad + \int_0^t \int_{-\infty}^{+\infty} |e_{yt}(y,t-s)||(N(v) + \dot\alpha v)(y,s)|\, dy\, ds \\ &\le |e_t|_{L^\infty}|v_0|_{L^1} + C\zeta(t)^2 \int_0^t |e_{yt}|_{L^2}(t-s)|(N(v)+\dot\alpha v)|_{L^2}(s) ds \\ &\le E_0(1+t)^{-\frac12} + C\zeta(t)^2 \int_0^t (t-s)^{-\frac34}(1+s)^{-\frac34} ds \\ &\le C(E_0 + \zeta(t)^2)(1+t)^{-\frac12}. \end{aligned} \tag{8.23}$$

By Lemma 8.2: $|z|_{H^2}(t) \le C|w|^2_{H^2}(t) \le C\zeta(t)^2$, so in particular, $|z|_{L^2}(t) \le C\zeta(t)^2(1+t)^{-\frac12}$. Applying Proposition 8.1, (8.22) and (8.23), we obtain:

$$|w|_{H^2}(t) \le C(E_0 + \zeta(t)^2)(1+t)^{-\frac14}. \tag{8.24}$$

Combining (8.22), (8.23), and (8.24), we obtain (8.20) as claimed. Finally, a computation parallel to (8.23) (see [MaZ3, Z2]) yields $|\alpha(t)| \leq C(E_0 + \zeta(t)^2)$, from which the last remaining bound on $|\alpha(t)|$ follows.

9. Cellular instability for flow in a duct [Z9]. We conclude by briefly discussing multidimensional stability and bifurcation of flow in a finite cross-sectional duct, or "shock tube", extending infinitely in the axial direction [Z9]. It is well known both experimentally and numerically [BE, MT, BMR, FW, MT, ALT, AT, F1, F2, KS] that shock and detonation waves propagating in a finite cross-section duct can exhibit time-oscillatory or "cellular" instabilities, in which the initially nearly planar shock takes on nontrivial transverse geometry.

In this section, following [Z9], we combine the analyses of [BSZ, TZ4, Z4] to make an explicit connection between stability of planar shocks on the whole space, and Hopf bifurcation in a finite cross-section duct. We point out that violation of the *refined stability condition* of [ZS, Z1, BSZ], a viscous correction of the inviscid planar stability condition of Majda [M1]–[M4], is generically associated with Hopf bifurcation corresponding to the observed cellular instability, for cross-section M sufficiently large. Indeed, we show more, that this is associated with a cascade of bifurcations to higher and higher wave numbers and more and more complicated solutions, with features on finer and finer length/time scales.

Consider a planar viscous shock solution:

$$u(x,t) = \bar{u}(x_1 - st) \tag{9.1}$$

of a two-dimensional system of viscous conservation laws:

$$u_t + \sum f^j(u)_{x_j} = \Delta_x u, \qquad u \in \mathbb{R}^n,\ x \in \mathbb{R}^2,\ t \in \mathbb{R}^+. \tag{9.2}$$

on the whole space. This may be viewed alternatively as a planar traveling-wave solution on an infinite channel $\mathcal{C} := \{x : (x_1, x_2) \in \mathbb{R}^1 \times [-M, M]\}$, under periodic boundary conditions:

$$u(x_1, M) = u(x_1, -M). \tag{9.3}$$

We take this as a simplified mathematical model for compressible flow in a duct, in which we have neglected boundary-layer phenomena along the wall $\partial\Omega$ in order to isolate the oscillatory phenomena of our main interest. Following [TZ2], consider a one-parameter family of standing planar viscous shock solutions $\bar{u}^\varepsilon(x_1)$ of a smooth family od PDEs:

$$u_t = \mathcal{F}(\varepsilon, u) := \Delta_x u - \sum_{j=1}^{2} F^j(\varepsilon, u)_{x_j}, \qquad u \in \mathbb{R}^n \tag{9.4}$$

in the fixed channel $\mathcal{C}$, with periodic boundary conditions (typically shifts $\sum F^j(\varepsilon, u)_{x_j} := \sum f^j(u)_{x_j} - s(\varepsilon) u_{x_1}$ of a single equation (9.2) written in coordinates $x_1 \mapsto x_1 - s(\varepsilon)t$ moving with traveling-wave solutions of varying speeds $s(\varepsilon)$), with linearized operators $L(\varepsilon) := \partial\mathcal{F}/\partial u|_{u=\bar{u}^\varepsilon}$.

Profiles $\bar{u}^\varepsilon$ satisfy the standing-wave ODE:

$$u' = F^1(\varepsilon, u) - F^1(\varepsilon, u_-). \tag{9.5}$$

Let $A^1_\pm(\varepsilon) := \lim_{z\to\pm\infty} F^1_u(\varepsilon, \bar{u}^\varepsilon)$. Following [Z1, TZ2, Z4], we assume:

(H0) $F^j \in C^k$, $k \geq 2$.

(H1) $\sigma(A^1_\pm(\varepsilon))$ are real, distinct, nonzero, and $\sigma(\sum \xi_j A^j_\pm(\varepsilon))$ real and semisimple for $\xi \in \mathbb{R}^d$.

For most of our results, we require also:

(H2) Considered as connecting orbits of (9.5), $\bar{u}^\varepsilon$ are transverse and unique up to translation, with dimensions of stable subpace $S(A^1_+)$ and unstable subspace $U(A^1_-)$ summing, for each ε, to $n+1$.

(H3) $\det(r_1^-, \dots, r_c^-, r_{c+1}^+, \dots, r_n^+, u_+ - u_-) \neq 0$, where $r_1^-, \dots, r_c^-$ are eigenvectors of A^1_- with negative eigenvalues and $r_{c+1}^-, \dots, r_n^-$ are eigenvectors of A^1_+ with positive eigenvalues.

Hypothesis (H2) asserts in particular that $\bar{u}^\varepsilon$ is of standard *Lax type*, meaning that the axial hyperbolic convection matrices $A^1_+(\varepsilon)$ and $A^1_-(\varepsilon)$ at plus and minus spatial infinity have, respectively, $n-c$ positive and $c-1$ negative real eigenvalues for $1 \leq c \leq n$, where c is the characteristic family associated with the shock. Hypothesis (H3) is the *Liu–Majda condition* corresponding to one-dimensional stability of the associated inviscid shock. In the present, viscous, context, this, together with transverality, (H2), plays the role of a spectral nondegeneracy condition corresponding in a generalized sense [ZH, Z1] to simplicity of the embedded zero eigenvalue associated with eigenfunction $\partial_{x_1}\bar{u}$ and translational invariance.

9.1. Stability and bifurcation conditions. Our first set of results characterizes stability/instability of waves $\bar{u}^\varepsilon$ in terms of the spectrum of the linearized operator $L(\varepsilon)$. Fixing ε, we suppress the parameter ε. We start with the routine observation that the semilinear parabolic equation (9.2) has a center-stable manifold about the equilibrium solution $\bar{u}$.

PROPOSITION 9.1 ([Z9]). *Under assumptions (H0)–(H1), there exists in an H^2 neighborhood of the set of translates of $\bar{u}$ a codimension-p translation invariant C^k (with respect to H^2) center stable manifold $\mathcal{M}_{cs}$, tangent at $\bar{u}$ to the center stable subspace Σ_{cs} of L, that is (locally) invariant under the forward time-evolution of (9.2)–(9.3) and contains all solutions that remain bounded and sufficiently close to a translate of $\bar{u}$ in forward time, where p is the (necessarily finite) number of unstable, i.e., positive real part, eigenvalues of L.*

Introduce now the *nonbifurcation condition*:

(D1) L has no nonzero imaginary eigenvalues.

As discussed above, (H2)–(H3) correspond to a generalized notion of simplicity of the embedded eigenvalue $\lambda = 0$ of L. Thus, (D1) together with (H2)–(H3) correspond to the assumption that there are no additional (usual or generalized) eigenvalues on the imaginary axis other than the translational eigenvalue at $\lambda = 0$.

THEOREM 9.1 ([Z9]). *Under (H0)–(H3) and (D1), $\bar{u}$ is nonlinearly orbitally stable as a solution of* (9.2)–(9.3) *under sufficiently small perturbations of $\bar{u}$ in $L^1 \cap H^2$ lying on $\mathcal{M}_{cs}$ and its translates, in the sense that, for some $\alpha(\cdot)$, and all L^p:*

$$\begin{aligned} |u(x,t) - \bar{u}(x - \alpha(t))|_{L^p} &\le C(1+t)^{-\frac{1}{2}(1-\frac{1}{p})}|u(x,0) - \bar{u}(x)|_{L^1 \cap H^2}, \\ |u(x,t) - \bar{u}(x - \alpha(t))|_{H^2} &\le C(1+t)^{-\frac{1}{4}}|u(x,0) - \bar{u}(x)|_{L^1 \cap H^2}, \\ \dot{\alpha}(t) &\le C(1+t)^{-\frac{1}{2}}|u(x,0) - \bar{u}(x)|_{L^1 \cap H^2}, \\ \alpha(t) &\le C|u(x,0) - \bar{u}(x)|_{L^1 \cap H^2} . \end{aligned} \tag{9.6}$$

Moreover, $\bar{u}$ is orbitally unstable with respect to small H^2 perturbations not lying in $\mathcal{M}_{cs}$, in the sense that the corresponding solution leaves a fixed-radius neighborhood of the set of translates of $\bar{u}$ in finite time.

Proof. Observing that transverse modes $\xi \neq 0$ due to finite tranvserse cross-section are exponentially damped, reduces the analysis essentially to the one-dimensional case $\xi = 0$ treated in Theorem 8.2; see [Z9]. □

Define the *Hopf bifurcation condition*:

(D2) Outside the essential spectrum of $L(\varepsilon)$, for ε and $\delta > 0$ sufficiently small, the only eigenvalues of $L(\varepsilon)$ with real part of absolute value less than δ are a crossing conjugate pair $\lambda_{\pm}(\varepsilon) := \gamma(\varepsilon) \pm i\tau(\varepsilon)$ of $L(\varepsilon)$, with $\gamma(0) = 0$, $\partial_\varepsilon \gamma(0) > 0$, and $\tau(0) \neq 0$.

PROPOSITION 9.2 ([TZ2, TZ3]). *Let $\bar{u}^\varepsilon$,* (9.4) *be a family of traveling-waves and systems satisfying assumptions (H0)–(H3) and (D2), and $\eta > 0$ sufficiently small. Then, for $a \ge 0$ sufficiently small and $C > 0$ sufficiently large, there are C^1 functions $\varepsilon(a)$, $\varepsilon(0) = 0$, and $T^*(a)$, $T^*(0) = 2\pi/\tau(0)$, and a C^1 family of solutions $u^a(x_1,t)$ of* (9.4) *with $\varepsilon = \varepsilon(a)$, time-periodic of period $T^*(a)$, such that:*

$$C^{-1}a \le \sup_{x_1 \in \mathbb{R}} e^{\eta|x_1|} \left|u^a(x,t) - \bar{u}^{\varepsilon(a)}(x_1)\right| \le Ca \qquad \text{for all } t \ge 0. \tag{9.7}$$

Up to fixed translations in x, t, for ε sufficiently small, these are the only nearby solutions as measured in norm $\|f\|_{X_1} := \|(1+|x_1|)f(x)\|_{L^\infty(x)}$ that are time-periodic with period $T \in [T_0, T_1]$, for any fixed $0 < T_0 < T_1 < +\infty$. Indeed, they are the only nearby solutions of form $u^a(x,t) = \mathbf{u}^a(x - \sigma^a t, t)$ with $\mathbf{u}^a$ periodic in its second argument.

Proof. Nonstandard Lyapunov–Schmidt reduction making use of nonlinear cancellation estimates similar to those appearing in the stability theory [TZ2]. Exponential damping of transverse modes $\xi \neq 0$ again reduces the analysis essentially to that of the one-dimensional case $\xi = 0$. □

Together with Theorem 9.1, Proposition 9.2 implies that, under the Hopf bifurcation assumption (D2) together with the further assumption that $L(\varepsilon)$ have no strictly positive real part eigenvalues other than possibly $\lambda_\pm$, waves $\bar{u}^\varepsilon$ are linearly and nonlinearly *stable* for $\varepsilon < 0$ and *unstable* for $\varepsilon > 0$, with bifurcation/exchange of stability at $\varepsilon = 0$.

9.2. Longitudinal vs. transverse bifurcation. The analysis of [TZ2] in fact gives slightly more information. Denote by:

$$\Pi^\varepsilon f := \sum_{j=\pm} \phi_j^\varepsilon(x)\langle\tilde{\phi}_j^\varepsilon, f\rangle \tag{9.8}$$

the $L(\varepsilon)$-invariant projection onto oscillatory eigenspace $\Sigma^\varepsilon := \mathrm{Span}\{\phi_\pm^\varepsilon\}$, where $\phi_\pm^\varepsilon$ are the eigenfunctions associated with $\lambda_\pm(\varepsilon)$. Then, we have the following result:

PROPOSITION 9.3 ([TZ2]). *Under the assumptions of Proposition 9.2:*

$$\sup_{x_1} e^{\eta|x_1|}|u^a - \bar{u} - \Pi^\varepsilon(u^a - \bar{u})| \le Ca^2 \qquad \text{for all } t \ge 0. \tag{9.9}$$

Bounds (9.7) and (9.9) together yield the standard finite-dimensional property that bifurcating solutions lie to quadratic order in the direction of the oscillatory eigenspace of $L(\varepsilon)$. From this, we may draw the following additional conclusions about the structure of bifurcating waves. By separation of variables, and x_2-independence of the coefficients of $L(\varepsilon)$, we have that the eigenfunctions ψ of $L(\varepsilon)$ decompose into families: $e^{i\xi x_2}\psi(x_1)$, $\xi = \frac{\pi k}{M}$, associated with different integers k, where M is cross-sectional width. Thus, there are two very different cases: (i) (*longitudinal instability*) the bifurcating eigenvalues $\lambda_\pm(\varepsilon)$ are associated with wave-number $k = 0$, or (ii) (*transverse instability*) the bifurcating eigenvalues $\lambda_\pm(\varepsilon)$ are associated with wave-numbers $\pm k \neq 0$.

COROLLARY 9.1 ([Z9]). *Under the assumptions of Proposition 9.2, u^a depend nontrivially on x_2 iff the bifurcating eigenvalues $\lambda_\pm$ are associated with transverse wave-numbers $\pm k \neq 0$.*

Bifurcation through longitudinal instability corresponds to "galloping" or "pulsating" instabilities described in detonation literature, while symmetry-breaking bifurcation through transverse instability corresponds to "cellular" instabilities introducing nontrivial transverse geometry to the structure of the propagating wave.

9.3. Review of multidimensional stability conditions. Inviscid stability analysis for shocks centers about the *Lopatinski determinant*:

$$\Delta(\tilde{\xi},\lambda) := \begin{pmatrix} \mathcal{R}_1^- & \cdots & \mathcal{R}_{p-1}^- & \mathcal{R}_{p+1}^+ & \cdots & \mathcal{R}_n^+ & \lambda[u] + i\tilde{\xi}[f^2] \end{pmatrix}, \tag{9.10}$$

$\tilde{\xi} \in \mathbb{R}^1$, $\lambda = \gamma + i\tau \in \mathbb{C}$, $\tau > 0$, a spectral determinant whose zeroes correspond to normal modes $e^{\lambda t}e^{i\tilde{\xi}x_2}w(x_1)$ of the constant-coefficient linearized equations about the discontinuous shock solution. Here, $\{\mathcal{R}_{p+1}^+, \dots, \mathcal{R}_n^+\}$ and $\{\mathcal{R}_1^-, \dots, \mathcal{R}_{p-1}^-\}$ denote bases for the unstable/resp. stable subspaces of: $\mathcal{A}_+(\tilde\xi, \lambda) := (\lambda I + i\tilde\xi df^2(u_\pm))(df^1(u_\pm))^{-1}$.

Weak stability $|\Delta| > 0$ for $\tau > 0$ is clearly necessary for linearized stability, while strong, or uniform stability, $|\Delta|/|(\tilde\xi, \lambda)| \geq c_0 > 0$, is sufficient for nonlinear stability. Between strong instability, or failure of weak stability, and strong stability, there lies a region of neutral stability corresponding to the appearance of surface waves propagating along the shock front, for which Δ is nonvanishing for $\Re\lambda > 0$ but has one or more roots $(\tilde\xi_0, \lambda_0)$ with $\lambda_0 = \tau_0$ pure imaginary. This region of neutral inviscid stability typically occupies an open set in physical parameter space [M1, M2, M3, BRSZ, Z1, Z2]. For details, see, e.g., [Er1, M1, M2, M3, Me, Se1, Se2, Se3, ZS, Z1, Z2, Z3, BRSZ], and references therein.

Viscous stability analysis for shocks in the whole space centers about the *Evans function* $D(\tilde\xi, \lambda)$, $\tilde\xi \in \mathbb{R}^1$, $\lambda = \gamma + i\tau \in \mathbb{C}$, $\tau > 0$, a spectral determinant analogous to the Lopatinski determinant of the inviscid theory, whose zeroes correspond to normal modes $e^{\lambda t} e^{i\tilde\xi x_2} w(x_1)$, of the linearized equations about $\bar u$ (now variable-coefficient), or spectra of the linearized operator about the wave. The main result of [ZS], establishing a rigorous relation between viscous and inviscid stability, was the expansion:

$$D(\tilde\xi, \lambda) = \gamma\Delta(\tilde\xi, \lambda) + o(|(\tilde\xi, \lambda)|) \tag{9.11}$$

of D about the origin $(\tilde\xi, \lambda) = (0, 0)$, where γ is a constant measuring tranversality of $\bar u$ as a connecting orbit of the traveling-wave ODE. Equivalently, considering $D(\tilde\xi, \lambda) = D(\rho\tilde\xi_0, \rho\lambda_0)$ as a function of polar coordinates $(\rho, \tilde\xi_0, \lambda_0)$, we have:

$$D|_{\rho=0} = 0 \text{ and } (\partial/\partial\rho)|_{\rho=0} D = \gamma\Delta(\tilde\xi_0, \lambda_0). \tag{9.12}$$

An important consequence of (9.11) is that *weak inviscid stability,* $|\Delta| > 0$*, is necessary for weak viscous stability,* $|D| > 0$ (an evident necessary condition for linearized viscous stability). For, (9.11) implies that the zero set of D is tangent at the origin to the cone $\{\Delta = 0\}$ (recall, (9.10), that Δ is homogeneous, degree one), hence enters $\{\tau > 0\}$ if $\{\Delta = 0\}$ does. Moreover, in case of *neutral inviscid stability* $\Delta(\xi_0, i\tau_0) = 0$, $(\xi_0, i\tau_0) \neq (0, 0)$, one may extract a further, *refined stability condition*:

$$\beta := -D_{\rho\rho}/D_{\rho\lambda}|_{\rho=0} \geq 0 \tag{9.13}$$

necessary for weak viscous stability. For, (9.12) then implies $D_\rho|_{\rho=0} = \gamma\Delta(\xi_0, i\tau_0) = 0$, whence Taylor expansion of D yields that the zero level set of D is concave or convex toward $\tau > 0$ according to the sign of β [ZS]. As discussed in [ZS, Z1], the constant β has a heuristic interpretation as an effective diffusion coefficient for surface waves moving along the front.

As shown in [ZS, BSZ], the formula (9.13) is well-defined whenever Δ is analytic at $(\tilde\xi_0, i\tau)$, in which case D considered as a function of polar coordinates is analytic at $(0, \tilde\xi_0, i\tau_0)$, and $i\tau_0$ is a simple root of $\Delta(\tilde\xi_0, \cdot)$. The determinant Δ in turn is analytic at $(\tilde\xi_0, i\tau_0)$, for all except a finite set of branch singularities $\tau_0 = \tilde\xi_0\eta_j$. As discussed in [BSZ, Z2, Z3], the apparently nongeneric behavior that the family of holomorphic functions Δ^ε associated with shocks $(u_+^\varepsilon, u_-^\varepsilon)$ have roots $(\xi_0^\varepsilon, i\tau_0(\varepsilon))$ with $i\tau_0$ pure imaginary on an open set of ε is explained by the fact that, on certain components of the complement on the imaginary axis of this finite set of branch singularities, $\Delta^\varepsilon(\xi_0^\varepsilon, \cdot)$ takes the imaginary axis to itself. Thus, zeros of odd multiplicity persist on the imaginary axis, by consideration of the topological degree of Δ^ε as a map from the imaginary axis to itself.

Moreover, the same topological considerations show that a simple imaginary root of this type can only enter or leave the imaginary axis at a branch singularity of $\Delta^\varepsilon(\tilde\xi^\varepsilon, \cdot)$ or at infinity, which greatly aids in the computation of transition points for inviscid stability [BSZ, Z1, Z2, Z3]. As described in [Z2, Z3, Se1], escape to infinity is always associated with transition to strong instability. Indeed, using real homogeneity of Δ, we may rescale by $|\lambda|$ to find in the limit as $|\lambda| \to \infty$ that $0 = |\lambda_0|^{-1}\Delta(\tilde\xi_0, \lambda_0) = \Delta(\tilde\xi_0/|\lambda_0|, \lambda_0/|\lambda_0|) \to \Delta(0, i)$, which, by the complex homogeneity $\Delta(0, \lambda) \equiv \lambda\Delta(0, 1)$ of the one-dimensional Lopatinski determinant $\Delta(0, \cdot)$, yields *one-dimensional instability* $\Delta(0, 1) = 0$. As described in [Z1], Section 6.2, this is associated not with surface waves, but the more dramatic phenomenon of *wave-splitting*, in which the axial structure of the front bifurcates from a single shock to a more complicated multi-wave Riemann pattern.

We note in passing that one-dimensional inviscid stability $\Delta^\varepsilon(0, 1) \neq 0$ is equivalent to (H3) through the relation

$$\Delta^\varepsilon(0, \lambda) = \lambda \det(r_1^-, \dots, r_c^-, r_{c+1}^+, \dots, r_n^+, u_+ - u_-). \tag{9.14}$$

9.4. Transverse bifurcation of flow in a duct. We now make an elementary observation connecting cellular bifurcation of flow in a duct to stability of shocks in the whole space. Assume for the family $\bar u^\varepsilon$:

(B1) For ε sufficiently small, the inviscid shock $(u_+^\varepsilon, u_-^\varepsilon)$ is weakly stable; more precisely, $\Delta^\varepsilon(1, \lambda)$ has no roots $\Re\lambda \geq 0$ but a single simple pure imaginary root $\lambda(\varepsilon) = i\tau_*(\varepsilon) \neq 0$ lying away from the singularities of Δ^ε.

(B2) Coefficient $\beta(\varepsilon)$ defined in (9.13) satisfies $\Re\beta(0) = 0$, $\partial_\varepsilon \Re\beta(0) < 0$.

LEMMA 9.1 ([ZS, Z1]). *Let (H0)–(H2) and (B1). For ε, $\tilde\xi$ sufficiently small, there exist a smooth family of roots $(\tilde\xi, \lambda_*^\varepsilon(\tilde\xi))$ of $D(\tilde\xi, \lambda)$ with:*

$$\lambda_*^\varepsilon(\tilde\xi) = i\tilde\xi\tau_*(\varepsilon) - \tilde\xi^2\beta(\varepsilon) + \delta(\varepsilon)\tilde\xi^3 + r(\varepsilon, \tilde\xi)\tilde\xi^4, \qquad r \in C^1(\varepsilon, \tilde\xi). \tag{9.15}$$

Moreover, these are the unique roots of D satisfying $\Re\lambda \geq -|\tilde\xi|/C$ for some $C > 0$ and $\rho = |(\tilde\xi, \lambda)|$ sufficiently small.

Consequently, there is a unique C^1 function $\mathcal{E}(\tilde{\xi}) \neq 0$, $\mathcal{E}(0) = O$, such that $\Re\lambda_*^\varepsilon(\tilde{\xi}) = 0$ for $\varepsilon = \mathcal{E}(\tilde{\xi})$. In the generic case $\Re\delta(0) \neq 0$, moreover:

$$\mathcal{E}(\tilde{\xi}) \sim (\Re\delta(0)/\partial_\varepsilon\beta(0))\tilde{\xi}. \tag{9.16}$$

To (B1) and (B2), adjoin now the additional assumptions:

(B3) $\delta(0) \neq 0$.

(B4) At $\varepsilon = 0$, the Evans function $D(\tilde{\xi}, \lambda)$ has no roots $\tilde{\xi} \in \mathbb{R}$, $\Re\lambda \geq 0$ outside a sufficiently small ball about the origin.

Then, we have the following main result:

THEOREM 9.2 ([Z9]). *Assuming (H0)–(H2), (B1)-(B4), for $\varepsilon_{max} > 0$ sufficiently small and each cross-sectional width M sufficiently large, there is a finite sequence $0 < \varepsilon_1(M) < \cdots < \varepsilon_k(M) < \cdots \leq \varepsilon_{max}$, with $\varepsilon_k(M) \sim (\Re\delta(0)/\partial_\varepsilon\beta(0))\frac{\pi k}{M}$, such that, as ε crosses successive ε_k from the left, there occur a series of transverse (i.e., "cellular") Hopf bifurcations of $\bar{u}^\varepsilon$ associated with wave-numbers $\pm k$, with successively smaller periods $T_k(\varepsilon) \sim \tau_*(0)\frac{2M}{k}$.*

Proof. ([Z9]) By (B4), for $|\varepsilon| \leq \varepsilon_{max}$ sufficiently small, we have by continuity that there exist no roots of $D(\tilde{\xi}, \lambda)$ for $\Re\lambda \geq -1/C$, $C > 0$, outside a small ball about the origin. By Lemma 9.1, within this small ball, there are no roots other than possibly $(\tilde{\xi}, \lambda^\varepsilon(\tilde{\xi}))$ with $\Re\lambda \geq 0$: in particular, *no nonzero purely imaginary spectra are possible other than at values* $\lambda^\varepsilon(\tilde{\xi})$ for operator $L(\varepsilon)$ acting on functions on the whole space.

Considering L instead as an operator acting on functions on the channel $\mathcal{C}$, we find by discrete Fourier transform/separation of variables that its spectra are exactly the zeros of $D(\xi_k, \lambda)$, as $\xi_k = \frac{\pi k}{L}$ runs through all integer wave-numbers k. Applying Lemma 9.1 and using (B3), we find, therefore, that pure imaginary eigenvalues of $L(\varepsilon)$ with $|\varepsilon| \leq \varepsilon_{max}$ sufficiently small occur precisely at values $\varepsilon = \varepsilon_k$, and consist of crossing conjugate pairs $\lambda_\pm^k(\varepsilon)$ associated with wave-numbers $\pm k$, satisfying Hopf bifurcation condition (D2) with:

$$\Im\lambda_\pm^k(\varepsilon) \sim \tau_*(0)\Big)\pi k/L.$$

Applying Proposition 9.2, the result follows. □

10. Discussion and open problems. We have described a comprehensive program, centered around the Evans function, for the study of stability and dynamics of viscous shock waves and related nonlinear phenomena: in particular, determination of stability, invariant manifolds, and bifurcation/exchange of stability. A novel feature of our approach is the systematic incorporation in our numerical studies of rigorous asymptotic analysis in high-frequency and other limits, allowing us to carry out global studies in parameter space, as in the arbitrary-amplitude study of Sec. 7.

The particular analyses presented here are special to their contexts. However, the basic approach is quite general. For large-amplitude/model parameter stability studies in the more complicated settings of MHD and nonisentropic gas dynamics, see [BHZ, BLZ, HLyZ1, HLyZ2]. For applications to viscoelastic shocks, see [BLeZ]. Similar programs have been carried out for noncharacteristic boundary layers [CHNZ, NZ1, NZ2, GMWZ2, Z3], detonation waves [LyZ1, LyZ2, LRTZ, TZ4, G1, W, Z12, Z13, HuZ1, BZ, HLyZ3], and solitary waves in shallow-water flow [BJRZ].

Extensions to spatially periodic traveling waves have been carried out in [G2, OZ1, OZ2, OZ3, OZ4, BJNRZ1, BJNRZ2, JZ1, JZ2, JZ3, JZ4, JZ5, JZB, JZN], and to temporally periodic waves in [BeSZ]. Other technically substantial extensions include the case of coupled hyperbolic-elliptic (e.g., radiative) shocks [LMNPZ, NPZ] and shocks of semidiscrete [B, BHu, BHuR, BeHSZ] and discrete [God] numerical schemes. Pointwise bounds sharpening Theorem 5.1 are given in [HZ, RZ, HR, HRZ].

Recall from the discussion at the end of Section 5.1, that inviscid shock stability is *necessary* for viscous stability [ZS]; this has the somewhat counterintuitive consequence that the inclusion of viscosity *may destabilize but never stabilize* an inviscid shock wave. A fundamental open question is whether in practice such destabilization can in fact occur; that is, to find, presumably numerically, a physical example for which viscous effects hasten the onset of instability as model parameters (say, shock amplitude) are varied, or to show analytically that this cannot occur.

Likewise, it would be very interesting to further investigate numerically the mechanism for cellular instabilities proposed in Section 9, in particular, to compare to alternative predictions of Majda et al [MR1, MR2, AM] based on hyperbolic effects (nonlinear rather than viscous corrrection), obtained by weakly nonlinear geometric optics.

More generally, the larger themes connecting the various topics presented here are stability analysis of solutions of systems of pde with large numbers of variables, and or about which the linearized equations have no spectral gap between decaying and neutral modes. As the main directions for the future we see the analysis of complex systems possessing orders of magnitude more variables, and of more complicated, genuinely multi-dimensional hdyrodynamic flows in unbounded domains. The recent studies [GLZ, Z12] represent first, preliminary, steps in these directions.

Extension to time-varying flows such as Riemann patterns is another important direction, as is the extension to infinite-dimensional kinetic systems such as Boltzmann's equation; see [MeZ3] and references therein.

Finally, we mention the problems of numerical proof and of purely analytical verification of stability in important cases. Though the basic ingredients (rigorous convergence theory, numerical well-conditioning, and mature algorithmic development) are present for numerical proof, this is of course a separate problem with its own challenges. However, the problems considered seem sufficiently physically fundamental to warrant such an investigation.

Regarding analytical proof, we point to the surprising recent result of Matsumura and Wang [MW], in which the authors establish spectral stability of arbitrary amplitude shock waves of the isentropic gas dynamics equations with density-dependent viscosity of form $\mu(v) = Cv^{-(\gamma-1)\beta}$, $\beta \geq 1/2$ including that ($\beta = 1/2$) predicted by statistical mechanics/Boltzmann's equation in the rarefied gas limit,[3] by an elegant energy estimate quite similar to that of Proposition 7.2. This demonstrates by example that, with sufficient skill and understanding, analytical verification of spectral stability may yet be possible.

Acknowledgement. Thanks to Olivier Lafitte and Benjamin Texier for their interest and encouragement in this work and in an earlier version of the course presented at University of Paris 13 in Summer 2008 focused on the topic of numerical Evans function evaluation. Thanks to the University of Paris 13 for their hospitality during the visit in which that course was developed and presented, and to the I.M.A. and organizers for their hospitality during the Summer 2009 workshop in which the current much extended course was presented. Finally, warm thanks to Editor Marta Lewicka for her generous and skillful contribution in converting the full course notes into the condensed form appearing here.

REFERENCES

[AT] G. Abouseif and T.Y. Toong, *Theory of unstable one-dimensional detonations,* Combust. Flame **45** (1982) 67–94.

[AGJ] J. Alexander, R. Gardner, and C.K.R.T. Jones, *A topological invariant arising in the analysis of traveling waves*, J. Reine Angew. Math. **410** (1990) 167–212.

[AMPZ] A. Azevedo, D. Marchesin, B. Plohr, and K. Zumbrun, *Nonuniqueness of solutions of Riemann problems.* Z. Angew. Math. Phys. **47** (1996), 977–998.

[AS] J.C. Alexander and R. Sachs, *Linear instability of solitary waves of a Boussinesq-type equation: a computer assisted computation,* Nonlinear World **2** (1995) 471–507.

[AlB] L. Allen and T.J. Bridges, *Numerical exterior algebra and the compound matrix method,* Numer. Math. **92** (2002) 197–232.

[AlT] R.L. Alpert and T.Y. Toong, *Periodicity in exothermic hypersonic flows about blunt projectiles,* Acta Astron. **17** (1972) 538–560.

[A] J. Anderson, *Magnetohydrodynamics Shock Waves*, MIT Press, Cambridge, MA (1963).

[AM] M. Artola and A. Majda, *Nonlinear kink modes for supersonic vortex sheets,* Phys. Fluids A **1** (1989), no. 3, 583–596.

[ALT] R.L. Alpert and T.Y. Toong, *Periodicity in exothermic hypersonic flows about blunt projectiles,* Acta Astron. **17** (1972) 538–560.

[3]Chapman's law $\mu(T) = CT^{1/2}$ [Ba], together with the isentropic approximation $T \sim v^{-(\gamma-1)}$, where T is temperature.

[ACR] U.M. Ascher, H. Chin, and S. Reich, *Stabilization of DAEs and invariant manifolds,* Numer. Math. **67** (1994) 131–149.

[BHRZ] B. Barker, J. Humpherys, K. Rudd, and K. Zumbrun, *Stability of viscous shocks in isentropic gas dynamics*, Comm. Math. Phys. **281** (2008), no. 1, 231–249.

[BHZ] B. Barker, J. Humpherys, and K. Zumbrun, *One-dimensional stability of parallel shock layers in isentropic magnetohydrodynamics,* to appear, J. Diff. Eq.

[BJNRZ1] B. Barker, M. Johnson, P. Noble, M. Rodrigues, and K. Zumbrun, *Spectral stability of periodic viscous roll waves*, in preparation:

[BJNRZ2] B. Barker, M. Johnson, P. Noble, M. Rodrigues, and K. Zumbrun, *Witham averaged equations and modulational stability of periodic solutions of hyperbolic-parabolic balance laws,* Proceedings, French GDR meeting on EDP, Port D'Albret, France; to appear.

[BJRZ] B. Barker, M. Johnson, M. Rodrigues, and K. Zumbrun, *Metastability of solitary roll wave solutions of the St. Venant equations with viscosity,* preprint (2010).

[BLZ] B. Barker, O. Lafitte, and K. Zumbrun, *Existence and stability of viscous shock profiles for 2-D isentropic MHD with infinite electrical resistivity,* to appear, J. Diff. Eq.

[BLeZ] B. Barker, M. Lewicka, and K. Zumbrun, *Existence and stability of viscoelastic shock profiles,* to appear, Arch. Ration. Mech. Anal.

[BZ] B. Barker and K. Zumbrun, *A numerical stability investigation of strong ZND detonations for Majda's model,* preprint (2010).

[Ba] G.K. Batchelor, *An introduction to fluid dynamics*, Cambridge Mathematical Library. Cambridge University Press, Cambridge, paperback edition, 1999.

[BE] A.A. Barmin and S.A. Egorushkin, *Stability of shock waves*, Adv. Mech. **15** (1992) no. 1–2, 3–37.

[BeSZ] M. Beck, B. Sandstede, and K. Zumbrun, *Nonlinear stability of time-periodic shock waves,* Arch. Rat. Mechanics and Anal. **196** (2010) no. 3, 1011–1076.

[BeHSZ] M. Beck, H.J. Hupkes, B. Sandstede, and K. Zumbrun, *Nonlinear stability of semi-discrete shocks for general numerical schemes,* SIAM J. Math. Anal. **42** (2010), no. 2, 857–903.

[B] S. Benzoni-Gavage, *Semi-discrete shock profiles for hyperbolic systems of conservation laws,* Phys. D **115** (1998) 109–123.

[BHu] S. Benzoni-Gavage and P. Huot, *Existence of semi-discrete shocks,* Discrete Contin. Dyn. Syst. **8** (2002) 163–190.

[BHuR] S. Benzoni–Gavage, P. Huot, and F. Rousset, *Nonlinear stability of semidiscrete shock waves,* SIAM J. Math. Anal. **35** (2003) 639–707.

[BRSZ] S. Benzoni–Gavage, F. Rousset, D. Serre, and K. Zumbrun, *Generic types and transitions in hyperbolic initial-boundary-value problems,* Proc. Roy. Soc. Edinburgh Sect. A **132** (2002) 1073–1104.

[BSZ] S. Benzoni-Gavage, D. Serre, and K. Zumbrun, *Transition to instability of planar viscous shock fronts,* to appear, Z.A.A.

[Be] W.-J. Beyn, *The numerical computation of connecting orbits in dynamical systems,* IMA J. Numer. Analysis **9** (1990) 379–405.

[BMR] A. Bourlioux, A. Majda, and V. Roytburd, *Theoretical and numerical structure for unstable one-dimensional detonations.* SIAM J. Appl. Math. **51** (1991) 303–343.

[BDG] T.J. Bridges, G. Derks, and G. Gottwald, *Stability and instability of solitary waves of the fifth-order KdV equation: a numerical framework.* Phys. D **172** (2002), no. 1–4, 190–216.

[Br1] L. Brin, *Numerical testing of the stability of viscous shock waves.* Doctoral thesis, Indiana University (1998).

[Br2] L.Q. BRIN, *Numerical testing of the stability of viscous shock waves.* Math. Comp. **70** (2001) 235, 1071–1088.

[BrZ] L. BRIN AND K. ZUMBRUN, *Analytically varying eigenvectors and the stability of viscous shock waves.* Seventh Workshop on Partial Differential Equations, Part I (Rio de Janeiro, 2001). Mat. Contemp. **22** (2002), 19–32.

[C] H. CABANNES, *Theoretical Magnetofluiddynamics,* Academic PRess, New York (1970).

[CHNZ] N. COSTANZINO, J. HUMPHERYS, T. NGUYEN, AND K. ZUMBRUN, *Spectral stability of noncharacteristic boundary layers of isentropic Navier–Stokes equations,* Arch. Rat. Mechanics and Anal. **192** (2009), no. 1–3, 119–136.

[Da] A. DAVEY, *An automatic orthonormalization method for solving stiff boundary value problems,* J. Comput. Phys. **51** (1983) 343–356.

[DDF] J.W. DEMMEL, L. DIECI, AND M.J. FRIEDMAN, *Computing connecting orbits via an improved algorithm for continuing invariant subspaces.* SIAM J. Sci. Comput. **22** (2000) 81–94.

[DE1] L. DIECI AND T. EIROLA, *Applications of smooth orthogonal factorizations of matrices.* In Numerical methods for bifurcation problems and large-scale dynamical systems (Minneapolis, MN, 1997). Springer, IMA Vol. Math. Appl. **119** (2000) 141–162.

[DE2] L. DIECI AND T. EIROLA, *On smooth decompositions of matrices.* SIAM J. Matrix Anal. Appl. **20** (1999) 800–819.

[DF] L. DIECI AND M.J. FRIEDMAN, *Continuation of invariant subspaces.* Numer. Lin. Alg. Appl. **8** (2001) 317–327.

[Dr] L.O. DRURY, *Numerical solution of Orr-Sommerfeld-type equations,* J. Comput. Phys. **37** (1980) 133–139.

[Er1] J.J. ERPENBECK, *Stability of steady-state equilibrium detonations,* Phys. Fluids **5** (1962), 604–614.

[Er2] J.J. ERPENBECK, *Stability of idealized one-reaction detonations,* Phys. Fluids **7** (1964).

[Er3] J.J. ERPENBECK, *Stability of step shocks.* Phys. Fluids **5** (1962) no. 10, 1181–1187.

[E1] J.W. EVANS, *Nerve axon equations: I. Linear approximations.* Ind. Univ. Math. J. **21** (1972) 877–885.

[E2] J.W. EVANS, *Nerve axon equations: II. Stability at rest.* Ind. Univ. Math. J. **22** (1972) 75–90.

[E3] J.W. EVANS, *Nerve axon equations: III. Stability of the nerve impulse.* Ind. Univ. Math. J. **22** (1972) 577–593.

[E4] J.W. EVANS, *Nerve axon equations: IV. The stable and the unstable impulse.* Ind. Univ. Math. J. **24** (1975) 1169–1190.

[EF] J.W. EVANS AND J.A. FEROE, *Traveling waves of infinitely many pulses in nerve equations,* Math. Biosci., **37** (1977) 23–50.

[F1] W. FICKETT, *Stability of the square wave detonation in a model system.* Physica **16**D (1985) 358–370.

[F2] W. FICKETT, *Detonation in miniature,* 133–182, in *The mathematics of combustion,* Frontiers in App. Math. (1985) SIAM, Philadelphia ISBN: 0-89871-053-7.

[FD] W. FICKETT AND W.C. DAVIS, *Detonation,* University of California Press, Berkeley, CA (1979): reissued as *Detonation: Theory and experiment,* Dover Press, Mineola, New York (2000), ISBN 0-486-41456-6.

[FW] FICKETT AND WOOD, *Flow calculations for pulsating one-dimensional detonations.* Phys. Fluids **9** (1966) 903–916.

[FS] H. FREISTÜHLER AND P. SZMOLYAN, *Spectral stability of small shock waves,* Arch. Ration. Mech. Anal. **164** (2002) 287–309.

[G1] R. GARDNER, *On the structure of the spectra of periodic traveling waves,* J. Math. Pures Appl. **72** (1993), 415–439.

[G2] R. GARDNER, *On the detonation of a combustible gas,* Trans. Amer. Math. Soc. **277** (1983), no. 2, 431–468.

[GJ1] R. GARDNER AND C.K.R.T. JONES, *A stability index for steady state solutions of boundary value problems for parabolic systems,* J. Diff. Eqs. **91** (1991), no. 2, 181–203.

[GJ2] R. GARDNER AND C.K.R.T. JONES, *Traveling waves of a perturbed diffusion equation arising in a phase field model,* Ind. Univ. Math. J. **38** (1989), no. 4, 1197–1222.

[GZ] R. A. GARDNER AND K. ZUMBRUN, *The gap lemma and geometric criteria for instability of viscous shock profiles,* Comm. Pure Appl. Math. **51** (1998), no. 7, 797–855.

[GLZ] F. GESZTESY, Y. LATUSHKIN, AND K. ZUMBRUN, *Derivatives of (Modified) Fredholm Determinants and Stability of Standing and Traveling Waves,* J. Math. Pures Appl. (9) **90** (2008), no. 2, 160–200.

[GB] F. GILBERT AND G. BACKUS, *Propagator matrices in elastic wave and vibration problems,* Geophysics, **31** (1966) 326–332.

[God] P. GODILLON, *Green's function pointwise estimates for the modified Lax-Friedrichs scheme,* M2AN Math. Model. Numer. Anal.**37** (2003) 1–39.

[GvL] G.H. GOLUB AND C.F. VAN LOAN, *Matrix computations,* Johns Hopkins University Press, Baltimore (1996).

[Go] J. GOODMAN, *Remarks on the stability of viscous shock waves,* in: Viscous profiles and numerical methods for shock waves (Raleigh, NC, 1990), 66–72, SIAM, Philadelphia, PA, (1991).

[GMWZ1] O. GUES, G. METIVIER, M. WILLIAMS, AND K. ZUMBRUN, *Multidimensional viscous shocks I: degenerate symmetrizers and long time stability,* Journal of the Amer. Math. Soc. **18** (2005), 61–120.

[GMWZ2] O. GUÈS, G.MÉTIVIER, M. WILLIAMS, AND K. ZUMBRUN, *Existence and stability of noncharacteristic hyperbolic-parabolic boundary-layers,* Arch. for Ration. Mech. Anal. **197** (2010), no. 1, 1–87.

[He] D. HENRY, *Geometric theory of semilinear parabolic equations.* Lecture Notes in Mathematics, Springer–Verlag, Berlin (1981), iv + 348 pp.

[Hol] OLGA HOLTZ, *Hermite–Biehler, Routh–Hurwitz, and total positivity,* Linear algebra and its applications **372** (2003) 105–110.

[H1] E. HOPF, *Abszweigung einer periodischen Lösung von einer stationaeren Lösung eines Differentialsystems,* Akad. Wiss. (Leipzig) **94** (1942), 2–11.

[H2] E. HOPF, *Selected works of Eberhard Hopf with commentaries,* Edited by Cathleen S. Morawetz, James B. Serrin and Yakov G. Sinai. American Mathematical Society, Providence, RI, 2002. xxiv+386 pp. ISBN: 0-8218-2077-X. (Commentary by M. Golubitsky and P.H. Rabinowitz.)

[HR] P. HOWARD AND M. RAOOFI, *Pointwise asymptotic behavior of perturbed viscous shock profiles,* Adv. Differential Equations (2006) 1031–1080.

[HRZ] P. HOWARD, M. RAOOFI, AND K. ZUMBRUN, *Sharp pointwise bounds for perturbed viscous shock waves,* J. Hyperbolic Differ. Equ. (2006) 297–373.

[HZ] P. HOWARD AND K. ZUMBRUN, *Stability of undercompressive viscous shock waves,* in press, J. Differential Equations **225** (2006), no. 1, 308–360.

[HLZ] J. HUMPHERYS, O. LAFITTE, AND K. ZUMBRUN, *Stability of viscous shock profiles in the high Mach number limit,* CMP **293** (2010), no. 1, 1–36.

[HLyZ1] J. HUMPHERYS, G. LYNG, AND K. ZUMBRUN, *Spectral stability of ideal gas shock layers,* Arch. for Rat. Mech. Anal. **194** (2009), no. 3, 1029–1079.

[HLyZ2] J. HUMPHERYS, G. LYNG, AND K. ZUMBRUN, *Multidimensional spectral stability of large-amplitude Navier-Stokes shocks,* in preparation.

[HLyZ3] J. HUMPHERYS, G. LYNG, AND K. ZUMBRUN, *Spectral stability of combustion waves: the Majda model,* in preparation.

[HuZ1] J. HUMPHERYS AND K. ZUMBRUN, *Spectral stability of small amplitude shock profiles for dissipative symmetric hyperbolic–parabolic systems.* Z. Angew. Math. Phys. **53** (2002) 20–34.

[HuZ2] J. HUMPHERYS AND K. ZUMBRUN, *An efficient shooting algorithm for Evans function calculations in large systems,* Phys. D **220** (2006), no. 2, 116–126.

[HuZ3] J. HUMPHERYS AND K. ZUMBRUN, *Efficient numerical stability analysis of detonation waves in ZND*, in preparation.

[HSZ] J. HUMPHERYS, K. ZUMBRUN, AND BJÖRN SANDSTEDE, *Efficient computation of analytic bases in Evans function analysis of large systems,* Numer. Math. **103** (2006), no. 4, 631–642.

[J] A. JEFFREY, *Magnetohydrodynamics,* University Mathematical Texts, no. 33 Oliver & Boyd, Edinburgh-London; Interscience Publishers Inc. John Wiley & Sons, Inc., New York 1966 viii+252 pp.

[JZ1] M. JOHNSON AND K. ZUMBRUN, *Rigorous Justification of the Whitham Modulation Equations for the Generalized Korteweg-de Vries Equation,* to appear, Studies in Appl. Math.

[JZ2] M. JOHNSON AND K. ZUMBRUN, *Nonlinear stability of periodic traveling wave solutions of systems of viscous conservation laws in the generic case*, J. Diff. Eq. **249** (2010), no. 5, 1213–1240.

[JZ3] M. JOHNSON AND K. ZUMBRUN, *Nonlinear stability of spatially-periodic traveling-wave solutions of systems of reaction diffusion equations,* preprint (2010).

[JZ4] M. JOHNSON AND K. ZUMBRUN, *Nonlinear stability and asymptotic behavior of periodic traveling waves of multidimensional viscous conservation laws in dimensions one and two*, preprint (2009).

[JZ5] M. JOHNSON AND K. ZUMBRUN, *Stability of KP waves,* to appear, SIAM J. for Math. Anal.

[JZB] M. JOHNSON, K. ZUMBRUN, AND J.C. BRONSKI, *On the modulation equations and stability of periodic GKdV waves via bloch decompositions,* to appear Physica D.

[JZN] M. JOHNSON, K. ZUMBRUN, AND P. NOBLE, *Nonlinear stability of viscous roll waves,* preprint (2010).

[K] T. KATO, *Perturbation theory for linear operators.* Springer–Verlag, Berlin Heidelberg (1985).

[KS] A.R. KASIMOV AND D.S. STEWART, *Spinning instability of gaseous detonations.* J. Fluid Mech. **466** (2002), 179–203.

[Kaw] S. KAWASHIMA, *Systems of a hyperbolic–parabolic composite type, with applications to the equations of magnetohydrodynamics.* thesis, Kyoto University (1983).

[KSh] S. KAWASHIMA AND Y. SHIZUTA, *On the normal form of the symmetric hyperbolic-parabolic systems associated with the conservation laws.* Tohoku Math. J. **40** (1988) 449–464.

[L] P.D. LAX, *Hyperbolic systems of conservation laws and the mathematical theory of shock waves*, Conference Board of the Mathematical Sciences Regional Conference Series in Applied Mathematics, no. 11. Society for Industrial and Applied Mathematics, Philadelphia, Pa., 1973. v+48 pp.

[LS] H.I. LEE AND D.S. STEWART, *Calculation of linear detonation instability: one-dimensional instability of plane detonation,* J. Fluid Mech., **216** (1990) 103–132.

[LyZ1] G. LYNG AND K. ZUMBRUN, *One-dimensional stability of viscous strong detonation waves,* Arch. Ration. Mech. Anal. **173** (2004), no. 2, 213–277.

[LyZ2] G. Lyng and K. Zumbrun, *A stability index for detonation waves in Majda's model for reacting flow*, Physica D, **194** (2004), 1–29.

[LRTZ] G. Lyng, M. Raoofi, B. Texier, and K. Zumbrun, *Pointwise Green Function Bounds and stability of combustion waves*, J. Differential Equations **233** (2007) 654–698.

[M1] A. Majda, *The stability of multi-dimensional shock fronts – a new problem for linear hyperbolic equations,* Mem. Amer. Math. Soc. **275** (1983).

[M2] A. Majda, *The existence of multi-dimensional shock fronts,* Mem. Amer. Math. Soc. **281** (1983).

[M3] A. Majda, *Compressible fluid flow and systems of conservation laws in several space variables,* Springer-Verlag, New York (1984), viii+ 159 pp.

[M4] A. Majda, *A qualitative model for dynamic combustion*, SIAM J. Appl. Math., **41** (1981), 70–91.

[MP] A. Majda and R. Pego, *Stable viscosity matrices for systems of conservation laws.* J. Diff. Eqs. **56** (1985) 229–262.

[MR1] A. Majda and R. Rosales, *A theory for the spontaneous formation of Mach stems in reacting shock fronts, I. The basic perturbation analysis,* SIAM J. Appl. Math. **43** (1983), no. 6, 1310–1334.

[MR2] A. Majda and R. Rosales, *A theory for spontaneous Mach-stem formation in reacting shock fronts. II. Steady-wave bifurcations and the evidence for breakdown,* Stud. Appl. Math. **71** (1984), no. 2, 117–148.

[LMNPZ] C. Lattanzio, C. Mascia, T. Nguyen, R. Plaza, and K. Zumbrun, *Stability of scalar radiative shock profiles*, SIAM J. for Math. Anal. **41** (2009/2010), no. 6, 2165–2206.

[MaZ1] C. Mascia and K. Zumbrun, *Pointwise Green's function bounds and stability of relaxation shocks.* Indiana Univ. Math. J. **51** (2002), no. 4, 773–904.

[MaZ2] C. Mascia and K. Zumbrun, *Stability of shock profiles of dissipative symmetric hyperbolic–parabolic systems*, preprint (2001): Comm. Pure Appl. Math. **57** (2004), no. 7, 841–876.

[MaZ3] C. Mascia and K. Zumbrun, *Pointwise Green function bounds for shock profiles of systems with real viscosity.* Arch. Ration. Mech. Anal. **169** (2003), no. 3, 177–263.

[MaZ4] C. Mascia and K. Zumbrun, *Stability of large-amplitude viscous shock profiles of hyperbolic-parabolic systems,* Arch. Ration. Mech. Anal. **172** (2004), no. 1, 93–131.

[MN] A. Matsumura and K. Nishihara, *On the stability of travelling wave solutions of a one-dimensional model system for compressible viscous gas,* Japan J. Appl. Math. **2** (1985) 17–25.

[MW] A. Matsumura and Y. Wang, *Asymptotic stability of viscous shock wave for a one-dimensional isentropic model of viscous gas with density dependent viscosity,* preprint (2010).

[NPZ] T. Nguyen, R. Plaza, and K. Zumbrun, *Stability of radiative shock profiles for hyperbolic-elliptic coupled systems,* Physica D. **239** (2010), no. 8, 428–453.

[MT] U.B. McVey and T.Y. Toong, *Mechanism of instabilities in exothermic blunt-body flows,* Combus. Sci. Tech. **3** (1971) 63–76.

[Me] G. Métivier, *Stability of multidimensional shocks.* Advances in the theory of shock waves, 25–103, Progr. Nonlinear Differential Equations Appl., **47**, Birkhäuser Boston, Boston, MA, 2001.

[MeZ1] G. Métivier and K. Zumbrun, *Large viscous boundary layers for noncharacteristic nonlinear hyperbolic problems,* Mem. Amer. Math. Soc. **175** (2005), no. 826, vi+107 pp.

[MeZ2] G. Métivier and K. Zumbrun, *Existence of semilinear relaxation shocks,* J. Math. Pures. Appl. (9) **92** (2009), no. 3, 209–231.

[MeZ3] G. MÉTIVIER AND K. ZUMBRUN, *Existence and sharp localization in velocity of small-amplitude Boltzmann shocks*, Kinet. Relat. Models **2** (2009), no. 4, 667–705.

[NR1] B.S. NG AND W.H. REID, *An initial value method for eigenvalue problems using compound matrices,* J. Comput. Phys. **30** (1979) 125–136.

[NR2] B.S. NG AND W.H. REID, *A numerical method for linear two-point boundary value problems using compound matrices,* J. Comput. Phys. **33** (1979) 70–85.

[NR3] B.S. NG AND W.H. REID, *On the numerical solution of the Orr-Sommerfeld problem: asymptotic initial conditions for shooting methods,* J. Comput. Phys., **38** (1980) 275–293.

[NR4] B.S. NG AND W.H. REID, *The compound matrix method for ordinary differential systems,* J. Comput. Phys. **58** (1985) 209–228.

[OZ1] M. OH AND K. ZUMBRUN, *Stability of periodic solutions of viscous conservation laws with viscosity: Analysis of the Evans function*, Arch. Ration. Mech. Anal. **166** (2003), no. 2, 99–166.

[OZ2] M. OH AND K. ZUMBRUN, *Stability of periodic solutions of viscous conservation laws: pointwise bounds on the Green function*, Arch. Ration. Mech. Anal. (2002), no. 2, 167–196.

[OZ3] M. OH AND K. ZUMBRUN, *Low-frequency stability analysis of periodic traveling-wave solutions of viscous conservation laws in several dimensions,* Z. Anal. Anwend. **25** (2006), no. 1, 1–21.

[OZ4] M. OH AND K. ZUMBRUN, *Stability and asymptotic behavior of periodic traveling wave solutions of viscous conservation laws in several dimensions,* Arch. for Rat. Mech. Anal. **196** (2010), no. 1, 1–20; Errata, Arch. for Rat. Mech. Anal. **196** (2010), no. 1, 21–23.

[NZ1] T. NGUYEN AND K. ZUMBRUN, *Long-time stability of large-amplitude noncharacteristic boundary layers for hyperbolic parabolic systems*, J. Math. Pures Appl. (9) **92** (2009), no. 6, 547–598.

[NZ2] T. NGUYEN AND K. ZUMBRUN, Long-time stability of multi-dimensional noncharacteristic viscous boundary layers, to appear, Comm. Math. Phys.

[PZ] R. PLAZA AND K. ZUMBRUN, *An Evans function approach to spectral stability of small-amplitude shock profiles*, J. Disc. and Cont. Dyn. Sys. **10** (2004), 885–924.

[Pe] R.L. PEGO, *Stable viscosities and shock profiles for systems of conservation laws.* Trans. Amer. Math. Soc. **282** (1984) 749–763.

[PW] R.L. PEGO AND M.I. WEINSTEIN, *Eigenvalues, and instabilities of solitary waves.* Philos. Trans. Roy. Soc. London Ser. A **340** (1992), 47–94.

[Sa] D. SATTINGER, *On the stability of waves of nonlinear parabolic systems*, Adv. Math. **22** (1976) 312–355.

[R] M. RAOOFI, *L^p asymptotic behavior of perturbed viscous shock profiles,* J. Hyperbolic Differ. Equ. **2** (2005), no. 3, 595–644.

[RZ] M. RAOOFI AND K. ZUMBRUN, *Stability of undercompressive viscous shock profiles of hyperbolic-parabolic systems,* J. Differential Equations (2009) 1539–1567.

[Se1] D. SERRE, *La transition vers l'instabilité pour les ondes de chocs multidimensionnelles,* Trans. Amer. Math. Soc. **353** (2001) 5071–5093.

[Se2] D. SERRE, *Systems of conservation laws. 1. Hyperbolicity, entropies, shock waves,* Translated from the 1996 French original by I. N. Sneddon. Cambridge University Press, Cambridge, 1999. xxii+263 pp. ISBN: 0-521-58233-4.

[Se3] D. SERRE, *Systems of conservation laws. 2. Geometric structures, oscillations, and initial-boundary value problems,* Translated from the 1996 French original by I. N. Sneddon. Cambridge University Press, Cambridge, 2000. xii+269 pp. ISBN: 0-521-63330-3.

[Sm] J. Smoller, *Shock waves and reaction–diffusion equations,* Second edition, Grundlehren der Mathematischen Wissenschaften [Fundamental Principles of Mathematical Sciences], **258**, Springer-Verlag, New York, 1994. xxiv+632 pp. ISBN: 0-387-94259-9.

[SB] J. Stoer and R. Bulirsch, *Introduction to numerical analysis,* Springer-Verlag, New York (2002).

[Tay] G.I. Taylor, Proceedings of the Royal Society, 1910 Volume.

[TZ1] B. Texier and K. Zumbrun, *Relative Poincaré–Hopf bifurcation and galloping instability of traveling waves,* Methods Anal. and Appl. **12** (2005), no. 4, 349–380.

[TZ2] B. Texier and K. Zumbrun, *Galloping instability of viscous shock waves,* Physica D. **237** (2008) 1553-1601.

[TZ3] B. Texier and K. Zumbrun, *Hopf bifurcation of viscous shock waves in gas dynamics and MHD*, Arch. Ration. Mech. Anal. **190** (2008) 107–140.

[TZ4] B. Texier and K. Zumbrun, *Transition to longitudinal instability of detonation waves is generically associated with Hopf bifurcation to time-periodic galloping solutions*, preprint (2008).

[W] M. Williams, *Heteroclinic orbits with fast transitions: a new construction of detonation profiles,* to appear, Indiana Mathematics Journal.

[Z1] K. Zumbrun, *Multidimensional stability of planar viscous shock waves,* Advances in the theory of shock waves, 307–516, Progr. Nonlinear Differential Equations Appl., **47**, Birkhäuser Boston, Boston, MA, 2001.

[Z2] K. Zumbrun, *Stability of large-amplitude shock waves of compressible Navier–Stokes equations,* with an appendix by Helge Kristian Jenssen and Gregory Lyng, in Handbook of mathematical fluid dynamics. Vol. III, 311–533, North-Holland, Amsterdam (2004).

[Z3] K. Zumbrun, *Stability of noncharacteristic boundary layers in the standing shock limit,* to appear, Trans. Amer. Math. Soc.

[Z4] K. Zumbrun, *Conditional stability of unstable viscous shock waves in compressible gas dynamics and MHD,* to appear, Arch. Ration. Mech. Anal.

[Z5] K. Zumbrun, *Planar stability criteria for viscous shock waves of systems with real viscosity,* in Hyperbolic Systems of Balance Laws, CIME School lectures notes, P. Marcati ed., Lecture Note in Mathematics **1911**, Springer (2004).

[Z6] K. Zumbrun, *Dynamical stability of phase transitions in the p-system with viscosity-capillarity,* SIAM J. Appl. Math. **60** (2000), no. 6, 1913–1924 (electronic).

[Z7] K. Zumbrun, *Conditional stability of unstable viscous shocks,* J. Differential Equations **247** (2009), no. 2, 648–671.

[Z8] K. Zumbrun, *Center stable manifolds for quasilinear parabolic pde and conditional stability of nonclassical viscous shock waves,* preprint (2008).

[Z9] K. Zumbrun, *The refined inviscid stability condition and cellular instability of viscous shock waves,* to appear, Physica D.

[Z10] K. Zumbrun, *Numerical error analysis for Evans function computations: a numerical gap lemma, centered-coordinate methods, and the unreasonable effectiveness of continuous orthogonalization*, preprint (2009).

[Z11] K. Zumbrun, *A local greedy algorithm and higher-order extensions for global continuation of analytically varying subspaces,* To appear, Quarterly Appl. Math.

[Z12] K. Zumbrun, *Stability of detonation waves in the ZND limit*, to appear, Arch. Ration. Mech. Anal.

[Z13] K. Zumbrun, *High-frequency asymptotics and stability of ZND detonations in the high-overdrive and small-heat release limits,* preprint (2010).

[ZH] K. Zumbrun and P. Howard, *Pointwise semigroup methods and stability of viscous shock waves.* Indiana Mathematics Journal V47 (1998), 741–871; Errata, Indiana Univ. Math. J. **51** (2002), no. 4, 1017–1021.

[ZS] K. Zumbrun and D. Serre, *Viscous and inviscid stability of multidimensional planar shock fronts,* Indiana Univ. Math. J. **48** (1999) 937–992.

Part II: Specialized Research Lectures

MATHEMATICAL ASPECTS OF A MODEL FOR GRANULAR FLOW

DEBORA AMADORI* AND WEN SHEN†

Abstract. The model for granular flow being studied by the authors was proposed by Hadeler and Kuttler in [21]. In one space dimension, by a change of variable, the system can be written as a 2×2 hyperbolic system of balance laws.

Various results are obtained for this system, under suitable assumptions on initial data which leads to a strictly hyperbolic system. For suitably small initial data, the solution remains smooth globally. Furthermore, the global existence of large BV solutions for Cauchy problem is established for initial data with small height of moving layer. Finally, at the slow erosion limit as the height of moving layer tends to zero, the slope of the mountain provides the unique entropy solution to a scalar integro-differential conservation law, implying that the profile of the standing layer depends only on the total mass of the avalanche flowing downhill.

Various open problems and further research topics related to this model are discussed at the end of the paper.

Key words. Granular matter, balance laws, weakly linearly degenerate systems, global large BV solutions, slow erosion.

AMS(MOS) subject classifications. Primary 35L45, 35L50, 35L60, 35L65; Secondary 35L40, 58J45.

1. Introduction. In [21] the following model was proposed to describe granular flows

$$\begin{cases} h_t &= \operatorname{div}(h\nabla u) - (1 - |\nabla u|)h\,, \\ u_t &= \big(1 - |\nabla u|\big)h\,. \end{cases} \tag{1.1}$$

These equations describe conservation of masses. The material is divided in two parts: a moving layer with height h on top and a standing layer with height u at the bottom. The moving layer slides downhill, in the direction of steepest descent, with speed proportional to the slope of the standing layer. If the slope $|\nabla u| > 1$ then grains initially at rest are hit by rolling grains of the moving layer and start moving as well. Hence the moving layer gets bigger. On the other hand, if $|\nabla u| < 1$, grains which are rolling can be deposited on the bed. Hence the moving layer becomes smaller.

This model is studied in one space dimension by the authors [30, 2, 3]. Define $p \doteq u_x$, and assume $p \geq 0$, one can rewrite (1.1) into the following 2×2 system of balance laws

$$\begin{cases} h_t - (hp)_x &= (p-1)h\,, \\ p_t + \big((p-1)h\big)_x &= 0\,. \end{cases} \tag{1.2}$$

*Dipartimento di Matematica Pura ed Applicata, University of L'Aquila, Italy (amadori@univaq.it).

†Department of Mathematics, Penn State University, University Park, PA 16802 (shen_w@math.psu.edu). The work of the second author is partially supported by NSF grant DMS-0908047.

Writing the system of balance laws (1.2) in quasilinear form, the corresponding Jacobian matrix is computed as

$$A(h,p) = \begin{pmatrix} -p & -h \\ p-1 & h \end{pmatrix}.$$

For $h \geq 0$ and $p > 0$, one finds two real distinct eigenvalues $\lambda_1 < 0 \leq \lambda_2$, with r_1, r_2 the corresponding eigenvectors. Denote "$\bullet$" as the directional derivative, a direct computation gives

$$r_1 \bullet \lambda_1 = -\frac{2(\lambda_1+1)}{\lambda_2-\lambda_1} \approx \frac{2(p-1)}{p}, \qquad r_2 \bullet \lambda_2 = -\frac{2\lambda_2}{\lambda_2-\lambda_1} \approx -2\frac{h}{p^2}.$$

This shows the fact that the first characteristic field is genuinely nonlinear away from the line $p = 1$ and the second field is genuinely nonlinear away from the line $h = 0$, therefore the system is weakly linearly degenerate at the point $(h,p) = (0,1)$. Also, the direction of increasing eigenvalues, for the first family, changes with the sign of $p - 1$. The lines $p = 1$, $h = 0$ are characteristic curves of the first, second family respectively, along which the system becomes separated and linearly degenerate:

$$p = 1, \quad h_t - h_x = 0; \qquad h = 0, \quad p_t = 0.$$

See Figure 1 for the characteristic curves.

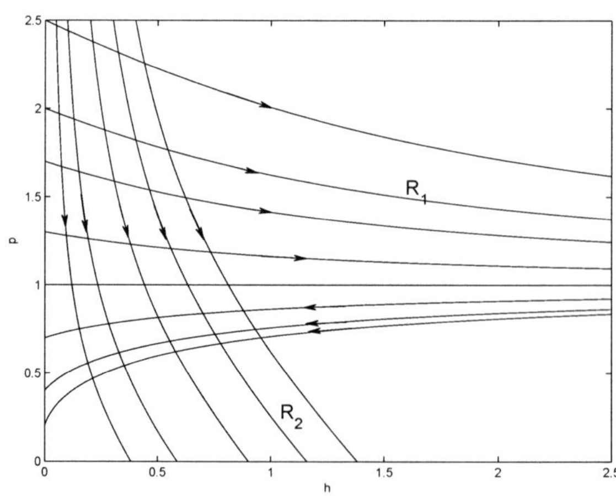

FIG. 1. *Characteristic curves of the two families in the h-p plane. The arrows point in the direction of increasing eigenvalues.*

In this paper we review some recent results about the existence of solutions for system (1.2), that are shown to exist globally in time for suitable classes of initial data. For systems of conservation laws with source term, some dissipation conditions are known in the literature that ensure the global in time existence of (smooth or weak) solutions; we refer to [23], to Kawashima–Shizuta condition (see [22]) for smooth solutions and to [15, 25] for the weak solutions. These conditions exploit a suitable balance between the differential terms and the source term that enable to control the nonlinearity of the system. It is interesting to remark that system (1.2) does not satisfy any of these conditions, nevertheless it admits global in time solutions.

For a derivation of the model (1.1) of granular flow we refer to [21]. A mathematical analysis of steady state solutions for (1.1) was carried out in [10, 11]; see also [17] for a numerical approach in one and two space dimensions.

Other models are presented in [7, 16] and in [28, 20]. We refer to [31] for a numerical study on the Savage-Hutter model [28], which describes a one-layer flow down an inclined bed in terms of its thickness h and of its velocity (or momentum).

2. Global smooth solutions. The global existence of smooth solutions is established in [30], under suitable assumptions on the initial data.

Let's first define the **decoupled initial data**

$$h(0,x) = \phi(x) \qquad\qquad p(0,x) = 1 + \psi(x) \tag{2.1}$$

with ϕ, ψ satisfying

$$\begin{cases} \phi(x) = 0 & \text{if } \ x \notin [a,b]\,, \\ \psi(x) = 0 & \text{if } \ x \notin [c,d]\,. \end{cases}$$

The intervals are disjoint, i.e., $a < b < c < d$. Moreover we assume $\psi(x) > -1$ for all x. For decoupled initial data, a global solution of the Cauchy problem can be explicitly given, namely

$$h(t,x) = \phi(x+t)\,, \qquad p(t,x) = 1 + \psi(x)\,, \qquad x \in I\!R\,, \quad t \geq 0\,.$$

Our first result provides the stability of these decoupled solutions. More precisely, every sufficiently small, compactly supported perturbation of a Lipschitz continuous decoupled solution eventually becomes decoupled. Moreover, no gradient catastrophe occurs, i.e., solutions remain smooth for all time.

THEOREM 2.1. *Let $a < b < c < d$ be given, together with Lipschitz continuous, decoupled initial data as in (2.1). Then there exists $\delta > 0$ such that the following holds. For every perturbations $\tilde{\phi}, \tilde{\psi}$, satisfying*

$$\tilde{\phi}(x) = \tilde{\psi}(x) = 0 \qquad \textit{if } \ x \notin [a,d]\,, \qquad \left|\tilde{\phi}'(x)\right| \leq \delta\,, \left|\tilde{\psi}'(x)\right| \leq \delta\,, \tag{2.2}$$

the Cauchy problem for (1.2) with initial data

$$h(0,x) = \phi(x) + \tilde{\phi}(x)\,, \qquad p(0,x) = 1 + \psi(x) + \tilde{\psi}(x)\,, \tag{2.3}$$

has a unique solution, defined for all $t \geq 0$ and globally Lipschitz continuous. Moreover, this solution becomes decoupled in finite time.

The proof replies on the method of characteristics [23]. One must bound the $\mathbf{L}^\infty$ and $\mathbf{L}^1$ norms of h_x and p_x. For details, we refer to [30].

3. Global existence of large BV solutions. For more general initial data, due to the nonlinearity of the flux, the solutions will develop discontinuities (shocks) in finite time. Solutions should be defined in the space of BV functions. Assuming the height of the moving layer h sufficiently small, in [2] we prove the global existence of large BV solutions, for a class of initial data with bounded but possibly large total variation.

More precisely, consider initial data of the form

$$h(0,x) = \bar{h}(x) \geq 0\,, \qquad p(0,x) = \bar{p}(x) > 0\,, \tag{3.1}$$

which satisfy the following properties:

$$\text{Tot.Var.}\{\bar{p}\} \leq M\,, \qquad \text{Tot.Var.}\{\bar{h}\} \leq M\,, \tag{3.2}$$

$$\|\bar{h}\|_{\mathbf{L}^1} \leq M\,, \qquad \|\bar{p}-1\|_{\mathbf{L}^1} \leq M\,, \qquad \bar{p}(x) \geq p_0 > 0\,, \tag{3.3}$$

for some constants M (possibly large) and p_0. The following theorem is proved in [2].

THEOREM 3.1. *For any constants M, $p_0 > 0$, there exists $\delta > 0$ small enough such that, if (3.2)–(3.3) hold together with*

$$\|\bar{h}\|_{\mathbf{L}^\infty} \leq \delta\,, \tag{3.4}$$

then the Cauchy problem (1.2)–(3.1) has an entropy weak solution, defined for all $t \geq 0$, with uniformly bounded total variation.

Compared with previous literature, the main novelty of this result stems from the fact that:

(i) We have arbitrarily large BV data;
(ii) We assume a small $\mathbf{L}^\infty$ bound on $\bar{h}$, but not on both initial data;
(iii) The system is strictly hyperbolic, but one of the characteristic fields is neither genuinely nonlinear nor linearly degenerate;
(iv) The system (1.2) contains source terms.

In the literature, for systems without source terms and small BV data, the global existence and uniqueness of entropy-weak solutions to the Cauchy problem are well known, using techniques such as the Glimm scheme [18, 24, 26], front tracking approximations [9, 5, 6], and vanishing viscosity approximations [8]. In some special cases, one has the existence and uniqueness of global solutions in the presence of a source term [15, 25, 12, 14, 1, 13].

However, global existence of solutions to hyperbolic systems with large BV data is a more difficult, still largely open problem. In addition to the special system [27], two main cases are known in the literature, where global existence of large BV solutions is achieved.

One is the case of Temple class systems [29], where one can measure the wave strengths in terms of Riemann invariants, so that the total strength of all wave fronts does not increase in time, across each interaction. A second major result [19] refers to general 2×2 systems, where again we can measure wave strengths in terms of Riemann coordinates; here, if the $\mathbf{L}^\infty$ norm of the solution is sufficiently small, the increase of total variation produced by the interaction is very small, and a global existence result of large BV solutions can then be established.

The validity of Theorem 3.1 relies heavily on some special properties of the hyperbolic system (1.2). First, the system is linearly degenerate along the straight line where $h = 0$. In the region where h is very small, the second field of the system is "almost-Temple class". Rarefaction curve and shock curve through the same point are very close to each other. This allows us to deduce refined interaction estimates, in which the effect of the nonlinearity is controlled by the quantity $\|h\|_{\mathbf{L}^\infty}$.

Second, the source term involves the quadratic form $h(p-1)$. Here the quantities h and $p-1$ have large, but bounded $\mathbf{L}^1$ norms. Moreover, they are transported with strictly different speeds. The total strength of the source term is thus expected to be $\mathcal{O}(1) \cdot \|h\|_{\mathbf{L}^1} \cdot \|p-1\|_{\mathbf{L}^1}$. In addition, since h itself is a factor in the source term, one can obtain a uniform bound on the norm $\|h\|_{\mathbf{L}^\infty}$, valid for all times $t \geq 0$.

For details of the proof of Theorem 3.1 we refer to [2].

4. Global large BV solutions of an initial boundary value problem. Next, we study how the mountain profile evolves when the thickness of the moving layer approaches zero, but the total mass of sliding material remains positive. This result is best formulated in connection with an initial-boundary value problem. On $\mathbb{R}_- \doteq \{x < 0\}$, consider the initial-boundary value problem for (1.2), with initial data (3.1) and the following boundary condition at $x = 0$

$$p(t,0)\,h(t,0) \;=\; F(t)\,. \tag{4.1}$$

Notice that here we prescribe the incoming flux $F(t)$ of the moving material, through the point $x = 0$. We assume

$$F(t) \geq 0\,, \qquad \text{Tot.Var}\{F\} \leq M\,, \qquad 0 < M' < \int_0^\infty F(\tau)\,d\tau \leq M\,. \tag{4.2}$$

As a partial step toward the slow erosion limit, we prove in [3] next theorem on the global existence of large BV solutions to this initial-boundary value problem, provided that $\|\bar{h}\|_{\mathbf{L}^\infty}$ and $\|F\|_{\mathbf{L}^\infty}$ are sufficiently small.

THEOREM 4.1. *Given M, $p_0 > 0$, there exists $\delta > 0$ such that the assumptions (3.2)–(3.3) and (4.2), together with*

$$\|\bar{h}\|_{\mathbf{L}^\infty} \leq \delta\,, \qquad \|F\|_{\mathbf{L}^\infty} \leq \delta\,, \tag{4.3}$$

imply that the initial-boundary value problem (1.2)-(3.1), (4.1) has a global solution, with uniformly bounded total variation for all $t \geq 0$.

The proof for Theorem 4.1 follows a similar setting as for Theorem 3.1. The additional difficulty lies in the treatment of the boundary condition at $x = 0$. Fortunately, the additional waves generated at the boundary (such as reflection waves and new entering waves) all contain a factor of $\|h\|_{\mathbf{L}^\infty}$ or the term $\|F\|_{\mathbf{L}^\infty}$, which are arbitrarily small. Therefore, same global a priori estimates as for Theorem 3.1 can be established, proving the global existence of large BV solution. For details, see [3].

5. Slow erosion limit. We now study the slow erosion limit. Numerical simulations in [30] show the following observation. When the height of the moving layer h is very small, the profile of the standing layer depends only on the total mass of the avalanche flowing downhill, not on the time-law describing at which rate the material slides down. This observation is proved rigorously in [3].

We define a new variable which measures the total mass of avalanche flowing down:

$$\mu(t) \doteq \int_0^t F(\tau)\, d\tau\,.$$

Recalling that $F(t) \geq 0$, the above function is monotone non-decreasing. Let $t(\mu)$ be its generalized inverse, and reparametrize the solution in terms of μ:

$$(\tilde{h}, \tilde{p})(\mu, x) = (h, p)\,(t(\mu), x)\,.$$

The last Theorem gives the slow erosion limit.

THEOREM 5.1. *Assume all the assumption in Theorem 4.1 hold. Then, as* $\|\bar{h}\|_{\mathbf{L}^\infty} \to 0$ *and* $\|F\|_{\mathbf{L}^\infty} \to 0$, *the rescaled* p *component of the solutions to the initial boundary value problem (1.2)–(3.1)–(4.1) converges to a limit function* $\hat{p}$, *which provides the unique entropy solution to the scalar integro-differential conservation law*

$$p_\mu + \left(\frac{p-1}{p} \cdot \exp \int_x^0 \frac{p(\mu, y) - 1}{p(\mu, y)}\, dy\right)_x = 0\,, \tag{5.1}$$

with initial data $\hat{p}(0, x) = \bar{p}(x)$ *for* $x < 0$.

A formal derivation is obtained as follows. For simplicity, assume $\bar{h}(x) = 0$ for $x < 0$. We stretch a given boundary data $\bar{F}(t) > 0$ by defining $F^\varepsilon(t) = \varepsilon\, \bar{F}(\varepsilon\, t)$. Introduce the new variable $\mu = \mu^\varepsilon(t)$,

$$\mu^\varepsilon(t) \doteq \int_0^t F^\varepsilon(s)\, ds = \int_0^{\varepsilon t} \bar{F}(s)\, ds\,, \qquad \text{so } \mu'(t) = F^\varepsilon(t) = \varepsilon \bar{F}(\varepsilon t)\,.$$

Using $\mu = \mu^\varepsilon$ as a rescaled time variable, the equations in (1.1) can now be rewritten as

$$\begin{cases} \mu' h^\varepsilon_\mu - (h^\varepsilon p^\varepsilon)_x & = (p^\varepsilon - 1)h^\varepsilon \,, \\ \mu' p^\varepsilon_\mu + \big((p^\varepsilon - 1)h^\varepsilon\big)_x & = 0 \,. \end{cases} \tag{5.2}$$

Taking the limit as $\varepsilon \to 0$ in the first equation, the term $\mu' h^\varepsilon_\mu$ turns out to be of higher order w.r.t. ε, while $h = \mathcal{O}(\varepsilon)$ and $p = \mathcal{O}(1)$.

Introducing the new variable $m \doteq hp/\varepsilon$, from (5.2) we formally obtain

$$-m_x = \frac{p-1}{p}\, m \,, \tag{5.3}$$

$$\bar{F}(\mu) p_\mu + \left(\frac{p-1}{p}\, m\right)_x = 0 \,. \tag{5.4}$$

Integrating the equation in (5.3) with proper boundary conditions, one obtains

$$m(\mu, x) = \exp\left(\int_x^\infty \frac{p(\mu, y) - 1}{p(\mu, y)}\, dy\right) \bar{F}(\mu) \,. \tag{5.5}$$

Roughly speaking, this is the size of the avalanche at the time when it crosses a given point $x < 0$. By inserting (5.5) in (5.4), and dividing both terms by the common factor $\bar{F}(\mu)$, we obtain (5.1).

The key point in the proof of Theorem 5.1 is to show that, taking a converging sequence of p-component of solutions to the initial boundary value problem (1.2)-(3.1)-(4.1), the limit $\hat{p}$ is a weak solution to the conservation law (5.1). This is achieved by passing to the limit in the corresponding weak formulations. Here one needs the weak convergence of the flux hp and the strong convergence of the function $\frac{p-1}{p}$

$$\tilde{p}(\mu, x)\, \tilde{h}(\mu, x) \rightharpoonup \exp \int_x^0 \frac{\hat{p}(\mu, \zeta) - 1}{\hat{p}(\mu, \zeta)}\, d\zeta \,, \qquad \frac{\tilde{p}(\mu, x) - 1}{\tilde{p}(\mu, x)} \to \frac{\hat{p}(\mu, x) - 1}{\hat{p}(\mu, x)} \,.$$

The above convergence is obtained in [3] using a compactness argument. By showing that the limiting integro-differential equation (5.1) is well posed, the convergence is extended to the whole sequence. The well-posedness of (5.1) is non-trivial because the flux is a global function. In the forthcoming work [4], we prove that the flow generated by the integro-differential equation (5.1) is Lipschitz continuous restricted to the domain of functions satisfying the bounds:

$$\inf_{x<0} p(x,t) \geq p_o > 0 \,, \quad \text{Tot.Var. } p(\cdot, t) \leq M \,, \quad \|p(\cdot, t) - 1\|_{\mathbf{L}^1(\mathbb{R}_-)} \leq M \,.$$

6. Further discussion. In this final section we discuss various interesting open problems related to this model.

(A). Uniqueness of entropy weak solutions for (1.2). After having established the global existence of large BV solutions to (1.2), it remains open the question of uniqueness of entropy-weak solutions. This does not immediately follow from the known results [14], because one of the characteristic fields is neither genuinely nonlinear, nor linearly degenerate. Uniqueness and continuous dependence of large BV solutions may be approached using the techniques in [5, 14].

(B). The Cauchy problem with a source term $f = f(t,x)$. The original model in [21] takes into account also precipitation effects. This corresponds to supplementing the first equation in (1.1) by an extra term $f = f(t,x) \geq 0$ in the right hand side, representing additional material that increments the size of the moving layer. In one dimension we are led to

$$\begin{cases} h_t - (hp)_x &= (p-1)h + f\,, \\ p_t + \big((p-1)h\big)_x &= 0\,. \end{cases} \tag{6.1}$$

Notice that this system still decouples for $p \equiv 1$: in this case, h satisfies the scalar equation $h_t - h_x = f(t,x)$.

For system (6.1), can one still establish the global existence result? What will be the proper assumptions on $f(t,\cdot)$? To start, one can assume that the support of $f(t,\cdot)$ is uniformly bounded, that

$$\int_0^\infty [\text{Tot.Var.}\, f(t,\cdot)]\; dt \leq C\,,$$

and that the norm $\|f\|_{\mathbf{L}^\infty}$ is sufficiently small, depending on the above constant C.

(C). Mountain slope changes sign. It is interesting to check whether the equations (1.1) have meaningful solutions also when the slope $p = u_x$ changes sign. In this case, the one-dimensional model takes the form

$$\begin{cases} h_t - (hp)_x &= (|p|-1)h + f\,, \\ p_t + \big((|p|-1)h\big)_x &= 0\,. \end{cases} \tag{6.2}$$

Since all these models do not account for the conservation of momentum, one might wonder if the predictions are realistic, also in cases where $p \approx 0$ and the actual motion of the granular matter may be dominated by inertial forces. We remark that, when p is allowed to change sign, the flux in (6.2) is no longer smooth but only Lipschitz continuous. From the point of view of basic theory, this is a situation not covered by standard existence and uniqueness results, and should be examined specifically.

(D). Radially symmetric solutions. As an intermediate step toward the fully two-dimensional model, one can study radially symmetric solutions in $\mathbb{R}^2$. By writing the system (1.1) into polar coordinates (r,θ) and looking for solutions $(h,u)(t,r)$, one reduces to

$$\begin{cases} h_t - (hp)_r &= (|p|-1)h + \frac{p}{r}h\,, \\ p_t + ((|p|-1)h)_r &= 0 \end{cases} \tag{6.3}$$

where $p \doteq u_r$ and $r > 0$. One may reduce to consider a conic-shaped mountain, therefore assuming $p < 0$. The resulting system is quite similar to (1.2), but differs for the additional source term ph/r, whose dependence on the space variable r is not integrable on the half line.

Notice that, in the simple case $p \equiv -1$, (6.3) reduces to the equation $h_t + h_r = -h/r$. Assuming that

$$h(0,r) \;=\; h_0(r) = \frac{\Phi(r)}{r}$$

with $h_0(r) \to 0$ as $r \to 0+$, then the solution is given by

$$h(r,t) = \begin{cases} \dfrac{\Phi(r-t)}{r} & \text{if } r > t \\ 0 & \text{if } 0 < r < t\,. \end{cases}$$

Here, due to the spreading of the mass, h becomes smaller as r grows.

A very interesting problem is to reach a priori BV bounds for the system (6.3), for suitable boundary conditions at $r = r_0 > 0$. For this case, the BV bound estimates run into several difficulties: (i) the source term depends on the space variable r in a non-integrable way; and (ii) the characteristic speed is not strictly bounded away from 0, unless p is proved to remain strictly different from 0.

(E). The two dimensional case. To present date, the mathematical analysis of two-dimensional granular flow model (1.1) has been mainly concerned with steady state solutions [10, 11] and has been approached numerically in [17]. Furthermore, the general existence-uniqueness result for a two-dimensional hyperbolic system such as (1.1) is not yet available. However, the special structure of this system suggests a possible line of attack, based on a multi-dimensional extension of the integro-differential formula (5.1).

For example, consider the case where we initially have a steady sandpile of height $u_0(x)$ on a table Ω and $h_0 \equiv 0$. Start to pour more sand on top of it at a very slow rate $f = \varepsilon \hat{f}(x)$; then the sand falls off as it reaches the edge of the table.

As $\varepsilon \to 0$, a formal extension of the formula (5.1) leads to the following approximate evolution equation:

$$u_t(t,x) = (1 - |\nabla u|)\, H(t,x)\,,$$

where H solves the following linear equation for every fixed t

$$\nabla u \cdot \nabla H + \big(|\nabla u| - 1 + \Delta u\big) H + \hat{f} = 0\,,$$

with the initial and boundary data

$$u(0,x) = u_0(x)\,,\;\; h(0,x) \equiv 0 \;\; \text{for} \;\; x \in \Omega\,, \qquad u(x) \equiv 0\,, \;\; \text{for} \;\; x \in \partial\Omega\,.$$

REFERENCES

[1] D. Amadori and G. Guerra, *Uniqueness and continuous dependence for systems of balance laws with dissipation.* Nonlinear Anal. TMA **49** (2002), 987–1014.

[2] D. Amadori and W. Shen, *Global Existence of large BV solutions in a Model of Granular Flow.* Comm. Partial Differential Equations **34** (2009), 1003–1040.

[3] D. Amadori and W. Shen, *The Slow Erosion Limit for a Model of Granular Flow.* Arch. Ration. Mech. Anal. (2010), to appear.

[4] D. Amadori and W. Shen, *An Integro-Differential Conservation Law arising in a Model of Granular Flow.* In preparation.

[5] F. Ancona and A. Marson, *Well-posedness for general* 2×2 *systems of conservation laws.* Mem. Amer. Math. Soc. **169** (2004), no. 801.

[6] F. Ancona and A. Marson, *Existence Theory by Front Tracking for General Nonlinear Hyperbolic Systems.* Arch. Ration. Mech. Anal. **185** (2007), 287–340.

[7] T. Boutreux and P.-G. de Gennes, *Surface flows of granular mixtures, I. General principles and minimal model.* J. Phys. I France **6** (1996), 1295–1304.

[8] S. Bianchini and A. Bressan, *Vanishing viscosity solutions to nonlinear hyperbolic systems. Ann. of Math.* **161** (2005), 223–342.

[9] A. Bressan, *Hyperbolic Systems of Conservation Laws. The One Dimensional Cauchy Problem*, Oxford University Press, 2000.

[10] P. Cannarsa and P. Cardaliaguet, *Representation of equilibrium solutions to the table problem for growing sandpiles.* J. Eur. Math. Soc. **6** (2004), 435–464.

[11] P. Cannarsa, P. Cardaliaguet, G. Crasta, and E. Giorgieri, *A boundary value problem for a PDE model in mass transfer theory: representation of solutions and applications.* Calc. Var. PDE. **24** (2005), 431–457.

[12] C. Christoforou, *Hyperbolic systems of balance laws via vanishing viscosity.* J. Differential Equations **221** (2006), 470–541.

[13] C. Christoforou, *Uniqueness and sharp estimates on solutions to hyperbolic systems with dissipative source.* Comm. Partial Differential Equations **31** (2006), 1825–1839.

[14] G. Crasta and B. Piccoli, *Viscosity solutions and uniqueness for systems of inhomogeneous balance laws.* Discrete Contin. Dyn. Systems **3** (1997), 477–502.

[15] C.M. Dafermos and L. Hsiao, *Hyperbolic systems of balance laws with inhomogeneity and dissipation.* Indiana U. Math. J. **31** (1982), 471–491.

[16] J. Duran, *Sands, Powders, and Grains: An Introduction to the Physics of Granular Materials*, Springer-Verlag, 2000.

[17] M. Falcone and S. Finzi Vita, *A finite-difference approximation of a two-layer system for growing sandpiles.* SIAM J. Sci. Comput. **28** (2006), 1120–1132.

[18] J. Glimm, *Solutions in the large for nonlinear hyperbolic systems of equations.* Comm. Pure Appl. Math. **18** (1965), 697–715.

[19] J. Glimm and P. Lax, *Decay of solutions of systems of nonlinear hyperbolic conservation laws*, Mem. Amer. Math. Soc. **101** (1970).

[20] J.M.N.T. Gray, Y.-C. Tai, and S. Noelle, *Shock waves, dead zones and particle-free regions in rapid granular free-surface flows.* J. Fluid Mech. **491** (2003), 161–181.

[21] K.P. Hadeler and C. Kuttler, *Dynamical models for granular matter.* Granular Matter **2** (1999), 9–18.

[22] B. Hanouzet and R. Natalini, *Global existence of smooth solutions for partially dissipative hyperbolic systems with a convex entropy.* Arch. Ration. Mech. Anal. **169** (2003), 89–117.

[23] T.T. Li, *Global classical solutions for quasilinear hyperbolic systems*, John Wiley & Sons, Chichester, 1994.

[24] T.P. Liu, *Admissible solutions of hyperbolic conservation laws.* Mem. Amer. Math. Soc. **30** (1981), no. 240.

[25] T.P. LIU, *Quasilinear hyperbolic systems.* Comm. Math. Phys. **68** (1979), 141–172.

[26] T.P. LIU AND T. YANG, *Weak Solutions of General Systems of Hyperbolic Conservation Laws.* Comm. Math. Phys. **230** (2002), 289–327.

[27] T. NISHIDA AND J. SMOLLER, *Solutions in the large for some nonlinear conservation laws.* Comm. Pure Appl. Math. **26** (1973), 183–200.

[28] S.B. SAVAGE AND K. HUTTER, *The dynamics of avalanches of granular materials from initiation to runout. Part I: Analysis.* Acta Mech. **86** (1991), 201–223.

[29] D. SERRE, *Systems of Conservation Laws I, II,* Cambridge University Press, 2000.

[30] W. SHEN, *On the Shape of Avalanches.* J. Math. Anal. Appl. **339** (2008), 828–838.

[31] Y.-C. TAI, S. NOELLE, J.M.N.T. GRAY, AND K. HUTTER, *Shock capturing and front tracking methods for granular avalanches.* J. Comput. Phys. **175** (2002), 269–301.

THE FLOW ASSOCIATED TO WEAKLY DIFFERENTIABLE VECTOR FIELDS: RECENT RESULTS AND OPEN PROBLEMS

LUIGI AMBROSIO*

Abstract. We illustrate some recent developments of the theory of flows associated to weakly differentiable vector fields, listing the regularity/structural conditions considered so far, extensions to state spaces more general than Euclidean and open problems.

Key words. Continuity equation, transport equation, flow.

1. Introduction. In the last few years quite some progress has been made on the well-posedness of the continuity and transport equation

$$\partial_t w_t + \nabla \cdot (\boldsymbol{b}_t w_t) = 0, \tag{1.1}$$

$$\partial_t f_t + \boldsymbol{b}_t \cdot \nabla f_t = 0 \tag{1.2}$$

and the relation of these well-posedness results with the existence and stability of the flow $\boldsymbol{X}(t,x)$ associated to $\boldsymbol{b}$, namely the family of solutions to

$$\dot{x}(t) = \boldsymbol{b}_t(x(t)) \qquad \text{for a.e. } t \in (0,T). \tag{1.3}$$

Here $\boldsymbol{b}(t,x) = \boldsymbol{b}_t(x) : (0,T) \times \mathbb{R}^d \to \mathbb{R}^d$ is a possibly nonautonomous Borel velocity field in $\mathbb{R}^d$. In order to fix the ideas and to avoid global issues, I will make the standing assumption that $\boldsymbol{b}$ is globally bounded and I will focus, in the same spirit of [4], on the continuity equation (1.1) only, departing a bit from the seminal paper [42]: the conservative form is more amenable for nonsmooth vector fields, possibly having also an unbounded divergence (and in the case of bounded divergence there is no essential difference between (1.1) and (1.2), provided we allow right hand sides of the form cw_t, cf_t).

Let us recall the definition of Regular Lagrangian Flow (RLF in short) associated to $\boldsymbol{b}$:

DEFINITION 1.1 ($\mathscr{L}^d$-RLF in $\mathbb{R}^d$). *Let $\boldsymbol{X}(t,x) : [0,T] \times \mathbb{R}^d \to \mathbb{R}^d$. We say that $\boldsymbol{X}(t,x)$ is a $\mathscr{L}^d$-RLF in $\mathbb{R}^d$ (relative to $\boldsymbol{b}$) if the following two conditions are fulfilled:*

(i) for $\mathscr{L}^d$-a.e. x, the function $t \mapsto \boldsymbol{X}(t,x)$ is an absolutely continuous integral solution to the ODE (1.3) in $[0,T]$ with $\boldsymbol{X}(0,x) = x$;

(ii) $\boldsymbol{X}(t,\cdot)_\#\mathscr{L}^d \le C\mathscr{L}^d$ for all $t \in [0,T]$, for some constant C independent of t.

*Scuola Normale Superiore, Piazza dei Cavalieri 7, 56126 Pisa, Italy (l.ambrosio@sns.it).

Here and in the sequel I use the notation $\mathscr{L}^d$ for the Lebesgue measure in $\mathbb{R}^d$ and I use $f_\sharp$ to denote the push forward operator between measures induced by f, namely $f_\sharp\mu(E) = \mu(f^{-1}(E))$.

Notice that, while (i) imposes a natural condition on the *individual* paths $\boldsymbol{X}(\cdot, x)$, (ii) should be understood as a *global* condition on this family of paths: heuristically it means that we are selecting paths which do not concentrate too much, and we don't rule out the possibility of the existence of concentrating paths (see in particular the illustration of the square root example in [4]). The following result is proved in [4, Theorem 19] for the part concerning existence and in [4, Theorem 16, Remark 17] for the part concerning uniqueness.

THEOREM 1.1 (Existence and uniqueness of the $\mathscr{L}^d$-RLF). *Assume that* (1.1) *has (forward) existence and uniqueness in* $L^\infty_+([0,T]; L^1 \cap L^\infty(\mathbb{R}^d))$. *Then the* $\mathscr{L}^d$*-RLF* $\boldsymbol{X}$ *exists and is unique.*

Here uniqueness is understood in the following sense: if $\boldsymbol{X}$ and $\boldsymbol{Y}$ are $\mathscr{L}^d$-RLF's, then $\boldsymbol{X}(\cdot, x) = \boldsymbol{Y}(\cdot, x)$ for $\mathscr{L}^d$-a.e. $x \in \mathbb{R}^d$.

In the next sections I will address several questions relative to Theorem 1.1 and some recent progress made in this area. The main question is: which assumptions on $\boldsymbol{b}$ ensure that (1.1) has existence and uniqueness in $L^\infty_+([0,T]; L^1 \cap L^\infty(\mathbb{R}^d))$?

2. Regularity of the vector field. DiPerna and Lions proved in [42], among many other things, that (1.1) has existence and uniqueness in the class

$$L^\infty([0,T]; L^1 \cap L^\infty(\mathbb{R}^d))$$

provided $\boldsymbol{b}_t \in W^{1,1}_{\mathrm{loc}}$ for a.e. $t \in (0,T)$, with:

(i) $\int_{B_R} |\nabla \boldsymbol{b}_t|\,dx$ integrable in $(0,T)$ for all $R > 0$;

(ii) $\|\nabla \cdot \boldsymbol{b}_t\|_\infty \in L^1(0,T)$.

Actually an inspection of the proof in [42] (or the stability under time reversal of the assumptions) shows that (ii) ensures both forward and backward uniqueness. On the other hand, bounds on the negative part of the divergence suffice to obtain forward well-posedness, and it was pointed out in [2] that only forward uniqueness is needed for the uniqueness of the (forward) $\mathscr{L}^d$-RLF $\boldsymbol{X}$. Also, existence and uniqueness in the smaller class $L^\infty_+([0,T]; L^1 \cap L^\infty(\mathbb{R}^d))$ of nonnegative solutions is sufficient to that purpose. We shall comment more on the bounds on divergence in the next section, and focus here on the regularity of $\boldsymbol{b}$:

- (Vector fields in LD) It was noticed in [26] that the *isotropic* smoothing scheme of [42], on which the uniqueness proof relies, works under the only assumption that the symmetric part $Du + {}^tDu$ of the distributional derivative is absolutely continuous. This vector space, usually denoted by LD in the theory of linear elasticity [60], can be strictly larger than $W^{1,1}$. Notice however that $Du + {}^tDu \in L^p_{\mathrm{loc}}$ for some $p > 1$ implies $u \in W^{1,p}_{\mathrm{loc}}$ by a local version of Korn's inequality.

• (Vector fields in BV) Bouchut has been the first one to achieve in [18] this extension, for Hamiltonian vector fields of the form $\boldsymbol{b}_t(x,p) = (p, V_t(x))$, with $V_t \in BV_{\mathrm{loc}}$. The proof uses a clever *anisotropic* smoothing scheme, where mollification in the "bad" variables x occurs at a faster rate. In [2] (see also [30], [29] for intermediate results) the scheme has been improved and extended to all BV_{loc} vector fields (with global integrability in time of the total variations $|D\boldsymbol{b}_t|(B_R)$ for all $R > 0$), under the assumptions that $D \cdot \boldsymbol{b}_t \ll \mathscr{L}^d$ and that the density $\mathrm{div}\boldsymbol{b}_t$ satisfies

$$\| [\mathrm{div}\boldsymbol{b}_t]^- \|_\infty \in L^1(0,T). \tag{2.1}$$

• (Vector fields in BD) Recall that BD consists in the space of functions u such that $(Du + {}^t Du)$ are representable by measures. The extension to BD vector fields is still an open problem: indeed, one is tempted to use symmetric mollifiers as in [26]. But, we know that even for BV vector fields anisotropic mollifiers are needed to get the result. In [6] we follow a different path and we achieve the result for SBD vector fields $\boldsymbol{b}_t$ (namely we need that $(D\boldsymbol{b}_t + {}^t D\boldsymbol{b}_t)$ has no "Cantor" part).

• (Vector fields representable as singular integrals) Recently Bouchut and Crippa achieved in [24] a very nice extension of the theory to vector fields $\boldsymbol{b}_t$ that can be represented as a singular integral

$$\boldsymbol{b}_t(x) = \int_{\mathbb{R}^d} K(x-y) F_t(y)\, dy \tag{2.2}$$

with $F_t \in [L^1(\mathbb{R}^d)]^d$. Here K is a matrix-valued map whose components satisfy the standard assumption of the theory of singular integral operators (in particular $|K(x)| \sim |x|^{1-d}$ as $|x| \to 0$), so that weak L^1 estimates are available. The proof uses a very clever improvement of the maximal estimates used in [10] and [33] (see Section 4 below) to obtain regularity properties of the flow and effective stability estimates: the main new ingredient is the use of a suitable family of maximal operators, instead of the standard maximal operator on balls.

Notice that the class (2.2) includes $W^{1,1}$ functions g (and vector fields), because the solution to $\Delta w = \nabla \cdot F$ is representable by (with appropriate boundary conditions)

$$w(x) = c(d) \int_{\mathbb{R}^d} |y-x|^{2-d} \nabla \cdot F(y)\, dy = c(d)(d-2) \int_{\mathbb{R}^d} \frac{y-x}{|y-x|^d} F(y)\, dy.$$

Choosing $F = \nabla g$ gives $w = g$ and hence the desired representation of g.

It is also easily seen that the class of vector fields (2.2) is not contained in $W^{1,1}$ or in BV, so that definitely [24] provides new results (and also new applications to stability of solutions to incompressible Euler equations with vorticity in L^1). In this direction, it would be very nice to have an extension of this result to the case when F_t is just a measure, and not necessarily an L^1 function. This would include BV and even BD vectorfields, by the same elliptic PDE argument illustrated above.

• (Vector fields having a special structure) In this research field it is difficult to imagine a result better than the others. Indeed, sometimes the special structure of the vector field of $\boldsymbol{b}$ helps a lot in getting well-posedness, under very mild regularity conditions. I will illustrate this by two examples.

The first example concerns bounded, divergence-free autonomous vector fields $\boldsymbol{b}$ in the plane; they can obviously can be represented as rotations of ∇H, for some Lipschitz potential function H, and solutions to the ODE should preserve H. This suggests a factorization of the dynamics on the level sets of H. This argument has been used by Bouchut and Desvillettes in [22] (see also [28], [45] for related results) to obtain well-posedness under an additional local "regularity" assumption on $\boldsymbol{b}$, and the assumption that H maps the critical set $\Sigma := \{\nabla H = 0\}$ into a $\mathscr{L}^1$-negligible set (notice that by Sard's theorem this condition always holds if $\boldsymbol{b}$ is smooth). A much more detailed analysis, performed by Alberti, Bianchini and Crippa in [1], reveals that no extra regularity of $\boldsymbol{b}$ is needed, and that the "weak Sard" condition

$$H_\sharp(\chi_\Sigma \mathscr{L}^2) \ll \mathscr{L}^1$$

suffices for well-posedness of (1.1). Also, it turns out that a further refinement of this condition, involving also the topology of the level sets of H, is *necessary and sufficient* for well-posedness.

The second example concerns vector fields $\boldsymbol{B}_t$ of the form

$$\boldsymbol{B}_t(x,y) := \big(\boldsymbol{b}_t(x), \nabla \boldsymbol{b}_t(x) y\big) \tag{2.3}$$

with $\boldsymbol{b}_t \in W^{1,1}_{\rm loc}$. Here we see that the last d components of $\boldsymbol{B}_t$ have no regularity in x, but they are very regular in y (on the other hand the divergence of $\boldsymbol{B}_t$ is nice as long as the divergence of $\boldsymbol{b}_t$ is nice, since $\nabla \cdot \boldsymbol{B}_t = 2\nabla \cdot \boldsymbol{b}_t$). Lions and Le Bris used in [48] this particular structure and adapted Bouchut's scheme [18] to obtain well-posedness of (1.1) (see also [49] for related results in a BV framework). Here the function space where well-posedness occurs is adapted to $\boldsymbol{B}_t$:

$$L^\infty_+\left([0,T]; L^1 \cap L^\infty(\mathbb{R}^{2d})\right) \cap L^\infty\left([0,T]; L^\infty_{\rm loc}(\mathbb{R}^d_x; L^1(\mathbb{R}^d_y))\right).$$

The reason for the restriction to this smaller space is the fact that $|\boldsymbol{B}_t|/(1+|x|+|y|)$ in general does not belong to

$$L^1\left([0,T]; L^1(\mathbb{R}^d_x \times \mathbb{R}^d_y)\right) + L^1\left([0,T]; L^\infty(\mathbb{R}^d_x \times \mathbb{R}^d_y)\right),$$

because the last d components do not tend to 0 as $|y| \to \infty$ while x is kept fixed (and their limit is possibly unbounded as a function of x). For this reason, a weaker growth condition on $\boldsymbol{B}_t$ turns into a stronger growth condition on w, still compatible with existence of solutions.

As we will see later on, the vector fields (2.3) occur in the study of the differentiability properties of the $\mathscr{L}^d$-RLF flow $\boldsymbol{X}$ associated to $\boldsymbol{b}_t$.

3. Bounds on divergence. The bound (2.1) on the divergence might be too restrictive for some applications, where even a singular divergence might appear. As an application, let us consider the multidimensional version of the Keyfitz-Kranzer system considered in [25], [7], [5]:

$$\partial_t u + \sum_{i=1}^{d} \partial_i\big(\boldsymbol{f}(|u|)u\big) = 0$$

with $u : (0,+\infty) \times \mathbb{R}^d \to \mathbb{R}^k$ and $\boldsymbol{f} : \mathbb{R}_+ \to \mathbb{R}^k$ smooth. The system can formally be decoupled into a scalar conservation law and a transport equation, in the polar variables $u = \rho\theta$, $\rho = |u|$:

$$\partial_t \rho + \nabla \cdot \big(\boldsymbol{f}(\rho)\rho\big) = 0, \qquad \partial_t \theta + f(\rho) \cdot \nabla\theta = 0.$$

If the initial condition $\bar{\rho} = |\bar{u}|$ is sufficiently nice, say BV_{loc} and L^∞, then Kruzkhov theory provides us with the unique entropy solution $t \mapsto \rho_t$ of the scalar conservation law, and this solution is locally BV on bounded sets of $(0,+\infty) \times \mathbb{R}^d$. The vector field $\boldsymbol{b}_t := \boldsymbol{f}(\rho_t)$ appearing in the transport equation for θ is bounded and BV, but its distributional divergence need not be bounded or even absolutely continuous (for instance this can be seen by computing the divergence on shocks, where ρ is discontinuous, if f is injective). In [7], [5] this difficulty has been bypassed by considering the autonomous, divergence-free and $(d+1)$-dimensional vector field

$$\boldsymbol{B}(t,x) := \big(\rho_t(x), f(\rho_t(x))\rho_t(x)\big)$$

and building from the flow of $\boldsymbol{B}$ a "natural" flow of $\boldsymbol{b}_t$ (by a reparameterization).

More generally, we may think that whenever we have a nonnegative function ρ satisfying

$$\partial_t \rho_t + \nabla \cdot (\boldsymbol{b}_t \rho_t) = 0 \tag{3.1}$$

then we should think to $\rho\mathscr{L}^d$ as our new reference measure and try to obtain well-posedness, at least in the regions where ρ does not vanish. This is particularly clear if ρ is independent of time: in this case $\nabla \cdot (\boldsymbol{b}_t\rho) = 0$ precisely corresponds to the fact that the ρ-divergence of $\boldsymbol{b}_t$, namely the $L^2(\rho)$ adjoint of the gradient, vanishes.

This point of view has been used in [8] (and more recently in [13]). The uniqueness of the flow and the well-posedness of the PDE when $\boldsymbol{b}_t \in BV$ and the function ρ in (3.1) belongs to L^∞ are still open problems: the main result of [8] is to answer these questions affirmatively when $\boldsymbol{b}_t \in SBV$, and to relate this problem to a compactness conjecture of Bressan [25]: we prove that if the limit field in Bressan's conjecture is BV, then compactness hold (see also [32] for a positive answer to the conjecture under $W^{1,p}$ bounds, $p > 1$).

4. Differentiability of the flow and effective stability results. In this section we start from the discussion of the differentiability properties of the $\mathscr{L}^d$-RLF associated to $W^{1,1}$ vector fields, briefly describing the results obtained in [48]. On the other hand, in the $W^{1,p}$ case, with $p > 1$, much stronger results are available [33], as we will see. The main theorem in [48] is the following:

THEOREM 4.1. *Let $\boldsymbol{b} \in L^1\Big((0,T); W^{1,1}_{\rm loc}(\mathbb{R}^d;\mathbb{R}^d)\Big)$ be satisfying*

(i) $\dfrac{|\boldsymbol{b}|}{1+|x|} \in L^1\left((0,T); L^1(\mathbb{R}^d)\right) + L^1\left((0,T); L^\infty(\mathbb{R}^d)\right)$;

(ii) $[\mathrm{div}\boldsymbol{b}_t] \in L^1\left((0,T); L^\infty(\mathbb{R}^d)\right)$;

and let $\boldsymbol{X}(t,x)$ be the corresponding $\mathscr{L}^d$-Lagrangian flow. Then for a.e. $t \in (0,T)$ there exists a measurable function $\boldsymbol{W}_t : \mathbb{R}^d \to M^{d\times d}$ such that

$$\frac{\boldsymbol{X}(t,x+\varepsilon y)-\boldsymbol{X}(t,x)}{\varepsilon} \to \boldsymbol{W}_t(x)y \quad \textit{locally in measure in } \mathbb{R}^d_x \times \mathbb{R}^d_y \tag{4.1}$$

as $\varepsilon \downarrow 0$, uniformly in time.

Notice that, in general, $\boldsymbol{W}_t$ need not be locally integrable. Actually in [48] it is not proved, or stated, that the limit of difference quotients is linear in y, but this can be proved by an approximation argument, using the stability properties of flows: indeed, it turns out that the difference quotients of $\boldsymbol{X}$, together with $\boldsymbol{X}$, are solutions to the ODE relative to

$$B^\varepsilon_t(x,y) := \big(\boldsymbol{b}_t(x), \frac{\boldsymbol{b}_t(x+\varepsilon y)-\boldsymbol{b}_t(x)}{\varepsilon}\big)$$

whose limit is precisely the vector field $\boldsymbol{B}_t$ in (2.3). Hence, $\boldsymbol{W}_t(x)y$ can be recovered from the flow of $\boldsymbol{B}_t$ and it is clear that the linear structure is preserved by a smooth approximation of $\boldsymbol{b}_t$ (and of $\boldsymbol{B}_t$ as well). In [12] we called $\boldsymbol{W}_t(x)$, as defined by (4.1), "derivative in measure": though strictly weaker than many other differentiability concept (even approximate differentiability, as proved in [12]), (4.1) is the only differentiability property known in the $W^{1,1}$ case, and unknown in the BV case.

On the other hand, in the $W^{1,p}$ case, $p > 1$, stronger results are available. The first results relative to the approximative differentiability of the flow have been obtained in [10], and then improved substantially in [33]. An important fact is that the approach is completely different from the one of [48]: the main idea is to estimate the difference quotient of $\boldsymbol{b}_t$ in terms of maximal functions of the modulus of $\nabla\boldsymbol{b}_t$. More precisely, for all $g \in W^{1,1}_{\rm loc}$ we have the *pointwise* inequality

$$|g(x)-g(y)| \le c_d|x-y|\left(M|\nabla g|(x) + M|\nabla g|(y)\right)$$

at all Lebesgue points x, y of g. The heuristic idea is to use this, in conjunction with maximal inequalities and the no-concentration property ((ii) in Definition 1.1) to get regularity properties of the flow.

The starting point of the estimates given in [33] is already present in [10] (and, at least in a formal way, in the introduction of [42]): in a smooth context, since we can differentiate in space the ODE to get

$$\frac{d}{dt}\nabla\boldsymbol{X}(t,x) = \nabla\boldsymbol{b}_t\big(\boldsymbol{X}(t,x)\big)\nabla\boldsymbol{X}(t,x)$$

(i.e. the spatial gradient satisfies a *linear* ODE), so that we can control from above the time derivative $\frac{d}{dt}\log\left(|\nabla\boldsymbol{X}|\right)$ with $|\nabla\boldsymbol{b}_t|(\boldsymbol{X})$. The latter quantity belongs to L^p thanks to the regularity of the flow. The strategy of [10] allows to make this remark rigorous: it is possible to consider some integral quantities which contain a discretization of the space gradient of the flow, more stable by approximation (by the concavity of the logarithm, which results in a lack of lower semicontinuity, there is no way to pass to the limit in the differential inequality, as stated above, from smooth to nonsmooth flows).

Now we state simplified versions of the results of [33], referring to that paper for the most general statements.

THEOREM 4.2 (Lipschitz estimates). *Let $p > 1$ and let $\boldsymbol{b}_t \in W^{1,p}_{\mathrm{loc}}$ be divergence-free, uniformly bounded, with $\int_0^T \int_{B_R} |\nabla\boldsymbol{b}_t|^p \,dxdt < \infty$ for all $R > 0$. Then, for every ε, $R > 0$, we can find a compact set $K \subset B_R(0)$ such that*

(i) $\mathscr{L}^d(B_R(0)\setminus K) \leq \varepsilon$;

(ii) the restriction of $\boldsymbol{X}$ to $[0,T]\times K$ is Lipschitz continuous.

This result is remarkable: it shows that we can recover somehow the standard Cauchy-Lipschitz theory provided we remove sets of small measure (the optimal statement, not given here, is quantitative). We conclude this section showing another result obtained in [33] with techniques which are very similar to the ones described so far: it provides a logarithmic error estimate, implying in particular an effective stability result for RLF's. Of course this result yields uniqueness of the flow as a consequence, in this respect see also [46].

THEOREM 4.3 (Quantitative stability). *Let $\boldsymbol{b}$ and $\tilde{\boldsymbol{b}}$ be vector fields as in the previous theorem, and let $\boldsymbol{X}$ and $\tilde{\boldsymbol{X}}$ be the respective flows. Then, for every $\tau \in [0,T]$, we have*

$$\|\boldsymbol{X}(\tau,\cdot) - \tilde{\boldsymbol{X}}(\tau,\cdot)\|_{L^1(B_r(0))} \leq C\left|\log\left(\|\boldsymbol{b}-\tilde{\boldsymbol{b}}\|_{L^1([0,\tau]\times B_R(0))}\right)\right|^{-1},$$

where $R - r > 0$ and C depend only on the supremum and Sobolev bounds on $\boldsymbol{b}$ and $\tilde{\boldsymbol{b}}$.

5. Infinite-dimensional spaces. In this final section I will illustrate two examples of extension of the theory to infinite-dimensional spaces. Here the main difficulty is that requiring that a reference measure ($\mathscr{L}^d$ in the Euclidean case) is invariant or quasi-invariant under the flow leads to a severe restriction on the direction of the vector field. This seems to be a limitation in the attempt to apply this theory to PDE's (viewed as ODE's in an infinite-dimensional space): the only result I am aware of is [57]. In [16], see also the announcement [15], in the case when the state space is $\mathscr{P}(\mathbb{R}^d)$ (Borel probability measures in $\mathbb{R}^d$) we have been able to use a weaker regularity condition on the flow; however for the moment our results apply only to the *linear* continuity equation

$$\partial_t \mu_t + \nabla \cdot (\boldsymbol{b}_t \mu_t) = 0 \tag{5.1}$$

which, in this abstract perspective, should be seen as a *constant coefficient* ODE in $\mathscr{P}(\mathbb{R}^d)$ (with rough coefficients if $\boldsymbol{b}_t$ is rough).

Flows in Wiener spaces. Let us first introduce, briefly, the structure of Gaussian Wiener space. We consider a separable Banach space X and a nondegenerate, centered, Gaussian measure $\gamma \in \mathscr{P}(X)$. The Cameron-Martin space $H \subset X$ is the vector subspace of all $h \in H$ such that $(\tau_h)_\sharp \gamma \ll \gamma$, where $\tau_h(x) = x + h$. It turns out that $\gamma(H) = 0$ whenever X is infinite-dimensional. The maps

$$R^*(x^*) := \langle x^*, \cdot \rangle_{X^*,X}, \qquad R(f) := \int_X f(x) x \, d\gamma(x)$$

(here the integral is understood in Bochner's sense) provide canonical embeddings of X^* in $L^2(X,\gamma)$ and of $L^2(X,\gamma)$ in X respectively, and it can be proved that, denoting by $\mathscr{H}$ the closure of R^*X^* in $L^2(X,\gamma)$, the restriction of R to $\mathscr{H}$ is injective and $R\mathscr{H} = H$. With this notation we can endow H with the L^2 distance inherited from $\mathscr{H}$ and we have the integration by parts formula (for sufficiently nice f and g)

$$\int_X f \partial_h g \, d\gamma = -\int_X g \partial_h f \, d\gamma + \int_X \hat{h} f g \, d\gamma \qquad \forall h = R\hat{h} \in H. \tag{5.2}$$

This formula corresponds precisely, in the standard (product) Gaussian space $(\mathbb{R}^d, \gamma_d)$ with variance 1 in all coordinates, to

$$\int_{\mathbb{R}^d} f \partial_i g \, d\gamma_d = -\int_{\mathbb{R}^d} g \partial_i f \, d\gamma_d + \int_{\mathbb{R}^d} x_i f g \, d\gamma_d,$$

easy to obtain because γ_d is a constant multiple of $e^{-|x|^2/2}\mathscr{L}^d$.

Using (5.2) it is not hard to define Sobolev spaces $W^{1,p}$ and we may expect that a theory analogous to the finite-dimensional one could be developed for vector fields $\boldsymbol{b}_t : X \to H$ (the restriction on the target of $\boldsymbol{b}_t$ being due to quasi-invariance). This is the goal that we achieve in [14], assuming bounds on the (intrinsic) divergence and on the Hilbert-Schmidt norm of the gradient, now a linear operator from H to H. It is not easy to compare the well-posedness results with those obtained, for instance, in [17] (see also [34], [35], [36], [56]): therein the much weaker operator norm is considered, but the norm is assumed to be *exponentially* integrable. So, the result in [14] seem closer to the finite-dimensional theory and are perfectly consistent with it (changing the reference measure from $\mathscr{L}^d$ to γ_d in the same spirit of Section 3).

Flows in $\mathscr{P}(\mathbb{R}^d)$. As I anticipated, we may view (5.1) as an infinite-dimensional ODE in $\mathscr{P}(\mathbb{R}^d)$ and try to obtain existence and uniqueness results for (5.1) in the same spirit of the finite-dimensional theory, starting from the simple observation that $t \mapsto \delta_{\boldsymbol{X}(t,x)}$ solves (5.1) whenever $t \mapsto \boldsymbol{X}(t,x)$ solves (1.3). We may expect that, if we fix a "good" measure $\boldsymbol{\nu}$ in the space $\mathscr{P}(\mathbb{R}^d)$ of initial data, then existence, uniqueness $\boldsymbol{\nu}$-a.e. and stability hold. Moreover, for $\boldsymbol{\nu}$-a.e. μ, the unique and stable solution of (5.1) starting from μ should be given by

$$\boldsymbol{\mu}(t,\mu) := \int \delta_{\boldsymbol{X}(t,x)}\,d\mu(x) \qquad \forall\, t \in [0,T],\ \mu \in \mathscr{P}(\mathbb{R}^d). \tag{5.3}$$

Let us start with a notation and a definition (I use $\mathscr{M}_+(X)$ for the space of positive finite Borel measures in X). Given a nonnegative σ-finite measure $\boldsymbol{\nu} \in \mathscr{M}_+(\mathscr{P}(\mathbb{R}^d))$, I denote by $\mathbb{E}\boldsymbol{\nu} \in \mathscr{M}_+(\mathbb{R}^d)$ its expectation, namely

$$\int_{\mathbb{R}^d} \phi\, d\mathbb{E}\boldsymbol{\nu} = \int_{\mathscr{P}(\mathbb{R}^d)} \int_{\mathbb{R}^d} \phi\, d\mu\, d\boldsymbol{\nu}(\mu) \qquad \text{for all } \phi \text{ bounded Borel.}$$

DEFINITION 5.1 (Regular measures in $\mathscr{M}_+(\mathscr{P}(\mathbb{R}^d))$). *We say that $\boldsymbol{\nu} \in \mathscr{M}_+(\mathscr{P}(\mathbb{R}^d))$ is* regular *if $\mathbb{E}\boldsymbol{\nu} \le C\mathscr{L}^d$ for some constant C.*

We now observe that Definition 1.1 has a natural (but not perfect) transposition to flows in $\mathscr{P}(\mathbb{R}^d)$:

DEFINITION 5.2 (Regular Lagrangian flow in $\mathscr{P}(\mathbb{R}^d)$). *Let $\boldsymbol{\mu} : [0,T] \times \mathscr{P}(\mathbb{R}^d) \to \mathscr{P}(\mathbb{R}^d)$ and $\boldsymbol{\nu} \in \mathscr{M}_+(\mathscr{P}(\mathbb{R}^d))$. We say that $\boldsymbol{\mu}$ is a $\boldsymbol{\nu}$-RLF in $\mathscr{P}(\mathbb{R}^d)$ (relative to $\boldsymbol{b}$) if*

(i) for $\boldsymbol{\nu}$-a.e. μ, $t \mapsto \mu_t := \boldsymbol{\mu}(t,\mu)$ is continuous from $[0,1]$ to $\mathscr{P}(\mathbb{R}^d)$ with $\boldsymbol{\mu}(0,\mu) = \mu$, $|\boldsymbol{b}| \in L^1_{\rm loc}(\mu_t dt)$ and μ_t solves (5.1) in the sense of distributions;

(ii) $\mathbb{E}(\boldsymbol{\mu}(t,\cdot)_\sharp\boldsymbol{\nu}) \le C\mathscr{L}^d$ for all $t \in [0,T]$, for some constant C independent of t.

Notice that condition (ii) is weaker than $\boldsymbol{\mu}(t,\cdot)_\sharp\boldsymbol{\nu} \le C\boldsymbol{\nu}$, which would be the analogue of (ii) in Definition 1.1, and it is actually sufficient (at least for the special ODE (5.1) in $\mathscr{P}(\mathbb{R}^d)$) and much more flexible for many purposes.

THEOREM 5.1 (Existence and uniqueness of the $\boldsymbol{\nu}$-RLF in $\mathscr{P}(\mathbb{R}^d)$). *Assume that* (1.1) *has uniqueness in $L^\infty_+\big([0,T];L^1(\mathbb{R}^d)\cap L^\infty(\mathbb{R}^d)\big)$. Then, for all $\boldsymbol{\nu} \in \mathscr{M}_+(\mathscr{P}(\mathbb{R}^d))$ regular, there exists at most one $\boldsymbol{\nu}$-RLF in $\mathscr{P}(\mathbb{R}^d)$. If* (1.1) *has existence in $L^\infty_+\big([0,T];L^1(\mathbb{R}^d)\cap L^\infty(\mathbb{R}^d)\big)$, this unique flow is given by*

$$\boldsymbol{\mu}(t,\mu) := \int_{\mathbb{R}^d} \delta_{\boldsymbol{X}(t,x)}\,d\mu(x).,\tag{5.4}$$

where $\boldsymbol{X}(t,x)$ denotes the unique $\mathscr{L}^d$-RLF.

Of course the main point here is about uniqueness rather than existence, since the linear formula (5.3) provides a natural recipe for the solution: it is remarkable that the same condition used for the uniqueness of the $\mathscr{L}^d$-RLF in $\mathbb{R}^d$ provides also uniqueness of the "higher level" flow in $\mathscr{P}(\mathbb{R}^d)$. At the level of existence, on the other hand, one may speculate about situations where a $\boldsymbol{\nu}$-RLF $\boldsymbol{\mu}$ exists in $\mathscr{P}(\mathbb{R}^d)$, with $\boldsymbol{\nu}$ regular, but the flow $\boldsymbol{\mu}$ is not induced by any $\mathscr{L}^d$-RLF $\boldsymbol{X}$ as in (5.3).

The main motivation for the development of the theory in [16] has been an application to semiclassical limits, where not only uniqueness, but also stability of the $\boldsymbol{\nu}$-RLF is relevant. In order to describe this application briefly, let $\alpha\in(0,1)$ and let $\psi^\varepsilon_{x_0,p_0,t}:\mathbb{R}^n\to\mathbb{C}$ be a family of solutions to the Schrödinger equation:

$$\begin{cases} i\varepsilon\partial_t\psi^\varepsilon_{x_0,p_0,t} = -\frac{\varepsilon^2}{2}\Delta\psi^\varepsilon_{x_0,p_0,t}+U\psi^\varepsilon_{x_0,p_0,t}\\ \psi^\varepsilon_{x_0,p_0,0}=\varepsilon^{-n\alpha/2}\phi_0\Big(\frac{x-x_0}{\varepsilon^\alpha}\Big)e^{i(x\cdot p_0)/\varepsilon}. \end{cases} \tag{5.5}$$

Here $\phi_0\in C^2_c(\mathbb{R}^n)$ and $\int|\phi_0|^2\,dx=1$. When the potential U is of class C^2, it was proven in [53] that for every (x_0,p_0) the Wigner transforms

$$W_\varepsilon\psi^\varepsilon_{x_0,p_0,t}(x,p):=\frac{1}{(2\pi)^n}\int_{\mathbb{R}^n}\psi^\varepsilon_{x_0,p_0,t}(x+\frac{\varepsilon}{2}y)\overline{\psi^\varepsilon_{x_0,p_0,t}(x-\frac{\varepsilon}{2}y)}e^{-ipy}dy$$

converge, in the natural dual space $\mathcal{A}'$ for the Wigner transforms, to $\delta_{\boldsymbol{X}(t,x_0,p_0)}$ as $\varepsilon\downarrow 0$. Here $\boldsymbol{X}(t,x,p)$ if the unique flow in $\mathbb{R}^{2n}$ associated to the Liouville equation

$$\partial_t W+p\cdot\nabla_x W-\nabla U(x)\cdot\nabla_p W=0. \tag{5.6}$$

In [16] we are able to consider a potential U which can be written as the sum of a repulsive Coulomb potential U_s plus a bounded Lipschitz interation term U_b with $\nabla U_b\in BV_{\rm loc}$. We observe that in this case the equation (5.6) does not even make sense for measure initial data, as ∇U is not continuous (so the product $\nabla U(x)\cdot\nabla_p W$ is not a well-defined distribution, if W is just a measure). Still, we can prove *full* convergence as $\varepsilon\downarrow 0$, namely

$$\lim_{\varepsilon\downarrow 0}\int_{\mathbb{R}^d}\rho(x_0,p_0)\sup_{t\in[-T,T]}d_{\mathcal{A}'}\big(W_\varepsilon\psi^\varepsilon_{x_0,p_0}(t),\delta_{\boldsymbol{X}(t,x_0,p_0)}\big)dx_0dp_0=0\quad\forall T>0 \tag{5.7}$$

for all $\rho\in L^1(\mathbb{R}^{2n})\cap L^\infty(\mathbb{R}^{2n})$ nonnegative, where $\boldsymbol{X}(t,x,p)$ if the unique $\mathscr{L}^{2n}$-RLF associated to (5.6) and $d_{\mathcal{A}'}$ is a bounded distance inducing the weak* topology in the unit ball of $\mathcal{A}'$.

The proof of (5.7) relies on an application of the stability properties of the flow in $\mathscr{P}(\mathbb{R}^d)$ to the Husimi transforms of $\psi^\varepsilon_{x_0,p_0}(t)$, namely the convolutions with a Gaussian kernel with variance $\varepsilon/2$ of the Wigner transforms.

The scheme is sufficiently flexible to allow more general families of initial conditions: for instance, the limiting case $\alpha = 1$ in (5.5) leads to

$$\lim_{\varepsilon \downarrow 0} \int_{\mathbb{R}^d} \rho(x_0) \sup_{t \in [-T,T]} d_{\mathcal{A}'}\big(W_\varepsilon \psi^\varepsilon_{x_0,p_0}(t), \boldsymbol{\mu}(t, \mu(x_0,p_0))\big)dx_0 = 0$$

for all $T > 0$, $p_0 \in \mathbb{R}^n$, $\rho \in L^1(\mathbb{R}^n) \cap L^\infty(\mathbb{R}^n)$ nonnegative, with $\boldsymbol{\mu}(t,\mu)$ given by (5.3) and $\mu(x_0,p_0) = \delta_{x_0} \times |\hat{\phi}_0|^2(\cdot - p_0)\mathscr{L}^n$.

REFERENCES

[1] G.Alberti, S.Bianchini, and G.Crippa, Work in progress.

[2] L.Ambrosio, *Transport equation and Cauchy problem for BV vector fields.* Inventiones Mathematicae, **158** (2004), 227–260.

[3] L.Ambrosio, *Lecture notes on transport equation and Cauchy problem for BV vector fields and applications.* Unpublished preprint, 2004 (available at http://cvgmt.sns.it).

[4] L.Ambrosio, *Lecture notes on transport equation and Cauchy problem for non-smooth vector fields and applications.* Lecture Notes in Mathematics "Calculus of Variations and Non-Linear Partial Differential Equations" (CIME Series, Cetraro, 2005) **1927**, B. Dacorogna, P. Marcellini eds., 2–41, 2008.

[5] L.Ambrosio, F.Bouchut, and C.De Lellis, *Well-posedness for a class of hyperbolic systems of conservation laws in several space dimensions.* Comm. PDE, **29** (2004), 1635–1651.

[6] L.Ambrosio, G.Crippa, and S.Maniglia, *Traces and fine properties of a BD class of vector fields and applications.* Ann. Sci. Toulouse, **XIV** (4) (2005), 527–561.

[7] L.Ambrosio and C.De Lellis, *Existence of solutions for a class of hyperbolic systems of conservation laws in several space dimensions.* International Mathematical Research Notices, **41** (2003), 2205–2220.

[8] L.Ambrosio, C.De Lellis, and J.Malý, *On the chain rule for the divergence of BV like vector fields: applications, partial results, open problems.* In "Perspectives in Nonlinear Partial Differential Equations: in honour of Haim Brezis", Contemporary Mathematics, **446** (2007), 31–67, AMS.

[9] L.Ambrosio, N.Fusco, and D.Pallara, *Functions of bounded variation and free discontinuity problems.* Oxford Mathematical Monographs, 2000.

[10] L.Ambrosio, M.Lecumberry, and S.Maniglia, *Lipschitz regularity and approximate differentiability of the DiPerna-Lions flow.* Rendiconti del Seminario Fisico Matematico di Padova, **114** (2005), 29–50.

[11] L.Ambrosio, S.Lisini, and G.Savaré, *Stability of flows associated to gradient vector fields and convergence of iterated transport maps.* Manuscripta Math., **121** (2006), 1–50.

[12] L.Ambrosio and J.Malý, *Very weak notions of differentiability.* Proceedings of the Royal Society of Edinburgh, **137 A** (2007), 447–455.

[13] L.Ambrosio, A.Figalli, G.Crippa, and L.Spinolo, *Some new well-posedness results for continuity and transport equations, and applications to the chromatography system.* Preprint 2009, to appear on SIAM J. Math. Anal.

[14] L.Ambrosio and A.Figalli, *On flows associated to Sobolev vector fields in Wiener spaces: an approach à la DiPerna-Lions.* J. Funct. Anal., **256** (2009), 179–214.

[15] L.Ambrosio and A.Figalli, *Almost everywhere well-posedness of continuity equations with measure initial data.* Preprint, 2009.

[16] L.Ambrosio, A.Figalli, G.Friesecke, and J.Giannoulis, *Well posedness of transport equations with measure initial data and convergence of Wigner measures.* Work in progress.

[17] V.Bogachev and E.M.Wolf, *Absolutely continuous flows generated by Sobolev class vector fields in finite and infinite dimensions.* J. Funct. Anal., **167** (1999), 1–68.

[18] F.Bouchut, *Renormalized solutions to the Vlasov equation with coefficients of bounded variation.* Arch. Rational Mech. Anal., **157** (2001), 75–90.

[19] F.Bouchut and G.Crippa, *Uniqueness, Renormalization, and Smooth Approximations for Linear Transport Equations.* SIAM J. Math. Anal., **38** (2006), 1316–1328.

[20] F.Bouchut, F.Golse, and M.Pulvirenti, *Kinetic equations and asymptotic theory.* Series in Appl. Math., Gauthiers-Villars, 2000.

[21] F.Bouchut and F.James, *One dimensional transport equation with discontinuous coefficients.* Nonlinear Analysis, **32** (1998), 891–933.

[22] F.Bouchut and L.Desvillettes, *On two-dimensional Hamiltonian transport equations with continuous coefficients.* Differential Integral Equations, **14** (2001), 1015–1024.

[23] F.Bouchut, F.James, and S.Mancini, *Uniqueness and weak stability for multi-dimensional transport equations with one-sided Lipschitz coefficients.* Annali Scuola Normale Superiore, Ser. 5, **4** (2005), 1–25.

[24] F.Bouchut and G.Crippa, *Equations de transport à coefficient dont le gradient est donné par une intégrale singuliére.* Séminaire Équations aux dérivées partielles. Exp. No. 1 2007-2008 (2009), 1–13.

[25] A.Bressan, *An ill posed Cauchy problem for a hyperbolic system in two space dimensions.* Rend. Sem. Mat. Univ. Padova, **110** (2003), 103–117.

[26] I.Capuzzo Dolcetta and B.Perthame, *On some analogy between different approaches to first order PDE's with nonsmooth coefficients.* Adv. Math. Sci Appl., **6** (1996), 689–703.

[27] A.Cellina, *On uniqueness almost everywhere for monotonic differential inclusions.* Nonlinear Analysis, TMA, **25** (1995), 899–903.

[28] F.Colombini, G.Crippa, and J.Rauch, *A note on two dimensional transport with bounded divergence.* Comm. Partial Differential Equations, **31** (2006), 1109–1115.

[29] F.Colombini and N.Lerner, *Uniqueness of continuous solutions for BV vector fields.* Duke Math. J., **111** (2002), 357–384.

[30] F.Colombini and N.Lerner, *Uniqueness of L^∞ solutions for a class of conormal BV vector fields.* Contemp. Math. **368** (2005), 133–156.

[31] F.Colombini, T.Luo, and J.Rauch, *Nearly Lipschitzean diverge free transport propagates neither continuity nor BV regularity.* Commun. Math. Sci., **2** (2004), 207–212.

[32] G.Crippa and C.De Lellis, *Oscillatory solutions to transport equations.* Indiana Univ. Math. J., **55** (2006), 1–13.

[33] G.Crippa and C.De Lellis, *Estimates and regularity results for the DiPerna–Lions flow.* Preprint, 2006 (available at http://cvgmt.sns.it). Accepted by J. Reine Angew. Math.

[34] A.B.Cruzeiro, *Équations différentielles ordinaires: non explosion et mesures quasi-invariantes.* J. Funct. Anal., **54** (1983), 193–205.

[35] A.B.Cruzeiro, *Équations différentielles sur l'espace de Wiener et formules de Cameron-Martin non linéaires.* J. Funct. Anal., **54** (1983), 206–227.

[36] A.B.Cruzeiro, *Unicité de solutions d'équations différentielles sur l'espace de Wiener.* J. Funct. Anal., **58** (1984), 335–347.

[37] M.Cullen and M.Feldman, *Lagrangian solutions of semigeostrophic equations in physical space.* SIAM J. Math. Anal., **37** (2006), 1371–1395.

[38] M.Cullen and W.Gangbo, *A variational approach for the 2-dimensional semi-geostrophic shallow water equations.* Arch. Rational Mech. Anal., **156** (2001), 241–273.

[39] C.De Lellis, *Blow-up of the BV norm in the multidimensional Keyfitz and Kranzer system.* Duke Math. J., **127** (2004), 313–339.

[40] N.Depauw, *Non unicité des solutions bornées pour un champ de vecteurs BV en dehors d'un hyperplan.* C.R. Math. Sci. Acad. Paris, **337** (2003), 249–252.

[41] R.J.DiPerna, *Measure-valued solutions to conservation laws.* Arch. Rational Mech. Anal., **88** (1985), 223–270.

[42] R.J.DiPerna and P.L.Lions, *Ordinary differential equations, transport theory and Sobolev spaces.* Invent. Math., **98** (1989), 511–547.

[43] R.J.DiPerna and P.L.Lions, *On the Cauchy problem for the Boltzmann equation: global existence and weak stability.* Ann. of Math., **130** (1989), 312–366.

[44] M.Hauray, *On Liouville transport equation with potential in $BV_{\rm loc}$.* Comm. Partial Differential Equations, **29** (2004), 207–217.

[45] M.Hauray, *On two-dimensional Hamiltonian transport equations with $L^p_{\rm loc}$ coefficients.* Ann. IHP Nonlinear Anal. Non Linéaire, **20** (2003), 625–644.

[46] M.Hauray, C.Lebris, P.L.Lions, *Two remarks on generalized flows for ordinary differential equations.* C. R. Acad. Sci. Paris Sér. I, **344** (2007), 759–764.

[47] B.L.Keyfitz and H.C.Kranzer, *A system of nonstrictly hyperbolic conservation laws arising in elasticity theory.* Arch. Rational Mech. Anal. 1980, **72**, 219–241.

[48] C.Le Bris and P.L.Lions, *Renormalized solutions of some transport equations with partially $W^{1,1}$ velocities and applications.* Annali di Matematica, **183** (2003), 97–130.

[49] N.Lerner, *Transport equations with partially BV velocities.* Ann. Sc. Norm. Super. Pisa Cl. Sci., **3** (2004), 681–703.

[50] P.L.Lions, *Mathematical topics in fluid mechanics, Vol. I: incompressible models.* Oxford Lecture Series in Mathematics and its applications, **3** (1996), Oxford University Press.

[51] P.L.Lions, *Mathematical topics in fluid mechanics, Vol. II: compressible models.* Oxford Lecture Series in Mathematics and its applications, **10** (1998), Oxford University Press.

[52] P.L.Lions, *Sur les équations différentielles ordinaires et les équations de transport.* C. R. Acad. Sci. Paris Sér. I, **326** (1998), 833–838.

[53] P.L.Lions and T.Paul, *Sur les mesures de Wigner.* Rev. Mat. Iberoamericana, **9** (1993), 553–618.

[54] S.Maniglia, *Probabilistic representation and uniqueness results for measure-valued solutions of transport equations.* Preprint, 2005 (available at http://cvgmt.sns.it). Accepted by J. Math. Pures Appl.

[55] E.Y.Panov, *On strong precompactness of bounded sets of measure-valued solutions of a first order quasilinear equation.* Math. Sb., **186** (1995), 729–740.

[56] G.Peters, *Anticipating flows on the Wiener space generated by vector fields of low regularity.* J. Funct. Anal., **142** (1996), 129–192.

[57] G.Peters, *Symplectic geometry on an abstract Wiener space and the sine-Gordon equation.* C.R. Acad. Sci. Paris Sér I Math., **320** (1995), 1123-1128.

[58] G.Petrova and B.Popov, *Linear transport equation with discontinuous coefficients.* Comm. PDE, **24** (1999), 1849–1873.

[59] F.Poupaud and M.Rascle, *Measure solutions to the liner multidimensional transport equation with non-smooth coefficients.* Comm. PDE, **22** (1997), 337–358.

[60] R.Temam, *Problémes mathématiques en plasticité.* Gauthier-Villars, Paris, 1983.

EXISTENCE AND UNIQUENESS RESULTS FOR THE CONTINUITY EQUATION AND APPLICATIONS TO THE CHROMATOGRAPHY SYSTEM

LUIGI AMBROSIO*, GIANLUCA CRIPPA†, ALESSIO FIGALLI‡, AND LAURA V. SPINOLO§

Abstract. We discuss some new well-posedness results for the continuity equation in arbitrary space dimension and we then illustrate applications to a system of conservation laws in one space dimension known as the chromatography system. In the last section, we discuss some related open problems.

Key words. Continuity equation, chromatography system, BV functions.

AMS(MOS) subject classifications. 35F10, 35L65.

1. Introduction. We are concerned with the Cauchy problem obtained by coupling the continuity equation

$$\partial_t u + \mathrm{div}_x(bu) = 0 \qquad (t,x) \in [0,+\infty[\times\mathbb{R}^d,\ b(t,x) \in \mathbb{R}^d,\ u \in \mathbb{R} \tag{1.1}$$

with the initial datum

$$u(0,x) = \bar{u}(x). \tag{1.2}$$

In particular, we focus on the case when b has low regularity with respect to the space variable. In this context, DiPerna and Lions [23] established existence and uniqueness results under the hypothesis that b has Sobolev regularity with respect to the space variable. More recently, Ambrosio [2] proved well-posedness results in the BV (bounded total variation) case: for a discussion concerning these topics, we refer for example to the lecture notes by Ambrosio [3] and Ambrosio and Crippa [5], and to Crippa [17].

In both [2, 23], the analysis required some mild regularity assumptions on the divergence $\mathrm{div}_x\, b$. These hypotheses, however, are not completely natural in view of the applications, provided by Ambrosio, Bouchut and DeLellis [4, 7], to the Keyfitz and Kranzer system [25] in several space variables. This was a motivation for introducing the notion of *nearly incompressible vector field* (see Definition 2.1 in Section 2.1): the related theory is discussed in the lecture notes by De Lellis [21] and can be more

*Scuola Normale Superiore, Piazza dei Cavalieri 7, 56126 Pisa, Italy (l.ambrosio@sns.it).

†Dipartimento di Matematica, Università degli Studi di Parma, viale G.P. Usberti 53/A (Campus), 43100 Parma, Italy (gianluca.crippa@unipr.it).

‡Department of Mathematics, The University of Texas at Austin, 1 University Station, C1200, Austin, Texas 78712-1082, USA (figalli@math.utexas.edu).

§Centro De Giorgi, Collegio Puteano, Scuola Normale Superiore, Piazza dei Cavalieri 3, 56126 Pisa, Italy (laura.spinolo@sns.it).

directly applied to the analysis of the Keyfitz and Kranzer system. As a side remark, we point out that a counter-example due to Crippa and De Lellis [18] rules out the possibility of extending to general systems of conservation laws in several space dimensions the approach used in [4] to tackle the Keyfitz and Kranzer system.

The exposition in here is organized as follows. In Section 2 we describe new well-posedness results for nearly incompressible vector fields, handling in particular the case when the BV norm of $b(t,\cdot)$ blows up as $t \to 0^+$. In Section 3 we illustrate applications to the chromatography system and in Section 4 we discuss some related open problems.

2. Well-posedness results for the continuity equation.

2.1. Nearly incompressible vector fields. In this section we recall the definition of near incompressibility and a related well-posedness result concerning the Cauchy problem (1.1), (1.2). As mentioned in the introduction, a detailed discussion on these topics can be found in the lecture notes by De Lellis [21].

DEFINITION 2.1. *The bounded vector field b is* nearly incompressible *if there exists a measurable function ρ and a constant $C > 0$ such that*

$$\partial_t \rho + \operatorname{div}_x (b\rho) = 0 \tag{2.1}$$

and

$$0 < \frac{1}{C} \leq \rho(t,x) \leq C \quad \text{for } \mathscr{L}^{d+1} - \text{a.e. } (t,x) \in]0,+\infty[\times\mathbb{R}^d. \tag{2.2}$$

By combining Corollary 3.14 and Lemma 5.10 in De Lellis [21] one obtains

THEOREM 2.1. *Let $b \in L^\infty_{\mathrm{loc}} \cap BV_{\mathrm{loc}}\big([0,+\infty[\times\mathbb{R}^d;\mathbb{R}^d\big)$ be a nearly incompressible vector field and assume that the function ρ in equation (2.1), besides satisfying the bound (2.2), enjoys the following regularity:*

$$\rho \in BV_{\mathrm{loc}}\big([0,+\infty[\times\mathbb{R}^d\big), \quad \text{and} \quad \rho(0,\cdot) \in BV_{\mathrm{loc}}(\mathbb{R}^d).$$

Then, for any $\bar{u} \in L^\infty(\mathbb{R}^d)$, there exists a unique bounded, distributional solution of the Cauchy problem (1.1), (1.2).

In the statement of the previous theorem uniqueness can be interpreted as follows. Classical results (see e.g. Dafermos [20, Lemma 1.3.3]) ensure that any bounded, distributional solution of the continuity equation (1.1) admits a representative such that the map $t \mapsto u(t,\cdot)$ is continuous from $\mathbb{R}$ to $L^\infty(\mathbb{R}^d)$ provided that $L^\infty(\mathbb{R}^d)$ is endowed with the w^*-topology. Hence the statement of Theorem 2.1 can be reformulated by saying that the Cauchy problem (1.1), (1.2) admits a unique solution in the space of bounded functions u such that the map $t \mapsto u(t,\cdot)$ is w^*-continuous. In the following, we will always identify a bounded, distributional solution of (1.1) with its w^*-continuous representative.

2.2. A well-posedness result in the class of strongly continuous solutions. Before introducing our first well-posedness result (Theorem 2.2), we make some further preliminary considerations.

An example due to Depauw [22] ensures that if in the statement of Theorem 2.1 one removes the hypothesis that b has BV regularity then the uniqueness result fails. More precisely, Depauw exhibited a time dependent, bounded vector field b taking values in $\mathbb{R}^2$ and satisfying the following conditions. The vector field $b(t,\cdot)$ has locally bounded total variation for every $t>0$, but, for any compact set $K \subseteq \mathbb{R}^2$, the norm $\|b(t,\cdot)\|_{BV(K)}$ blows up as $t \to 0^+$ and the function $t \mapsto \|b(t,\cdot)\|_{BV(K)}$ is not integrable up to $t=0$. Also, the vector field b is divergence free and hence nearly incompressible (one can take $\rho(t,x) \equiv 1$ in Definition 2.1). In this case the uniqueness part in the statement of Theorem 2.1 fails because Depauw exhibits a non trivial solution of (1.1) satisfying the Cauchy datum $u(0,x) \equiv 0$, namely

$$\lim_{t \to 0^+} u(t,\cdot) = 0 \quad \text{weakly}^* \text{ in } L^\infty(\mathbb{R}^d). \tag{2.3}$$

Loosely speaking, our first new well-posedness result for the Cauchy problem (1.1)-(1.2) states that, for vector fields b like the one in Depauw's example, uniqueness can be restored provided one imposes additional regularity requirements on the solution u. The point is the following: the space $L^\infty(\mathbb{R}^d)$ can be endowed with the strong topology of $L^1_{\mathrm{loc}}(\mathbb{R}^d)$. As a matter of fact, one can verify that the non trivial solution u in Depauw's example is highly oscillating as $t \to 0^+$ and hence the limit (2.3) is *not* satisfied if, instead of the w^* topology, we consider the strong topology of $L^1_{\mathrm{loc}}(\mathbb{R}^2)$. Hence, there is hope of obtaining a uniqueness result for the Cauchy problem (1.1)-(1.2) in the class of bounded, distributional solutions u such that the map $t \mapsto u(t,\cdot)$ is continuous with respect to the strong topology in $L^1_{\mathrm{loc}}(\mathbb{R}^d)$.

Before stating our precise well-posedness result, we have to introduce some notations: we denote by B_R the ball of radius R and center at $\vec{0}$ in $\mathbb{R}^d$. Also, assume that the function $\rho(t,x)$ has, for a given t, locally bounded total variation with respect to the x variable, then the distributional derivative $|D\rho(t,\cdot)|$ is a Radon measure and we denote by $|D\rho(t,\cdot)|(B_R)$ the value assumed on the set B_R.

THEOREM 2.2. *Let $b \in L^\infty \cap BV_{\mathrm{loc}}(]0,+\infty[\times\mathbb{R}^d;\mathbb{R}^d)$ be a nearly incompressible vector field and assume that the function ρ in equation (2.1), besides fulfilling the bound (2.2), satisfies $\rho \in BV_{\mathrm{loc}}\big(]0,+\infty[\times\mathbb{R}^d\big)$ and enjoys the following regularity properties: the map $t \mapsto \rho(t,\cdot)$ is continuous from $[0,+\infty[$ in the space $L^1_{\mathrm{loc}}(\mathbb{R}^d)$ endowed with the strong topology. Also, the map $t \mapsto |D\rho(t,\cdot)|(B_R)$ belongs to $L^\infty_{\mathrm{loc}}([0,+\infty[)$ for every $R>0$.*

Then, for any $\bar{u} \in L^\infty(\mathbb{R}^d)$, there exists a unique bounded, distributional solution of the Cauchy problem (1.1)-(1.2) such that the map $t \mapsto u(t,\cdot)$ is continuous with respect to the strong topology in $L^1_{\mathrm{loc}}(\mathbb{R}^d)$.

The proof of Theorem 2.2 is provided in [6]: the main issue is actually establishing the existence of a strongly continuous solutions, which is achieved by approximating b by a sequence of smooth, nearly incompressible vector fields.

2.3. A new well-posedness result for weakly continuous solutions. In this section we discuss Theorem 2.3, another well-posedness result for the Cauchy problem (1.1)-(1.2) which, as Theorem 2.2, can be regarded as an extension of Theorem 2.1. More precisely, in the statement of Theorem 2.3 we keep the same regularity assumptions on b and ρ as in Theorem 2.1, but we relax the assumption that the function ρ in Definition 2.1 is bounded away from 0 by handling the case when ρ attains the value zero.

THEOREM 2.3. *Let $b \in BV_{\mathrm{loc}}\big([0,+\infty[\times\mathbb{R}^d;\mathbb{R}^d\big)$ be a bounded vector field and assume that there exists a nonnegative, locally bounded function $\rho \in BV_{\mathrm{loc}}\big([0,+\infty[\times\mathbb{R}^d\big)$ satisfying the following assumptions: first, the map $t \mapsto |D\rho(t,\cdot)|(B_R)$ belongs to $L^\infty_{\mathrm{loc}}\big([0,+\infty[\big)$ for all $R>0$. Also, ρ is a distributional solution of*

$$\partial_t \rho + \mathrm{div}_x\,(b\rho) = 0.$$

If, for some positive constant $K>0$, the initial datum $\bar{u}$ satisfies $|\bar{u}(x)| \leq K\rho(0,x)$ for a.e. $x \in \mathbb{R}^d$, then there exists a unique bounded, distributional solution of the Cauchy problem (1.1), (1.2) such that

$$|u(t,x)| \leq K\rho(t,x) \quad \text{for every } t \geq 0 \text{ and for a.e. } x \in \mathbb{R}^d. \tag{2.4}$$

Note that, as mentioned before, we are identifying any solution of the continuity equation (1.1) with its w^*-continuous representative and hence the inequality (2.4) is well-defined for every $t \geq 0$.

3. Applications to the chromatography system. In this section we discuss how Theorems 2.2 and 2.3 can be applied to the analysis of the so-called chromatography system, namely

$$\begin{cases} \partial_t u_1 + \partial_x \left(\dfrac{u_1}{1+u_1+u_2} \right) = 0 \\[2ex] \partial_t u_2 + \partial_x \left(\dfrac{u_2}{1+u_1+u_2} \right) = 0. \end{cases} \tag{3.1}$$

In the previous expression, $(t,x) \in [0,+\infty[\times\mathbb{R}$ and u_1 and u_2 are both real valued functions. The analysis of (3.1) is motivated by the study of the two-component chromatography (see e.g. Bressan [12, page 102]) and hence one is usually interested in finding nonnegative solutions, $u_1 \geq 0$ and $u_2 \geq 0$. As a matter of fact, both Theorems 3.1 and 3.2 can be easily extended to the case of k components,

$$\partial_t u_i + \partial_x \left(\frac{u_i}{1+\sum_{j=1}^k u_j} \right) = 0, \quad u_i \geq 0 \qquad \text{for } i = 1, \ldots, k.$$

However, to simplify the exposition in the following we focus on the case (3.1). Also, we are concerned with the Cauchy problem obtained by coupling (3.1) with the initial data

$$u_1(0, x) = \bar{u}_1(x) \quad \text{and} \quad u_2(0, x) = \bar{u}_2(x). \tag{3.2}$$

3.1. The chromatography system. The chromatography system (3.1) belongs to the so-called Temple class [31], and hence well-posedness results are available even for initial data that violate the classical hypothesis of small total variation (see e.g. the book by Bressan [12] for an overview of the theory of one-dimensional systems of conservation laws with Cauchy data having small total variation). Concerning Temple systems, Serre [30] established existence results under the hypothesis that the total variation of the initial data is bounded but arbitrarily large, while later on Bianchini [10] obtained existence and uniqueness results for initial data that are merely bounded. See also Baiti and Bressan [9] and Bressan and Goatin [14]. The well-posedness results in both Serre [30] and Bianchini [10] apply to general Temple systems, while here we restrict to the chromatography system. However, in [10, 30] the analysis requires a standard hypothesis of so-called *strict hyperbolicity* (see e.g. Dafermos [20, Chapter 7] for the exact definition), which is violated by the chromatography system if the sum $u_1 + u_2$ attains the value 0. For general systems of conservation laws, when strict hyperbolicity is violated even the existence of bounded, distributional solutions may fail (see again Dafermos [20, Section 9.6] for a related discussion). However, in the case of the chromatography system well-posedness results can be extended to the case the sum $u_1 + u_2$ attains the value 0, see Theorem 3.2.

The main difference between the well-posedness results in Serre [30] and Bianchini [10] and Theorems 3.1 and 3.2 lies in their approach: indeed, the analysis in both [10, 30] is based on the introduction of suitable approximation schemes. Conversely, Theorems 3.1 and 3.2 are obtained by introducing a change of variables which transforms (3.1) in the coupling between a scalar conservation law and a continuity equation with a weakly differentiable vector field (see Section 3.2 in here for more details). This is similar to the approach in Bressan and Shen [15] and Panov [28, Remark 4, page 140]. See also the analysis by Ambrosio, Bouchut and De Lellis [4, 7] concerning the so-called Keyfitz and Kranzer system [25], which was inspired by considerations in Bressan [13].

3.2. Well-posedness results for the chromatography system. In order to apply Theorems 2.2 and 2.3 to the analysis of the chromatography system (3.1) we first introduce the change of variables

$$v := u_1 + u_2 \qquad w := u_1 - u_2, \tag{3.3}$$

which transforms (3.1) in the coupling between the scalar conservation law

$$\partial_t v + \partial_x \left(\frac{v}{1+v} \right) = 0 \tag{3.4}$$

and the continuity equation

$$\partial_t w + \partial_x \left(\frac{w}{1+w} \right) = 0. \tag{3.5}$$

By applying Kružkov's theory [26], we get that, for any bounded initial datum, the Cauchy problem for (3.4) admits a unique bounded, entropy admissible, distributional solution. For the definition of the entropy admissibility criterion for conservation laws see e.g. Dafermos [20, Section 3.2].

We can then plug the solution v into (3.5), thus obtaining a continuity equation in the form (1.1) with $b(t,x) := 1/[1+v(t,x)]$. However, it is well known (see again Dafermos [20, Section 4.3]) that, even if the initial datum is extremely regular, in general classical solutions of scalar conservation laws break down in finite time. Hence, when handling (3.5) we are forced to take into account the possibility that the vector field $1/(1+v)$ has low regularity. Nevertheless, note that Kružkov's theory [26] ensures that equation (3.4) preserves the BV regularity, namely $v(t,\cdot) \in BV(\mathbb{R})$ for every $t > 0$ if the initial datum $v(0,\cdot) \in BV(\mathbb{R})$. Also, in the case of equation (3.4) the flux $v \mapsto 1/(1+v)$ is strictly concave, hence we can apply Oleĭnik's estimate [27] and get that, even if the initial datum is merely bounded, the entropy admissible solution of (3.4) satisfies $v(t,\cdot) \in BV(\mathbb{R})$ for every $t > 0$. Note, however, that in this case the BV norm of $v(t,\cdot)$ blows up as $t \to 0^+$.

Our first well-posedness result for the Cauchy problem (3.1)-(3.2) is the following.

THEOREM 3.1. *Assume that the initial data (3.2) satisfy the following conditions:*

- $\bar{u}_1,\ \bar{u}_2 \in L^\infty_{\text{loc}}(\mathbb{R})$;
- $\bar{u}_1(x) \geq 0$ *and* $\bar{u}_2(x) \geq 0$ *for a.e.* $x \in \mathbb{R}$;
- *for every* $R > 0$ *there exists* $\delta_R > 0$ *such that* $\bar{u}_1 + \bar{u}_2 \geq \delta_R$ *for a.e.* $x \in]-R, R[$.

Then there exists a unique vector valued function (u_1, u_2) *satisfying*

1. (u_1, u_2) *is an entropy admissible, distributional solution of the Cauchy problem (3.1)-(3.2);*
2. *the maps* $t \mapsto u_1(t,\cdot)$ *and* $t \mapsto u_2(t,\cdot)$ *are both continuous with respect to the strong topology of* $L^1_{\text{loc}}(\mathbb{R})$.

Also, the unique vector valued function (u_1, u_2) *satisfying 1. and 2. above enjoys the additional property that for every* $t \geq 0$ *and for a.e* $x \in \mathbb{R}$ $u_1(t,x) \geq 0$ *and* $u_2(t,x) \geq 0$.

For the proof of Theorem 3.1, we refer to [6], however the main ingredients are the changes of variables (3.3), Kružkov's theory [26], Theorem 2.2,

Oleĭnik's estimate [27], a regularity result due to Chen and Rascle [16] and a result which classifies the entropies of (3.1) and ensures that the entropy admissible solutions of (3.1) are exactly those such that the sum $v = u_1 + u_2$ is an entropy admissible solution of (3.4) and the difference $w = u_1 - u_2$ is a distributional soluton of (3.5) (see Lemma 4.1 and Proposition 4.2 in [6] for a more precise statement).

In our second well-posedness theorem we remove the assumption that the sum $\bar{u}_1 + \bar{u}_2$ is locally bounded away from 0, but the price we have to pay is that now we impose that $\bar{u}_1 + \bar{u}_2$ has BV regularity.

THEOREM 3.2. *Assume that the initial data (3.2) satisfy the following conditions:*

- $\bar{u}_1,\ \bar{u}_2 \in L^\infty_{\mathrm{loc}}(\mathbb{R})$;
- $\bar{u}_1(x) \geq 0$ *and* $\bar{u}_2(x) \geq 0$ *for a.e.* $x \in \mathbb{R}$;
- $(\bar{u}_1 + \bar{u}_2) \in BV_{\mathrm{loc}}(\mathbb{R})$.

Then the Cauchy problem (3.1), (3.2) admits a unique entropy admissible, distributional solution (u_1, u_2) *such that*

1. *the maps* $t \mapsto u_1(t, \cdot)$ *and* $t \mapsto u_2(t, \cdot)$ *are both continuous with respect to the* w^* *topology of* $L^\infty_{\mathrm{loc}}(\mathbb{R})$;
2. *for every* $t > 0$, $u_1(t, \cdot)$, $u_2(t, \cdot) \in L^\infty_{\mathrm{loc}}(\mathbb{R})$ *and* $u_1(t,x)$, $u_2(t,x) \geq 0$ *for a.e.* $x \in \mathbb{R}$.

Also, this unique solution satisfies the additional regularity property that the maps $t \mapsto u_1(t, \cdot)$ *and* $t \mapsto u_2(t, \cdot)$ *are both continuous with respect to the strong topology of* $L^1_{\mathrm{loc}}(\mathbb{R})$.

The proof of Theorem 3.2 is given in [6] and relies on Theorem 2.3 and, again, on Kružkov's theory [26], on the regularity result by Chen and Rascle [16] and on the classification of the entropies of (3.1).

Note that by applying Theorems 2.2 and 2.3 one can also get well-posedness results for the so-called Keyfitz and Kranzer system [25], in the same spirit as those in Ambrosio, Bouchut and De Lellis [4] (multi-dimensional case with BV data), Freistühler [24] and Panov [28] (one-dimensional case with bounded initial data). See [6, Section 5] for the details.

REMARK 3.1. Theorems 2.2 and 2.3 are valid in any space dimension, but the applications in Theorems 3.1 and 3.2 are concerned with the one-dimensional case only. In the one-dimensional case, well-posedness results for the continuity equation (1.1) are available under much weaker assumption than those required to deal with higher dimensions: see for example Panov [29]. By combining the classification of the entropies provided by Lemma 4.1 in [6] with the results in [29] one can establish well-posedness results for the Cauchy problem (3.1), (3.2) under the assumption that the initial data $\bar{u}_1$ and $\bar{u}_2$ are merely bounded and nonnegative.

4. Related open problems. For a recent discussion on open problems concerning the continuity equation (1.1) and its extensions, see the contribution by Ambrosio [1] to this volume. Here we focus on the connection between the issues discussed in Section 2 and the following compactness conjecture, proposed by Bressan [13].

CONJECTURE 4.1. *Let* $b_k : [0,+\infty[\times\mathbb{R}^d \to \mathbb{R}^d$, $k \in \mathbb{N}$, *be a sequence of smooth vector fields and denote by* X_k *the solution of the Cauchy problem*

$$\begin{cases} \dfrac{dX_k}{dt} = b_k\big[t, X_k(t,x)\big] \\ \\ X_k(0,x) = x. \end{cases}$$

Assume that $\|b_k\|_\infty + \|\nabla b_k\|_{L^1}$ *is uniformly bounded and that the flows* X_k *satisfy, for some constant* $C > 0$,

$$\frac{1}{C} \leq \det(\nabla_x X_k(t,x)) \leq C.$$

Then the sequence $\{X_k\}$ *is precompact with respect to the strong topology of* $L^1_{\rm loc}([0,+\infty[\times\mathbb{R}^d)$.

As a matter of fact, uniqueness results for the Cauchy problem (1.1), (1.2) are strongly related to the so-called renormalization properties: indeed, these properties can be used (see e.g. Di Perna and Lions [23] and Ambrosio [2]) to establish a comparison principle. See also Bouchut and Crippa [11] for a characterization of the relation between uniqueness and renormalization. Ambrosio, Bouchut and De Lellis [4] proposed the following renormalization conjecture (note that no regularity assumption is imposed on the function ρ).

CONJECTURE 4.2. *Let* $b \in L^\infty_{\rm loc} \cap BV_{\rm loc}\big([0,+\infty[\times\mathbb{R}^d;\mathbb{R}^d\big)$ *be a nearly incompressible vector field, and let* ρ *be the function appearing in Definition 2.1. Then, for every bounded function* u *solving*

$$\partial_t(\rho u) + \mathrm{div}_x\,(b\rho u) = 0$$

in the sense of distributions and for every $\beta \in C^1(\mathbb{R})$, *the function* $\beta(u)$ *satisfies*

$$\partial_t\big[\rho\beta(u)\big] + \mathrm{div}_x\,\big[b\rho\beta(u)\big] = 0$$

in the sense of distributions.

Moreover, in [4] the authors show that, if Conjecture 4.2 is true, then also Conjecture 4.1 is verified. A proof of Conjecture 4.2, under additional assumptions, is due to Ambrosio, De Lellis and Malý [8]. See also Crippa and De Lellis [19] for results concerning Bressan's conjecture.

REFERENCES

[1] L. AMBROSIO, *The flow associated to weakly differentiable vector fields: recent results and open problems*, in Nonlinear Conservation Laws and Applications, IMA Vol. Math. Appl., Springer. To appear. Available at http://cvgmt.sns.it/cgi/get.cgi/papers/amb09/.

[2] ———, *Transport equation and Cauchy problem for BV vector fields*, Invent. Math., **158** (2004), pp. 227–260.

[3] ———, *Transport equation and Cauchy problem for non-smooth vector fields*, in Calculus of variations and nonlinear partial differential equations, Vol. **1927** of Lecture Notes in Math., Springer, Berlin, 2008, pp. 1–41.

[4] L. Ambrosio, F. Bouchut, and C. De Lellis, *Well-posedness for a class of hyperbolic systems of conservation laws in several space dimensions*, Comm. Partial Differential Equations, **29** (2004), pp. 1635–1651.

[5] L. Ambrosio and G. Crippa, *Existence, uniqueness, stability and differentiability properties of the flow associated to weakly differentiable vector fields*, In: Transport Equations and Multi-D Hyperbolic Conservation Laws, Lecture Notes of the Unione Matematica Italiana, **5** (2008).

[6] L. Ambrosio, G. Crippa, A. Figalli, and L.V. Spinolo, *Some new well-posedness results for continuity and transport equations, and applications to the chromatography system*, SIAM J. Math. Anal., **41** (2009), pp. 1890–1920.

[7] L. Ambrosio and C. De Lellis, *Existence of solutions for a class of hyperbolic systems of conservation laws in several space dimensions*, Int. Math. Res. Not. (2003), pp. 2205–2220.

[8] L. Ambrosio, C. De Lellis, and J. Malý, *On the chain rule for the divergence of BV-like vector fields: applications, partial results, open problems*, in Perspectives in nonlinear partial differential equations, Vol. **446** of Contemp. Math., Amer. Math. Soc., Providence, RI, 2007, pp. 31–67.

[9] P. Baiti and A. Bressan, *The semigroup generated by a Temple class system with large data*, Differential Integral Equations, **10** (1997), pp. 401–418.

[10] S. Bianchini, *Stability of L^∞ solutions for hyperbolic systems with coinciding shocks and rarefactions*, SIAM J. Math. Anal., **33** (2001), pp. 959–981 (electronic).

[11] F. Bouchut and G. Crippa, *Uniqueness, renormalization, and smooth approximations for linear transport equations*, SIAM J. Math. Anal., **38** (2006), pp. 1316–1328 (electronic).

[12] A. Bressan, *Hyperbolic systems of conservation laws. The one-dimensional Cauchy problem*, Vol. 20 of Oxford Lecture Series in Mathematics and its Applications, Oxford University Press, Oxford, 2000.

[13] ———, *An ill posed Cauchy problem for a hyperbolic system in two space dimensions*, Rend. Sem. Mat. Univ. Padova, **110** (2003), pp. 103–117.

[14] A. Bressan and P. Goatin, *Stability of L^∞ solutions of Temple class systems*, Differential Integral Equations, **13** (2000), pp. 1503–1528.

[15] A. Bressan and W. Shen, *Uniqueness for discontinuous ODE and conservation laws*, Nonlinear Anal., **34** (1998), pp. 637–652.

[16] G.-Q. Chen and M. Rascle, *Initial layers and uniqueness of weak entropy solutions to hyperbolic conservation laws*, Arch. Ration. Mech. Anal., **153** (2000), pp. 205–220.

[17] G. Crippa, *The flow associated to weakly differentiable vector fields*, Vol. **12** of Tesi. Scuola Normale Superiore di Pisa (Nuova Serie) [Theses of Scuola Normale Superiore di Pisa (New Series)], Edizioni della Normale, Pisa, 2009.

[18] G. Crippa and C. De Lellis, *Oscillatory solutions to transport equations*, Indiana Univ. Math. J., **55** (2006), pp. 1–13.

[19] ———, *Estimates and regularity results for the DiPerna-Lions flow*, J. Reine Angew. Math., **616** (2008), pp. 15–46.

[20] C.M. Dafermos, *Hyperbolic Conservation Laws in Continuum Physics*, Springer-Verlag, Berlin, Second ed., 2005.

[21] C. De Lellis, *Notes on hyperbolic systems of conservation laws and transport equations*, Handbook of Differential Equations: Evolutionary Differential Equations, III (2006), pp. 277–383.

[22] N. Depauw, *Non unicité des solutions bornées pour un champ de vecteurs BV en dehors d'un hyperplan*, C. R. Math. Acad. Sci. Paris, **337** (2003), pp. 249–252.

[23] R.J. DiPerna and P.L. Lions, *Ordinary differential equations, transport theory and Sobolev spaces*, Invent. Math., **98** (1989), pp. 511–547.

[24] H. Freistühler, *On the Cauchy problem for a class of hyperbolic systems of conservation laws*, J. Differential Equations, **112** (1994), pp. 170–178.

[25] B.L. Keyfitz and H.C. Kranzer, *A system of nonstrictly hyperbolic conservation laws arising in elasticity theory*, Arch. Rational Mech. Anal., **72** (1979/80), pp. 219–241.

[26] S.N. Kružkov, *First order quasilinear equations with several independent variables.*, Mat. Sb. (N.S.), **81** (123) (1970), pp. 228–255.

[27] O.A. Oleĭnik, *Discontinuous solutions of non-linear differential equations*, Amer. Math. Soc. Transl. (2), **26** (1963), pp. 95–172.

[28] E.Y. Panov, *On the theory of generalized entropy solutions of the cauchy problem for a class of non-strictly hyperbolic systems of conservation laws*, SB MATH, **191** (2000), pp. 121–150.

[29] E.Y. Panov, *Generalized Solutions of the Cauchy Problem for a Transport Equation with Discontinuous Coefficients*, in Instability in Models Connected with Fluid Flows II. Edited by Claude Bardos and Andrei Fursikov, Springer New York, 2008, pp. 23–84.

[30] D. Serre, *Solutions à variations bornées pour certains systèmes hyperboliques de lois de conservation*, J. Differential Equations, **68** (1987), pp. 137–168.

[31] B. Temple, *Systems of conservation laws with invariant submanifolds*, Trans. Amer. Math. Soc., **280** (1983), pp. 781–795.

FINITE ENERGY WEAK SOLUTIONS TO THE QUANTUM HYDRODYNAMICS SYSTEM

PAOLO ANTONELLI* AND PIERANGELO MARCATI†

Abstract. In this paper we consider the global existence of weak solutions to a class of Quantum Hydrodynamics (QHD) systems with initial data, arbitrarily large in the energy norm. These type of models, initially proposed by Madelung [24], have been extensively used in Physics to investigate Supefluidity and Superconductivity phenomena [10], [19] and more recently in the modeling of semiconductor devices [11] . Our approach is based on various tools, namely the wave functions polar decomposition, the construction of approximate solution via a fractional steps method which iterates a Schrödinger Madelung picture with a suitable wave function updating mechanism. Therefore several *a priori* bounds of energy, dispersive and local smoothing type allow us to prove the compactness of the approximating sequences. No uniqueness result is provided. A more detailed exposition of the results is given in [2].

Key words. Analysis of PDEs, mathematical physics.

AMS(MOS) subject classifications. 35Q40 (Primary); 35Q55, 35Q35, 82D37, 82C10, 76Y05 (Secondary).

1. Introduction. In this paper we study the Cauchy problem for the Quantum Hydrodynamics (QHD) system:

$$\begin{cases} \partial_t \rho + \operatorname{div} J = 0 \\ \partial_t J + \operatorname{div}\left(\dfrac{J \otimes J}{\rho}\right) + \nabla P(\rho) + \rho \nabla V + J = \dfrac{\hbar^2}{2}\rho \nabla\left(\dfrac{\Delta\sqrt{\rho}}{\sqrt{\rho}}\right) \\ -\Delta V = \rho \end{cases} \tag{1.1}$$

with initial data

$$\rho(0) = \rho_0, \quad J(0) = J_0. \tag{1.2}$$

We are interested to study the global existence in the class of finite energy initial data without higher regularity hypotheses or smallness assumptions. There is an extensive literature (see for example [7], [19], [15], [17], and references therein) where superfluidity phenomena are described by means of quantum hydrodynamic systems. Another example is provided by the so called Schrödinger-Langevin equation in chemical physics [18]. Furthermore, the Quantum Hydrodynamics system is well known in literature since it has been used in modeling semiconductor devices at nanometric scales (see [11]). For a derivation of the QHD system we refer to [12], [9]. The unknowns ρ, J represent the charge and the current densities respectively, $P(\rho)$ the classical pressure which we assume to satisfy $P(\rho) = \frac{p-1}{p+1}\rho^{\frac{p+1}{2}}$

*Department of Applied Mathematics and Theoretical Physics, CMS, Wilberforce Road, Cambridge CB3 0WA, England.

†Department of Pure and Applied Mathematics, University of L'Aquila, Italy.

(here and throughout the paper we assume $1 \leq p < 5$). The function V is the self-consistent electric potential, given by the Poisson equation.

The term $\frac{\hbar^2}{2}\rho\nabla\left(\frac{\Delta\sqrt{\rho}}{\sqrt{\rho}}\right)$ can be interpreted as the quantum Bohm potential, or as a quantum correction to the pressure, indeed with some regularity assumptions we can write the dispersive term in different ways:

$$\begin{aligned} \frac{\hbar^2}{2}\rho\nabla\left(\frac{\Delta\sqrt{\rho}}{\sqrt{\rho}}\right) &= \frac{\hbar^2}{4}\operatorname{div}(\rho\nabla^2\log\rho) \\ &= \frac{\hbar^2}{4}\Delta\nabla\rho - \hbar^2\operatorname{div}(\nabla\sqrt{\rho}\otimes\nabla\sqrt{\rho}). \end{aligned} \tag{1.3}$$

There is a formal equivalence between the system (1.1) and the following nonlinear Schrödinger-Poisson system:

$$\begin{cases} i\hbar\partial_t\psi + \dfrac{\hbar^2}{2}\Delta\psi = |\psi|^{p-1}\psi + V\psi + \tilde{V}\psi \\ -\Delta V = |\psi|^2 \end{cases} \tag{1.4}$$

where $\tilde{V} = \frac{1}{2i}\log\left(\frac{\psi}{\overline{\psi}}\right)$, in particular the hydrodynamic system (1.1) can be obtained by defining $\rho = |\psi|^2$, $J = \hbar\text{Im}\left(\overline{\psi}\nabla\psi\right)$ and by computing the related balance laws.

This problem has to face a serious mathematical difficulty connected with the need to solve (1.4) with the ill-posed potential $\tilde{V}$. Presently there are no mathematical results concerning the solutions to (1.4), except small perturbations around constant plane waves or local existence results under various severe restrictions (see [14], [20]). In the paper by Li and the second author [21] there is a global existence result for the system (1.1), regarding small perturbations in higher Sobolev norms of subsonic stationary solutions, with periodic boundary conditions.

Another nontrivial problem concerning the derivation of solutions to (1.1) starting from the solutions to (1.4), regards the reconstruction of the initial datum $\psi(0)$ in terms of the observables $\rho(0), J(0)$. Actually this is a case of a more general important problem in physics, pointed out by Weigert in [27]. He named it the *Pauli problem*, and it regards the possibility of reconstructing a pure quantum state, just by knowing a finite set of measurements of the state (in our case, the mass and current densities). Here the possible existence of nodal regions, or vacuum in fluid terms, namely where $\rho = 0$, forbids in general this reconstruction in a classical way. In any case, various authors (see [27] and references therein) showed that knowledge of only position and momentum distribution does not specify any single state.

The opposite direction, namely the derivation of solutions of (1.4) starting from solutions of (1.1) also can face severe mathematical difficulties in various points. In particular if we prescribe $\psi(0)$, we can define $\rho(0)$ and $J(0)$, however from the evolution of the quantities $\rho(t)$ and $J(t)$, we cannot reconstruct the wave function $\psi(t)$. Furthermore, from the moment equation in (1.1) we cannot derive the quantum eikonal equation

$$\partial_t S + \frac{1}{2}|\nabla S|^2 + h(\rho) + V + S = \frac{\hbar^2}{2}\frac{\Delta\sqrt{\rho}}{\sqrt{\rho}} \tag{1.5}$$

which is the key element to reconstruct a solution of (1.4) via the WKB ansatz for the wave function, $\psi = \sqrt{\rho}e^{iS/\hbar}$. A related problem arises in the study of Nelson stochastic mechanics (see for instance Nelson [22], Guerra and Morato [13]), where the mathematical theory is based on the analysis of the velocity fields (see Carlen [4]). Unfortunately the lack of regularity of the solutions does not allow to apply this apporach, or even the more recent advances of the theory of transport equations [8], [1] or the somehow related approach developed by Teufel and Tumulka [26] in the study of trajectories of Bohmian mechanics.

Similar difficulties arise when approaching with Wigner functions, which has been recently quite popular to deduce quantum fluid systems in a kinetic way. Even in the case we know the initial data $\rho(0), J(0)$ to be originated from a wave function $\psi(0)$, it is very difficult to show that the solutions $\rho(t), J(t)$ coincide for all times with the first and second momenta of the Wigner function obtained by solving the Wigner quantum transport equation. Moreover in our case there is also the difficulty due to the non-classical potential $\tilde{V}$.

A natural framework to study the existence of the weak solutions to (1.1) is given by the space of finite energy states. Here the energy associated to the system (1.1) is given by

$$E(t) := \int_{\mathbb{R}^3} \frac{\hbar^2}{2}|\nabla\sqrt{\rho(t)}|^2 + \frac{1}{2}|\Lambda(t)|^2 + f(\rho(t)) + \frac{1}{2}|\nabla V(t)|^2 \mathrm{d}x, \tag{1.6}$$

where $\Lambda := J/\sqrt{\rho}$, and $f(\rho) = \frac{2}{p+1}\rho^{\frac{p+1}{2}}$. The function $f(\rho)$ denotes the internal energy, which is related to the pressure through the identity $P(\rho) = \rho f'(\rho) - f(\rho)$.

Therefore our initial data are required to satisfy $E_0 < \infty$, or equivalently (if we have $1 \le p < 5$), $\sqrt{\rho_0} \in H^1(\mathbb{R}^3), \Lambda_0 \in L^2(\mathbb{R}^3)$.

DEFINITION 1.1. *We say the pair* (ρ, J) *is a* weak solution *of the Cauchy problem* (1.1), (1.2) *in* $[0,T) \times \mathbb{R}^3$ *with Cauchy data* $(\rho_0, J_0) \in L^2(\mathbb{R}^3)$, *if there exist locally integrable functions* $\sqrt{\rho}, \Lambda$, *such that* $\sqrt{\rho} \in L^2_{loc}([0,T); H^1_{loc}(\mathbb{R}^3))$, $\Lambda \in L^2_{loc}([0,T); L^2_{loc}(\mathbb{R}^3))$ *and by defining* $\rho := (\sqrt{\rho})^2$, $J := \sqrt{\rho}\Lambda$, *one has*

- *for any test function* $\eta \in \mathcal{C}_0^\infty([0,T) \times \mathbb{R}^3)$ *we have*

$$\int_0^T \int_{\mathbb{R}^3} \rho\partial_t\eta + J\cdot\nabla\eta \mathrm{d}x\mathrm{d}t + \int_{\mathbb{R}^3} \rho_0\eta(0)\mathrm{d}x = 0; \tag{1.7}$$

- *for any test function* $\zeta \in \mathcal{C}_0^\infty([0,T) \times \mathbb{R}^3; \mathbb{R}^3)$

$$\begin{aligned}\int_0^T \int_{\mathbb{R}^3} & J \cdot \partial_t \zeta + \Lambda \otimes \Lambda : \nabla \zeta + P(\rho) \operatorname{div} \zeta \\ & - \rho \nabla V \cdot \zeta - J \cdot \zeta + \hbar^2 \nabla \sqrt{\rho} \otimes \nabla \sqrt{\rho} : \nabla \zeta \\ & - \frac{\hbar^2}{4} \rho \Delta \operatorname{div} \zeta \mathrm{d}x \mathrm{d}t + \int_{\mathbb{R}^3} J_0 \cdot \zeta(0) \mathrm{d}x = 0;\end{aligned} \tag{1.8}$$

- (generalized irrotationality condition) *for almost every* $t \in (0,T)$,

$$\nabla \wedge J = 2 \nabla \sqrt{\rho} \wedge \Lambda \tag{1.9}$$

holds in the sense of distributions.

We say that the weak solution (ρ, J) *to the Cauchy problem* (1.1), (1.2) *is a* finite energy weak solution *(FEWS) in* $[0,T) \times \mathbb{R}^3$, *if in addition for almost every* $t \in [0,T)$, *the energy* (1.6) *is finite.*

REMARK 1.1. Suppose we are in the smooth case, so that we can factorize $J = \rho u$, for some current velocity field u, then the last condition (1.9) simply means $\rho \nabla \wedge u = 0$, the current velocity u is irrotational in $\rho \mathrm{d}x$. This is why we will call it *generalized irrotationality condition.* The hydrodynamic structure of the system (1.1), (1.2) should not lead to conclude that the solutions behave like classical fluids. Indeed the connection with Schrödinger equations suggests that ρ_0, J_0 should in any case be seen as momenta related to some wave function ψ_0. The main result of this paper is the existence of FEWS by assuming the initial data ρ_0, J_0 are momenta of some wave function $\psi_0 \in H^1(\mathbb{R}^3)$.

THEOREM 1.1 (Main Theorem). *Let* $\psi_0 \in H^1(\mathbb{R}^3)$ *and let us define*

$$\rho_0 := |\psi_0|^2, \qquad J_0 := \hbar \operatorname{Im}(\overline{\psi_0} \nabla \psi_0).$$

Then, for each $0 < T < \infty$, *there exists a finite energy weak solution to the QHD system* (1.1) *in* $[0,T) \times \mathbb{R}^3$, *with initial data* (ρ_0, J_0) *defined as above.*

2. Polar decomposition. Let us notice that the measure $\psi \mathrm{d}x$ is absolutely continuous with respect to the measure $\sqrt{\rho}\mathrm{d}x$, where $\sqrt{\rho} := |\psi|$. Hence by the Radon-Nykodim theorem there exists ϕ, uniquely defined in $\sqrt{\rho}\mathrm{d}x$, such that $\psi = \sqrt{\rho}\phi$ and $|\phi| = 1$ $\sqrt{\rho}\mathrm{d}x$-a.e. We will call ϕ the *unitary factor* of ψ. In this section we will establish some nice properties for ϕ, which will be useful later on to determine the hydrodynamical quantities.

LEMMA 2.1. *Let* $\psi \in H^1(\mathbb{R}^3)$, $\sqrt{\rho} := |\psi|$, *then there exists* $\phi \in L^\infty(\mathbb{R}^3)$ *such that* $\psi = \sqrt{\rho}\phi$ *a.e. in* $\mathbb{R}^3$, $\sqrt{\rho} \in H^1(\mathbb{R}^3)$, $\nabla \sqrt{\rho} = \operatorname{Re}(\overline{\phi} \nabla \psi)$. *If we set* $\Lambda := \hbar \operatorname{Im}(\overline{\phi} \nabla \psi)$, *one has* $\Lambda \in L^2(\mathbb{R}^3)$ *and moreover the following identity holds*

$$\hbar^2 \operatorname{Re}(\partial_j \overline{\psi} \partial_k \psi) = \hbar^2 \partial_j \sqrt{\rho} \partial_k \sqrt{\rho} + \Lambda^{(j)} \Lambda^{(k)}. \tag{2.1}$$

Furthermore, let $\psi_n \to \psi$ strongly in $H^1(\mathbb{R}^3)$, then it follows

$$\nabla\sqrt{\rho_n} \to \nabla\sqrt{\rho}, \quad \Lambda_n \to \Lambda \qquad \text{in } L^2(\mathbb{R}^3), \tag{2.2}$$

where $\Lambda_n := \hbar\mathrm{Im}(\overline{\phi_n}\nabla\psi_n)$.

Proof. Let us consider a sequence $\{\psi_n\} \subset \mathcal{C}^\infty(\mathbb{R}^3)$, $\psi_n \to \psi$ in $H^1(\mathbb{R}^3)$, define

$$\phi_n(x) = \begin{cases} \dfrac{\psi_n(x)}{|\psi_n(x)|} & \text{if } \psi_n(x) \neq 0 \\ 0 & \text{if } \psi(x) = 0. \end{cases} \tag{2.3}$$

Then, there exists $\phi \in L^\infty(\mathbb{R}^3)$ such that $\phi_n \overset{*}{\rightharpoonup} \phi$ in $L^\infty(\mathbb{R}^3)$, hence by (2.3)

$$\nabla|\psi_n| = \mathrm{Re}\left(\overline{\phi_n}\nabla\psi_n\right) \rightharpoonup \mathrm{Re}\left(\overline{\phi}\nabla\psi\right) \qquad \text{in } L^2(\mathbb{R}^3).$$

Moreover, one has $\nabla\sqrt{\rho_n} \rightharpoonup \nabla\sqrt{\rho}$ in $L^2(\mathbb{R}^3)$, therefore $\nabla\sqrt{\rho} = \mathrm{Re}\left(\overline{\phi}\nabla\psi\right)$, where ϕ is a unitary factor of ψ. The identity (2.1) follows immediately from the following

$$\hbar^2\mathrm{Re}(\partial_j\overline{\psi}\partial_k\psi) = \hbar^2\mathrm{Re}((\phi\partial_j\overline{\psi})(\overline{\phi}\partial_k\psi)) = \hbar^2\partial_j\sqrt{\rho}\partial_k\sqrt{\rho} + \Lambda^{(j)}\Lambda^{(k)}.$$

Now we prove (2.2). Let us consider $\nabla\sqrt{\rho_n} = \mathrm{Re}\left(\overline{\phi_n}\nabla\psi_n\right)$, $\Lambda_n := \hbar\mathrm{Im}\left(\overline{\phi_n}\nabla\psi_n\right)$. As before, $\phi_n \overset{*}{\rightharpoonup} \phi$ in $L^\infty(\mathbb{R}^3)$; then $\nabla\sqrt{\rho_n} \rightharpoonup \nabla\sqrt{\rho}$, $\mathrm{Re}\left(\overline{\phi_n}\nabla\psi_n\right) \rightharpoonup \mathrm{Re}\left(\overline{\phi}\nabla\psi\right)$, and $\nabla\sqrt{\rho} = \mathrm{Re}\left(\overline{\phi}\nabla\psi\right)$. Moreover, $\Lambda_n := \hbar\mathrm{Im}\left(\overline{\phi_n}\nabla\psi_n\right) \rightharpoonup \hbar\mathrm{Im}\left(\overline{\phi}\nabla\psi\right) =: \Lambda$. To upgrade the weak convergence into the strong one, simply notice that by (2.1) one has

$$\begin{aligned}\hbar^2\|\nabla\psi\|_{L^2}^2 &= \hbar^2\|\nabla\sqrt{\rho}\|_{L^2}^2 + \|\Lambda\|_{L^2}^2 \leq \liminf_{n\to\infty}\left(\hbar^2\|\nabla\sqrt{\rho_n}\|_{L^2}^2 + \|\Lambda_n\|_{L^2}^2\right) \\ &= \hbar^2\|\nabla\psi\|_{L^2}^2.\end{aligned}$$

□

COROLLARY 2.1. *Let $\psi \in H^1(\mathbb{R}^3)$, then*

$$\hbar\nabla\overline{\psi} \wedge \nabla\psi = 2i\nabla\sqrt{\rho} \wedge \Lambda. \tag{2.4}$$

LEMMA 2.2. *Let $\psi \in H^1(\mathbb{R}^3)$, and let $\tau, \varepsilon > 0$ be two arbitrary (small) real numbers. Then there exists $\tilde{\psi} \in H^1(\mathbb{R}^3)$ such that*

$$\tilde{\rho} = \rho, \quad \tilde{\Lambda} = (1-\tau)\Lambda + r_\varepsilon,$$

where $\sqrt{\rho} := |\psi|, \sqrt{\tilde{\rho}} := |\tilde{\psi}|, \Lambda := \hbar\mathrm{Im}(\overline{\phi}\nabla\psi), \tilde{\Lambda} := \hbar\mathrm{Im}(\overline{\tilde{\phi}}\nabla\tilde{\psi})$, and $\|r_\varepsilon\|_{L^2(\mathbb{R}^3)} \leq \varepsilon$. Furthermore we have

$$\nabla\tilde{\psi} = \nabla\psi - i\frac{\tau}{\hbar}\phi^\star\Lambda + r_{\varepsilon,\tau}, \tag{2.5}$$

where $\|\phi^\star\|_{L^\infty(\mathbb{R}^3)} \leq 1$ and $\|r_{\varepsilon,\tau}\|_{L^2(\mathbb{R}^3)} \leq C(\tau\|\nabla\psi\|_{L^2(\mathbb{R}^3)} + \varepsilon)$.

3. QHD without collisions. Now let us summarize some key points regarding the existence of weak solutions to the Quantum Hydrodynamic system, in the collisionless case.

The balance equations can be written in the following way:

$$\begin{cases} \partial_t \rho + \operatorname{div} J = 0 \\ \partial_t J + \operatorname{div}\left(\dfrac{J \otimes J}{\rho}\right) + \nabla P(\rho) + \rho \nabla V = \dfrac{\hbar^2}{2}\rho\nabla\left(\dfrac{\Delta\sqrt{\rho}}{\sqrt{\rho}}\right) \\ -\Delta V = \rho, \end{cases} \tag{3.1}$$

where $P(\rho) = \frac{p-1}{p+1}\rho^{(p+1)/2}$, $1 \leq p < 5$. The next existence result roughly speaking shows how to get a weak solution to the system (3.1) out of a strong solution to the Schrödinger-Poisson system

$$\begin{cases} i\hbar\partial_t\psi + \dfrac{\hbar^2}{2}\Delta\psi = |\psi|^{p-1}\psi + V\psi \\ -\Delta V = |\psi|^2 \,. \end{cases} \tag{3.2}$$

Clearly the definition of finite energy weak solutions to (3.1) is completely similar to Definition 1.1. The quadratic nonlinearities in (3.1) are originated by a term of the form $\mathrm{Re}\left(\nabla\overline{\psi} \otimes \nabla\psi\right)$ since formally

$$\hbar^2\mathrm{Re}\left(\nabla\overline{\psi} \otimes \nabla\psi\right) = \hbar^2\nabla\sqrt{\rho} \otimes \nabla\sqrt{\rho} + \frac{J \otimes J}{\rho}.$$

However this identity can be justified in the nodal region $\{\rho = 0\}$ only by means of the polar factorization discussed in the previous section (see lemma 2.1). Furthermore, the study of the existence of weak solutions of (3.1) is done with Cauchy data of the form $(\rho_0, J_0) = (|\psi_0|^2, \hbar\mathrm{Im}(\overline{\psi_0}\nabla\psi_0))$, for some $\psi_0 \in H^1(\mathbb{R}^3;\mathbb{C})$.

PROPOSITION 3.1. *Let $0 < T < \infty$, let $\psi_0 \in H^1(\mathbb{R}^3)$ and define the initial data for (3.1), $(\rho_0, J_0) := (|\psi_0|^2, \hbar\mathrm{Im}(\overline{\psi_0}\nabla\psi_0))$. Then there exists a finite energy weak solution (ρ, J) to the Cauchy problem (3.1) in the space-time slab $[0,T) \times \mathbb{R}^3$. Furthermore the energy $E(t)$ defined in (1.6) is conserved for all times $t \in [0,T)$.* The idea behind the proof of this Proposition is the following. Let us consider the Cauchy problem for (3.2) with $\psi(0) = \psi_0$; then it is globally well-posed for initial data in $H^1(\mathbb{R}^3)$ (see [5]), and the solution satisfies $\psi \in \mathcal{C}^0(\mathbb{R}; H^1(\mathbb{R}^3))$. It thus makes sense to define for each time $t \in [0,T)$ the quantities $\rho(t) := |\psi(t)|^2, J(t) := \hbar\mathrm{Im}(\overline{\psi(t)}\nabla\psi(t))$ and we can see that (ρ, J) is a finite energy weak solution of (3.1): it is sufficient to differentiate with respect to time the quantities (ρ, J) and to use the equation (3.2). We only remark that we need the Lemma 2.1 to write

$$\hbar^2\mathrm{Re}\left(\nabla\overline{\psi} \otimes \nabla\psi\right) = \hbar^2\nabla\sqrt{\rho} \otimes \nabla\sqrt{\rho} + \Lambda \otimes \Lambda$$

and to obtain the following formal identity:

$$\partial_t J + \operatorname{div}(\Lambda \otimes \Lambda) + \nabla P(\rho) + \rho \nabla V = \frac{\hbar^2}{4} \Delta \nabla \rho - \hbar^2 \operatorname{div}(\nabla \sqrt{\rho} \otimes \nabla \sqrt{\rho}). \quad (3.3)$$

Of course these calculations are just formal, since ψ doesn't have the necessary regularity to implement them, but we can proceed with some straightforward regularizing argument.

4. The fractional step: Definitions and consistency. In this section we make use of the results of the previous sections to construct a sequence of approximate solutions of the QHD system

$$\begin{cases} \partial_t \rho + \operatorname{div} J = 0 \\ \partial_t J + \operatorname{div}\left(\dfrac{J \otimes J}{\rho}\right) + \nabla P(\rho) + \rho \nabla V + J = \dfrac{\hbar^2}{2} \rho \nabla \left(\dfrac{\Delta \sqrt{\rho}}{\sqrt{\rho}}\right) \\ -\Delta V = \rho \end{cases} \quad (4.1)$$

with Cauchy data

$$(\rho(0), J(0)) = (\rho_0, J_0). \quad (4.2)$$

DEFINITION 4.1. *Let $0 < T < \infty$. We say $\{(\rho^\tau, J^\tau)\}$ is a* sequence of approximate solutions *to the system* (1.1) *in $[0,T) \times \mathbb{R}^3$, with initial data $(\rho_0, J_0) \in L^1_{loc}(\mathbb{R}^3)$, if there exist locally integrable functions $\sqrt{\rho^\tau}, \Lambda^\tau$, such that $\sqrt{\rho^\tau} \in L^2_{loc}([0,T); H^1_{loc}(\mathbb{R}^3)), \Lambda^\tau \in L^2_{loc}([0,T); L^2_{loc}(\mathbb{R}^3))$ and if we define $\rho^\tau := (\sqrt{\rho^\tau})^2$, $J^\tau := \sqrt{\rho^\tau}\Lambda^\tau$, then*

- *for any test function $\eta \in C_0^\infty([0,T) \times \mathbb{R}^3)$ one has*

$$\int_0^T \int_{\mathbb{R}^3} \rho^\tau \partial_t \eta + J^\tau \cdot \nabla \eta \mathrm{d}x\mathrm{d}t + \int_{\mathbb{R}^3} \rho_0 \eta(0) \mathrm{d}x = o(1) \quad (4.3)$$

  as $\tau \to 0$,
- *for any test function $\zeta \in C_0^\infty([0,T) \times \mathbb{R}^3; \mathbb{R}^3)$ we have*

$$\begin{aligned} &\int_0^T \int_{\mathbb{R}^3} J^\tau \cdot \partial_t \zeta + \Lambda^\tau \otimes \Lambda^\tau : \nabla \zeta + P(\rho^\tau) \operatorname{div} \zeta - \rho^\tau \nabla V^\tau \cdot \zeta \\ &- J^\tau \cdot \zeta + \hbar^2 \nabla \sqrt{\rho^\tau} \otimes \nabla \sqrt{\rho^\tau} : \nabla \zeta - \frac{\hbar^2}{4} \rho^\tau \Delta \operatorname{div} \zeta \mathrm{d}x\mathrm{d}t \quad (4.4) \\ &+ \int_{\mathbb{R}^3} J_0 \cdot \zeta(0) \mathrm{d}x = o(1) \end{aligned}$$

  as $\tau \to 0$;
- (generalized irrotationality condition) *for almost every $t \in (0,T)$ we have*

$$\nabla \wedge J^\tau = 2 \nabla \sqrt{\rho^\tau} \wedge \Lambda^\tau.$$

Our fractional step method is based on the following simple idea. We split the evolution of our problem into two separate steps. Let us fix a (small) parameter $\tau > 0$ then in the former step we solve a non-collisional QHD problem, while in the latter one we solve the collisional problem without QHD, and at this point we can start again with the non-collisional QHD problem, the main difficulty being the updating of the initial data at each time step. Indeed, as remarked in the previous section we are able to solve the non-collisional QHD only in the case of Cauchy data compatible with the Schrödinger picture. This restriction impose to reconstruct a wave function at each time step. The iteration procedure can be defined in this following way. Let us take $\tau > 0$; we therefore define the approximate solutions in each time interval $[k\tau, (k+1)\tau)$, for any integer $k \geq 0$. At the first step, $k = 0$, we solve in $[0, \tau) \times \mathbb{R}^3$ the Cauchy problem for the Schrödinger-Poisson system

$$\begin{cases} i\hbar\partial_t\psi^\tau + \dfrac{\hbar^2}{2}\Delta\psi^\tau = |\psi^\tau|^{p-1}\psi^\tau + V^\tau\psi^\tau \\ -\Delta V^\tau = |\psi^\tau|^2 \\ \psi^\tau(0) = \psi_0 \ . \end{cases} \tag{4.5}$$

Let us define $\rho^\tau := |\psi^\tau|^2, J^\tau := \hbar\mathrm{Im}(\overline{\psi^\tau}\nabla\psi^\tau)$. Then (ρ^τ, J^τ) is a weak solution to the non-collisional QHD system. Let us assume that we know ψ^τ in the space-time slab $[(k-1)\tau, k\tau) \times \mathbb{R}^3$, we want how to define $\psi^\tau, \rho^\tau, J^\tau$ in the strip $[k\tau, (k+1)\tau)$. In order to take into account the presence of the collisional term we update ψ in $t = k\tau$, namely we define $\psi^\tau(k\tau+)$. Let us apply the Lemma 2.2, with $\psi = \psi^\tau(k\tau-)$, $\varepsilon = \tau 2^{-k}\|\psi_0\|_{H^1(\mathbb{R}^3)}$, then we can define

$$\psi^\tau(k\tau+) = \tilde{\psi}, \tag{4.6}$$

by using the wave function $\tilde{\psi}$ defined in the Lemma 2.2. Therefore we have

$$\rho^\tau(k\tau+) = \rho^\tau(k\tau-), \qquad \Lambda^\tau(k\tau+) = (1-\tau)\Lambda^\tau(k\tau-) + R_k \tag{4.7}$$

where $\|R_k\|_{L^2(\mathbb{R}^3)} \leq \tau 2^{-k}\|\psi_0\|_{H^1(\mathbb{R}^3)}$ and

$$\nabla\psi^\tau(k\tau+) = \nabla\psi^\tau(k\tau-) - i\frac{\tau}{\hbar}\phi^*\Lambda^\tau(k\tau-) + r_{k,\tau}, \tag{4.8}$$

with $\|\phi^*\|_{L^\infty} \leq 1$ and

$$\|r_{k,\tau}\|_{L^2} \leq C(\tau\|\nabla\psi^\tau(k\tau-)\| + \tau 2^{-k}\|\psi_0\|_{H^1(\mathbb{R}^3)}) \lesssim \tau E_0^{\frac{1}{2}}.$$

We then solve the Schrödinger-Poisson system with initial data $\psi(0) = \psi^\tau(k\tau+)$. We define ψ^τ in the time strip $[k\tau, (k+1)\tau)$ as the restriction of the Schrödinger-Poisson solution just found in $[0, \tau)$, furthermore, we define $\rho^\tau := |\psi^\tau|^2, J^\tau := \hbar\mathrm{Im}(\overline{\psi^\tau}\nabla\psi^\tau)$ and we go on for each time strip.

THEOREM 4.1 (Consistency of the approximate solutions). *Let us consider the sequence of approximate solutions* $\{(\rho^\tau, J^\tau)\}_{\tau>0}$ *constructed via the fractional step method, and assume there exists* $0 < T < \infty$, $\sqrt{\rho} \in L^2_{loc}([0,T); H^1_{loc}(\mathbb{R}^3))$ *and* $\Lambda \in L^2_{loc}([0,T); L^2_{loc}(\mathbb{R}^3))$, *such that*

$$\sqrt{\rho^\tau} \to \sqrt{\rho} \qquad \text{in } L^2([0,T); H^1_{loc}(\mathbb{R}^3)) \tag{4.9}$$

$$\Lambda^\tau \to \Lambda \qquad \text{in } L^2([0,T); L^2_{loc}(\mathbb{R}^3)). \tag{4.10}$$

Then the limit function (ρ, J), *where as before* $J = \sqrt{\rho}\Lambda$, *is a finite energy weak solution of the QHD system, with Cauchy data* (ρ_0, J_0).

5. A priori estimates and convergence. In this section we obtain various a priori estimates necessary to show the strong convergence of (a subsequence of) $\{\sqrt{\rho^\tau}\}$ in $L^2_{loc}([0,T); H^1_{loc}(\mathbb{R}^3))$ of $\{\Lambda^\tau\}$ in $L^2_{loc}([0,T); L^2_{loc}(\mathbb{R}^3))$, in order to apply theorem 4.1 to find out a FEWS to (1.1). The plan of this section is the following, first of all we get a discrete version of the (dissipative) energy inequality for the system (1.1), later we use the Strichartz estimates (see [16]) for $\nabla\psi^\tau$ by means of the formula (5.2) below. Consequently via the Strichartz estimates and by using the local smoothing results due to Constantin and Saut [6], we deduce some further regularity properties of $\{\nabla\psi^\tau\}$.

Let us begin with the energy inequality.

PROPOSITION 5.1 (Discrete energy inequality). *Let* (ρ^τ, J^τ) *be an approximate solution of the QHD system, with* $0 < \tau < 1$. *Then, for* $t \in [N\tau, (N+1)\tau)$ *we have*

$$E^\tau(t) \le -\frac{\tau}{2}\sum_{k=1}^{N} \|\Lambda(k\tau-)\|_{L^2(\mathbb{R}^3)} + (1+\tau)E_0. \tag{5.1}$$

Unfortunately the energy estimates are not sufficient to get enough compactness to show the convergence of the sequence of the approximate solutions. Indeed from the discrete energy inequality, we get only the weak convergence of $\nabla\psi^\tau$ in $L^\infty([0,\infty); H^1(\mathbb{R}^3))$, and therefore the quadratic terms in (4.4) could exhibit some concentrations phenomena in the limit.

More precisely, from energy inequality we get the sequence $\{\psi^\tau\}$ is uniformly bounded in $L^\infty([0,\infty); H^1(\mathbb{R}^3))$, hence there exists $\psi \in L^\infty([0,\infty); H^1(\mathbb{R}^3))$, such that

$$\psi^\tau \rightharpoonup \psi \qquad \text{in } L^\infty([0,\infty); H^1(\mathbb{R}^3)).$$

The need to pass into the limit the quadratic expressions leads us to look for a priori estimates in stronger norms. The relationships with the Schrödinger equation brings naturally into this search the Strichartz-type estimates (see [16]). The following results are concerned with these estimates. However they are *not* an immediate consequence of the Strichartz estimates for the Schrödinger equation since we have to take into account the effects of the updating procedure which we implement at each time step.

LEMMA 5.1. *Let ψ^τ be the wave function defined by the fractional step method and let $t \in [N\tau, (N+1)\tau)$. Then we have*

$$\begin{aligned}\nabla\psi^\tau(t) = U(t)\nabla\psi_0 - i\frac{\tau}{\hbar}\sum_{k=1}^{N} U(t-k\tau)\left(\phi_k^\tau\Lambda^\tau(k\tau-)\right)\\ -i\int_0^t U(t-s)F(s)\mathrm{d}s + \sum_{k=1}^{N} U(t-k\tau)r_k^\tau,\end{aligned} \tag{5.2}$$

where $U(t)$ is the free Schrödinger group, $F = \nabla(|\psi^\tau|^{p-1}\psi^\tau + V^\tau\psi^\tau)$ and

$$\|\phi_k^\tau\|_{L^\infty(\mathbb{R}^3)} \leq 1, \quad \|r_k^\tau\|_{L^2(\mathbb{R}^3)} \leq \tau\|\psi_0\|_{H^1(\mathbb{R}^3)}. \tag{5.3}$$

Proof. Since ψ^τ is solution of the Schrödinger-Poisson system in the space-time slab $[N\tau, (N+1)\tau) \times \mathbb{R}^3$, then we can write

$$\nabla\psi^\tau(t) = U(t-N\tau)\nabla\psi^\tau(N\tau+) - i\int_{N\tau}^t U(t-s)F(s)\mathrm{d}s. \tag{5.4}$$

As specified in the proof of the Lemma 2.2, we have

$$\psi(N\tau+) = e^{i(1-\tau)\theta_N}\sqrt{\rho_n}.$$

Therefore, from (5.4)

$$\begin{aligned}\nabla\psi^\tau(t) = U(t-N\tau)\nabla\psi^\tau(N\tau-) - i\frac{\tau}{\hbar}U(t-N\tau)(e^{i(1-\tau)\theta_N}\Lambda^\tau(N\tau-))\\ + U(t-N\tau)r_N^\tau - i\int_{N\tau}^t U(t-s)F(s)\mathrm{d}s.\end{aligned}$$

By iterating this formula we get (5.2). □

We can now apply the Strichartz estimates in [16] to (5.2) to get:

PROPOSITION 5.2 (Strichartz estimates for $\nabla\psi^\tau$). *Let, $0 < T < \infty$, let ψ^τ be as in the previous section, then one has*

$$\|\nabla\psi^\tau\|_{L_t^q L_x^r([0,T]\times\mathbb{R}^3)} \leq C(E_0^{\frac{1}{2}}, \|\rho_0\|_{L^1(\mathbb{R}^3)}, T) \tag{5.5}$$

for each admissible pair of exponents (q, r). By using the Strichartz estimates we just obtained, we apply some results concerning local smoothing due to Constantin, Saut [6].

PROPOSITION 5.3 (Local smoothing for $\nabla\psi^\tau$). *Let $0 < T < \infty$ and let ψ^τ be defined as in the previous section. Then one has*

$$\|\nabla\psi^\tau\|_{L^2([0,T];H^{1/2}_{loc}(\mathbb{R}^3))} \leq C(E_0, T, \|\rho_0\|_{L^1}). \tag{5.6}$$

Since $H^{1/2}_{loc}$ is compactly embedded in L^2_{loc}, we can apply some Aubin-Lions type lemma, namely the Theorem due to Rakotoson, Temam [25], to get the strong convergence of (a subsequence of) $\{\nabla\psi^\tau\}$.

PROPOSITION 5.4. *The sequence $\{\nabla\psi^{\tau}\}$ is strongly convergent in $L^2([0,T];L^2_{loc}(\mathbb{R}^3))$, namely*

$$\nabla\psi := s - \lim_{k\to\infty} \nabla\psi^{\tau_k} \qquad in\ L^2([0,T];L^2_{loc}(\mathbb{R}^3)). \tag{5.7}$$

In particular, one has $\nabla\sqrt{\rho^{\tau}} \to \nabla\sqrt{\rho}$ and $\Lambda^{\tau} \to \Lambda$ in $L^2([0,T];L^2_{loc}(\mathbb{R}^3))$.

PROPOSITION 5.5. *(ρ, J) is a weak solution to the Cauchy problem* (1.1), (1.2).

Proof. It follows directly by combining the Theorem 4.1 in the section 4 and the Proposition 5.4. As for the collisionless QHD system we should note that the generalized irrotationality condition holds by the definition of the current density and Corollary 2.1. □

6. Further extensions ([3]).

THEOREM 6.1 (2D case). *Let us consider the same Cauchy problem* (1.1), (1.2) *in two space dimensions and let us assume furthermore that*

$$\int_{\mathbb{R}^2} \rho_0 \log \rho_0 \mathrm{d}x < \infty, \tag{6.1}$$

and $V(0,x) := -\frac{1}{2\pi}\int_{\mathbb{R}^2} \log|x-y|\rho_0(y)\mathrm{d}y$ satisfies $V(0,\cdot) \in L^r(\mathbb{R}^2)$, for some $2 < r < \infty$. Then, for $0 < T < \infty$ there exists a finite energy weak solution of (1.1), (1.2), *in $[0,T) \times \mathbb{R}^2$.*

THEOREM 6.2 (Non-trivial doping profile). *Let us consider the same Cauchy problem as in Theorem 1.1, but with the Poisson equation for the electrostatic potential replaced by*

$$-\Delta V = \rho - C(x), \tag{6.2}$$

where C is a (given) background doping, such that $C \in L^{p_1}(\mathbb{R}^3) + L^{p_2}(\mathbb{R}^3)$. Then, for $0 < T < \infty$ there exists a finite energy weak solution in $[0,T) \times \mathbb{R}^3$.

REFERENCES

[1] AMBROSIO L., *Transport Equations and Cauchy Problem for BV Vector Fields*, Inventiones Mathematicae **158**: 227–260 (2004).

[2] P. ANTONELLI AND P. MARCATI, *On the Finite Energy Weak Solutions to a System in Quantum Fluid Dynamics*, Comm. Math. Phys., **287**(2): 657–686 (2009).

[3] ANTONELLI P. AND MARCATI P., *to appear.*

[4] CARLEN E., *Conservative Diffusions*, Comm. Math. Phys. **94**: 293–315 (1984).

[5] CAZENAVE T., *Semilinear Schrödinger Equations*, Courant Lecture Notes in Mathematics, **10**. New York University, Courant Institute of Mathematical Sciences, AMS, 2003.

[6] CONSTANTIN P. AND SAUT J.C., *Local Smoothing Properties of Dispersive Equations*, J. Amer. Math. Soc. **1**: 413–439 (1988).

[7] DALFOVO F., GIORGINI S., PITAEVSKII L., AND STRINGARI S., *Theory of Bose-Einstein condensation in trapped gases*, Rev. Mod. Phys., **71**: 463–512 (1999).

[8] DI PERNA R.J. AND LIONS P.-L., *Ordinary differential equations, transport theory and Sobolev spaces*, Invent. Math. **98**: 511–547 (1989).

[9] DEGOND P. AND RINGHOFER C., *Quantum moment hydrodynamics entropy principle*, J. Stat. Phys. **112**: 587–628 (2003).

[10] FEYNMAN R.P., *Superfluidity and Superconductivity*, Rev. Mod. Phys., **29**(2): 205 (1957).

[11] GARDNER C., *The Quantum Hydrodynamic Model for Semiconductor Devices* SIAM J. Appl. Math. **54**: 409–427 (1994).

[12] GASSER I. AND MARKOWICH P., *Quantum hydrodynamics, Wigner transforms and the classical limit*, Asymptot. Anal. **14**(2): 97–116 (1997).

[13] GUERRA F. AND MORATO L., *Quantization of Dynamical Systems and Stochastic Control Theory*, Phys. Rev. D **27**: 1771–1786 (1983).

[14] JÜNGEL A., MARIANI M.C., AND RIAL D., *Local Existence of Solutions to the Transient Quantum Hydrodynamic Equations*, Math. Models Methods Appl. Sci. **12**(4): 485–495 (2002).

[15] KHALATNIKOV I.M., *An introduction to the Theory of Superfluidity*, Benjamin N.Y., 1965.

[16] KEEL M. AND TAO T., *Endpoint Strichartz Estimates*, Amer. J. Math. **120**: 955–980 (1998).

[17] KIRKPATRICK T.R. AND DORFMAN J.R., *Transport theory for a weakly interacting condensed Bose gas*, Phys. Rev. A **28**: 2576 (1983).

[18] KOSTIN M., *On the Schrödinger-Langevin equation*, J. Chem. Phys. **57**: 3589–3591 (1972).

[19] LANDAU L.D., *Theory of the Superfluidity of Helium II*, Phys. Rev. **60**: 356 (1941).

[20] LI H. AND LIN C.-K., *Semiclassical limit and well-posedness of nonlinear Schrödinger-Poisson systems*, EJDE **93**: 1–17 (2003).

[21] LI H. AND MARCATI P., *Existence and asymptotic behavior of multi-dimensional quanntum hydrodynamic model for semiconductors*, Comm. Math. Phys. **245**(2): 215–247 (2004).

[22] NELSON E., *Quantum Fluctuations*, Princeton University Press, 1984.

[23] LIFSHITZ E.M. AND PITAEVSKII L., *Physical Kinetics*, Pergamon, Oxford (1981).

[24] MADELUNG E., *Quantentheorie in hydrodynamischer form*, Z. Physik, **40**: 322 (1927).

[25] RAKOTOSON J.M. AND TEMAM R., *An Optimal Compactness Theorem and Application to Elliptic-Parabolic Systems*, Appl. Math. Letters **14**: 303–306 (2001).

[26] TEUFEL S. AND TUMULKA R., *Simple proof for global existence of bohmian trajectories*, Comm. Math. Phys. **258**: 349–365 (2005).

[27] WEIGERT S., *How to determine a quantum state by measurements: the Pauli problem for a particle with arbitrary potential*, Phys. Rev. A **53**: 4, 2078–2083 (1996).

THE MONGE PROBLEM IN GEODESIC SPACES

STEFANO BIANCHINI* AND FABIO CAVALLETTI†

Abstract. We address the Monge problem in metric spaces with a geodesic distance: (X, d) is a Polish non branching geodesic space. We show that we can reduce the transport problem to 1-dimensional transport problems along geodesics. We introduce an assumption on the transport problem π which implies that the conditional probabilities of the first marginal on each geodesic are continuous. It is known that this regularity is sufficient for the construction of an optimal transport map.

AMS(MOS) subject classifications. 28-06.

1. Introduction. This paper concerns the Monge transportation problem in geodesic spaces, i.e. metric spaces with a geodesic structure. Given two Borel probability measures $\mu, \nu \in \mathcal{P}(X)$, where (X, d) is a locally compact Polish space, i.e. a separable complete locally compact metric space, we study the minimization of the functional

$$\mathcal{I}(T) = \int d(x, T(x))\mu(dy)$$

where T varies over all Borel maps $T : X \to X$ such that $T_\sharp\mu = \nu$ and d is a distance that makes (X, d) a non branching geodesic space.

Before giving an overview of the paper and of the existence result, we recall which are the main results concerning the Monge problem.

In the original formulation given by Monge in 1781 the problem was settled in $\mathbb{R}^n$, with the cost given by the Euclidean norm and the measures μ, ν were supposed to be absolutely continuous and supported on two disjoint compact sets. The original problem remained unsolved for a long time. In 1978 Sudakov [13] claimed to have a solution for any distance cost function induced by a norm: an essential ingredient in the proof was that if $\mu \ll \mathcal{L}^d$ and $\mathcal{L}^d$-a.e. $\mathbb{R}^d$ can be decomposed into convex sets of dimension k, then the conditional probabilities are absolutely continuous with respect to the $\mathcal{H}^k$ measure of the correct dimension. But it turns out that when $d > 2$, $0 < k < d - 1$ the property claimed by Sudakov is not true. An example with $d = 3$, $k = 1$ can be found in [11] and [1].

The Euclidean case has been correctly solved only during the last decade. L. C. Evans and W. Gangbo in [8] solved the problem under the assumptions that $\operatorname{spt}\mu \cap \operatorname{spt}\nu = \emptyset$, $\mu, \nu \ll \mathcal{L}^d$ and their densities are Lipschitz function with compact support. The first existence results for general absolutely continuous measures μ, ν with compact support have been independently obtained by L. Caffarelli, M. Feldman and R.J. McCann in [5] and by N. Trudinger and X.J. Wang in [14]. Afterwards M. Feldman and

*SISSA, via Beirut 2, IT-34014 Trieste, Italy (bianchin@sissa.it).

†SISSA, via Beirut 2, IT-34014 Trieste, Italy (cavallet@sissa.it).

R.J. McCann [9] extended the results to manifolds with geodesic cost. The case of a general norm as cost function on $\mathbb{R}^d$, including also the case with non strictly convex unitary ball, has been solved first in the particular case of crystalline norm by L. Ambrosio, B. Kirchheim and A. Pratelli in [1], and then in fully generality independently by L. Caravenna in [6] and by T. Champion and L. De Pascale in [7].

1.1. Overview of the paper. The presence of 1-dimensional sets (the geodesics) along which the cost is linear is a strong degeneracy for transport problems. This degeneracy is equivalent to the following problem in $\mathbb{R}$: if μ is concentrated on $(-\infty, 0]$, and ν is concentrated on $[0, +\infty)$, then every transference plan is optimal for the 1-dimensional distance cost $|\cdot|$. In fact, every $\pi \in \Pi(\mu, \nu)$ is supported on the set $(-\infty, 0] \times [0, +\infty)$, on which $|x - y| = y - x$ and thus

$$\int |x - y|\pi(dxdy) = -\int x\mu(dx) + \int y\nu(dy).$$

Nevertheless, for this easy case an explicit map $T : \mathbb{R} \to \mathbb{R}$ can be constructed if μ is non atomic: the easiest choice is the monotone map, a minimizer of the quadratic cost $|\cdot|^2$.

The approach suggested by the above simple case is the following:

1. reduce the problem to transportation problems along distinct geodesics;
2. show that the disintegration of the marginal μ on each geodesic is continuous;
3. find a transport map on each geodesic and piece them together.

While the last point can be seen as an application of selection principles in Polish spaces, the first two points are more subtle.

The geodesics used by a given transference plan π to transport mass can be obtained from a set Γ on which π is concentrated. If π wants to be a minimizer, then it certainly chooses the shortest paths: however the metric space can be branching, i.e. geodesics can bifurcate. In this paper we assume that *the space is non branching.*

Under this assumption, a cyclically monotone plan π yields a natural partition R of the transport set $\mathcal{T}_e$, i.e. the set of the geodesics used by π:

- the set $\mathcal{T}$ made of inner points of geodesics,
- the set $a \cup b := \mathcal{T}_e \setminus \mathcal{T}$ of initial points a and end points b.

The non branching assumption and the cyclical monotonicity of Γ imply that the geodesics used by π are a partition on $\mathcal{T}$, but no other conditions can be obtained on $a \cup b$: one can think to the one dimensional torus $\mathbb{T}^1$ with $\mu = \delta_0$ and $\nu = \delta_{1/2}$. We note here that π gives also a direction along each component of R, as the one dimensional example above shows.

Even if we have a natural partition R in $\mathcal{T}$ and $\mu(a \cup b) = 0$, we cannot reduce the transport problem to one dimensional problems: a necessary and sufficient condition is that the disintegration of the measure μ is strongly consistent, which is equivalent to the fact that there exists a μ-measurable quotient map $f : \mathcal{T} \to \mathcal{T}$. In this case, one can write

$$m := f_\sharp \mu, \quad \mu = \int \mu_y m(dy), \quad \mu_y(f^{-1}(y)) = 1,$$

i.e. the conditional probabilities μ_y are concentrated on the counterimages $f^{-1}(y)$ (which are single geodesics). At this point we can obtain the one dimensional problems by partition π w.r.t. the partition $R \times (X \times X)$,

$$\pi = \int \pi_y m(dy), \quad \nu = \int \nu_y m(dy) \quad \nu_y := (P_2)_\sharp \pi_y,$$

and considering the one dimensional problems along the geodesic $R(y)$ with marginals μ_y, ν_y and cost $|\cdot|$, the length on the geodesic. At this point we can study the problem of the regularity of the conditional probabilities μ_y.

The existence of a strongly consistent disintegration relies only on the properties of geodesics in Polish spaces. Moreover, a natural operation on sets can be considered: the translation along geodesics. If A is a subset of $\mathcal{T}$, we denote by A_t the set translated by t in the direction determined by π.

It turns out that the fact that $\mu(a \cup b) = 0$ and the measures μ_y are continuous depends on how the function $t \mapsto \mu(A_t)$ behaves. We can now state the main result.

THEOREM 1.1 (Lemma 5.3 and Proposition 5.1). *If $\sharp\{t > 0 : \mu(A_t) > 0\}$ is uncountable for all A Borel such that $\mu(A) > 0$, then $\mu(a \cup b) = 0$ and m-a.e. conditional probability μ_y is continuous.*

This is sufficient to solve the Monge problem, i.e. to find a transport map which has the same cost as π. For a more general setting we refer to [3].

2. Preliminaries. In this section we recall some general facts about projective classes, the Disintegration Theorem for measures, measurable selection principles, geodesic spaces and optimal transportation problems.

2.1. Borel, projective and universally measurable sets. The *projective class* $\Sigma^1_1(X)$ is the family of subsets A of the Polish space X for which there exist Y Polish and $B \in \mathcal{B}(X \times Y)$ such that $A = P_1(B)$. The *coprojective class* $\Pi^1_1(X)$ is the complement in X of the class $\Sigma^1_1(X)$. The class Σ^1_1 is called *the class of analytic sets*, and Π^1_1 are the *coanalytic sets.*

We will denote by $\mathcal{A}$ the σ-algebra generated by Σ^1_1.

We recall that a subset of X Polish is *universally measurable* if it belongs to all completed σ-algebras of all Borel measures on X: it can be proved that every set in $\mathcal{A}$ is universally measurable.

2.2. Disintegration of measures. Given a measurable space $(R, \mathscr{R})$ and a function $r : R \to S$, with S generic set, we can endow S with the *push forward σ-algebra* $\mathscr{S}$ of $\mathscr{R}$:

$$Q \in \mathscr{S} \iff r^{-1}(Q) \in \mathscr{R},$$

which could be also defined as the biggest σ-algebra on S such that r is measurable. Moreover, given a measure space $(R, \mathscr{R}, \rho)$, the *push forward measure* η is then defined as $\eta := (r_\sharp \rho)$.

Consider a probability space $(R, \mathscr{R}, \rho)$ and its push forward measure space $(S, \mathscr{S}, \eta)$ induced by a map r. From the above definition the map r is clearly measurable and inverse measure preserving.

DEFINITION 2.1. *A disintegration of ρ consistent with r is a map $\rho : \mathscr{R} \times S \to [0,1]$ such that*

1. *$\rho_s(\cdot)$ is a probability measure on $(R, \mathscr{R})$, for all $s \in S$,*
2. *$\rho_\cdot(B)$ is η-measurable for all $B \in \mathscr{R}$,*

and satisfies for all $B \in \mathscr{R}, C \in \mathscr{S}$ the consistency condition

$$\rho\left(B \cap r^{-1}(C)\right) = \int_C \rho_s(B)\eta(ds).$$

A disintegration is strongly consistent with r if for all s we have $\rho_s(r^{-1}(s)) = 1$.

We say that a σ-algebra $\mathcal{A}$ is *essentially countably generated* with respect to a measure m, if there exists a countably generated σ-algebra $\hat{\mathcal{A}}$ such that for all $A \in \mathcal{A}$ there exists $\hat{A} \in \hat{\mathcal{A}}$ such that $m(A \vartriangle \hat{A}) = 0$.

We recall the following version of the theorem of disintegration of measure that can be found on [10], Section 452.

THEOREM 2.1 (Disintegration of measure). *Assume that $(R, \mathscr{R}, \rho)$ is a countably generated probability space, $R = \cup_{s \in S} R_s$ a decomposition of R, $r : R \to S$ the quotient map and $(S, \mathscr{S}, \eta)$ the quotient measure space. Then $\mathscr{S}$ is essentially countably generated w.r.t. η and there exists a unique disintegration $s \to \rho_s$ in the following sense: if ρ_1, ρ_2 are two consistent disintegration then $\rho_{1,s}(\cdot) = \rho_{2,s}(\cdot)$ for $\eta-$a.e. s.*

If $\{S_n\}_{n \in \mathbb{N}}$ is a family essentially generating $\mathscr{S}$ define the equivalence relation:

$$s \sim s' \iff \{s \in S_n \iff s' \in S_n,\ \forall n \in \mathbb{N}\}.$$

Denoting with p the quotient map associated to the above equivalence relation and with $(L, \mathscr{L}, \lambda)$ the quotient measure space, the following properties hold:

- *$R_l := \cup_{s \in p^{-1}(l)} R_s = (p \circ r)^{-1}(l)$ is ρ-measurable and $R = \cup_{l \in L} R_l$;*
- *the disintegration $\rho = \int_L \rho_l \lambda(dl)$ satisfies $\rho_l(R_l) = 1$, for λ-a.e. l. In particular there exists a strongly consistent disintegration w.r.t. $p \circ r$;*

- *the disintegration* $\rho = \int_S \rho_s \eta(ds)$ *satisfies* $\rho_s = \rho_{p(s)}$, *for* η*-a.e.* s.

In particular we will use the following corollary.

COROLLARY 2.1. *If* $(S, \mathscr{S}) = (X, \mathcal{B}(X))$ *with* X *Polish space, then the disintegration is strongly consistent.*

2.3. Selection principles. Given a multivalued function $F : X \to Y$, X, Y metric spaces, the *graph* of F is the set

$$\mathrm{graph}(F) := \big\{(x, y) : y \in F(x)\big\}. \tag{2.1}$$

The *inverse image* of a set $S \subset Y$ is defined as:

$$F^{-1}(S) := \big\{x \in X \ : \ F(x) \cap S \neq \emptyset\big\}. \tag{2.2}$$

For $F \subset X \times Y$, we denote also the sets

$$F_x := F \cap \{x\} \times Y, \quad F^y := F \cap X \times \{y\}. \tag{2.3}$$

In particular, $F(x) = P_2(\mathrm{graph}(F)_x)$, $F^{-1}(y) = P_1(\mathrm{graph}(F)^y)$. We denote by F^{-1} the graph of the inverse function

$$F^{-1} := \big\{(x, y) : (y, x) \in F\big\}. \tag{2.4}$$

We say that F is $\mathcal{R}$*-measurable* if $F^{-1}(B) \in \mathcal{R}$ for all B open. We say that F is *strongly Borel measurable* if inverse images of closed sets are Borel. A multivalued function is called *upper-semicontinuous* if the preimage of every closed set is closed: in particular u.s.c. maps are strongly Borel measurable.

In the following we will not distinguish between a multifunction and its graph. Note that the *domain of* F (i.e. the set $P_1(F)$) is in general a subset of X. The same convention will be used for functions, in the sense that their domain may be a subset of X.

Given $F \subset X \times Y$, a *section* u *of* F is a function from $P_1(F)$ to Y such that $\mathrm{graph}(u) \subset F$. A *cross-section of the equivalence relation* E is a set $S \subset E$ such that the intersection of S with each equivalence class is a singleton. We recall that a set $A \subset X$ is saturated for the equivalence relation $E \subset X \times X$ if $A = \cup_{x \in A} E(x)$.

We recall the following selection principle, Theorem 5.2.1 of [12].

THEOREM 2.2. *Let* Y *be a Polish space,* X *a nonempty set, and* $\mathcal{L}$ *a* σ*-algebra of subset of* X*. Every* $\mathcal{L}$*-measurable, closed value multifunction* $F : X \to Y$ *admits an* $\mathcal{L}$*-measurable selection.*

2.4. Metric setting. In this section we refer to [4].

DEFINITION 2.2. *A length structure on a topological space* X *is a class* $\mathtt{A}$ *of admissible paths, which is a subset of all continuous paths in*

X, together with a map $L : \mathtt{A} \to [0, +\infty]$: the map L is called length of path.

The class $\mathtt{A}$ and the map L must satisfy the natural assumptions that one expects (for shortness we do not list them here).

Given a length structure, we can define a distance

$$d_L(x, y) = \inf \Big\{ L(\gamma) : \gamma : [a, b] \to X, \gamma \in A, \gamma(a) = x, \gamma(b) = y \Big\},$$

that makes (X, d_L) a metric space (allowing d_L to be $+\infty$). The metric d_L is called *intrinsic*.

DEFINITION 2.3. *A length structure is said to be* complete *if for every two points x, y there exists an admissible path joining them whose length $L(\gamma)$ is equal to $d_L(x, y) < +\infty$.*

In other words, a length structure is complete if there exists a shortest path between two points with finite length.

Intrinsic metrics associated with complete length structure are said to be *strictly intrinsic*. The metric space (X, d) with d strictly intrinsic is called a *geodesic space*. A curve whose length equals the distance between its end points is called *geodesic*.

It follows from Proposition 2.5.9 of [4] that every admissible curve of finite length admits a constant speed parametrization, i.e. γ defined on $[0, 1]$ and $L(\gamma\llcorner[t, t']) = v(t' - t)$, with v velocity. Hence from now on geodesics when parametrized are understood as constant speed geodesics.

DEFINITION 2.4. *Let (X, d_L) be a metric space. The distance d_L is said to be* strictly convex *if, for all $r \geq 0$, $d_L(x, y) = r/2$ implies that*

$$\{z : d_L(x, z) = r\} \cap \{z : d_L(y, z) = r/2\}$$

is a singleton.

The definition can be restated as: geodesics cannot branch in the interior. More precisely: let $\gamma_1, \gamma_2 : [0, 1] \to X$ be geodesics such that $\gamma_1(I_1) = \gamma_2(I_2)$ for some intervals $I_1, I_2 \subset [0, 1]$. Then either $\gamma_1 \subset \gamma_2$ or $\gamma_2 \subset \gamma_1$. An equivalent requirement is that if $\gamma_1(0) = \gamma_2(0)$, $\gamma_1(1) = \gamma_2(1)$, then such geodesics do not admit a geodesic extension i.e. they are not a part of a longer one. The metric space (X, d) is said non-branching.

From now on

(X, d) is a non-branching geodesic locally compact Polish space.

2.5. General facts about optimal transportation. Let (X, Ω, μ) and (Y, Σ, ν) be two probability spaces and $c : X \times Y \to \mathbb{R}^+$ be a $\Omega \times \Sigma$ measurable function. Consider the set of transference plans

$$\Pi(\mu, \nu) := \Big\{ \pi \in \mathcal{P}(X \times Y) : (P_1)_\sharp \pi = \mu, (P_2)_\sharp \pi = \nu \Big\},$$

where $P_i(x_1,x_2) = x_i, i = 1,2$ and $\mathcal{P}(X \times Y)$ is the set of probability measure on $X \times Y$. Define the functional

$$\begin{aligned} \mathcal{I} \ : \ \Pi(\mu,\nu) &\longrightarrow \mathbb{R}^+ \\ \pi &\longmapsto \mathcal{I}(\pi) := \int_{X\times Y} c\pi. \end{aligned} \tag{2.5}$$

The *Monge-Kantorovich minimization problem* is to find the minimum of $\mathcal{I}$ over all transference plans.

If we consider a map $T : X \to Y$ such that $T_\sharp\mu = \nu$, the functional 2.5 becomes

$$\mathcal{I}(T) := \mathcal{I}((Id \times T)_\sharp\mu) = \int_X c(x,T(x))\mu(dx).$$

The minimum problem over all T is called *Monge minimization problem.*

The Kantorovich problem admits a (pre) dual formulation: before stating it, we give a definition.

DEFINITION 2.5. *A set* $\Gamma \subset X \times Y$ *is said to be* c-cyclically monotone *if, for any* $n \in \mathbb{N}$ *and for any family* $(x_1,y_1),\dots,(x_n,y_n)$ *of points of* Γ, *the following inequality holds*

$$\sum_{i=0}^n c(x_i,y_i) \le \sum_{i=0}^n c(x_{i+1},y_i),$$

with $x_{n+1} = x_1$. *A transference plan is said to be* c-cyclically monotone *if it is concentrated on a c-cyclically monotone set.*

Consider the set

$$\Phi_c := \Big\{(\varphi,\psi) \in L^1(\mu) \times L^1(\nu) : \varphi(x) + \psi(y) \le c(x,y)\Big\}. \tag{2.6}$$

Define for all $(\varphi,\psi) \in \Phi_c$ the functional

$$J(\varphi,\psi) := \int \varphi\mu + \int \psi\nu. \tag{2.7}$$

The following is a well known result (see Theorem 5.10 of [16]).

THEOREM 2.3 (Kantorovich Duality). *Let* X *and* Y *be Polish spaces, let* $\mu \in \mathcal{P}(X)$ *and* $\nu \in \mathcal{P}(Y)$, *and let* $c : X \times Y \to \mathbb{R}^+ \cup \{+\infty\}$ *be lower semicontinuous. Then the following holds:*

1. Kantorovich duality:

$$\inf_{\pi\in\Pi[\mu,\nu]} \mathcal{I}(\pi) = \sup_{(\varphi,\psi)\in\Phi_c} J(\varphi,\psi).$$

Moreover, the infimum on the left-hand side is attained and the right-hand side is also equal to

$$\sup_{(\varphi,\psi)\in\Phi_c\cap C_b} J(\varphi,\psi),$$

where $C_b = C_b(X,\mathbb{R}) \times C_b(Y,\mathbb{R})$.

2. *If* c *is real valued and the optimal cost*

$$C(\mu,\nu) := \inf_{\pi\in\Pi(\mu,\nu)} I(\pi)$$

is finite, then there is a measurable c*-cyclically monotone set* $\Gamma \subset X \times Y$*, closed if* c *is continuous, such that for any* $\pi \in \Pi(\mu,\nu)$ *the following statements are equivalent:*
   - *(a)* π *is optimal;*
   - *(b)* π *is* c*-cyclically monotone;*
   - *(c)* π *is concentrated on* Γ*;*
   - *(d) there exists a* c*-concave function* φ *such that* π*-a.s.* $\varphi(x) + \varphi^c(y) = c(x,y)$.

3. *If moreover*

$$c(x,y) \leq c_X(x) + c_Y(y), \qquad (c_X, c_Y) \in L^1(\mu) \times L^1(\nu),$$

then there exist a couple of potentials and the optimal transference plan π *is concentrated on the set*

$$\big\{(x,y) \in X \times Y \,|\, \varphi(x) + \psi(y) = c(x,y)\big\}.$$

Finally if $(c_X, c_Y) \in \mathcal{L}^1(\mu) \times \mathcal{L}^1(\nu)$ *then the supremum is attained*

$$\sup_{\Phi_c} J = J(\varphi, \varphi^c).$$

We recall also that if c is Borel, then every optimal transference plan π is concentrated on a c-cyclically monotone set [2].

3. Optimal transportation in geodesic spaces. Take $\mu,\nu \in \mathcal{P}(X)$ and consider the optimal transportation problem with cost $c(x,y) = d(x,y)$. In our setting the following holds.

1. It is possible to restrict the Kantorovich duality just to 1-Lipschitz functions.
2. For a 1-Lipschitz map the d-transform has a particular form $\varphi^d(x) = -\varphi(x)$.
3. It follows that the support of the minimizing measures is the *transport set*

$$\Gamma := \Big\{(x,y) \in X \times X : \varphi(x) - \varphi(y) = d(x,y)\Big\}, \tag{3.1}$$

for any potential $\varphi \in \mathrm{Lip}_1(X)$.

4. The distance cost allows to assume $\mu \perp \nu$ because of the triangular inequality.

In this section we study the transport set Γ. Note that Γ is closed, hence σ-compact.

DEFINITION 3.1 (Transport rays). *Define for $x \in X$ the outgoing transport ray*

$$G(x) := \{y \in X : \varphi(x) - \varphi(y) = d(x, y)\} \tag{3.2}$$

and the incoming transport ray

$$G^{-1}(x) := \{y \in X : \varphi(y) - \varphi(x) = d(x, y)\}. \tag{3.3}$$

Define the set of transport rays through x as the set

$$R(x) := G(x) \cup G^{-1}(x). \tag{3.4}$$

Observe that the multivalued maps G, G^{-1}, and R have σ-compact graph.

DEFINITION 3.2. *Define the transport sets*

$$\mathcal{T} := P_1\big(\mathrm{graph}(G^{-1}) \setminus \{x = y\}\big) \cap P_1\big(\mathrm{graph}(G) \setminus \{x = y\}\big), \tag{3.5}$$

$$\mathcal{T}_e := P_1\big(\mathrm{graph}(G^{-1}) \setminus \{x = y\}\big) \cup P_1\big(\mathrm{graph}(G) \setminus \{x = y\}\big). \tag{3.6}$$

From the definition of G it is fairly easy to prove that $\mathcal{T}$, $\mathcal{T}_e$ are σ-compact sets. The subscript e refers to the endpoints of the geodesics: clearly we have

$$\mathcal{T}_e = P_1(R \setminus \{x = y\}). \tag{3.7}$$

Since $\pi(\Gamma) = 1$, it is fairly easy to prove that $\pi(\mathcal{T}_e \times \mathcal{T}_e \cup \{x = y\}) = 1$. As a consequence, $\mu(\mathcal{T}_e) = \nu(\mathcal{T}_e)$ and any maps T such that for $\nu \llcorner_{\mathcal{T}_e} = T_\sharp \mu \llcorner_{\mathcal{T}_e}$ can be extended to a map T' such that $\nu = T_\sharp \mu$ with the same cost by setting

$$T'(x) = \begin{cases} T(x) & x \in \mathcal{T}_e \\ x & x \notin \mathcal{T}_e \end{cases}. \tag{3.8}$$

Therefore we have only to study the Monge problem in $\mathcal{T}_e$.

REMARK 3.1. Take $y \in G(x)$, and take the points z such that

$$d(x, y) = d(x, z) + d(z, y).$$

From the definition of G, $z \in G(x)$ and $y \in G(z)$ or, equivalently, $z \in G^{-1}(y)$. Similarly if we take $y \in G(x)$ and $z \in G(y)$ we get $z \in G(x)$ and y is on the geodesic from x to z. So we can say that if $y \in G(x)$, the set

$$G(x) \cap G^{-1}(y)$$

is the union of the minimizing geodesic connecting x to y. Therefore, by the non-branching assumption, if $x \in \mathcal{T}$, then $R(x)$ is a single geodesic.

DEFINITION 3.3. *Define the multivalued endpoint maps* $a, b : \mathcal{T}_e \to \mathcal{T}_e$ *by:*

$$a(y) := \big\{z \in G^{-1}(y) : \nexists x \in X \setminus \{z\}, x \in G^{-1}(z)\big\}, \tag{3.9}$$

$$b(y) := \big\{z \in G(y) : \nexists x \in X \setminus \{z\}, x \in G(z)\big\}. \tag{3.10}$$

We call $a(\mathcal{T}_e)$ *the set of initial points and* $b(\mathcal{T}_e)$ *the set of final points.*

Observe that the multivalued maps $a, b : \mathcal{T}_e \longrightarrow \mathcal{T}_e$ have σ-compact graph. Other properties of the end point maps:

1. $a(x)$, $b(x)$ are singleton or empty when $x \in \mathcal{T}$;
2. $a(\mathcal{T}) = a(\mathcal{T}_e)$, $b(\mathcal{T}) = b(\mathcal{T}_e)$;
3. $\mathcal{T}_e = \mathcal{T} \cup a(\mathcal{T}) \cup b(\mathcal{T})$, $\mathcal{T} \cap (a(\mathcal{T}) \cup b(\mathcal{T})) = \emptyset$.

W.l.o.g. we can assume that the μ-measure of final points and the ν-measure of the initial points are 0.

4. Partition of the transport set. Let $\{x_i\}_{i\in\mathbb{N}}$ be a dense sequence in (X, d).

LEMMA 4.1. *The sets*

$$\mathcal{Z}_{ijk} := \Big\{x \in \mathcal{T} \cap \bar{B}_{2^{-j}}(x_i) : L(G(x)), L(G^{-1}(x)) \geq 2^{2-k}\Big\}$$

form a countable covering of $\mathcal{T}$ *of* σ*-compact sets.*

Proof. We first prove the measurability. We consider separately the conditions defining Z_{ijk}.

Point 1. The set

$$A_{ij} := \mathcal{T} \cap \bar{B}_{2^{-j}}(x_i)$$

is clearly σ-compact.

Point 2. The set

$$B_k := \big\{x \in \mathcal{T} : L(G(x)) \geq 2^{2-k}\big\} = P_1\Big(G \cap \big\{d(x, y) \geq 2^{2-k}\big\}\Big)$$

is σ-compact, being the projection of a σ-compact set. Similarly, the set

$$C_k := \{x \in \mathcal{T} : L(G^{-1}(x)) \geq 2^{2-k}\} = P_1\Big(G^{-1} \cap \{d(x,y) \geq 2^{2-k}\}\Big)$$

is again σ-compact.

We finally can write

$$\mathcal{Z}_{ijk} = A_{ij} \cap B_k \cap C_k.$$

To show that it is a covering, notice that for all $x \in \mathcal{T}$ it holds

$$\min\{L(G(x)), L(G^{-1}(x))\} \geq 2^{2-k}$$

for some $k \in \mathbb{N}$. □

From $\mathcal{Z}_{ijk}$ we can define a countable covering of $\mathcal{T}_e$ of σ-compact saturated sets just taking

$$\mathcal{T}_{ijk} := R^{-1}(\mathcal{Z}_{ijk}).$$

In the natural way, we can find a countable partition into σ-compact saturated sets by defining

$$\mathcal{Z}_{m,e} := \mathcal{T}_{i_m j_m k_m} \setminus \bigcup_{m'=1}^{m-1} \mathcal{T}_{i_{m'} j_{m'} k_{m'}}, \quad \mathcal{Z}_{0,e} := \mathcal{T}_e \setminus \bigcup_{m \in \mathbb{N}} \mathcal{Z}_{m,e}, \tag{4.1}$$

where

$$\mathbb{N} \ni m \mapsto (i_m, j_m, k_m) \in \mathbb{N}^3$$

is a bijective map. Intersecting the above sets with $\mathcal{T}$, we obtain the countable partition of $\mathcal{T}$ in σ-compact sets

$$\mathcal{Z}_m := \mathcal{Z}_{m,e} \cap \mathcal{T}, \quad m \in \mathbb{N}_0. \tag{4.2}$$

Since

$$R = \Big\{(x,y) \in X \times X : |\varphi(x) - \varphi(y)| = d(x,y)\Big\}$$

is the graph of an equivalence relation on $\mathcal{T}$, we use this partition to prove the strong consistency of the disintegration induced by R.

On $\mathcal{Z}_m$, $m > 0$, we define the closed values map

$$\mathcal{Z}_m \ni x \mapsto F(x) := R(x) \cap \bar{B}_{2^{-j_m}}(x_{i_m}) \subset \mathcal{Z}_m. \tag{4.3}$$

LEMMA 4.2. *The equivalence relation R admits a Borel section: there exists a Borel map $f : \mathcal{T} \to \mathcal{T}$ such that*

1. $xRf(x)$,
2. xRy *implies* $f(x) = f(y)$.

Proof. It is enough to consider just one $\mathcal{Z}_m$.

Step 1. First we show that F has σ-compact graph:

$$\mathrm{graph}(F) = \mathcal{Z}_m \times X \cap \big(R \cap X \times B_{2^{-j_m}}(x_{i_m})\big)$$

and $F(x)$ is clearly compact.

Step 2. Let $\mathcal{L}$ the family of saturated closed sets w.r.t. F, i.e. inverse images of closed sets, and let $\mathcal{L}_\sigma$ be the smallest σ-algebra containing $\mathcal{L}$. Note that $\mathcal{L}_\sigma$ is a subset of the Borel σ-algebra. Clearly, by construction, F is $\mathcal{L}_\sigma$-measurable.

Step 3. From Theorem 2.2 there exists a $\mathcal{L}_\sigma$-measurable selection f of F. Clearly the atoms of $\mathcal{L}_\sigma$ are $R(x) \setminus \{a(x), b(x)\}$ for $x \in \mathcal{Z}_m$ and f is constant along $R(x) \setminus \{a(x), b(x)\}$ being $\mathcal{L}_\sigma$-measurable. Moreover $f(x) \in R(x) \setminus \{a(x), b(x)\}$. Hence f is a Borel section. □

The set $S = f(\mathcal{T})$ is a Borel cross-section of R restricted to $\mathcal{T}$: indeed

$$S = \{x : d(f(x), x) = 0\}$$

and $R(x) \cap S = \{f(x)\}$. Having a measurable cross-section we can define the parametrization of $\mathcal{T}$, $\mathcal{T}_e$ by geodesics.

$$S \times \mathbb{R} \ni (y,t) \mapsto g(y,t) := \big\{x : \varphi(y) - \varphi(x) = t\big\}. \tag{4.4}$$

We summarize the properties of the set graph$g$

1. The set graph$g$ is Borel.
2. It is the graph of a map with range $\mathcal{T}_e$.
3. $t \mapsto g(y,t)$ is a d_L 1-Lipschitz G-order preserving for $y \in \mathcal{T}$.

5. Regularity of the disintegration. Let μ be a probability measure on (X,d). This section is divided in two parts.

In the first one we consider the translation of Borel sets by the optimal geodesic flow, we introduce the fundamental regularity assumption (Assumption 1) on the measure η and we show that an immediate consequence is that the set of initial points is negligible. A second consequence is that the disintegration of η w.r.t. the R has continuous conditional probabilities.

5.1. Evolution of Borel sets. Let $A \subset \mathcal{T}_e$ be an analytic set and define for $t \in \mathbb{R}$ the *t-evolution A_t of A* by

$$A_t := g\big(g^{-1}(A) + (0,t)\big). \tag{5.1}$$

LEMMA 5.1. *The set A_t is analytic.*

We can show that $t \mapsto \mu(A_t)$ is measurable.

LEMMA 5.2. *Let A be analytic. The function $t \mapsto \mu(A_t)$ is universally measurable.*

Proof. Since A is analytic, then $g^{-1}(A)$ is analytic, and the set

$$\tilde{A} := \{(y,t,\tau) : (y,t-\tau) \in g^{-1}(A)\}$$

is easily seen to be again analytic. From Fubini theorem applied to the measure $\mu \times \eta$, $\eta \in \mathcal{P}(\mathbb{R})$, it follows that $t \mapsto \mu((A)_t)$ is η-integrable. Since this holds for all η, by definition $t \mapsto \mu((A)_t)$ is universally measurable for $t \in \mathbb{R}$. □

The next assumption is the fundamental assumption of the paper.

ASSUMPTION 1 (Non-degeneracy assumption). *For all compact sets A such that $\mu(A) > 0$ the set $\{t \in \mathbb{R}^+ : \mu(A_t) > 0\}$ has cardinality $> \aleph_0$.*

An immediate consequence of the Assumption 1 is that the measure μ is concentrated on $\mathcal{T}$.

LEMMA 5.3. *If μ satisfies Assumption 1 then*

$$\mu(\mathcal{T}_e \setminus \mathcal{T}) = 0.$$

Proof. If $A \subset a(X)$ is compact, then $A_t \cap A_s = \emptyset$ for $0 \leq s < t$. Hence

$$\sharp\{t \in \mathbb{R}^+ : \mu(A_t) > 0\} \leq \aleph_0,$$

because of the boundedness of μ. This contradicts the assumption unless $\mu(A) = 0$. □

Once we know that $\mu(\mathcal{T}) = 1$, we can use the Disintegration Theorem 2.1 to write

$$\mu = \int_S \mu_y m(dy), \quad m = f_\sharp \mu, \ \mu_y \in \mathcal{P}(R(y)). \tag{5.2}$$

The disintegration is strongly consistent since the quotient map $f : \mathcal{T} \to \mathcal{T}$ is μ-measurable and $(\mathcal{T}, \mathcal{B}(\mathcal{T}))$ is countably generated.

The second consequence of Assumption 1 is that μ_y is continuous, i.e. $\mu_y(\{x\}) = 0$ for all $x \in X$.

PROPOSITION 5.1. *The conditional probabilities μ_y are continuous for m-a.e. $y \in S$.*

Proof. From the regularity of the disintegration and the fact that $m(S) = 1$, we can assume that the map $y \mapsto \mu_y$ is weakly continuous on a compact set $K \subset S$ of comeasure $< \epsilon$ such that $L(R(y)) > \epsilon$ for all $y \in K$. It is enough to prove the proposition on K.

Step 1. From the continuity of $K \ni y \mapsto \mu_y \in \mathcal{P}(X)$ w.r.t. the weak topology, it follows that the map

$$y \mapsto A(y) := \big\{x \in R(y) : \mu_y(\{x\}) > 0\big\} = \cup_n \big\{x \in R(y) : \mu_y(\{x\}) \geq 2^{-n}\big\}$$

is σ-closed: in fact, if $(y_m, x_m) \to (y, x)$ and $\mu_{y_m}(\{x_m\}) \geq 2^{-n}$, then $\mu_y(\{x\}) \geq 2^{-n}$ by u.s.c. on compact sets.

Hence it is Borel, and by Lusin Theorem (Theorem 5.8.11 of [12]) it is the countable union of Borel graphs: setting in case $c_i(y) = 0$, we can consider them as Borel functions on S and order w.r.t. G,

$$\mu_{y,\text{atomic}} = \sum_{i \in \mathbb{Z}} c_i(y) \delta_{x_i(y)}, \quad x_{i+1}(y) \in G(x_i(y)), \ i \in \mathbb{Z}.$$

Step 2. Define the sets

$$S_{ij}(t) := \Big\{y \in K : x_i(y) = g\big(g^{-1}(x_j(y)) + t\big)\Big\} \cap \mathcal{T}.$$

Since $K \subset S$, to define S_{ij} we are using the graph $g \cap S \times \mathbb{R} \times \mathcal{T}$, which is analytic: hence $S_{ij} \in \Sigma^1_1$.

For $A_j := \{x_j(y), y \in K\}$ and $t \in \mathbb{R}^+$ we have that

$$\begin{aligned} \mu((A_j)_t) &= \int_K \mu_y((A_j)_t) m(dy) = \int_K \mu_{y,\text{atomic}}((A_j)_t) m(dy) \\ &= \sum_{i \in \mathbb{Z}} \int_K c_i(y) \delta_{x_i(y)} \big(g(g^{-1}(x_j(y)) + t)\big) m(dy) = \sum_{i \in \mathbb{Z}} \int_{S_{ij}(t)} c_i(y) m(dy). \end{aligned}$$

We have used the fact that $A_j \cap R(y)$ is a singleton.

Step 3. For fixed $i, j \in \mathbb{N}$, again from the fact that $A_j \cap R(y)$ is a singleton

$$S_{ij}(t) \cap S_{ij}(t') = \begin{cases} S_{ij}(t) & t = t' \\ \emptyset & t \neq t' \end{cases}$$

so that

$$\sharp\big\{t : m(S_{ij}(t)) > 0\big\} \leq \aleph_0.$$

Finally

$$\mu((A_j)_t) > 0 \quad \Longrightarrow \quad t \in \bigcup_i \big\{t : m(S_{ij}(t)) > 0\big\},$$

whose cardinality is $\leq \aleph_0$, contradicting Assumption 1. □

6. Solution to the Monge problem. In this section we show that Proposition 5.1 allows to construct an optimal map T. We recall the one dimensional result for the Monge problem that can be found on [15].

THEOREM 6.1. *Let μ, ν be probability measures on $\mathbb{R}$, μ with no atoms, and let*

$$G(x) = \mu((-\infty; x)), \quad F(x) = \nu((-\infty; x)),$$

be the distribution functions of μ and ν respectively. Then

1. *The non decreasing function $T : \mathbb{R} \to \mathbb{R} \cup \{\pm\infty\}$ defined by*

$$T(x) = \sup \{y \in \mathbb{R} : F(y) \leq G(x)\},$$

with the convention $\sup \emptyset = -\infty$, maps μ to ν. Moreover any other non decreasing map T' such that $T'_\sharp\mu = \nu$ coincides with T on the support of μ up to a countable set.

2. *If $\phi : [0, +\infty] \to \mathbb{R}$ is non decreasing and convex, then T is an optimal transport relative to the cost $c(x, y) = \phi(|x - y|)$. Moreover T is the unique optimal transference map if ϕ is strictly convex.*

Assume that μ satisfies Assumption 1. Then we can disintegrate μ and the optimal transference plan π respect to the ray equivalence relation R and $R \times X$ as in equation 5.2,

$$\mu = \int \mu_y m(dy), \ \pi = \int \pi_y m(dy), \quad \mu_y \text{ continuous}, \ (P_1)_\sharp \pi_y = \mu_y. \tag{6.1}$$

We write moreover

$$\nu = \int \nu_y m(dy) = \int (P_2)_\sharp \pi_y m(dy). \tag{6.2}$$

Note that $\pi_y \in \Pi(\mu_y, \nu_y)$ is d-monotone (and hence optimal, because $R(y)$ is one dimensional) for m-a.e. y. If $\nu(\mathcal{T}) = 1$, then (6.2) is the disintegration of ν w.r.t. R.

THEOREM 6.2. *Let $\pi \in \Pi(\mu, \nu)$ be an optimal transference plan, and assume that Assumption 1 holds. Then there exists a Borel map $T : X \to X$ with the same transport cost as π.*

Proof. By means of the map g^{-1}, we reduce to a transport problem on $S \times \mathbb{R}$, with cost

$$c((y, s), (y', t)) = \begin{cases} |t - s| & y = y' \\ +\infty & y \neq y'. \end{cases}$$

It is enough to prove the theorem in this setting under the following assumptions: S compact and $S \ni y \mapsto (\mu_y, \nu_y)$ weakly continuous. We consider here the probabilities μ_y, ν_y on $\mathbb{R}$.

Step 1. From the weak continuity of the map $y \mapsto (\mu_y, \nu_y)$, it follows that for all t the map

$$(y,t) \mapsto H(y,t) := \mu_y((-\infty,t)),$$

is continuous in t and l.s.c. in y, hence l.s.c.. Similarly, the map

$$(y,t) \mapsto F(y,t) := \nu_y((-\infty,t))$$

is easily seen to be l.s.c.. Both are clearly increasing in t.

Step 2. The map T defined as Theorem 6.1 by

$$T(y,s) := \Big(y, \sup\big\{t : F(y,t) \le H(y,s)\big\}\Big)$$

is Borel. In fact, for A Borel,

$$T^{-1}(A \times [t,+\infty)) = \Big\{(y,s) : y \in A, H(y,s) \ge F(y,t)\Big\} \in \mathcal{B}(S \times \mathbb{R}).$$

Step 3. By the definition of the set G, it follows that along each geodesic $\mu_y(g(y,(-\infty,t))) \ge \nu_y(g(y,(-\infty,t)))$, because in the opposite case G is not d-monotone. Hence we can conclude that $T(s) \ge s$, and $c((y,s),T(y,s)) = P_2(T(y,s)) - s$. □

6.1. Final Remarks. So from Theorem 6.2 we know that to solve the Monge problem we need Assumption 1 to hold. As it is clear from its formulation, Assumption 1 is not a direct hypothesis on the geometry of the metric space (X,d). Indeed the evolution defined in the paper is induced by the d-monotone set Γ that is deeply related to the problem.

However it is well known that in a finite dimensional manifold a lower bound on Ricci curvature implies an estimates of how the mass is moved from a point by the exponential map. Similar notion of curvature are present also in metric spaces. In [3] we prove that if the metric measure space (X,d,μ) satisfies the *measure contraction property* (MCP) then Assumption 1 is satisfied and therefore the Monge problem is solved.

A possible direction of a future research is to remove the hypothesis of strictly convexity on the distance cost d. The main problem will be that, since bifurcation of geodesics must be taken into account, the reduction to the one-dimensional case, that is the main ingredient we use to solve the Monge problem, will not hold anymore.

REFERENCES

[1] L. Ambrosio, B. Kirchheim, and A. Pratelli, *Existence of optimal transport maps for crystalline norms*, Duke Math. J., **125**(2) (2004), pp. 207–241.

[2] S. Bianchini and L. Caravenna, *On the extremality, uniqueness and optimality of transference plans.* preprint.

[3] S. Bianchini and F. Cavalletti, *The Monge problem for distance cost in geodesic spaces*, preprint (2009).

[4] D. Burago, Y. Burago, and S. Ivanov, *A course in metric geometry*, Graduate studies in mathematics, AMS, 2001.

[5] L. Caffarelli, M. Feldman, and R. McCann, *Constructing optimal maps for Monge's transport problem as a limit of strictly convex costs*, J. Amer. Math. Soc., **15** (2002), pp. 1–26.

[6] L. Caravenna, *An existence result for the Monge problem in $\mathbb{R}^n$ with norm cost function*, preprint (2009).

[7] T. Champion and L.D. Pascale, *The Monge problem in $\mathbb{R}^d$*, preprint (2009).

[8] L. Evans and W. Gangbo, *Differential equations methods for the Monge-Kantorovich mass transfer problem*, Current Developments in Mathematics (1997), pp. 65–126.

[9] M. Feldman and R. McCann, *Monge's transport problem on a Riemannian manifold.*, Trans. Amer. Math. Soc., **354** (2002), pp. 1667–1697.

[10] D.H. Fremlin, *Measure Theory*, Vol. **4**, Torres Fremlin, 2002.

[11] D. Larman, *A compact set of disjoint line segments in $\mathbb{R}^3$ whose end set has positive measure*, Mathematika, **18** (1971), pp. 112–125.

[12] A.M. Srivastava, *A course on Borel sets*, Springer, 1998.

[13] V. Sudakov, *Geometric problems in the theory of dimensional distributions*, Proc. Steklov Inst. Math., **141** (1979), pp. 1–178.

[14] N. Trudinger and X. Wang, *On the Monge mass transfer problem*, Calc. Var. PDE, **13** (2001), pp. 19–31.

[15] C. Villani, *Topics in optimal transportation*, Graduate studies in mathematics, AMS, 2003.

[16] ———, *Optimal transport, old and new*, Springer, 2008.

EXISTENCE OF A UNIQUE SOLUTION TO A NONLINEAR MOVING-BOUNDARY PROBLEM OF MIXED TYPE ARISING IN MODELING BLOOD FLOW

SUNČICA ČANIĆ*, ANDRO MIKELIĆ†, TAE-BEOM KIM‡, AND GIOVANNA GUIDOBONI§

Abstract. We prove the (local) existence of a unique mild solution to a nonlinear moving-boundary problem of a mixed hyperbolic-degenerate parabolic type arising in modeling blood flow through compliant (viscoelastic) arteries.

Key words. Moving boundary problem, PDE of mixed type, Hyperbolic-degenerate parabolic problem, Blood flow, Compliant arteries.

AMS(MOS) subject classifications. 35M10, 35G25, 35K65, 35Q99, 76D03, 76D08, 76D99.

1. Introduction. This work was motivated by a study of blood flow in compliant arteries. In medium-to-large arteries blood can be modeled by the Navier-Stokes equations for an incompressible, viscous Newtonian fluid, while the arterial walls behave as a viscoelastic material [2, 3, 4]. To study the coupled fluid-structure interaction (FSI) problem we derived in [6] a leading-order, closed, effective model for the benchmark problem in blood flow: the pressure-driven FSI problem defined on a time-dependent cylindrical domain $\Omega(t)$ with small aspect ratio $\varepsilon = R/L$ (see Fig. 1) and axially symmetric flow. Kelvin-Voigt linearly viscoelastic cylindrical membrane and Kelvin-Voigt linearly viscoelastic cylindrical Koiter shell were used in [6] to model the viscoelastic behavior of arterial walls.

The leading-order problem derived in [6] defines a nonlinear, moving-boundary problem for a system of partial differential equations of mixed hyperbolic-parabolic type in two space dimensions. The resulting problem is a hydrostatic approximation of the full FSI problem between the Navier-Stokes equations for an incompressible, viscous Newtonian fluid and the Kelvin-Voigt linearly viscoelastic cylindrical membrane or Koiter shell

* Department of Mathematics, University of Houston, Houston, Texas 77204-3476. Research supported by NSF under grant DMS-0806941, by NSF/NIH grant number DMS-0443826, and by the Texas Higher Education Board under ARP grant 003652-0051-2006.

† Université de Lyon, Lyon, F-69003, Université Lyon 1, Institut Camille Jordan, UMR 5208, Département de Mathématiques, 43, Bd du 11 nov. 1918, 69622 Villeurbanne Cedex, France (Andro.Mikelic@univ-lyon1.fr). Research supported by NSF/NIH under grant DMS-0443826.

‡ Department of Mathematics, University of Houston, Houston, Texas 77204-3476. Graduate Student Support by the NSF/NIGMS under grant 0443826, and by the Texas Higher Education Board under ARP grant 003652-0051-2006.

§ Department of Mathematics, University of Houston, Houston, Texas 77204-3476. Research supported by the NSF under grant DMS-0811138, and by the Texas Higher Education Board under ARP grant 003652-0051-2006.

model. The problem is given in terms of two unknown functions: the axial component of the fluid velocity, $v_z = v_z(r,z,t)$, and the radial displacement of the arterial wall, $\eta = \eta(z,t)$:

$$\frac{\partial (R+\eta)^2}{\partial t} + \frac{\partial}{\partial z}\int_0^{R+\eta} 2rv_z dr = 0, \quad 0<z<L,\ 0<t<T, \quad (1.1)$$

$$\varrho_F \frac{\partial v_z}{\partial t} - \mu_F \frac{1}{r}\frac{\partial}{\partial r}\left(r\frac{\partial v_z}{\partial r}\right) = -\frac{\partial p}{\partial z}, \quad (r,z)\in\Omega(t),\ 0<t<T, \quad (1.2)$$

with

$$p - p_{\text{ref}} = \left(\frac{hE}{R(1-\sigma^2)}K + p_{\text{ref}}\right)\frac{\eta}{R} + \frac{hC_v}{R^2}K\frac{\partial \eta}{\partial t}, \quad 0<z<L,\ 0<t<T, \quad (1.3)$$

where $K = 1$ for the membrane and $K = 1 + \frac{h^2}{12R^2}$ for the Koiter shell.

Here ϱ_F is the fluid density, μ_F is the fluid dynamic viscosity coefficient, p is the fluid pressure with p_{ref} denoting the pressure at which the domain reference configuration is assumed (straight cylinder of radius R). The constants describing the structure properties are the Young's modulus of elasticity E, the Poisson ratio σ, the wall thickness h, and the structure viscoelasticity constant C_v. Typical values for these constants can be found, e.g., in [6, 8]. Problem (1.1)–(1.3) is defined on $\Omega(t)$, $0 < t < T$, where

$$\Omega(t) = \{(r,z):\ 0\le r < R+\eta(z,t),\ 0<z<L, \text{ so that } (r\cos\vartheta, r\sin\vartheta, z)\in I\!R^3 \text{ for } \vartheta\in[0,2\pi) \text{ defines a cylinder in } I\!R^3\}. \quad (1.4)$$

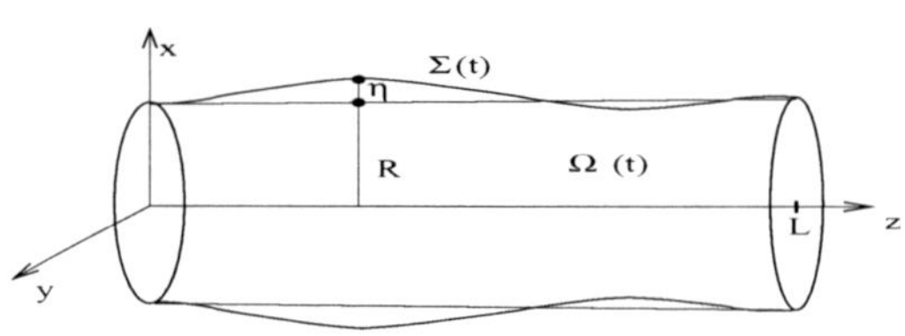

FIG. 1. *Deformed domain* $\Omega(t)$.

Problem (1.1)–(1.3) is supplemented by the following initial and boundary conditions

$$v_z(0,z,t) - \text{ bounded}, v_z(R+\eta(z,t),z,t) = 0, v_z(r,z,0) = v_z^0(r,z), \quad (1.5)$$

$$\eta(z,0) = \eta^0(z), p(0,t) = P_0(t), p(L,t) = P_L(t), \quad (1.6)$$

describing pressure-driven fluid flow in a compliant cylinder $\Omega(t)$, with no-slip boundary conditions at the lateral boundary of $\Omega(t)$.

Equation (1.1) is derived from the conservation of mass condition for incompressible fluids, while equation (1.2) is the leading-order hydrostatic approximation of the balance of axial momentum. It was shown in [7] that (1.1)–(1.3) plus its ε-correction (not shown here) satisfy the original fluid-structure interaction problem to the ε^2 accuracy.

Problem (1.1)–(1.6) in a nonlinear, initial-boundary-value problem of hyperbolic-parabolic type. The hyperbolic waves described by the quasi-linear transport equation (1.1) may develop shock waves in the axial direction z giving rise to discontinuities in the displacement of the arterial wall (lateral boundary), not typically observed in cardiovascular flow of healthy humans [5]. The second equation, (1.2), could potentially smooth out the steep wave fronts due to the fluid viscosity effects. Unfortunately, the fluid viscosity in the axial direction in not present in (1.2) since the corresponding terms are negligible (higher order) in comparison with the diffusion in the radial direction [6]. This gives rise to a problem with degenerate/anisotropic diffusion in the momentum equation (1.2) which presents various difficulties in the proof of the existence of a solution. However, due to the time-differentiated term in equation (1.3) coming from the viscoelasticity of arterial walls, the sharp wave fronts in the displacement of the arterial walls will be smoothed out, giving rise to a solution of (1.1)–(1.6) which is physiologically reasonable. More precisely, we will prove in this manuscript the existence of a unique **mild solution** to problem (1.1)–(1.6) with sufficient regularity in the axial direction allowing solutions with no shock formation. As we shall see in the proof, the dominant smoothing of shock fronts in the displacement of the arterial walls is provided by the viscoelastic term in the pressure-displacement relationship (1.3) describing the arterial wall properties.

This reduced problem has many interesting features. It captures the main properties of fluid-structure interaction in blood flow with physiologically reasonable equations and data [8], while allowing fast numerical computations and a relatively simple analysis related to its well-posedness.

Within the past ten years there has been considerable progress in the analysis of fluid-structure interaction problems between an incompressible, viscous fluid and an elastic or viscoelastic structure. All the results that are related to an elastic structure interacting with a viscous, incompressible fluid have been obtained under the assumption that the structure is entirely immersed in the fluid, see e.g., [9, 10, 12]. To our knowledge, there have been no results showing existence of a solution to a fluid-structure interaction problem where an elastic structure is a part of the fluid boundary, which is the case, for example, in modeling blood flow through elastic arteries. Often times additional ad hoc terms of viscoelastic nature are added to the vessel wall model to provide stability and convergence of the underlying numerical algorithm [16, 18], or to provide enough regularity in the proof of the existence of a solution as in [11, 10, 14, 21]. In [11, 9] terms describing bending (flexion) rigidity were added to provide smoothing mechanisms for the evolution of the structure displacement.

The novelty of the present paper is in considering a problem with the viscoelastic smoothing in the structure equation described by the lowest possible time derivative appearing in the physiologically relevant equations allowing the use of measurements data to describe the viscoelastic arterial wall properties.

2. The Nonlinear Problem on a Fixed Domain. We begin by mapping the moving-boundary problem (1.1)–(1.6) onto a fixed domain. At the same time we will be introducing the non-dimensional variables to derive the corresponding nonlinear problem defined on a fixed domain in non-dimensional form.

To simplify notation we introduce

$$\gamma(z,t) := R + \eta(z,t).$$

Introduce the mapping $r \mapsto \dfrac{r}{\gamma} =: \tilde{r}$ which maps $\Omega(t)$ onto the fixed domain $\Omega := (0,1) \times (0,L) \times (0,T)$. In addition, consider the following scalings of the independent and dependent variables

$$z = L\tilde{z}, t = \tau\tilde{t}, v_z = V\tilde{v}_z, \eta = N\tilde{\eta}, V = \frac{L}{\tau}, \gamma = R\tilde{\gamma} \text{ where } \tilde{\gamma} = 1 + \frac{N}{R}\tilde{\eta}.$$

Also, denote $\tilde{T} = T/\tau$. With these transformations, the problem is now defined on the scaled fixed domain

$$\begin{aligned}\tilde{\Omega} =& \big\{(\tilde{r},\tilde{z}) : \tilde{r} \in (0,1), \tilde{z} \in (0,1), \text{ so that } (\tilde{r}\cos\vartheta, \tilde{r}\sin\vartheta, \tilde{z}),\\ &\vartheta \in [0,2\pi), \text{defines a cylinder in } I\!R^3\big\}.\end{aligned} \tag{2.1}$$

The corresponding **nonlinear, fixed-boundary problem** in non-dimensional form then reads: for $0 < \tilde{t} < \tilde{T}$ find $\tilde{\gamma}(\tilde{z},\tilde{t})$ and $\tilde{v}_z(\tilde{r},\tilde{z},\tilde{t})$ so that

$$\tilde{\gamma}\frac{\partial\tilde{\gamma}}{\partial\tilde{t}} + \frac{\partial}{\partial\tilde{z}}\int_0^1 \tilde{\gamma}^2\tilde{v}_z\tilde{r}\, d\tilde{r} = 0, \quad 0 < \tilde{z} < 1, \tag{2.2}$$

$$\begin{aligned}\frac{\partial\tilde{v}_z}{\partial\tilde{t}} - C_1\frac{1}{\tilde{\gamma}^2}\frac{1}{\tilde{r}}\frac{\partial}{\partial\tilde{r}}\left(\tilde{r}\frac{\partial\tilde{v}_z}{\partial\tilde{r}}\right) - \frac{\tilde{r}}{\tilde{\gamma}}\frac{\partial\tilde{\gamma}}{\partial\tilde{t}}\frac{\partial\tilde{v}_z}{\partial\tilde{r}} &= -C_2\frac{\partial\tilde{\gamma}}{\partial\tilde{z}} - C_3\frac{\partial^2\tilde{\gamma}}{\partial\tilde{z}\partial\tilde{t}},\\ &(\tilde{r},\tilde{z}) \in \tilde{\Omega},\end{aligned} \tag{2.3}$$

with

$$\begin{cases}\tilde{\gamma}(0,\tilde{t}) = \tilde{\gamma}_0(\tilde{t}),\ \tilde{\gamma}(1,\tilde{t}) = \tilde{\gamma}_L(\tilde{t}),\ \tilde{\gamma}(\tilde{z},0) = \tilde{\gamma}^0(\tilde{z}),\\ \tilde{v}_z(1,\tilde{z},\tilde{t}) = 0, \tilde{v}_z(\tilde{r},\tilde{z},\tilde{t}=0) = \tilde{v}_z^0(\tilde{r},\tilde{z}),\ \left|\tilde{v}_z(0,\tilde{z},\tilde{t})\right| < +\infty,\end{cases} \tag{2.4}$$

where

$$C_1 = \frac{\mu_F \tau}{\rho_F R^2},\ C_2 = \left(\frac{Eh}{(1-\sigma^2)R}K + p_{ref}\right)\frac{1}{V^2\rho_F},\ C_3 = \frac{hC_v K}{RLV\rho_F}. \quad (2.5)$$

The inlet and outlet data $\tilde{\gamma}_0$ and $\tilde{\gamma}_L$ are obtained from the pressure data $P_0(t)$ and $P_L(t)$ given in (1.6), by integrating the pressure-displacement relationship (1.3) with respect to t, and then transforming the result into the non-dimensional form. Thus, $\tilde{\gamma}_0$ and $\tilde{\gamma}_L$ are scaled by R. They describe the inlet and outlet fluctuations of the domain radius around the "reference" domain radius $\tilde{r} = 1$.

PROPOSITION 2.1. *For the initial data* $\tilde{\gamma}^0 = R = 1$, $\tilde{v}_z^0 = 0$ *and boundary data* $\tilde{\gamma}_0 = \tilde{\gamma}_L = R = 1$ *problem* (2.2)–(2.5) *has a solution* $\tilde{\gamma} = R = 1$, $\tilde{v}_z = 0$.

We will show below, by using the Implicit Function Theorem, that problem (2.2)–(2.5) has a unique "mild" solution whenever the initial and boundary data are "close" to those listed in Proposition 2.1, namely, whenever the initial and boundary displacement from the reference radius $r = R = 1$ is small and whenever the initial velocity v_z^0 is close to zero.

In the rest of the manuscript we will be working with the non-dimensional form of the problem. To simplify notation, superscript "wiggle" that denotes the non-dimensional variables, will now be dropped, and this nomenclature will continue throughout the rest of the manuscript. Also, whenever R is used in the remainder of the paper, it refers to $R = 1$. Domain Ω below corresponds to the fixed, scaled domain, defined in (2.2).

3. Mild solution of the nonlinear problem. We will consider solutions of problem (2.2)–(2.5) with the initial and boundary data corresponding to the following function spaces:

$$\gamma^0 \in H^1(0,1),\ v_z^0 \in H^1_{0,0}(\Omega, r),\ \gamma_0, \gamma_L \in H^2(0,T). \quad (3.1)$$

Here $H^1_{0,0}(\Omega, r) = \{w \in L^2(\Omega, r) : \frac{\partial w}{\partial r} \in L^2(\Omega, r),\ < w > \in H^1(0,1),$ $w|_{r=1} = 0,\ |\ w|_{r=0}\ | < +\infty\}$, where $< w > := \int_0^1 wrdr$. The norm on $H^1_{0,0}(\Omega, r)$ is given by:

$$\|w\|^2_{H^1_{0,0}(\Omega,r)} = \int_\Omega \left(|w|^2 + \left|\frac{\partial w}{\partial r}\right|^2\right) rdrdz + \int_0^1 \left|\frac{\partial}{\partial z}\int_0^1 w\ rdr\right|^2 dz.$$

(The norm on $L^2(\Omega, r)$ is given by $\|f\|^2_{L^2(\Omega,r)} = \int_\Omega f\ rdrdz$.)

Thus, we define the space of data Λ to be

$$\Lambda = H^1(0,1) \times H^1_{0,0}(\Omega, r) \times (H^2(0,T))^2. \quad (3.2)$$

In order to define mild solution of problem (2.2)–(2.5) we introduce the following solution spaces

$$X_v := \{v \in L^2(0,T; H^1_{0,0}(\Omega, r)) \cap L^\infty(0,T; L^2(\Omega, r)) \mid \partial_t v \in L^2(0,T; L^2(\Omega, r)),$$
$$\Delta_r v \in L^2(0,T; L^2(\Omega, r)), \partial^2_{z,z} < v > \in L^2(0,T; L^2(0,1))\},$$

corresponding to the velocity space, and

$$X_\gamma := \left\{ \gamma \in H^1(0,T;H^1(0,1)) \mid \partial_t \gamma \in L^\infty(0,T;L^2(0,1)) \right\},$$

corresponding to the space of displacements.

DEFINITION 3.1. *Suppose that the initial data* $\gamma^0 \in H^1(0,1)$, $v_z^0 \in H^1_{0,0}(\Omega, r)$ *and that the boundary data* $(\gamma_0, \gamma_L) \in (H^2(0,T))^2$. *Function* $(\gamma, v_z) \in X_\gamma \times X_v$ *is called a mild solution of problem* (2.2)–(2.5) *if* (2.2)–(2.5) *holds for a.a.* $z \in (0,1), r \in [0,1)$ *and* $t \in (0,T)$.

4. Existence of a mild solution.

4.1. The framework. We aim at using the Implicit Function Theorem of Hildebrandt and Graves [22] to prove the (local) existence of a mild solution in a neighborhood of the solution stated in Proposition 2.1.

THEOREM 4.1. **(Implicit Function Theorem [22])** *Suppose that:*

- $F : U(\lambda_0, x_0) \subset \Lambda \times X \to Z$ *is defined on an open neighborhood* $U(\lambda_0, x_0)$ *and* $F(\lambda_0, x_0) = 0$, *where* Λ, X, Z *are Banach spaces.*
- F_x *exists as a Frechét partial derivative on* $U(\lambda_0, x_0)$ *and* $F_x(\lambda_0, x_0) : X \to Z$ *is bijective,*
- F *and* F_x *are continuous at* (λ_0, x_0).

Then the following are true:

- *Existence and uniqueness: There exist positive numbers* δ_0 *and* δ *such that for every* $\lambda \in \Lambda$ *satisfying* $\|\lambda - \lambda_0\| \le \delta_0$ *there is exactly one* $x \in X$ *for which* $\|x - x_0\| \le \delta$ *and* $F(\lambda, x(\lambda)) = 0$.
- *Continuity: If* F *is continuous in a neighborhood of* (λ_0, x_0), *than* x *is continuous in a neighborhood of* λ_0.

4.2. The mapping F**.** To define F we first remark that we will consider the conservation of mass equation (2.2) as a condition which will be satisfied for all possible solution candidates (γ, v_z). More precisely, when considering the continuity of F and F_x and when showing the bijective property of F_x will be "perturbing" our function F by a small source term f only in the balance of momentum equation, and not in the conservation of mass equation, preserving the conservation of mass property identically for all possible solutions, which is physically reasonable.

First notice that the conservation of mass equation (2.2) can be rewritten, after dividing (2.2) by γ, as a linear operator in γ, $\mathcal{L}_{<v_z>}(\gamma^0, \gamma_0, \gamma_L)$, which to each given $< v_z >$ and initial and boundary data γ^0, γ_0 and γ_L associates the (unique) solution $\gamma \in X_\gamma$ of the following linear transport problem:

$$\frac{\partial \gamma}{\partial t} + 2 < v_z > \frac{\partial \gamma}{\partial z} + \gamma \frac{\partial < v_z >}{\partial z} = 0, \tag{4.1}$$

with $\gamma(0,t) = \gamma_0(t)$ whenever $< v_z >$ is positive, $\gamma(1,t) = \gamma_L(t)$ whenever $< v_z >$ is negative, and $\gamma(z,0) = \gamma^0(z)$.

Define F via the momentum equation (2.3) where $\gamma \in X_\gamma$ in equation (2.3) is obtained from the conservation of mass "condition" (4.1).

DEFINITION 4.1. **(Mapping F)** *Let $Z := L^2(0,T;L^2(\Omega,r))$. Define mapping $F : U((R,0,R,R),0) \subset \Lambda \times X_v \to Z$, which associates to each $((\gamma^0, v_z^0, \gamma_0, \gamma_L), v_z) \in U((R,0,R,R),0)$ an $f \in Z$*

$$F : ((\gamma^0, v_z^0, \gamma_0, \gamma_L), v_z) \mapsto f \tag{4.2}$$

such that

$$\begin{cases} F((\gamma^0, v_z^0, \gamma_0, \gamma_L), v_z) := \dfrac{\partial v_z}{\partial t} - C_1 \dfrac{1}{\gamma^2}\Delta_r v_z - \dfrac{r}{\gamma}\dfrac{\partial \gamma}{\partial t}\dfrac{\partial v_z}{\partial r} \\ \qquad\qquad\qquad\qquad\qquad + C_2 \dfrac{\partial \gamma}{\partial z} + C_3 \dfrac{\partial^2 \gamma}{\partial z \partial t}, \\ v_z(1,z,t) = 0, v_z(r,z,t=0) = v_z^0(r,z), \ |v_z(0,z,t)| < +\infty, \end{cases} \tag{4.3}$$

where $\gamma \in X_\gamma$ depends on v_z and is given as a solution of

$$\begin{cases} \dfrac{\partial \gamma}{\partial t} + 2 < v_z > \dfrac{\partial \gamma}{\partial z} + \gamma \dfrac{\partial < v_z >}{\partial z} = 0, \\ \gamma(0,t) = \gamma_0(t), \ \gamma(1,t) = \gamma_L(t), \ \gamma(z,0) = \gamma^0(z). \end{cases} \tag{4.4}$$

Denote by $(\lambda_0, x_0) = ((R,0,R,R),0)$. Then we see, by Proposition 2.1, that $F(\lambda_0, x_0) = 0$.

We will be using the Implicit Function Theorem to show the existence of a unique mild solution $(\gamma, v_z) \in X_\gamma \times X_v$ of (2.2)–(2.5) for each set of data $\lambda = (\gamma^0, v_z^0, \gamma_0, \gamma_L)$ in a neighborhood of $\lambda_0 = (R,0,R,R)$, by considering small perturbations (λ, v_z) of the zero set $(\lambda_0, 0)$ of the mapping F, given by the balance of momentum equation (2.3), in which $\gamma \in X_\gamma$ satisfies the mass conservation condition (2.2).

PROPOSITION 4.1. *Mapping F is continuous at (λ_0, x_0).*

The proof is a direct consequence of the form of (4.3) and of the continuous dependence of the solution γ of (4.1) on the coefficients depending on $< v_z >$ and on the initial and boundary data.

4.3. The Frechét Derivative of F. Introduce perturbation of v_z around $\hat{v}_z$ as follows:

$$v_z = \hat{v}_z + \delta w_z, \quad \delta > 0.$$

Define $\hat{\gamma}$ via $\hat{v}_z$ as the solution of (4.4) corresponding to $v_z = \hat{v}_z$. Then the Frechét derivative of F with respect to $x = v_z$, evaluated at $((\hat{\gamma}^0, \hat{v}_z^0, \hat{\gamma}_0, \hat{\gamma}_L), \hat{v}_z)$ is a mapping

$$F_x((\hat{\gamma}^0, \hat{v}_z^0, \hat{\gamma}_0, \hat{\gamma}_L), \hat{v}_z) : X_v \to Z$$

defined by

$$F_x((\hat{\gamma}^0, \hat{v}_z^0, \hat{\gamma}_0, \hat{\gamma}_L), \hat{v}_z)w_z := \frac{\partial w_z}{\partial t} - C_1 \frac{1}{\hat{\gamma}^2}\Delta_r w_z + C_1 \frac{2}{\hat{\gamma}^3}\eta \Delta_r \hat{v}_z$$
$$- \frac{r}{\hat{\gamma}}\frac{\partial \eta}{\partial t}\frac{\partial \hat{v}_z}{\partial r} - \frac{r}{\hat{\gamma}}\frac{\partial \hat{\gamma}}{\partial t}\frac{\partial w_z}{\partial r} + \frac{r}{\hat{\gamma}^2}\eta \frac{\partial \hat{\gamma}}{\partial t}\frac{\partial \hat{v}_z}{\partial r} + C_2 \frac{\partial \eta}{\partial z} + C_3 \frac{\partial^2 \eta}{\partial z \partial t}, \quad (4.5)$$

where η is given as a solution of

$$\begin{cases} \dfrac{\partial \eta}{\partial t} + 2 < \hat{v}_z > \dfrac{\partial \eta}{\partial z} + 2 < w_z > \dfrac{\partial \hat{\gamma}}{\partial z} + \hat{\gamma} \dfrac{\partial < w_z >}{\partial z} + \eta \dfrac{\partial < \hat{v}_z >}{\partial z} = 0, \\ \eta(0,t) = 0, \ \eta(1,t) = 0, \ \eta(z,0) = 0, \end{cases} \quad (4.6)$$

with

$$w_z(1,z,t) = 0, w_z(r,z,0) = 0, \ w_z(0,z,t) - \textit{bounded}. \quad (4.7)$$

By the similar arguments as those used for continuity of the mapping F one can see that the following is true.

THEOREM 4.2. *The Frechét derivative F_x is a continuous mapping from X_v to Z.*

Next we need to show that the Frechét derivative, evaluated at (λ_0, x_0), is a bijection. From (4.6)–(4.7) we see that the Frechét derivative evaluated at $(\lambda_0, x_0) = ((R, 0, R, R), 0)$ is given by the following

$$\begin{cases} F_x((R,0,R,R),0)w_z := \dfrac{\partial w_z}{\partial t} - C_1 \dfrac{1}{R^2}\Delta_r w_z + C_2 \dfrac{\partial \eta}{\partial z} + C_3 \dfrac{\partial^2 \eta}{\partial z \partial t}, \\ w_z(1,z,t) = 0, w_z(r,z,0) = 0, \ w_z(0,z,t) - \textit{bounded}, \end{cases} \quad (4.8)$$

where η, which depends on w_z, satisfies

$$\begin{cases} \dfrac{\partial \eta}{\partial t} + R \dfrac{\partial}{\partial z} < w_z > = 0, \\ \eta(0,t) = 0 \text{ whenever } < w_z > \text{ positive}, \\ \eta(1,t) = 0 \text{ whenever } < w_z > \text{ negative}, \\ \eta(z,0) = 0. \end{cases} \quad (4.9)$$

THEOREM 4.3. *The Frechét derivative defined by* (4.8)–(4.9) *is a bijection from X_v to Z.*

Theorem 4.3 is a consequence of the following result: for every $f \in L^2(0,T; L^2(\Omega, r))$ and $(\eta^0, w_z^0, \eta_0, \eta_L) \in \Lambda$ there exists a unique function $(\eta, w_z) \in X_\gamma \times X_v$ satisfying for a.e. $0 < z < 1, 0 \le r < 1, 0 \le t \le T$

$$\frac{\partial \eta}{\partial t} + R \frac{\partial}{\partial z} < w_z > = 0, \quad (4.10)$$
$$\frac{\partial w_z}{\partial t} - \frac{C_1}{R^2}\Delta_r \, w_z + C_2 \frac{\partial \eta}{\partial z} + C_3 \frac{\partial^2 \eta}{\partial z \partial t} = f(r,z,t), \quad (4.11)$$

with

$$\begin{cases} \eta(0,t) = 0, \ \eta(1,t) = 0, \ \eta(z,0) = 0 \\ w_z(1,z,t) = 0, \ w_z(0,z,t) - \text{bounded}, \ w_z(r,z,0) = 0. \end{cases} \tag{4.12}$$

(Equation (4.10) implies that the boundary conditions for η are equivalent to the homogeneous Neumann condition for $< w_z >$ at $z = 0, 1$.) In fact, we will show a slightly more general result (general data):

THEOREM 4.4. *Let $f \in L^2(0,T; L^2(\Omega, r))$ and $\eta_0, \eta_L \in H^2(0,T)$, $\eta^0 \in H^1(0,1)$ and $w_z^0 \in H^1_{0,0}(\Omega, r)$. Then, there exists a unique function $(\eta, w_z) \in X_\gamma \times X_v$ satisfying for a.e. $0 < z < 1, 0 \le r < 1, 0 < t \le T$*

$$\frac{\partial \eta}{\partial t} + R\frac{\partial}{\partial z} < w_z >= 0, \tag{4.13}$$

$$\frac{\partial w_z}{\partial t} - \frac{C_1}{R^2} \Delta_r \, w_z + C_2\frac{\partial \eta}{\partial z} + C_3\frac{\partial^2 \eta}{\partial z \partial t} = f(r,z,t), \tag{4.14}$$

with

$$\begin{cases} \eta(0,t) = \eta_0(t), \ \eta(1,t) = \eta_L(t), \ \eta(z,0) = \eta^0(z) \\ w_z(1,z,t) = 0, \ w_z(0,z,t) - \text{bounded}, \ w_z(r,z,0) = w_z^0(r,z). \end{cases} \tag{4.15}$$

This result motivated the choice of the parameter space Λ for the existence of a unique mild solution to the nonlinear problem (2.2)–(2.4).

To prove this result we proceed in two steps:

1. Show the existence of a unique *weak solution* to (4.13), (4.14) and (4.15).
2. Obtain energy estimates which provide higher regularity of the weak solution, giving rise to the *mild solution* $(\eta, w_z) \in X_\gamma \times X_v$.

STEP 1. **Existence of a unique weak solution of (4.13)–(4.15).**

Introduce the function $\bar\eta$ which satisfies the homogeneous boundary data at $z = 0$ and $z = 1$: $\bar\eta = \eta(z,t) - ((\eta_L(t) - \eta_0(t))z + \eta_0(t))$. Problem (4.13)–(4.15) can then be rewritten in terms of $\bar\eta$ as follows

$$\frac{\partial \bar\eta}{\partial t} + R\frac{\partial}{\partial z} < w_z >= -g_1, \tag{4.16}$$

$$\frac{\partial w_z}{\partial t} - \frac{C_1}{R^2} \Delta_r \, w_z + C_2\frac{\partial \bar\eta}{\partial z} + C_3\frac{\partial^2 \bar\eta}{\partial z \partial t} = f - g_2. \tag{4.17}$$

with

$$\begin{cases} \bar\eta(0,t)=0, \ \bar\eta(1,t)=0, \ \bar\eta(z,0)=(\eta_L(0)-\eta_0(0))z + \eta_0(0)=\bar\eta^0(z), \\ w_z(1,z,t) = 0, \ w_z(0,z,t) - \text{bounded}, \ w_z(r,z,0) = w^0(r,z), \end{cases} \tag{4.18}$$

where

$$\begin{cases} g_1(z,t) = ((\eta_L'(t) - \eta_0'(t))z + \eta_0'(t)), \\ g_2(r,z,t) = C_2(\eta_L(t) - \eta_0(t)) + C_3(\eta_L'(t) - \eta_0'(t)). \end{cases} \tag{4.19}$$

To define a weak solution introduce the following function spaces

$$\Gamma = H^1(0,T : L^2(0,1)), \tag{4.20}$$

$$V = \{w \in L^2(0,T : H^1_{0,0}(\Omega, r)) : \frac{\partial w}{\partial t} \in L^2(0,T : H^{-1}_{0,0}(\Omega, r))\}. \tag{4.21}$$

DEFINITION 4.2. *We say that* $(\bar{\eta}, w_z) \in \Gamma \times V$ *is a weak solution of* (4.16)–(4.19) *provided that for all* $\varphi \in H^1_0(0,1)$ *and* $\xi \in H^1_{0,0}(\Omega, r)$

$$\int_0^1 \frac{\partial \bar{\eta}}{\partial t}\varphi \, dz - R\int_0^1 < w_z > \frac{\partial \varphi}{\partial z}\, dz = -\int_0^1 g_1\varphi \, dz \tag{4.22}$$

$$\begin{aligned}\int_\Omega \frac{\partial w_z}{\partial t}\xi \, rdrdz + \frac{C_1}{R^2}\int_\Omega \frac{\partial w_z}{\partial r}\frac{\partial \xi}{\partial r}\, rdrdz - C_2\int_0^1 \bar{\eta}\, \frac{\partial}{\partial z} < \xi > \, dz \\ -C_3\int_0^1 \frac{\partial \bar{\eta}}{\partial t}\frac{\partial}{\partial z} < \xi > \, dz = \int_0^1 f\varphi \, dz - \int_0^1 g_2\varphi \, dz,\end{aligned} \tag{4.23}$$

for a.e. $0 \le t \le T$, *and satisfying* $\bar{\eta}(z,0) = \bar{\eta}^0(z)$, $w_z(r,z,0) = w_z^0(r,z)$.

We first show that for the boundary data η_0 and η_L in $H^1(0,T)$ and for the initial data $\eta^0 \in L^2(0,1)$, $w_z^0 \in L^2(\Omega, r)$, there exists a unique weak solution of (4.16)–(4.19).

Notice that the weak formulation of the problem reflects lack of regularity in the z-direction due to the parabolic degeneracy in the momentum equation (4.23) and due to the hyperbolic nature of the averaged conservation of mass equation (4.22). This will introduce various difficulties in the proof of the existence of a unique weak solution which we state next.

THEOREM 4.5. *Let* $f \in L^2(0,T; L^2(\Omega, r))$. *Assume that the initial data* $\bar{\eta}^0$ *and* w_z^0 *satisfy* $\bar{\eta}^0 \in L^2(0,1)$ *and* $w_z^0 \in L^2(\Omega, r)$ *and that the boundary data* $\eta_0(t)$ *and* $\eta_L(t)$ *satisfy* $\eta_0, \eta_1 \in H^1(0,T)$. *Then there exists a unique weak solution* $(\bar{\eta}, w_z) \in \Gamma \times V$ *of* (4.16)–(4.19).

Proof. The proof is an application of the Galerkin method combined with the nontrivial energy estimates to deal with the lack of regularity in the z-direction. We present the proof in the following four steps:

1. Construction of the Galerkin approximations.
2. Uniform energy estimates.
3. Weak convergence of a sub-sequence of Galerkin approximations to a solution using compactness arguments.
4. Uniqueness of the weak solution.

CONSTRUCTION OF THE GALERKIN APPROXIMATIONS: Let $\{\phi_k\}_{k=1}^\infty$ be the smooth functions which are orthogonal in $H^1_0(0,1)$, orthonormal in $L^2(0,1)$ and span the solution space for $\bar{\eta}$. Furthermore, let $\{w_k\}_{k=1}^\infty$ be the smooth functions which satisfy $w_k|_{r=1} = 0$, and are orthonormal in $L^2(\Omega, r)$ and span the solution space for the velocity w_z. Introduce the function space $C^k_{0,0}(\Omega) = \{v \in C^k(\Omega) : v|_{r=1} = 0\}$, for any $k = 0, 1, ..., \infty$.

Fix positive integers m and n. We look for the functions $\bar{\eta}_m : [0,T] \to C_0^\infty(0,1)$ and $w_{z_n} : [0,T] \to C_{0,0}^\infty(\Omega)$ of the form

$$\bar{\eta}_m(t) = \sum_{i=1}^{m} d_i^m(t)\,\phi_i, \quad w_{z_n}(t) = \sum_{j=1}^{n} l_j^n(t)\,w_j, \tag{4.24}$$

where the coefficient functions d_h^m and l_k^n are chosen so that the functions $\bar{\eta}_m$ and w_{z_n} satisfy the weak formulation (4.22)–(4.23) of the linear problem (4.16)–(4.19), projected onto the finite dimensional subspaces spanned by $\{\phi_i\}$ and $\{w_j\}$ respectively:

$$\int_0^1 \frac{\partial \bar{\eta}_m}{\partial t}\phi_h\,dz - R\int_0^1 < w_{z_n} > \frac{\partial \phi_h}{\partial z}\,dz = -\int_0^1 g_1\phi_h\,dz \tag{4.25}$$

$$\begin{aligned}\int_\Omega \frac{\partial w_{z_n}}{\partial t} w_k\,rdrdz + \frac{C_1}{R^2}\int_\Omega \frac{\partial w_{z_n}}{\partial r}\frac{\partial w_k}{\partial r} rdrdz - C_2\int_0^1 \bar{\eta}_m \frac{\partial}{\partial z} < w_k > dz \\ - C_3\int_0^1 \frac{\partial \bar{\eta}_m}{\partial t}\frac{\partial}{\partial z} < w_k > \,dz = \int_\Omega f w_k\,dz - \int_\Omega g_2 w_k\,rdrdz\end{aligned} \tag{4.26}$$

for a.e $0 \le t \le T$, $h = 1, \cdots, m$ and $k = 1, \cdots, n$, and

$$\begin{cases} d_h^m(0) = \displaystyle\int_0^1 \bar{\eta}^0(z)\phi_h(z)dz, \\ l_k^n(0) = \displaystyle\int_\Omega w_z^0 w_k r dr dz. \end{cases} \tag{4.27}$$

The existence of the coefficient functions satisfying these requirements is guaranteed by the following Lemma.

LEMMA 4.1. *Assume that* $f \in L^2(0,T;L^2(\Omega,r))$. *For each* $m = 1,2,...$ *and* $n = 1,2,...$ *there exist unique functions* $\bar{\eta}_m$ *and* w_{z_n} *of the form (4.24), satisfying* (4.25)–(4.27). *Moreover*

$$(\bar{\eta}_m, v_{w_n}) \in C^1(0,T : C_0^\infty(0,1)) \times C^1(0,T : C_{0,0}^\infty(\Omega)).$$

Proof. To simplify notation, let us first introduce the following vector functions:

$$d^m(t) = \begin{pmatrix} d_1^m(t) \\ \vdots \\ d_m^m(t) \end{pmatrix}, \; l^n(t) = \begin{pmatrix} l_1^n(t) \\ \vdots \\ l_n^n(t) \end{pmatrix}, \; Y(t) = \begin{pmatrix} d^m(t) \\ l^n(t)) \end{pmatrix} \tag{4.28}$$

Then, equation (4.25) written in matrix form reads:

$$A_1 d^{m'}(t) + A_2 l^n(t) = S_1(t), \tag{4.29}$$

where A_1 is an $m \times m$ matrix, A_2 an $m \times n$ matrix and S_1 an $m \times 1$ matrix defined by the following: $[A_1]_{h,i} = (\phi_i, \phi_h)_{L^2(0,1)} = \delta_{h,i}$, $[A_2]_{h,i} = -R\left(< w_j >, \frac{\partial \phi_h}{\partial z}\right)_{L^2(0,1)}$, $[S_1(t)]_{h,1} = (g_1, \phi_h)_{L^2(0,1)}$, where $h, i = 1, ..., m$ and $j = 1, ..., n$. Similarly, equation (4.26) written in matrix form reads:

$$B_1 l^{n'}(t) + B_2 l^n(t) - B_3 d^m(t) - B_4 d^{m'}(t) = S_2(t), \tag{4.30}$$

where B_1 and B_2 are $n \times n$ matrices, B_3 and B_4 are $n \times m$ matrices, and $S_2(t)$ is an $n \times 1$ matrix defined by the following: $[B_1]_{k,j} = (w_j, w_k)_{L^2(\Omega,r)} = \delta_{k,j}$, $[B_2]_{k,j} = \frac{C_1}{R^2}\left(\frac{\partial w_j}{\partial r}, \frac{\partial w_k}{\partial r}\right)_{L^2(\Omega,r)}$, $[B_3]_{k,i} = C_2\left(\frac{\partial < w_k >}{\partial z}, \phi_i\right)_{L^2(0,1)}$, $[B_4]_{k,i} = C_3\left(\frac{\partial < w_k >}{\partial z}, \phi_i\right)_{L^2(0,1)}$, $[S_2(t)]_{k,1} = (f - g_2, w_k)_{L^2(\Omega,r)}$, where $k, j = 1, ..., n$ and $i = 1, ..., m$.

Equations (4.29) and (4.30) can be written together as the following system

$$\begin{cases} AY'(t) + BY(t) = S(t), \\ Y(0) = \begin{pmatrix} d^m(0) \\ l^n_k(0) \end{pmatrix}, \end{cases} \tag{4.31}$$

where Y is defined in (4.28) and

$$A = \begin{pmatrix} A_1^{m\times m} & 0^{m\times n} \\ -B_4^{n\times m} & B_1^{n\times n} \end{pmatrix}_{(m+n)\times(m+n)},$$

$$B = \begin{pmatrix} 0^{m\times m} & A_2^{m\times n} \\ -B_3^{n\times m} & B_2^{n\times n} \end{pmatrix}_{(m+n)\times(m+n)}.$$

Function S is an $(m+n) \times 1$ matrix which incorporates the initial and boundary data obtained from the right hand-sides of (4.29) and (4.30).

To guarantee the existence of a solution $Y(t)$ of appropriate regularity first notice that linear independence of the sets $\{\phi_1, \cdots, \phi_m\}$ and $\{w_1, \cdots, w_n\}$ guarantees that the matrix A is nonsingular. Additionally, since the coefficient matrices are constant, there exists a unique C^1 function $Y(t) = (d^m(t), l^n(t))$ satisfying (4.31). Moreover $(\bar{\eta}_m, w_{z_n})$, defined via $d^m(t)$ and $l^n(t)$ in (4.24), respectively, solves (4.25)–(4.27) for all $0 \le t \le T$, thus $(\bar{\eta}_m, w_{z_n}) \in C^1(0, T : C_0^\infty(0,1)) \times C^1(0, T : C_{0,0}^\infty(\Omega))$. This completes the proof of Lemma 4.1. ◻

Energy Estimate: We continue our proof of the existence of a weak solution to (4.16)–(4.19) by obtaining an energy estimate for $\bar{\eta}_m$ and w_{z_n} which is uniform in m and n. The estimate will bound the L^2-norms of $\bar{\eta}_m$ and w_{z_n}, the L^2-norms of $\frac{\partial w_{z_n}}{\partial r}$ and $\frac{\partial \bar{\eta}_m}{\partial t}$, and the $L^2(0, T; H_{0,0}^{-1}(\Omega, r))$-norm of $\frac{\partial w_{z_n}}{\partial t}$, in terms of the initial and boundary data and the coefficients of (4.16)–(4.18). Notice again the lack of information about the smoothness in z of the functions $\bar{\eta}$ and w_z.

THEOREM 4.6. *There exists a constant C depending on $1/R, T, C_2$ and C_3, such that*

$$\sup_{0\leq t\leq T}\left[\|w_{z_n}\|^2_{L^2(\Omega,r)}+\frac{C_2}{R}\|\bar\eta_m\|^2_{L^2(0,1)}\right]+\frac{2C_1}{R^2}\left\|\frac{\partial w_{z_n}}{\partial r}\right\|^2_{L^2(0,T;L^2(\Omega,r))}$$
$$+\frac{C_3}{R}\left\|\frac{\partial\bar\eta_m}{\partial t}\right\|^2_{L^2(0,T;L^2(0,1))}+\left\|\frac{\partial w_{z_n}}{\partial t}\right\|^2_{L^2(0,T;H^{-1}_{0,0}(\Omega,r))}\leq C\Big[\|\bar\eta^0\|^2_{L^2(0,1)}$$
$$+\|w_z^0\|^2_{L^2(\Omega,r)}+\|f\|^2_{L^2(0,T;L^2(\Omega,r))}+\|\eta_L-\eta_0\|^2_{H^1(0,T)}+\|\eta_0\|^2_{H^1(0,T)}\Big]$$

Furthermore, $\frac{\partial}{\partial z}\int_0^1 w_{z_n}rdr\in L^2(0,T;L^2(0,1))$, *and its* $L^2(0,T;L^2(0,1))$-*norm is bounded by the right hand-side of the above energy estimate.*

Proof. We aim at using the Gronwall's inequality. However, due to the lack of smoothness in z, it is impossible to control the terms with the z-derivative of $\bar\eta_m$. To deal with this problem, we manipulate the conservation of mass and balance of momentum equations in order to cancel the unwanted terms. The remaining terms, which we will estimate in terms of the data, will be those appearing in the estimate above.

We begin by first multiplying (4.26) by l_k^n and summing $k=1,\cdots,n$ to find

$$\frac{1}{2}\frac{d}{dt}\int_\Omega|w_{z_n}|^2rdrdz+\frac{C_1}{R^2}\int_\Omega\left|\frac{\partial w_{z_n}}{\partial r}\right|^2rdrdz-\underbrace{C_2\int_0^1\bar\eta_m\frac{\partial}{\partial z}<w_{z_n}>dz}_{(i)}$$
$$-\underbrace{C_3\int_0^1\frac{\partial\bar\eta_m}{\partial t}\frac{\partial}{\partial z}<w_{z_n}>dz}_{(ii)}=\int_\Omega fw_{z_n}rdrdz-\int_\Omega g_2w_{z_n}rdrdz. \tag{4.32}$$

Multiply (4.25) by d_h^m and sum $h=1,\cdots,m$ to find

$$\frac{1}{2}\frac{d}{dt}\int_0^1|\bar\eta_m|^2\,dz-\underbrace{R\int_0^1<w_{z_n}>\frac{\partial\bar\eta_m}{\partial z}\,dz}_{(i)}=-\int_0^1g_1\bar\eta_m\,dz. \tag{4.33}$$

Multiply (4.25) by $\dot d_h^m$ and sum $h=1,\cdots,m$ to find

$$\int_0^1\left|\frac{\partial\bar\eta_m}{\partial t}\right|^2dz-\underbrace{R\int_0^1<w_{z_n}>\frac{\partial^2\bar\eta_m}{\partial t\partial z}\,dz}_{(ii)}=-\int_0^1g_1\frac{\partial\bar\eta_m}{\partial t}\,dz. \tag{4.34}$$

Multiply equation (4.33) by $\frac{C_2}{R}$ and (4.34) by $\frac{C_3}{R}$ and add the two resulting equations to equation (4.32) to obtain

$$\frac{1}{2}\frac{d}{dt}\left[\int_\Omega |w_{z_n}|^2\, rdrdz + \frac{C_2}{R}\int_0^1 |\bar{\eta}_m|^2\, dz\right] + \frac{C_1}{R^2}\int_\Omega \left|\frac{\partial w_{z_n}}{\partial r}\right|^2 rdrdz$$
$$+ \frac{C_3}{R}\int_0^1 \left|\frac{\partial \bar{\eta}_m}{\partial t}\right|^2 dz = \int_\Omega f w_{z_n}\, rdrdz - \int_\Omega g_2 w_{z_n}\, rdrdz \qquad (4.35)$$
$$- \frac{C_2}{R}\int_0^1 g_1 \bar{\eta}_m\, dz - \frac{C_3}{R}\int_0^1 g_1 \frac{\partial \bar{\eta}_m}{\partial t}\, dz.$$

We can see that the terms denoted by (i) and (ii), which we cannot control, canceled out. By using the Cauchy inequality to estimate the right hand-side of (4.35) we obtain

$$\frac{1}{2}\frac{d}{dt}\left[\|w_{z_n}\|^2_{L^2(\Omega,r)} + \frac{C_2}{R}\|\bar{\eta}_m\|^2_{L^2(0,1)}\right] + \frac{C_1}{R^2}\left\|\frac{\partial w_{z_n}}{\partial r}\right\|^2_{L^2(\Omega,r)}$$
$$+ \frac{C_3}{2R}\left\|\frac{\partial \bar{\eta}_m}{\partial t}\right\|^2_{L^2(0,1)} \le \left\|f_2 + \frac{g_2}{4} + \frac{C_2+C_3}{2R} g_1\right\|^2_{L^2(\Omega,r)}$$
$$+ \frac{1}{2}\|w_{z_n}\|^2_{L^2(\Omega,r)} + \frac{C_2}{2R}\|\bar{\eta}_m\|^2_{L^2(0,1)}.$$

We are now in a position to apply the differential form of the Gronwall's inequality to conclude that there exists a constant $C > 0$ depending on T, C_2, C_3 and $1/R$ such that

$$\sup_{0\le t\le T}\left[\|w_{z_n}\|^2_{L^2(\Omega,r)} + \frac{C_2}{R}\|\bar{\eta}_m\|^2_{L^2(0,1)}\right] + \frac{2C_1}{R^2}\int_0^T \left\|\frac{\partial w_{z_n}}{\partial r}\right\|^2_{L^2(\Omega,r)} dt$$
$$+ \frac{C_3}{2R}\int_0^T \left\|\frac{\partial \bar{\eta}_m}{\partial t}\right\|^2_{L^2(0,1)} dt \le C\left[\|\bar{\eta}^0\|^2_{L^2(0,1)} + \|w_z^0\|^2_{L^2(\Omega,r)}\right. \qquad (4.36)$$
$$\left. + \|f\|^2_{L^2(0,T:L^2(\Omega,r))} + \|\eta_1 - \eta_0\|^2_{H^1(0,T)} + \|\eta_0\|_{H^1_0(0,T)}\right].$$

We conclude the proof by showing that $\dfrac{\partial w_{z_n}}{\partial t} \in L^2(0,T; H^{-1}_{0,0}(\Omega,r))$, and that $\partial v_{z_n}/\partial t$ satisfies the estimate stated in Theorem 4.6.

Fix $\nu \in H^1_{0,0}(\Omega,r)$ such that $\|\nu\|_{H^1_{0,0}(\Omega,r)} \le 1$. Since $C^\infty_{0,0}(\Omega)$ is dense in $H^1_{0,0}(\Omega,r)$, we can write $\nu = \nu_1 + \nu_2$, where $\nu_1 \in \text{span }\{w_j\}_{j=1}^n$ and $(\nu_2, w_j)_{L^2(\Omega,r)} = 0$ for $j = 1, \cdots, n$. Then (4.24) and (4.26) imply

$$\left|\int_\Omega \frac{\partial w_{z_n}}{\partial t}\nu\, rdrdz\right| \le \left[\frac{C_1}{R}\left\|\frac{\partial w_n}{\partial r}\right\|_{L^2(\Omega,r)} + C_2\|\bar{\eta}_m\|_{L^2}\right.$$
$$\left. + C_3\left\|\frac{\partial \bar{\eta}_m}{\partial t}\right\|_{L^2} + \|f\|_{L^2} + \|g_2\|_{L^2}\right]\|\nu\|_{H^1_{0,0}(\Omega,r)},\ a.e.\ 0 \le t \le T.$$

Thus, since $\|\nu_1\|_{H^1_{0,0}(\Omega,r)} \le 1$, by using the energy estimate (4.36), we find that there exists a constant $\tilde{C}$ depending on $T, 1/R, C_2, C_3$ such that

$$
\begin{aligned}
\int_0^T \left\| \frac{\partial w_{z_n}}{\partial t} \right\|^2_{H^{-1}_{0,0}(\Omega,r)} dt \leq \tilde{C} \Big[& \|\bar{\eta}^0\|^2_{L^2(0,1)} + \left\| w_z^0 \right\|^2_{L^2(\Omega,r)} \\
& + \|f\|^2_{L^2(0,T;L^2(\Omega,r))} + \|\eta_1 - \eta_0\|^2_{H^1(0,T)} + \|\eta_0\|^2_{H^1_0(0,T)} \Big] .
\end{aligned}
\tag{4.37}
$$

This concludes the proof of Theorem 4.6. □

It is interesting to notice that the coefficient of the vessel wall viscosity, C_3, governs the estimate for the time-derivative of the structure displacement, which is to be expected. Thus, our estimate shows how the structure viscoelasticity regularizes the time evolution of the structure.

Also, notice that the right hand-side of the energy estimate incorporates the initial data for both the structure displacement and the structure velocity, but the boundary data for only the structure displacement. This is a consequence of the parabolic degeneracy in the balance of momentum equation and is an interesting feature of this reduced, effective model.

WEAK CONVERGENCE TO A SOLUTION: We use the uniform energy estimate, presented in Theorem 4.6, to conclude that there exist convergent subsequences that converge weakly to the functions which satisfy (4.16)–(4.19) in the weak sense. This is a standard approach except for the fact that we need to deal with the weighted L^2-norms in Ω, with the singular weight r that is present due to the axial symmetry of the problem. We deal with this technical obstacle by using Lemma 4.58 on page 120 in [1], with $p = 2$ and $\nu = 1$.

By the energy estimate stated in Theorem 4.6 we see that the sequence $\{\bar{\eta}_m\}_{m=1}^{\infty}$ is bounded in $H^1(0,T;L^2(0,1))$. Similarly, $\{w_{z_n}\}_{n=1}^{\infty}$ is bounded in $L^2(0,T;H^1_{0,0}(\Omega,r))$ and that $\partial w_{z_n}/\partial t$ is bounded in $L^2(0,T;H^{-1}_{0,0}(\Omega,r))$. Therefore, there exist convergent subsequences $\{\bar{\eta}_{m_j}\}_{m_j=1}^{\infty}$ and $\left\{w_{z_{n_j}}\right\}_{n_j=1}^{\infty}$ such that

$$
\begin{cases}
\eta_{m_j} \rightharpoonup \eta & \text{weakly in } H^1(0,T;L^2(0,1)),\\
w_{z_{n_j}} \rightharpoonup w_z & \text{weakly in } L^2(0,T;L^2(\Omega,r)),\\
\dfrac{\partial w_{z_{n_j}}}{\partial r} \rightharpoonup \dfrac{\partial w_z}{\partial r} & \text{weakly in } L^2(0,T;L^2(\Omega,r)),\\
\dfrac{\partial w_{z_{n_j}}}{\partial t} \rightharpoonup \dfrac{\partial w_z}{\partial t} & \text{weakly in } L^2(0,T;H^{-1}_{0,0}(\Omega,r)).
\end{cases}
\tag{4.38}
$$

What is left to show is that the limiting functions satisfy (4.16)–(4.19) in the weak sense and that the limiting functions satisfy the initial data. This requires relatively standard arguments which can be found in, e.g., [13]. For details of this calculation, please see [19]. Similar arguments have been used also in [20].

UNIQUENESS: Uniqueness of the weak solution is a direct consequence of the linearity of the problem.

This completes the proof of Theorem 4.5. □

This proof completes the first step in the proof of Theorem 4.4. What is left to show is that the weak solution of (4.13) and (4.15) has higher regularity and that, in fact, it belongs to the space $X_\gamma \times X_v$.

COROLLARY 4.1. *The energy estimate stated in Theorem 4.6 implies that, in fact,* $\bar{\eta} \in L^\infty(0,T;L^2(0,1)) \cap H^1(0,T;L^2(0,1))$, $w_z \in L^2(0,T; H^1_{0,0}(\Omega,r)) \cap L^\infty(0,T;L^2(\Omega,r))$ *with* $\frac{\partial w_z}{\partial t} \in L^2(0,T;H^{-1}_{0,0}(\Omega,r))$.

STEP 2. **Higher regularity of the weak solution to (4.13)-(4.15).**

To show that our weak solution $(\bar{\eta}, w_z)$ is actually in $X_\gamma \times X_v$ we proceed in two steps. First we show that the sequence $\{\frac{\partial w_{z_n}}{\partial t}\}_{n=1}^\infty$ is bounded in $L^2(0,T;L^2(\Omega,r))$, and then, using this information, we show that $(\bar{\eta}, w_z) \in X_\gamma \times X_v$. To show this improved regularity property of our weak solution we need to assume, as usual, some higher regularity of the initial and boundary data. The precise assumptions are given below.

THEOREM 4.7. *(Improved Regularity: Part I) Suppose that the boundary data* $\eta_1, \eta_0 \in H^2(0,T)$ *and the initial data* $\bar{\eta}^0 \in L^2(0,1)$, $w_z^0 \in H^1_{0,0}(\Omega,r)$. *Then the weak solution* $(\bar{\eta}, w_z) \in \Gamma \times V$ *satisfies* $\dfrac{\partial \bar{\eta}}{\partial t} \in L^\infty(0,T:L^2(0,1))$, $\dfrac{\partial w_z}{\partial r} \in L^\infty(0,T:L^2(\Omega,r))$, $\dfrac{\partial w_z}{\partial t} \in L^2(0,T:L^2(\Omega,r))$.

Moreover, there exists a $C>0$, *depending on* $1/R, C_2, C_3, T$, *such that*

$$\sup_{0\le t\le T}\left[\frac{C_3}{R}\left\|\frac{\partial\bar{\eta}}{\partial t}\right\|^2_{L^2(0,1)} + \frac{2C_1}{R^2}\left\|\frac{\partial w_z}{\partial r}\right\|^2_{L^2(\Omega,r)}\right]$$
$$+2\int_0^T\left\|\frac{\partial w_z}{\partial t}\right\|^2_{L^2(\Omega,r)} ds \le C\Big(\|f\|^2_{L^2(0,T:L^2(\Omega,r))} \tag{4.39}$$
$$+\|\eta_1-\eta_0\|^2_{H^2(0,T)} + \|\eta_0\|^2_{H^2(0,T)} + \left\|\bar{\eta}^0\right\|^2_{L^2(0,1)} + \left\|w_z^0\right\|^2_{H^1_{0,0}(\Omega,r)}\Big).$$

Proof. Again, we need to deal with the lack of regularity in the z direction by canceling the terms which we cannot control at this point. As before, we need to manipulate the conservation of mass equation and the conservation of momentum equation in such as way that, when they are added up, the unwanted terms cancel out and produce an equation whose terms on the right hand-side can be estimated using the Cauchy's and Young's inequalities. The energy estimate will then follow by an application of the Gronwall's inequality.

As in the previous proof, we begin by multiplying equation (4.26) by $\dot{l}_k^n(t)$, and sum over $k=1,\cdots,n$. Then we differentiate (4.25) with respect to t, multiply by $\dot{d}_h^m(t)$ and by $\frac{C_3}{R}$, and sum over $h=1,\cdots,m$. Finally, we differentiate (4.25) with respect to t, multiply by $d_h^m(t)$ and by $\frac{C_2}{R}$, and sum over $h=1,\cdots,m$. The resulting equations contain the unwanted terms which, when added up, cancel and produce the following equality:

$$\frac{1}{2}\frac{d}{dt}\left[\frac{C_3}{R}\left\|\frac{\partial\bar{\eta}_m}{\partial t}\right\|^2_{L^2(0,1)}+\frac{C_1}{R^2}\left\|\frac{\partial w_{z_n}}{\partial r}\right\|^2_{L^2(\Omega,r)}\right]+\left\|\frac{\partial w_{z_n}}{\partial t}\right\|^2_{L^2(\Omega,r)}$$

$$=\int_\Omega f\frac{\partial w_{z_n}}{\partial t}\,rdrdz-\int_\Omega g_2\frac{\partial w_{z_n}}{\partial t}\,rdrdz-\frac{C_2}{R}\int_0^1\frac{\partial g_1}{\partial t}\bar{\eta}_m\,dz \tag{4.40}$$

$$-\frac{C_3}{R}\int_0^1\frac{\partial g_1}{\partial t}\frac{\partial\bar{\eta}_m}{\partial t}\,dz+\frac{C_2}{R}\int_0^1\frac{\partial^2\bar{\eta}_m}{\partial t^2}\bar{\eta}_m\,dz.$$

Before we estimate the right hand-side of this equation, we will integrate the entire equation with respect to t in order to deal with the term on the right hand-side which contains the second derivative with respect to t of $\bar{\eta}_m$. We obtain

$$\frac{1}{2}\left[\frac{C_3}{R}\left\|\frac{\partial\bar{\eta}_m(t)}{\partial t}\right\|^2_{L^2(0,1)}+\frac{C_1}{R^2}\left\|\frac{\partial w_{z_n}(t)}{\partial r}\right\|^2_{L^2(\Omega,r)}\right]+\int_0^t\left\|\frac{\partial w_{z_n}}{\partial s}\right\|^2_{L^2(\Omega,r)}ds$$

$$=\int_0^t\int_\Omega(f-g_2)\frac{\partial w_{z_n}}{\partial s}\,rdrdzds-\frac{C_2}{R}\int_0^t\int_0^1\frac{\partial g_1}{\partial s}\bar{\eta}_m\,dzds$$

$$-\frac{C_3}{R}\int_0^t\int_0^1\frac{\partial g_1}{\partial s}\frac{\partial\bar{\eta}_m}{\partial s}\,dzds+\frac{C_2}{R}\int_0^t\int_0^1\frac{\partial^2\bar{\eta}_m}{\partial s^2}\bar{\eta}_m\,dzds \tag{4.41}$$

$$+\frac{1}{2}\left[\frac{C_3}{R}\left\|\frac{\partial\bar{\eta}_m(0)}{\partial t}\right\|^2_{L^2(0,1)}+\frac{C_1}{R^2}\left\|\frac{\partial w_{z_n}(0)}{\partial r}\right\|^2_{L^2(\Omega,r)}\right].$$

The first three terms on the right hand-side can be estimated by using the Cauchy inequality. To estimate the fourth term, we use integration by parts with respect to s to obtain

$$-\frac{C_2}{R}\int_0^t\int_0^1\frac{\partial^2\bar{\eta}_m}{\partial s^2}\bar{\eta}_m\,dzds=\frac{C_2}{R}\int_0^t\int_0^1\left|\frac{\partial\bar{\eta}_m}{\partial s}\right|^2\,dzds$$

$$-\frac{C_2}{R}\int_0^1\frac{\partial\bar{\eta}_m(z,t)}{\partial t}\bar{\eta}_m(z,t)\,dz+\frac{C_2}{R}\int_0^1\frac{\partial\bar{\eta}_m(z,0)}{\partial t}\bar{\eta}_m(z,0)\,dz$$

where $\displaystyle\int_0^1\frac{\partial\bar{\eta}_m(z,0)}{\partial t}\bar{\eta}_m(z,0)dz=-\int_0^1\left(g_1+R^2\frac{\partial<w_{z_n}>}{\partial z}\right)_{t=0}\bar{\eta}_m(z,0)dz.$
This implies

$$\left|\frac{C_2}{R}\int_0^t\int_0^1\frac{\partial^2\bar{\eta}_m}{\partial s^2}\bar{\eta}_m\,dzds\right|\le\tilde{K}\left(\left\|\frac{\partial\bar{\eta}_m}{\partial s}\right\|^2_{L^2(0,T:L^2(0,1))}+\|\bar{\eta}_m(t)\|^2_{L^2(0,1)}\right.$$

$$\left.+\left\|\bar{\eta}^0\right\|^2_{H^1(0,1)}+\left\|w_z^0\right\|^2_{H^1_{0,0}(\Omega,r)}+\|g_1(0)\|^2_{L^2(0,1)}\right)+\frac{C_3}{4R}\left\|\frac{\partial\bar{\eta}_m(t)}{\partial t}\right\|^2_{L^2(0,1)}$$

where $\tilde{K}>0$ depends on $C_2, C_3, 1/C_3, 1/R, R$.

The last two terms in (4.41) can be estimated by first using the conservation of mass equation (4.25) to obtain

$$\int_0^1 \left|\frac{\partial \bar{\eta}_m(0)}{\partial t}\right|^2 dz = -\int_0^1 \frac{\partial < w_{z_n}(0) >}{\partial z}\frac{\partial \bar{\eta}_m(0)}{\partial t} dz - \int_0^1 g_1(0)\frac{\partial \bar{\eta}_m(0)}{\partial t} dz.$$

and then the Cauchy's inequality so that:

$$\int_0^1 \left|\frac{\partial \bar{\eta}_m(0)}{\partial t}\right|^2 dz \leq \left\|\frac{\partial}{\partial z} < w_z^0 >\right\|^2_{L^2(0,1)} + \|g_1(0)\|^2_{L^2(0,1)}$$

to obtain

$$\frac{1}{2}\left[\frac{C_3}{R}\left\|\frac{\partial \bar{\eta}_m(0)}{\partial t}\right\|^2_{L^2(0,1)} + \frac{C_1}{R^2}\left\|\frac{\partial w_{z_n}(0)}{\partial r}\right\|^2_{L^2(\Omega,r)}\right]$$
$$\leq \tilde{C}\left(\|w_z^0\|^2_{H^1_{0,0}(\Omega,r)} + (\eta_1'(0) - \eta_0'(0))^2 + (\eta_0'(0))^2\right)$$

where $\tilde{C} > 0$ depends on $C_1, C_3, 1/R$.

By combining these estimates and by using the energy estimate stated in Theorem 4.6 we see that there exists a constant $C > 0$ depending on $C_1, C_2, C_3, 1/R, R$ such that

$$\frac{C_3}{4R}\left\|\frac{\partial \bar{\eta}_m(t)}{\partial t}\right\|^2_{L^2(0,1)} + \frac{C_1}{2R^2}\left\|\frac{\partial w_{z_n}(t)}{\partial r}\right\|^2_{L^2(\Omega,r)} + \frac{1}{2}\int_0^t \left\|\frac{\partial w_{z_n}}{\partial s}\right\|^2_{L^2(\Omega,r)} ds$$
$$\leq C(\|f\|^2_{L^2(0,T:L^2(\Omega,r))} + \|\eta_1 - \eta_0\|^2_{H^2(0,T)} + \|\eta_0\|^2_{H^2(0,T)} + \left\|\bar{\eta}^0\right\|^2_{H^1(0,1)}$$
$$+ \left\|w_z^0\right\|^2_{L^2_{0,0}(\Omega,r)} + \left|\eta_1'(0) - \eta_0'(0)\right|^2 + (\eta_0'(0))^2)$$

for a.e $0 \leq t \leq T$. Passing to the limit as $m \to \infty$ and $n \to \infty$ we recover the estimate (4.39). This completes the proof of Theorem 4.7. □

Next we show that the weak solution $(\bar{\eta}, w_z)$ is, in fact, a *mild solution*, namely, that $(\bar{\eta}, w_z) \in X_\gamma \times X_v$ under some additional assumptions on the smoothness of the initial data. It is in this step that we can finally take control over certain derivatives with respect to z of our solution.

THEOREM 4.8 (Improved regularity: Part II). *Assume, in addition to the assumptions of Theorem 4.7, that the initial data $\bar{\eta}^0 \in H^1(0,1)$. Then the weak solution $\bar{\eta}$ satisfies $\bar{\eta} \in H^1(0,T; H^1(0,1))$. Furthermore, the following estimate holds:*

$$\sup_{0\leq t\leq T} \frac{C_2 C_3}{12}\left\|\frac{\partial \bar{\eta}}{\partial z}\right\|^2_{L^2(0,1)} + \frac{C_3^2}{12}\left\|\frac{\partial^2 \bar{\eta}}{\partial t \partial z}\right\|^2_{L^2(0,T:L^2(0,1))}$$
$$\leq C\Big(\|f\|^2_{L^2(0,T:L^2(\Omega,r))} + \|\eta_1 - \eta_0\|^2_{H^2(0,T)} \quad (4.42)$$
$$+ \|\eta_0\|^2_{H^2(0,T)} + \left\|\bar{\eta}^0\right\|^2_{H^1(0,1)} + \left\|w_z^0\right\|^2_{H^1_{0,0}(\Omega)}\Big),$$

where C depends on $1/R, C_2, C_3, T$. This implies that, in fact,

$$\frac{\partial^2 < w_z >}{\partial z^2} \in L^2(0,T : L^2(0,1)), \quad \Delta_r w_z \in L^2(0,T : L^2(\Omega,r)). \quad (4.43)$$

Proof. The proof is based on the following idea. We will use the weak form of the momentum equation (4.26) to estimate $\partial\bar{\eta}/\partial z$ and $\partial^2\bar{\eta}/\partial z\partial t$. In order to obtain the desired estimate, we will substitute the test function w_k in the weak form of the momentum equation (4.26) by $(1-r)\frac{\partial\phi_k(z)}{\partial z} \in C^1_{0,0}(\Omega)$. We will then use the fact that

$$(1-r)\frac{\partial^2\bar{\eta}_m}{\partial z\partial t} = (1-r)\sum_{k=1}^{m}\dot{d}_k^m(t)\frac{\partial\phi_k(z)}{\partial z} = \sum_{k=1}^{m}\dot{d}_k^m(t)\underbrace{(1-r)\frac{\partial\phi_k(z)}{\partial z}}_{w_k(r,z)},$$

where

$$w_k(r,z) = (1-r)\frac{\partial\phi_k(z)}{\partial z} \in C^1_{0,0}(\Omega,r). \quad (4.44)$$

Notice that, without loss of generality, we could have used the space $C^1_{0,0}$ in the definition of the Galerkin approximation for the velocity, instead of the space $C^\infty_{0,0}$. Thus, everything obtained will hold assuming $w_k \in C^1_{0,0}$. This relaxed choice of the space for w_k is now important to obtain improved regularity.

We now proceed by substituting w_k in (4.26) with (4.44) and by multiplying equation (4.26) by $\dot{d}_k^m(t)$ and summing over $k = 1,...m$ to obtain

$$\begin{aligned}
&\int_\Omega \frac{\partial w_{z_n}}{\partial t}\frac{\partial^2\bar{\eta}_m}{\partial z\partial t}(1-r)rdrdz - \frac{C_1}{R^2}\int_\Omega \frac{\partial w_{z_n}}{\partial r}\frac{\partial^2\bar{\eta}_m}{\partial z\partial t}rdrdz \\
&+ C_2\int_\Omega \frac{\partial\bar{\eta}_m}{\partial z}\frac{\partial^2\bar{\eta}_m}{\partial z\partial t}(1-r)rdrdz + C_3\int_\Omega \left|\frac{\partial\bar{\eta}_m}{\partial z\partial t}\right|^2(1-r)rdrdz \quad (4.45)\\
&= \int_\Omega f\frac{\partial^2\bar{\eta}_m}{\partial z\partial t}(1-r)rdrdz - \int_\Omega g_2\frac{\partial^2\bar{\eta}_m}{\partial z\partial t}(1-r)rdrdz.
\end{aligned}$$

Multiplying (4.45) by C_3 and integrating with respect to r where possible, we get

$$\begin{aligned}
&\frac{C_2C_3}{12}\frac{d}{dt}\left\|\frac{\partial\bar{\eta}_m}{\partial z}\right\|^2_{L^2(0,1)} + \frac{C_3^2}{6}\left\|\frac{\partial^2\bar{\eta}_m}{\partial t\partial z}\right\|^2_{L^2(0,1)} \\
&= -C_3\int_\Omega \frac{\partial w_{z_n}}{\partial t}\frac{\partial^2\bar{\eta}_m}{\partial t\partial z}(1-r)rdrdz + \frac{C_1C_3}{R^2}\int_\Omega \frac{\partial w_{z_n}}{\partial r}\frac{\partial^2\bar{\eta}_m}{\partial t\partial z}rdrdz \\
&+ C_3\int_\Omega (f-g_2)\frac{\partial^2\bar{\eta}_m}{\partial t\partial z}(1-r)rdrdz.
\end{aligned}$$

By applying the Cauchy inequality to the right hand-side, and then using the differential form of Gronwall's inequality, and by employing the improved regularity estimate (4.39) we obtain

$$\frac{C_2C_3}{12}\sup_{0\le t\le T}\left\|\frac{\partial\bar{\eta}_m}{\partial z}\right\|^2_{L^2(0,1)}+\frac{C_3^2}{12}\int_0^T\left\|\frac{\partial^2\bar{\eta}_m}{\partial t\partial z}\right\|^2_{L^2(0,1)}dt$$
$$\le C(\|f\|^2_{L^2(0,T:L^2(\Omega,r))}+\|\eta_1-\eta_0\|^2_{H^2(0,T)}+\|\eta_0\|^2_{H^2(0,T)}+\left\|\bar{\eta}^0\right\|^2_{H^1(0,1)}$$
$$+\left\|w_z^0\right\|^2_{H^1_{0,0}(\Omega)}+\left|\eta_1'(0)-\eta_0'(0)\right|^2+(\eta_0'(0))^2).$$

Passing to limit we recover the desired estimate (4.42). Moreover since $\bar{\eta}\in H^1(0,T;H^1(0,1))$, from equations (4.16) and (4.17), we conclude (4.43). For details please see [19]. This concludes the proof of Theorem 4.8. □

With this proof we have completed the second step in showing that problem (4.13)–(4.15) has a unique mild solution. This result implies, in particular, that the Frechét derivative is a bijection from X_v to Z and, thus, completes the proof of Theorem 4.3.

REMARK. An alternate proof for the problem with zero initial and boundary data can be obtained by using a (distributional) Laplace transform approach, the (complex) Lax-Milgram Lemma, the Paley-Wiener Theorem, and abstract elliptic regularity theory [15].)

The Implicit Function Theorem 4.1 now implies existence of a unique, mild solution to the nonlinear, moving boundary problem (1.1)–(1.6).

In order to state this result in terms of the pressure inlet and outlet boundary data as formulated in (1.6) we remark that the condition on the boundary data $\eta_0,\eta_L\in H^2(0,T)$ translates into the following condition in terms of the pressure data $P_0,P_L\in H^1(0,T)$. This is due to the pressure-displacement relationship (1.3). Thus, the parameter space Λ in terms of the pressure boundary data becomes $\tilde{\Lambda}:=H^1(0,L)\times H^1_{0,0}(\Omega,r)\times(H^1(0,T))^2$. We can now state our main result in terms of the pressure data:

THEOREM 4.9 (**Main Result**). *Assume that the initial data η^0 for the displacement η from the reference cylinder of radius R, is in $H^1(0,L)$, and that the initial data v_z^0 for the axial component of the velocity is in $H^1_{0,0}(\Omega,r)$. Furthermore, suppose that the inlet and outlet pressure data $P_0(t)$ and $P_L(t)$ which correspond to the fluctuations around the reference pressure $p_{\rm ref}$, are such that $P_0,P_L\in H^1(0,T)$. Then, there exists a neighborhood $S\subset X_\gamma\times X_v$ around the solution $\eta=0,v_z=0$, and a neighborhood $D\subset\tilde{\Lambda}$ around the initial and boundary data $\eta^0=0,v_z^0=0,P_0=p_{\rm ref},P_L=p_{\rm ref}$ such that there exists exactly one mild solution $(\eta,v_z)\in S\subset X_\gamma\times X_v$ of (1.1)–(1.6) for each choice of the initial and boundary data $(\eta^0,v_z^0,P_0,P_L)\in D\subset\tilde{\Lambda}$.*

5. Conclusions. In this manuscript we proved the existence of a unique mild solution to a nonlinear moving-boundary problem of mixed hyperbolic-parabolic type arising in modeling blood flow through viscoelastic arteries. The result holds for small perturbations of the data around the reference cylinder of radius R and axial velocity equal to zero. Future research in this direction includes an extension of this result to the solutions obtained as small perturbations of flow in a cylinder of radius R with the axial velocity corresponding to the Womersley profile, assumed for time-periodic pressure gradients. This scenario corresponds more closely to the physiologically relevant blood flow conditions.

Acknowledgement. The authors would like to thank the referee for her/his insightful comments.

REFERENCES

[1] R. Adams. Sobolev. Spaces. Second Edition. Academic Press, New York 2008.

[2] R.L. Armentano, J.G. Barra, J. Levenson, A. Simon, and R.H. weak Pichel. *Arterial wall mechanics in conscious dogs: assessment of viscous, inertial, and elastic moduli to characterize aortic wall behavior.* Circ. Res. **76** (1995), pp. 468–478.

[3] R. L. Armentano, J.L. Megnien, A. Simon, F. Bellenfant, J.G. Barra, and J. Levenson. *Effects of hypertension on viscoelasticity of carotid and femoral arteries in humans.* Hypertension **26** (1995), pp. 48–54.

[4] R.D. Bauer, R. Busse, A. Shabert, Y. Summa, and E. Wetterer. *Separate determination of the pulsatile elastic and viscous forces developed in the arterial wall in vivo.* Pflugers Arch. **380** (1979), pp. 221–226.

[5] S. Čanić and E.-H. Kim. *Mathematical analysis of the quasilinear effects in a hyperbolic model of blood flow through compliant axi–symmetric vessels,* Mathematical Methods in the Applied Sciences, **26**(14) (2003), pp. 1161–1186.

[6] S. Čanić, J. Tambača, G. Guidobini, A. Mikelić, C.J. Hartley, and D. Rosenstrauch. *Modeling viscoelastic behavior of arterial walls and their interaction with pulsatile blood flow* SIAM J. App. Math., V 67-1 (2006), 164–193.

[7] S. Čanić, A. Mikelić, D. Lamponi, and J. Tambača. *Self-Consistent Effective Equations Modeling Blood Flow in Medium-to-Large Compliant Arteries.* SIAM J. Multiscale Analysis and Simulation **3**(3) (2005), pp. 559–596.

[8] S. Čanić, C.J. Hartley, D. Rosenstrauch, J. Tambača, G. Guidoboni, and A. Mikelić. *Blood Flow in Compliant Arteries: An Effective Viscoelastic Reduced Model, Numerics and Experimental Validation.* Annals of Biomedical Engineering. **34** (2006), pp. 575–592.

[9] D. Coutand and S. Shkoller. *On the motion of an elastic solid inside of an incompressible viscous fluid.* Archive for rational mechanics and analysis, Vol. **176**, pp. 25–102, 2005.

[10] D. Coutand and S. Shkoller. *On the interaction between quasilinear elastodynamics and the Navier-Stokes equations* Archive for Rational Mechanics and Analysis, Vol. **179**, pp. 303–352, 2006.

[11] A. Chambolle, B. Desjardins, M. Esteban, and C. Grandmont. *Existence of weak solutions for an unsteady fluid-plate interaction problem.* J Math. Fluid Mech. **7** (2005), pp. 368–404.

[12] B. Desjardin, M.J. Esteban, C. Grandmont, and P. Le Tallec. *Weak solutions for a fluid-elastic structure interaction model.* Revista Mathem'atica complutense, Vol. XIV, num. 2, 523–538, 2001.

[13] L.C. Evans, *Parrial differential equations.* Graduate Studies in Mathematics, (19), American Mathematical Society, RI 2002.

[14] M. Guidorzi, M. Padula, and P. Plotnikov. *Galerkin method for fuids in domains with elastic walls.* University of Ferrara, Preprint.

[15] A. Mikelić. *Bijectivity of the Frechét derivative for a Biot problem in blood flow.* Private communication. Dec 19, 2009.

[16] F. Nobile, *Numerical Approximation of Fluid-Structure Interaction Problems with Application to Haemodynamics*, Ph.D. Thesis, EPFL, Lausanne, 2001.

[17] G. Pontrelli. *A mathematical model of flow through a viscoelastic tube.* Med. Biol. Eng. Comput, 2002.

[18] A. Quarteroni, M. Tuveri, and A. Veneziani. *Computational vascular fluid dynamics: problems, models and methods. Survey article*, Comput. Visual. Sci. **2** (2000), pp. 163–197.

[19] T.-B. Kim. *Some mathematical issues in blood flow problems.* Ph.D. Thesis. University of Houston 2009.

[20] T.-B. Kim, S. Čanić, G. Guidoboni. *Existence and uniqueness of a solution to a three-dimensional axially symmetric Biot problem arising in modeling blood flow.* Communications on Pure and Applied Analysis. To appear (2009).

[21] B. da Veiga. *On the existence of strong solutions to a coupled fluid-structure evolution problem.* Journal of Mathematical Fluid Mechanics, Vol. **6**, pp. 1422–6928 (Print), pp. 1422–6952 (Online), 2004.

[22] E. Zeidler. *Nonlinear Functional Analysis and its Applications I. (Fixed Point Theorems)* 1986, Springer-Verlag New York, Inc.

TRANSONIC FLOWS AND ISOMETRIC EMBEDDINGS

GUI-QIANG G. CHEN*, MARSHALL SLEMROD†, AND DEHUA WANG‡

Abstract. Transonic flows past an obstacle such as an airfoil are first considered. A viscous approximation to the steady transonic flow problem is presented, and its convergence is obtained by the method of compensated compactness. Then the isometric embedding problem in geometry is discussed. A fluid dynamic formulation of the Gauss-Codazzi system for the isometric embedding of two-dimensional Riemannian manifolds is provided, and an existence result of isometric immersions with negative Gauss curvature is given.

Key words. Transonic flow, viscosity method, Euler equations, gas dynamics, compensated compactness, entropy solutions, isometric embedding, two-dimensional Riemannian manifold, Gauss-Codazzi system, negative Gauss curvature.

AMS(MOS) subject classifications. 76H05, 35M10, 35A35, 76N10, 53C42.

1. Introduction. Transonic flows past an obstacle arise in many applications of gas dynamics (cf. [8, 13, 15]). The mathematical analysis of transonic flows is "exceedingly challenging" (quote from Bers [1]). Generally, the transonic flow problem is of hyperbolic-elliptic mixed type. Only when the flow is subsonic, the governing equations are elliptic. In this case, Shiffman [16] and Bers [1] established the existence of classical subsonic solutions. Morawetz in [11, 12] tackled the problem of transonic flows in a channel or exterior to an airfoil by the method of vanishing viscosity. Following Morawetz [11, 12], in Chen-Slemrod-Wang [3], we introduced a new viscous approximation to the transonic flow problem, constructed the invariant regions, and proved the convergence by the method of compensated compactness ([17]). The idea in [3] was applied to study the isometric embedding problem in [4]. More specifically, in Chen-Slemrod-Wang [4], we introduced a general approach to deal with the isometric immersion problem involving nonlinear partial differential equations of mixed hyperbolic-elliptic type by combining a fluid dynamic formulation of balance laws with

*Department of Mathematics, Northwestern University, Evanston, IL 60208, USA; Mathematical Institute, University of Oxford, Oxford, OX1 3LB, UK; School of Mathematical Sciences, Fudan University, Shanghai 200433, China. G.-Q. Chen's research was supported in part by the National Science Foundation under Grants DMS-0807551, DMS-0720925, and DMS-0505473, the Natural Science Foundation of China under Grant NSFC-10728101, and the Royal Society-Wolfson Research Merit Award (UK).

†Department of Mathematics, University of Wisconsin, Madison, WI 53706, USA; Department of Mathematics, University of Pittsburgh, Pittsburgh, PA 15260, USA; Department of Mathematical Sciences, Korean Advanced Institute for Science and Technology, Daejeon, Republic of Korea. M. Slemrod's research was supported in part by the National Science Foundation under Grant DMS-0647554.

‡Department of Mathematics, University of Pittsburgh, Pittsburgh, PA 15260, USA. D. Wang's research was supported in part by the National Science Foundation under Grants DMS-0604362 and DMS-0906160, and by the Office of Naval Research under Grant N00014-07-1-0668.

a compensated compactness framework. As an example of direct applications of this approach, we show how this idea can be applied to establish an isometric immersion of a two-dimensional Riemannian manifold with negative Gauss curvature in $\mathbb{R}^3$. Of course, the isometric embedding is a classical problem in differential geometry. The first global existence of smooth C^∞ embeddings was given by Nash [14]. Many other important results have been achieved for the embedding of surfaces with positive Gauss curvature, which can be formulated as an elliptic boundary value problem. For the case of surfaces of negative Gauss curvature where the underlying partial differential equations are hyperbolic, the problem is more difficult. See Han-Hong [7] and Yau [19] for related results. In this paper, we present a viscous approximation for the transonic flow problem in Section 2, and a fluid dynamic approach of the isometric immersion problem for the two-dimensional surfaces in Section 3. Finally, in Section 4, we discuss some related open problems.

2. Transonic flows past an obstacle. In this section, we discuss the Euler equations in compressible fluid dynamics for transonic flows past an obstacle such as an airfoil.

The two-dimensional steady, isentropic transonic flow is governed by the following steady Euler equations of conservation of mass and momentum in gas dynamics:

$$\begin{cases} (\rho u)_x + (\rho v)_y = 0, \\ (\rho u^2 + p)_x + (\rho uv)_y = 0, \\ (\rho uv)_x + (\rho v^2 + p)_y = 0, \end{cases} \tag{2.1}$$

where (x, y) are the spatial variables, ρ is the density, (u, v) is the velocity, and $p = \rho^\gamma/\gamma$, $\gamma \geq 1$, is the pressure. Under the assumption that the flow is irrotational, (2.1) can be reduced to the following system of irrotationality and conservation of mass:

$$\begin{cases} v_x - u_y = 0, \\ (\rho u)_x + (\rho v)_y = 0, \end{cases} \tag{2.2}$$

together with Bernoulli's law:

$$\rho = \left(1 - \frac{\gamma - 1}{2} q^2\right)^{\frac{1}{\gamma - 1}}, \quad \gamma > 1, \tag{2.3}$$

where q is the flow speed defined by $q^2 = u^2 + v^2$. The sound speed c is defined as

$$c^2 = p'(\rho) = 1 - \frac{\gamma - 1}{2} q^2.$$

At the cavitation point $\rho = 0$,

$$q = q_{cav} := \sqrt{\frac{2}{\gamma - 1}}.$$

At the stagnation point $q = 0$, the density reaches its maximum $\rho = 1$. Bernoulli's law (2.3) is valid for $0 \le q \le q_{cav}$. Define the critical speed q_{cr} as

$$q_{cr} := \sqrt{\frac{2}{\gamma+1}}.$$

We rewrite Bernoulli's law (2.3) in the form

$$q^2 - q_{cr}^2 = \frac{2}{\gamma+1}\left(q^2 - c^2\right). \tag{2.4}$$

We see that the flow is *subsonic when* $q < q_{cr}$, *sonic when* $q = q_{cr}$, *and supersonic when* $q > q_{cr}$. Thus, the problem of transonic flows is of mixed type. The isothermal case ($\gamma = 1$) is similar, which is omitted here.

When the upstream speed is sufficiently small, the flow remains subsonic and the governing equations are elliptic. Shiffman [16] and Bers [1] proved that, for a given upstream speed $w_\infty = (u_\infty, v_\infty)$, there exists a number (determined by the flow geometry) $\hat{q} < q_{cr}$ such that the problem has a unique subsonic solution (u,v) for $q_\infty := |w_\infty| < \hat{q}$, and the maximum speed $q_m \to q_{cr}$ as $q_\infty \to \hat{q}$. See [2] for other related results and references.

It is also indicated in [1, 16] that the maximum speed $q_m \to q_{cr}$ as $q_\infty \nearrow \hat{q}$, that is, some sonic points appear and the flow becomes subsonic-sonic. In Chen-Dafermos-Slemrod-Wang [2], we proved the following: Let $q_\infty^\varepsilon < \hat{q}$ be a sequence of speeds at infinity, and let $(u^\varepsilon, v^\varepsilon)$ be the corresponding subsonic solutions. Then, as $q_\infty^\varepsilon \nearrow \hat{q}$, the sequence $(u^\varepsilon, v^\varepsilon)$ possesses a subsequence that converges strongly to a weak solution (u,v) with $q = |(u,v)| \le q_{cr}$. In this case, we obtained a subsonic-sonic solution by the method of compensated compactness.

When $q_\infty > \hat{q}$, the flow becomes transonic and shock waves form. Morawetz presented a program in her two papers [11, 12] for proving the existence of weak solutions to the transonic flow in a channel or exterior to an airfoil. She proved that, if the solutions of some viscous approximation problem satisfy the compensated compactness framework, then there is a convergent subsequence whose limit is a solution of the transonic flow problem. The viscous approximation problem was not identified in [11, 12]. In Chen-Slemrod-Wang [3], we introduced the polar coordinates in the phase plane:

$$u = q\cos\theta, \qquad v = q\sin\theta,$$

with

$$\theta = \tan^{-1}\left(\frac{v}{u}\right)$$

being the flow angle, and provided such an viscous approximation problem:

$$\begin{cases} v_x - u_y = R_1 := \varepsilon\Delta\theta, \\ (\rho u)_x + (\rho v)_y = R_2 := \varepsilon\nabla\cdot(\sigma(\rho)\nabla\rho), \end{cases} \tag{2.5}$$

where $\sigma(\rho)$ is any positive smooth function satisfying

$$\sigma(\rho) = 1 - \frac{c^2}{q^2} \qquad \text{for } q > c.$$

In terms of (q, θ), we obtain from (2.5)

$$\begin{bmatrix} -\sin\theta & -q\cos\theta \\ \frac{c^2-q^2}{c^2 q}\cos\theta & -\sin\theta \end{bmatrix} \begin{bmatrix} q \\ \theta \end{bmatrix}_x + \begin{bmatrix} \cos\theta & -q\sin\theta \\ \frac{c^2-q^2}{c^2 q}\sin\theta & \cos\theta \end{bmatrix} \begin{bmatrix} q \\ \theta \end{bmatrix}_y = \begin{bmatrix} -R_1 \\ \frac{1}{\rho q} R_2 \end{bmatrix}. \quad (2.6)$$

The two matrices in (2.6) commute, so do their transposes. Thus, they have common eigenvectors. Their eigenvalues are

$$\lambda_\pm = -\sin\theta \pm \frac{\sqrt{q^2-c^2}}{c}\cos\theta, \quad \mu_\pm = \cos\theta \pm \frac{\sqrt{q^2-c^2}}{c}\sin\theta,$$

and the left eigenvectors of the first matrix are

$$(\mp\frac{\sqrt{q^2-c^2}}{qc}, 1).$$

Thus the Riemann invariants $W_\pm$ defined by

$$\frac{\partial W_\pm}{\partial\theta} = 1, \quad \frac{\partial W_\pm}{\partial q} = \mp\frac{\sqrt{q^2-c^2}}{qc} \qquad \text{for } q \geq c \quad (2.7)$$

satisfy

$$\lambda_\pm \frac{\partial W_\pm}{\partial x} + \mu_\pm \frac{\partial W_\pm}{\partial y} = -\frac{\partial W_\pm}{\partial q} R_1 + \frac{1}{\rho q}\frac{\partial W_\pm}{\partial\theta} R_2. \quad (2.8)$$

For a bounded domain Ω with the boundary $\partial\Omega_1$ of the obstacle and the far field boundary $\partial\Omega_2$, $\partial\Omega = \partial\Omega_1 \cup \partial\Omega_2$. In particular, we limit ourselves to the case $v_\infty = 0$ and the two types of domains Ω in Fig. 1 (a)–(b), where $\partial\Omega_1$ (the solid curve in (a) and the solid closed curve in (b)) is the boundary of the obstacle, $\partial\Omega_2$ (dashed line segments in both (a) and (b)) is the far field boundary, and Ω is the domain bounded by $\partial\Omega_1$ and $\partial\Omega_2$. This assumption implies $\theta = 0$ on $\partial\Omega_2$. In case (b), the circulation about the boundary $\partial\Omega_2$ is zero.

Consider the viscous problem (2.5) with the boundary conditions:

$$\begin{cases} \nabla\theta \cdot \mathbf{n} = 0 \qquad \text{on } \partial\Omega_1, \\ \varepsilon\sigma_2 \nabla\rho \cdot \mathbf{n} = -|\rho\,(u, v) \cdot \mathbf{n}| \qquad \text{on } \partial\Omega_1, \\ (u, v) - (u_\infty, v_\infty) = 0 \qquad \text{on } \partial\Omega_2 \text{ with } q_\infty < q_{cav}. \end{cases} \quad (2.9)$$

Then all solutions to (2.5) and (2.9) are bounded on the domain Ω staying uniformly in ε away from cavitation, i.e., there exists $q^* < q_{cav}$ such that

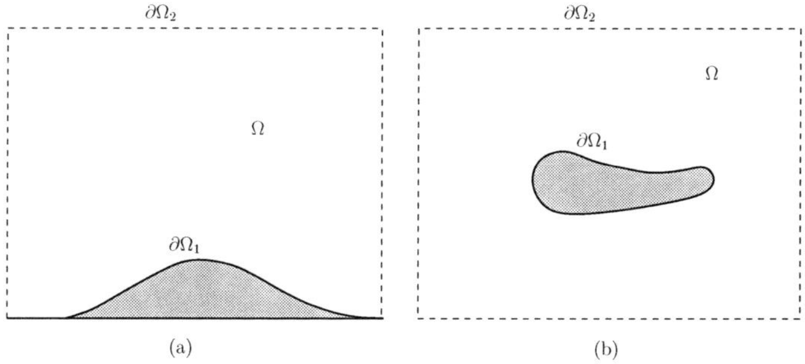

FIG. 1. *Domains.*

$q^\varepsilon \leq q^*$ that is equivalent to $\rho(q^\varepsilon) \geq \underline{\rho} > 0$ for some $\underline{\rho} = \underline{\rho}(q^*) > 0$. More specifically, when $1 \leq \gamma < 3$, the viscous approximate solutions stay in a family of "apple" shaped invariant regions. Fig. 2 shows two examples of the invariant regions.

To show the convergence of the viscous approximate solutions, we introduce the entropy pairs (Q_1, Q_2) given by the Loewner-Morawetz relation:

$$Q_1 = \rho q H_\mu \cos\theta - qH_\theta \sin\theta, \qquad Q_2 = \rho q H_\mu \sin\theta + qH_\theta \cos\theta. \tag{2.10}$$

Here H is the generator satisfying the generalized Tricomi equation:

$$H_{\mu\mu} + \frac{1}{\rho^2}(1 - M^2)H_{\theta\theta} = 0, \tag{2.11}$$

where $\mu = \mu(\rho)$ is defined by $\mu'(\rho) = c^2/q^2$, and $M = q/c$ is the Mach number. Then the corresponding entropy dissipation measures are analyzed so that the viscous approximate solutions satisfy the compensated compactness framework. Finally, the argument of Morawetz [12] leads the convergence. More precisely, in Chen-Slemrod-Wang [3], we proved the following: Let $v_\infty = 0$, $|u_\infty| < q_{cav}$, $1 \leq \gamma < 3$, and $\|\theta^\varepsilon\|_{L^\infty} \leq C$ for some C independent of ε, and assume that there exists a positive ε-independent $\alpha(\delta) \to 0$ as $\delta \to 0$ such that $q^\varepsilon(x, y) \geq \alpha(\delta)$ for any $(x, y) \in \Omega_\delta := \{(x, y) \in \Omega : dist((x, y), \partial\Omega_1) \geq \delta > 0\}$. Then the support of the Young measure $\nu_{x,y}$ strictly excludes the stagnation point $q = 0$ and reduces to a point for a.e. $(x, y) \in \Omega_\delta$ for any $\delta > 0$. Hence, the Young measure is a Dirac mass, and the sequence $(u^\varepsilon, v^\varepsilon)$ has a subsequence converging strongly in $L^2_{loc}(\Omega)^2$ to an entropy solution of the inviscid transonic flow problem (2.2).

3. Isometric embedding. We now consider the isometric embedding problem of differential geometry and show a way of using the approach for the transonic flow problem in gas dynamics to study the problem of isometric embedding.

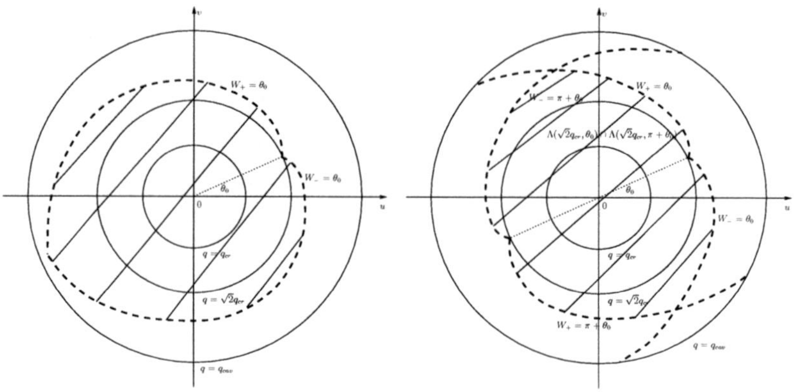

FIG. 2. *Examples of invariant regions.*

The isometric embedding of a two-dimensional Riemannian manifold into $\mathbb{R}^3$ can be studied through the Gauss-Codazzi system described as follows. Let $g_{ij}, i, j = 1, 2$, be the given metric of a two-dimensional Riemannian manifold $\mathcal{M}$ parameterized on an open set $\Omega \subset \mathbb{R}^2$. The first fundamental form I for $\mathcal{M}$ on Ω is

$$I := g_{11}(dx)^2 + 2g_{12}dxdy + g_{22}(dy)^2,$$

and the isometric embedding problem is to seek a map $\mathbf{r} : \Omega \to \mathbb{R}^3$ such that $d\mathbf{r} \cdot d\mathbf{r} = I$, that is,

$$\partial_x \mathbf{r} \cdot \partial_x \mathbf{r} = g_{11}, \quad \partial_x \mathbf{r} \cdot \partial_y \mathbf{r} = g_{12}, \quad \partial_y \mathbf{r} \cdot \partial_y \mathbf{r} = g_{22},$$

so that $\{\partial_x \mathbf{r}, \partial_y \mathbf{r}\}$ in $\mathbb{R}^3$ are linearly independent. The corresponding second fundamental form is

$$II := h_{11}(dx)^2 + 2h_{12}dxdy + h_{22}(dy)^2.$$

The fundamental theorem of surface theory (cf. [6, 7]) indicates that there exists a surface in $\mathbb{R}^3$ whose first and second fundamental forms are I and II if the coefficients (g_{ij}) and (h_{ij}) of the two given quadratic forms I and II with $(g_{ij}) > 0$ satisfy the Gauss-Codazzi system. It is indicated in Mardare [10] (Theorem 9) that this theorem holds even when (h_{ij}) is only in L^∞ for given (g_{ij}) in $C^{1,1}$, for which the immersion surface is $C^{1,1}$. This shows that, for the realization of a two-dimensional Riemannian manifold in $\mathbb{R}^3$ with given metric $(g_{ij}) > 0$, it suffices to solve $(h_{ij}) \in L^\infty$ determined by the Gauss-Codazzi system to recover $\mathbf{r}$ a posteriori. The Gauss-Codazzi system (cf. [6, 7]) can be written as:

$$\begin{cases} \partial_x M - \partial_y L = \Gamma_{22}^{(2)} L - 2\Gamma_{12}^{(2)} M + \Gamma_{11}^{(2)} N, \\ \partial_x N - \partial_y M = -\Gamma_{22}^{(1)} L + 2\Gamma_{12}^{(1)} M - \Gamma_{11}^{(1)} N, \end{cases} \tag{3.1}$$

with

$$LN - M^2 = \kappa, \tag{3.2}$$

where

$$L = \frac{h_{11}}{\sqrt{|g|}}, \quad M = \frac{h_{12}}{\sqrt{|g|}}, \quad N = \frac{h_{22}}{\sqrt{|g|}}, \quad |g| = det(g_{ij}) = g_{11}g_{22} - g_{12}^2,$$

$\kappa(x, y)$ is the Gauss curvature that is determined by the relation:

$$\kappa(x, y) = \frac{R_{1212}}{|g|}, \; R_{ijkl} = g_{lm}\left(\partial_k \Gamma_{ij}^{(m)} - \partial_j \Gamma_{ik}^{(m)} + \Gamma_{ij}^{(n)}\Gamma_{nk}^{(m)} - \Gamma_{ik}^{(n)}\Gamma_{nj}^{(m)}\right),$$

R_{ijkl} is the curvature tensor and depends on (g_{ij}) and its first and second derivatives, and

$$\Gamma_{ij}^{(k)} = \frac{1}{2} g^{kl}\left(\partial_j g_{il} + \partial_i g_{jl} - \partial_l g_{ij}\right)$$

is the Christoffel symbol and depends on the first derivatives of (g_{ij}), where the summation convention is used, (g^{kl}) denotes the inverse of (g_{ij}), and $(\partial_1, \partial_2) = (\partial_x, \partial_y)$. Therefore, given a positive definite metric $(g_{ij}) \in C^{1,1}$, the Gauss-Codazzi system gives us three equations for the three unknowns (L, M, N) determining the second fundamental form II. Note that, the Gauss curvature κ may change sign, thus the Gauss-Codazzi system (3.1)–(3.2) generically is of mixed hyperbolic-elliptic type, as in transonic flow (cf. [3]).

Although, from the viewpoint of geometry, the Gauss equation (3.2) is a constraint condition and the Codazzi equations in (3.1) are integrability relations, we can put the problem into a fluid dynamic formulation so that the isometric immersion problem may be solved via the approaches for transonic flows of fluid dynamics in Chen-Slemrod-Wang [3]. We now describe the fluid dynamic formulation of the Gauss-Codazzi system (3.1)–(3.2) in Chen-Slemrod-Wang [4]. To do this, we set

$$L = \rho v^2 + p, \quad M = -\rho uv, \quad N = \rho u^2 + p,$$

and set $q^2 = u^2 + v^2$ as usual. Then the Codazzi equations in (3.1) become the familiar balance laws of momentum:

$$\begin{cases} \partial_x(\rho uv) + \partial_y(\rho v^2 + p) = -(\rho v^2 + p)\Gamma_{22}^{(2)} - 2\rho uv\Gamma_{12}^{(2)} - (\rho u^2 + p)\Gamma_{11}^{(2)}, \\ \partial_x(\rho u^2 + p) + \partial_y(\rho uv) = -(\rho v^2 + p)\Gamma_{22}^{(1)} - 2\rho uv\Gamma_{12}^{(1)} - (\rho u^2 + p)\Gamma_{11}^{(1)}, \end{cases} \tag{3.3}$$

and the Gauss equation (3.2) becomes

$$\rho p q^2 + p^2 = \kappa. \tag{3.4}$$

We choose the pressure p as for the Chaplygin-type gas:

$$p = -\frac{1}{\rho}.$$

Then, from (3.4), we have the "Bernoulli" relation:

$$\rho = \frac{1}{\sqrt{q^2 + \kappa}}. \tag{3.5}$$

This yields

$$p = -\sqrt{q^2 + \kappa}, \tag{3.6}$$

and

$$u^2 = p(p - M), \quad v^2 = p(p - L), \quad M^2 = (N - p)(L - p),$$

which lead to the relations for (u, v) in terms of (L, M, N).

We define the "sound" speed as

$$c^2 = p'(\rho) = \frac{1}{\rho^2},$$

then, from our "Bernoulli" relation (3.5), we see

$$c^2 = q^2 + \kappa. \tag{3.7}$$

Hence, under this formulation,
(i) when $\kappa > 0$, the "flow" is subsonic, and system (3.3)–(3.4) is elliptic;
(ii) when $\kappa < 0$, the "flow" is supersonic, and (3.3)–(3.4) is hyperbolic;
(iii) when $\kappa = 0$, the "flow" is sonic, and (3.3)–(3.4) is degenerate.
In general, system (3.3)–(3.4) is of mixed hyperbolic-elliptic type.

We rewrite (3.3) as

$$\begin{cases} \partial_x(\rho uv) + \partial_y(\rho v^2 + p) = R_1, \\ \partial_x(\rho u^2 + p) + \partial_y(\rho uv) = R_2, \end{cases} \tag{3.8}$$

where R_1 and R_2 denote the right-hand sides of (3.3). Then we can write down our "rotationality-continuity equations" as

$$\begin{cases} \partial_x v - \partial_y u = \frac{1}{\rho q^2}\Big(u\big(\frac{1}{2}\rho\partial_y\kappa + R_1\big) - v\big(\frac{1}{2}\rho\partial_x\kappa + R_2\big)\Big), \\ \partial_x(\rho u) + \partial_y(\rho v) = \frac{1}{2}\frac{\rho u}{q^2}\partial_x\kappa + \frac{1}{2}\frac{\rho v}{q^2}\partial_y\kappa + \frac{v}{q^2}R_1 + \frac{u}{q^2}R_2. \end{cases} \tag{3.9}$$

In summary, the Gauss-Codazzi system (3.1)–(3.2), the momentum equations (3.3)–(3.6), and the rotationality-continuity equations (3.5) and (3.9) are all formally equivalent. It is the Gauss-Codazzi system that must be

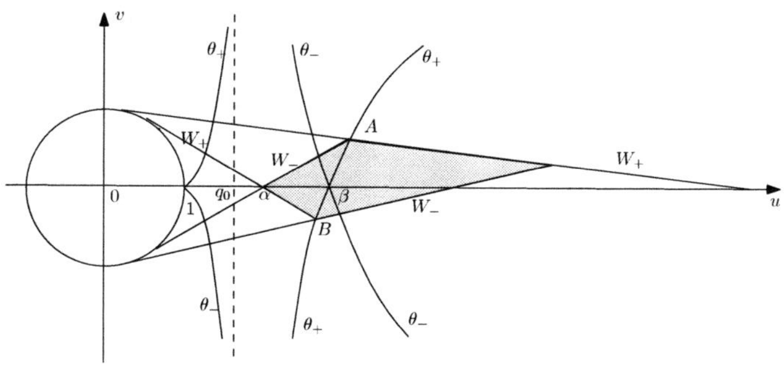

FIG. 3. *An invariant region for isometric immersion.*

solved exactly and hence the rotationality-continuity equations will become "entropy" inequalities for weak solutions.

In Chen-Slemrod-Wang [4], we considered one of the variables, x, as time-like and the other variable y as space-like, introduced a vanishing viscosity method via parabolic regularization to obtain the uniform L^∞ estimate by identifying invariant regions for the approximate solutions, and showed that the H^{-1}_{loc}–compactness can be achieved for the viscous approximate solutions. Then, as in Chen-Slemrod-Wang [3], the compensated compactness framework yields a weak solution to the initial value problem of system (3.3)–(3.4) when the initial data lies in the diamond-shaped invariant region in Fig. 3, for the catenoid with the given metric:

$$g_{11} = g_{22} = (cosh(cx))^{\frac{2}{\beta^2-1}}, \quad g_{12} = 0, \quad \kappa(x) = -\kappa_0 E(x)^{-\beta^2},$$

with $c \neq 0, \kappa_0 > 0, \beta > \sqrt{2}$. This establishes a $C^{1,1}(\mathbb{R}^2)$ immersion of the Riemannian manifold into $\mathbb{R}^3$. In particular, our existence result asserts the existence of a $C^{1,1}$-surface for the associated metric for a class of noncircular cross-sections prescribed at $x = x_0$ ($\neq 0$) for a catenoid. Possible implication of our approach may be in existence theorems for equilibrium configurations of a catenoidal shell as detailed in Vaziri-Mahedevan [18].

4. Discussions. We introduced a viscous approximation to the transonic flow problem in [3] and proved the convergence of the approximate solutions through the compensated compactness framework. In [3], we showed that the flow is away from the vacuum, but assumed that the flow angles remain bounded and the flow is away from the stagnation point. It is still open to remove these two assumptions. One may try to obtain better estimates for the viscous approximation problem in [3] or improve the viscous approximation.

In [4], we established the existence of $C^{1,1}$ isometric immersions for a class of initial data where the given metric is that of a catenoid. The existence of a global isometric embedding or immersion of a general surface with negative Gauss curvature is still open and is of course generally not possible due to Efimov's theorem (cf. [7]). When the Gauss curvature κ changes sign, the problem becomes transonic and thus mixed hyperbolic-elliptic type. In this mixed-type problem, only special local solutions are known to exist for special data (cf. [9, 7]), and the existence of global solutions is a significantly difficult open problem. An extremely challenging direction is the higher dimensional isometric embedding, and all major problems are open. See the discussion and results in Chen-Slemrod-Wang [5].

REFERENCES

[1] L. Bers, *Mathematical Aspects of Subsonic and Transonic Gas Dynamics*, John Wiley & Sons, Inc.: New York; Chapman & Hall, Ltd.: London, 1958.

[2] G.-Q. Chen, C. Dafermos, M. Slemrod, and D. Wang, *On two-dimensional sonic-subsonic flow*, Commun. Math. Phys. **271** (2007), 635–647.

[3] G.-Q. Chen, M. Slemrod, and D. Wang, *Vanishing viscosity method for transonic flow*, Arch. Rational Mech. Anal. **189** (2008), 159–188.

[4] G.-Q. Chen, M. Slemrod, and D. Wang, *Isometric immersions and compensated compactness*, Commun. Math. Phys. **294** (2010), 411–437.

[5] G.-Q. Chen, M. Slemrod, and D. Wang, *Weak continuity of the Gauss-Codazzi-Ricci system for isometric embedding*, Proc. Amer. Math. Soc. **138** (2010), 1843–1852.

[6] M. P. do Carmo, *Riemannian Geometry*, Transl. by F. Flaherty, Birkhäuser: Boston, MA, 1992.

[7] Q. Han, and J.-X. Hong, *Isometric embedding of Riemannian manifolds in Euclidean spaces*, AMS: Providence, RI, 2006.

[8] A. G. Kuz'min, *Boundary Value Problems for Transonic Flow*, Wiley, 2002.

[9] C.-S. Lin, *The local isometric embedding in $\mathbb{R}^3$ of 2-dimensional Riemannian manifolds with Gaussian curvature changing sign cleanly*, Comm. Pure Appl. Math. **39** (1986), 867–887.

[10] S. Mardare, *On Pfaff systems with L^p coefficients and their applications in differential geometry*, J. Math. Pure Appl. **84** (2005), 1659–1692.

[11] C. S. Morawetz, *On a weak solution for a transonic flow problem*, Comm. Pure Appl. Math. **38** (1985), 797–818.

[12] C. S. Morawetz, *On steady transonic flow by compensated compactness,* Methods Appl. Anal. **2** (1995), 257–268.

[13] T. H. Moulden, *Fundamentals of Transonic Flow*, Krieger Pub Co, 1990.

[14] J. Nash, *The imbedding problem for Riemannian manifolds*, Ann. Math. (2), **63** (1956), 20–63.

[15] H. J. Ramm, *Fluid Dynamics for the Study of Transonic Flow*, Oxford Engineering Science Series, Oxford University Press, 1990.

[16] M. Shiffman, *On the existence of subsonic flows of a compressible fluid*, J. Rational Mech. Anal. **1** (1952). 605–652.

[17] L. Tartar, *Compensated compactness and applications to partial differential equations. In, Nonlinear Analysis and Mechanics, Heriot-Watt Symposium IV*, Res. Notes in Math. **39**, 136–212, Pitman: Boston-London, 1979.

[18] A. Vaziri and L. Mahedevan, *Localized and extended deformations of elastic shells*, Proc. National Acad. Sci. **105** (2008), 7913–7918.

[19] S.-T. Yau, *Review of geometry and analysis*, In: Mathematics: Frontiers and Perspectives, pp. 353–401, International Mathematics Union, Eds. V. Arnold, M. Atiyah, P. Lax, and B. Mazur, AMS: Providence, 2000.

WELL POSEDNESS AND CONTROL IN MODELS BASED ON CONSERVATION LAWS

RINALDO M. COLOMBO*

Abstract. Consider a control problem based on a balance law, where a given cost, say an integral functional, needs to be minimized. Once suitable well posedness results are available, the existence of optimal controls easily follows. This presentation overviews several examples of this problem.

Key words. Hyperbolic conservation laws, optimal control of conservation laws.

AMS(MOS) subject classifications. 35L65, 49J20.

1. Introduction. This presentation is devoted to control problems of the type: maximize $\mathcal{J}(u)$ where u solves

$$\begin{cases} \partial_t u + \operatorname{div}_x f(t,x,u) = g(t,x,u) & (t,x) \in \mathbb{R}^+ \times \Omega \\ b\left(u(t,x)\right) = \psi(t) & (t,x) \in \mathbb{R}^+ \times \partial\Omega \\ u(0,x) = \bar{u}(x) & x \in \Omega\,. \end{cases} \tag{1.1}$$

$\mathcal{J}$ is an integral functional, the flow f and the source term g are sufficiently smooth, $\Omega \subseteq \mathbb{R}^N$ and $u \in \mathbb{R}^n$. The examples below show that control parameters may enter f, g, ψ or $\bar{u}$. As it is well known, a basic analytical theory for (1.1) is available only when the number n of equations and the space dimension N are in one of the two cases

$$n \geq 1 \text{ and } N = 1 \qquad \text{or} \qquad n = 1 \text{ and } N \geq 1\,.$$

2. The Case of junctions. Note preliminarily that, when $N = 1$, the general setting of (1.1) comprises also the case of junctions. Consider for instance a traffic light, say sited at $x = 0$, separating an incoming road, $x < 0$, from the outgoing one, $x > 0$. Then, the traffic densities u_1, before the traffic light, and u_2, after it, can be assumed to solve the following conservation law at a junction:

$$\begin{cases} \partial_t u_i + \partial_x f_i(u_i) = 0 & (t,x) \in \mathbb{R}^+ \times \mathbb{R},\ i = 1,2 \\ \Psi\left(u_1(t,0-), u_2(t,0+)\right) = \psi(t) & t \in \mathbb{R}^+ \\ u(0,x) = \bar{u}(x) & x \in \mathbb{R} \end{cases} \tag{2.1}$$

where f is the traffic flow, see [36, 53]. The condition Ψ at the junction prescribes the conservation of vehicles as well as other conditions, such as priority rules or the maximization of the junction efficiency, see [53]. The framework of (2.1) comprises also other situations. For instance, u_1 might

*Dept. of Mathematics, Brescia University, Italy (rinaldo@ing.unibs.it).

be the vector of the densities of mass and linear momentum of a fluid in a pipe entering a junction or an elbow, sited at $x = 0$, while u_2 is the analogous vector for the fluid in the pipe exiting the junction, with f being the flow of the p-system, see for instance [33, formula (1.1)] or [31, 32, 58]. A key role is played by the choice of the condition Ψ: depending on the specific problem under consideration, it constrains the solution to (2.1) to satisfy specific conditions. In the case of fluid dynamics, Ψ ensures the conservation of mass and that of a component of the linear momentum, see [10, 11, 31, 32, 33], or it may also describe the effect of a compressor at $x = 0$, see [40, 43].

Problem (2.1) is equivalent to (1.1) with $N = 1$, $\Omega = \mathbb{R}^+$, $g = 0$, $u = (u_1, u_2)$, $f = (f_1, f_2)$ and $b(u) = \Psi(u_1, u_2)$, see [40, Proposition 4.2]. More general junctions can be treated similarly, as well as general networks with arcs of finite length. Below, we indifferently refer to initial–boundary value problems like (1.1) or to problems at junctions, like (2.1).

3. Examples based on the p-system. Consider a fluid flowing in pipes with constant sections having a common origin at a junction or at an elbow, for instance as one of the three below. In the isentropic, or isothermal, approximation the flow along each pipe is described through the p-system:

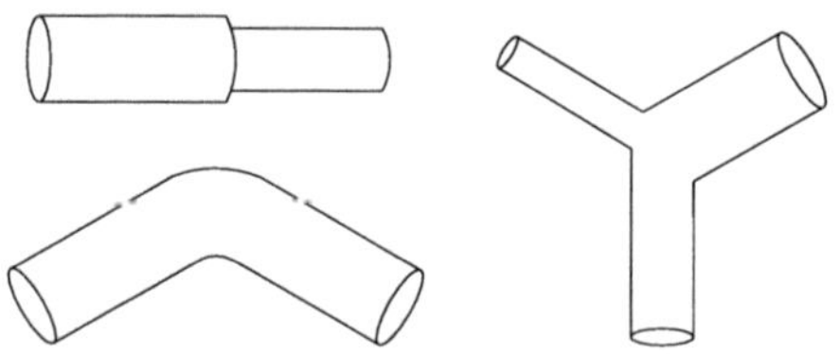

$$\begin{cases} \partial_t \rho_i + \partial_x q_i = 0 \\ \partial_t q_i + \partial_x P(\rho_i, q_i) = 0 \end{cases} \qquad P(\rho, q) = \frac{q^2}{\rho} + p(\rho) \tag{3.1}$$

where (ρ_i, q_i) are the density of mass and of linear momentum in the i-th pipe. The pressure is assumed to satisfy standard assumptions, such as [33, **(P)**], typically satisfied by the usual γ-law $p(\rho) = k\,\rho^\gamma$. Remark that q_i is the component of the linear momentum density along the axis of the i-th pipe, so that part of the geometry of the junction can be recovered in (3.1). However, the flow of gas in junctions such as those above is clearly an intrinsically $3D$ phenomenon. Nevertheless, providing a good $1D$ description may significantly shorten numerical integrations. Besides, a full analytical treatment of the p-system in $3D$ is now not available.

Physically, the core of the present $1D$ description lies in the choice of the condition to be imposed on the traces at the junction of the solutions to (3.1), namely

$$\Psi\left((\rho_1, q_1)(t, 0-); (\rho_2, q_2)(t, 0+)\right) = 0\,.$$

Here, $\Psi\colon (\mathbb{R}^+ \times \mathbb{R})^2 \to \mathbb{R}^2$ is sufficiently smooth. The first component of Ψ ensures the conservation of mass, i.e. $\Psi_1\left((\rho_1, q_1); (\rho_2, q_2)\right) = a_1 q_1 - a_2 q_2$,

a_i being the section of the i-th tube. We collect some of the choices found in the current literature for the case of 2 pipes in the table below, from [34], where qualitative properties of the solutions to (3.1) with different conditions at the junction are compared.

Ψ_2	Meaning
$a_1 P(\rho_1, q_1) - a_2 P(\rho_2, q_2)$	Partial conservation of linear momentum, see [32]
$p(\rho_r) - p(\rho_l)$	Equal pressure, typically justified at the static equilibrium, see [10, 11]
$P(\rho_1, q_1) - P(\rho_2, q_2)$	Equal dynamic pressure, see [31, 33]
$a_1 P(\rho_1, q_1) - a_2 P(\rho_2, q_2) + $ $+ \int_{a_1}^{a_2} p\left(R(\alpha; \rho_l, q_l)\right) \, d\alpha$	For two parallel pipes, limit of the condition for smooth variations of the pipes' sections, see [45, 55].

We refer also to [58] for a treatment specific of kinks and to [46, 44] for the full 3×3 system of Euler equations. Above, R is the ρ component of the stationary solution to (3.1), see [34] or [45, Proposition 2.7].

The above structure is relevant from the application point of view, due to its applicability to gas networks and pipelines, see for instance [56, 66, 67, 68, 72]. We consider now in more detail a compressor sited at a junction joining two pipes. Its role is, for example, to pump the fluid up along an inclined pipe as in the figure below.

$$\begin{cases} \partial_t \rho_i + \partial_x q_i = 0 \\ \partial_t q_i + \partial_x P_i = -\frac{\nu q_i |q_i|}{\rho_i} - \rho_i g \sin \alpha_i \end{cases}$$
$$P_i = \frac{(q_i)^2}{\rho_i} + p(\rho_i)$$

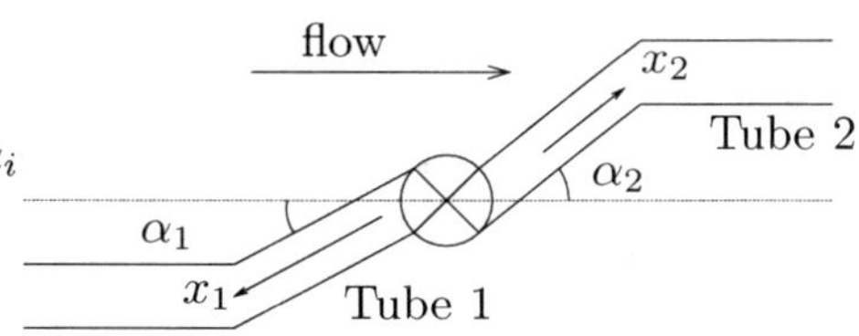

Here, α_i is the slope of the pipe, ν accounts for friction along the pipe's walls and g is gravity.

As a condition at the junction, a typical choice is that in [72, § 2.2]:

$$q_2(t, 0+) \left(\left(\frac{p\left(\rho_2(t, 0+)\right)}{p\left(\rho_1(t, 0+)\right)} \right)^{(\gamma-1)/\gamma} - 1 \right) = \Pi(t)$$

where the control Π is the power exerted from the compressor. A reasonable lower semicontinuous cost functional on the time interval $[0, T]$ is then

$$\mathcal{J}(\Pi) = \mathrm{TV}\left(\Pi; [0, T]\right) + \|\Pi\|_{\mathbf{L}^\infty([0,T];\mathbb{R})} + \int_0^T \int_a^b \left| p(\rho_2(t, x; \Pi) - \bar{p} \right| \, dx \, dt$$

where $\bar p$ is the desired gas pressure along the stretch $[a,b]$ of the second pipe. Here, as usual, $\mathrm{TV}\left(\Pi;[0,T]\right)$ denotes the total variation of the function Π over the time interval $[0,T]$. The first two terms in the right hand side above tend to penalize variations in the compressor power and to minimize its consumption, see [40, 43].

Along a river or a canal, an underflow gate as in the figure below can also be described by the p-system, see [51]:

$$\begin{cases} \partial_t H_i + \partial_x Q_i = 0 \\ \partial_t Q_i + \partial_x P_i = -\frac{\nu Q_i |Q_i|}{H_i} - H_i g \sin\alpha_i \end{cases}$$
$$P_i = (Q_i)^2/H_i + (g/2)(H_i)^2$$

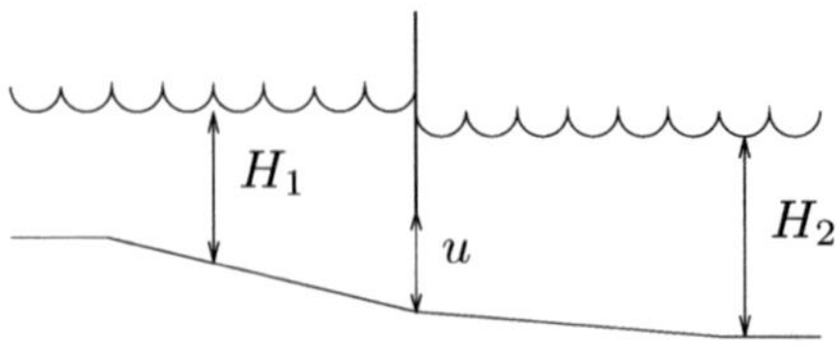

Here, H_i is the height of water in the i-th part of the canal; g, ν and α_i are as above. The system is controlled through u, which is the height of the opening in the gate.

The condition at the junction consists of an equation for the conservation of water together with

$$\left(H_1(t,0-) - H_2(t,0+)\right)\, u(t) = \left(Q_1(t,0-)\right)^2 .$$

Typically, one wants to minimize variations in the water level downstream the gate while ensuring a suitable through flow $\bar Q$. Therefore, a reasonable cost functional is

$$\mathcal{J}(u) = \int_0^T \int_a^b \left|Q_2(t,x;u) - \bar Q\right| \mathrm{d}x\, \mathrm{d}t + \int_0^T \int_{\mathbb{R}+} w(x) |\partial_x H_2| .$$

Above, $|\partial_x H_2|$ is the measure theoretic total variation of the space derivative of H_2 and $w \in \mathbf{C}_\mathbf{c}^\infty(\mathbb{R}^+;\mathbb{R}^+)$ is a suitable weight. The cost $\mathcal{J}$ is lower semicontinuous as a function of the solution (H,Q), see [37, Lemma 2.1 and Theorem 2.2].

Several other applications of $1D$ systems of conservation laws are found in the literature. We refer to [21] for a model related to blood flow; to [9, 13, 25, 36] for traffic flow models and to [53] for the treatment of road networks; to [3, 4, 41, 57, 70] for systems describing the movement of granular matter. A wide literature is concerned with the propagation of phase boundaries in fluids, see for instance [1, 26, 64, 71, 74]. In this context, the continuous dependence from the kinetic relation was proved in [29]. This structure also applies to detonation and deflagration phenomena, see for instance [27, 75].

In the case $N = 1$ and $n \geq 1$, the existence of an optimal control in the examples above readily follows, as soon as the solution to (1.1) is proved to depend continuously from the various control parameters. The global in time existence of **BV** solutions to $1D$ hyperbolic systems of conservation laws was proved in [54].

The $\mathbf{L^1}$ Lipschitz continuous dependence of the solution from the initial data was proved in the 2×2 case in [18] and in the general case in [14, 19, 20], see also [17, 50].

The $\mathbf{L^1}$ Lipschitz continuous dependence of the solutions from the $\mathbf{L^1}$ distance between boundary data and from the $\mathbf{C^0}$ distance between the boundary profiles was obtained in the 2×2 case in [2]. In the case of a non-characteristic boundary, this result was extended to the $n \times n$ case in [39, 52]. The case of junctions was specifically investigated in [40, 43].

In the case of general hyperbolic conservation laws generating Standard Riemann Semigroups [17, Definition 9.1], the $\mathbf{L^1}$ Lipschitz dependence of the solutions from the $\mathbf{C^0}$ distance between the Jacobians of the flows was obtained in [15].

Most of the above results rely on proving suitable estimates, first on approximate solutions and then passing to the limit. Wave Front Tracking proved to be a very effective procedure to construct approximate solutions to conservation laws, see [17, 50, 59]. Other tools are Glimm scheme, see [54], and vanishing viscosity, see [14]. To pass to balance laws, a standard strategy is based on operator splitting, see [38, § 3.3] or [23, 28, 73]. A general framework that allows to comprise problems with local/non-local sources and/or boundary and/or junctions is in [38, § 3.1] and [39].

4. 1D Conservation laws with unilateral constraints. The traffic flow along a rectilinear one-way road is often described through the Lighthill-Whitham [65] and Richards [69] model

$$\partial_t \rho + \partial_x f(\rho) = 0 \qquad\qquad f(\rho) = \rho\, v(\rho)\,,$$

where the traffic speed v is assumed to be a known function of the traffic density ρ. A typical choice can be $v(\rho) = V \cdot \big(1 - \rho/R\big)$, V being the maximal speed and R the maximal density, see also [35, (R1)] for a more general condition on the flow f.

The effect of a toll gate sited at, say, x_r is to limit the flow of traffic below a threshold $q_r = q_r(t)$. We thus obtain the Cauchy problem

$$\left\{\begin{array}{l} \partial_t \rho + \partial_x f(\rho) = 0 \\ \rho(0,x) = \bar\rho(x) \\ f\left(\rho(t,x_r)\right) \leq q_r(t) \end{array}\right.$$

$$f(\rho) = V\,\rho\,(1 - \rho/R)$$

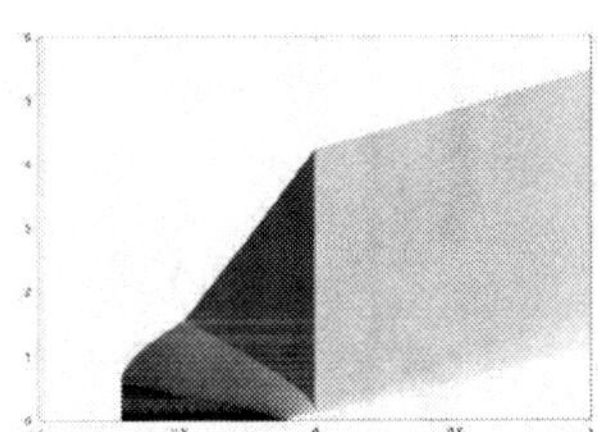

where the (x,t) diagram above on the right corresponds to $x_r = 0$, $\bar\rho = \chi_{[-0.7,-0.1]}$, $R = 1$ and $V = 1$. From the traffic point of view, it is more realistic to consider the initial boundary value problem with both the inflow and the constraint q_r periodic in time. We thus obtain the following situation:

$$\left\{\begin{array}{l} \partial_t \rho + \partial_x f(\rho) = 0 \\ \rho(0,x) = 0 \\ f\left(\rho(t,0)\right) = q_o(t) \\ f\left(\rho(t,x_r)\right) \leq q_r(t) \end{array}\right.$$
$$f(\rho) = V\,\rho\,(1-\rho/R).$$

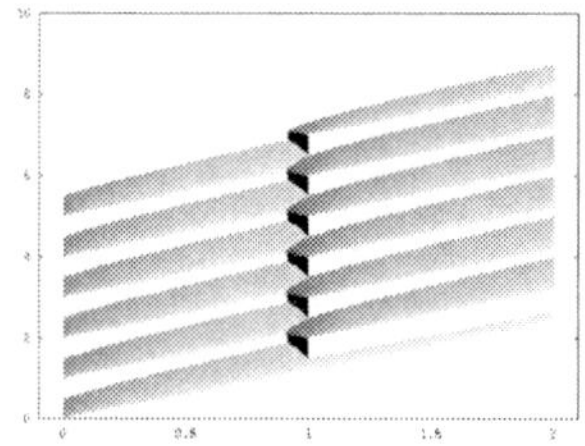

Here, both q_o and q_r vary between 0 and the maximal flow 1/4.

The basic analytical issues concerning the well posedness of these problems are solved in [6, 35]. Various control problems can now be posed, such as minimizing the travel time, see [5] or the variations in the traffic speed, see [37]. Furthermore, assume that road constructions hinder the flow of traffic at x_c, with $x_c > x_r$, see the figure below. Then, describing the effects of the road construction by means of a further unilateral constraint, we are lead to

$$\left\{\begin{array}{l} \partial_t \rho + \partial_x f(\rho) = 0 \\ \rho(0,x) = 0 \\ f\left(\rho(t,0)\right) = q_o(t) \\ f\left(\rho(t,x_r)\right) \leq q_r(t) \\ f\left(\rho(t,x_c)\right) \leq q_c \end{array}\right.$$
$$f(\rho) = V\,\rho\,(1-\rho/R).$$

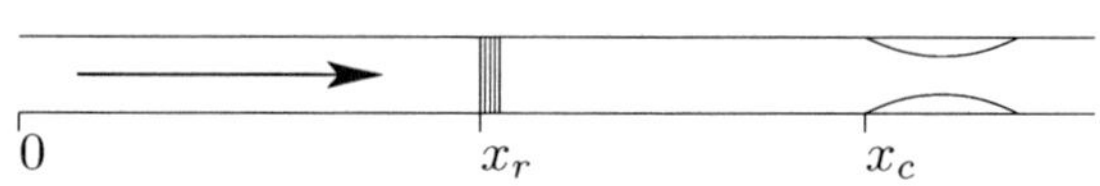

An analytical and numerical study of this problem is in [36].

In the study of scalar conservation laws with unilateral constraints, a suitable mixture of Kružkov technique, see [63], with wave front tracking, see [17, 49, 50], proved to be effective, see [6, 35, 36].

5. MultiD scalar conservation laws. A natural application of scalar conservation laws in $2D$ is provided by the modeling of crowd dynamics, see [24, 30, 60, 61]. In this case, ρ is the pedestrian density, which is assumed to solve a conservation law of the type, for instance,

$$\left\{\begin{array}{lrcl} \partial_t \rho + \operatorname{div} f(x,\rho) = 0 & (t,x) & \in & \mathbb{R}^+ \times \Omega \\ \rho(0,x) = \bar\rho(x) & x & \in & \Omega \\ f\left(x,\rho(t,x)\right)\cdot\nu(x) = \psi(t) & (t,x) & \in & \mathbb{R}^+ \times \mathcal{D} \\ f\left(x,\rho(t,x)\right)\cdot\nu(x) = 0 & (t,x) & \in & \mathbb{R}^+ \times \mathcal{W} \end{array}\right. \tag{5.1}$$

with $\Omega \subseteq \mathbb{R}^2$, $\nu(x)$ is the interior normal to $\partial\omega$ at x, $\mathcal{D} \subseteq \partial\Omega$ is the door and $\mathcal{W} \subseteq \partial\Omega$ is the wall, with $\mathcal{D}\cup\mathcal{W} = \partial\Omega$ and $\mathcal{D}\cap\mathcal{W} = \emptyset$. $\psi(t)$ is the flow of people entering Ω at time t. Note however that a unilateral constraint $f\left(x, \rho(t,x)\right) \cdot \nu(x) \leq \psi(t)$ might often be more suitable, in particular in the case of people exiting Ω.

Problems of this type apparently received little attention from the mathematical community. In the case of the Cauchy problem

$$\begin{cases} \partial_t \rho + \operatorname{div} f(x,\rho) = 0 \\ \rho(0,x) = \bar\rho(x) \end{cases}$$

the existence of solutions, as well as the fact that the resulting semigroup is non expansive in $\mathbf{L^1}$, was proved in the classical paper by Kružkov [63], whereas the case of bounded domains was dealt with in [12]. However, the stability of solutions with respect to the flow was proved only recently in [47]. Other results in the literature provided similar estimates in the case of particular flows, for instance $f(x,\rho) = a(x)\, b(\rho)$, or required *a priori* bounds on the total variation of the solution, see [16, 22, 62].

Recent results in this direction considers the case $N \geq 1$ and $n = 1$ of (1.1) with a *nonlocal* flow, see [42, 48]. This case is motivated by examples from pedestrian dynamics, see [42, § 4], as well as supply chain management, see [7, 8]. In [42, Theorem 2.10] it is proved that the Cauchy problem for the continuity equation

$$\begin{cases} \partial_t \rho + \operatorname{div}\left(\rho\, V(\rho)\right) = 0 \\ \rho(0,x) = \rho_o(x) \end{cases} \tag{5.2}$$

with $V \colon \mathbf{L^1} \mapsto \mathbf{C^2}$ being a *nonlocal* operator, generates a semigroup S in the sense that $t \mapsto S_t \rho_o$ is the solution to (5.2). Under strong regularity assumptions on v, [42, Theorem 2.10] proves that S_t is differentiable with respect to ρ_o and that its derivative computed at ρ_o in the direction r is characterized by $\left(DS_t(\rho_o)\right) r = \Sigma_t^{\rho_o} r$. Here Σ^{ρ_o} is the semigroup generated by the *linearized* equation

$$\begin{cases} \partial_t r + \operatorname{div}\left(r\, V(\rho) + \rho\left(DV(\rho)\right)(r)\right) = 0 \\ r(0,x) = r_o\,. \end{cases}$$

A necessary condition for the optimal control of integral functionals then follows. In [48] another existence result is proved by means also of $\mathbf{L^2}$ techniques.

6. Open problems. It is clear from the above presentation that several issues, to the present knowledge of the author, are still unanswered.

Concerning the dependence of the solution to (1.1) from the various quantities appearing therein, not all situations have been fully considered, in particular in the case of (5.1).

Concerning optimal control problems, all the results above ensure the existence of such a control. Finding it, either through suitable necessary conditions or through approximate constructive procedures, is a problem still open in many cases when non smooth solutions arise and may develop interacting shocks. The existence of closed–loop, or feedback, controls is also of great interest in most of the examples cited above.

Two other research areas seem worth being considered: the inverse problem and stochastic evolutions. The former is of interest in particular in those situations, such as traffic modeling, in which the various parameters entering the equations are not motivated *a priori* from physics. The latter seems unavoidable when trying to provide a macroscopic description of phenomena that are inherently microscopic, an example being nucleation in phase transitions.

REFERENCES

[1] R. Abeyaratne and J.K. Knowles, *Kinetic relations and the propagation of phase boundaries in solids*, Arch. Rat. Mech. Anal., **114** (1991), pp. 119–154.

[2] D. Amadori and R.M. Colombo, *Continuous dependence for* 2×2 *conservation laws with boundary*, J. Differential Equations, **138** (1997), pp. 229–266.

[3] D. Amadori and W. Shen, *Global existence of large bv solutions in a model of granular flow*, Submitted (2009).

[4] ———, *Slow erosion limit in model of granular flow*, Submitted (2009).

[5] F. Ancona and A. Marson, *Scalar non-linear conservation laws with integrable boundary data*, Nonlinear Anal., **35** (1999), pp. 687–710.

[6] B. Andreianov, P. Goatin, and N. Seguin, *Finite volume schemes for locally constrained conservation laws.*, Submitted (2009).

[7] D. Armbruster, P. Degond, and C. Ringhofer, *A model for the dynamics of large queuing networks and supply chains*, SIAM J. Appl. Math., **66** (2006), pp. 896–920 (electronic).

[8] D. Armbruster, D.E. Marthaler, C. Ringhofer, K. Kempf, and T.-C. Jo, *A continuum model for a re-entrant factory*, Oper. Res., **54** (2006), pp. 933–950.

[9] A. Aw and M. Rascle, *Resurrection of "second order" models of traffic flow*, SIAM J. Appl. Math., **60** (2000), pp. 916–938 (electronic).

[10] M.K. Banda, M. Herty, and A. Klar, *Coupling conditions for gas networks governed by the isothermal Euler equations*, Netw. Heterog. Media, 1 (2006), pp. 295–314 (electronic).

[11] ———, *Gas flow in pipeline networks*, Netw. Heterog. Media, **1** (2006), pp. 41–56 (electronic).

[12] C. Bardos, A.Y. le Roux, and J.-C. Nédélec, *First order quasilinear equations with boundary conditions*, Comm. Partial Differential Equations, **4** (1979), pp. 1017–1034.

[13] S. Benzoni-Gavage and R.M. Colombo, *An n-populations model for traffic flow*, European J. Appl. Math., **14** (2003), pp. 587–612.

[14] S. Bianchini and A. Bressan, *Vanishing viscosity solutions of nonlinear hyperbolic systems*, Ann. of Math. (2), **161** (2005), pp. 223–342.

[15] S. Bianchini and R.M. Colombo, *On the stability of the standard Riemann semigroup*, Proc. Amer. Math. Soc., **130** (2002), pp. 1961–1973 (electronic).

[16] F. Bouchut and B. Perthame, *Kružkov's estimates for scalar conservation laws revisited*, Trans. Amer. Math. Soc., **350** (1998), pp. 2847–2870.

[17] A. Bressan, *Hyperbolic systems of conservation laws*, Vol. **20** of Oxford Lecture Series in Mathematics and its Applications, Oxford University Press, Oxford, 2000. The one-dimensional Cauchy problem.

[18] A. Bressan and R.M. Colombo, *The semigroup generated by 2×2 conservation laws*, Arch. Rat. Mech. Anal., **133** (1995), pp. 1–75.

[19] A. Bressan, G. Crasta, and B. Piccoli, *Well-posedness of the Cauchy problem for $n \times n$ systems of conservation laws*, Mem. Amer. Math. Soc., **146** (2000).

[20] A. Bressan, T.-P. Liu, and T. Yang, *L^1 stability estimates for $n \times n$ conservation laws*, Arch. Rat. Mech. Anal., **149** (1999), pp. 1–22.

[21] S. Čanić and E.H. Kim, *Mathematical analysis of the quasilinear effects in a hyperbolic model blood flow through compliant axi-symmetric vessels*, Math. Methods Appl. Sci., **26** (2003), pp. 1161–1186.

[22] G.-Q. Chen and K.H. Karlsen, *Quasilinear anisotropic degenerate parabolic equations with time-space dependent diffusion coefficients*, Comm. Pure Appl. Anal., **4** (2005), pp. 241–266.

[23] P.R. Chernoff, *Product formulas, nonlinear semigroups, and addition of unbounded operators*, American Mathematical Society, Providence, R. I., 1974. Memoirs of the American Mathematical Society, No. 140.

[24] R.R. Clements and R.L. Hughes, *Mathematical modelling of a mediaeval battle: the Battle of Agincourt, 1415*, Math. Comput. Simul., **64** (2004), pp. 259–269.

[25] R.M. Colombo, *Hyperbolic phase transitions in traffic flow.*, SIAM J. Appl. Math., **63** (2002), pp. 708–721.

[26] R.M. Colombo and A. Corli, *Continuous dependence in conservation laws with phase transitions*, SIAM Journal Math. Anal., **31** (1999), pp. 34–62.

[27] ———, *Sonic hyperbolic phase transitions and Chapman-Jouguet detonations*, J. Differential Equations, **184** (2002), pp. 321–347.

[28] ———, *A semilinear structure on semigroups in a metric space*, Semigroup Forum, **68** (2004), pp. 419–444.

[29] ———, *Stability of the Riemann semigroup with respect to the kinetic condition*, Quart. Appl. Math., **62** (2004), pp. 541–551.

[30] R.M. Colombo, G. Facchi, G. Maternini, and M.D. Rosini, *On the continuum modeling of crowds*, in Hyperbolic Problems: Theory, Numerics and Applications, E. Tadmor, L. Jian-Guo, and A.E. Tzavaras, eds., Vol. **67**, June 2010.

[31] R.M. Colombo and M. Garavello, *A well posed Riemann problem for the p-system at a junction*, Netw. Heterog. Media, **1** (2006), pp. 495–511 (electronic).

[32] ———, *On the p-system at a junction*, in Control methods in PDE-dynamical systems, Vol. **426** of Contemp. Math., Amer. Math. Soc., Providence, RI, 2007, pp. 193–217.

[33] ———, *On the Cauchy problem for the p-system at a junction*, SIAM J. Math. Anal., **39** (2008), pp. 1456–1471.

[34] ———, *On the 1d modeling of fluid flowing through a junction*, Subitted (2009).

[35] R.M. Colombo and P. Goatin, *A well posed conservation law with a variable unilateral constraint*, J. Differential Equations, **234** (2007), pp. 654–675.

[36] R.M. Colombo, P. Goatin, and M.D. Rosini, *On the modeling and management of traffic*, submitted (2009).

[37] R.M. Colombo and A. Groli, *Minimising stop and go waves to optimise traffic flow*, Appl. Math. Lett., **17** (2004), pp. 697–701.

[38] R.M. Colombo and G. Guerra, *Hyperbolic balance laws with a dissipative non local source*, Comm. Pure Appl. Anal., **7** (2008), pp. 1077–1090.

[39] ———, *On general balance laws with boundary*, J. Differential Equations, **248**, 1017–1043, 2010.

[40] R.M. Colombo, G. Guerra, M. Herty, and V. Sachers, *Modeling and optimal control of networks of pipes and canals*, SIAM J. Math. Anal., **48** (2009), pp. 2032–2050.

[41] R.M. Colombo, G. Guerra, and F. Monti, *Modeling the dynamics of granular matter*. Preprint, 2009.

[42] R.M. Colombo, M. Herty, and M. Mercier, *Control of the continuity equation with a non local flow*. To appear on ESAIM, COCV.

[43] R.M. Colombo, M. Herty, and V. Sachers, *On* 2×2 *conservation laws at a junction*, SIAM J. Math. Anal., **40** (2008), pp. 605–622.

[44] R.M. Colombo and F. Marcellini, *Coupling conditions for the* 3×3 *Euler system.* To appear on Networks and Heterogeneous Media.

[45] ———, *Smooth and discontinuous junctions in the p-system*, J. Math. Anal. Appl., **361** (2010), pp. 440–456.

[46] R.M. Colombo and C. Mauri, *Euler system for compressible fluids at a junction*, J. Hyperbolic Differ. Equ., **5** (2008), pp. 547–568.

[47] R.M. Colombo, M. Mercier, and M.D. Rosini, *Stability and total variation estimates on general scalar balance laws*, Comm. Math. Sci., **7** (2009), pp. 37–65.

[48] J.-M. Coron, M. kawski, and Z. Wang, *Analysis of a conservation law modeling a highly re-entrant manufacturing system*, 2009.

[49] C.M. Dafermos, *Polygonal approximations of solutions of the initial value problem for a conservation law*, J. Math. Anal. Appl., **38** (1972), pp. 33–41.

[50] ———, *Hyperbolic conservation laws in continuum physics*, Vol. **325** of Grundlehren der Mathematischen Wissenschaften [Fundamental Principles of Mathematical Sciences], Springer-Verlag, Berlin, second ed., 2005.

[51] J. de Halleux, C. Prieur, J.-M. Coron, B. d'Andréa Novel, and G. Bastin, *Boundary feedback control in networks of open channels*, Automatica J. IFAC, **39** (2003), pp. 1365–1376.

[52] C. Donadello and A. Marson, *Stability of front tracking solutions to the initial and boundary value problem for systems of conservation laws*, NoDEA Nonlinear Differential Equations Appl., **14** (2007), pp. 569–592.

[53] M. Garavello and B. Piccoli, *Traffic flow on networks*, Vol. **1** of AIMS Series on Applied Mathematics, American Institute of Mathematical Sciences (AIMS), Springfield, MO, 2006. Conservation laws models.

[54] J. Glimm, *Solutions in the large for nonlinear hyperbolic systems of equations*, Comm. Pure Appl. Math., **18** (1965), pp. 697–715.

[55] G. Guerra, F. Marcellini, and V. Schleper, *Balance laws with integrable unbounded source*, SIAM J. Math. Anal. (to appear).

[56] M. Gugat, G. Leugering, and E.J.P. Georg Schmidt, *Global controllability between steady supercritical flows in channel networks*, Math. Methods Appl. Sci., **27** (2004), pp. 781–802.

[57] K. Hadeler and C. Kuttler, *Dynamical models for granular matter*, Granular Matter, **2** (1999).

[58] H. Holden and N.H. Risebro, *Riemann problems with a kink*, SIAM J. Math. Anal., **30** (1999), pp. 497–515 (electronic).

[59] ———, *Front tracking for hyperbolic conservation laws*, Vol. **152** of Applied Mathematical Sciences, Springer-Verlag, New York, 2002.

[60] R.L. Hughes, *A continuum theory for the flow of pedestrians*, Transportation Research Part B, **36** (2002), pp. 507–535.

[61] ———, *The flow of human crowds*, in Annual review of fluid mechanics, Vol. **35**, Vol. **35** of Annu. Rev. Fluid Mech., Palo Alto, CA, 2003, pp. 169–182.

[62] K.H. Karlsen and N.H. Risebro, *On the uniqueness and stability of entropy solutions of nonlinear degenerate parabolic equations with rough coefficients*, Discrete Cont. Dyn. Syst., **9** (2003), pp. 1081–1104.

[63] S.N. Kružhkov, *First order quasilinear equations with several independent variables.*, Mat. Sb. (N.S.), **81** (123) (1970), pp. 228–255.

[64] P. LeFloch, *Propagating phase boundaries: formulation of the problem and existence via the Glimm method*, Arch. Rat. Mech. Anal., **123** (1993), pp. 153–197.

[65] M.J. Lighthill and G.B. Whitham, *On kinematic waves. II. A theory of traffic flow on long crowded roads*, Proc. Roy. Soc. London. Ser. A., **229** (1955), pp. 317–345.

[66] M. Luskin, *An approximation procedure for nonsymmetric, nonlinear hyperbolic systems with integral boundary conditions*, SIAM J. Numer. Anal., **16** (1979), pp. 145–164.

[67] ———, *On the existence of global smooth solutions for a model equation for fluid flow in a pipe*, J. Math. Anal. Appl., **84** (1981), pp. 614–630.

[68] M. Luskin and B. Temple, *The existence of a global weak solution to the nonlinear waterhammer problem*, Comm. Pure Appl. Math., **35** (1982), pp. 697–735.

[69] P.I. Richards, *Shock waves on the highway*, Operations Res., **4** (1956), pp. 42–51.

[70] W. Shen, *On the shape of avalanches*, J. Math. Anal. Appl., **339** (2008), pp. 828–838.

[71] M. Slemrod, *A limiting "viscosity" approach to the Riemann problem for materials exhibiting change of phase*, Arch. Rat. Mech. Anal., **105** (1989), pp. 327–365.

[72] M.C. Steinbach, *On PDE solution in transient optimization of gas networks*, J. Comput. Appl. Math., **203** (2007), pp. 345–361.

[73] H.F. Trotter, *On the product of semi-groups of operators*, Proc. Amer. Math. Soc., **10** (1959), pp. 545–551.

[74] L. Truskinovsky, *About the "normal growth" approximation in the dynamical theory of phase transitions*, Cont. Mech. Thermodyn., **6** (1994), pp. 185–208.

[75] L. Truskinovsky, *Transition to detonation in dynamic phase changes*, Arch. Rat. Mech. Anal., **125** (1994), pp. 375–397.

HOMOGENIZATION OF NONLINEAR PARTIAL DIFFERENTIAL EQUATIONS IN THE CONTEXT OF ERGODIC ALGEBRAS: RECENT RESULTS AND OPEN PROBLEMS

HERMANO FRID* AND JEAN SILVA†

Abstract. We review recent results on homogenization of nonlinear partial differential equations in ergodic algebras. We describe some open problems concerning the general theory of with mean value (algebras w.m.v., in short). We also prove a new result establishing the invariance of any algebra w.m.v. under the flow of a Lipschitz continuous divergence-free vector field whose components belong to the corresponding algebra w.m.v., thus solving a problem left open by the authors in a previous paper.

Key words. Homogenization, Young measures, two-scale, almost periodic functions.

1. Introduction. The purpose of this paper is to review some recent results on homogenization of nonlinear differential equations, specially the degenerate parabolic-hyperbolic equations, in ergodic algebras, and to describe some open problems. We also prove a new result (see Theorem 7.1) establishing the invariance of any algebra w.m.v. under the flow of a Lipschitz continuous divergence-free vector field whose components belong to the corresponding algebra w.m.v. This closes a question left open by the authors in [9]. Algebras w.m.v. and ergodic algebras were introduced in [15]. The main tool for the homogenization results described in this paper is the parametrized family of two-scale Young measures. The latter was introduced in homogenization problems in the periodic setting in [7], as a nonlinear extension of the concept of two-scale convergence introduced in [13] and further developed in [1].

2. Ergodic algebras. In this section we recall some basic facts about algebras with mean values and ergodic algebras that will be needed for the purposes of this paper. To begin with, we recall the notion of mean value for functions defined in $\mathbb{R}^n$.

DEFINITION 2.1. *Let $g \in L^1_{loc}(\mathbb{R}^n)$. A number $M(g)$ is called the* mean value of g *if for all Borel set A, with n-dimensional Lebesgue measure $|A| \neq 0$,*

$$\lim_{t\to\infty} \frac{1}{t^n|A|} \int_{A_t} g(x)\,dx = M(g), \tag{2.1}$$

*Instituto de Matemática Pura e Aplicada - IMPA Estrada Dona Castorina, 110 Rio de Janeiro, RJ, 22460-320, Brazil (hermano@impa.br). H. Frid gratefully acknowledges the support of CNPq, grant 306137/2006-2 , and FAPERJ, grant E-26/152.192-2002.

†Instituto de Matemática Pura e Aplicada - IMPA Estrada Dona Castorina, 110 Rio de Janeiro, RJ, 22460-320, Brazil (jcsmat@impa.br).

where $A_t := \{x \in \mathbb{R}^n : t^{-1}x \in A\}$ *for* $t > 0$. *We also use the notation* $⨍_{\mathbb{R}^n} g\,dx$ *for* $M(g)$.

As usual, we denote by BUC($\mathbb{R}^n$) the space of the bounded uniformly continuous real-valued functions in $\mathbb{R}^n$.

We recall now the definition of algebra with mean value introduced in [15].

DEFINITION 2.2. *Let* $\mathcal{A}$ *be a linear subspace of* BUC($\mathbb{R}^n$). *We say that* $\mathcal{A}$ *is an* algebra with mean value *(or* algebra w.m.v., *in short), if the following conditions are satisfied:*

(A) If f *and* g *belong to* $\mathcal{A}$, *then the product* fg *belongs to* $\mathcal{A}$.

(B) $\mathcal{A}$ *is invariant under the translations* $\tau_y : \mathbb{R}^n \to \mathbb{R}^n$, $x \mapsto x + y$, $y \in \mathbb{R}^n$, *that is, if* $f \in \mathcal{A}$, *then* $\tau_y f \in \mathcal{A}$, *for all* $y \in \mathbb{R}^n$, *where* $\tau_y f := f \circ \tau_y$, $f \in \mathcal{A}$.

(C) Any $f \in \mathcal{A}$ *possesses a mean value.*

(D) $\mathcal{A}$ *is closed in* BUC($\mathbb{R}^n$) *and contains the unity, i.e., the function* $e(x) := 1$ *for* $x \in \mathbb{R}^n$.

For $1 \le p < \infty$, the generalized Besicovitch space $\mathcal{B}^p$ is defined as the abstract completion of the algebra $\mathcal{A}$ with respect to the Besicovitch seminorm

$$|f|_p := \left(⨍_{\mathbb{R}^n} |f|^p\,dx \right)^{1/p}.$$

Both the action of translations and the mean value extend by continuity to $\mathcal{B}^p$, and we will keep using the notation $\tau_y f$ and $M(f)$ even when $f \in \mathcal{B}^p$. Furthermore, for $p > 1$ the product in the algebra extends to a bilinear operator from $\mathcal{B}^p \times \mathcal{B}^q$ into $\mathcal{B}^1$, with q equal to the dual exponent of p, satisfying

$$|fg|_1 \le |f|_p |g|_q.$$

In particular, the operator $M(fg)$ provides a nonnegative definite bilinear form on $\mathcal{B}^2$.

Since there is an obvious inclusion between elements of this family of spaces, we may define the space $\mathcal{B}^\infty$ as follows:

$$\mathcal{B}^\infty = \{f \in \bigcap_{1 \le p < \infty} \mathcal{B}^p \,:\, \sup_{1 \le p < \infty} |f|_p < \infty\}.$$

We endow $\mathcal{B}^\infty$ with the (semi)norm

$$|f|_\infty := \sup_{1 \le p < \infty} |f|_p.$$

Obviously the corresponding quotient spaces for all these spaces (with respect to the null space of the seminorms) are Banach spaces, and in the case $p = 2$ we obtain a Hilbert space. We denote by $\stackrel{\mathcal{B}^p}{=}$, the equivalence relation given by the equality in the sense of the $\mathcal{B}^p$ semi-norm. We will keep the notation $\mathcal{B}^p$ also for the corresponding quotient spaces.

REMARK 2.1. A classical argument going back to Besicovitch [4] (see also [12], p.239) shows that the elements of $\mathcal{B}^p$ can be represented by functions in $L^p_{\mathrm{loc}}(\mathbb{R}^n)$, $1 \le p < \infty$.

We next recall a result established in [3] which provides a connection between algebras with mean value and the algebra $C(\mathcal{K})$ of continuous functions on a certain compact (Hausdorff) topological space. We state here only the parts of the corresponding result in [3] that will be used in this paper.

THEOREM 2.1 (cf. [3]). *For an algebra $\mathcal{A}$, we have:*

(i) There exist a compact space $\mathcal{K}$ and an isometric isomorphism i identifying $\mathcal{A}$ with the algebra $C(\mathcal{K})$ of continuous functions on $\mathcal{K}$.

(ii) The translations $\tau_y : \mathbb{R}^n \to \mathbb{R}^n$, $\tau_y x = x + y$, induce a family of homeomorphisms $T(y) : \mathcal{K} \to \mathcal{K}$, $y \in \mathbb{R}^n$, satisfying the group properties $T(0) = I$, $T(x+y) = T(x) \circ T(y)$, such that the mapping $T : \mathbb{R}^n \times \mathcal{K} \to \mathcal{K}$, $T(y, z) := T(y)z$, is continuous.

(iii) The mean value on $\mathcal{A}$ extends to a Radon probability measure $\mathfrak{m}$ on $\mathcal{K}$ defined by

$$\int_{\mathcal{K}} i(f)\, d\mathfrak{m} := \fint_{\mathbb{R}^n} f\, dx, \qquad f \in \mathcal{A},$$

which is invariant by the group of homeomorphisms $T(y) : \mathcal{K} \to \mathcal{K}$, $y \in \mathbb{R}^n$, that is, $\mathfrak{m}(T(y)E) = \mathfrak{m}(E)$ for all Borel sets $E \subseteq \mathcal{K}$.

(iv) For $1 \le p \le \infty$, the Besicovitch space $\mathcal{B}^p / \stackrel{\mathcal{B}^p}{=}$ is isometrically isomorphic to $L^p(\mathcal{K}, \mathfrak{m})$.

A function $f \in \mathcal{B}^2$ is said to be *invariant* if $\tau_y f \stackrel{\mathcal{B}^2}{=} f$, for all $y \in \mathbb{R}^n$. More clearly, $f \in \mathcal{B}^2$ is invariant if

$$M\big(|\tau_y f - f|^2\big) = 0, \qquad \text{for all } y \in \mathbb{R}^n. \tag{2.2}$$

The concept of ergodic algebra is then introduced as follows.

DEFINITION 2.3. *An algebra w.m.v. $\mathcal{A}$ is called an* ergodic algebra *if any invariant function f belonging to the corresponding space $\mathcal{B}^2$ is equivalent (in $\mathcal{B}^2$) to a constant.*

A very useful alternative definition of ergodic algebra is also given in [12], p. 247, and shown therein to be equivalent to Definition 2.3. We state that as the following lemma.

LEMMA 2.1 (cf. [12]). *Let $\mathcal{A} \subseteq \mathrm{BUC}(\mathbb{R}^n)$ be an algebra w.m.v.. Then $\mathcal{A}$ is ergodic if and only if*

$$\lim_{t\to\infty} M_y \left(\left| \frac{1}{|B(0;t)|} \int_{B(0;t)} f(x+y)\, dx - M(f) \right|^2 \right) = 0 \qquad \forall f \in \mathcal{A}. \tag{2.3}$$

3. Regular algebras w.m.v. and the Fourier-Stieltjes space $\mathrm{FS}(\mathbb{R}^n)$**.** In this section we recall the definition of regular algebra w.m.v. introduced by the authors in [10], and the definition and some basic properties of the Fourier-Stieltjes space introduced by the authors in [9], which is, to the best of our knowledge, the largest known example of a regular algebra w.m.v..

For any $f \in L^\infty(\mathbb{R}^n)$, let us denote by $\hat{f}$ the Fourier transform of f defined as the following distribution

$$\langle \hat{f}, \phi \rangle := \int f(x)\check{\phi}(x)\,dx, \qquad \text{for all } \phi \in C_c^\infty(\mathbb{R}^n),$$

where $\check{\phi}$ denotes the usual inverse Fourier transform of ϕ, i.e.,

$$\check{\phi}(x) = \frac{1}{(2\pi)^{\frac{n}{2}}} \int \phi(y) e^{iy\cdot x}\,dx.$$

Given an algebra w.m.v. $\mathcal{A}$, let us denote by $V(\mathcal{A})$ the subspace formed by the elements $f \in \mathcal{A}$ such that $M(f) = 0$, namely,

$$V(\mathcal{A}) := \{f \in \mathcal{A} \,:\, M(f) = 0\}.$$

Also, let us denote by $Z(\mathcal{A})$ the subset of those $f \in \mathcal{A}$ such that the distribution $\hat{f}$ has compact support not containing the origin 0, that is,

$$Z(\mathcal{A}) := \{f \in \mathcal{A} \,:\, \operatorname{supp}(\hat{f}) \text{ is compact and } 0 \notin \operatorname{supp}(\hat{f})\}.$$

We collect in the following lemma some useful properties of the functions in $Z(\mathcal{A})$, whose proof is found in [12], p. 246.

LEMMA 3.1 (cf. [12]). *Let $\mathcal{A}$ be an algebra w.m.v. in $\mathbb{R}^n$ and $f \in Z(\mathcal{A})$. Then:*

(i) There exists $u \in C^\infty(\mathbb{R}^n) \cap Z(\mathcal{A})$ such that $\Delta u = f$, where Δ is the usual Laplace operator in $\mathbb{R}^n$; $u = f * \zeta$ for certain smooth ζ, fast decaying together with all its derivatives, satisfying $\hat{\zeta} \in C_c^\infty(\mathbb{R}^n)$ and $0 \notin \operatorname{supp}(\hat{f})$.

(ii) For any Borelian $Q \subseteq \mathbb{R}^n$, with $|Q| > 0$, we have

$$\lim_{t\to\infty} \frac{1}{t^n|Q|} \int_{Q_t} f(x+y)\,dx = 0, \qquad \text{uniformly in } y \in \mathbb{R}^n, \quad (3.1)$$

where $|Q|$ denotes the n-dimensional Lebesgue measure of Q and $Q_t = \{x \in \mathbb{R}^n \,:\, t^{-1}x \in Q\}$.

The fundamental result about ergodic algebras, proved by Zhikov and Krivenko [15], is the following.

THEOREM 3.1 (cf. [15]). *If $\mathcal{A}$ is an ergodic algebra, then $Z(\mathcal{A})$ is dense in $V(\mathcal{A})$ in the topology of the corresponding space $\mathcal{B}^2$.*

The following useful result is an immediate consequence of the last theorem and was stated and applied in [3] (see also [10]) in connection with the homogenization problem for a porous medium type equation.

LEMMA 3.2 (cf. [3]). *Let $\mathcal{A}$ be an ergodic algebra in* BUC($\mathbb{R}^n$) *and $h \in \mathcal{B}^2$ such that $M(h\Delta f) = 0$ for all $f \in \mathcal{A}$ such that $\Delta f \in \mathcal{A}$. Then h is $\mathcal{B}^2$-equivalent to a constant.*

Theorem 3.1 also motivates the following definition.

DEFINITION 3.1. *An algebra w.m.v. $\mathcal{A}$ is said to be* regular *if $Z(\mathcal{A})$ is dense in $V(\mathcal{A})$ in the topology of the* sup*-norm.*

We have the following important fact about regular algebras w.m.v..

PROPOSITION 3.1 (cf. [10]). *If $\mathcal{A}$ is a regular algebra w.m.v., then $\mathcal{A}$ is ergodic.*

The next property of regular algebras w.m.v. has been used by the authors in application to homogenization of porous medium type equations on bounded domains in [9, 10].

LEMMA 3.3. *Let $\mathcal{A}$ be a regular algebra w.m.v. If $f \in V(\mathcal{A})$, then for any $\varepsilon > 0$ there exists a function $u_\varepsilon \in Z(\mathcal{A})$ satisfying the inequalities*

$$f - \varepsilon \leq \Delta u_\varepsilon \leq f + \varepsilon. \tag{3.2}$$

The space FS($\mathbb{R}^n$) introduced in [9] provides a very encompassing example of a regular algebra w.m.v..

DEFINITION 3.2. *The Fourier-Stieltjes space, denoted by* FS($\mathbb{R}^n$), *is the completion relatively to the* sup*-norm of the space of functions* $\mathrm{FS}_*(\mathbb{R}^n)$ *defined by*

$$\mathrm{FS}_*(\mathbb{R}^n) := \left\{ f : \mathbb{R}^n \to \mathbb{R} \, : \, f(x) = \int_{\mathbb{R}^n} e^{ix\cdot y}\, d\nu(y) \text{ for some } \nu \in \mathcal{M}_*(\mathbb{R}^n) \right\}, \tag{3.3}$$

where by $\mathcal{M}_*(\mathbb{R}^n)$ we denote the space of complex-valued measures μ with finite total variation, i.e., $|\mu|(\mathbb{R}^n) < \infty$.

Recall that a subalgebra $\mathcal{B} \subseteq \mathcal{A}$ is called an ideal of $\mathcal{A}$ if for any $f \in \mathcal{A}$ and $g \in \mathcal{B}$ we have $fg \in \mathcal{B}$. Let $C_0(\mathbb{R}^n)$ denote the closure of $C_c^\infty(\mathbb{R}^n)$ with respect to the sup norm. The following result was established in [9].

PROPOSITION 3.2 (cf. [9]). FS($\mathbb{R}^n$) $\subseteq$ BUC($\mathbb{R}^n$) *and it is an algebra w.m.v. containing $C_0(\mathbb{R}^n)$ as an ideal. Moreover,* FS($\mathbb{R}^n$) *is a regular algebra w.m.v. and the space* PAP($\mathbb{R}^n$) *of the perturbed almost periodic functions, defined as*

$$\mathrm{PAP}(\mathbb{R}^n) := \{ f \in \mathrm{BUC}(\mathbb{R}^n) \, : \, f = g + \psi, \; g \in \mathrm{AP}(\mathbb{R}^n), \; \psi \in C_0(\mathbb{R}^n) \},$$

is a closed strict subalgebra of FS($\mathbb{R}^n$).

4. Open problems concerning algebras w.m.v. The following questions, listed in order of difficulty, concerning the general theory of algebras w.m.v., briefly recalled in the last section, remain open:

1. Are there regular algebras w.m.v. in $\mathbb{R}^n$ that are not subalgebras of FS($\mathbb{R}^n$)?

2. Are there ergodic algebras that are not regular algebras w.m.v.?
3. Are there ergodic algebras that are not subalgebras of $\mathrm{FS}(\mathbb{R}^n)$?

5. Two-scale Young Measures. In this section we recall the theorem giving the existence of two-scale Young measures established in [3]. We begin by recalling the concept of vector-valued algebra with mean value.

Given a Banach space E and an algebra w.m.v. $\mathcal{A}$, we denote by $\mathcal{A}(\mathbb{R}^n; E)$ the space of functions $f \in \mathrm{BUC}(\mathbb{R}^n; E)$ satisfying the following:

(i) $L_f := \langle L, f\rangle$ belongs to $\mathcal{A}$ for all $L \in E^*$;

(ii) The family $\{L_f : L \in E^*,\ \|L\| \le 1\}$ is relatively compact in $\mathcal{A}$.

THEOREM 5.1 (cf. [3]). *Let E be a Banach space, $\mathcal{A}$ an algebra w.m.v. and $\mathcal{K}$ be the compact associated with $\mathcal{A}$. There is an isometric isomorphism between $\mathcal{A}(\mathbb{R}^n; E)$ and $C(\mathcal{K}; E)$. Denoting by $g \mapsto \underline{g}$ the canonical map from $\mathcal{A}$ to $C(\mathcal{K})$, the isomorphism associates to $f \in \mathcal{A}(\mathbb{R}^n; E)$ the map $\tilde f \in C(\mathcal{K}; E)$ satisfying*

$$\underline{\langle L, f\rangle} = \langle L, \tilde f\rangle \in C(\mathcal{K}) \qquad \forall L \in E^*. \tag{5.1}$$

In particular, for each $f \in \mathcal{A}(\mathbb{R}^n; E)$, $\|f\|_E \in \mathcal{A}$.

For $1 \le p < \infty$, we define the space $L^p(\mathcal{K}; E)$ as the completion of $C(\mathcal{K}; E)$ with respect to the norm $\|\cdot\|_p$, defined as usual,

$$\|f\|_p := \left(\int_{\mathcal{K}} \|f\|_E^p \, d\mathfrak{m}\right)^{1/p}.$$

As a standard procedure, we identify functions in L^p that coincide $\mathfrak{m}$-a.e. in $\mathcal{K}$.

Similarly, we define the space $\mathcal{B}^p(\mathbb{R}^n; E)$ as the completion of $\mathcal{A}(\mathbb{R}^n; E)$ with respect to the seminorm

$$|f|_p := \left(⨍_{\mathbb{R}^n} \|f\|_E^p \, dx\right)^{1/p},$$

identifying functions in the same equivalence class determined by the seminorm $|\cdot|_p$. Clearly, the isometric isomorphism given by Theorem 5.1 extends to an isometric isomorphism between $\mathcal{B}^p(\mathbb{R}^n; E)$ and $L^p(\mathcal{K}; E)$.

The next theorem gives the existence of two-scale Young measures associated with an algebra $\mathcal{A}$. For the proof, we again refer to [3].

Let $\Omega \subseteq \mathbb{R}^n$ be a bounded open set and $\{u_\varepsilon(x)\}_{\varepsilon>0}$ be a family of functions in $L^\infty(\Omega; K)$, for some compact metric space K.

THEOREM 5.2. *Given any infinitesimal sequence $\{\varepsilon_i\}_{i\in\mathbb{N}}$ there exist a subnet $\{u_{\varepsilon_{i(d)}}\}_{d\in D}$, indexed by a certain directed set D, and a family of probability measures on K, $\{\nu_{z,x}\}_{z\in\mathcal{K}, x\in\Omega}$, weakly measurable with respect to the product of the Borel σ-algebras in $\mathcal{K}$ and $\mathbb{R}^n$, such that*

$$\lim_D \int_\Omega \Phi\Big(\frac{x}{\varepsilon_{i(d)}}, x, u_{\varepsilon_{i(d)}}(x)\Big)\, dx = \int_\Omega \int_{\mathcal{K}} \langle \nu_{z,x}, \underline{\Phi}(z, x, \cdot)\rangle \, d\mathfrak{m}(z)\, dx \qquad \forall \Phi \in \mathcal{A}\left(\mathbb{R}^n; C_0(\Omega \times K)\right). \tag{5.2}$$

Here $\underline{\Phi} \in C(\mathcal{K}; C_0(\Omega \times K))$ *denotes the unique extension of* Φ. *Moreover, equality* (5.2) *still holds for functions* Φ *in the following function spaces:*

1. $\mathcal{B}^1(\mathbb{R}^n; C_0(\Omega \times K))$;
2. $\mathcal{B}^p(\mathbb{R}^n; C(\bar{\Omega} \times K))$ *with* $p > 1$;
3. $L^1(\Omega; \mathcal{A}(\mathbb{R}^n; C(K)))$.

As in the classical theory of Young measures we have the following consequence of Theorem 5.2.

THEOREM 5.3. *Let* $\Omega \subseteq \mathbb{R}^n$ *be a bounded open set, let* $\{u_\varepsilon\} \subseteq L^\infty(\Omega; \mathbb{R}^m)$ *be uniformly bounded and let* $\nu_{z,x}$ *be a two-scale Young measure generated by a subnet* $\{u_{\varepsilon(d)}\}_{d \in D}$, *according to Theorem 5.2. Assume that* U *belongs either to* $L^1(\Omega; \mathcal{A}(\mathbb{R}^n; \mathbb{R}^m)))$ *or to* $\mathcal{B}^p(\mathbb{R}^n; C(\bar{\Omega}; \mathbb{R}^m))$ *for some* $p > 1$. *Then*

$$\nu_{z,x} = \delta_{\underline{U}(z,x)} \text{ if and only if } \lim_D \left\| u_{\varepsilon(d)}(x) - U\left(\frac{x}{\varepsilon(d)}, x\right) \right\|_{L^1(\Omega)} = 0. \quad (5.3)$$

6. Applications and ongoing projects. Two-scale Young measures have been used in the analysis of the homogenization problem in ergodic algebras in [2, 3, 9, 10] for particular cases of degenerate hyperbolic-parabolic equations of the general form

$$u_t + D \cdot A\left(x, \frac{x}{\varepsilon}, u\right) = D^2 \cdot B\left(x, \frac{x}{\varepsilon}, u\right), \quad (6.1)$$

where $A(x,y,u) = (a_1(x,y,u), \dots, a_n(x,y,u))$, $B(x,y,u) = (b_{ij}(x,y,u))_{i,j=1}^N$, $D = (\partial_{x_1}, \dots, \partial_{x_n})$ and $D^2 = (\partial^2_{x_i x_j})_{i,j=1}^N$. Namely, in [2] and [9], there have been considered the equations

$$u_t + a\left(\frac{x}{\varepsilon}\right) \cdot Df(u) = 0, \quad (6.2)$$

where $a = (a_1, \dots, a_n)$ is a divergence-free field with components in an ergodic algebra, and

$$u_t + \left(f_1(u) - V\left(\frac{x}{\varepsilon}\right)\right)_{x_1} + \sum_{k=2}^n f_k(u)_{x_k} = 0, \quad (6.3)$$

with $V(y)$ belonging to an ergodic algebra, while [3] and [9] considered the equation

$$u_t = \Delta\left(f(u) - V\left(x, \frac{x}{\varepsilon}, u\right)\right), \quad (6.4)$$

and, more generally, [10] addresses the equation

$$u_t = \Delta f\left(x, \frac{x}{\varepsilon}, u\right), \quad (6.5)$$

where $f(x,y,\cdot)$ is increasing but may degenerate in the sense that $f_u(x,y,\cdot)$ may vanish in a set with empty interior.

In the periodic case, the equation was considered by W. E [7], and, more generally, Dalibard [5] establishes a theory for general scalar conservation laws of the form

$$u_t + D\cdot A\Big(\frac{x}{\varepsilon}, u\Big) = 0. \tag{6.6}$$

A framework for the analysis of the homogenization problem for the general degenerate parabolic-hyperbolic equation (6.1) in ergodic algebras is currently in preparation [11].

7. Flows generated by Lipschitz vector fields in an algebra w.m.v. The problem of the invariance of a given algebra w.m.v. under the flow generated by a Lipschitz continuous vector field whose components belong the corresponding algebra w.m.v. was first addressed for the algebra of almost periodic functions in [2], and then in [9] for the algebra $\mathrm{FS}(\mathbb{R}^n)$. In [9], we left open the question whether this invariance property could be extended to more general algebras w.m.v. This problem is completely solved in this section (see Theorem 7.1).

Let $a\in\mathcal{A}\cap W^{1,\infty}(\mathbb{R}^n;\mathbb{R}^n)$, and let us assume that a is incompressible, i.e.

$$\nabla_z\cdot a = 0. \tag{7.1}$$

Let us consider the Cauchy problem

$$\begin{cases}\dfrac{dX}{dt}(z,t) = a(X(z,t)),\\ X(z,0)=z.\end{cases} \tag{7.2}$$

The map $t\mapsto X(z,t)$ will be denoted by $X_t(z)$. We will now use the caracterization given by Theorem 2.1 for to show that the map $\boldsymbol{X}_t : \mathrm{BUC}(\mathbb{R}^n)\to \mathrm{BUC}(\mathbb{R}^n)$ defined by $g\mapsto g\circ X_t$ is invariant under the flow of Lipschtiz algebra w.m.v. fields. Moreover, we will see that the property above is false if we delete the condition of the field belongs to some algebra w.m.v..

THEOREM 7.1. *$\varphi\circ X_t\in\mathcal{A}$ for any $\varphi\in\mathcal{A}$ and*

$$⨍_{\mathbb{R}^n} |\varphi(X(z,t))|^2\,dz = ⨍_{\mathbb{R}^n} |\varphi(z)|^2\,dz. \tag{7.3}$$

Therefore $\boldsymbol{X}_t$ can be extended to an operator in $\mathcal{B}^2$ satisfying

$$⨍_{\mathbb{R}^n} |\boldsymbol{X}_t(\varphi)|^2\,dz = ⨍_{\mathbb{R}^n} |\varphi(z)|^2\,dz \qquad \forall\varphi\in\mathcal{B}^2. \tag{7.4}$$

Proof. 1. Denoting by $f \mapsto \underline{f}$ the canonical map i from $\mathcal{A}$ to $C(\mathcal{K})$, we begin by proving that if $b \in \mathcal{A}(\mathbb{R}^n;\mathbb{R}^n)$, then $\varphi(\cdot + b(\cdot)) \in \mathcal{A}$ for all $\varphi \in \mathcal{A}$. For this end, consider in $\mathbb{R}^n$ the equivalence relation $x \sim y$ iff $f(x) = f(y)$ for all $f \in \mathcal{A}$. As in the proof of Theorem 2.1, $\mathbb{R}^n/\sim$ can be naturally embedded in the compact $\mathcal{K}$ as a dense subset such that $\underline{f}(\bar{x}) = f(x)$ and $T(y)\bar{x} = \overline{x+y}$ for any $x, y \in \mathbb{R}^n$, where $\bar{x}$ denotes the equivalence class of x. If $b = (b^1, \cdots, b^n)$, denote $\underline{b}$ by $(\underline{b^1}, \cdots, \underline{b^n})$. Since the functions $z \in \mathcal{K} \mapsto (\underline{b}(z), z) \in \mathbb{R}^n \times \mathcal{K}$ and $(y, z) \in \mathbb{R}^n \times \mathcal{K} \mapsto T(y)z \in \mathcal{K}$ are continuous, we have that the function $\psi(z) := \underline{\varphi}(T(\underline{b}(z))z)$ belongs to $C(\mathcal{K})$. Thus, by Theorem 2.1, there exist a function $g \in \mathcal{A}$ such that $\underline{g} = \psi$. Therefore,

$$\begin{aligned} g(x) &= \underline{g}(\bar{x}) = \psi(\bar{x}) = \underline{\varphi}(T(\underline{b}(\bar{x}))\bar{x}) \\ &= \underline{\varphi}(T(b(x))\bar{x}) = \underline{\varphi}\left(\overline{x + b(x)}\right) = \varphi(x + b(x)), \end{aligned}$$

thus proving that $\varphi(\cdot + b(\cdot)) \in \mathcal{A}$.

2. Define $Y := \{f \in C(\mathbb{R}^n;\mathbb{R}^n);\ f(x) - x \in L^\infty(\mathbb{R}^n;\mathbb{R}^n)\}$. Note that if $f, g \in Y$, then, $f - g \in L^\infty(\mathbb{R}^n;\mathbb{R}^n)$. We define in Y the metric $d_Y(f,g) := \|f - g\|_\infty$ and observe that (Y, d_Y) is a complete metric space. Fix $T > 0$ and let $X := C([-T,T];Y)$ with the metric $d(\varphi_1, \varphi_2) := \sup_{t\in[-T,T]} d_Y(\varphi_1(t), \varphi_2(t))$. The space (X, d) is a complete metric space. Define $F : X \to X$ by:

$$F(\varphi)(t,z) := z + \int_0^t a(\varphi(s,z))ds.$$

Using a standard argument, we see that there exists a single fixed point $X_t(z)$ of F which is the unique solution of (7.2). Moreover, $\forall\, \Phi \in X$, we have that $F^{(j)}(\Phi) \to X_t(z)$ in X. Therefore, for each fixed t, $F^{(j)}(\Phi)(t) \to X_t$ uniformly in z. Now, we take $\Phi(t,z) = z$ and $X^{(j)} := F^{(j)}(\Phi) = F(F^{(j-1)}(\Phi)) = F(X^{(j-1)})$ and note that $X^{(1)}(t,z) = z + ta(z)$ and it is uniformly continuous in $[-T,T] \times \mathbb{R}^n$. By step 1, $\varphi \circ X^{(1)}(t,\cdot) = \varphi(\cdot + ta(\cdot)) \in \mathcal{A}$ for any $\varphi \in \mathcal{A}$.

3. Suppose that for all $\varphi \in \mathcal{A}$, we have $\varphi(X^{(j-1)}(t,\cdot)) \in \mathcal{A}$ for each fixed $t \in [-T,T]$ and $X^{(j-1)}$ is uniformly continuous in $[-T,T] \times \mathbb{R}^n$. Observe that

$$\varphi(X^{(j)}(z,t)) = \varphi(F(X^{(j-1)})(z,t)) = \varphi\Big(z + \int_0^t a(X^{(j-1)}(z,s))\,ds\Big).$$

Since $X^{(j-1)}$ is uniformly continuous, then the Riemann sums of $\int_0^t a(X^{(j-1)}(z,s))\,ds$ uniformly converge in z. Therefore, $\int_0^t a(X^{(j-1)}(z,s))\,ds \in \mathcal{A}(\mathbb{R}^n;\mathbb{R}^n)$ and by step 1, $\varphi(X^{(j)}(t,\cdot)) \in \mathcal{A}$. Moreover, it is easy to see that $X^{(j)}$ is uniformly continuous in $[-T,T] \times \mathbb{R}^n$. Thus, we have proved, by induction, that $\varphi(X^{(j)}(t,\cdot)) \in \mathcal{A}$

for all j. Therefore, the uniform convergence of $X^{(j)}(t,\cdot)$ to $X_t(\cdot)$ provides the proof of the first part of the theorem.

4. Now we prove (7.3). The incompressibility assumption (7.1) implies that the Jacobian determinant of X_t is a.e. equal to 1, and we have

$$\begin{aligned}\frac{1}{L^n}\int_{[0,L]^n}|\varphi(X(z,t))|^2\,dz &= \frac{1}{L^n}\int_{X_t([0,L]^n)}|\varphi(w)|^2\,dw\\ = \frac{1}{L^n}\int_{[0,L]^n}|\varphi(w)|^2\,dw &- \frac{1}{L^n}\int_{[0,L]^n\setminus X_t([0,L]^N)}|\varphi(w)|^2\,dw\\ &+ \frac{1}{L^n}\int_{X_t([0,L]^n)\setminus[0,L]^n}|\varphi(w)|^2\,dw.\end{aligned}$$

Take the limit as $L\to\infty$ observing that the two last terms on the right-hand side of the last equality above go to 0 as $L\to\infty$ since

$$[\|a\|_\infty t, L-\|a\|_\infty t]^n \subseteq X_t([0,L]^n) \subseteq [-\|a\|_\infty t, L+\|a\|_\infty t]^n.$$

We then obtain (7.3). Relation (7.3) immediately implies that $\boldsymbol{X}_t$ can be extended to an operator in $\mathcal{B}^2$, and that $\boldsymbol{X}_t$ fulfills (7.4). □

REMARK 7.1. We observe that the Theorem 7.1 is, in general, false if the vector field only belongs to $W^{1,\infty}(\mathbb{R}^n;\mathbb{R}^n)$. The following provides a simple counter-example. Let

$$a(x) := \begin{cases} 1, & \text{for } x>1,\\ x, & \text{for } |x|\le 1,\\ -1, & \text{for } x<-1,\end{cases}$$

in the Cauchy problem 7.2 and let $\mathcal{A}$ be the space of continuous periodic functions in $\mathbb{R}$ with period 2π. In this case, the validity of the assertions in Theorem 7.1 would mean, in particular, that $\cos(X_t(\cdot))$ should be 2π-periodic. But then, for all $t>0$,

$$1=\cos(X_t(0)) = \cos(X_t(2\pi k) = \cos(2\pi k+t) = \cos(t),$$

which gives a contradiction.

REMARK 7.2. It is somehow amazing the generality of Theorem 7.1, in the sense that it applies to any algebra w.m.v., with no ergodicity assumption. We recall that an example presented in [12] allows to construct an algebra w.m.v. which is not ergodic. Namely, we take the closed algebra with unity in $\mathrm{BUC}(\mathbb{R})$ generated by the function $\cos\sqrt[3]{x}$ and its translates $\cos\sqrt[3]{x+t}$, $t\in\mathbb{R}$. Indeed, $\cos\sqrt[3]{x}-\cos\sqrt[3]{x+t}\to 0$ as $|x|\to\infty$ for all $t\in\mathbb{R}$, and each of the functions $\cos^k\sqrt[3]{x}$, $k\in\mathbb{N}$, possesses a mean value M_k. Clearly, any product of translates $\cos\sqrt[3]{x+t_1}\cdots\cos\sqrt[3]{x+t_k}$ is $\mathcal{B}^2$-equivalent to $\cos^k\sqrt[3]{x}$, $k\in\mathbb{N}$. Moreover, $M_1=0$ and $M_2=\frac{1}{2}$. Hence, this algebra is an algebra w.m.v. which is not ergodic since the function $\cos\sqrt[3]{x}$ is invariant and is not $\mathcal{B}^2$-equivalent to a constant.

Let $\mathcal{A} = \mathcal{A}(\mathbb{R}^n)$ be an algebra w.m.v. and $a \in \mathcal{A} \bigcap W^{1,\infty}(\mathbb{R}^n; \mathbb{R}^n)$ satisfying (7.1) as above. Let $\mathcal{S}$ be the closed subspace of $\mathcal{B}^2$ defined as follows. Let us consider the equation

$$\nabla \cdot (av) = 0. \tag{7.5}$$

We define a class of asymptotic solutions of (7.5) as follows. Let us define the space of test functions

$$\mathcal{T} := \{v \in \mathcal{A}(\mathbb{R}^n) \cap W^{1,\infty}(\mathbb{R}^n) \,:\, \nabla_a v := a \cdot \nabla v \in \mathcal{A}(\mathbb{R}^n)\}. \tag{7.6}$$

We then define

$$\mathcal{S} := \left\{ v \in \mathcal{B}^2 \,:\, \int\!\!\!\!\!\!-_{\mathbb{R}^n} v(z) \nabla_a \varphi(z)\, dz = 0, \text{ for all } \varphi \in \mathcal{T} \right\}. \tag{7.7}$$

We also consider the following subspaces of $\mathcal{S}$:

$$\mathcal{S}^* := \left\{ v \in \mathcal{A}(\mathbb{R}^n) \cap W^{1,\infty}(\mathbb{R}^n) :\ \nabla_a v = 0 \quad \text{a.e.} \right\}, \tag{7.8}$$

$$\mathcal{S}^\dagger := \left\{ v \in \mathcal{S} \,:\, \exists\, (v_k)_{k \in \mathbb{N}} \subseteq \mathcal{T}, v_k \overset{\mathcal{B}^2 \cap L^2_{\mathrm{loc}}}{\longrightarrow} v \text{ and } \nabla_a v_k \overset{\mathcal{B}^2 \cap L^2_{\mathrm{loc}}}{\longrightarrow} 0 \right\}. \tag{7.9}$$

and

$$\mathcal{S}^\flat := \left\{ v \in \mathcal{S} \,:\, \exists\, (v_k)_{k \in \mathbb{N}} \subseteq \mathcal{T}, v_k \overset{\mathcal{B}^2}{\longrightarrow} v \text{ and } \nabla_a v_k \overset{\mathcal{B}^2}{\longrightarrow} 0 \right\}. \tag{7.10}$$

Clearly, we have $\mathcal{S}^* \subseteq \mathcal{S}^\dagger \subseteq \mathcal{S}^\flat$. For the study of the homogenization problem for the nonlinear transport equation (6.2), it is relevant the question whether $\mathcal{S}^\dagger$ is dense in $\mathcal{S}$. In the case of periodic functions, the analogues of $\mathcal{S}^\dagger$ and $\mathcal{S}^\flat$ coincide since convergences in L^2_{loc} and $\mathcal{B}^2$ are equivalent in that case. We will see in the next proposition that in fact $\mathcal{S}^\flat = \mathcal{S}$. The analogous result in the periodic case implies $\mathcal{S}^\dagger = \mathcal{S}$. The latter was given another proof in [5] by a standard argument using convolutions with approximations of the Dirac measure. Nevertheless, this argument is heavily supported on the equivalence between L^2_{loc} and $\mathcal{B}^2$ convergences in the periodic case and so cannot be extended even to the case of almost periodic functions. Therefore, here, as in [2], instead of considering the question of the density of $\mathcal{S}^\dagger$ in $\mathcal{S}$, we will address the question concerning the stronger property of the density of $\mathcal{S}^*$ in $\mathcal{S}$.

Let us consider for a moment the one-parameter group of unitary operators $\boldsymbol{X}_t : \mathcal{B}^2 \to \mathcal{B}^2$ defined by $\boldsymbol{X}_t v = v \circ X_t$, where X_t is the flow generated by the vector field a. Let us denote also by $\mathcal{B}^2$ its standard complex extension and consider the natural extension of $\boldsymbol{X}_t$ to the complexification of $\mathcal{B}^2$. By Stone's Theorem (see, e.g., [14], p. 266) there is a self-adjoint operator A on $\mathcal{B}^2$ so that $\boldsymbol{X}_t = e^{itA}$. We have the following fact.

PROPOSITION 7.1. *The self-adjoint operator A such that $\boldsymbol{X}_t = e^{itA}$ is essentially self-adjoint on $\mathcal{T}$ and $A|\mathcal{T} = \frac{1}{i}\nabla_a$. Moreover, $\mathcal{S}$ is the invariant space under $\boldsymbol{X}_t$ and $\mathcal{S}^\flat = \mathcal{S}$.*

The next result improves a similar proposition in [2] for the case of almost periodic functions. The proof is similar to the corresponding one in [2] but now we can dispense with the density of $\mathcal{S}^*$ in $\mathcal{S}$ by using the equality $\mathcal{S}^\flat = \mathcal{S}$ which holds in general by Proposition 7.1.

PROPOSITION 7.2. *$\mathcal{S} \cap L^\infty(\mathcal{K})$ is an algebra and*

$$\widetilde{gr} = g\tilde{r} \qquad \forall g \in \mathcal{S} \cap L^\infty(\mathcal{K}),\ r \in L^2(\mathcal{K}). \tag{7.11}$$

We remark that the Mean Ergodic Theorem (see [6], Theorem VIII.7.1), which is applicable due to Theorem 7.1 and to the fact that $\mathcal{S}$ is the invariant space of $\boldsymbol{X}_t$ (see Proposition 7.1 above) implies that for $g \in \mathcal{B}^2$ we have

$$\lim_{t\to\infty} \frac{1}{t}\int_0^t \boldsymbol{X}_s g(z)\,ds = \tilde{g}(z) \qquad \forall g \in \mathcal{B}^2, \tag{7.12}$$

in the sense of convergence in $\mathcal{B}^2$, and one can use this formula to link in a more explicit way $\tilde{g}$ to g.

So far, the most general results concerning the density $\mathcal{S}^*$ in $\mathcal{S}$, which obviously implies the density of $\mathcal{S}^\dagger$ in $\mathcal{S}$, are restricted to the case when $\mathcal{A}$ is $\mathrm{FS}(\mathbb{R}^n)$ or one of its subalgebras. We recall here some results of [9] concerning this question.

LEMMA 7.1 (cf. [9]). *If $a = (a_1, \cdots, a_n)$ is a vector field and $x = (x_1, \cdots, x_n)$, let us use the notation $\bar{a} = (a_1, \cdots, a_{n-1})$ and $\bar{x} = (x_1, \cdots, x_{n-1})$. Suppose the following assumptions hold:*

(A1) $a \in \mathrm{FS} \cap W^{1,\infty}(\mathbb{R}^n; \mathbb{R}^n)$, $\operatorname{div} a = 0$ *and* $|a_n| \geq \delta > 0$;

(A2) The function defined by $\Phi(x) := (\Phi_{x_n}(\bar{x}), x_n)$, where $\Phi_{x_n}(\bar{x})$ is the flow associated with the Cauchy problem

$$\begin{cases} \dfrac{dX}{dx_n}(\bar{x}, x_n) = \dfrac{\bar{a}(X(\bar{x}, x_n), x_n)}{a_n(X(\bar{x}, x_n), x_n)}, \\ X(\bar{x}, 0) = \bar{x}, \end{cases} \tag{7.13}$$

is such that $\Phi : \mathbb{R}^n \mapsto \mathbb{R}^n$ is a bi-Lipschitz map and $g \circ \Phi$, $g \circ \Phi^{-1} \in \mathrm{FS}(\mathbb{R}^n)$ for any $g \in \mathrm{FS}(\mathbb{R}^n)$.

Then, $\mathcal{S}^*$ is dense in $\mathcal{S}$.

The next lemma is an important tool for the purpose of obtaining examples of vector fields satisfying the assumptions (A1) and (A2) of Lemma 7.1.

LEMMA 7.2 (cf. [9]). *Let us consider the Cauchy problem*

$$\begin{cases} \frac{dX}{dt}(z, t) = \bar{\alpha}(X(z, t))\theta(t), \\ X(z, 0) = z. \end{cases} \tag{7.14}$$

where $\bar{\alpha} \in \mathrm{FS} \cap W^{1,\infty}(\mathbb{R}^{n-1};\mathbb{R}^{n-1})$ *and* $\theta \in \mathrm{FS}(\mathbb{R})$ *is such that* $\varrho(t) := \int_0^t \theta(s)\,ds \in \mathrm{FS}(\mathbb{R})$. *Let* $X_t(x)$ *be the flow generated by the vector field* $\bar{\alpha}(x)$. *Then,* $\Phi(x,t) := (X_{\varrho(t)}(x),t)$ *satisfies* $\varphi \circ \Phi$, $\varphi \circ \Phi^{-1} \in \mathrm{FS}(\mathbb{R}^n)$ *for any* $\varphi \in \mathrm{FS}(\mathbb{R}^n)$.

COROLLARY 7.1 (cf. [9]). *Let* a *be a vector field of the form* $a(\bar{x},x_n) := \big(\alpha_1(\bar{x})\theta(x_n), \cdots, \alpha_{n-1}(\bar{x})\theta(x_n), \alpha_n(\bar{x})\big)$, *where* $\alpha_i \in \mathrm{FS} \cap W^{1,\infty}(\mathbb{R}^{n-1})$, $i = 1,\dots,n$, $|\alpha_n| \geq \delta > 0$, $\theta \in \mathrm{FS} \cap W^{1,\infty}(\mathbb{R})$, $\int_0^{x_n} \theta(y)\,dy \in \mathrm{FS}(\mathbb{R})$ *and* $\mathrm{div}_{\bar{x}}\,\bar{\alpha} = 0$. *Then,* $\mathcal{S}^*$ *is dense in* $\mathcal{S}$.

REMARK 7.3. We observe that if $\theta \in \mathrm{FS}_*(\mathbb{R})$ is such that $\frac{1}{y} \in L^1_{|\hat{\theta}|}(\mathbb{R})$, then the indefinite integral of θ belongs to $\mathrm{FS}(\mathbb{R})$.

REFERENCES

[1] G. ALLAIRE. *Homogenization and two-scale convergence.* SIAM J. Math. Anal. **23** (1992), No. 6, 1482–1518.

[2] L. AMBROSIO AND H. FRID. *Multiscale Young measures in almost periodic homogenization and applications.* Archive for Rational Mechanics and Analysis **192** (2009), 37–85.

[3] L. AMBROSIO, H. FRID, AND JEAN SILVA. *Multiscale Young Measures in Homogenization of Continuous Stationary Processes in Compact Spaces and Applications.* Journal of Functional Analysis **256** (2009), 1962–1997.

[4] A.S. BESICOVITCH. Almost Periodic Functions. Cambridge University Press, 1932.

[5] A.-L. DALIBARD. *Homogenization of non-linear scalar conservation laws.* Arch. Rational Mech. Anal. **192** (2009), no. 1, 117–164.

[6] N. DUNFORD AND J.T. SCHWARTZ. Linear Operators. Parts I and II. Interscience Publishers, Inc., New York, 1958, 1963.

[7] W. E. *Homogenization of linear and nonlinear transport equations.* Comm. Pure and Appl. Math. **45** (1992), 301–326.

[8] H. FEDERER. Geometric Measure Theory. Springer-Verlag, New York, 1969.

[9] H. FRID AND J. SILVA. *Homogenization of Nonlinear PDE's in the Fourier-Stieltjes Algebras.* SIAM Journal on Mathematical Analysis, Vol. **41**, 1589–1620, 2009.

[10] H. FRID AND J. SILVA. *Homogenization of degenerate porous medium type equation in Ergodic algebras.* Submitted, 2010.

[11] H. FRID, E. PANOV, AND J. SILVA. *Homogenization of degenerate parabolic-hyperbolic equations in ergodic algebras.* In preparation.

[12] V.V. JIKOV, S.M. KOZLOV, AND O.A. OLEINIK. Homogenization of Differential Operators and Integral Functionals. Springer-Verlag, Berlin Heidelberg, 1994.

[13] G. NGUETSENG. *A general convergence result for a functional related to the theory of homogenization.* SIAM J. Math. Anal. **20** (1989), no. 3, 608–623.

[14] M. REED AND B. SIMON. *Methods of Modern Mathematical Physics I: Functional Analysis.* Academic Press, New York, 1972.

[15] V.V. ZHIKOV AND E.V. KRIVENKO. *Homogenization of singularly perturbed elliptic operators.* Matem. Zametki **33** (1983), 571–582. (English transl.: Math. Notes **33** (1983), 294–300).

CONSERVATION LAWS AT A NODE

MAURO GARAVELLO*

Abstract. This survey paper deals with a system of conservation laws on a network composed by a single node with n incoming and m outgoing arcs. We analyze the Cauchy problem in the case of the Lighthill-Whitham-Richards (LWR) model for traffic and in the case of the p-system. Some open problems are presented.

Key words. Conservation laws, networks, car traffic, gas pipeline.

AMS(MOS) subject classifications. Primary 90B20, 35L65.

1. Introduction. Partial differential equations on networks attract lot of attention. The main motivation is that there are different possible applications: vehicular traffic, data networks, irrigation channels, gas pipelines, supply chains, blood circulation and so on (see [4, 9, 10, 11, 12, 13, 14, 16, 21, 22]). A network is simply a finite collection of arcs connected together by vertices (called also nodes or junctions). On each arc of the network we consider a system of partial differential equation of hyperbolic type in conservation form. Here we consider just the case of a network, composed by a single node with n incoming and m outgoing arcs.

The paper is a review of conservation laws at a junction. More precisely we treat just the scalar case (the LWR model for vehicular traffic) and the case of the p-system, used to describe the evolution of a gas in a tube. We present some results about existence, uniqueness and well-posedness for solutions to Cauchy problems at the node. To describe the dynamics, it is sufficient to define solutions to Riemann problems at the node, which are Cauchy problems with constant initial conditions on the arcs meeting at the node. The maps providing such solutions are called Riemann solvers. The main technique for constructing solutions to Cauchy problem is the wave-front tracking method, which consists in approximating the exact solution by piecewise constant functions, obtained by solving Riemann problems inside the arcs and at the node.

The paper is organized as follows. In Section 2 we introduce the basic definitions about system of conservation laws on networks and the Riemann and Cauchy problem at a node. Section 3 deals with the scalar case, in which we consider the model for traffic, introduced by Lighthill, Whitham [23] and independently by Richards [25]. We present the state of art in this setting and some open problems conclude the section. Section 4 contains the case of the p-system at a node: a definition of solution at the junction and a result about the existence of a semigroup of solutions for

*Dipartimento di Scienze e Tecnologie Avanzate, University of Eastern Piedmont "A. Avogadro", viale Teresa Michel 11, 15121 Alessandria, Italy (mauro.garavello@mfn.unipmn.it). Partially supported by Dipartimento di Matematica e Applicazioni, University of Milano-Bicocca.

the Cauchy problem. Subsection 4.1 deals with some open problems for the p-system at the node.

2. Basic definitions. Consider a node or a vertex J with n incoming arcs $I_1, \dots, I_n$ and m outgoing ones $I_{n+1}, \dots, I_{n+m}$. We model each incoming arc I_i ($i \in \{1, \dots, n\}$) of the node with the real interval $I_i = \mathbb{R}^- :=]-\infty, 0]$. Similarly we model each outgoing arc I_j ($j \in \{n+1, \dots, n+m\}$) of the node with the real interval $I_j = \mathbb{R}^+ := [0, +\infty[$.

On each arc I_l ($l \in \{1, \dots, n+m\}$) we consider the system of conservation laws

$$(u_l)_t + f(u_l)_x = 0, \tag{2.1}$$

where $u_l = u_l(t,x) \in \Omega$ is the *conserved quantity*, $f : \Omega \to \mathbb{R}^N$ is the *flux* and Ω is an open and connected subset of $\mathbb{R}^N$. Hence the datum is given by a finite collection of functions u_l defined on $[0, +\infty[\times I_l$. Suppose that the flux f is a smooth function and that the Jacobian matrix $A(u) = Df(u)$ has N real and distinct eigenvalues $\lambda_1(u) < \cdots < \lambda_N(u)$, e.g. the system (2.1) is strictly hyperbolic; see [5]. Denote by $r_h(u)$ ($h \in \{1, \dots, N\}$) the right eigenvalues of $A(u)$. If $N \geq 2$, then we assume that, for every $h \in \{1, \dots, N\}$, either $\nabla\lambda_h(u) \cdot r_h(u) = 0$ for every $u \in \Omega$ or $\nabla\lambda_h(u) \cdot r_h(u) > 0$ for every $u \in \Omega$. This means that the h-th characteristic field is respectively linearly degenerate or genuinely nonlinear; see [5].

Below we introduce the definition of a Riemann problem and of weak solution to the Riemann problem at J.

DEFINITION 2.1. *A Riemann problem at the node J is the following Cauchy problem*

$$\begin{cases} (u_1)_t + f(u_1)_x = 0, & t > 0, x \in I_1 \\ \vdots & \vdots \\ (u_{n+m})_t + f(u_{n+m})_x = 0, & t > 0, x \in I_{n+m} \\ u_1(0,x) = u_{1,0}, & x \in I_1 \\ \vdots & \vdots \\ u_{n+m}(0,x) = u_{n+m,0}, & x \in I_{n+m}, \end{cases} \tag{2.2}$$

where $u_{1,0}, \dots, u_{n+m,0} \in \Omega$ are constant.

DEFINITION 2.2. *A weak solution to the Riemann problem (2.2) is a vector $(u_1(t,x), \dots, u_{n+m}(t,x))$, whose components are functions $u_l :]0, +\infty[\times I_l \to \Omega$ satisfying the following properties:*

1. *for every $i \in \{1, \dots, n\}$, u_i is the restriction to $]0, +\infty[\times I_i$ of the solution to the classical Riemann problem*

$$\begin{cases} (u_i)_t + f(u_i)_x = 0, & t > 0, x \in \mathbb{R} \\ u_i(0,x) = u_{i,0}, & x < 0, \\ u_i(0,x) = u_i(1,0-), & x > 0; \end{cases}$$

2. *for every* $j \in \{n+1,\ldots,n+m\}$, u_j *is the restriction to* $]0,+\infty[\times I_j$ *of the solution to the classical Riemann problem*

$$\begin{cases} (u_j)_t + f(u_j)_x = 0, & t>0, x\in\mathbb{R} \\ u_j(0,x) = u_{j,0}, & x>0, \\ u_j(0,x) = u_j(1,0+), & x<0; \end{cases}$$

3. *for every* $l \in \{1,\ldots,n+m\}$, $\lim_{t\to 0^+} u_l(t,\cdot) = u_{l,0}$ *with respect to the* L^1 *topology.*

REMARK 2.1. Note that giving the trace at the node J of a weak solution to a Riemann problem is completely equivalent to give the weak solution itself.

Note also that conditions 1. and 2. of Definition 2.2 imply that the waves generated by a weak solution to the Riemann problem (2.2) have negative speed in incoming arcs and positive speed in outgoing arcs.

According to Remark 2.1, we introduce, for every $l \in \{1,\ldots,n+m\}$, the set Ω_l, composed by all the possible traces at $x=0$ in the arc I_l.

DEFINITION 2.3. *For a given Riemann problem (2.2) at* J *we define the following subsets of* Ω.

1. *For* $i \in \{1,\ldots,n\}$, *the set* Ω_i *consists of all the elements* $\tilde{u} \in \Omega$ *such that the classical Riemann problem with initial condition* $(u_{i,0},\tilde{u})$ *is solved with waves with negative speed.*
2. *For* $j \in \{n+1,\ldots,n+m\}$, *the set* Ω_j *consists of all the elements* $\tilde{u} \in \Omega$ *such that the classical Riemann problem with initial condition* $(\tilde{u},u_{j,0})$ *is solved with waves with positive speed.*

Finally, let us consider the following Cauchy problem at J:

$$\begin{cases} (u_1)_t + f(u_1)_x = 0, & t>0, x\in I_1 \\ \vdots & \vdots \\ (u_{n+m})_t + f(u_{n+m})_x = 0, & t>0, x\in I_{n+m} \\ u_1(0,x) = u_{1,0}, & x\in I_1 \\ \vdots & \vdots \\ u_{n+m}(0,x) = u_{n+m,0}, & x \in I_{n+m}, \end{cases} \tag{2.3}$$

where, for every $l \in \{1,\ldots,n+m\}$, $u_{l,0} \in L^1(I_l)$ is a function with finite total variation.

3. The scalar case. In this section we consider the scalar model, introduced by Lighthill, Whitham and Richards; see [23, 25]. The model is based on the conservation of the number of cars. In this setting we have $N=1$, $f(u)=uv$, where v is the average speed of cars, and $\Omega=[0,\rho_{max}]$, where ρ_{max} is the maximum density of cars. Without loss of generality we assume that $\rho_{max}=1$. The main assumption of this model is that v is a given function depending only on the density in a decreasing way, i.e. $v=v(u)$ and $v'(u)\le 0$. Therefore the model is described by the scalar

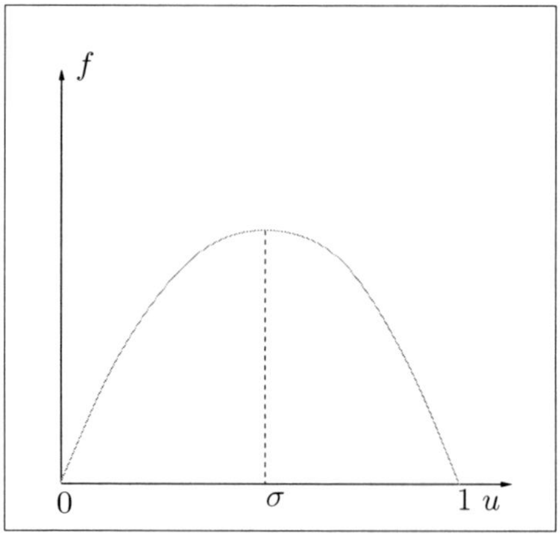

FIG. 1. *The flux function for the LWR model.*

equation $u_t + (uv(u))_x = 0$. We also assume that the flux f is a smooth function such that:

1. $f(0) = f(1) = 0$;
2. f is strictly concave;
3. there exists a unique $0 < \sigma < 1$ such that $f'(\sigma) = 0$; see Figure 1.

In this setting the sets Ω_i and Ω_j, introduced in Definition 2.3, are explicitly given in the next lemma. For a proof see [17].

LEMMA 3.1. *For $i \in \{1, \ldots, n\}$ and $j \in \{n+1, \ldots, n+m\}$ we have that*

$$\Omega_i = \begin{cases} \{u_{i,0}\} \cup]\max f^{-1}(f(u_{i,0})), 1], & \text{if } u_{i,0} < \sigma, \\ [\sigma, 1], & \text{if } u_{i,0} \geq \sigma, \end{cases}$$

$$\Omega_j = \begin{cases} \{u_{j,0}\} \cup [0, \min f^{-1}(f(u_{j,0}))[, & \text{if } u_{j,0} > \sigma, \\ [0, \sigma], & \text{if } u_{j,0} \leq \sigma. \end{cases}$$

We need to introduce some notation. For a collection of functions $u_l \in C([0, +\infty[; L^1_{loc}(I_l))$ ($l \in \{1, \ldots, n+m\}$) such that, for every $l \in \{1, \ldots, n+m\}$ and a.e. $t > 0$ the map $x \mapsto u_l(t, x)$ has a version with bounded total variation, we define the functionals

$$\Gamma(t) := \sum_{i=1}^{n} f(u_i(t, 0-)), \quad \text{Tot.Var.}_f(t) := \sum_{l=1}^{n+m} \text{Tot.Var.}\left(f\left(u_l(t, \cdot)\right)\right).$$

DEFINITION 3.1. *A Riemann solver $\mathcal{RS}$ is a function*

$$\begin{array}{cccc} \mathcal{RS}: & [0,1]^{n+m} & \longrightarrow & \Omega_1 \times \cdots \times \Omega_{n+m} \\ & (u_{1,0}, \ldots, u_{n+m,0}) & \longmapsto & (\bar{u}_1, \ldots, \bar{u}_{n+m}) \end{array}$$

such that $\sum_{i=1}^{n} f(\bar{u}_i) = \sum_{j=n+1}^{n+m} f(\bar{u}_j)$.

DEFINITION 3.2. *We say that a Riemann solver* $\mathcal{RS}$ *satisfies the consistency condition if* $\mathcal{RS}(\mathcal{RS}(u_{1,0},\ldots,u_{n+m,0})) = \mathcal{RS}(u_{1,0},\ldots,u_{n+m,0})$ *for every* $(u_{1,0},\ldots,u_{n+m,0}) \in [0,1]^{n+m}$.

DEFINITION 3.3. *We say that* $(u_{1,0},\ldots,u_{n+m,0})$ *is an equilibrium for the Riemann solver* $\mathcal{RS}$ *if* $\mathcal{RS}(u_{1,0},\ldots,u_{n+m,0}) = (u_{1,0},\ldots,u_{n+m,0})$.

DEFINITION 3.4. *We say that a datum* $u_i \in [0,1]$ *in an incoming arc is a good datum if* $u_i \in [\sigma,1]$ *and a bad datum otherwise.*

We say that a datum $u_j \in [0,1]$ *in an outgoing arc is a good datum if* $u_i \in [0,\sigma]$ *and a bad datum otherwise.*

Now we can state the three key properties of a Riemann solver, which ensure the necessary bounds for the existence of solutions to Cauchy problems.

DEFINITION 3.5. *We say that a Riemann solver* $\mathcal{RS}$ *has the property (P1) if the following condition holds. Given* $(u_{1,0},\ldots,u_{n+m,0})$ *and* $(u'_{1,0},\ldots,u'_{n+m,0})$ *two initial data such that* $u_{l,0} = u'_{l,0}$ *whenever either* $u_{l,0}$ *or* $u'_{l,0}$ *is a bad datum, then* $\mathcal{RS}(u_{1,0},\ldots,u_{n+m,0}) = \mathcal{RS}(u'_{1,0},\ldots,u'_{n+m,0})$.

DEFINITION 3.6. *We say that a Riemann solver* $\mathcal{RS}$ *has the property (P2) if there exists a constant* $C \geq 1$ *such that the following condition holds. For every equilibrium* $(u_{1,0},\ldots,u_{n+m,0})$ *of* $\mathcal{RS}$ *and for every wave* $(u_{l,0},u_l)$ $(l \in \{1,\ldots,n+m\})$ *interacting with* J *at time* $\bar{t} > 0$ *and producing waves in the arcs according to* $\mathcal{RS}$, *we have*

$$\begin{aligned} &Tot.Var._f(\bar{t}+) - Tot.Var._f(\bar{t}-) \\ &\leq C \min\{|f(u_{l,0}) - f(u_l)|, |\Gamma(\bar{t}+) - \Gamma(\bar{t}-)|\}. \end{aligned}$$

DEFINITION 3.7. *We say that a Riemann solver* $\mathcal{RS}$ *has the property (P3) if, for every equilibrium* $(u_{1,0},\ldots,u_{n+m,0})$ *of* $\mathcal{RS}$ *and for every wave* $(u_{l,0},u_l)$ $(l \in \{1,\ldots,n+m\})$ *with* $f(u_l) < f(u_{l,0})$ *interacting with* J *at time* $\bar{t} > 0$ *and producing waves in the arcs according to* $\mathcal{RS}$, *we have* $\Gamma(\bar{t}+) \leq \Gamma(\bar{t}-)$.

The following theorem deals with existence of solution for the Cauchy problem (2.3) at J. A proof, based on the wave-front tracking technique, is contained in [18].

THEOREM 3.1. *Consider the Cauchy problem (2.3) and a Riemann solver* $\mathcal{RS}$ *satisfying the consistency condition and the properties (P1), (P2) and (P3). Then there exists a weak solution* $(u_1(t,x),\ldots,u_{n+m}(t,x))$ *such that*

1. *for every* $l \in \{1,\ldots,n+m\}$, $u_l(0,x) = u_{0,l}(x)$ *for a.e.* $x \in I_l$;
2. *for a.e.* $t > 0$,

$$\mathcal{RS}(u_1(t,0-),\ldots,u_{n+m}(t,0+)) = (u_1(t,0-),\ldots,u_{n+m}(t,0+)).$$

REMARK 3.1. An similar result about existence of solutions for the Cauchy problem (2.3) holds also in the case of a Riemann solver, which varies in time. The exact formulation of the statement is contained in [20].

This generalization can be seen as a first step for controllability problems on networks.

For uniqueness and well posedness of solutions to the Cauchy problem (2.3), the following theorem holds.

THEOREM 3.2. *Consider the Cauchy problem (2.3) and a Riemann solver $\mathcal{RS}$ satisfying the consistency condition and the properties (P1), (P2) and (P3). Assume that the functional Γ does not increase when a wave reaches the junction J. Then there exists a unique weak solution $(u_1(t,x),\ldots,u_{n+m}(t,x))$ such that*

1. *for every $l \in \{1,\ldots,n+m\}$, $u_l(0,x) = u_{0,l}(x)$ for a.e. $x \in I_l$;*
2. *for a.e. $t > 0$,*

$$\mathcal{RS}(u_1(t,0-),\ldots,u_{n+m}(t,0+)) = (u_1(t,0-),\ldots,u_{n+m}(t,0+)).$$

Moreover the solution depends in a Lipschitz continuous way on the initial datum with respect to the L^1-topology.

The proof of the theorem is contained in [18]. It is based on the concept of generalized tangent vectors, introduced in the papers [6, 7].

3.1. Open problems. In this subsection we describe some open problems in the scalar case.

Continuous dependence. In Theorem 3.2 the hypothesis that the functional Γ does not increase when waves interact with the junction is fundamental. More precisely, in [17, Chapter 5.4] there is a counterexample to Theorem 3.2 in the case of the Riemann solver $\mathcal{RS}$ (introduced for traffic in [11]) satisfying the consistency conditions and (P1)-(P2)-(P3) but not the hypothesis on the functional Γ. The example shows that, fixed a junction J with 2 incoming and 2 outgoing arcs and a constant $C > 0$, one can construct two different initial conditions in such a way the corresponding two solutions at a positive time differ in L^1-norm more than C times the L^1-norm of the difference of initial conditions. The example does not produce a blow up of the L^1-distance of the solutions, so it is not a counterexample for the continuous dependence with respect the initial condition, which remains an open problem.

Control problems. Controllability for hyperbolic conservation laws is a difficult issue, since solutions can develop discontinuities in finite time; see [1, 2, 3]. In the case of networks, one possibility is modifying the solution controlling the fluxes at the junctions, i.e. controlling the Riemann solvers at nodes; the papers [15, 20] are a first step in this directions.

Entropy conditions at junctions. In [19] two different entropy conditions for solutions at junctions were introduced. They are generalizations of the classical Kružkov entropy condition. Unfortunately, there are infinitely many Riemann solvers satisfying these entropy conditions. Is it possible to define an entropy-type condition at junctions such that there is a unique Riemann solver satisfying this condition? If so, does this Riemann solver satisfy the hypotheses of Theorem 3.1 and/or Theorem 3.2?

4. The system case. Here we present an example of a system of conservation laws on a vertex. More precisely we consider the p-system, which describes the evolution of the gas in a pipe. Therefore we have $N = 2$, $\Omega = \{(\rho, q) \in \mathbb{R} \times \mathbb{R} : \rho > 0\}$ and the flux $f(u) = f(\rho, q) = \begin{pmatrix} q \\ \frac{q^2}{\rho} + p(\rho) \end{pmatrix}$, where ρ and q represent respectively the density and the linear momentum of the gas, while the pressure p is a given increasing, smooth and convex function. A typical example is the γ-pressure law $p(\rho) = k\rho^\gamma$ with $k > 0$ and $\gamma \geq 1$. Thus the p-system can be written in the form

$$\begin{cases} \partial_t \rho + \partial_x q = 0, & t \in [0, +\infty[, \\ \partial_t q + \partial_x \left(\dfrac{q^2}{\rho} + p(\rho) \right) = 0, & x \in I_l. \end{cases} \tag{4.1}$$

For a later use, we introduce the *dynamic pressure*, i.e. the flow of the linear momentum $P(u) = \frac{q^2}{\rho} + p(\rho)$ and the entropy flow $F(u) = q\left(\frac{q^2}{2\rho^2} + \int_1^\rho \frac{p'(r)}{r} dr\right)$.

PROPOSITION 4.1. *In the set Ω, system (4.1) is strictly hyperbolic. The eigenvalues of the Jacobian matrix of the flux are $\lambda_1(u) = \frac{q}{\rho} - \sqrt{p'(\rho)}$ and $\lambda_2(u) = \frac{q}{\rho} + \sqrt{p'(\rho)}$. The two characteristic speeds are both genuinely nonlinear. Finally, in the (ρ, q) plane, the Lax curves of the first family are concave, while the Lax curves of the second family are convex.*

Differently from the previous cases, we do not distinguish between incoming and outgoing arcs. Therefore we assume that all the arcs are modeled by the real interval $\mathbb{R}^+$, i.e. $n = 0$, and, for every $l \in \{1, \ldots, m\}$, we define the vector ν_l such that it is parallel to the l-tube and the norm $\|\nu_l\|$ is the section of the duct.

LEMMA 4.1. *Define the following sets*

$$\mathcal{R}_1 = \begin{cases} \left\{ (\rho, L_2^-(\rho; \bar\rho_j, \bar q_j)) : \begin{array}{c} \rho > \bar\rho_j \\ L_2^-(\rho; \bar\rho_j, \bar q_j) > \bar q_j \end{array} \right\} & \text{if } \lambda_2(\bar\rho_j, \bar q_j) < 0 \\ \left\{ (\rho, L_2^-(\rho; \bar\rho_j, \bar q_j)) : \lambda_2\left(\rho, L_2^-(\rho; \bar\rho_j, \bar q_j)\right) > 0 \right\} & \text{if } \lambda_2(\bar\rho_j, \bar q_j) \geq 0 \end{cases}$$

$$\mathcal{R}_2 = \left\{ (\rho, q) \colon \lambda_1(\rho, q) \geq 0,\ q > q^+(\rho; \bar\rho_j, \bar q_j) \right\}$$

where $L_2^-(\rho; \bar\rho_j, \bar q_j)$ denotes the reverse Lax curve of second family through the point $(\bar\rho_j, \bar q_j)$ and $q^+(\rho; \bar\rho_j, \bar q_j)$ is a suitable function such that the eigenvalue $\lambda_1(\rho, q^+(\rho; \bar\rho_j, \bar q_j))$ is positive for every $\rho > 0$. It holds that $\Omega_j = \mathcal{R}_1 \cup \mathcal{R}_2$; see Figure 2.

We propose to solve the Riemann problem (2.2) according to the following definition.

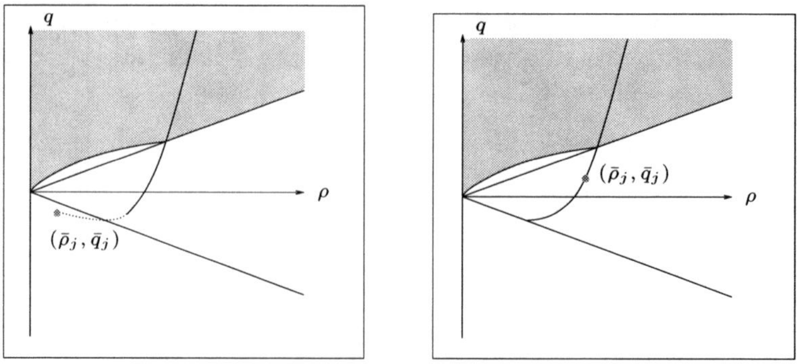

FIG. 2. *The sets Ω_j in the cases $\lambda_2(\bar\rho_j, \bar q_j) < 0$ (left) and $\lambda_2(\bar\rho_j, \bar q_j) \geq 0$ (right).*

DEFINITION 4.1. *A solution to the Riemann problem (2.2) is a weak solution*

$$(u_1(t,x), \dots, u_{n+m}(t,x)) = ((\rho_1, q_1)(t,x), \dots, (\rho_{n+m}, q_{n+m})(t,x)),$$

such that

1. *mass is conserved at J, i.e.*

$$\sum_{l=1}^{m} \|\nu_l\|\, q_l(t,0) = 0; \tag{4.2}$$

2. *the linear momentum is conserved at J, i.e.*

$$P((\rho_l, q_l)(t,0)) = P((\rho_j, q_j)(t,0)) \tag{4.3}$$

for every $l, j \in \{1, \dots, m\}$;

3. *entropy may not decrease, i.e.*

$$\sum_{l=1}^{m} F((\rho_l, q_l)(t,0)) \leq 0. \tag{4.4}$$

We have the following result about the existence and well posedness of solutions to the Cauchy problem (2.3).

THEOREM 4.1. *Fix an m-tuple of states*

$$(\hat\rho, \hat q) = ((\hat\rho_1, \hat q_1), \dots, (\hat\rho_m, \hat q_m)) \in \Omega^m$$

such that

$$\begin{cases} \lambda_1(\hat\rho_l, \hat q_l) < 0 < \lambda_2(\hat\rho_l, \hat q_l), & \forall l \in \{1, \dots, m\}, \\ \sum_{l=1}^{m} \|\nu_l\|\, \hat q_l = 0, & \\ P(\hat\rho_l, \hat q_l) = P(\hat\rho_j, \hat q_j), & \forall j, l \in \{1, \dots, m\}, \\ \sum_{l=1}^{m} F(\hat\rho_l, \hat q_l) < 0. & \end{cases}$$

Then, there exist positive constants δ_0, L and a map $S\colon [0,+\infty[\times \mathcal{D} \to \mathcal{D}$, with the following properties:

1. $\mathcal{D} \supseteq \{(\rho,q) \in (\hat{\rho},\hat{q}) + L^1(\mathbb{R}^+; (\mathbb{R}^+ \times \mathbb{R})^n) \colon$ *Tot. Var.*$(\rho,q) \leq \delta_0\}$.
2. *For* $(\rho,q) \in \mathcal{D}$, $S_0(\rho,q) = (\rho,q)$ *and for* $s,t \geq 0$, $S_s S_t(\rho,q) = S_{s+t}(\rho,q)$.
3. *For* $(\rho,q),(\rho',q') \in \mathcal{D}$ *and* $s,t \geq 0$,

$$\|S_t(\rho,q) - S_s(\rho',q')\|_{L^1} \leq L \cdot (\|(\rho,q) - (\rho',q')\|_{L^1} + |t-s|).$$

4. *If* $(\rho,q) \in \mathcal{D}$ *is piecewise constant, then for* $t > 0$ *sufficiently small,* $S_t(\rho,q)$ *coincides with the juxtaposition of the solutions to Riemann problems centered at the points of jumps or at the junction.*

Moreover, for every $(\rho,q) \in \mathcal{D}$, *the map* $t \to S_t(\rho,q)$ *is a solution to the Cauchy problem (2.3).*

The proof of this theorem is contained in [14]. It is based on the wave-front tracking technique and on the Liu-Yang functional; see [8, 24].

4.1. Open problems. In this subsection we describe some open problems for the p-system at a junction.

Case $\gamma = 1$. If the pressure law $p(\rho)$ is equal to $k\rho$ with k constant, then the classical Riemann problem in a tube admits a unique solution for each initial condition. Moreover, the classical Cauchy problem in a tube has a solution for initial data with finite total variation. At a junction with $m \geq 3$ tubes, both problems are open.

Solution at junctions. In literature there are other definitions of solutions at junctions; see [4, 13]. In [4], the authors proposed a solution which has the same pressure at J and so (4.3) is not necessary satisfied. This kind of solution does not have the property of continuous dependence with respect to the initial condition; see [12]. In [13] a solution depending on the geometry of the junction was introduced. Is it possible to define a solution, which permits to describe in a better way the real behavior of a gas at a junction?

REFERENCES

[1] F. Ancona, A. Bressan, and G.M. Coclite, *Some results on the boundary control of systems of conservation laws*, in Hyperbolic problems: theory, numerics, applications, Springer, Berlin, 2003, pp. 255–264.

[2] F. Ancona and G.M. Coclite, *On the attainable set for Temple class systems with boundary controls*, SIAM J. Control Optim., **43** (2005), pp. 2166–2190 (electronic).

[3] F. Ancona and A. Marson, *Asymptotic stabilization of systems of conservation laws by controls acting at a single boundary point*, in Control methods in PDE-dynamical systems, Vol. **426** of Contemp. Math., Amer. Math. Soc., Providence, RI, 2007, pp. 1–43.

[4] M.K. Banda, M. Herty, and A. Klar, *Gas flow in pipeline networks*, Netw. Heterog. Media, **1** (2006), pp. 41–56 (electronic).

[5] A. Bressan, *Hyperbolic systems of conservation laws*, Vol. **20** of Oxford Lecture Series in Mathematics and its Applications, Oxford University Press, Oxford, 2000. The one-dimensional Cauchy problem.

[6] A. Bressan and R.M. Colombo, *The semigroup generated by* 2×2 *conservation laws*, Arch. Rational Mech. Anal., **133** (1995), pp. 1–75.

[7] A. Bressan, G. Crasta, and B. Piccoli, *Well-posedness of the Cauchy problem for* $n \times n$ *systems of conservation laws*, Mem. Amer. Math. Soc., **146** (2000), pp. viii+134.

[8] A. Bressan, T.-P. Liu, and T. Yang, L^1 *stability estimates for* $n \times n$ *conservation laws*, Arch. Ration. Mech. Anal., **149** (1999), pp. 1–22.

[9] G. Bretti, C. D'Apice, R. Manzo, and B. Piccoli, *A continuum-discrete model for supply chains dynamics*, Netw. Heterog. Media, **2** (2007), pp. 661–694 (electronic).

[10] S. Čanić and E.H. Kim, *Mathematical analysis of the quasilinear effects in a hyperbolic model blood flow through compliant axi-symmetric vessels*, Math. Methods Appl. Sci., **26** (2003), pp. 1161–1186.

[11] G.M. Coclite, M. Garavello, and B. Piccoli, *Traffic flow on a road network*, SIAM J. Math. Anal., **36** (2005), pp. 1862–1886 (electronic).

[12] R.M. Colombo and M. Garavello, *A well posed Riemann problem for the p-system at a junction*, Netw. Heterog. Media, **1** (2006), pp. 495–511 (electronic).

[13] ———, *On the p-system at a junction*, in Control methods in PDE-dynamical systems, Vol. **426** of Contemp. Math., Amer. Math. Soc., Providence, RI, 2007, pp. 193–217.

[14] ———, *On the Cauchy problem for the p-system at a junction*, SIAM J. Math. Anal., **39** (2008), pp. 1456–1471.

[15] R.M. Colombo and P. Goatin, *A well posed conservation law with a variable unilateral constraint*, J. Differential Equations, **234** (2007), pp. 654–675.

[16] C. D'apice, R. Manzo, and B. Piccoli, *Packet flow on telecommunication networks*, SIAM J. Math. Anal., **38** (2006), pp. 717–740 (electronic).

[17] M. Garavello and B. Piccoli, *Traffic flow on networks*, Vol. **1** of AIMS Series on Applied Mathematics, American Institute of Mathematical Sciences (AIMS), Springfield, MO, 2006. Conservation laws models.

[18] ———, *Conservation laws on complex networks*, Ann. H. Poincaré, doi:10.1016/j.anihpc.2009.04.001, **26** (2009), pp. 1925–1951.

[19] ———, *Entropy type conditions for riemann solvers at nodes*, preprint, (2009).

[20] ———, *Time-varying Riemann solvers for conservation laws on networks*, J. Differential Equations, **247** (2009), pp. 447–464.

[21] S. Göttlich, M. Herty, and A. Klar, *Modelling and optimization of supply chains on complex networks*, Commun. Math. Sci., **4** (2006), pp. 315–330.

[22] G. Leugering and E.J.P.G. Schmidt, *On the modelling and stabilization of flows in networks of open canals*, SIAM J. Control Optim., **41** (2002), pp. 164–180 (electronic).

[23] M.J. Lighthill and G.B. Whitham, *On kinematic waves. II. A theory of traffic flow on long crowded roads*, Proc. Roy. Soc. London. Ser. A., **229** (1955), pp. 317–345.

[24] T.-P. Liu and T. Yang, L_1 *stability of conservation laws with coinciding Hugoniot and characteristic curves*, Indiana Univ. Math. J., **48** (1999), pp. 237–247.

[25] P.I. Richards, *Shock waves on the highway*, Operations Res., **4** (1956), pp. 42–51.

NONLINEAR HYPERBOLIC SURFACE WAVES

JOHN K. HUNTER*

Abstract. We describe examples of hyperbolic surface waves and discuss their connection with initial boundary value and discontinuity problems for hyperbolic systems of PDEs that are weakly but not uniformly stable.

Key words. Nonlinear waves, conservation laws, initial-boundary value problems.

AMS(MOS) subject classifications. Primary 35L50, 35L65, 74J15, 74J30.

1. Introduction. Hyperbolic surface waves are described by initial boundary value problems (IBVPs) and discontinuity problems for hyperbolic systems of conservation laws that are weakly stable but not uniformly stable. Our aim here is to give an informal overview of such surface waves, with an emphasis on 'genuine' surface waves, or surface waves 'of finite energy,' that are localized near the boundary or discontinuity and decay exponentially into the interior. In particular, we are interested in understanding the effect of nonlinearity on such waves.

A well-known example of a genuine hyperbolic surface wave is the Rayleigh wave, or surface acoustic wave (SAW), that propagates on the stress-free boundary of an elastic half space. Rayleigh waves are important in seismology and electronics, where ultrasonic SAW devices are widely used as components in cell phones and other systems (although, at present, nearly always in a linear regime).

Another example of a genuine hyperbolic surface wave occurs in magnetohydrodynamics (MHD). We consider incompressible MHD for simplicity. A tangential discontinuity consists of a magnetic field and fluid velocity that are tangent to the discontinuity and jump across it. If the jump in the velocity is nonzero and the magnetic field is zero, then the tangential discontinuity is a vortex sheet, which is subject to the Kelvin-Helmholtz instability. If the magnetic field is large enough compared with the jump in the velocity, then it stabilizes the discontinuity. In that case, the discontinuity is weakly but not strongly stable and genuine surface waves propagate along it.

In a unidirectional surface wave, the displacement $y = \phi(x,t)$ of a weakly stable tangential discontinuity satisfies the following quadratically nonlinear, nonlocal asymptotic equation [1]

$$\phi_t + \mathbf{H}\left[\psi\psi_x\right]_x + \psi\phi_{xx} = 0 \qquad \psi = \mathbf{H}[\phi] \tag{1.1}$$

where $\mathbf{H}$ is the spatial Hilbert transform, defined by

$$\mathbf{H}[e^{ikx}] = -i(\operatorname{sgn} k)e^{ikx}.$$

---

*Department of Mathematics, University of California, Davis, CA 95616. Partially supported by NSF grant DMS-0604947.

Equation (1.1) provides a model equation for nonlinear hyperbolic genuine surface waves that plays an analogous role to the inviscid Burgers equation for bulk waves. It was first proposed by Hamilton, Il'insky, and Zabolotskaya [14] as a simplification of asymptotic equations for SAWs, so we refer to it as the HIZ equation for short.

Hyperbolic surface waves are nondispersive. As a result, their nonlinear behavior differs qualitatively from that of dispersive water waves that propagate on the free surface of a fluid. Even shallow water waves, which are nondispersive when their slope is sufficiently small, become dispersive once they steepen; by contrast, hyperbolic surface waves remain nondispersive however steep they become. Similarly, the Benjamin-Ono equation $\phi_t + \phi\phi_x + \mathbf{H}[\phi]_{xx} = 0$ has local nonlinearity and nonlocal linear dispersion, whereas the HIZ equation (1.1) has nonlocal nonlinearity and no dispersion.

2. Half-space hyperbolic IBVPs. The well-posedness of IBVPs for hyperbolic PDEs was studied by Kreiss [18] for systems and Sakamoto [26, 27] for wave equations. Majda [21, 22] extended this analysis to discontinuity problems for hyperbolic systems of conservation laws and used the results to study the stability of shock waves.

Consider a half-space IBVP in d-space dimensions that consists of a hyperbolic system of conservation laws for $u(x,t) \in \mathbb{R}^n$

$$u_t + \sum_{j=1}^{d} f^j(u)_{x_j} = 0 \qquad x_d > 0,\, t > 0 \tag{2.1}$$

with initial and boundary conditions

$$u = u_0 \quad \text{on } t = 0, \qquad h(u) = 0 \quad \text{on } x_d = 0. \tag{2.2}$$

Here, $x = (x_1, \dots, x_d) \in \mathbb{R}^d$, $f^j : \mathbb{R}^n \to \mathbb{R}^n$ is the flux vector in the jth direction, and $h : \mathbb{R}^n \to \mathbb{R}^m$ defines the boundary conditions. Nonhyperbolic systems of conservation laws, such as the incompressible Euler or MHD equations, may be regarded as a limit of the corresponding compressible equations. We consider only inviscid equations here; for viscous problems, see *e.g.* [30].

Neither the geometry nor the PDE define length or time parameters, and the IBVP is invariant under rescalings $x \mapsto cx$, $t \mapsto ct$ for any $c > 0$. Thus, the existence of a solution that is bounded in space and grows in time implies the existence of spatially-bounded solutions that grow arbitrarily quickly in time. As a result, there is a sharp division between weakly or strongly stable problems without growing modes and 'violently' unstable problems with growing modes. Unstable problems typically arise when the boundary conditions couple positive and negative energy bulk waves [19].

The linearization of (2.1)–(2.2) about a constant solution has the form

$$u_t + \sum_{j=1}^{d} A^j u_{x^j} = 0 \qquad \text{for } x_d,\ t > 0,$$
$$Cu = 0 \quad \text{on } x_d = 0, \qquad u = u_0 \quad \text{on } t = 0$$

where $A^j \in \mathbb{R}^{n\times n}$, $C \in \mathbb{R}^{m\times n}$ are constant matrices. For definiteness, We suppose that $x_d = 0$ is non-characteristic, meaning that A^d is non-singular.[1] Then a necessary condition for well-posedness is that the number of BCs m must equal the number of positive eigenvalues of A^d and the BCs must determine the associated wave-components that propagate into the half-space in terms of the wave-components that propagate out of the half-space.

A sufficient condition for the strong L^2-well-posedness of the IBVP in any number of space dimensions is provided by a uniform Kreiss-Sakamoto-Lopatinski condition, or uniform Lopatinski condition for short. To obtain this condition, we look for Fourier solutions of the IBVP that oscillate tangent to the boundary and decay away from it of the form

$$u(x,t) = e^{i(k_1x_1+\cdots+k_{d-1}x_{d-1})-i\omega t}U(x_d)$$

where $k = (k_1,\ldots,k_{d-1}) \in \mathbb{R}^{d-1}$, $\omega \in \mathbb{C}$, and $U(x_d) \to 0$ as $x_d \to +\infty$. Typically, U decays exponentially and given by

$$U(x_d) = \sum_{j=1}^{m} c_j e^{-\beta_j(\omega,k)x_d} r_j(\omega,k)$$

where $r_j(\omega,k) \in \mathbb{C}^n$ and $\beta_j(\omega,k) \in \mathbb{C}$ with $\operatorname{Re}\beta_j > 0$. In the case of repeated roots, we may obtain additional polynomial factors in x_d.

There are three alternatives that correspond in a rough sense (made precise by the Kreiss theory) to the growth, oscillation, or decay in time of appropriate Fourier solutions. (a) If there is a solution of the PDE and BC that grows in time and decays in space away from boundary, meaning that $\operatorname{Im}\omega > 0$ and $\operatorname{Re}\beta_j > 0$ for every $1 \le j \le m$ for some $k \in \mathbb{R}^{d-1}$, then the IBVP is subject to Hadamard instability and is ill-posed in any Sobolev space. (b) If there is a solution of the PDE and BC that oscillates in time and does not grow away from boundary, meaning that $\operatorname{Im}\omega = 0$ and $\operatorname{Re}\beta_j \ge 0$ for some $k \in \mathbb{R}^{d-1}$, *and* this solution is a limit of solutions of the PDE that grow in time and decay in space away from boundary, then the linearized problem is weakly stable in L^2-Sobolev spaces, but there is a 'loss' of derivatives in the energy estimates. (c) If there are no such

[1] Characteristic IBVPs arise in many applications, for example in the study of discontinuities that move with a fluid such as the tangential discontinuities considered below, but it is more difficult to develop a general theory for them *c.f.* [7] for further discussion.

solutions of the PDE and BC that grow or oscillate in time, then there are good L^2-energy estimates and the problem is strongly stable. This case occurs when the norm of a suitable $m \times m$ matrix $L(\omega, k)$ is uniformly bounded away from zero in $\operatorname{Im}\omega > 0$ and $k \in \mathbb{S}^{d-2}$. We then say that the IBVP satisfies a uniform Lopatinski condition. See [7] for a detailed description and derivation of these results.

In the weakly stable case (b) there are two alternatives for the spatial behavior of the surface waves in the limit that they are approximated by spatially decaying solutions of the PDE with $\operatorname{Im}\omega \to 0^+$. (b1) If $\lim \operatorname{Re}\beta_j > 0$ for every j, then we get genuine surface waves that decay away from the boundary. (b2) If $\lim \operatorname{Re}\beta_j = 0$ for some j, then we get 'radiative' or 'leaky' surface waves that generate bulk waves which propagate away from the boundary into the interior.

As an illustration, consider solutions of the scalar wave equation

$$u_{tt} = u_{xx} + u_{yy} \tag{2.3}$$

that propagate with speed λ along the boundary $y = 0$. If $|\lambda| < 1$, then the corresponding Fourier solution is given by

$$u(x, y, t) = e^{i(x-\lambda t) - \sqrt{1-\lambda^2}\, y}.$$

Thus, a surface wave that is 'subsonic' with respect to the speed of bulk waves is genuine. For example, the speed of a Rayleigh wave in isotropic elasticity is less than both the transverse and longitudinal bulk wave speeds. If $|\lambda| > 1$, then the Fourier solution is

$$u(x, y, t) = e^{i(x-\lambda t) + i\sqrt{\lambda^2-1}\, y},$$

and a 'supersonic' surface wave is radiative. The difference between 'subsonic' genuine waves and 'supersonic' radiative waves is a consequence of the geometrical effect that the phase speed along the boundary of a plane bulk wave propagating at an angle θ to a boundary is *greater* than its normal phase speed by a factor of $\sec\theta$.

Next, consider (2.3) in the half-space $y > 0$ with the initial and boundary conditions [23]

$$u = u_0, \quad u_t = v_0 \quad \text{on } t = 0, \qquad \gamma u_t + u_y = 0 \quad \text{on } y = 0 \tag{2.4}$$

where $\gamma \in \mathbb{R}$ is a parameter. The BC in (2.4) is invariant under a simultaneous spatial reflection and time reversal $y \mapsto -y$, $t \mapsto -t$, but not under $t \mapsto -t$ alone. Thus, unlike the free space problem, the IBVP is not reversible in time and IBVPs that are stable forward in time may be unstable backward in time.

The IBVP (2.3)–(2.4) is uniformly stable if $\gamma < 0$, weakly stable if $0 \le \gamma \le 1$, and unstable if $\gamma > 1$. For $\gamma > 1$, the mode that grows in t and decays in y is

$$u(x,y,t) = \exp\left[ix + \sigma(t - \gamma y)\right], \qquad \sigma = \frac{1}{\sqrt{\gamma^2 - 1}}.$$

For $\gamma < -1$, this mode grows in t only if it grows in y, consistent with the uniform stability. The weak stability for $0 < \gamma < 1$ is a consequence of the existence of surface waves of the form

$$u(x,y,t) = \exp\left[ix + i\lambda(\gamma y - t)\right], \qquad \lambda = \pm\frac{1}{\sqrt{1 - \gamma^2}}$$

that radiate bulk waves into the interior $y > 0$. These solutions also exist for $-1 < \gamma < 0$, but in that case they are not the limit of of solutions of the PDE that grow in time and decay in space, so their existence does not violate the uniform Lopatinski condition. Equivalently, the waves are coupled with bulk waves whose group velocity is directed toward the boundary, not away from it as in the weakly stable case [15].

This IBVP for a scalar wave equation leads to radiative surface waves in the weakly stable case, but at least two wave equations are required to obtain genuine surface waves. Thus, genuine surface waves do not arise in gas dynamics, only in larger systems such as elasticity or MHD.

3. Discontinuity problems. Discontinuity problems for hyperbolic system of conservation laws lead to analogous equations to the ones for IBVPs in which the location of the discontinuity appears as an additional variable. Consider a solution of (2.1) that contains a discontinuity — such as a shock wave, vortex sheet, or contact discontinuity — located at

$$x_d = \Phi\left(x_1, \ldots, x_{d-1}, t\right).$$

The PDE for $u = u^+$ in $x_d > \Phi$ and $u = u^-$ in $x_d < \Phi$ can be regarded as giving a single half-space PDE in twice the number of dependent variables. The jump conditions for (2.1) have the form

$$\Phi_t + W\left(u^+, u^-, \nabla\Phi\right) = 0, \quad h\left(u^+, u^-, \nabla\Phi\right) = 0 \qquad \text{on } x_d = \Phi,$$

which gives as an equation for the motion of the discontinuity as well as BCs on the discontinuity. Linearization of these equations about a planar discontinuity leads to similar equations to the linearized half-space equations.

If the unperturbed discontinuity is a Lax shock, then the boundary is non-characteristic. The Lax shock condition (that characteristics in one family enter the shock from both sides and characteristics in the other families cross the shock) implies that we have the correct number of boundary conditions: n jump conditions give $(n-1)$ BCs for the waves that propagate into the half-spaces on either side of shock and one condition for the shock location. We also require an 'evolutionary' condition, meaning that the jump conditions determine the wave components that are outgoing from the shock in terms of the incoming wave components.

Overcompressive shocks give linearized IBVPs with too many BCs, and such shocks typically split into multiple Lax shocks under arbitrary perturbations. Undercompressive shocks give linearized IBVPs with too few BCs, and extra conditions, or 'kinetic relations,' are required to determine their motion.

An evolutionary Lax shock is strongly stable in several space dimensions if it satisfies an analog of the uniform Lopatinski condition [21]. If the Lopatinski condition fails entirely, then there are modes that grow arbitrarily quickly, and the discontinuity is violently unstable as in the Kelvin-Helmholtz instability of a vortex sheet. If the discontinuity is weakly but not strongly stable, then surface waves propagate along the discontinuity.

Surface waves on a discontinuity may be radiative or genuine. Examples of discontinuities that support radiative surface waves include strong shocks in the compressible Euler equations for some equations of state [23], supersonic vortex sheets in a compressible fluid [3, 10], and detonation waves in combustion [24]. Examples of discontinuities that support genuine surface waves are tangential discontinuities in MHD [8, 29] and propagating phase boundaries in a van der Waals fluid [5, 6].

Short-time existence results have been obtained for various weakly stable problems, typically by the use of a Nash-Moser scheme to compensate for the loss of derivatives in the linearized energy estimates *e.g.* [8, 10, 25, 29]. The Nash-Moser scheme is a powerful method but because of its generality it makes relatively little use of the specific features of these problems, such as the distinction between genuine and radiative surface waves or the structure of the nonlinearity.

4. Magnetohydrodynamics. The non-conservative form of the system of MHD equations for a conducting, incompressible fluid with velocity $\mathbf{u}$, magnetic field $\mathbf{B}$, and total pressure p is

$$\begin{aligned} &\mathbf{u}_t + \mathbf{u}\cdot\nabla\mathbf{u} - \mathbf{B}\cdot\nabla\mathbf{B} + \nabla p = 0, \qquad && \operatorname{div}\mathbf{u} = 0,\\ &\mathbf{B}_t + \mathbf{u}\cdot\nabla\mathbf{B} - \mathbf{B}\cdot\nabla\mathbf{u} = 0, && \operatorname{div}\mathbf{B} = 0. \end{aligned}$$

The jump conditions for a tangential discontinuity located at $\Phi(\mathbf{x}, t) = 0$ are

$$\Phi_t + \mathbf{u}\cdot\nabla\Phi = 0, \quad \mathbf{B}\cdot\nabla\Phi = 0, \quad [p] = 0 \qquad \text{on } \Phi = 0$$

where $[p]$ denotes the jump in p across the discontinuity.

A planar tangential discontinuity located at $y = 0$, say, consists of arbitrary velocities and magnetic fields that are constant in each half-space $y > 0$ and $y < 0$, tangent to the discontinuity, and jump across it:

$$\mathbf{u} = \begin{cases} \mathbf{U}_+ & y > 0 \\ \mathbf{U}_- & y < 0 \end{cases}, \qquad \mathbf{B} = \begin{cases} \mathbf{B}_+ & y > 0 \\ \mathbf{B}_- & y < 0 \end{cases}.$$

If $\mathbf{B}_\pm = 0$ and $\mathbf{U}_+ \neq \mathbf{U}_-$, then the tangential discontinuity is an unstable vortex sheet. If the magnetic field components in the direction of the discontinuous velocity components are strong enough, specifically if [20]

$$|\mathbf{B}_+|^2 + |\mathbf{B}_-|^2 > \frac{1}{2}|\mathbf{U}_+ - \mathbf{U}_-|^2,$$
$$|\mathbf{B}_+ \times \mathbf{B}_-|^2 \geq \frac{1}{2}\left(|\mathbf{B}_+ \times \mathbf{U}_+|^2 + |\mathbf{B}_- \times \mathbf{U}_-|^2\right),$$

then the tangential discontinuity is weakly stable, and genuine surface waves propagate along it.

For times that are not too long compared with a typical period, which we nondimensionalize to be of the order one, a unidirectional surface wave with small slope of the order ε consists of an arbitrary profile that propagates without change of shape at the linearized surface wave speed λ. Over longer times, of the order ε^{-1}, nonlinear effects distort the profile. Considering planar flows for simplicity, one finds that the weakly nonlinear solution for the location of the tangential discontinuity in a surface wave is given by [1]

$$y = \varepsilon\phi(x - \lambda t, \varepsilon t) + O(\varepsilon^2) \qquad \text{as } \varepsilon \to 0$$

where $\phi(x, t)$ satisfies (1.1) after the change of variables $x - \lambda t \mapsto x$, $\varepsilon t \mapsto t$ and a normalization.

It is interesting to consider the dynamics of a tangential discontinuity in MHD near the onset of Kelvin-Helmholtz instability. This bifurcation problem differs from standard ones because all of the spatial modes of the solution become unstable simultaneously.

In a reference frame moving with the linearized surface wave speed at the bifurcation point where the left and right wave velocities coalesce and become complex, the location $y = \phi(x, t)$ of the discontinuity satisfies the normalized asymptotic equation [17]

$$\phi_{tt} + \left(\mathbf{H}\left[\psi\psi_x\right]_x + \psi\phi_{xx}\right)_x = \mu\phi_{xx} \qquad \psi = \mathbf{H}[\phi]. \tag{4.1}$$

Here, $\mu = \mathbf{B}_+^2 + \mathbf{B}_-^2 - \frac{1}{2}\left(\mathbf{U}_+ - \mathbf{U}_-\right)^2$ is the bifurcation parameter. If $\mu > 0$, then the tangential discontinuity is linearly stable and the asymptotic equation is a wave equation. For large μ, it reduces in a unidirectional approximation $\phi_t \sim \pm\sqrt{\mu}\phi_x$ to decoupled equations of the form (1.1). If $\mu < 0$, then the tangential discontinuity is linearly unstable, and the linearized asymptotic equation is a Laplace equation that is subject to Hadamard instability, corresponding to the Kelvin-Helmholtz instability of the discontinuity.

One might hope that nonlinearity has a regularizing effect on the Hadamard instability (as in the model equations studied in [4]). Preliminary numerical solutions of (4.1), however, suggest the reverse: finite amplitude effects appear to trigger a Kelvin-Helmholtz stability on the less stable side of the discontinuity where the magnetic field is weaker.

The apparent ill-posedness in any Sobolev space of vortex sheet solutions of the Euler and MHD equations raises questions about the validity of these equations as general physical models, especially for compressible flows where vortex sheets can be generated spontaneously from smooth initial data by the formation of shocks and triple points. Perhaps related to this instability is the apparent nonuniqueness of numerical solutions of the compressible Euler equations containing vortex sheets [13] and the nonuniqueness of low-regularity weak solutions [12].

5. The HIZ equation. Consider, for definiteness, solutions of the HIZ equation (1.1) that are 2π-periodic in space with the Fourier expansion

$$\phi(x,t) = \sum_{k\in\mathbb{Z}} \hat{\phi}(k,t)e^{ikx}. \tag{5.1}$$

The spectral form of (1.1) is

$$\hat{\phi}_t(k,t) + i(\operatorname{sgn} k)\sum_{n\in\mathbb{Z}} T(-k,k-n,n)\phi(k-n,t)\phi(n,t) = 0 \tag{5.2}$$

where the interaction coefficient $T : \mathbb{Z}\times\mathbb{Z}\times\mathbb{Z}\to\mathbb{C}$ is given by

$$T(k,m,n) = \frac{2|kmn|}{|k|+|m|+|n|}. \tag{5.3}$$

The coefficient $T(k,m,n)$ describes the strength of the coupling between Fourier modes in the resonant three-wave interaction $m+n\to -k$. Only the values of $T(k,m,n)$ on $k+m+n=0$ appear in (5.2), but it is convenient to show all three wave numbers explicitly.

The symmetry of $T(k,m,n)$ in (k,m,n) is equivalent to the existence of a canonical Hamiltonian structure for the HIZ equation; its spatial form is

$$\phi_t = \mathbf{H}\left[\frac{\delta\mathcal{H}}{\delta\phi}\right], \qquad \mathcal{H}(\phi) = \int \psi\phi_x\psi_x\,dx, \qquad \psi = \mathbf{H}[\phi]. \tag{5.4}$$

The homogeneity of T of degree two is implied by a dimensional analysis of quadratically nonlinear nondispersive Hamiltonian waves, depending only on velocity parameters, that propagate on a boundary with codimension one [2].

It follows from (5.3) that

$$|T(k,m,n)| \le |kmn|^{1/2}\min\left\{|k|^{1/2}, |m|^{1/2}, |n|^{1/2}\right\} \tag{5.5}$$

on $k+m+n=0$. This estimate implies a limit on the rate at which energy can be transferred from low to high wavenumbers by three-wave interactions, since the interaction coefficient is bounded by a factor that depends on the *lowest* wavenumber that participates in the interaction. Related to this spectral property is the following spatial form of (1.1)

$$\phi_t + [\mathbf{H},\psi]\,\psi_{xx} + \mathbf{H}\left[\psi_x^2\right] = 0 \tag{5.6}$$

where $[\mathbf{H}, \psi] = \mathbf{H}\psi - \psi\mathbf{H}$ is the commutator of $\mathbf{H}$ with multiplication by ψ. As (5.6) shows, there is a cancelation of the second order spatial derivatives appearing in (1.1). Generalizing (5.5), we make the following definition.

DEFINITION 5.1. *An interaction coefficient $T : \mathbb{Z} \times \mathbb{Z} \times \mathbb{Z} \to \mathbb{C}$ has minimal spectral growth with exponent $\mu \geq 0$ if there is a constant C such that*

$$|T(k, m, n)| \leq C|kmn|^{1/2} \min\{|k|^\mu, |m|^\mu, |n|^\mu\}$$

for every $k, m, n \in \mathbb{Z}$ such that $k + m + n = 0$.

For example, the HIZ-coefficient (5.3) has minimal spectral growth with exponent $\mu = 1/2$. We then have the following local existence result [16].

THEOREM 5.1. *Suppose that $T : \mathbb{Z} \times \mathbb{Z} \times \mathbb{Z} \to \mathbb{C}$ is a symmetric interaction coefficient that has minimal spectral growth with exponent $\mu \geq 0$. Let $s > \mu + 2$ and $\phi_0 \in H^s(\mathbb{T})$. Then there exists $t_* > 0$, depending only on T and $\|\phi_0\|_{H^s}$, and a unique local solution*

$$\phi \in C([-t_*, t_*]; H^s(\mathbb{T})) \cap C^1\left([-t_*, t_*]; H^{s-1}(\mathbb{T})\right)$$

of (5.1)–(5.2) with $\phi(0) = \phi_0$.

Thus, we get the short-time existence of smooth H^s-valued solutions of (1.1) for $s > 5/2$. The proof uses a standard energy method in Fourier space, and the main point of this result is the identification of the minimal spectral growth condition as a sufficient condition for well-posedness. This condition allows one to estimate a term such as $|k|^\alpha |m|^\beta T(k, m, n)$ by a term proportional to $|k|^\alpha |m|^\beta |n|^\mu$, thus avoiding a loss of derivatives. This corresponds to the use of integration by parts to eliminate higher derivatives in spatial estimates, although the nonlocality of the spatial form of (5.2) makes it more difficult to carry out the spatial estimates directly. If T does not satisfy the minimal spectral growth condition, then there appears to be a loss of derivatives in the energy estimates for solutions of (5.1)–(5.2), and it is not clear that local H^s-solutions exist for any $s \in \mathbb{R}$.

An interesting question is to characterize which boundary conditions in weakly stable IBVP problems with genuine surface waves for Hamiltonian or variational systems of PDEs lead to asymptotic equations whose interaction coefficients satisfy the minimal spectral growth condition. This would provide a possible criterion for the identification of weakly stable problems with a good nonlinear theory.

Numerical results obtained using a spectral viscosity method show the formation of singularities in smooth solutions of (1.1) in finite time (see Figure 1). The solution appears to remain continuous while its derivative blows up. Furthermore, solutions of a viscous regularization of (1.1) appear to converge to a weak solution of (1.1) in the zero-viscosity limit.

Using the distributional formula for the Hilbert transform,

$$\mathbf{H}[|x|^\alpha] = -c_\alpha \operatorname{sgn} x |x|^\alpha \quad c_\alpha = \tan\frac{\pi\alpha}{2} \qquad \text{for } \alpha \notin 2\mathbb{Z} + 1,$$

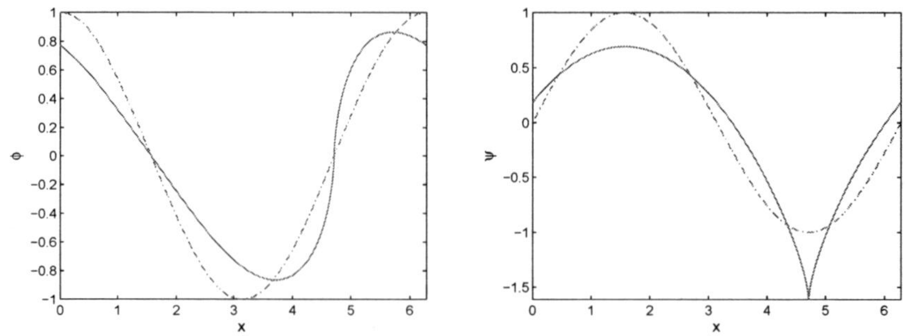

FIG. 1. *Formation of a singularity in the solution of (1.1) with initial data* $\phi(x, 0) = \cos x$. *The dashed lines are the initial data and the solid lines are the solution at time* $t = 0.8$. *The plot on the left is* ϕ*; the plot on the right is* $\psi = \mathbf{H}\phi$.

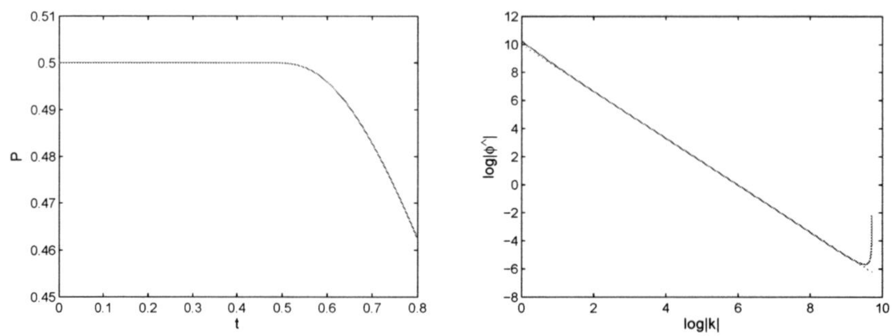

FIG. 2. *The plot on the left is the momentum* $P = (2\pi)^{-1} \int_0^{2\pi} \phi|\partial_x|\phi \, dx$ *as a function of time for the solution shown in Figure 1;* P *is conserved until the singularity develops at* $t \approx 0.5$. *The plot on the right is the spectrum for the numerical solution with* 32768 *modes at* $t = 0.8$. *The dotted line is a best linear fit to the power-law range of the spectrum, which gives* $|\hat{\phi}(k)| \sim C|k|^{-\alpha}$ *with* $\alpha \approx 1.66657$ *and* $\log C \approx 9.96648$.

we find that an exact steady distributional solution of (1.1) is

$$\phi(x) = A \operatorname{sgn} x \, |x|^{\alpha} \tag{5.7}$$

where A is an arbitrary constant and α satisfies $(2\alpha - 1)c_{2\alpha} = (\alpha - 1)c_{\alpha}$. The unique solution of this equation with $\alpha, 2\alpha \notin 2\mathbb{Z} + 1$ is $\alpha = 2/3$. As shown in Figure 2, numerical solutions have a power-law spectral decay with exponent close to $5/3$, consistent with this analytical solution. Only singularities with $A > 0$ appear, presumably as the result of a viscous admissibility condition.

The validity of the asymptotic solution for MHD once the derivative of ϕ blows up is open to question. In particular, although the asymptotic solutions for the fluid velocity and magnetic field are smooth in the interior, they are unbounded near a singularity in the tangential discontinuity. Nevertheless, analogous weakly nonlinear asymptotic solutions for bulk waves correctly describe the formation and propagation of shocks and it is plausible that the asymptotic solution for surface waves remains valid in a weak sense. The asymptotic analysis therefore predicts the formation of singularities with Hölder-exponent 2/3 in the displacement of a weakly stable tangential discontinuity in incompressible MHD. This mechanism for singularity formation on the boundary in weakly stable IBVPs and discontinuities differs qualitatively from shock formation in the interior.

The numerical results suggest that (1.1) has global admissible weak solutions, but a proof of this conjecture is open. Apart from the energy $\int \psi\phi_x\psi_x\, dx$, which is not positive definite, the only globally conserved quantity for smooth solutions of (1.1) that we know of is the momentum $\int \phi|\partial_x|\phi\, dx$, which is the homogeneous $H^{1/2}$-norm of ϕ. This quantity is conserved for smooth solutions and decreases for weak solutions that satisfy a viscous admissibility condition, but it does not seem to be strong enough to imply the global existence of weak solutions.

6. Conclusion. Despite significant recent progress in the analysis of weakly stable IBVPs and discontinuity problems in hyperbolic conservation laws, much remains to be understood. It seems likely that one will need to identify and use crucial features of the nonlinear structure of these problems including, perhaps, their Hamiltonian or variational structure [28] and their Lagrangian formulation [9, 11]. There appears to be little hope of a reasonable nonlinear theory for 'violently' unstable IBVPs or discontinuities, such as vortex sheets. This raises serious questions about the consistency of hyperbolic conservation laws, which neglect all small-scale regularizing effects, as mathematical models for such problems.

REFERENCES

[1] G. Alì and J.K. Hunter, *Nonlinear surface waves on a tangential discontinuity in magnetohydrodynamics*, Quart. Appl. Math., **61** (2003), 451–474.

[2] G. Alì, J.K. Hunter, and D.F. Parker, *Hamiltonian equations for nonlinear scale-invariant waves*, Stud. Appl. Math., **108** (2002) 305–321.

[3] M. Artola and A.J. Majda, *Nonlinear development of instabilities in supersonic vortex sheets I. The basic kink modes*, Physica D, **28** (1987), 253–281.

[4] J.T. Beale and D. Schaeffer, *Nonlinear behavior of model equations which are linearly ill-posed*, Comm. Partial Differential Equations, **13** (1988), 423–467.

[5] S. Benzoni-Gavage, *Stability of subsonic planar phase boundaries in a van der Waals fluid*, Arch. Rat. Mech. Anal. **150** (1999), 23–55.

[6] S. Benzoni-Gavage and H. Freistühler, *Effects of surface tension on the stability of dynamical liquid-vapor interfaces*, Arch. Ration. Mech. Anal., **174** (2004), 111–150.

[7] S. Benzoni-Gavage and D. Serre, *Multi-dimensional Hyperbolic Partial Differential Equations*, Oxford University Press, Oxford, 2007.

[8] G.-Q. Chen and Y.-G. Wang, *Existence and stability of compressible current-vortex sheets in three-dimensional magnetohydrodynamics*, Arch. Rational Mech. Anal., **187** (2008), 369–408.

[9] A. Cheng, D. Coutand, and S. Shkoller, *On the Motion of Vortex Sheets with Surface Tension in the 3D Euler Equations with Vorticity*, Comm. Pure Appl. Math., **61** (2008), 1715–1752.

[10] J.F. Coulombel and P. Secchi, *Stability of compressible vortex sheet in two space dimensions*, Indiana Univ. Math. J. **53** (2004), 941-1012.

[11] D. Coutand and S. Shkoller, *Well-posedness of the free-surface incompressible Euler equations with or without surface tension*, J. Amer. Math. Soc., **20** (2007), 829–930.

[12] C. de Lellis and L. Székelyhidi, *The Euler equations from the point of view of differential inclusions*, Mat. Ital. **9** (2008), 873–879.

[13] V. Elling, *A possible counter example to well posedness of entropy solutions and to Godunov scheme convergence*, Math. Comp., **75** (2006), 1721–1733.

[14] M.F. Hamilton, Yu. A. Il'insky, and E.A. Zabolotskaya, *Evolution equations for nonlinear Rayleigh waves*, J. Acoust. Soc. Am., **97** (1995), 891–897.

[15] R.L. Higdon, *Initial-boundary value problems for linear hyperbolic systems*, SIAM Review, **28** (1986), 177–217.

[16] J.K. Hunter, *Short-time existence for scale-invariant Hamiltonian waves*, J. Hyperbolic Differential Equations, **3** (2006), 247–267.

[17] J.K. Hunter and J.B. Thoo, *A weakly nonlinear Kelvin-Helmholtz instability in MHD*, in preparation.

[18] H.-O. Kreiss, *Initial boundary value problems for hyperbolic systems*, Comm. Pure Appl. Math., **23** (1970), 277–298.

[19] S.C. Lashmore-Davies, *Negative energy waves*, J. Plasma Physics, **71** (2005), 101–109.

[20] L.D. Landau, L.P. Pitaevskii, and E.M. Lifschitz, *The Electrodynamics of Continuous Media*, Pergamon Press, New York, 2nd ed., 1984.

[21] A. J. Majda, *The stability of multidimensional shock fronts*, Mem. Amer. Math. Soc., **41** (1983), no. 275.

[22] A. J. Majda, *The existence of multi-dimensional shock fronts*, Mem. Amer. Math. Soc., **43** (1983), no. 281, AMS, Providence, 1983.

[23] A. J. Majda, *Compressible Fluid Flow and Systems of Conservation Laws in Several Space Dimensions*, Springer-Verlag, 1984.

[24] A. J. Majda, and R. R. Rosales, *A theory for spontaneous Mach stem formation in reacting shock fronts. I. The basic perturbation analysis*, SIAM J. Appl. Math., **43** (1983), 1310–1334.

[25] M. Sable-Tougeron, *Existence pour un problème de l'élastodynamique Neumann non linéaire en dimension* 2, Arch. Rational Mech. Anal., **101** (1988), 261–292.

[26] R. Sakamoto, *Mixed problems for hyperbolic equations. I. Energy inequalities*, J. Math. Kyoto Univ., **10** (1970), 349–373.

[27] R. Sakamoto, *Mixed problems for hyperbolic equations. II. Existence theorems with zero initial datas and energy inequalities with initial datas, J. Math. Kyoto Univ.*, **10** (1970), 403–417.

[28] D. Serre, *Second order initial boundary-value problems of variational type*, J. Functional Anal., **236** (2006), 409–446.

[29] Y. Trakhinin, *Existence of compressible current-vortex sheets: variable coefficients linear analysis*, Arch. Ration. Mech. Anal. **177** (2005), 331-366.

[30] K. Zumbrun, *Multidimensional stability of planar viscous shock waves*, Advances in the theory of shock waves, 307–516, Progr. Nonlinear Differential Equations Appl., **47**, Birkhäuser Boston, Boston, MA, 2001.

VACUUM IN GAS AND FLUID DYNAMICS

JUHI JANG* AND NADER MASMOUDI†

Abstract. In this paper, we review some interesting problems of vacuum states arising in hyperbolic conservations laws with applications to gas and fluid dynamics. We present the current status of the understanding of compressible Euler flows near vacuum and discuss related open problems.

Key words. Compressible Euler equations, physical vacuum, free boundary, energy estimates.

1. Introduction. A system of conservation laws is the collection of partial differential equations in divergence form, which describe the dynamics of continua such as fluids and plasmas, by means of the physical principles of conservation of mass, momentum, and energy with constitutive relations encoding the material properties of the medium. Systems of conservation laws identify the theory of mechanics, thermodynamics, electrodynamics and so on. Its canonical form is $\partial_t U + \sum_{j=1}^d \partial_{x_j} F_j(U) = 0$ where $U = U(t, x_1, \ldots, x_d) \in \mathbb{R}^n$ is a vector of conserved quantities and $F_j(U) \in \mathbb{R}^n$ is a flux function. This system is called hyperbolic in the t-direction if for any fixed U and $\nu \in S^{d-1}$, the $n \times n$ matrix $\sum_{j=1}^d \nu_j D_U F_j(U)$ has only real eigenvalues $\lambda_1, \ldots, \lambda_n$, called characteristic speeds, and it is diagonalizable [15].

One of the most fundamental examples of a system of hyperbolic conservation laws is the compressible Euler system of isentropic, ideal gas dynamics. In Eulerian coordinates, it takes the form:

$$\begin{aligned} \partial_t \rho + \operatorname{div}(\rho u) &= 0\,, \\ (\rho u)_t + \operatorname{div}(\rho u \otimes u) + \operatorname{grad} p &= 0\,. \end{aligned} \tag{1.1}$$

Here ρ, u and p denote respectively the density, velocity, and pressure of the gas. The first and second equations express respectively conservation of mass and momentum. In considering the polytropic gases, the constitutive relation which is also called the equation of state is given by

$$p = K\rho^\gamma \tag{1.2}$$

where K is an entropy constant and $\gamma > 1$ is the adiabatic gas exponent. The case of $\gamma = 1$ corresponds to the isothermal gas flow. In this article, we discuss isentropic compressible Euler equations (1.1) with (1.2) rather than general hyperbolic conservation laws. The main interest is to study vacuum states in the framework of classical solutions.

*Courant Institute, New York University, New York, NY 10012 (juhijang@cims.nyu.edu).

†Courant Institute, New York University, New York, NY 10012 (masmoudi@cims.nyu.edu).

When the initial density function contains a vacuum, the vacuum boundary Γ is defined as

$$\Gamma = cl\{(t,x) : \rho(t,x) > 0\} \cap cl\{(t,x) : \rho(t,x) = 0\}$$

where cl denotes the closure. We also introduce the sound speed c of Euler equations (1.1)

$$c = \sqrt{\frac{d}{d\rho}p(\rho)} \quad \Big(= \sqrt{A\gamma}\rho^{\frac{\gamma-1}{2}} \text{ for polytropic gases}\Big).$$

We recall that for one-dimensional Euler flows, $u \pm c$ are the characteristic speeds. If Γ is nonempty, Euler equations become degenerate along Γ, namely degenerate hyperbolic.

We briefly review some existence theories of compressible flows. In the absence of vacuum, namely if the density is bounded below from zero everywhere, then one can use the theory of symmetric hyperbolic systems developed by Friedrichs-Lax-Kato [19, 31, 33]; for instance, see Majda [46]. The breakdown of classical solutions was demonstrated by Sideris [61].

When the initial datum is compactly supported, there are at least three ways of looking at the problem. The first consists in solving the Euler equations in the whole space and requiring that the system (1.1) holds in the sense of distribution for all $x \in \mathbb{R}^d$ and $t \in [0,T]$. This is in particular the strategy used to construct global weak solutions (see for instance Diperna [16] and [10, 39]). The second way consists in symmetrizing the system first and then solving it using the theory of symmetric hyperbolic system. Again the symmetrized form has to be solved in the whole space. The third way is to require the Euler equations to hold on the set $\{(t,x) : \rho(t,x) > 0\}$ and write an equation for Γ. Here, the vacuum boundary Γ is part of the unknown: this is a free boundary problem and in this case, an appropriate boundary condition at vacuum is necessary.

In the first and second ways (called later the Cauchy problem), there is no need of knowing exactly the position of the vacuum boundary. DiPerna used the theory of compensated compactness to pass to the limit weakly in a parabolic approximation of the system and recovered a weak solution of the Euler system (see also [39] where a kinetic formulation of the system was also used). Makino, Ukai and Kawashima [51] wrote the system in a symmetric hyperbolic form which allows the density to vanish. The system they get is not equivalent to the Euler equations when the density vanishes. This special symmetrization was also used for the Euler-Poisson system. This formulation was also used by Chemin [9] to prove the local existence of regular solutions in the sense that $c, u \in C([0,T); H^m(\mathbb{R}^d))$ for some $m > 1 + d/2$ and d is the space dimension (see also Serre [59] and Grassin [22], for some global existence result of classical solutions under some special conditions on the initial data, by extracting a dispersive effect after some invariant transformation). However, it was noted in [47, 48] that the requirement that c is continuously differentiable excludes many interesting solutions such as the stationary solutions of the Euler-Poisson system which have a behavior of the type $\rho \sim |x - x_0|^{\frac{1}{\gamma-1}}$, namely $c^2 \sim |x - x_0|$ near the vacuum boundary. Indeed, Nishida in [55] suggested to consider a free boundary problem which includes this kind of singularity caused by vacuum, not shock wave singularity.

For the third way (called later the free boundary problem), we divide into a few cases according to the initial behavior of the sound speed c. For simplicity, let the origin be the initial vacuum contact point ($x_0 = 0$). And let $c \sim |x|^\alpha$. When $\alpha \geq 1$, namely initial contact to vacuum is smooth enough, Liu and Yang [43] constructed the local-in-time solutions to one-dimensional Euler equations with damping by using the energy method based on the adaptation of the theory of symmetric hyperbolic system and characteristic method. They also prove that c^2 can not be smooth across Γ after a finite time. We note that in these regimes there is no acceleration along the vacuum boundary.

For $0 < \alpha < 1$, the initial contact to vacuum is only Holder continuous. In particular, the corresponding behavior to $\alpha = 1/2$ can be realized by some self-similar solutions and stationary solutions for different physical systems such as Euler equations with damping, Navier-Stokes-Poisson or Euler-Poisson equations for gaseous stars [27, 28, 44, 64]. This motivates the following definition: A vacuum boundary Γ is called *physical* if the acceleration is bounded from below and above, namely

$$-\infty < \frac{\partial c^2}{\partial n} < 0 \tag{1.3}$$

in a small neighborhood of the boundary, where n is the outward unit normal to Γ. Despite its physical importance, the local existence theory of smooth solutions featuring the physical vacuum boundary even for one-dimensional flows was established only recently. This is because if the physical vacuum boundary condition (1.3) is assumed, the classical theory of hyperbolic systems can not be applied [44, 64]: the characteristic speeds $u \pm c$ become singular with infinite spatial derivatives at the vacuum boundary and this singularity creates an analytical difficulty. The first rigorous result regarding the physical vacuum was given by the authors [29] for one-dimensional Euler equations in mass Lagrangian coordinates based on extracting a new structure lying upon the physical vacuum. Recently Coutand and Shkoller [12] constructed more regular solutions in Lagrangian coordinates. For multi-dimensional case, Coutand, Lindblad and Shkoller [14] have a priori estimates by assuming smoothness of solutions. More recently, Coutand and Shkoller [13] extended their 1D methodology to construct solutions in 3D with physical vacuum and independently of this work, the authors [30] have established the well-posedness of 3D case. The methods are very different.

For $0 < \alpha < 1/2$ or $1/2 < \alpha < 1$, the corresponding boundary behavior is believed to be ill-posed and indeed, we think that it should instantaneously change into the physical vacuum. However, there is no mathematical justification available so far.

The case $\alpha = 0$ is when there is no continuous initial contact of the density with vacuum. It can be considered as either Cauchy problem or free boundary problem. An example of Cauchy problem when $\alpha = 0$ is the Riemann problem for genuinely discontinuous initial datum. For instance, see [24], which was brought up by Hunter [1] and Bouchut [7]. An example of a free boundary problem when $\alpha = 0$ is the work by Lindblad [36] where the density is positive at the vacuum boundary.

The paper proceeds as follows. In Section 2, we discuss the well-posedness of vacuum free boundary problems in detail. In Section 3, the comparison between Euler equations with damping and porus media equations and some open problems are presented. In Section 4, we present some analogues of vacuum singularity in other areas and further discuss other research directions.

2. Euler equations with vacuum free boundary.

2.1. Mass Lagrangian formulation in 1D flows and $\alpha \geq 1$. Let the initial density $\rho_0(0, x)$ for $a \leq x \leq b$ be given so that $\rho_0(a) = 0$ and $\rho_0(x) > 0$ for $a < x \leq b$. Let $a(t)$ be the particle path from $x = a$. We seek $\rho(t, x)$, $u(t, x)$, and $a(t)$ for $t \in [0, T]$, $T > 0$ and $x \in [a(t), b]$ so that for such t and x, $\rho(t, x)$ and $u(t, x)$ satisfy Euler equations (1.1) with boundary conditions

$$\rho(t, a(t)) = 0\,,\ u(t, b) = 0\,,\ 0 < \frac{\partial}{\partial x}\rho^{\frac{\gamma-1}{2\alpha}}\big|_{x=a(t)} < \infty\,.$$

In the following, we present the work of Liu and Yang [43] which concerns the well-posedness of the boundary condition of the case $\alpha \geq 1$. For one-dimensional Euler equations, there is a natural transformation which fixes the free boundary, mass Lagrangian coordinates:

$$y \equiv \int_{a(t)}^{x} \rho(t, z)dz, \quad x \in [a(t), b].$$

Note that the vacuum free boundary $x = a(t)$ corresponds to $y = 0$, and $x = b$ to $y = M$ where M is the total mass of the gas, and thus both boundaries are fixed in (t, y). The Euler equations (1.1) can be rewritten as a symmetric hyperbolic system

$$\phi_t + \mu u_y = 0\,; \quad u_t + \mu\phi_y = 0 \tag{2.1}$$

in the variables $\phi \sim \rho^{(\gamma-1)/2}$ and $\mu \sim \rho^{(\gamma+1)/2}$. Since $\phi \sim y^{\alpha(\gamma-1)/(2\alpha+\gamma-1)}$ vanishes algebraically at $y = 0$, one can try to find another change of variables so that (2.1) reduces to the system in non-vanishing unknown and the corresponding propagation speed becomes smooth in a new variable. To this end, let $z \equiv y^q$, $\phi \equiv z^\beta \eta$ and $u \equiv C + z^\beta \zeta$ where C is a given constant, $\beta = \alpha(\gamma-1)/q(2\alpha+\gamma-1)$ and q is a constant to be determined. Now the interest is when $\eta > 0$. Write (2.1) for η, ζ:

$$\begin{aligned} \eta_t + dz^{(\alpha-1)(\gamma-1)/q(2\alpha+\gamma-1)}\eta^{(\gamma+1)/(\gamma-1)}(z\zeta_z + \beta\zeta) = 0 \\ \zeta_t + dz^{(\alpha-1)(\gamma-1)/q(2\alpha+\gamma-1)}\eta^{(\gamma+1)/(\gamma-1)}(z\eta_z + \beta\eta) = 0 \end{aligned} \tag{2.2}$$

where d is a constant depending on q, γ. We now state the result in [43].

THEOREM 2.1. *Let $\alpha \geq 1$. There exist solutions for the system (2.1) of the following form locally in time $\phi(t,y) = y^{\alpha(\gamma-1)/(2\alpha+\gamma-1)}\eta$, $u(t,y) = C + y^{\alpha(\gamma-1)/(2\alpha+\gamma-1)}\zeta$, where $\eta > 0$, ζ satisfies* (2.2).

Here the point is that for $\alpha \geq 1$, one can choose a positive constant q so that the propagation speed $z^{(\alpha-1)(\gamma-1)/q(2\alpha+\gamma-1)+1}$ of (2.2) is smooth, indeed either z or z^2 and thus Kato's theorem [31] on a symmetric quasi-linear hyperbolic system for η, ζ can be applied. However, this method is not applied to other singular cases $0 < \alpha < 1$.

2.2. Well-posedness of physical vacuum. We fix $\alpha = 1/2$. Let us start with (2.1). Note that the degeneracy for the physical singularity is given by $\phi \sim y^{(\gamma-1)/2\gamma}$ and $\mu \sim y^{(\gamma+1)/2\gamma}$. In order to get around this degeneracy, the following change of variables was introduced by Liu and Yang [44]:

$$\xi \equiv \frac{2\gamma}{\gamma-1} y^{\frac{\gamma-1}{2\gamma}} \text{ such that } \partial_y = y^{-\frac{\gamma+1}{2\gamma}}\partial_\xi \,.$$

After normalizing A and M, the equations (2.1) take the form:

$$\phi_t + (\frac{\phi}{\xi})^{2k} u_\xi = 0\,; \quad u_t + (\frac{\phi}{\xi})^{2k}\phi_\xi = 0\,, \quad k \equiv \frac{1}{2}\frac{\gamma+1}{\gamma-1} \tag{2.3}$$

for $t \geq 0$ and $0 \leq \xi \leq 1$. The physical singularity condition (1.3) is written as $0 < |\phi_\xi| < \infty$. Thus we expect $\phi \sim \xi$ for a short time near 0. The propagation speed is now non-degenerate. However, its behavior is different in the interior and on the boundary, which makes it hard to apply any standard energy method to construct solutions in the current formulation.

In [29], we proposed a new formulation to (2.3) so that some energy estimates can be closed in the appropriate energy space. As a preparation, first define the operators V and V^* associated to (2.3) as follows:

$$V(f) \equiv \frac{1}{\xi^k}\partial_\xi\left[\frac{\phi^{2k}}{\xi^k} f\right], \quad V^*(g) \equiv -\frac{\phi^{2k}}{\xi^k}\partial_\xi\left[\frac{1}{\xi^k} g\right] \tag{2.4}$$

for $f,\ g \in L^2_\xi$. In terms of V and V^*, the Euler equations (2.3) can be rewritten as follows:

$$\partial_t(\xi^k\phi) - V^*(\xi^k u) = 0\,;\ \ \partial_t(\xi^k u) + V(\xi^k\phi)/(2k+1) = 0 \tag{2.5}$$

with the boundary conditions $\phi(t,0) = 0$ and $u(t,1) = 0$. Equations (2.5) look like a symmetric linear hyperbolic system with respect to nonlinear operators V, V^*. Define the associated energy

$$\mathcal{E}(\phi,u) \equiv \sum_{i=0}^{\lceil k\rceil+3} \int |(V)^i(\xi^k\phi)|^2 + |(V^*)^i(\xi^k u)|^2 d\xi$$

where $\lceil k\rceil = \min\{n \in \mathbb{Z} : k \leq n\}$. The study of V, V^* operators and V, V^* energy estimates, which require Hardy-type inequalities, leads to the well-posedness of physical vacuum [29].

THEOREM 2.2. *Fix k, $1/2 < k < \infty$. Suppose given initial data ϕ_0 and u_0 have finite energy $\mathcal{E}(\phi_0,u_0) < \infty$ and ϕ_0 satisfies the physical vacuum singularity condition: $\phi_0/\xi \sim 1$ near $\xi \sim 0$. Then there exist a time $T > 0$ only depending on the initial data, and a unique solution (ϕ,u) to the reformulated Euler equations* (2.5) *on the time interval* $[0,T]$ *so that $\mathcal{E}(\phi,u) < \infty$ and $\phi/\xi \sim 1$ near $\xi \sim 0$.*

We remark that the above energy $\mathcal{E}$ is designed to guarantee such *minimal* regularity that $\frac{\phi}{\xi}$ and ϕ_ξ are bounded away from zero and from above, and continuous. In particular, since $\frac{\partial c^2}{\partial x} = (\frac{\phi}{\xi})^{2k}\phi_\xi$, up to constant, the solution constructed in Theorem 2.2 satisfies the physical vacuum boundary condition (1.3). The energy $\mathcal{E}$ is a ρ-weighted energy in Eulerian coordinates where the weights and the derivatives are carefully combined and the smaller γ, the more derivatives are needed. And the energy may be finite for initial data which is regular inside but not very regular at the boundary. The evolution of the vacuum boundary $x = a(t)$ is given by $\dot{a}(t) = u(t,\xi = 0)$. By Theorem 2.2, we deduce that the vacuum interface is well-defined.

2.3. Lagrangian formulation and physical vacuum. Parallel to the recent progress in free surface boundary problems with geometry involved (see for instance [2, 3, 4, 11, 60, 67]), one-dimensional result is expected to be generalized to multi-dimensional case, since the difficulty of the physical singularity lies in how the solution behaves with respect to the normal direction to the boundary. Indeed, by assuming to have smooth solutions in hand, Coutand, Shkoller and Lindblad [14] have a priori estimates for 3D compressible Euler equations for $\gamma = 2$. In this section, we discuss their result established in real Lagrangian coordinates. This Lagrangian approach leads to the study of Lagrangian velocity and flow map only, since the density is constructed from the initial datum and the Jacobian determinant of the deformation gradient.

Let $\eta(t,x)$ be the position of the gas particle x at time t so that

$$\eta_t = u(t,\eta(t,x)) \text{ for } t>0 \text{ and } \eta(0,x)=x \text{ in } \Omega\,.$$

Defining the following Lagrangian quantities

$$v(t,x)\equiv u(t,\eta(t,x))\,,\ f(t,x)\equiv \rho(t,\eta(t,x))\,,$$
$$A\equiv [D\eta]^{-1}\,,\ J\equiv \det D\eta\,,\ a\equiv JA\,,$$

and using Einstein's summation convention and the notation $F_{,k}$ to denote the kth-partial derivative of F, Euler equations (1.1) for $\gamma=2$ read as follows:

$$f_t + fA_i^j v^i{}_{,j}=0\,;\ \ f v_t^i + A_i^k f^2{}_{,k}=0\,. \tag{2.6}$$

Since $J_t = JA_i^j v^i{}_{,j}$ and $J(0)=1$, together the equation for f, we find that $fJ=\rho_0$ where ρ_0 is given initial density function. Thus, using $A_i^k = J^{-1}a_i^k$, (2.6) reduce to the following:

$$\rho_0 v_t^i + a_i^k({\rho_0}^2 J^{-2})_{,k}=0. \tag{2.7}$$

For simplicity, $\Omega=\mathbb{T}^2\times(0,1)$, $\Gamma(0)=\{x_3=1\}$ as the reference vacuum boundary and $\rho_0 = 1-x_3$ are considered. The moving vacuum boundary is given by $\Gamma(t)=\eta(t)(\Gamma(0))$. Define the higher-order energy function

$$\begin{aligned}E(t) &= \sum_{i=0}^{4}\|\partial_t^{2i}\eta(t)\|_{4-i}^2 + \sum_{i=0}^{4}\left\{\|\rho_0\bar\partial^{4-i}\partial_t^{2i}D\eta(t)\|_0^2 + \|\sqrt{\rho_0}\bar\partial^{4-i}\partial_t^{2i}v(t)\|_0^2\right\}\\ &\quad+\sum_{i=0}^{3}\|\rho_0\partial_t^{2i}J^{-2}(t)\|_{4-i}^2 + \|\mathrm{curl}_\eta v(t)\|_3^2 + \|\rho_0\bar\partial^4\mathrm{curl}_\eta v(t)\|_0^2\end{aligned}$$

where $\bar\partial=(\partial_{x_1},\partial_{x_2})$ tangential derivatives. We now state the result [14].

THEOREM 2.3. *Suppose $\eta(t)$ is a smooth solution of (2.7). Then there exists a sufficiently small $T_0>0$ depending on $E(0)$ such that for $0<T<T_0$, the energy function $E(t)$ constructed from $\eta(t)$ satisfies the a priori estimate* $\sup_{t\in[0,T]}E(t)\le M_0$ *where M_0 depends on $E(0)$.*

The proof consists of a few different key ingredients such as weighted Sobolev embedding, curl estimates, *time differentiated* energy estimates, and elliptic estimates for normal derivatives. The additional estimates for normal derivatives are obtained by inverting time derivatives via the equation. As noted in [14], in order to do so legally, sufficient smoothness of solutions was assumed and its justification by their energy should require additional work.

Recently Coutand and Shkoller [12] constructed H^2-type solutions for one-dimensional Euler equations based on the a priori estimates given in Theorem 2.3 and Hardy inequalities by degenerate parabolic regularization. This work provides an answer to the regularity question; these solutions are smoother than the solutions constructed in [29], of course, with smoother initial data. Due to the regularized approximations, the initial data need to have more regularity than the solutions to guarantee the uniqueness.

More recently, there are many works trying to prove local existence in 3D: Coutand and Shkoller [13] have a way of constructing solutions in 3D with the same method used in 1D case. Also, Lindblad [37] has a similar result using the linearized compressible Euler equations with a Nash-Moser iteration. Independently of these works, the authors [30] have constructed solutions to 3D compressible Euler equations in Lagrangian coordinates by a new analysis of physical vacuum. We briefly discuss the analysis of [30] for general γ.

Let $w \equiv K\rho_0^{\gamma-1}$ satisfying (1.3). For instance, one can take $w = x_3(1-x_3)$ for the domain $\Omega = \mathbb{T}^2 \times (0,1)$. Note that the equations (1.1) or (2.7) can be viewed as a degenerate nonlinear acoustic equation for η:

$$w^\alpha \eta^i_{tt} + (w^{1+\alpha}\, A^k_i J^{-1/\alpha})_{,k} = 0 \quad \text{and} \quad \eta^i_t = v^i \tag{2.8}$$

where $\alpha \equiv 1/(\gamma-1)$. Define the instant energy and the total energy:

$$\mathcal{E}^N \equiv \sum_{m+n=0}^{N} \frac{1}{2}\int_\Omega w^{\alpha+n}|\overline{\partial}^m \partial_3^n v|^2 dx + \frac{1}{2}\int_\Omega w^{1+\alpha+n} J^{-1/\alpha}|D_\eta \overline{\partial}^m \partial_3^n \eta|^2 dx$$

$$\mathcal{TE}^N \equiv \mathcal{E}^N + \sum_{m+n=1}^{N} \frac{1}{2}\int_\Omega w^{1+\alpha+n} J^{-1/\alpha}|\mathrm{curl}_\eta \overline{\partial}^m \partial_3^n v|^2 dx.$$

THEOREM 2.4 ([30]). *Let $\alpha > 0$ be fixed and $N \geq 2[\alpha] + 9$ be given. Let $\mathcal{TE}^N(0) < \infty$. Then there exist a time $T > 0$ depending only on $\mathcal{TE}^N(0)$ and a unique solution (η, v) to the Euler equation (2.8) on the time interval $[0,T]$ satisfying $\mathcal{TE}^N(\eta, v) \leq 2\mathcal{TE}^N(0)$ and $\|A-I\|_\infty \leq 1/8$.*

The proof is based on a hyperbolic type of new energy estimates which consist in the instant energy estimates and the curl estimates. The new key is to extract *right algebraic weighted structure* for the linearization in the normal direction such that one can directly estimate normal derivatives via the energy estimates and thus the energy estimates provide a unified, systematic way of treating all the spatial derivatives. The solutions are constructed by the duality argument and the degenerate elliptic regularity.

Theorem 2.4 indicates that the minimal number of derivatives needed to capture the physical vacuum (1.3) depends on γ as captured in the 1D result [29]. Besides boundary geometry, a critical difference between 1D case and the new analysis is that while in V, V^* framework, the energy space is nonlinear and the number of V, V^* is rigid, the energy spaces given in the above are indeed equivalent to the standard linear weighted Sobolev spaces and also the higher regularity can be readily established.

3. Open problems.

3.1. Long time behavior with or without damping. Having the local existence theory of vacuum states, the next important question is whether such a local solution exists globally in time or how it breaks down. We note that study of vacuum free boundary problems automatically excludes the breakdown of solutions caused by vacuum, which is one possible senario for positive solutions to compressible Euler equations (1.1). It was shown in [41] that the shock waves vanish at the vacuum and the singular behavior is similar to the behavior of the centered rarefaction waves corresponding to the case when c is regular [44], which indicates that vacuum has a regularizing effect. Therefore it would be interesting to investigate the long time behavior of vacuum states.

When there is damping, based on self-similar behavior, Liu conjectured [40] that time asymptotically, solutions to Euler equations with damping

$$\rho_t + (\rho u)_x = 0\,; \quad \rho(u_t + uu_x) + (A\rho^\gamma)_x = -\rho u$$

should behave like, via Darcy's law, the ones to the porus media equation:

$$\rho_t - (A\rho^\gamma)_{xx} = 0 \tag{3.1}$$

where the canonical boundary is characterized by the physical vacuum condition (1.3). The statement on the canonical boundary for porus media equations will be clear in Section 3.2. This conjecture was established by Huang, Marcati and Pan [26] in the framework of the entropy solution where the method of compensated compactness yields a global weak solution in L^∞. But in their work, there is no way of tracking the vacuum boundary. Thus it would be interesting to investigate the asymptotic relationship between regular solutions of physical vacuum and regular solutions to the porus media equation. Of course, prior to it, one should have global solutions in hand.

3.2. Ill-posedness and change of behavior. We now go back to other vacuum states of compressible Euler equations: $c \sim x^\alpha$ for $0 < \alpha < 1/2$ or $1/2 < \alpha < 1$ described in the introduction. Recall that the physical vacuum corresponds to $\alpha = 1/2$. The question is whether the boundary condition $c \sim x^\alpha$ is well-posed or not for Euler equations. Indeed, for such a fixed α, if assuming that it were well-posed for a short time and tracking the behavior $c \sim x^\alpha$ near vacuum boundary through both equations in (2.3) within that time, one can see that always the more singular mode will be created. The conjecture is that the behavior should instantaneously change into the physical vacuum, namely from $\alpha \in (0, 1/2) \cup (1/2, 1)$ to $\alpha = 1/2$, but its mathematical justification is an open problem.

In Section 3.1, we discussed some connection between Euler equations with damping and porus media equations. For the porus media equations (3.1), which are nonlinear degenerate parabolic equations, the initial boundary value problem including the behavior of solutions and the regularity of the vacuum free boundary has been well studied. In particular, in the work of Knerr, Cafarelli and Friedman [32, 8], it was shown that if the initial data $\rho_0^{\gamma-1} \leq Cx^2$, then there exists a waiting time $t^* > 0$ such that the boundary starts moving after this waiting time and $(\rho^{\gamma-1})_x$ is bounded away from zero and infinity for $t > t^*$; on the other hand, if the initial data $\rho_0^{\gamma-1} \geq Cx^{2\alpha}$ where $0 < \alpha < 1$, the boundary moves instantaneously and $(\rho^{\gamma-1})_x$ is bounded away from zero and infinity for $t > 0$. We note that the canonical boundary behavior $\rho^{\gamma-1} \sim x$ for the porus media equations corresponds to the physical vacuum $c^2 \sim x$ for Euler equations. One expects to have this kind of waiting time behavior or instantaneous change of behavior for Euler equations with damping, but again it is an open question.

3.3. Relativistic fluids. The relativistic Euler equations for a perfect fluid in four-dimensional Minkowski spacetime are given by

$$\partial_t\left(\frac{\rho+\varepsilon^2 p}{1-\varepsilon^2u^2} - \varepsilon^2 p\right) + \partial_x\left(\frac{\rho+\varepsilon^2 p}{1-\varepsilon^2u^2}u\right) = 0\,,$$
$$\partial_t\left(\frac{\rho+\varepsilon^2 p}{1-\varepsilon^2u^2}u\right) + \partial_x\left(\frac{\rho+\varepsilon^2 p}{1-\varepsilon^2u^2}u^2 + p\right) = 0\,,$$

where the parameter $1/\varepsilon$ represents the speed of light. The range of physical interest is $\rho \geq 0$, $|u| < 1/\varepsilon$, and $c < 1/\varepsilon$. Letting $\varepsilon \to 0$, non-relativistic Euler equations (1.1) are formally recovered. For the physical background, see [23, 34, 49, 50, 57, 58] and the references therein.

The relativistic Euler equations is known to be symmetric hyperbolic in the so-called entropy variables introduced by Makino and Ukai [49, 50] away from vacuum. A particular interest is compactly supported relativistic flows which for instance can be applied to the dynamics of stars in the context of special relativity. LeFloch and Ukai [34] proposed a new symmetrization, which makes sense for solutions containing vacuum states and generalizes the theory in [51] for non-relativistic Euler equations as Cauchy problem. Whether one can extend the theory of free boundary problems including physical vacuum developed for non-relativistic Euler equations to relativistic case is an open problem.

4. Further discussions.

4.1. Viscous flows and vacuum. Compressible Navier-Stokes equations

$$\partial_t\rho + \operatorname{div}(\rho u) = 0\,,$$
$$(\rho u)_t + \operatorname{div}(\rho u \otimes u) + \nabla p = \operatorname{div}(\mu[\nabla u + \nabla u^T]) + \nabla(\delta \operatorname{div} u)\,,$$

where the viscosity coefficients are assumed to satisfy $\mu \geq 0$ and $2\mu/d + \delta \geq 0$, describe the dynamics of viscous gases and fluids. Often $\operatorname{div}(\mu\nabla u)$ is used as the viscosity. There is huge literature on the studies of the existence and the behavior of solutions; we will not attempt to address exhaustive references: for classical works, see [18, 38]. One of the major difficulties when trying to prove global existence of weak solutions and strong regularity results is the possible appearance of vacuum.

When vacuum is allowed to appear initially, studying Cauchy problems for compressible Navier-Stokes equations with constant viscosity coefficients yields somewhat negative results: for instance, a finite time blow-up for nontrivial compactly supported initial density [63] and a failure of continuous dependence on initial data [25]. There are some existence theories available with the physical vacuum boundary: the vacuum interface behavior as well as the regularity to one-dimensional Navier-Stokes free boundary problems were investigated in [45]. And the local-in-time well-posedness of Navier-Stokes-Poisson equations in three dimensions with radial symmetry featuring the physical vacuum boundary was established in [28]. Global existence of such strong solutions is an open problem. See also [17, 53, 56] for related models. On the other hand, to resolve the issue of no continuous dependence on initial data in [25] for constant viscosity coefficient, a density-dependent viscosity coefficient was introduced in [42]. Since then, there has been a lot of studies on global weak solutions for various models and stabilization results under gravitation and external forces: see [35, 54, 65, 66] and the references therein. Despite the significant progress over the years, many interesting and important questions are still unanswered especially for general multi-dimensional flows.

4.2. Magnetohydrodynamics and vacuum. The theory of magnetohydrodynamics (MHD), another interesting system of hyperbolic conservation laws arising in electromechanical phenomena, describes the interaction of a magnetic field with an electrically conducting thermoelastic fluid. The equations, also called Lundquist's equations, read as follows:

$$
\begin{aligned}
&\partial_t \rho + \operatorname{div}(\rho u) = 0\,,\\
&(\rho u)_t + \operatorname{div}(\rho u \otimes u - H \otimes H) + \operatorname{grad}\left(p + \frac{|H|^2}{2}\right) = 0\,,\\
&\partial_t H - \operatorname{curl}(u \times H) = 0\,, \quad \operatorname{div} H = 0\,,
\end{aligned}
$$

where H is the magnetic field and the electric field is given by $E = H \times u$. See [15, 21] for more physical background. Vacuum states of MHD can be realized by seeking equilibria for the system and for example, static axisymmetric equilibria are obtained by solving the Grad-Shafranov equation which is a nonlinear elliptic partial differential equation [6]. Due to the interplay between the scalar pressure of the fluid and the anisotropic magnetic stress, vacuum states are richer than in hydrodynamics and their rigorous study in the context of nonlinear partial differential equations seems to be widely open.

4.3. Degenerate elliptic equations. One of the main difficulty of studying the Euler system with a free boundary is that it leads to a degenerate hyperbolic equation due to the fact that the density vanishes at the free boundary. For this we need techniques coming from elliptic degenerate equations (see for instance Baouendi [5]). Also, similar problems arise in degenerate parabolic equations (in porous medium equations [8], in thin film equations [20], in the study of polymeric flows [52] and so on).

4.4. Validity of vacuum in hydrodynamics and kinetic theory. The following Von Neumann's comment on vacuum of hydrodynamics in 1949 [62] was brought up and read by Serre at the meeting [1].

There is a further difficulty in the expansion case considered by Burgers. It was accepted that the front advances into a vacuum. It is evident that you cannot get the normal conditions of kinetic theory here either, because the density of the gas goes to zero at the front, which means that the mean free path of the molecules will go to infinity. This means that if we are in the expanding gas and approach the (theoretical) front, we will necessarily come to regions where the mean free path is larger than the distance from the front. In such regions one cannot use the hydrodynamical equations. But, as in the case of the shock wave, where ordinary conditions are reached at a distance of a few mean free paths from the shock itself, so in the case of expansion into a vacuum, at a short distance from the theoretical front, one comes into regions where the mean free path is considerably smaller than the distance from the front, and where again the classical hydrodynamical equations can be applied. If this is applied to expanding interstellar clouds, I think that in order that the classical theory be true down to 1/1000 of the density of the clouds, it is necessary that the distance from the theoretical front should be of the order of a percent of a parsec.

As to the issue of validity of vacuum in hydrodynamics, Serre [1] suggested to consider ρ^ε problem for Euler equations (1.1) where $\inf \rho^\varepsilon \geq \varepsilon$ and to study the limit $\varepsilon \to 0$ to test the stability of vacuum.

The above comment also suggests the importance of the modeling of the vacuum boundary. It suggests the existence of a layer where we need to use kinetic theory, in particular Boltzmann equations. This poses an other question, namely the boundary condition between the region where the kinetic theory is necessary and the region where hydrodynamic equations are used. This study would not be only important for the mathematical theory of vacuum but it can also be applicable to other physical problems for instance the modeling of stellar structure by using both kinetic equations and hydrodynamical equations in rarefied gas region and vacuum region.

REFERENCES

[1] Video of Discussion: "Free boundary problems related to water waves," Summer program: Nonlinear Conservation Laws and Applications, July 13–21, 2009, Institute for Mathematics and its Applications, http://www.ima.umn.edu/videos/?id=915.

[2] B. Alvarez-Samaniego and D. Lannes. Large time existence for 3D water-waves and asymptotics. *Invent. Math.*, **171**(3):485–541, 2008.

[3] D.M. Ambrose and N. Masmoudi. The zero surface tension limit of two-dimensional water waves. *Comm. Pure Appl. Math.*, **58**(10):1287–1315, 2005.

[4] D. Ambrose and N. Masmoudi. Well-posedness of 3D vortex sheets with surface tension, *Commun. Math. Sci.*, **5**:391–430, 2007.

[5] M.S. Baouendi. Sur une classe d'opérateurs elliptiques dégénérés. *Bull. Soc. Math. France*, **95**:45–87, 1967.

[6] G. Bateman. MHD Instabilities, The MIT Press, 1978.

[7] F. Bouchut. *Nonlinear stability of finite volume methods for hyperbolic conservation laws and well-balanced schemes for sources.* Frontiers in Mathematics. Birkhäuser Verlag, Basel, 2004.

[8] L. Caffarelli and A. Friedman. Regularity of the free boundary for the one-dimensional flow of gas in a porous medium. *Amer. J. Math.*, **101**(6): 1193–1218, 1979.

[9] J.-Y. Chemin. Dynamique des gaz à masse totale finie. *Asymptotic Anal.*, **3**:215–220, 1990.

[10] G.-Q. Chen. Remarks on R.J. DiPerna's paper: "Convergence of the viscosity method for isentropic gas dynamics" [Comm. Math. Phys., **91**(1):1–30, 1983; MR0719807 (85i:35118)]. *Proc. Amer. Math. Soc.*, **125**(10):2981–2986, 1997.

[11] D. Coutand and S. Shkoller. Well-posedness of the free-surface incompressible Euler equations with or without surface tension. *J. Amer. Math. Soc.*, **20**(3):e829–e930 (electronic), 2007.

[12] D. Coutand and S. Shkoller. Well-posedness in smooth function spaces for the moving-boundary 1-D compressible Euler equations in physical vacuum *preprint*, 2009, available at arXiv:0910.3136.

[13] D. Coutand and S. Shkoller. Well-posedness in smooth function spaces for the moving-boundary 3-D compressible Euler equations in physical vacuum *preprint*, 2010, available at arXiv:1003.4721.

[14] D. Coutand, S. Shkoller, and H. Lindblad. A priori estimates for the free-boundary 3D compressible Euler equations in physical vacuum *Commun. Math. Phys.*, **296**:559–587, 2010.

[15] C. Dafermos. *Hyperbolic conservation laws in continuum physics.* Second edition. Grundlehren der Mathematischen Wissenschaften [Fundamental Principles of Mathematical Sciences], **325**. Springer-Verlag, Berlin, 2005.

[16] R.J. DiPerna. Convergence of the viscosity method for isentropic gas dynamics. *Comm. Math. Phys.*, **91**(1):1–30, 1983.

[17] B. Ducomet and A. Zlotnik. Stabilization and stability for the spherically symmetric Navier-Stokes-Poisson system, *Appl. Math. Lett.*, **18**:1190–1198, 2005.

[18] E. Feireisl. *Dynamics of viscous compressible fluids. Oxford Lecture Series in Mathematics and its Applications*, **26**. Oxford University Press, 2004.

[19] K. Friedrichs. Symmetric hyperbolic linear differential equations. *Comm. Pure Appl. Math.*, **7**:345–392, 1954.

[20] L. Giacomelli, H. Knüpfer, and F. Otto. Smooth zero-contact-angle solutions to a thin-film equation around the steady state. *J. Differential Equations*, **245**(6):1454–1506, 2008.

[21] H. Grad. Mathematical problems in magneto-fluid dynamics and plasma physics. 1963 *Proc. Internat. Congr. Mathematicians (Stockholm, 1962)* pp. 560–583 *Inst. Mittag-Leffler, Djursholm.*

[22] M. GRASSIN. Global smooth solutions to Euler equations for a perfect gas. *Indiana Univ. Math. J.*, **47**:1397–1432, 1998.

[23] Y. GUO AND S. TAHVILDAR-ZADEH. Formation of singularities in relativistic fluid dynamics and in spherically symmetric plasma dynamics. *Nonlinear partial differential equations (Evanston, IL, 1998)*, 151–161, Contemp. Math., **238**, *Amer. Math. Soc., Providence, RI*, 1999.

[24] H.P. GREENSPAN AND D.S. BUTLER. On the expansion of a gas into vacuum, *J. Fluid Mech.*, **13**:101–119, 1962.

[25] D. HOFF AND D. SERRE. The failure of continuous dependence on initial data for the Navier-Stokes equations of compressible flow. *SIAM J. Appl. Math.*, **51**(4):887–898, 1991.

[26] F. HUANG, P. MARCATI, AND R. PAN: Convergence to the Barenblatt solution for the compressible Euler equations with damping and vacuum, *Arch. Ration. Mech. Anal.*, **176**:1–24, 2005.

[27] J. JANG. Nonlinear Instability in Gravitational Euler-Poisson system for $\gamma = 6/5$, *Arch. Ration. Mech. Anal.*, **188**:265–307, 2008.

[28] J. JANG. Local wellposedness of viscous gaseous stars, *Arch. Ration. Mech. Anal.*, **195** (2010), no. 3, 797–863.

[29] J. JANG AND N. MASMOUDI. Well-posedness for compressible Euler equations with physical vacuum singularity, *Comm. Pure Appl. Math.*, **62**:1327–1385, 2009.

[30] J. JANG AND N. MASMOUDI. Well-posedness of compressible Euler equations in a physical vacuum, *preprint* 2010, available at arXiv:1005.4441.

[31] T. KATO. The Cauchy problem for quasi-linear symmetric hyperbolic systems. *Arch. Rational Mech. Anal.*, **58**:181–205, 1975.

[32] B. KNERR. The porous medium equation in one dimension. *Trans. Amer. Math. Soc.*, **234** (1977), no. 2, 381–415.

[33] P. LAX. Weak solutions of nonlinear hyperbolic equations and their numerical computation, *Comm. Pure Appl. Math.*, **7** (1954), 159–193.

[34] P.G. LEFLOCH AND S. UKAI. A symmetrization of the relativistic Euler equations with several spatial variables. *Kinet. Relat. Models*, **2** (2009), no. 2, 275–292.

[35] H-L. LI, J. LI, AND Z. XIN. Vanishing of vacuum states and blow-up phenomena of the compressible Navier-Stokes equations. *Comm. Math. Phys.*, **281** (2008), no. 2, 401–444.

[36] H. LINDBLAD. Well posedness for the motion of a compressible liquid with free surface boundary, *Comm. Math. Phys.*, **260**:319–392, 2005).

[37] H. LINDBLAD. Personal communication, 2009.

[38] P.-L. LIONS. *Mathematical topis in fluid mechanics*, Vol. 2. Compressible Models. New York: Oxford University Press, 1998.

[39] P.-L. LIONS, B. PERTHAME, AND P.E. SOUGANIDIS. Existence and stability of entropy solutions for the hyperbolic systems of isentropic gas dynamics in Eulerian and Lagrangian coordinates. *Comm. Pure Appl. Math.*, **49**(6): 599–638, 1996.

[40] T.-P. LIU. Compressible flow with damping and vacuum, *Japan J. Appl. Math*, **13**:25–32 (1996).

[41] T.-P. LIU AND J. SMOLLER: On the vacuum state for isentropic gas dynamics equations, *Advances in Math.*, **1**:345–359 (1980).

[42] T.-P. LIU, Z. XIN, AND T. YANG: Vacuum states for compressible flow. *Discrete Contin. Dynam. Systems*, **4** (1998), no. 1, 1–32.

[43] T.-P. LIU AND T. YANG. Compressible Euler equations with vacuum, *J. Differential Equations*, **140**:223–237, 1997.

[44] T.-P. LIU AND T. YANG. Compressible flow with vacuum and physical singularity, *Methods Appl. Anal.*, **7**:495–509, 2000.

[45] T. LUO, Z. XIN, AND T. YANG. Interface behavior of compressible Navier-Stokes equations with vacuum, *SIAM J. Math. Anal.*, **31**:1175–1191, 2000.

[46] A. MAJDA. *Compressible fluid flow and systems of conservation laws in several space variables*, Vol. **53** of *Applied Mathematical Sciences*. Springer-Verlag, New York (1984).

[47] T. Makino. Blowing up solutions of the Euler-Poisson equation for the evolution of gaseous stars. In *Proceedings of the Fourth International Workshop on Mathematical Aspects of Fluid and Plasma Dynamics (Kyoto, 1991)*, **21**:615–624, 1992.

[48] T. Makino and S. Ukai. Sur l'existence des solutions locales de l'équation d'Euler-Poisson pour l'évolution d'étoiles gazeuses. *J. Math. Kyoto Univ.*, **27**:387–399, 1987.

[49] T. Makino and S. Ukai. Local smooth solutions of the relativistic Euler equation. II. *Kodai Math. J.*, **18** (1995), no. 2, 365–375.

[50] T. Makino and S. Ukai. Local smooth solutions of the relativistic Euler equation. *J. Math. Kyoto Univ.*, **35** (1995), no. 1, 105–114.

[51] T. Makino, S. Ukai, and S. Kawashima. Sur la solution à support compact de l'équations d'Euler compressible. *Japan J. Appl. Math.*, **3**:249–257, 1986.

[52] N. Masmoudi. Well-posedness for the FENE dumbbell model of polymeric flows. *Comm. Pure Appl. Math.*, **61** (2008), no. 12, 1685–1714.

[53] S. Matusu-Necasova, M. Okada, and T. Makino. Free boundary problem for the equation of spherically symmetric motion of viscous gas III, *Japan J. Indust. Appl. Math.*, **14**:199–213, 1997.

[54] A. Mellet and A. Vasseur: On the barotropic compressible Navier-Stokes equations. *Comm. Partial Differential Equations*, **32** (2007), no. 1–3, 431–452.

[55] T. Nishida: Equations of fluid dynamics—free surface problems. Frontiers of the mathematical sciences: 1985 (New York, 1985). *Comm. Pure Appl. Math.*, **39** (1986), no. S, suppl., S221–S238.

[56] M. Okada and T. Makino. Free boundary problem for the equation of spherically symmetric motion of viscous gas, *Japan J. Indust. Appl. Math.*, **10**:219–235, 1993.

[57] R. Pan and J. Smoller. Blowup of smooth solutions for relativistic Euler equations. *Comm. Math. Phys.*, **262** (2006), no. 3, 729–755.

[58] A. Rendall. The initial value problem for self-gravitating fluid bodies. *Mathematical physics, X (Leipzig, 1991)*, 470–474, Springer, Berlin, 1992.

[59] D. Serre. Solutions classiques globales des équations d'Euler pour un fluide parfait compressible, *Ann. Inst. Fourier (Grenoble)*, **47**, 139–153 (1997).

[60] J. Shatah and C. Zeng. Geometry and a priori estimates for free boundary problems of the Euler equation. *Comm. Pure Appl. Math.*, **61**(5):698–744, 2008.

[61] T.C. Sideris. Formation of singularities in three-dimensional compressible fluids. *Comm. Math. Phys.*, **101**:475–485, 1985.

[62] Chairman: J. von Neumann. Discussion on the existence and uniqueness or multiplicity of solutions of the aerodynamical equations, Proceedings of the Symposium on the Motion of Gaseous Masses of Cosmical Dimensions held at Paris, August 16–19, 1949.

[63] Z. Xin. Blowup of smooth solutions to the compressible Navier-Stokes equation with compact density. *Comm. Pure Appl. Math.*, **51** (1998), no. 3, 229–240.

[64] T. Yang. Singular behavior of vacuum states for compressible fluids, *Comput. Appl. Math.*, **190**:211–231, 2006.

[65] T. Zhang and D. Fang. Global behavior of compressible Navier-Stokes equations with a degenerate viscosity coefficient. *Arch. Ration. Mech. Anal.*, **182** (2006), no. 2, 223–253.

[66] T. Zhang and D. Fang. Global behavior of spherically symmetric Navier-Stokes-Poisson system with degenerate viscosity coefficients. *Arch. Ration. Mech. Anal.*, **191** (2009), no. 2, 195–243.

[67] P. Zhang and Z. Zhang. On the free boundary problem of three-dimensional incompressible Euler equations. *Comm. Pure Appl. Math.*, **61**(7):877–940, 2008.

ON RADIALLY SYMMETRIC SOLUTIONS TO CONSERVATION LAWS

HELGE KRISTIAN JENSSEN*

Abstract. Radially symmetric solutions to multi-dimensional systems of conservations laws are important in applications and computations, as well as in the general theory of conservative systems. Notwithstanding their one-dimensional nature they are poorly understood. In particular this is true for the Euler equations in gas-dynamics. After a short review of symmetric solutions to the Euler system, we introduce a scalar model for symmetric waves. The model is contrasted with standard multi-dimensional conservation laws, both scalar equations and systems. We give explicit examples of focusing nonlinear waves that blow up in amplitude. Based on the examples a suitable solution concept is introduced and compared to related treatments in the literature.

Key words. Conservation laws, existence theory, radial symmetry, nonlinear waves.

AMS(MOS) subject classifications. 35L65, 35L67, 76L05, 76N10.

1. Introduction. We consider radially symmetric solutions to multi-dimensional (multi-D) conservation laws. Although such solutions are essentially one-dimensional (1-D), the only spatial dependence being through the radial distance to the origin, little is rigorously known about their existence, uniqueness, and qualitative behavior. When formulated in polar form, i.e. in terms of time t and the radial variable r, the equations typically contain lower order geometric source terms. These source terms are singular at $r = 0$ and this rules out any straightforward application of techniques that work in standard 1-D theory.

We are motivated by the compressible Euler system and Section 2 contains a review of radially symmetric gas flow. It is noteworthy that even the basic question of whether converging waves can/must lead to blowup at the origin, is open.

Any attempt to study general Cauchy problems for conservation laws in several space dimensions must necessarily also address the case of radially symmetric solutions. Given the analytic obstacles one encounters for symmetric solutions to systems, it is natural, as a preliminary step, to consider scalar models that captures converging and diverging non-linear waves. In Section 3 we present one such model, see (3.3). The type of scalar equations we consider are tailor-made to generate symmetric solutions, and they are not covered by the standard Vol′pert-Kružkov theory [14,24,53] for scalar conservation laws. The model captures the phenomena

*Department of Mathematics, Penn State University, University Park, PA 16801.
Part of the results in this paper were obtained during the international research program on Nonlinear Partial Differential Equations at the Centre for Advanced Study (CAS) at the Norwegian Academy of Science and Letters in Oslo, 2008-09. Part of the present work will also appear in a contribution to a book from that program. The work was partially supported by NSF grant DMS-0539549.

of imploding shocks, along which the solution may blow up in amplitude. On the other hand, as a scalar model it does not capture reflection of waves off the origin.

2. Radially symmetric Euler solutions. The Euler system for compressible fluids plays a distinguished role in the theory of conservation laws. Solutions describing radially symmetric gas flow are important in applications as well in theory, and they also serve to validate numerical codes. Particular types of symmetric solutions, such as single, expanding and contracting shock waves or detonations, received a great deal of attention during World War II. Below we recall some of these findings.

Before considering time-varying solutions we mention the case of *stationary* symmetric Euler flows. Such flows may be set up as follows: in the region between two concentric spheres (or cylinders) we inject gas at one of the boundaries while extracting it at the other. By carefully adjusting the rates of injection and extraction one may build up (non-constant) stationary profiles which may or may not contain shocks. The case of smooth, inviscid flow without swirl of an ideal polytropic gas in a cone was analyzed in [13]. Chen and Glimm [8,9], in their work on flow in the exterior of a sphere, have performed a detailed local analysis of stationary shocks for isentropic symmetric flow. In [16,17] stationary shock profiles were constructed for general barotropic flow, as well as for the full Euler system. It was also shown that these solutions may be realized as limits of Navier-Stokes profiles as viscosity and heat conductivity tend to zero. Let us note here that no non-trivial stationary symmetric solution to the Euler system can be defined on a whole ball about the origin, [16,17]. Starting from the exterior boundary and moving towards the origin, any stationary symmetric flow will necessarily become sonic at a radius $r^* > 0$, beyond which the flow cannot be continued in a stationary manner.

The much harder problem of constructing dynamic (non-stationary) symmetric solutions to the compressible Euler system remains to a large extent open. The Holy Grail would be global (in time) existence results for weak solutions with radially symmetric initial data. However, very few rigorous results have been established for such flows, and we currently lack a proper understanding of even the most basic cases, such as converging shocks. It is not known if blowup of some of the physical variables (pressure, temperature, particle velocity) may occur in this case. Even less is known about the ensuing flow where, presumably, a reflected outward moving shock wave interacts with the converging flow. We will return to these issues below in connection with single-shock solutions of the Euler system.

We also mention two other special types flows that arise naturally in applications. A closely related system describes flow in ducts with varying cross-section. In a similar fashion this yields lower order source terms that are given in terms of the variable cross-section. This case is somewhat more tractable as the source terms are now non-singular. For various results on duct flows see [7, 10, 30, 31] and references therein.

Another central source of multi-D solutions is given by multi-D *Riemann problems*. In particular, 2-D Riemann problems, where the data are assumed to be constant on sectors in the (x_1, x_2)-plane, have been extensively studied. We refer to Zheng [56] for a comprehensive account. When the data prescribe the same constant thermodynamic state in all of space, together with a radially directed velocity field of constant magnitude, then the resulting flow will be radially symmetric. In this way one can construct examples of explicit solutions, at least for isentropic flows. E.g., an initially inward directed flow generates an immediately reflected shock which leaves the fluid at rest behind it. Similarly, an initially outward directed flow generates a rarefaction wave which may leave a growing circular vacuum region behind. See [56] for details as well as further examples of flows with swirl.

Concerning general spherical symmetric solutions, to the best of our knowledge, the only result that deals with solutions defined on all of space, i.e. where the origin is part of the flow domain, is the recent work by LeFloch & Westdickenberg [29]. This work builds on the earlier work by Chen & Glimm [8] and employs the method of compensated compactness. Global existence of finite energy weak solutions is demonstrated through a detailed analysis. The authors explicitly points out that it is unknown whether such solutions may blow up in L^∞ or not.

For completeness let us also consider *exterior* problems where one considers radially directed flow of a compressible fluid outside of a fixed ball, $\{r > 1\}$ say, with zero velocity at the boundary, $u\big|_{r=1} \equiv 0$. For the case of isothermal flow (pressure p proportional to density ρ) such flows were constructed by Makino & Mizohata & Ukai [35, 36]. For isentropic flow ($p \propto \rho^\gamma$, $\gamma \in (1, \frac{5}{3}]$) Makino & Takeno [37] constructed local-in-time solutions, while Chen & Glimm [8, 9] provided global (in time) weak solutions for arbitrary L^∞ data. Similar results hold for 1-d nozzle flow, see e.g. [23]. As far as we know the exterior problem for the full system remains open.

2.1. Particular types of solutions. Confronted with the considerable technical issues involved in constructing compressible Euler flows with general symmetric initial data, it is natural to search for particular types of such flows. There are several approaches to this in the literature, although the arguments have a common structure. One starts out by making an ansatz about the form of the solutions. For example, in addition to requiring radial symmetry of the solution in physical space, one may posit that the solution only depends on certain types of similarity variables. These may be derived from a consideration of the symmetry-group of the system; see [1, 6, 41] for classifications of symmetries of the Euler system. A simpler system of equations are obtained for the dependent variables in the new variables. It should be stressed that even when the reduced system can be solved, it still remains to decide whether or not the resulting solutions describe physically attainable flow. E.g., it is a separate issue to decide what the possible initial and boundary conditions for such flows are, and whether entropy admissible shocks can be built from them.

An important example concerns the so-called progressing wave solutions of the full Euler system [13]. It appears that these were first considered by Guderley [19] for an ideal, polytropic gas. Written in non-conservative form in terms of density (ρ), radial particle speed (u), and pressure (p), the Euler system for an ideal polytropic gas takes the form

$$\rho_t + u\rho_r + \rho\big[u_r + \frac{mu}{r}\big] = 0 \tag{2.1}$$

$$u_t + uu_r + \frac{p_r}{\rho} = 0 \tag{2.2}$$

$$\big(p\rho^{-\gamma}\big)_t + u\big(p\rho^{-\gamma}\big)_r = 0\,, \tag{2.3}$$

where $p = (\gamma - 1)\rho e$, e being the specific internal energy (proportional to temperature). The cases $m = 0,\ 1,\ 2$ correspond to flows with planar, cylindrical, and spherical symmetry, respectively. Defining the similarity variables

$$\xi := \frac{r}{\lambda t}\,, \qquad \eta := \frac{t}{r^\lambda}\,, \tag{2.4}$$

we make the following ansatz about how the flow variables depend on ξ and η:

$$\begin{aligned} \rho &= r^\kappa \Omega(\eta)\,, & u &= \xi U(\eta)\,, \\ c &= \xi C(\eta) \quad \text{(sound speed)}, & p &= r^\kappa \xi^2 P(\eta)\,, \end{aligned} \tag{2.5}$$

for unknown functions Ω, U, C, and P. Here λ and κ are constants whose values need to be assigned to produce particular flows. As the local sound speed is given by $c = \sqrt{\gamma p/\rho}$ we must require that

$$\Omega = \frac{\gamma P}{C^2}$$

for consistency. Upon substitution into the field equations (2.1)-(2.3), the Euler system reduces to ODE system for U, C, P:

$$U_\eta = \frac{A(U,C)}{\eta\Delta(U,C)} \tag{2.6}$$

$$C_\eta = \frac{B(U,C)C}{\eta\Delta(U,C)} \tag{2.7}$$

$$P_\eta = \frac{E(U,C)P}{\eta\Delta(U,C)}\,. \tag{2.8}$$

The functions Δ, A, B, and E are given, rational functions of U and C that also involve the parameters γ, κ, and λ (see Section 160 in [13] for details). We note that (2.6) and (2.7) provide a single ODE for C in terms of U.

It turns out that several types of flows may be realized as solutions of the ODE system (2.6)-(2.8), and that it also covers solutions in the limiting regime of strong shocks for which the upstream pressure in front of the shock is set to zero. E.g., the blast wave solutions of Taylor and Primakoff fit this setup. See [13, 43, 46, 54] for further details; in particular [43] contains a comprehensive discussion as well as an updated bibliography. We comment on the case of a single, focusing shock in Section 2.2.

An alternative approach to building classes of spherically symmetric solutions of the full compressible Euler system is provided by the work of McVittie [39]. In this method one considers a "potential" $\varphi(t,r)$ which is used to express the physical flow variables ρ, u, p (notation as above) according to

$$\rho = -\Delta\varphi\,, \qquad u = -\frac{\varphi_{rt}}{\Delta\varphi}\,, \qquad p = \Pi(t) - \varphi_{tt} + 2\int \frac{I}{r}\,dr + I\,.$$

Here Π is an arbitrary function of time, and

$$I = \frac{\varphi_{rt}^2}{\Delta\varphi}\,.$$

A direct substitution shows that any potential φ in this way yields a (possibly) formal solution to the mass equation (2.1) as well as the momentum equation (2.2). (This applies to symmetric 3-D flow, i.e. $m = 2$). In order to provide a complete solution one also needs to verify conservation of energy, which for an ideal polytropic fluid amounts to requiring that (2.3) holds, i.e. the entropy remains constant along fluid parcels: $\frac{d}{dt}(p\rho^{-\gamma}) = 0$. In the presence of shocks one should also require that the physical entropy of fluid particles increases across shock waves.

In [39] McVittie considered "separated" potentials of the form $\varphi(t,r) = f(t)w(\eta)$, where η is given by (2.4) (keeping λ as a parameter in the solution). It is further required that the particle speed u be proportional to the radial distance to the origin at every fixed moment in time. More recently this technique has been revisited by Sachdev, Joseph & Haque [44] who also produce flows velocity profiles that are non-linear in r at each instant in time.

A third method is due to Keller [22] who studied the equation of motion for fluid particles. Introducing the Lagrangian mass-coordinate h, the particle trajectories (i.e., their radii) $r = y(t,h)$ satisfy a non-linear wave equation in t and h whose coefficients depend on the entropy profile of the flow, see [13, 22]. Again, by making the ansatz that the solution is of a separated form $y(t,h) = f(h)j(t)$ one obtains two decoupled, second order, non-linear ODEs for h and j. Keller builds various types of flows from these special solutions. In particular he provides expanding shock solutions as well expansions of a gas into vacuum. There is, in fact, a close relationship between the solutions built by Keller and the ones built by McVittie. For a discussion of this point see Section 5.2 in Sachdev's monograph [43].

Apparently, no example of an *exact*, imploding shock solution has been constructed with these methods.

2.2. Solutions with a single shock. Both for theoretical reasons and for applications the case of solutions to the compressible Euler system with a single (converging or diverging) shock is of particular interest. Due to its relevance to detonations this problem was studied intensively during World War II. In particular, techniques for approximate descriptions of such waves were derived by groups in Germany, the UK, the USA, and in the USSR.

A presentation of the seminal results of von Neumann [52], Guderley [19] and Taylor [50] is contained in the monograph by Courant & Friedrichs [13]. In particular they treat the problem of expanding shock waves and build various exact and approximate solutions of such.

They also consider the more challenging case of *converging* shocks, and their reflection at the origin. We briefly describe a possible physical situation for this type of solutions. One way to generate such a shock wave is to set up a "Riemann problem" where initially a quiescent inner state (of zero velocity and constant density and pressure) is separated from a (not necessarily quiescent) fluid by a spherical membrane of radius $\bar{r}$. At time zero the membrane is removed and an ensuing spherically symmetric flow develops. The initial state immediately on the outside of the membrane may be chosen so that the Rankine-Hugoniot jump relations are satisfied, and with a denser state at the outside. The initial state of the fluid for $r > \bar{r}$ is assumed to be such that (at least until collapse) it results in smooth flow everywhere on the outside of the contracting shock. A fluid parcel in the quiescent fluid in the region interior to the shock remains at rest, with its initial density and pressure, until the focusing shock passes across it. Under most circumstances we expect the shock to accelerate and strengthen as it approaches the center: the closer a fluid parcel is to the origin, the more violent change it will suffer as the shock passes.

To construct and analyze such a flow is a challenging flow problem, which in turn results in the still more complicated issue of describing the interaction between the (presumably) reflected shock and the converging wake of the incoming shock. To the best of our knowledge it remains an open problem to construct an *exact* solution of this type for an an ideal, polytropic gas. In the case of isentropic flow the result in [29] provides existence of the flow resulting from the "Riemann-data" described above, provided suitable decay is ensured as $r \to \infty$. However, its qualitative behavior is not clear and the question remains: can or must a focusing gas-dynamical shock result in blowup of some of the flow variables?

Particular flows of this type may be treated in an approximate manner by exploiting the special type of progressing wave solutions described above. Guderley [19] assumed that the flow in the wake of the imploding shock can be described in terms of a solution to (2.6)-(2.8), and that the shock is strong enough to justify a strong shock approximation (upstream pressure is neglected). This simplifies the Rankine-Hugoniot jump conditions enough to yield a tractable system of equations. With $\kappa = 0$ and for a particular value of $\lambda > 1$ (see (2.5)) one can construct *approximate* solutions where a shock contracts and hits the origin with infinite speed. The shock is immediately reflected off the origin, again with infinite speed, after which it decelerates. Along this type of solutions also the pressure tends to infinity in the immediate wake of the incoming shock, while the density remains finite. A comprehensive study of this approach, including numerical results, appears in [25].

Notwithstanding the lack of analytic results, converging shock waves and their approximations continue to generate interest. Some more recent works can be found in [3,11,20,25,26,42,48,51,55]. For the extensive USSR literature, mostly in Russian, we refer to [46, 49]. Much of these works concerns various approximate models (weak or strong shock regimes), and is often accompanied by numerical results.

We finally note that very little seems to be known about the *stability* of converging shocks; see [26] for a discussion and partial results.

3. A scalar model for focusing shocks. Since the construction and analysis of focusing shocks for the full compressible Euler system is a highly challenging analytical task, it seems reasonable as a first step to consider simplified models. The goal is to capture and study in isolation some facet of the complicated behavior of the full model.

In this section we consider a severe reduction in complexity by studying a scalar toy model. The purpose is to generate a situation which is simple enough that we can readily produce exact solutions and study their properties. As mentioned earlier our model captures the phenomena of amplitude blowup due to focusing of non-linear waves. However, as a scalar model it does not capture the reflection of waves off the center of motion. To formulate a relevant existence theory for the simplified scalar model requires some care. We provide a first step in this direction by defining and discussing a relevant notion of radially symmetric weak solutions.

To motivate the model let's recall the spatial formulation of the Euler system for radially directed compressible barotropic flow:

$$\rho_t + \mathrm{div}_x(\rho u \vec{e}_r) = 0 \tag{3.1}$$

$$(\rho u)_t + \mathrm{div}_x(\rho u^2 \vec{e}_r) + \mathrm{grad}(p(\rho)) = 0\,, \tag{3.2}$$

where the notation is as above and $\vec{e}_r$ is the radially outward pointing unit vector. Somewhat artificially we may now set $\rho \equiv u$ in the continuity equation (3.1), and disregard (3.2). Alternatively we may set the density to be constant in the momentum equation (3.2), and disregard (3.1). Either reduction results in a single scalar equation for $u = u(t, r)$:

$$u_t + \mathrm{div}_x(u^2 \vec{e}_r) = 0\,.$$

Generalizing this Burgers-type equation we let $f : \mathbb{R} \to \mathbb{R}$ be a smooth function, and consider the equation

$$u_t + \mathrm{div}_x\big[f(u)\vec{e}_r\big] = 0\,. \tag{3.3}$$

We will only consider radially symmetric solutions of (3.3), i.e. we assume (with a slight abuse of notation) that $u(t, x) = u(t, |x|)$ for all $t \geq 0$, $x \in \mathbb{R}^n$. Rewritten in polar form the equation becomes

$$(r^m u)_t + \big(r^m f(u)\big)_r = 0 \qquad\qquad r = |x|\,, \tag{3.4}$$

or

$$u_t + f(u)_r + \frac{mf(u)}{r} = 0 \qquad \text{for } r > 0, \text{ with } m := n - 1. \tag{3.5}$$

In what follows we restrict attention to the multi-D case $n \geq 2$, whence $m \geq 1$.

Before continuing with the analysis of (3.3) let us contrast this model with the case of a "standard" scalar conservation law of the form

$$u_t + \mathrm{div}_x g(u) = 0\,, \qquad\qquad \text{with initial data } u(0, x) = u_0(x), \tag{3.6}$$

where $g : \mathbb{R} \to \mathbb{R}^n$. As is well-known, under mild smoothness conditions on the flux g, the classical Vol′pert-Kružkov theory [14, 24, 53] applies and guarantees the existence of a unique entropy solution to (3.6) for any data $u_0 \in L^\infty(\mathbb{R}^n_x)$. In particular, the solutions obey a maximum principle which precludes blowup of the amplitude. In fact, as far as modeling of radially symmetric solutions is concerned, the standard case (3.6) is irrelevant:

PROPOSITION 3.1. *Assume that the scalar conservation law (3.6) preserves radial symmetry, i.e., whenever the initial data $u_0(x)$ depend on x only through its modulus $r = |x|$, then the same holds for $u(t, x)$ at all times $t \geq 0$. Then the flux g is necessarily trivial in the sense that $g(u)$ is constant on each connected component of its domain of definition.*

Proof. Consider smooth, radially symmetric initial data u_0 with values in the domain of definition of g. Assume that $r \mapsto u_0(r)$ is one-to-one. Then, until a gradient catastrophe occurs, the solution $u(t, x)$ is smooth and

$$u(t, x) = u_0\big(x - tg'(u(t, x))\big)\,.$$

Thus, if the solution operator preserves radial symmetry, it follows that for fixed t, $r > 0$ the map

$$x \mapsto \left|x - tg'(u(t,r))\right|$$

is constant along $\{|x| = r\}$, where we have written $u(t,r)$ for $u(t,x)$ when $|x| = r$. Writing $\bar{u}$ for $u(t,r)$, we have that

$$\left|x - tg'(u(t,r))\right|^2 = r^2 + t^2|g'(\bar{u})|^2 - 2tx \cdot g'(\bar{u}) .$$

It follows that $x \mapsto x \cdot g'(\bar{u})$ is constant along $\{|x| = r\}$, whence $g'(\bar{u}) = 0$. As we can prescribe smooth and radially symmetric data taking any value in the domain of definition of g, it follows that this conclusion applies to all values $\bar{u}$ in the domain of definition of g. □

We note that models closely related to the (3.3) have been considered in the literature. In particular the polar forms (3.4), (3.5) have been studied by several authors. The equation

$$u_t + (c_0 + c_1 u)u_r + \frac{\beta c_0 u}{r} = 0 \qquad r > 0 \text{ and } c_0,\, c_1,\, \beta \text{ constants}, \tag{3.7}$$

and the more general version

$$u_t + f(u)_r + \frac{\phi(u)}{r} = 0 \qquad r > 0,\, f \text{ convex/concave}, \tag{3.8}$$

were studied by Whitham [54] and Schonbeck [45], respectively. While Whitham introduced (3.7) as a test-case for the so-called non-linearization technique, Schonbeck considered the initial-boundary value problem in the first quadrant with Dirichlet boundary conditions. The particular case with $f(u) = \frac{u^2}{2}$, $\phi(u) = u$ was treated in [15]. LeFloch & Nedelec [28] and LeFloch [27] used a generalization of Lax' formula to study the initial-boundary value problem for weighted equations of the form

$$\big[W(r)u\big]_t + \big[W(r)f(u)\big]_r = 0 \qquad r > 0,\, f \text{ convex}. \tag{3.9}$$

The weight function $W(r)$ is assumed positive and may be any power function $W(r) = r^\alpha$, $\alpha \in \mathbb{R}$. Equation (3.9) thus covers (3.4) for convex f.

REMARK 3.1. When considering the Cauchy problem for the polar forms (3.4), (3.5) one also needs to prescribe appropriate boundary conditions at $r = 0$ to obtain a solution, say $\tilde{u}(t,r)$, for $r \geq 0$. We stress the fact that defining $u(t,x) := \tilde{u}(t,|x|)$ does not necessarily yield a *conservative* solution of the original, multi-D equation (3.3) in all of $\mathbb{R}^n$, see Example 1. We elaborate on this issue below in connection with the weak formulation of (3.3).

Next, returning to the multi-D model (3.3), we shall apply the method of characteristics to construct a few explicit examples of solutions. In particular we consider cases where the singular geometric source term yields finite-time blowup of the solution itself through focusing at the origin. We shall also observe how the lack of reflection at the origin, a consequence of the fact that we consider only scalar equations, motivates a non-standard weak formulation for (3.3). Finally we consider how the resulting solution concept relates to the other works mentioned above.

3.1. Explicit solutions. For this we consider the polar form (3.5) and let $u(t,r)$ be a smooth solution. The associated characteristic curve $R(t)$ is given by

$$\begin{cases} \dot{R}(t) = f'(u(t,R(t))) \\ R(0) = R_0 > 0\,. \end{cases} \tag{3.10}$$

Letting $U(t) = U(t;R_0) := u(t,R(t))$ we thus have

$$\begin{cases} \dot{R} = f'(U) \\ \dot{U} = -m\frac{f(U)}{R} \end{cases}, \tag{3.11}$$

whence $r^m f(u)$ is constant along characteristics. The Rankine-Hugoniot relation for a solution of (3.5) with a discontinuity located along $r = X(t)$ is

$$\dot{X}(t)\,[u] = [f(u)]\,, \tag{3.12}$$

where $[\cdot]$ denotes the jump across $r = X(t)$. We observe that (3.5) admits self-similar solutions of the form $u(t,r) = \varphi(\frac{t}{r})$.

REMARK 3.2 (Stationary solutions). For the standard conservation law (3.6) the constants provide stationary solutions. Instead, the stationary solutions $\bar{u}(r)$ of (3.5) are given by

$$r^m f(\bar{u}(r)) = const.. \tag{3.13}$$

Recalling that we only consider the multi-D case $m \geq 1$, the only constant solutions to (3.5) are the roots of f. Also, as mentioned in Section 2, the compressible Euler system (barotropic or full case) has the feature that non-trivial stationary solutions are only defined strictly away from the origin. Whether this occurs for the scalar model (3.3) depends on the flux f. Indeed, in some cases (e.g. $f(u) = u^2$) all stationary solutions are defined for all $r > 0$; in other cases (e.g. $f(u) = \sin u$) no non-constant stationary solution can be continued down to $r = 0+$; finally, in other cases (e.g. $f(u) = u^2 - 1$) some stationary solutions are defined for all $r > 0$, while others are not.

We continue by exhibiting particular solutions built in a standard fashion by solving the characteristic equations (3.11) and using the Rankine-Hugoniot relation (3.12).

EXAMPLE 1 (Focusing shock with amplitude blowup). *Let $f(u) = \frac{u^2}{2}$ (Burgers) and $n = 3$:*

$$u_t + \left(\frac{u^2}{2}\right)_r + \frac{u^2}{r} = 0\,. \tag{3.14}$$

In this case the stationary solutions are $\bar{u}(r) = \frac{C}{r}$ and we can use these to give an easy example of an imploding shock along which the solution blows up. For this we choose initial data of the form

$$u_0(r) = \left\{ \begin{array}{ll} 0 & r < X_0 \\ -\frac{C}{r} & r > X_0 \end{array} \right\}, \qquad \textit{where } X_0,\, C > 0\,. \tag{3.15}$$

The two parts of the initial data are themselves parts of two stationary solutions. Without going into details about entropy admissible solutions, we observe that the only reasonable solution to (3.14)-(3.15) consists of a single shock moving towards the origin. Letting $r = X(t)$, with $X(0) = X_0$, denote its location at time t, the Rankine-Hugoniot condition yields the shock path

$$X(t) = \sqrt{X_0^2 - Ct}\,.$$

Hence, the shock accelerates inward, progressively "reveals" more and more of the stationary solution $-C/r$ in its wake, and hits the origin with infinite speed at time $t^* = \frac{X_0^2}{C}$.

Next, consider instead the following initial data:

$$u_0(r) = \left\{ \begin{array}{ll} -\frac{C}{r} & 0 < r \leq Y_0^2 \\ 0 & r > Y_0\,. \end{array} \right. \tag{3.16}$$

These data produce a rarefaction wave which rarefies toward the origin. For times $t \in (0, \frac{Y_0^2}{2Ct})$ the solution is obtained by solving along characteristics and is given by

$$u(t,r) = \left\{ \begin{array}{ll} -\frac{C}{r} & 0 < r \leq \sqrt{Y_0^2 - 2Ct} \\ -\frac{(Y_0^2 - r^2)}{2tr} & \sqrt{Y_0^2 - 2Ct} \leq r \leq Y_0 \\ 0 & r > Y_0\,. \end{array} \right. \tag{3.17}$$

We note that the rarefaction reaches the origin in finite time and that the total "mass"

$$4\pi \int_0^\infty u(t,r) r^2\, dr$$

is finite and decreases to zero as time increases.

Two remarks are in order at this point. First, it is not immediately clear how to continue the first solution in Example 1 beyond the time of collapse. Secondly, it is clear from the second part of the example that solutions obtained by simply solving the polar form of the equation may not provide *conservative* solutions to the original equation (3.3). We shall return to these issues in connection with the weak formulation of (3.3) in Section 3.2.

We also note that, due to the non-existence of stationary solutions defined in all of $\{r > 0\}$ for the Euler system (see comments in Section 2), there is no corresponding simple way of constructing focusing shocks for the Euler system in a similar manner.

We next consider an analogue of the so-called *strong shock approximation* for converging gas dynamical shocks. As mentioned in Section 2.2 this approximation amounts to neglecting the upstream pressure (i.e., near the origin) for sufficiently violent shocks. Since the solutions of the Euler system are hard to estimate it is unclear exactly how good an approximation this is. The following example indicates that a strong shock approximation in the case of the simple scalar model (3.3) yields (exactly) the correct profile at time of collapse. On the other hand it predicts a time of collapse that differs from the true one by O(max.initial perturbation).

EXAMPLE 2 (Strong shock approximation). *We consider the same Burgers equation as in Example 1 but now with initial data of the form*

$$u_0(r) = \left\{ \begin{array}{ll} \frac{r}{2\alpha} & r < X_0 \\ \\ -\frac{C}{r} & r > X_0 \end{array} \right\}, \qquad \textit{where } \alpha,\, X_0,\, C > 0\,. \tag{3.18}$$

Again we take it for granted that the relevant solution in this case is given by a single shock, the position of which is to be determined from the jump relation. The inner part of the data is chosen such that

$$u(t,x) := \frac{r}{2(t+\alpha)}\,, \tag{3.19}$$

is a rarefaction solution in the inner region $r < X(t)$. *The inner and outer states at the shock are therefore* $\frac{X(t)}{2(t+\alpha)}$ *and* $-\frac{C}{X(t)}$, *respectively. Again the Rankine-Hugoniot relation may be integrated explicitly and this yields the shock location*

$$X(t) = \Big[\frac{X_0^2 + 2\alpha C}{\sqrt{\alpha}}\sqrt{t+\alpha} - 2C(t+\alpha)\Big]^{\frac{1}{2}}\,,$$

from which it follows that the shock will initially expand or contract according to whether $X_0^2 \gtrless 2\alpha C$. *However, in either case the shock eventually ends up accelerating inward and hits the origin with infinite speed at time*

$$\tilde{t}^* = t^* + \frac{X_0^4}{4\alpha C^2}\,.$$

Here $t^* = \frac{X_0^2}{C}$ *is the time of collapse determined in Example 1, i.e. when the inner state is the constant* 0. *We interpret the solution in this latter case as the strong shock approximation to the "exact" solution of (3.14) with data (3.18). Thus, while the time of collapse in the exact solution with data (3.18) differs from* t^* *by* $\frac{O(1)X_0}{\alpha}$, *the final profiles are the same in both cases. We note that* $\frac{X_0}{2\alpha}$ *is the maximal amplitude of the initial perturbation from the inner* 0*-state.*

Before turning to the issue of a suitable weak solution concept we briefly review a solution procedure that applies to a restricted class of Burgers-type equations. Namely, consider (3.5) with flux $f(u) = \frac{u^p}{p}$ where p (to start with) is any positive number:

$$u_t + u^{p-1}u_r + \frac{mu^p}{pr} = 0 \qquad \text{for } t > 0 \text{ and } r > 0. \tag{3.20}$$

In this case the stationary solutions are

$$\bar{u}(r) = \frac{C}{r^{m/p}}, \qquad C = const., \tag{3.21}$$

and it is natural to scale the dependent variable according to

$$v(t,r) := r^{m/p}u(t,r). \tag{3.22}$$

The equation for v is

$$r^{m(1-1/p)}v_t + v^{p-1}v_r = 0, \tag{3.23}$$

which may be brought into conservative form by introducing the new spatial variable

$$s(r) := \frac{p}{np-m}r^{n-m/p}, \qquad \text{and setting} \qquad v(t,r) =: w(t,s(r)). \tag{3.24}$$

(For simplicity we assume that $p \neq \frac{m}{n}$.) This finally yields the equation

$$w_t + \left(\frac{w^p}{p}\right)_s = 0, \tag{3.25}$$

to be solved for $t > 0$ and $s > 0$. This reformulation was considered by LeFloch & Nedelec [28] and LeFloch [27] for the case $p = 2$ and $m \in \mathbb{R}$ arbitrary. It also appears as a special case of a more general procedure given by Joseph & Sachdev in [21]. Now, stationary solutions of (3.25) correspond to stationary solutions of (3.20): $w(t,s) \equiv C$ (constant) corresponds to the stationary solutions recorded in (3.21). Indeed, since $L^\infty(ds)$ is a "good" space for equation (3.25), it appears that an appropriate function space for (3.20) consists of those functions that remain bounded when weighted with the inverse of the stationary blowup solutions.

This approach may be further elaborated when p is an even and positive integer. In this particular case it turns out that the correspondence $u(t,r) \leftrightarrow w(t,s)$, together with odd extension about the origin in $\mathbb{R}_s$, allow us to generate a natural solution candidate to (3.20) for any data in $L^\infty(r^{m/p}dr)$. More precisely, after extending w to all of $\mathbb{R}_s$ by odd reflection about $s = 0$, we obtain a Cauchy problem for a standard 1-D conservation law with data in $L^\infty(ds)$. The Kružkov solution $w(t,s)$ may then be restricted to $\mathbb{R}_s^+$ and transformed back to a solution of (3.20) on $\mathbb{R}_r^+$. We refer to [12] for the details and consider a concrete example:

EXAMPLE 3 (Infinite amplitude shock and rarefaction for Burgers equation). *Recall the case (3.14) of Burgers flux in 3-D. We consider the two Cauchy problems with data $u_{0,1}(r) = -\frac{1}{r}$ and $u_{0,2}(r) := +\frac{1}{r}$, respectively. The solution process outlined above of rescaling and odd reflection yields, after transforming back to (t,r)-coordinates, the following solution candidates:*

$$u_1(t,r) = -\frac{1}{r}\,, \qquad \forall t,\, r > 0\,, \tag{3.26}$$

and

$$u_2(t,r) = \begin{cases} \frac{r}{2t} & 0 < r < \sqrt{2t}\,, \\ \\ \frac{1}{r} & r > \sqrt{2t}\,. \end{cases} \tag{3.27}$$

These are the obvious candidates for the "good" entropy solutions in the two cases. Letting $u_{0,1}$ and $u_{0,2}$ both be zero 0 at the origin, we may consider these as defining an infinitely strong down-jump and an infinitely strong up-jump, respectively, at $r = 0$. Since the Burgers flux is convex one should expect a shock to form immediately in the former case, while a rarefaction wave should open up in the latter case. Indeed, (3.26) provides a stationary "shock" connecting 0 to $-\infty$ at $r = 0$, and (3.27) consists of a rarefaction that invades a stationary solution.

3.2. Weak formulation of scalar radial model. In this section we discuss the weak formulation for (3.3) and a possible approach to an existence theory.

The purpose of introducing the scalar toy-model (3.3) is to have a simple, yet non-linear, model for focusing and blowup of radially symmetric waves. As demonstrated by the examples above, (3.3) does capture this type of behavior. Now, given the completely satisfactory theory for standard scalar conservation laws (3.6), it is natural to pursue a similar theory for (3.3). An immediate issue is how to continue solutions beyond blowup.

In this connection let's consider a special 2-D Riemann problem for isentropic gas dynamics [56]. The initial data consist of a constant density field and together with a velocity field of constant magnitude and directed radially towards the origin. In [56] a solution is constructed which consists of an immediately reflected, expanding shock wave which leaves the gas in its wake at rest and at constant density (in both space and time). Outside the expanding shock the solution is non-constant and determined by integrating an ODE from $r = \infty$.

It is natural to inquire whether a similar scenario may occur for the scalar case model (3.3). However, for a scalar model there is only one characteristic speed, or "mode of propagation", and this seems to prevent any natural mechanism by which waves can be reflected at the origin. The following example provides an illustration of this.

EXAMPLE 4 (Scalar reflection). *Consider again the 3-D Burgers model (3.14) with data $u_0(r) = -\frac{C}{r}$, $C > 0$. Motivated by the Euler solution described above we attempt to construct a reflected wave that consists of an (at least initially) outgoing shock, followed by a rarefaction wave. It seems reasonable to let the rarefaction wave be given by the following self-similar solution*

$$u(t,r) = \frac{r}{2t}\,. \tag{3.28}$$

We then insert a shock at the location $r = X(t)$, connecting the inner rarefaction (3.28) to the outer (stationary) solution $-\frac{C}{r}$. It turns out that the Rankine-Hugoniot relation yields a singular ODE for the shock location. This ODE has a one parameter family of solutions:

$$X(t) = X_a(t) = \sqrt{a\sqrt{t} - 2Ct}\,, \qquad a > 0\,.$$

These provide one solution to the initial value problem for each choice of $a > 0$. Each of these solutions consist of an shock that emerges from the origin, with infinite speed, at time zero, and then slows down. At time $t_a = (a/4C)^2$ the shock stops and begins to accelerate back towards the origin. As it reaches the origin again, we have returned to the original situation at time zero.

The last example shows that there may be no unique way to extend solutions to the scalar model (3.3) if we allow reflection of waves. We also note that, while there is no maximum principle available for (3.3) (due to the geometric source term), solutions with reflected waves also violate the weaker *comparison principle*, according to which ordered data $u_0(r) \leq v_0(r)$ should generate ordered solutions $u(t,r) \leq v(t,r)$. We thus abandon the idea of incorporating reflection into the scalar toy model.

Instead one may consider the weak formulation of the polar form, viz. (3.4) or (3.5), in the quarter-plane $\{r > 0, t > 0\}$. This approach then requires relevant boundary data along $r = 0$. This in turn raises two issues: the equation itself is singular at this point, and it is not immediately clear what type of boundary data one should prescribe in order to obtain a "good" solution of the original, multi-D equation (3.3). For the general theory of boundary conditions for standard scalar conservation laws (3.6) we refer to [4, 18, 38, 40, 47].

The prescription of boundary data at $r =$ has been considered in several works, but apparently not with the goal of constructing conservative solutions to equations of the form (3.3). Instead, these works consider the polar form (3.4) "in its own right", i.e., as an equation to be solved for $r \in \mathbb{R}^+$.

Schonbek [45] obtained a solution of equation (3.8) (when $u\phi(u) \geq 0$ for $|u|$ large enough) by the method of vanishing viscosity and regularization of the singular source term. The approximate solutions satisfy homogeneous Dirichlet condition at $r = 0$, while the boundary behavior of the limiting solution is not specified.

For the equation (3.9) LeFloch [27] and LeFloch & Nedelec [28] used convexity of the flux f to obtain a generalization of Lax' formula that applies to initial-boundary value problems. Building on the situation in 1-D [27] (see also [2]) one obtains a unique, entropy solution to the weighted Burgers equation $\left[r^m u\right]_t + \left[\frac{r^m}{2}u^2\right]_r = 0$. The boundary data assigned to u are obtained by insisting that the re-scaled solution w in (3.25) (for $p = 2$) takes on bounded boundary data: the trace of $r^{\frac{m}{2}}u(t,r)$ is bounded as $r \downarrow 0$. The method is extended further in LeFloch & Nedelec [28] where the more general case (3.9) is treated. In the latter case it is required that the scaled flux, viz. $W(r)f(u(t,r))$, has finite trace as $r \downarrow 0$.

Our main motivation for studying (3.3) derives from its use as a simple model for the more complicated case of systems. Of course, we are particularly interested in radially symmetric solutions to the Euler system. Now, for the latter the natural problem to study is the pure Cauchy problem without a priori specification of the behavior of the solution as one approaches the origin. A minimal requirement should be that the mass is conserved in time. While it seems natural to insist that the velocity field vanishes *at* $r = 0$, we want to keep open the possibility that there are solutions whose amplitudes may tend to infinity as $r \downarrow 0$ (at least at isolated times).

We are therefore interested in formulating a weak form of the scalar toy model (3.3) for which we can prove existence of *conservative* solutions. As already mentioned, this requires a more careful construction than simply extending the (weak) solutions of the polar form (3.4) to al of $\mathbb{R}^n$ by radial symmetry.

A reasonable weak formulation for (3.3) is obtained by multiplying (3.3) by a compactly supported Lipschitz function $\phi : \mathbb{R}_0^+ \times \mathbb{R}^n \to \mathbb{R}$, and integrating (formally) by parts to obtain

$$\int_0^\infty\int_{\mathbb{R}^n}\phi_t u+f(u)\,\mathrm{grad}_x\,\phi\cdot\vec{e}_r\,dxdt+\int_{\mathbb{R}^n}\phi_0u_0\,dx \\ =\int_0^\infty\int_{\mathbb{R}^n}\mathrm{div}_x\big[\phi f(u)\vec{e}_r\big]\,dxdt\,, \tag{3.29}$$

where a 0-subscript denotes evaluation at $t=0$. We write

$$\int_0^\infty\int_{\mathbb{R}^n}\mathrm{div}_x\big[\phi f(u)\vec{e}_r\big]\,dxdt=\lim_{\epsilon\downarrow 0}\int_0^\infty\int_{B_\epsilon^c}\mathrm{div}_x\big[\phi f(u)\vec{e}_r\big]\,dxdt \\ =-\omega_n\int_0^\infty\phi(t,0)\cdot\lim_{\epsilon\downarrow 0}\big\{\epsilon^m f(u(t,\epsilon,t))\big\}\,dt\,,$$

where B_ϵ is the ball of radius ϵ about the origin, and we have applied (formally) the Divergence Theorem. Example 1 shows that the last "flux-trace integral" may be non-zero: for the stationary solutions $\frac{C}{r}$ to the 3-D Burgers equation (3.14) we have

$$\int_0^\infty\phi(t,0)\cdot\lim_{\epsilon\downarrow 0}\big\{\epsilon^2 f(u(t,\epsilon,t))\big\}\,dt=\frac{C^2}{2}\int_0^\infty\phi(t,0)\,dt\,.$$

This type of unbounded solutions are bona fide solutions that we would like to cover with our concept of weak solutions. This motivates the following definition.

DEFINITION 3.1 (Weak formulation of (3.3)). *By a* radially symmetric weak solution *of equation (3.3) with radially symmetric initial data $u_0(x)$ we mean a function $u=u(t,x)$ which depends on x only through $|x|$ and satisfies:*

(i) the map $t\mapsto u(t,\cdot)$ is a continuous map from $\mathbb{R}_0^+$ into $L^1_{loc}(\mathbb{R}^n)$,
(ii) $u\in L^1_{loc}(\mathbb{R}_0^+\times\mathbb{R}^n)$,
(iii) $f(u)\in L^1_{loc}(\mathbb{R}_0^+\times\mathbb{R}^n)$,
(iv) the limit

$$M(t):=\lim_{\epsilon\downarrow 0}\epsilon^m f(u(t,\epsilon)) \qquad \text{exists for a.a. times } t\geq 0 \\ \text{and belongs to } L^1(\mathbb{R}^+),$$

(v) the following identity holds for all Lipschitz continuous test functions $\phi:[0,\infty)\times\mathbb{R}^n\to\mathbb{R}$ with compact support:

$$\int_{\mathbb{R}^n}u_0\phi_0\,dx+\int_0^\infty\int_{\mathbb{R}^n}u\phi_t+f(u)\,\vec{e}_r\cdot\mathrm{grad}_x\,\phi\,dxdt \\ =-\omega_n\int_0^\infty\phi(t,0)M(t)\,dt\,. \tag{3.30}$$

We thus consider a weak solution to (3.3) as consisting of a standard weak solution in $\mathbb{R}^n\setminus\{0\}$, together with a non-standard singular part in the form of a Dirac delta located at the origin. The solution for $r > 0$ is presumably obtainable from the polar form (3.5) of the equation, while the time-varying Dirac mass $M(t)$ at the origin $r = 0$ should account for the cumulative amount of u which enters (and "sticks" to) the origin. In order to obtain a conservative solution we must evidently require that the quantity

$$M(t) + \omega_n \int_0^\infty u(t,r) r^m \, dr \qquad \text{remains constant in time.}$$

The further issues of prescribing an appropriate selection criterion and to prove the existence of weak solutions according to Definition 3.1 will be addressed in a future work.

Acknowledgements. The author is indebted to Nick Costanzino and Mark Williams for discussions about the scalar model presented above.

REFERENCES

[1] W.F. Ames, R.L. Anderson, V.A. Dorodnitsyn, E.V. Ferapontov, R.K. Gazizov, N.H. Ibragimov, and S.R. Svirshchevskiĭ, *CRC handbook of Lie group analysis of differential equations. Vol.* **1**, CRC Press, Boca Raton, FL, 1994. Symmetries, exact solutions and conservation laws. MR **1278257 (95h:**58145)

[2] Fabio Ancona and Andrea Marson, *Scalar non-linear conservation laws with integrable boundary data*, Nonlinear Anal. **35** (1999), no. 6, Ser. A: Theory Methods, 687–710. MR **1663615 (99m:**35149)

[3] R.A. Axford and D.D. Holm, *Converging finite-strength shocks*, Phys. D **2** (1981), no. 1, 194–202.

[4] C. Bardos, A.Y. le Roux, and J.-C. Nédélec, *First order quasilinear equations with boundary conditions*, Comm. Partial Differential Equations **4** (1979), no. 9, 1017–1034. MR **542510 (81b:**35052)

[5] Sylvie Benzoni-Gavage and Denis Serre, *Multidimensional hyperbolic partial differential equations*, Oxford Mathematical Monographs, The Clarendon Press Oxford University Press, Oxford, 2007. First-order systems and applications. MR **2284507 (2008k:**35002)

[6] Brian J. Cantwell, *Introduction to symmetry analysis*, Cambridge Texts in Applied Mathematics, Cambridge University Press, Cambridge, 2002. With 1 CD-ROM. MR **1929892 (2003h:**35001)

[7] Gui-Qiang Chen and Mikhail Feldman, *Existence and stability of multidimensional transonic flows through an infinite nozzle of arbitrary cross-sections*, Arch. Ration. Mech. Anal. **184** (2007), no. 2, 185–242. MR **2299761 (2008k:**35508)

[8] Gui-Qiang Chen and James Glimm, *Global solutions to the compressible Euler equations with geometrical structure*, Comm. Math. Phys. **180** (1996), no. 1, 153–193. MR **1403862 (97j:**35120)

[9] Gui-Qiang Chen and James Glimm, *Global solutions to the cylindrically symmetric rotating motion of isentropic gases*, Z. Angew. Math. Phys. **47** (1996), no. 3, 353–372. MR **1394912 (97h:**76100)

[10] Shuxing Chen, *Transonic shocks in 3-D compressible flow passing a duct with a general section for Euler systems*, Trans. Amer. Math. Soc. **360** (2008), no. 10, 5265–5289. MR **2415074 (2009d:**35216)

[11] R.F. Chisnell, *An analytic description of converging shock waves*, J. Fluid Mech. **354** (1998), 357–375. MR **1606487 (99a:**76076)

[12] NICOLA COSTANZINO and H.K. JENSSEN, *Symmetric solutions to multi-dimensional conservation laws*, to appear in publication from the 2008-09 program "Nonlinear Partial Differential Equations" at the Centre for Advanced Study (CAS), Norwegian Academy of Science and Letters.

[13] R. COURANT and K.O. FRIEDRICHS, *Supersonic flow and shock waves*, Springer-Verlag, New York, 1976. Reprinting of the 1948 original; Applied Mathematical Sciences, Vol. **21**. MR 0421279 (54 #9284)

[14] CONSTANTINE M. DAFERMOS, *Hyperbolic conservation laws in continuum physics*, 2nd ed., Grundlehren der Mathematischen Wissenschaften [Fundamental Principles of Mathematical Sciences], vol. **325**, Springer-Verlag, Berlin, 2005. MR **2169977 (2006d:**35159)

[15] JOÃO-PAULO DIAS, MÁRIO FIGUEIRA, and LUIS SANCHEZ, *On the existence of shock fronts for a Burgers equation with a singular source term*, Math. Methods Appl. Sci. **21** (1998), no. 12, 1107–1113. MR **1637546 (99e:**35196)

[16] ERIK ENDRES, HELGE KRISTIAN JENSSEN, and MARK WILLIAMS, *Symmetric Euler and Navier-Stokes shocks in stationary barotropic flow on a bounded domain*, J. Differential Equations **245** (2008), no. 10, 3025–3067. MR 2454811

[17] ———, *Singularly perturbed ODEs and profiles for stationary symmetric Euler and Navier-Stokes shocks* (Submitted, 2009).

[18] EDWIGE GODLEWSKI and PIERRE-ARNAUD RAVIART, *Numerical approximation of hyperbolic systems of conservation laws*, Applied Mathematical Sciences, vol. 118, Springer-Verlag, New York, 1996. MR **1410987 (98d:**65109)

[19] K.G. GUDERLEY, *Starke kugelige und zylindrische Verdichtungsstgsse in der Niihe des Kugelmittelspunktes bzw. der Zylinderachse*, Luftfahrt-Forsch. **19** (1942), 302–312.

[20] Peter HAFNER, *Strong convergent shock waves near the center of convergence: a power series solution*, SIAM J. Appl. Math. **48** (1988), no. 6, 1244–1261. MR **968828 (90a:**76121)

[21] K.T. JOSEPH and P.L. SACHDEV, *On the solution of the equation* $u_t + u^n u_x + H(x,t,u) = 0$, Quart. Appl. Math. **52** (1994), no. 3, 519–527. MR **1292202 (95g:**35118)

[22] J.B. KELLER, *Spherical, cylindrical and one-dimensional gas flows*, Quart. Appl. Math. **14** (1956), 171–184. MR 0080481 (18,253c)

[23] CHRISTIAN KLINGENBERG and YUN-GUANG LU, *Existence of solutions to hyperbolic conservation laws with a source*, Comm. Math. Phys. **187** (1997), no. 2, 327–340. MR **1463831 (98d:**35139)

[24] S.N. KRUŽKOV, *First order quasilinear equations with several independent variables.*, Mat. Sb. (N.S.) **81 (123)** (1970), 228–255 (Russian). MR 0267257 (42 #2159)

[25] ROGER B. LAZARUS, *Self-similar solutions for converging shocks and collapsing cavities*, SIAM J. Numer. Anal. **18** (1981), no. 2, 316–371. MR **612145 (82i:**76054)

[26] ———, *One-dimensional stability of self-similar converging flows*, Phys. Fluids **25** (1982), no. 7, 1146–1155. MR **668286 (83h:**76054)

[27] PHILIPPE LEFLOCH, *Explicit formula for scalar nonlinear conservation laws with boundary condition*, Math. Methods Appl. Sci. **10** (1988), no. 3, 265–287. MR **949657 (90c:**35141)

[28] PHILIPPE LEFLOCH and JEAN-CLAUDE NÉDÉLEC, *Explicit formula for weighted scalar nonlinear hyperbolic conservation laws*, Trans. Amer. Math. Soc. **308** (1988), no. 2, 667–683. MR **951622 (90c:**35142)

[29] PHILIPPE G. LEFLOCH and MICHAEL WESTDICKENBERG, *Finite energy solutions to the isentropic Euler equations with geometric effects*, J. Math. Pures Appl. (9) **88** (2007), no. 5, 389–429 (English, with English and French summaries). MR **2369876 (2009f:**76144)

[30] TAI PING LIU, *Transonic gas flow in a duct of varying area*, Arch. Rational Mech. Anal. **80** (1982), no. 1, 1–18. MR **656799 (83h:**76050)

[31] ———, *Quasilinear hyperbolic systems*, Comm. Math. Phys. **68** (1979), no. 2, 141–172. MR **543196 (81b:**35051)

[32] ANDREW MAJDA, *The stability of multidimensional shock fronts*, Mem. Amer. Math. Soc. **41** (1983), no. 275, iv+95. MR **683422 (84e:**35100)

[33] ———, *The existence of multidimensional shock fronts*, Mem. Amer. Math. Soc. **43** (1983), no. 281, v+93. MR **699241 (85f:**35139)

[34] A. MAJDA, *Compressible fluid flow and systems of conservation laws in several space variables*, Applied Mathematical Sciences, vol. 53, Springer-Verlag, New York, 1984. MR **748308 (85e:**35077)

[35] TETU MAKINO, KIYOSHI MIZOHATA, and SEIJI UKAI, *The global weak solutions of compressible Euler equation with spherical symmetry*, Japan J. Indust. Appl. Math. **9** (1992), no. 3, 431–449. MR **1189949 (93k:**35205)

[36] ———, *Global weak solutions of the compressible Euler equation with spherical symmetry. II*, Japan J. Indust. Appl. Math. **11** (1994), no. 3, 417–426. MR **1299954 (95j:**35172)

[37] TETU MAKINO and SHIGEHARU TAKENO, *Initial-boundary value problem for the spherically symmetric motion of isentropic gas*, Japan J. Indust. Appl. Math. **11** (1994), no. 1, 171–183. MR **1266527 (95h:**76095)

[38] J. MÁLEK, J. NEČAS, M. ROKYTA, and M. RŮŽIČKA, *Weak and measure-valued solutions to evolutionary PDEs*, Applied Mathematics and Mathematical Computation, vol. **13**, Chapman & Hall, London, 1996. MR **1409366 (97g:**35002)

[39] G.C. MCVITTIE, *Spherically symmetric solutions of the equations of gas dynamics*, Proc. Roy. Soc. London. Ser. A. **220** (1953), 339–355. MR 0059719 (15,576a)

[40] FELIX OTTO, *Initial-boundary value problem for a scalar conservation law*, C. R. Acad. Sci. Paris Sér. I Math. **322** (1996), no. 8, 729–734 (English, with English and French summaries). MR **1387428 (97c:**35132)

[41] L.V. OVSIANNIKOV, *Group analysis of differential equations*, Academic Press Inc. [Harcourt Brace Jovanovich Publishers], New York, 1982. Translated from the Russian by Y. Chapovsky; Translation edited by William F. Ames. MR **668703 (83m:**58082)

[42] N.F. PONCHAUT, H.G. HORNUNG, D.I. PULLIN, and C.A. MOUTON, *On imploding cylindrical and spherical shock waves in a perfect gas*, J. Fluid Mech. **560** (2006), 103–122. MR **2265076 (2007g:**76127)

[43] P.L. SACHDEV, *Shock waves and explosions*, Chapman & Hall/CRC Monographs and Surveys in Pure and Applied Mathematics, vol. 132, Chapman & Hall/CRC, Boca Raton, FL, 2004. MR **2068608 (2006e:**76086)

[44] P.L. SACHDEV, K.T. JOSEPH, and M. EJANUL HAQUE, *Exact solutions of compressible flow equations with spherical symmetry*, Stud. Appl. Math. **114** (2005), no. 4, 325–342. MR **2131550 (2006c:**76099)

[45] MARIA ELENA SCHONBEK, *Existence of solutions to singular conservation laws*, SIAM J. Math. Anal. **15** (1984), no. 6, 1125–1139. MR **762969 (86c:**35099)

[46] L.I. SEDOV, *Similarity and dimensional methods in mechanics*, "Mir", Moscow, 1982. Translated from the Russian by V.I. Kisin. MR **693457 (84f:**00028)

[47] DENIS SERRE, *Systems of conservation laws. 2*, Cambridge University Press, Cambridge, 2000. Geometric structures, oscillations, and initial-boundary value problems; Translated from the 1996 French original by I.N. Sneddon. MR **1775057 (2001c:**35146)

[48] V.D. SHARMA and CH. RADHA, *Strong converging shock waves in a radiating gas*, Z. Angew. Math. Mech. **75** (1995), no. 11, 847–859 (English, with English and German summaries). MR **1362540 (96i:**76085)

[49] K.P. STANYUKOVICH, *Unsteady motion of continuous media*, Translation edited by Maurice Holt; literal translation by J. George Adashko, Pergamon Press, New York, 1960. MR 0114423 (22 #5244)

[50] GEOFFREY TAYLOR, *The Formation of a Blast Wave by a Very Intense Explosion. I. Theoretical Discussion*, Proc. R. Soc. Lond. Ser. A Math. Phys. Eng. Sci. **201** (1950), no. 1065, 159-174.

[51] MILTON VAN DYKE and A.J. GUTTMANN, *The converging shock wave from a spherical or cylindrical piston*, J. Fluid Mech. **120** (1982), 451–462. MR **669437 (83h:**76052)

[52] JOHN VON NEUMANN, *Collected works. Vol.* **VI***: Theory of games, astrophysics, hydrodynamics and meteorology*, General editor: A. H. Taub. A Pergamon Press Book, The Macmillan Co., New York, 1963. MR 0157876 (28 #1105)

[53] A.I. VOL'PERT, *Spaces* BV *and quasilinear equations*, Mat. Sb. (N.S.) **73 (115)** (1967), 255–302 (Russian). MR 0216338 (35 #7172)

[54] G.B. WHITHAM, *Linear and nonlinear waves*, Pure and Applied Mathematics (New York), John Wiley & Sons Inc., New York, 1999. Reprint of the 1974 original; A Wiley-Interscience Publication. MR **1699025 (2000c:**35001)

[55] M. YOUSAF, *Imploding spherical and cylindrical shocks*, Phys. Fluids **29** (1986), no. 3, 680–684. MR **828185 (87d:**76106)

[56] YUXI ZHENG, *Systems of conservation laws*, Progress in Nonlinear Differential Equations and their Applications, 38, Birkhäuser Boston Inc., Boston, MA, 2001. Two-dimensional Riemann problems. MR **1839813 (2002e:**35155)

CHARGE TRANSPORT IN AN INCOMPRESSIBLE FLUID: NEW DEVICES IN COMPUTATIONAL ELECTRONICS

JOSEPH W. JEROME*

Abstract. A model for electro-diffusion is discussed, characterized by the Navier-Stokes/Poisson-Nernst-Planck system. In particular, we emphasize: (i) significant applications; (ii) existence for the initial/boundary-value problem; (iii) aspects of the steady problem.

Key words. Navier-Stokes/Poisson-Nernst-Planck, charge transport, ionic channels, slip condition, initial/boundary-value problem, steady problem, EOSFET.

AMS(MOS) subject classifications. Primary 35Q30, 76D03, 76C05.

1. Introduction.

Comprehensive conservation law models began to be used in computational electronics for simulation in the latter 1980s, when the charged carriers were characterized as a compressible (energetic) fluid in a solid state semiconductor device. Earlier, Blotekjaer [1] had characterized energy valleys as a basis for carrier distinction in a hydrodynamic model, derived from the Boltzmann equation. However, the timely paper [37] serves as a marker for the confluence of emerging technology and numerical algorithms. It was shortly after this time that the International Workshops on Computational Electronics [16] began to be held, permitting the systematic presentation and publication of hydrodynamic model simulations and analysis for charge transport [14, 11, 7, 25, 4, 5, 19, 21] and those of the closely related energy transport model [24, 26, 27]. In addition, devices such as the resonant tunneling diode motivated the study of quantum hydrodynamic models, based upon Madelung's transformation [13, 6, 41, 27, 10]. The use of charge conservation as a legitimate, physics-based, modeling tool for electrons and holes dates to the period shortly after the invention of the transistor (see [39]). A recent survey [22] fills in this discussion.

1.1. New devices.

By the term, 'new devices', is meant those electronic devices that do not function exclusively via solid-state physics technology. Among these, organic devices and hybrid devices have recently emerged. Bio-chips (discussed below) and solar cells [2] are included, for example. As an illustration, we consider the study of charge transport in Voltage Operated ionic Channels (VOCs) for application in Bio–Electronics and biophysics more generally. VOCs are present on cellular membranes, where they regulate and maintain the dynamical electro-chemical equilibrium between the cell-surrounding environment and the intracellular en-

---

*Department of Mathematics, Northwestern University, Evanston, IL 60208-2730. The work of the author was supported in part by ONR/Darpa grant LLCN00014-05-C-0241 and in part by NSF grant DMS-0935967.

vironment. Single channel current recordings have been made possible by the development of the patch clamp by E. Neher and B. Sakmann. VOCs can be included in a more comprehensive model: Nanoscale Biological Chips (NBC). An example of such a device is the Electrolyte Oxide Silicon Field Effect Transistor (EOSFET) in which a transistor is gated by cell-generated ionic current via an electrolyte (see [40, 12] for studies and [32] for illuminating graphics and discussion). This cited work also includes a characterization of the electrolyte as an equivalent circuit model. In this paper we rely on biophysical conservation laws to define the appropriate system of partial differential equations. In particular, we discuss: (i) a Navier-Stokes/Poisson-Nernst-Planck model; (ii) the existence theory for the initial/boundary-value problem; (iii) the steady problem. These issues are characterized by electro-diffusion.

1.2. Related literature. The mathematical model presented in the next section is due to Rubinstein [36]. Analytical studies include local smooth solutions of the Cauchy problem [20], and global weak solutions of the initial/boundary-value problem [23, 38]. For simulations of the model presented here, see [30, 28, 29]. Since the determination of the current density was a prime goal of these studies, a critical byproduct was the conclusion that the fluid fluctuations in the ion channels decisively impacted the current. This conclusion mandates, in some sense, the analysis of the extended model. Another application is that of tissue engineering in bioreactors [9, 35]. A culture medium creates hydrodynamic stress within a scaffolded micro-architecture. Future models may permit charge flow. Growth of neuronal cells [33] and electro-hydrodynamics [15] provide additional applications. Finally, a mathematical model similar to the one employed here is employed for electro-osmotic flow [8].

2. The mathematical model. The equations are discussed in the first subsection, and a weak solution is defined following this.

2.1. The fluid/transport system. If $\vec{v}$ is the velocity of the electrolyte, and the ionic concentrations are denoted by n, p, the current densities are:

$$\vec{J_n} = eD_n \nabla n + e\mu_n n\vec{E} - e\vec{v}n,$$
$$\vec{J_p} = -eD_p \nabla p + e\mu_p p\vec{E} + e\vec{v}p.$$

$\vec{J_n}, \vec{J_p}$ are the anion and cation current densities, with corresponding diffusion and mobility coefficients, D_n, D_p, μ_n, μ_p. The charge modulus is given by e, and ϕ is the electric potential, with electric field $\vec{E} = -\nabla\phi$. The enhanced PNP system is, with ϵ the dielectric constant:

$$\frac{\partial n}{\partial t} - \frac{1}{e}\nabla\cdot \vec{J_n} = 0, \quad D_n = (kT_0/e)\mu_n,$$
$$\frac{\partial p}{\partial t} + \frac{1}{e}\nabla\cdot \vec{J_p} = 0, \quad D_p = (kT_0/e)\mu_p,$$

$$\nabla\cdot(\epsilon\nabla\phi) = e(n - p - d) \quad \text{(Poisson equation)}.$$

The Einstein relations have been used: T_0 is the ambient temperature; k denotes Boltzmann's constant; d is the non-mobile charge. The velocity of the electrolyte is determined by the Navier-Stokes system:

$$\rho(\vec{v}_t + \vec{v}\cdot\nabla\vec{v}) - \eta\Delta\vec{v} = -\nabla P_f + e(p - n)\vec{E},$$
$$\nabla\cdot\vec{v} = 0,$$

where ρ is the (mass) density of the electrolyte, P_f denotes fluid pressure, and η is the dynamic viscosity. We shall make use of the kinematic viscosity, $\nu_* = \eta/\rho$, which has a nominal value of $1\text{nm}^2/\text{ps}$ at room temperature.

2.2. Weak solution in the sense of Leray. We first introduce the required function spaces. Note that Σ_D denotes the Dirichlet boundary for the variables n, p, ϕ. For $\vec{v}$, the entire boundary of Ω is used for the specification of (possibly) slip boundary conditions. Let m denote the Euclidean space dimension, and let $s \geq m/2$ be prescribed ($H^{s-1} \subset L_m$). Denote by $\mathcal{H}$ the divergence free functions in the m-fold Cartesian product of $H^1(\Omega)$, by $\mathcal{H}^s$ the intersection of $\mathcal{H}$ with the m-fold Cartesian product of $H^s(\Omega)$; by $\mathcal{H}_0^s$ the zero trace (on $\partial\Omega$) subspace of $\mathcal{H}^s$. $(\mathcal{H}_0^s)^*$ is the corresponding dual. Analogously, H^s_{0,Σ_D} denotes those H^s functions with zero trace on Σ_D. The space-time domain is: $\mathcal{D} = \Omega \times [0, T]$. Define

$$\mathcal{C} = \{\vec{u} = (\vec{v}, \phi, n, p) : \vec{v} \in L_2((0,T);\mathcal{H}); n, p \in L_2((0,T); H^1);$$
$$\phi \in L_2((0,T); H^2);\ (\vec{v}_t, n_t, p_t) \in L_2((0,T); (\mathcal{H}_0^s)^* \times (H^s_{0,\Sigma_D})^* \times (H^s_{0,\Sigma_D})^*)\}.$$

A weak solution of the Navier-Stokes/PNP system is a vector $\vec{u} \in \mathcal{C}$ such that boundary conditions are satisfied, such that ϕ is related to n, p via the Poisson equation, and, for $a(\vec{v}, \vec{v}, \vec{\psi}) = \int_\Omega \vec{v}\cdot\nabla\vec{v}\vec{\psi}\, d\xi$, and for test functions $\vec{\psi}, \omega_n, \omega_p$, continuous from $[0, T]$ into the space $\mathcal{H}_0^s \times H^s_{0,\Sigma_D} \times H^s_{0,\Sigma_D}$, with time derivatives in $L^2(\mathcal{D})$, we have (the highlighted term is the NS-term):

$\vec{v}$ equation

$$\int_{\mathcal{D}} [\rho\vec{v}\frac{\partial\vec{\psi}}{\partial t} - \eta\nabla\vec{v}\cdot\nabla\vec{\psi}]d\xi dt \boxed{-\rho\int_0^T a(\vec{v},\vec{v},\vec{\psi})dt} - e\int_{\mathcal{D}} (p-n)\nabla\phi\cdot\vec{\psi}\, d\xi dt$$
$$= \int_{\Omega\times\{T\}} \vec{v}\cdot\vec{\psi}\, d\xi - \int_{\Omega\times\{0\}} \vec{v}_0\cdot\vec{\psi}\, d\xi$$

n equation

$$\int_{\mathcal{D}} [n\frac{\partial\omega_n}{\partial t} - D_n\nabla n\cdot\nabla\omega_n + \left(\frac{neD_n}{kT_0}\right)\nabla\phi\cdot\nabla\omega_n + \vec{v}n\cdot\nabla\omega_n]d\xi dt$$
$$= \int_{\Omega\times\{T\}} n\omega_n\, d\xi - \int_{\Omega\times\{0\}} n_0\omega_n\, d\xi$$

p equation

$$\int_{\mathcal{D}} [p\frac{\partial \omega_p}{\partial t} - D_p \nabla p \cdot \nabla \omega_p - \left(\frac{peD_p}{kT_0}\right) \nabla \phi \cdot \nabla \omega_p + \vec{v} p \cdot \nabla \omega_p] d\xi dt$$
$$= \int_{\Omega \times \{T\}} p \omega_p \, d\xi - \int_{\Omega \times \{0\}} p_0 \omega_p \, d\xi \ .$$

2.3. Multiscales in the EOSFET.

2.3.1. Brief description of the EOSFET. The EOSFET consists of a single cell immersed in a surrounding bath, connected to an electronic substrate: oxide/transistor. Gating of the transistor is achieved via electrolyte current. Vertical cross-sections can be used to define two-dimensional computational domains. The role of the boundary conditions becomes important here in the separation of the domain components. For example, Dirichlet boundary conditions are imposed on a portion of the top of the cell, to specify cell-reference ionic concentrations. Dirichlet conditions are again specified on a reference portion of the lower cell, near the cleft separating the cell from the electronic substrate. These conditions also contain information regarding potential differences in the device. The EOSFET has extraordinary long-term implications for use: (i) transduction of chemical signals generated by the biological component into an electronically identifiable signal; (ii) activation of the biological component by an applied electronic signal.

2.3.2. Disparity of scales. This model encompasses significant spatial and temporal scales. The ion channels in the cell membrane are nanometers in length, and the current is gauged on the nanosecond scale. The channel gating is on the millisecond scale. The cell is of micron dimensions, and the communication between cell and transistor is affected by gating. Moreover, experience in the case of the PNP model and also the hydrodynamic model has shown that the dynamic problem reaches a steady state much more rapidly than the gating time scale. This indicates the importance of the steady problem in the open channel. An early simulation of this case, via the full conservation law model, can be found in [3].

2.4. Assumptions. In this section, we summarize the assumptions.

2.4.1. Boundary conditions. Dirichlet boundary values for $\vec{v}$ are imposed on all of $\partial\Omega$ with the stipulation that a nonnegative normal fluid velocity component condition is imposed on $\partial\Omega$. For the remaining variables, Σ_D serves as the Dirichlet boundary, with homogeneous Neumann boundary conditions on the boundary complement. Thus, we assume the existence of functions $\vec{v}_B, \phi_B, n_B, p_B$, defined continuously from $[0, T]$ into appropriately smooth spaces, such that the following requirements hold for the solution vector: $\vec{v}_{|\partial\Omega} = \vec{v}_{B|\partial\Omega}$, $\phi_{|\Sigma_D} = \phi_{B|\Sigma_D}$, $n_{|\Sigma_D} = n_{B|\Sigma_D}$, $p_{|\Sigma_D} = p_{B|\Sigma_D}$. Moreover, it is assumed that $\vec{v}_B$ is divergence free on Ω.

2.4.2. Boundary regularity. We assume the boundary of Ω is sufficiently regular that the classical trace formulas and Green's integration by parts formulas are valid, and we assume that the Poisson solver is H^2 regularizing for L_2 data. Note that this implies that the mixed boundary conditions imposed by ϕ_B are not completely arbitrary.

2.4.3. Technical assumptions on given data. $\vec{v}_B(\cdot\,,t)$ has components in $\mathcal{H}^s \cap L^\infty$ $\forall t \in [0,T]$; $\phi_B(\cdot\,,t) \in H^{\max(s,2)}$; and $n_B(\cdot\,,t), p_B(\cdot\,,t) \in H^s$. The functions $\vec{v}_B, n_B, p_B$ have time derivatives in $L^2(\mathcal{D})$ and (trace time derivatives) in $L^2(\mathcal{D}_b)$. The initial data functions, $\vec{v}_0, n_0, p_0$, are of finite energy, with $\vec{v}_0$ divergence free. The non-mobile charge d is continuous from $[0,T]$ into L^2.

3. The existence theory. The essential approach of [23] is summarized; Rothe's method forms the basis for the existence proof. Because of this, we begin with the stationary case.

3.1. Abstract stationary result. The analysis is based upon finite-dimensional approximations, combined with an appropriate passage to the limit. Let X, Y be separable, reflexive Banach spaces, with Y a subspace, densely and continuously embedded in X; X is compactly embedded in the reflexive Banach space W. Consider $A : X \mapsto Y^*$,

$$A(u) = Lu + a(u,u,\cdot\,) + F(u,\cdot\,),$$

where $L : X \mapsto X^*$ is an isomorphism. The structure of L is induced by a continuous, coercive bilinear form $B(\cdot\,,\cdot\,)$ on $X \times X$:

$$\langle Lu, v\rangle = B(u,v),\ B(u,u) \geq c\|u\|^2.$$

We outline the general assumptions now.

- a is continuous on $X \times X \times Y$ and F is continuous on $X \times Y$.
- For each $u \in X$, $a(u,u,\cdot\,), F(u,\cdot\,)$ are continuous *linear* functionals on Y.
- The coerciveness property,

  $$\langle A(u), u\rangle / \|u\|_X \to \infty, \text{ as } \|u\|_X \to \infty, u \in Y,$$

  holds in the norm on X for elements in Y.
- If $u_k \rightharpoonup u$ (weakly) in X and $u_k \to u$ in W, then

  $$\begin{aligned} a(u_k,u_k,v) &\to a(u,u,v),\ \forall v \in Y,\\ F(u_k,v) &\to F(u,v),\ \forall v \in Y. \end{aligned}$$

The following theorem is proved in [23].

THEOREM 3.1. *Under the stated hypotheses, there is an element $u \in X$ satisfying $A(u) = f_{|Y}$ for any given $f \in X^*$.*

3.2. Rothe's method. Given a partition $\{t_k = k\Delta t : 0 \leq k \leq K\}$ of $[0,T]$, we define the following semidiscrete system:

$$\begin{aligned}
\frac{\vec{v}_k - \vec{v}_{k-1}}{\Delta t} + V(\vec{v}_k, \vec{v}_k, n_k, p_k, \cdot\,) &= 0, \\
\frac{n_k - n_{k-1}}{\Delta t} + N(\vec{v}_k, \vec{v}_k, n_k, p_k, \cdot\,) &= 0, \\
\frac{p_k - p_{k-1}}{\Delta t} - P(\vec{v}_k, \vec{v}_k, n_k, p_k, \cdot\,) &= 0, \\
\nabla\cdot(\epsilon\nabla\phi_k) &= e(n_k - p_k - d_k).
\end{aligned}$$

Here, d_k is the obvious trace of d on $t_k = k\Delta t$. Adjoined to these semidiscrete equations are the traces of the prescribed boundary data functions. An explanation of the mappings is as follows. A typical domain element is written $(\vec{v}, \vec{w}, n, p)$. V has range in the dual of divergence free functions in $\prod_1^m \mathcal{H}_0^s$. Thus, for $\vec{\psi}$ in $\mathcal{H}_0^s$, this can be written as follows.

$$V(\vec{v}, \vec{w}, n, p, \vec{\psi}) = \nu_*(\nabla\vec{v}, \nabla\vec{\psi})_{L_2} + (\vec{v}\cdot\nabla\vec{w}, \vec{\psi})_{L_2} + (e/\rho)((p-n)\nabla\phi, \vec{\psi})_{L_2}.$$

The mappings N, P are each defined into the dual of H^s_{0,Σ_D} as follows. For ω_n, ω_p in H^s_{0,Σ_D},

$$\begin{aligned}
N(\vec{v}, \vec{w}, n, p, \omega_n) &= \frac{1}{e}(\vec{J}_n, \nabla\omega_n)_{L_2}, \\
P(\vec{v}, \vec{w}, n, p, \omega_p) &= -\frac{1}{e}(\vec{J}_p, \nabla\omega_p)_{L_2}.
\end{aligned}$$

3.3. Analysis of the hypotheses for the semidiscrete system. We begin with the identifications of the spaces in the abstract theorem.

3.3.1. Identifications. The space X is a product of finite energy spaces, with appropriate zero boundary trace and divergence free first component. The spaces Y, W are now defined: $Y = \mathcal{H}_0^s \times H^s_{0,\Sigma_D} \times H^s_{0,\Sigma_D}$, and $W = W_0 \times L_2 \times L_2$, where $W_0 = \prod_1^m L_2$. We now make the identifications with mappings of the previous subsection. Because of the definition of X used here, where the homogeneous boundary conditions are employed, we necessarily carry this over to B, a, and F. We write: $\vec{u} = (\vec{v}, n, p) = (\vec{v}_B + \vec{\sigma}, n_B + \nu, p_B + \pi)$, and formulate the definitions in terms of $\vec{\zeta} = (\vec{\sigma}, \nu, \pi)$ in order to accommodate the choice of X. Thus,

$$\begin{aligned}
B(\vec{\zeta}, \vec{\zeta}) &= \nu_*(\nabla\vec{\sigma}, \nabla\vec{\sigma})_{L_2} + \Delta t^{-1}(\vec{\sigma}, \vec{\sigma})_{L_2} + D_n(\nabla\nu, \nabla\nu)_{L_2} \\
&\quad + \Delta t^{-1}(\nu, \nu)_{L_2} + D_p(\nabla\pi, \nabla\pi)_{L_2} + \Delta t^{-1}(\pi, \pi)_{L_2}, \\
a(\vec{\sigma}, \vec{\tau}, \cdot\,) &= ((\vec{v}_B + \vec{\sigma})\cdot\nabla(\vec{v}_B + \vec{\tau}), \cdot\,)_{L_2} = (\vec{v}\cdot\nabla\vec{w}, \cdot\,)_{L_2}.
\end{aligned}$$

For later reference, we write: $b(\vec{v}, \vec{w}, \cdot\,) := a(\vec{\sigma}, \vec{\tau}, \cdot\,)$. A technical issue in verifying (stationary) convergence is the requirement of employing truncation to modify F (and later passing to the limit). Truncation outside the interval $[-M, M]$ for each component in the case of $\vec{v}$, and for a scalar function otherwise, is indicated by τ_M. Set: $c_n = \frac{eD_n}{kT_0}, c_p = \frac{eD_p}{kT_0}$, and

$$\begin{aligned}
F_M(\vec{v}, n, p; \vec{\psi}, \omega_n, \omega_p) &= \frac{e}{\rho}\Big((\tau_M(p) - \tau_M(n))\nabla\phi, \vec{\psi}\Big)_{L^2} \\
&\quad -c_n(\tau_M(n)\nabla\phi, \nabla\omega_n)_{L^2} - (n\tau_M(\vec{v}), \nabla\omega_n)_{L^2} \\
&\quad +c_p(\tau_M(p)\nabla\phi, \nabla\omega_p)_{L^2} - (p\tau_M(\vec{v}), \nabla\omega_p)_{L^2} \\
f(\vec{\psi}, \omega_n, \omega_p) &= -\nu_*(\nabla\vec{v}_B, \nabla\vec{\psi})_{L^2} - D_n(\nabla n_B, \nabla\omega_n)_{L^2} \\
&\quad -D_p(\nabla p_B, \nabla\omega_p)_{L^2} + \Delta t^{-1}(\vec{u}_{k-1}, \vec{\psi})_{L^2} \\
&\quad +\Delta t^{-1}(n_{k-1}, \omega_n)_{L^2} + \Delta t^{-1}(p_{k-1}, \omega_p)_{L^2} \\
&\quad -\Delta t^{-1}(\vec{u}_B(t_k), \vec{\psi})_{L^2} - \Delta t^{-1}(n_B(t_k), \omega_n)_{L^2} \\
&\quad -\Delta t^{-1}(p_B(t_k), \omega_p)_{L^2}.
\end{aligned}$$

It is understood that ϕ is implicitly defined via the Poisson equation.

3.3.2. Continuity. It is not possible in this review to do more than indicate general techniques of proof. The analysis of the bilinear form B is standard. For a, continuity in the argument $(\vec{\sigma}, \vec{\tau}, \vec{\psi})$, (i. e., on $X \times X \times Y$) is equivalent to continuity of b in the argument $(\vec{v}, \vec{w}, \vec{\psi})$; for $s \geq m/2$:

$$|a(\vec{\sigma}, \vec{\tau}, \vec{\psi})| = |b(\vec{v}, \vec{w}, \vec{\psi})| \leq C\|\vec{v}\|_{L^2}\|\vec{w}\|_{H^1}\|\vec{\psi}\|_{H^s}.$$

This estimate uses $b(\vec{v}, \vec{w}, \vec{\psi}) = -b(\vec{v}, \vec{\psi}, \vec{w})$, followed by the Hölder inequality with reciprocal indices $1/2, 1/m, (m-2)/(2m)$, and the Sobolev inequality. Estimation of the other terms is similar.

3.3.3. Coerciveness. The bilinear form B is coercive by hypothesis. For a, we observe: $a(\vec{\sigma}, \vec{\sigma}, \vec{\sigma}) = b(\vec{v}, \vec{v}, \vec{v} - \vec{v}_B)$, and we estimate:

$$b(\vec{v}, \vec{v}, \vec{v} - \vec{v}_B) = b(\vec{v}, \vec{v}, \vec{v}) - b(\vec{v}, \vec{v}, \vec{v}_B) \geq -C\|\vec{v}\|_{L^2}\|\vec{v}\|_{\mathcal{H}}\|\vec{v}_B\|_{L^\infty}.$$

Notice that the inequality $b(\vec{v}, \vec{v}, \vec{v}) \geq 0$, has been used. The homogeneous trace terms may be absorbed into B for Δt sufficiently small; the remaining terms are controlled by problem dependent constants.

$b(\vec{v}, \vec{v}, \vec{v}) \geq 0$ follows from:

$$b(\vec{v}, \vec{v}, \vec{v}) = \frac{1}{2}\int_\Omega \vec{v}\cdot\nabla|\vec{v}|^2\, dx = \frac{1}{2}\int_\Omega \nabla\cdot(\vec{v}|\vec{v}|^2)\, dx,$$

which is then integrated by parts, and observed to have a nonnegative boundary integral by the assumption on the sign of the normal boundary component of $\vec{v}_B$. Notice that, prior to integration by parts, we have used the divergence free property of $\vec{v}$. Estimation of the other terms follows a similar line of argument. The truncation is used here in a fundamental way. The sequential hypothesis is routine.

One concludes: For each fixed M, solutions of the stationary problem $(F \mapsto F_M)$ exist for Δt sufficiently small.

4. Existence of weak solutions for the dynamic problem. We make use of the stationary solutions previously constructed.

4.1. Step function and piecewise linear sequences. We now define the sequences which permit weak compactness estimates. Indeed, $\vec{u}_{S,\Delta t}$ is the step function and $\vec{u}_{PL,\Delta t}$ the piecewise linear interpolant, both given explicitly: for $(k-1)\Delta t \leq t < k\Delta t$,

$$\vec{u}_{S,\Delta t}(\cdot\,,t) = \vec{u}_{k,\Delta t},\ \vec{u}_{PL,\Delta t}(\cdot\,,t) = \vec{u}_{k,\Delta t} + \frac{(t-k\Delta t)(\vec{u}_{k,\Delta t} - \vec{u}_{k-1,\Delta t})}{\Delta t}.$$

We have the following fundamental estimates.

LEMMA 4.1. *The sequences, $\vec{u}_{S,\Delta t}(x,t), \vec{u}_{PL,\Delta t}(x,t)$, are bounded in the topology of $L^2((0,T);X)$, and $\vec{u}_{PL,\Delta t}(x,t)$ is bounded in $H^1((0,T);Y^*)$. The bounds do not depend on the truncation parameter M.*

4.2. Weak solutions of the reduced system. The technique of proof is to extract weakly convergent subsequences of $\vec{u}_{S,\Delta t}$ and $\vec{u}_{PL,\Delta t}$ in $L^2((0,T);X)$; the Aubin lemma [18, Lemma 5.4.2], based upon Lemma 4.1, allows one to obtain a further $L^2(\mathcal{D})$-convergent subsequence of the piecewise linear sequence. Any limit can be identified with a weak solution, in the sense of Leray, of the modified (via truncation) evolution system.

4.3. Conclusion of the proof. It remains to conclude the proof. The same arguments used to obtain a solution of the reduced problem may be repeated, as applied to the sequence with $M \to \infty$. Similar weak limits, together with the Aubin lemma, are valid. We have the following theorem.

THEOREM 4.1. *Under the hypotheses stated earlier, there is a weak solution of the Navier-Stokes/PNP system. The requirement that the normal component of the velocity boundary data be nonnegative can be relaxed in the case of the Stokes system.*

REMARK 4.1. The question arises as to whether n, p can be chosen to be nonnegative. One can implement this argument at the level of the semidiscrete equations, via a discrete Gronwall inequality, as derived in [18]. Initial and boundary values are required to be nonnegative in this case, for the validity of the argument via the discrete Gronwall inequality. Altogether, one obtains nonnegativity for the modified system, and, finally, for the system of the definition.

5. Computational aspects and open problems.

5.1. Numerical approximation. A staggered algorithm is adopted for the successive solution of the PNP and NS subsystems, in the same fashion as in the treatment of fluid-structure interaction problems. For each time level t_k, a PNP system with a given velocity field $\mathbf{v}^{(k)}$ is solved using the Gummel Map typically employed in semiconductor device simulation [18]. This provides the updated concentrations $n^{(k+1)}$, $p^{(k+1)}$ and electric field $\mathbf{E}^{(k+1)}$. Then, the NS system is solved using a fixed point iteration based on Oseen subproblems [34]. This provides the updated velocity

$\mathbf{v}^{(k+1)}$ and pressure $P_f^{(k+1)}$. The process is repeated until self-consistency is achieved for the solution at the considered time level. Note here that the pressure is employed explicitly, indicating the gap between the theoretical Leray framework, and the implicit strong solution framework used for computation.

5.2. Open problems and promising research directions. The regularity presently demonstrated for the NS/PNP model, i.e., that associated with Leray solutions, is not appropriately aligned with the regularity of the assumptions, particularly regarding boundary data. Also, a more explicit treatment of the pressure, as is done in the numerical algorithm, is desirable. In brief, a strong solution theory would be desirable. It would be helpful if the role of the non-slip boundary condition could be clarified. A *challenging* open problem is the existence theory for the steady-state NS/PNP model.

In terms of scientific significance, a clearer understanding of the impact of the fluid fluctuations on the current density is highly desirable. Moreover, even if it is determined that these are significant, what is the physical role of the convective term in the NS/PNP system? Is this system actually required for modeling, or is the Stokes/PNP system adequate? Computationally, what are the precise time scale observations as the dynamic model approaches steady-state? Finally, this model is no longer out of reach of skilled computational science. Can the three-dimensional problem now be solved for a meaningful application?

REFERENCES

[1] K. Blotekjaer, *Transport equations for electrons in two-valley semiconductors*, IEEE Trans. Electron Devices ED-**17** (1970), pp. 38–47.

[2] G.A. Buxton and N. Clarke, *Computer simulation of polymer solar cells*, Modelling Simul. Mater. Sci. Eng. **15** (2007), pp. 13–26.

[3] D.-P. Chen, R.E. Eisenberg, J.W. Jerome, and C.-W. Shu, *A hydrodynamic model of temperature change in the open ionic channel*, Biophys. J. **69** (1995), pp. 2304–2322.

[4] G.-Q. Chen, J.W. Jerome, C.-W. Shu, and D. Wang, *Two -carrier semiconductor device models with geometric structure and symmetry properties*. In, Modelling and Computation for Applications in Mathematics, Science, and Engineering (J. Jerome, ed.), Oxford Univ. Press, pages 103–140, 1998.

[5] G.-Q. Chen, J.W. Jerome, and D. Wang *Compressible Euler-Maxwell equations*, Transport Theory and Statistical Physics **29** (2000), pp. 311–331.

[6] Z. Chen, B. Cockburn, C. Gardner, and J. Jerome, *Quantum hydrodynamic simulation of hysteresis in the resonant tunneling diode*, J. Comp. Phys. **117** (1995), pp. 274–280.

[7] Z. Chen, B. Cockburn, J. Jerome, and C.-W. Shu, *Mixed-RKDG finite element methods for the 2-D hydrodynamic model for semiconductor device simulation*, VLSI DESIGN **3** (1995), pp. 145–158.

[8] H.-W. Choi and M. Paraschivoiu, *Advanced hybrid-flux approach for output bounds of electro-osmotic flows: adaptive refinement and direct equilibrating strategies*, Microfluidics and Nanofluidics **2** (2006), pp. 154–170.

[9] C.A. Chung, C.W. Chen, C.P. Chen, and C.S. Tseung, *Enhancement of cell growth in tissue engineering constructs under direct perfusion: modeling and simulation*, Biotechnol. Bioeng. **97** (2007), pp. 1603–1616.

[10] P. Degond, F. Mehats, and C. Ringhofer, *Quantum hydrodynamic models derived from the entropy principle.* In, Nonlinear Partial Differential Equations and Related Analysis (G.-Q. Chen, G. Gasper, and J.W. Jerome, eds.), Contemporary Math. Vol. **371**, American Math. Soc., Providence, pages 107–131, 2005.

[11] E. Fatemi, J. Jerome, and S. Osher, *Solution of the hydrodynamic device model using high-order nonoscillatory shock capturing algorithms*, IEEE Trans. Computer-Aided Design of Integrated Circuits and Systems CAD-**10** (1991), pp. 232–244.

[12] P. Fromherz, *Neuroelectronics interfacing: Semiconductor chips with ion channels, cells, and brain*, In, Nanoelectronics and Information Technology (R. Weise, ed.), Wiley-VCH, Berlin, pages 781–810, 2003.

[13] C.L. Gardner, *The quantum hydrodynamic model for semiconductor devices*, SIAM J. Appl. Math. **54** (1994), pp. 409–427.

[14] C.L. Gardner, J.W. Jerome, and D.J. Rose, *Numerical methods for the hydrodynamic device model: Subsonic flow*, IEEE Trans. Computer-Aided Design of Integrated Circuits and Systems CAD-**8** (1989), pp. 501–507.

[15] Y. Hu, J. Lee, C. Werner, and D. Li, *Electrokinetically controlled concentration gradients in micro-chambers in microfluidic systems*, Microfluidics and Nanofluidics **2** (2006), pp. 141–153.

[16] *International Workshops on Computational Electronics* (held at eighteen month intervals), URL: `http://www.iwce.org`.

[17] J.W. Jerome, *Approximation of Nonlinear Evolution Systems*, Academic Press, New York, 1983

[18] ———*Analysis of Charge Transport*, Springer-Verlag, 1996.

[19] ———*An analytical study of smooth solutions of the Blotekjaer hydrodynamic model of electron transport*, VLSI Design **15** (2002), pp. 729–742.

[20] ———*An analytical approach to charge transport in a moving medium*, Transport Theory Statist. Phys. **31** (2002), pp. 333–366.

[21] ———*The Cauchy problem for compressible hydrodynamic-Maxwell systems: A local theory for smooth solutions*, Differential and Integral Equations **16** (2003), pp. 1345–1368.

[22] ———*Mathematical advances and horizons for classical and quantum-perturbed drift-diffusion systems: solid state devices and beyond*, Journal of Computational Electronics **8** (2009), pp.132–141.

[23] J.W. Jerome and R. Sacco, *Global weak solutions for an incompressible charged fluid with multi-scale couplings: Initial-boundary value problem*, Nonlinear Analysis, **71** (2009), e2487–e2497.

[24] J.W. Jerome and C.-W. Shu, *Energy models for one-carrier transport in semiconductor devices*, In *Semiconductors, Part II* (W.M. Coughran, J. Cole, P. Lloyd, and J.K. White eds.) IMA Volumes in Mathematics and its Applications, Vol. **59**, Springer, New York, pages 185–207, 1994.

[25] ———*Transport effects and characteristic modes in the modeling and simulation of submicron devices*, IEEE Trans. Computer-Aided Design of Integrated Circuits and Systems, CAD-**14** (1995), pp. 917–923.

[26] ———*Energy transport systems for semiconductors: Analysis and simulation*, In Proceedings, First World Congress of Nonlinear Analysts, Walter de Gruyter, Berlin, pages 3835–3846, 1995.

[27] A. Jüngel, *Quasi-hydrodynamic Semiconductor Equations*, Progress in Nonlinear Differential Equations and Their Applications, Birkhäuser, 2001.

[28] M. Longaretti, B. Chini, and J.W. Jerome, and R. Sacco, *Electrochemical modeling and characterization of voltage operated channels in nano-bio-electronics*, Sensor Letters **6** (2008), pp. 49–56.

[29] M. Longaretti, B. Chini, J.W. Jerome, and R. Sacco *Computational modeling and simulation of complex systems in bio-electronics*, Journal of Computational Electronics **7** (2008), pp. 10–13.

[30] M. Longaretti, G. Marino, B. Chini, J.W. Jerome, and R. Sacco *Computational models in nano-bio-electronics: simulation of ionic transport in voltage operated channels*, Journal of Nanoscience and Nanotechnology **8** (2008), pp. 3686–3694.

[31] Yoichiro Mori, Joseph W. Jerome, and Charles Peskin, *Three-dimensional model of electrical activity in biological cells*, Bulletin of the Mathematical Institute, Academia Sinica (Taiwan) **2** (2007), pp. 367–390.

[32] E. Neher, *Molecular biology meets microelectronics*, Nature Biotechnol. **19** (2001), pp. 121–124.

[33] A. Offenhäuser, J. Rühe, and W. Knoll, *Neuronal cells cultured on modified microelectronic device structures*, Journal of Vacuum Science and Technology A **13** (5) (1995), pp. 2606–2612.

[34] A. Quarteroni and A. Valli, *Domain Decomposition Methods for Partial Differential Equations*, Oxford Science Publications, New York, 1999.

[35] M.T. Raimondi, F. Boschetti, F. Migliavacca, M. Cioffi, and G. Dubini, *Micro-fluid dynamics in three-dimensional engineered cell systems in bioreactors*, In, Topics in Tissue Engineering (N. Ashammakhi and R.L. Reis, eds.), Vol. **2**, Chapter 9, 2005.

[36] I. Rubinstein, *Electro-Diffusion of Ions*, SIAM Studies in Applied Mathematics, SIAM, Philadelphia, 1990.

[37] M. Rudan and F. Odeh, *Multi-dimensional discretization scheme for the hydrodynamic model of semiconductor devices*, COMPEL **5** (1986), pp. 149–183.

[38] M. Schmuck, *Analysis of the Navier-Stokes-Nernst-Planck-Poisson system* Math. Models and Methods in Appl. Sci. **19** (2009), pp. 993–1015.

[39] W. Shockley, *Electrons and Holes in Semiconductors*, Van Nostrand, Princeton, N.J., 1950.

[40] B. Straub, E. Meyer, and P. Fromherz, *Recombinant maxi-K channels on transistor, a prototype of ionic-electronic interfacing* **19** (2001), pp. 121–124.

[41] B. Zhang and J. Jerome, *On a steady-state quantum hydrodynamic model for semiconductors*, Nonlinear Analysis **26** (1996), pp. 845–856.

LOCALIZATION AND SHEAR BANDS IN HIGH STRAIN-RATE PLASTICITY

THEODOROS KATSAOUNIS* AND ATHANASIOS TZAVARAS†

Abstract. This article is devoted to the explanation of the onset of localization and the formation of shear bands in high strain-rate plasticity of metals. We employ the Arrhenius constitutive model and show Hadamard instability for the linearized problem. For the nonlinear model, using an asymptotic procedure motivated by the theory of relaxation and the Chapman-Enskog expansion, we derive an effective equation for the evolution of the strain rate, which is backward parabolic with a small stabilizing fourth order correction. We construct self-similar solutions that describe the self-organization into a localized solution starting from well prepared data.

Key words. shear bands, localization, effective equations, self-similarity.

AMS(MOS) subject classifications. 74C20, 74H35, 35F55, 35K55, 35Q72.

1. Introduction. The phenomenon of shear strain localization appears in several instances of material instability in mechanics. It is associated with ill-posedness of an underlying initial value problem, what has coined the term Hadamard-instability for its description in the mechanics literature. It should however be noted that while Hadamard instability indicates the catastrophic growth of oscillations around a mean state, it does not by itself explain the formation of coherent structures typically observed in localization. The latter is a nonlinear effect that will be the center-stage of the present study.

The mathematical theory of localization in high strain rate plasticity aims to understand a destabilizing feedback mechanism proposed in [20, 3] and deemed responsible for formation of shear band. It poses analytical challenges at the interface of dynamical systems theory and (small viscosity) parabolic regularizations for ill-posed problems. The onset of localization may be decomposed into two stages. At the first stage (a) one aims to determine whether a Hadamard-type instability is at play for a linearized problem. The second stage (b) requires to understand how the Hadamard instability interacts with the nonlinear features of the problem to form a localized coherent state, associated to a shear band. The latter is a challenging problem in the realm of nonlinear analysis. In this survey we draw material from [10, 11, 1] and focus on the Arrhenius law as a paradigm to describe the sequence of events occurring in the process of localization.

*Department of Applied Mathematics, University of Crete, Heraklion, 71409, Greece and Institute of Applied and Computational Mathematics, FORTH, Heraklion 71110, Greece (thodoros@tem.uoc.gr).

†Department of Applied Mathematics, University of Crete, Heraklion, 71409, Greece; and Institute of Applied and Computational Mathematics, FORTH, Heraklion, 71110, Greece (tzavaras@tem.uoc.gr).

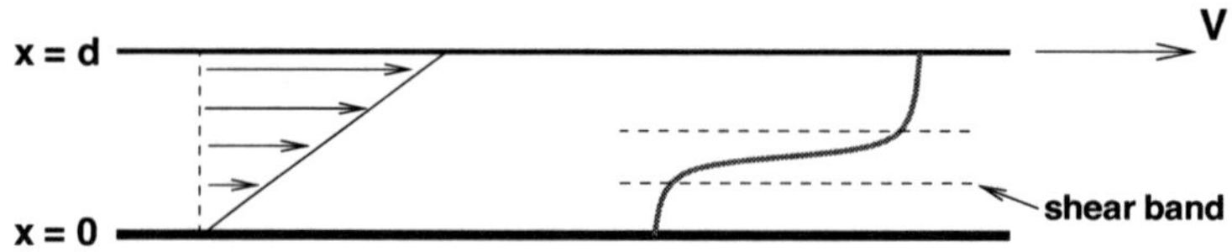

FIG. 1. *Uniform shear versus shear band.*

2. Description of the problem. The formation of shear bands during rapid shearing deformations of steels [4, 9, 20] is a striking instance of material instability in mechanics. The shear strain concentrates in a narrow band inside the specimen and concurrently an elevation of the temperature occurs in the interior of the band. A caricature of shear band forming is presented in Fig. 1. Shear bands are often precursors to rupture and their study has attracted considerable attention in the mechanics and mathematics literature including experimental works [4, 9], mechanical modeling and linearized analysis studies ([3, 8, 12, 17, 19] and references therein), asymptotic analysis [6, 18], nonlinear analysis [5, 13, 14, 2], and numerical investigations [16, 7, 1].

2.1. Modeling shear bands. Shear bands appear and propagate as one dimensional structures (up to interaction times), and most investigations have focused on the study of one-dimensional, simple shearing deformations. As a test problem we consider the simple shear between two parallel plates of a thermoviscoplastic material, Fig. 1. The associated model (see [3, 17, 10] for details) reads

$$\begin{aligned} v_t &= \frac{1}{r}\,\sigma_x, \\ \theta_t &= \kappa\theta_{xx} + \sigma\gamma_t, \\ \gamma_t &= v_x, \end{aligned} \tag{2.1}$$

where r, κ are non-dimensional constants. The variables, velocity in the shearing direction $v(x,t)$, the (plastic) shear strain $\gamma(x,t)$, the temperature $\theta(x,t)$, the heat flux $Q(x,t)$, and the shear stress $\sigma(x,t)$, are connected through the balance of linear momentum, kinematic compatibility, and the balance of energy equations.

It was recognized by Zener and Hollomon [20] that the effect of the deformation speed is twofold: First, an increase in the deformation speed changes the deformation conditions from isothermal to nearly adiabatic. Under such conditions the combined effect of thermal softening and strain hardening tends to produce net softening response. (Indeed, experimental observations of shear bands are typically associated with strain softening response – past a critical strain – of the measured stress-strain curve [3].) Second, strain rate has an effect *per se*, and needs to be included in the constitutive modeling. Accordingly, a model is employed where the shear stress is described by a constitutive law within the framework of thermoviscoplasticity: $\sigma = f(\theta, \gamma, \gamma_t)$ with $f_p > 0$. This may be viewed as a plastic flow rule or a yield surface, what suggests the terminology: the material exhibits thermal softening at state variables (θ, γ, p) where $f_\theta(\theta, \gamma, p) < 0$, strain hardening at state variables where $f_\gamma(\theta, \gamma, p) > 0$, and strain softening when $f_\gamma(\theta, \gamma, p) < 0$. The slopes of f measure the degree of thermal softening, strain hardening (or softening) and strain-rate sensitivity, respectively. The difficulty of performing high strain-rate experiments causes uncertainty as to the specific form of the constitutive form of the stress. In the literature two constitutive laws are widely used to describe the stress σ: the *power law* and the *Arrhenius law*,

$$\sigma = \theta^{-\alpha}\gamma^m\gamma_t^n, \qquad \text{Power Law}, \tag{2.2}$$

$$\sigma = e^{-\alpha\theta}v_x^n, \qquad \text{Arrhenius Law}. \tag{2.3}$$

The upper plate at $x = 1$ is subjected to a prescribed constant velocity $V = 1$ (in non-dimensional variables) while the lower plate at $x = 0$ is held at rest: $v(0, t) = 0$, $v(1, t) = 1$. It is further assumed that the plates are thermally insulated: $\theta_x(0, t) = 0$, $\theta_x(1, t) = 0$. For the heat flux Q one either uses the adiabatic assumption $Q = 0$ (equivalently $\kappa = 0$) or alternatively a Fourier law $Q = \kappa\theta_x$ with thermal diffusivity parameter κ. Imposing adiabatic conditions projects the belief that, at high strain rates, heat diffusion operates at a slower time scale than the one required for the development of a shear band. It appears a plausible assumption for the shear band initiation process, but not necessarily for the evolution of a developed band, due to the high temperature differences involved.

We summarize the equations describing the model. For the case of the power law the resulting system reads

$$\begin{aligned} v_t &= \frac{1}{r}\,\sigma_x, \\ \theta_t &= \kappa\theta_{xx} + \sigma\gamma_t, \\ \gamma_t &= v_x, \\ \sigma &= \theta^{-\alpha}\gamma^m\gamma_t^n. \end{aligned} \tag{2.4}$$

The Arrhenius law does not exhibit any strain hardening and thus the system decouples and leads to the form

$$\begin{aligned} v_t &= \frac{1}{r}\,\sigma_x, \\ \theta_t &= \kappa\theta_{xx} + \sigma v_x \\ \sigma &= e^{-\alpha\theta}v_x^n. \end{aligned} \tag{2.5}$$

In the sequel we focus on the Arrhenius model (2.5); we refer to [10, 11] for analogous results for the power law.

The models (2.4) and (2.5) admit a class of special solution describing uniform shearing. For the Arrhenius model (2.5), the uniform shear flow is

$$\begin{aligned} v_s &= x, \\ \theta_s &= \frac{1}{\alpha}\log(\alpha t + k_0), \qquad k_0 = e^{\alpha\theta_0}, \\ \sigma_s &= \frac{1}{\alpha t + k_0}. \end{aligned} \tag{2.6}$$

The graph $\sigma - t$ is viewed as describing the stress vs. average strain response. As the graph is always decreasing, it means that there is always net softening response.

2.2. Analytical challenges posed by the localization problem. In this review work we focus on the initiation stage of shear bands and seek to understand the mechanism of formation. There are the following tasks to be carried out :

- To define a notion of stability (or instability) for the uniform shear; this task is complicated by the time dependent nature of the base solutions.
- To derive quantitative criteria for stability (or instability) and a mathematical mechanism pinpointing the onset of localization.
- To describe the localization process and the formation of the shear band.

Naturally, one distinguishes two stages in the process: The initial stage of stability or instability of the uniform shearing solutions leads to questions in the realm of linearized analysis, but with certain special features. Because the uniform shearing solution is time dependent the notion of stability has to be specified. An efficient definition due to Molinari and Clifton [12, 8] espouses the idea of relative perturbation. That is, the uniform shear is stable if perturbations grow slower than the base solutions, and unstable if perturbations grow faster than the base solution. Following that framework leads to questions of linearized stability analysis for non-autonomous linear systems. A rigorous analysis for a simplified model illustrating the difficulties can be found in [15]. In section 3 we carry out this program for the Arrhenius model (2.5).

The second stage of localization lies within the realm of nonlinear analysis, and the focal issue is how the catastrophic growth of high frequency oscillations resulting from Hadamard instability interacts with the nonlinearity to form a coherent structure. A quantitative criterion accounting for the nonlinear aspects of localization was recently derived in [10], based on ideas from the theory of relaxation system and the Chapman Enskog expansion. In follow-up work [11], this analysis is supplemented with a study of self-similar solutions associated with a time-reversed problem with delta-mass initial data. Numerical investigations of the self-similar solutions indicate how the localized state evolves from well prepared initial data for a power law. The corresponding analysis for the Arrhenius system is presented in sections 5 and 6. We refer to [10] and [11] for analogous results for system (2.4).

3. Stability of the Arrhenius system. We introduce new variables $(V,\ \Theta,\ \Sigma)$ that describe the relative change of a solution to (2.5) with respect to the uniform shear flow (2.6):

$$\begin{aligned} v &= V(x,\tau(t)), \\ \theta &= \theta_s(t) + \Theta(x,\tau(t)), \\ \sigma &= \sigma_s(t)\Sigma(x,\tau(t))\,, \end{aligned} \tag{3.1}$$

where $\tau(t)$ is a rescaling of time, defined by

$$\dot{\tau}(t) = \sigma_s(t) = \frac{1}{\alpha t + k_0}, \quad \tau(0) = 0, \implies \tau(t) = \frac{1}{\alpha}\log\left(\frac{\alpha}{k_0}t + 1\right).$$

The new variables (u, Θ, Σ), where $u = V_x$ satisfy the system

$$\begin{aligned} u_\tau &= \frac{1}{r}\Sigma_{xx}, \\ \Theta_\tau &= k_0\kappa e^{\alpha\tau}\Theta_{xx} + (\Sigma u - 1)\,, \\ \Sigma &= e^{-\alpha\Theta}u^n. \end{aligned} \tag{3.2}$$

The uniform shear flow, via (3.1), is mapped to the point

$$u_0 = 1,\ \Theta_0 = 0,\ \Sigma_0 = 1\,,$$

which is an equilibrium of (3.2). Moreover, let us consider the ode system consisting of the second ($\kappa = 0$) and third equation in (3.2),

$$\begin{aligned} \Theta_\tau &= e^{-\alpha\Theta}u^{n+1} - 1, \\ \Sigma &= e^{-\alpha\Theta}u^n. \end{aligned}$$

This system has an equilibrium manifold parametrized by the moment u,

$$(\Theta, \Sigma) = \left(\frac{n+1}{\alpha}\log u, u^{-1}\right) \quad \text{or} \quad (\Theta, \Sigma) = \left(\Theta, e^{-\frac{\alpha}{n+1}\Theta}\right),$$

and its flow is globally attracted to the equilibrium manifold.

3.1. Linearized stability. We turn to linearized stability for the rescaled system (3.2). We will say that the uniform shear is stable if the perturbation decays and unstable if the perturbation grows. This notion of stability corresponds to the idea of relative perturbation stability analysis in [12] (see also [8, 15]). Consider a perturbation of the uniform shear $\Sigma_0 = 1$, $u_0 = 1$, $\Theta_0 = 0$, that is

$$\begin{aligned} u &= 1 + \delta\, u_1 + O(\delta^2), \\ \Sigma &= 1 + \delta\, \Sigma_1 + O(\delta^2), \\ \Theta &= 0 + \delta\, \Theta_1 + O(\delta^2), \end{aligned}$$

and obtain the linearized system

$$\begin{aligned} u_{1,\tau} &= \frac{1}{r}\, \Sigma_{1,xx}, \\ \Theta_{1,\tau} &= k_0 \kappa e^{\alpha\tau}\, \Theta_{1,xx} - \alpha\Theta_1 + (n+1)u_1, \\ \Sigma_1 &= -\alpha\Theta_1 + n u_1, \end{aligned} \tag{3.3}$$

with boundary conditions

$$\Theta_{1,x}(0,\tau) = \Theta_{1,x}(\pi,\tau) = 0, \quad u_{1,x}(0,\tau) = u_{1,x}(\pi,\tau) = 0.$$

The solutions of (3.3) can be expressed in terms of the Fourier modes

$$u_1(x,\tau) = \hat{v}_j(\tau)\cos(jx), \quad \Theta_1(x,\tau) = \hat{\xi}_j(\tau)\cos(jx),$$

and the Fourier coefficients $(\hat{v}_j, \hat{\xi}_j)$ satisfy the non-autonomous system of ordinary differential equations.

$$\begin{aligned} \frac{d\hat{v}_j}{d\tau} &= -j^2 \frac{n}{r} \hat{v}_j + j^2 \frac{\alpha}{r} \hat{\xi}_j, \\ \frac{d\hat{\xi}_j}{d\tau} &= -j^2 k_0 \kappa e^{\alpha\tau} \hat{\xi}_j - \alpha\hat{\xi}_j + (n+1)\hat{v}_j. \end{aligned} \tag{3.4}$$

We proceed to analyze the behavior of solutions to (3.4). When $\kappa = 0$ the system is autonomous and the analysis concludes with an examination of the eigenvalues. When $\kappa \neq 0$ we will consider the problem with the diffusion coefficient frozen in time, and the eigenvalue analysis gives an indication on what to expect (but is not rigorous). Assuming the diffusion coefficient to be frozen in time, consider the ode system:

$$\frac{d}{dt}\begin{pmatrix} \hat{v}_j \\ \hat{\xi}_j \end{pmatrix} = \begin{pmatrix} -j^2 \frac{n}{r} & j^2 \frac{\alpha}{r} \\ n+1 & -j^2\kappa - \alpha \end{pmatrix} \begin{pmatrix} \hat{v}_j \\ \hat{\xi}_j \end{pmatrix}.$$

The characteristic polynomial of the matrix is

$$\lambda^2 + \lambda\left(j^2 \frac{n}{r} + j^2\kappa + \alpha\right) + j^2\left[\frac{n}{r}(j^2\kappa + \alpha) - (n+1)\frac{\alpha}{r}\right] = 0,$$

and the discriminant

$$\Delta = j^4 (\frac{n}{r} - \kappa)^2 + 2j^2 (\frac{n}{r} + \kappa)\alpha + 4j^2 \frac{\alpha}{r} + \alpha^2 > 0\,.$$

The eigenvalues (characteristic speeds) are all real and

$$\lambda_1\lambda_2 = j^2\left(j^2\kappa\frac{n}{r} - \frac{\alpha}{r}\right) = \frac{j^2}{r}\left(j^2\kappa n - \alpha\right).$$

Hence, if $j^2\kappa n - \alpha < 0$ then $\lambda_1\lambda_2 < 0$ and there is one positive eigenvalue. In the adiabatic case $\kappa = 0$ we always have

$$\lambda_1 + \lambda_2 < 0 \quad \text{and} \quad \lambda_1\lambda_2 = -j^2\frac{\alpha}{r} < 0.$$

We conclude

- If $\kappa = 0$ then for every Fourier mode j there is one positive and one negative eigenvalue, what indicates Hadamard instability.
- If $\kappa \neq 0$ there is one positive eigenvalue for the j-the mode, $j = 1, 2, \ldots$ if and only if $j^2 n\kappa - \alpha < 0$. Therefore, the first few Fourier modes may well be unstable but for $\kappa \neq 0$ the high frequency modes will decay.

This analysis is rigorous for $\kappa = 0$ but the analysis of frozen coefficients is only suggestive of the actual behavior in the heat conducting case. An exact analysis of the linearized non-autonomous system is carried out in [11] and shows that for $\kappa \neq 0$ the Fourier modes grow initially but after a transient period the inhomogeneities of the state variables eventually dissipate. This suggests the possibility of metastable response even in the nonlinear regime. Indeed, this has been observed numerically [1] for the case of a power law (2.2).

4. Numerical results for the Arrhenius system. We present numerical results for the Arrhenius system obtained by discretizing (2.5). A small initial temperature perturbation of Gaussian form is introduced at the center of the slab, while the initial velocity profile is retained as the uniform shearing solution (2.6). The boundary conditions are

$$v(0,t) = 0, \quad v(1,t) = 1, \quad \theta_x(0,t) = \theta_x(1,t) = 0.$$

The spatial discretization of (2.5) is based on a finite element method using linear elements, while for the temporal discretization we use the implicit Euler method. Special adaptive techniques in space as well as in time are used in order to fully resolve the band, see [1]. We consider adiabatic deformations ($\kappa = 0$) with $\alpha = 0.25$, $n = 0.1$. The numerical solution of the system, up to time $t \sim 7.47$ is shown in Figures 2 and 3. As seen in Figure 3 stress diffusion collapses across the shear band.

5. Derivation of an effective equation. Next, we derive an effective equation describing the long-time behavior of the system (2.5). Our analysis follows [10] adapted here to the case of the Arrhenius model. When $\kappa = 0$ the system (3.2) takes the form

$$\begin{aligned} u_t &= \frac{1}{r}\Sigma_{xx}, \\ \Theta_t &= \Sigma u - 1, \\ \Sigma &= e^{-\alpha\Theta}u^n. \end{aligned} \tag{5.1}$$

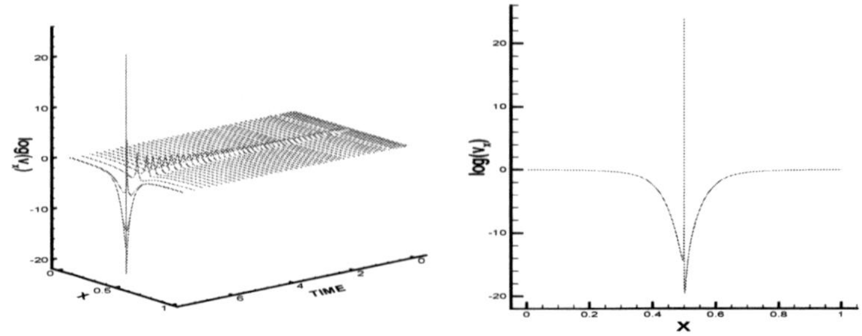

FIG. 2. $v_x(\gamma_t)$ *in log-scale: evolution (left), final time (right).*

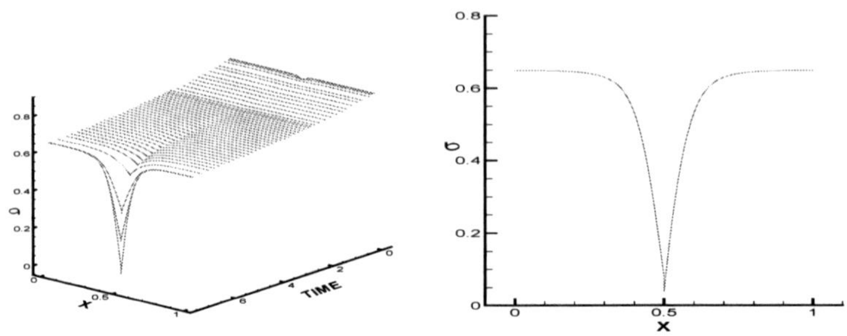

FIG. 3. σ *: evolution (left), final time (right).*

The forthcoming analysis expands on the following view of the problem. We expect that as $t \to \infty$ the first equation in (5.1) will act as a moment equation while the large time behavior of the remaining equations will be slaved into the equilibrium manifold, that is $\Sigma u \to 1$ as $t \to \infty$ and the limiting dynamics is captured by the effective equation

$$u_t = \frac{1}{r}(u^{-1})_{xx}, \tag{5.2}$$

which is a backward parabolic.

To capture this asymptotic behavior of (5.1) we introduce a parameter T thought of as an observational time scale. Consider the time rescaling $s(t) = t/T$, then

$$\begin{aligned} u_s &= \frac{T}{r}\Sigma_{xx}, \\ \frac{1}{T}\Theta_s &= (\Sigma u - 1), \\ \Sigma &= e^{-\alpha\Theta}u^n. \end{aligned} \tag{5.3}$$

Moreover, we either consider an inner expansion of the solution (see [11]) or consider a limit where r is also scaled with T and $r = O(T)$. This corresponds to looking at an asymptotic regime where inertia is dominant. We introduce the Hilbert expansion

$$\begin{aligned} u &= u_0 + \frac{1}{T} u_1 + O(\frac{1}{T^2}), \\ \Sigma &= \Sigma_0 + \frac{1}{T} \Sigma_1 + O(\frac{1}{T^2}), \\ \Theta &= \Theta_0 + \frac{1}{T} \Theta_1 + O(\frac{1}{T^2}), \end{aligned} \tag{5.4}$$

and compute the first few terms of the expansion. The $O(1)$ and $O(\frac{1}{T})$ terms give

$$\begin{aligned} &O(1): \quad \Sigma_0 u_0 = 1, \quad \Sigma_0 = e^{-\alpha\Theta_0} u_0^n \implies \Sigma_0 = \frac{1}{u_0}, \quad \Theta_0 = \frac{n+1}{\alpha} \log u_0, \\ &O(\frac{1}{T}): \quad \Theta_{0,s} = \Sigma_1 u_0 + \Sigma_0 u_1, \quad \Sigma_1 = -\alpha \Sigma_0 \Theta_1 + n \frac{\Sigma_0}{u_0} u_1, \\ &\qquad \implies \Sigma_1 = -\frac{u_1}{u_0^2} + \frac{1}{u_0} \partial_s \left(\frac{n+1}{\alpha} \log u_0 \right). \end{aligned}$$

Moreover, (5.3)(a) implies

$$u_{0,s} + \frac{1}{T} u_{1,s} + O(\frac{1}{T^2}) = \partial_{xx} \left(\Sigma_0 + \frac{1}{T} \Sigma_1 + O(\frac{1}{T^2}) \right)$$

which in turn yields

$$u_{0,s} = \partial_{xx} \Sigma_0 = \partial_{xx} (\frac{1}{u_0}).$$

Next, we reconstruct the effective equation up to order $O(\frac{1}{T^2})$ using a procedure analogous to the Chapman-Enskog expansion. From (5.3) we get

$$\begin{aligned} u_s &= \partial_{xx} \left(\frac{1}{u_0} - \frac{1}{T} \frac{u_1}{u_0^2} + \frac{1}{T} \frac{1}{u_0} \partial_s \left(\frac{n+1}{\alpha} \log u_0 \right) \right) + O(\frac{1}{T^2}), \\ &= \partial_{xx} \left((u_0 + \frac{1}{T} u_1)^{-1} + O(\frac{1}{T^2}) \right) + \frac{1}{T} \partial_{xx} \left(\frac{1}{u_0} \frac{n+1}{\alpha} \frac{1}{u_0} u_{0,s} \right) + O(\frac{1}{T^2}). \end{aligned}$$

Hence, u satisfies within $O(\frac{1}{T^2})$ the effective equation

$$u_s = \partial_{xx} \left(\frac{1}{u} \right) + \frac{1}{T} \frac{n+1}{\alpha} \partial_{xx} \left(\frac{1}{u^2} \partial_{xx} \left(\frac{1}{u} \right) \right). \tag{5.5}$$

The first term in the right hand of (5.5) is backward parabolic. Since the coefficient $\frac{n+1}{\alpha} > 0$, the fourth order term has a regularizing effect.

The uniform shear corresponds to $u_0 = 1$. Set $u = 1 + v$ and compute the linearized equation of (5.5) as follows

$$\begin{aligned}\partial_s v &= \partial_{xx}\left(\frac{1}{1+v} + \frac{1}{T}\frac{n+1}{\alpha}\frac{1}{(1+v)^2}\partial_{xx}\left(\frac{1}{1+v}\right)\right), \\ &= \partial_{xx}\left(1 - v + O(v^2) + \frac{n+1}{T\alpha}(1 - 2v + O(v^2))\partial_{xx}(1 - v + O(v^2))\right),\end{aligned}$$

and thus the linearized equation is

$$\partial_s v = -\partial_{xx} v - \frac{1}{T}\frac{n+1}{\alpha}\partial_{xxxx} v. \tag{5.6}$$

The Fourier transform $\hat{v}$ of v satisfies

$$\partial_s \hat{v} = \left(\xi^2 - \frac{1}{T}\frac{n+1}{\alpha}\xi^4\right)\hat{v},$$

and thus the low frequencies will grow but the high frequency modes still decay since $\frac{n+1}{\alpha} > 0$. Hence the linearized equation (5.6) is well posed.

6. Localization. In this section we show that the effective equation (5.5) admits solutions that develop coherent localized states for well-prepared data. Consider the equation

$$u_t = \partial_{xx}\left(\frac{1}{u}\right) + \varepsilon\partial_{xx}\left(\frac{1}{u^2}\partial_{xx}\left(\frac{1}{u}\right)\right). \tag{6.1}$$

Rescalings of the form

$$u_\lambda = \lambda^\alpha u(\lambda x, \lambda^\beta t) \tag{6.2}$$

leave (6.1) invariant whenever $2\alpha + \beta = 2$. If the scaling is also required to be consistent with conservation of the L^1 mass this fixes the parameters as $\alpha = 1$ and $\beta = 0$.

We seek solutions of (6.1) of the form

$$u = \frac{1}{R(t)}U\left(\frac{x}{R(t)}\right) \tag{6.3}$$

where $R(t)$ decays in time, so that the self-similar solution remains constant on lines $x = \xi R(t)$ focusing to $x = 0$ as $t \to \infty$. Introducing the *ansatz* (6.3) to (6.1), we obtain

$$-\frac{\dot{R}}{R}(\xi U)' = \left(\frac{1}{U} + \varepsilon\frac{1}{U^2}\left(\frac{1}{U}\right)''\right)''.$$

Set $R(t) = e^{-t}$ in which case the form of (6.3) becomes

$$u(x,t) = e^t\, U\left(x\, e^t\right). \tag{6.4}$$

We look to identify a symmetric profile $U(\xi)$, where $U(0) = U_0 > 0$ and the $U''(0) = -\zeta < 0$ are given as parameters. For an even profile $U'(0) = U'''(0) = 0$. Hence, U is selected by solving the initial value problem

$$\begin{gathered} \xi U = \left(\frac{1}{U} + \varepsilon \frac{1}{U^2} \left(\frac{1}{U} \right)'' \right)' \\ U(0) = U_0 \,, \quad U'(0) = 0 \,, \quad U''(0) = -\zeta \,, \end{gathered} \tag{6.5}$$

where $U_0 > 0$ and $\zeta > 0$ are parameters.

The limiting problem for $\varepsilon = 0$,

$$\begin{gathered} \xi U = \left(\frac{1}{U} \right)' \\ U(0) = U_0 > 0 \,, \end{gathered} \tag{6.6}$$

can be solved explicitly and its solution reads

$$U(\xi) = \frac{U_0}{(\xi^2 U_0^2 + 1)^{\frac{1}{2}}} \,. \tag{6.7}$$

In turn, it induces the following form for the associated solution u of (6.1) with $\varepsilon = 0$

$$u(x,t) = e^t U(x\, e^t) = \frac{U_0}{(U_0^2 x^2 + e^{-2t})^{\frac{1}{2}}} \,. \tag{6.8}$$

Observe that $u(0,t) = e^t$ grows exponentially at the origin $x = 0$, also that away from $|x| << 1$ the solution $u(x,t) \sim \frac{1}{|x|}$ for any $t > 1$. The solution is formally consistent with conservation of mass and is constant on lines $x = \xi e^{-t}$. This behavior indicates localization.

The problem (6.5) is solved numerically. We ask U to be an even function with normalized value $U(0) = 1$ and with $U''(0) = -\zeta$. We solve for various values of the parameter $\zeta > 0$. The solution of the o.d.e in the limiting $\varepsilon = 0$ case is $U_\infty(\xi) = 1/\sqrt{\xi^2 + 1}$. In Figure 4 the profiles of solutions U are presented for a fixed value of $\zeta = 10$ and three values of ϵ. The profile of U_∞ numerically coincides with the one of value $\epsilon = 0.001$. From the numerical results some observations can be made :

- The profile of U for large values of ξ behaves like $U(\xi) \sim 1/\xi$, independently of ϵ and ζ.
- Oscillations appear in the computed profile in accordance with expectations. The appearance and the frequency of oscillations is strongly related to the values of ε and ζ, but no definitive conclusion can be drawn from the numerical results.

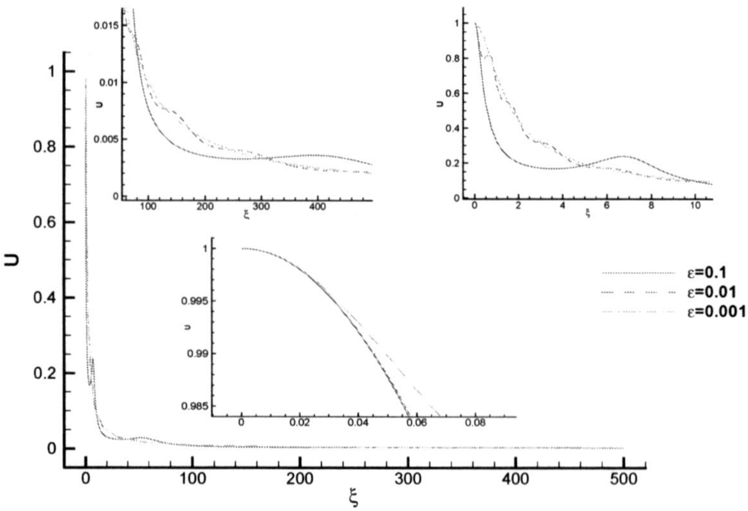

FIG. 4. $\zeta = 10,\ \epsilon \in [5 \cdot 10^{-4}, 5 \cdot 10^{-2}]$.

Acknowledgments. We thank Professor LEV TRUSKINOVSKY for very insightful remarks. AET partially supported by the National Science Foundation. Research partially supported by the European Union FP7 program Capacities (Regpot 2009-1).

REFERENCES

[1] TH. BAXEVANIS, TH. KATSAOUNIS, AND A.E. TZAVARAS, *Adaptive finite element computations of shear band formation*, Math. Models Meth. Appl. Sciences (2010), in press.

[2] M. BERTSCH, L.A. PELETIER, AND S.M. VERDUYN LUNEL, *The effect of temperature dependent viscosity on shear flow of incompressible fluids*, SIAM J. Math. Anal., **22** (1991), 328–343.

[3] R.J. CLIFTON, J. DUFFY, K.A. HARTLEY, AND T.G. SHAWKI, *On critical conditions for shear band formation at high strain rates*, Scripta Met., **18** (1984), pp. 443–448.

[4] L.S. COSTIN, E.E. CRISMAN, R.H. HAWLEY, AND J. DUFFY, *On the localization of plastic flow in mild steel tubes under dynamic torsional loading*, In "Proc. 2nd Conf. on the Mechanical Properties of Materials at high rates of strain", Inst. Phys. Conf. Ser. no 47, Oxford, **90**, 1979.

[5] C.M. DAFERMOS AND L. HSIAO, *Adiabatic shearing of incompressible fluids with temperature dependent viscosity*, Quart. Appl. Math., **41** (1983), 45–58.

[6] J.A. DILELLIO AND W.E. OLMSTEAD, *Shear band formation due to a thermal flux inhomogeneity*, SIAM J. Appl. Math., **57** (1997), 959–971.

[7] D.J. ESTEP, S.M.V. LUNEL, AND R.D. WILLIAMS, *Analysis of shear layers in a fluid with temperature-dependent viscosity*, Comp. Physics, **173** (2001), 17–60.

[8] C. FRESSENGEAS AND A. MOLINARI, *Instability and localization of plastic flow in shear at high strain rates*, J. Mech. Phys. Solids **35** (1987), pp. 185–211.

[9] K.A. Hartley, J. Duffy, and R.J. Hawley, *Measurement of the temperature profile during shear band formation in steels deforming at high-strain rates*, J. Mech. Physics Solids, **35** (1987), 283–301.

[10] Th. Katsaounis and A.E. Tzavaras, *Effective equations for localization and shear band formation*, SIAM J. of Applied Math., **69**, no. 6 (2009), pp. 1618–1643.

[11] Th. Katsaounis and A.E. Tzavaras, *Self-similar organization of localization in high strain-rate plasticity*, in preparation.

[12] A. Molinari and R.J. Clifton, *Analytical characterization of shear localization in thermoviscoplastic materials*, J. Appl. Mech., **54** (1987), 806–812.

[13] A.E. Tzavaras, *Effect of thermal softening in shearing of strain-rate dependent materials*, Arch. Rational Mech. Analysis, **99** (1987), 349–374.

[14] A.E. Tzavaras, *Strain softening in viscoelasticity of the rate type*, J. Integral Equations Appl., **3** (1991), 195–238.

[15] A.E. Tzavaras, *Nonlinear analysis techniques for shear band formation at high strain rates*, Appl. Mech. Reviews, **45** (1992), S82–S94.

[16] J.W. Walter, *Numerical experiments on adiabatic shear band in one space dimension*, Int. J. of Plasticity, **8** (1992), 657–693.

[17] T.W. Wright and J.W. Walter, *On stress collapse in adiabatic shear bands*, J. Mech. Phys. Solids, **35** (1988), 701–720.

[18] T.W.Wright and H. Ockendon, *A model for fully formed shear bands*, J. Mech. Phys. Solids, **40** (1992), 1217–1226.

[19] T.W. Wright, *The Physics and Mathematics of Shear Bands*, Cambridge University Press, 2002.

[20] C. Zener and J.H. Hollomon, *Effect of strain rate upon plastic flow of steel*, J. Appl. Physics, **15** (1944), 22–32.

HYPERBOLIC CONSERVATION LAWS ON SPACETIMES

PHILIPPE G. LEFLOCH*

Abstract. We present a generalization of Kruzkov's theory to manifolds. Nonlinear hyperbolic conservation laws are posed on a differential $(n + 1)$-manifold, called a spacetime, and the flux field is defined as a field of n-forms depending on a parameter. The entropy inequalities take a particularly simple form as the exterior derivative of a family of n-form fields. Under a global hyperbolicity condition on the spacetime, which allows arbitrary topology for the spacelike hypersurfaces of the foliation, we establish the existence and uniqueness of an entropy solution to the initial value problem, and we derive a geometric version of the standard L^1 semi-group property. We also discuss an alternative framework in which the flux field consists of a parametrized family of vector fields.

Key words. Hyperbolic conservation law, manifold, spacetime, entropy solution, well-posed theory, finite volume scheme.

AMS(MOS) subject classifications. Primary: 35L65. Secondary: 76L05, 76N.

1. Introduction. Hyperbolic conservation laws on manifolds arise in many applications to continuum physics. Specifically, the shallow water equations for perfect fluids posed on the sphere with prescribed topography provide an important model describing geophysical flows and exhibiting complex wave structures (Rosby waves, etc.) with several distinct scales. In [1]–[5] and [13, 18, 19, 20], the author together with several collaborators recently initiated a research program and posed the foundations for the analysis and numerical approximation of hyperbolic conservation laws on manifolds. These equations form a simplified, yet challenging, mathematical model for studying the propagation of nonlinear waves, including shock waves, and their interplay with the background geometry.

2. Hyperbolic conservation laws based on vector fields. The standard shock wave theory posed on the Euclidian space covers nonlinear hyperbolic equations of the form

$$\partial_t u + \sum_{j=1}^{n} \partial_j f^j(u) = 0, \qquad u = u(t,x) \in \mathbb{R},\ t \geq 0,\ x \in \mathbb{R}^n,$$

where $f^j : \mathbb{R} \to \mathbb{R}$ are given functions, and, more generally, applies to balance laws with variable coefficients and source-terms. Pioneering works by Lax, Oleinik, and others in the 50's and 60's set the foundations about discontinuous entropy solutions to such equations. Later, well-posedness

*Laboratoire Jacques-Louis Lions & Centre National de la Recherche Scientifique, Université Pierre et Marie Curie (Paris 6), 4 Place Jussieu, 75252 Paris, France (Blog: philippelefloch.wordpress.com).
The author was supported by the Agence Nationale de la Recherche via the grant ANR 06-2-134423.

theorems within the class of entropy solutions were established by Conway and Smoller (1966), Volpert (1967) (for solutions with bounded variation), Kruzkov (1970) (for bounded solutions), and finally DiPerna (1984) (for measure-valued solutions). Based on this mathematical theory, active research followed in the 80's and 90's that led to the development of robust and high-order accurate, shock-capturing schemes.

The generalization of the above theory to *manifolds* has only recently received particular attention, driven by problems arising in geophysical science and general relativity. In the present section, we consider a class of conservation laws based on parametrized vector fields, and we suppose that M is a compact, oriented n-dimensional manifold, endowed with an L^∞ volume form such that, in a finite atlas of local coordinate charts (x^j) and for uniform constants $c \le C$,

$$\omega = \overline{\omega}\, dx^1 \cdots dx^n, \qquad 0 < c \le \overline{\omega} \le C.$$

Recall that the divergence of an L^∞ vector field X is defined in the sense of distributions by

$$\langle \mathrm{div}_\omega X, \theta \rangle := \int_M (d\theta)(X)\, \omega$$

for all test-function $\theta : M \to \mathbb{R}$.

DEFINITION 2.1. *To any parameterized family of (smooth) vector fields $f = f_p(\overline{u}) \in T_pM$ depending upon $\overline{u} \in \mathbb{R}$, one associates the* **hyperbolic conservation law on the manifold** (M, ω)

$$\partial_t u + div_\omega\, (f(u)) = 0, \qquad u = u(t,p), \quad t \ge 0,\, p \in M. \tag{2.1}$$

The parametrized flux field f is called **geometry-compatible** *if constants are trivial solutions, that is*

$$(div_\omega\, f)_p(\overline{u}) = 0, \qquad \overline{u} \in \mathbb{R},\, p \in M \tag{2.2}$$

in the distribution sense. A **convex entropy pair** *is a convex function $U : \mathbb{R} \to \mathbb{R}$ together with a parametrized family of vector field*

$$F_p(\overline{u}) := \int^{\overline{u}} \partial_u U(v)\, \partial_u f_p(v)\, dv \in T_pM, \qquad \overline{u} \in \mathbb{R},\, p \in M.$$

The equation (2.1) is geometric in nature and does not depend on a particular choice of local coordinates on M. The flux field is a section of the tangent bundle TM and, in general, there is no concept of "spatially constant flux", unlike in the Euclidian case where one may arbitrarily choose a vector $f_p(\overline{u})$ at some point $p \in \mathbb{R}^n$ and parallel-transport it in $\mathbb{R}^n$, trivially generating a constant (and, therefore, smooth) vector field. The geometry-compatibility condition is natural if one observes that it is satisfied by the shallow water equation.

REMARK 2.1. An important class of geometry-compatible conservation laws is based on **projected gradient flux** fields, as it is called in [4], and is defined as follows. Consider the unit sphere $\mathbf{S}^2$ embedded in the Euclidian space $\mathbb{R}^3$ and a function $h : \mathbb{R}^3 \times \mathbb{R} \to \mathbb{R}$ being fixed, and define

$$f_p(\overline{u}) := p \wedge (\nabla_{\mathbb{R}^3} h)_p(\overline{u}) \in T_p\mathbf{S}^2, \qquad p \in \mathbf{S}^2 \subset \mathbb{R}^3.$$

Now, to formulate the initial value problem associated with (2.1) we are given an initial data $u_0 : M \to \mathbb{R}$ and we impose

$$u(0,p) = u_0(p), \quad p \in M. \tag{2.3}$$

It is well-known that entropy inequalities must be imposed on weak solutions to conservation laws. In this juncture, it is interesting to point out the following property of geometry-compatible flux field: If $u = u(t,p)$ is a smooth solution and $U = U(\overline{u})$ is an arbitrary function, then clearly

$$\partial_t U(u) + \partial_u U(u) \operatorname{div}_\omega (f_p(u)) = 0,$$

and it can then be checked that a necessary and sufficient condition for the existence of a vector field $F = F_p(\overline{u}) \in T_pM$ such that

$$\partial_u U(v) \operatorname{div}_\omega (f_p(v)) = \operatorname{div}_\omega (F_p(v))$$

for all functions $v = v(p)$ is precisely that the flux $f_p(\overline{u})$ be geometry-compatible. This observation motivates the following definition which impose the entropy inequality

$$\partial_t U(u) + \operatorname{div}_\omega (F(u)) \leq 0 \tag{2.4}$$

in the sense of distributions for all convex U.

DEFINITION 2.2. *An* **entropy solution to the conservation law** *(2.1), posed on the manifold (M,ω) and assuming the initial data $u_0 \in L^\infty(M)$, is a scalar field $u \in L^\infty\big(\mathbb{R}^+ \times M\big)$ satisfying the entropy inequalities*

$$\int_{\mathbb{R}^+} \int_M \big(U(u)\, \partial_t\theta + d\theta(F(u))\big)\, \omega dt + \int_M U(u_0)\, \theta(0,\cdot)\, \omega \geq 0$$

for all convex (U,F) and test-functions $\theta \geq 0$.

Note that integration is performed w.r.t. the volume form on M, and the differential $d\theta$ is a 1-form field acting on vector fields. Our main existence result is as follows.

THEOREM 2.1 (Well-posedness theory on a manifold). *Let (M,ω) be a compact n-manifold endowed with an L^∞ volume form, and $f = f_p(\overline{u}) \in T_pM$ be a geometry-compatible flux field.*

1. *Given any data $u_0 \in L^\infty(M)$, the initial value problem associated with the conservation law (2.1) admits a unique entropy solution $u \in L^\infty(\mathbb{R}^+ \times M)$.*
2. *Entropy solutions satisfy the following L^q stability property for all $q \in [1, \infty]$ and $t_2 \geq t_1$*

$$\|u(t_2)\|_{L^q_\omega(M)} \leq \|u(t_1)\|_{L^q_\omega(M)}. \tag{2.5}$$

3. *Moreover, the following L^1 semi-group property holds for any two entropy solutions u, v and $t_2 \geq t_1$*

$$\|v(t_2) - u(t_2)\|_{L^1_\omega(M)} \leq \|v(t_1) - u(t_1)\|_{L^1_\omega(M)}. \tag{2.6}$$

This theorem provides a generalization of Kruzkov's theory to a manifold, endowed with a volume form with possibly low regularity. The L^q stability and L^1 contraction properties provide *geometry-independent* bounds, which are useful to design approximation schemes. For further discussions on the well-posedness theory based on vector fields, we refer the reader to Ben-Artzi and LeFloch [4] and Amorim, Ben-Artzi and LeFloch [1], as well as Panov [21, 22].

Consider a geometry-compatible conservation law with flux field f together with some initial data $u_0 \in L^\infty(M)$. Then, an **entropy measure-valued solution** to (2.1)-(2.3), by definition, is a measure-valued field $(t, p) \in \mathbb{R}^+ \times M \mapsto \nu_{t,p} \in \text{Prob}(\mathbb{R})$ such that

$$\int_{\mathbb{R}^+} \int_M \Big(\langle \nu_{t,p}, U \rangle \, \partial_t \theta + \langle \langle \nu_{t,p}, F \rangle, d\theta \rangle \Big) \, \omega(p) dt + \int_M U(u_0) \, \theta(0, \cdot) \, \omega \geq 0$$

for all convex entropy pairs and test-functions $\theta = \theta(t, p) \geq 0$. Following DiPerna [11] we can establish the following uniqueness result.

THEOREM 2.2 (Uniqueness and compactness framework). *If ν is an entropy measure-valued solution to the conservation law (2.1) posed on the manifold (M, ω) and assuming some initial data $u_0 \in L^\infty(M)$ at $t = 0$, then*

$$\nu_{t,p} = \delta_{u(t,p)} \qquad a.e. \ p \in M.$$

where $u \in L^\infty(\mathbb{R}^+ \times M)$ denotes the unique entropy solution to the same problem.

This theorem is very useful to establish the convergence of approximation schemes since, in comparison with the standard technique based on a total variation bound, it requires weaker and *geometrically more natural* bounds on (the entropy dissipation of) the approximate solutions, as well as *weaker assumptions* on the manifold geometry.

In the Euclidian case, the total variation diminishing (TVD) property was found to be very useful for the development of high-order schemes. On manifolds, provided ω is sufficiently smooth and the initial data u_0 has bounded variation then $u(t)$ has bounded variation for all $t \geq 0$. In fact, there exists $C > 0$ depending on the geometry of M and $\|u_0\|_{L^\infty(M)}$ such that

$$TV(u(t)) \leq e^{C\,t}\,(1 + TV(u_0)),$$

and the total variation might be unbounded as $t \to \infty$.

Introduce the total variation along a given vector field X by

$$TV_X(u) := \sup_{\|\phi\|_{L^\infty}} \int_M u \operatorname{div}_\omega (\phi\, X)\, \omega. \tag{2.7}$$

THEOREM 2.3 (Total variation estimate). *Consider a geometry-compatible conservation law (2.1) posed on the manifold (M, ω) and suppose that the initial data have bounded variation. Then, the* **total variation diminishing property** *(TVD)*

$$TV_X(u(t_2)) \leq TV_X(u(t_1)), \qquad t_2 \geq t_1,$$

holds provided the flux field f and the given field X satisfy the compatibility condition

$$\mathcal{L}_X\big(\partial_u f(\overline{u})\big) := \big[\partial_u f(\overline{u}), X\big] = 0. \tag{2.8}$$

On the other hand, from the time-invariance of the equation and the L^1_ω contraction property, it follows that

$$\|div_\omega\,(f(u(t)))\|_{\mathcal{M}_\omega(M)} := \sup_{\|\theta\|_{C^0} \leq 1} \int_M d\theta(f(u(t)))\,\omega \tag{2.9}$$

is diminishing in time.

The condition (2.8) is usually not satisfied in the examples of interest, except when there exists some decoupling into a one-parameter family of independent, one-dimensional equations. (See below.) Hence, the curved geometry restricts the class of flux enjoying the TVD property, and we can not expect to rely on a TVD property to develop approximation schemes. In contrast, in the Euclidian case no restriction is implied on (p-independent) flux and the TVD property always holds.

REMARK 2.2. Returning to the projected gradient flux field on $\mathbf{S}^2 \subset \mathbb{R}^3$ introduced in Remark 2.1 and choosing a vector field X of the form

$$X_p := p \wedge (\nabla_{\mathbb{R}^3} k)_p, \qquad p \in \mathbb{R}^3,$$

then the condition $\mathcal{L}_X\big(\partial_u f(\overline{u})\big) = 0$ on $\mathbf{S}^2$ holds if and only if the vectors $\partial_u f_p(\overline{u})$ and X_p are parallel, and the proportionality factor $C(\overline{u}, p)$ is constant along the integral curves of X:

$$\partial_u f_p(\overline{u}) = \partial_u C(\overline{u}, p)\, X_p, \quad X(C(\overline{u}, p)) = 0, \qquad \overline{u} \in \mathbb{R},\ p \in M.$$

3. Hyperbolic conservation laws based on differential forms. In this section based on LeFloch and Okutmustur [19, 20], we consider hyperbolic conservation laws posed on an $(n+1)$-dimensional differentiable manifold M, referred to as a spacetime. We emphasize that, in the framework now developed, no geometric structure is a priori imposed on M.

DEFINITION 3.1. *The* **hyperbolic conservation law on the differentiable** $(n+1)$**-manifold** M *associated with a parametrized family of* L^∞ *n-form fields* $p \in M \mapsto \omega_p(\overline{u})$ *depending smoothly upon the parameter* $\overline{u}$*, by definition, reads*

$$d(\omega(u)) = 0, \tag{3.1}$$

whose unknown is the scalar field $u : M \to \mathbb{R}$*. The parametrized flux field* ω *is called* **geometry-compatible** *if it is closed, that is, for each* $\overline{u} \in \mathbb{R}$*,* $(d\omega)(\overline{u}) = 0$ *in the distribution sense.*

By definition, weak solutions $u \in L^\infty(M)$ must satisfy the following weak form of the conservation law

$$\int_M d\theta \wedge \omega(u) = 0 \qquad \text{for all test-functions } \theta : M \to \mathbb{R}. \tag{3.2}$$

Then, to pose the initial value problem we impose the following **global hyperbolicity condition:** M admits a smooth foliation by compact, oriented hypersurfaces, i.e.,

$$M = \bigcup_{t \geq 0} \mathcal{H}_t.$$

Furthermore, denoting by $i_{\mathcal{H}_t} : \mathcal{H}_t \to M$ the injection map, we impose that each hypersurface is **spacelike** in the sense that the n-forms

$$\partial_u \big(i^*_{\mathcal{H}_t} \omega(\overline{u}) \big) > 0 \tag{3.3}$$

are (positive) n-volume forms for each $t \geq 0$ and all relevant $\overline{u}$.

To formulate the initial value problem associated with (3.1) we are given some initial data $u_0 \in L^\infty(\mathcal{H}_0)$ and we search for a scalar field $u \in L^\infty(M)$ satisfying

$$u_{|\mathcal{H}_0} = u_0. \tag{3.4}$$

In the present framework, a **convex entropy flux** is now a parametrized family of n-forms $\Omega = \Omega_p(\overline{u})$ such that there exists a convex $U : \mathbb{R} \to \mathbb{R}$

$$\Omega_p(\overline{u}) := \int^{\overline{u}} \partial_u U \, \partial_u \omega_p \, du', \qquad p \in M, \, \overline{u} \in \mathbb{R}.$$

If the flux ω is geometry-compatible, then smooth solutions satisfy the additional conservation laws

$$d(\Omega(u)) = 0,$$

so that the entropy inequalities take the following form

$$d(\Omega(u)) \leq 0. \tag{3.5}$$

DEFINITION 3.2. *An* **entropy solution to the geometry-compatible conservation law** *(3.1) is a scalar field $u = u(p) \in L^\infty(M)$ satisfying for all convex entropy flux $\Omega = \Omega_p(\overline{u})$ and smooth functions $\theta \geq 0$*

$$\int_M d\theta \wedge \Omega(u) + \int_{\mathcal{H}_0} \theta_{\mathcal{H}_0} \, (i^*\Omega)(u_0) \geq 0.$$

Introduce also Kruzkov's entropy flux fields, defined here as parametrized fields of n-forms:

$$\Omega(u, v) := \text{sgn}\,(u - v)\big(\omega(u) - \omega(v)\big).$$

THEOREM 3.1 (Well-posedness theory on a spacetime. Geometry-compatible flux). *Consider a conservation law with a geometry-compatible flux $\omega = \omega(\overline{u})$, posed on a globally hyperbolic, spatially compact manifold $M = \bigcup_{t \geq 0} \mathcal{H}_t$. Then, the initial value problem with data $u_0 \in L^\infty(\mathcal{H}_0)$ admits a unique entropy solution $u \in L^\infty(M)$, and for any convex entropy flux Ω the function $t \mapsto \int_{\mathcal{H}_t} i^*_{\mathcal{H}_t}\Omega(u)$ is well-defined and non-increasing. Moreover, for any two entropy solutions u, v*

$$\int_{\mathcal{H}_t} i^*_{\mathcal{H}_t}\Omega(u, v)$$

is non-increasing in time.

THEOREM 3.2 (Well-posedness theory on a spacetime. General flux). *Consider a conservation law with general flux field $\omega = \omega(u)$, posed on globally hyperbolic, spatially compact M with (past) boundary $\mathcal{H}_0$. Consider initial data u_0 satisfying $\int_{\mathcal{H}_0} |i_{\mathcal{H}_0}\omega(u_0)| < \infty$. Then, there exists a semi-group of entropy solutions $u_0 \mapsto u = Su_0$ and for all smooth functions $\theta \geq 0$*

$$\int_M d\theta \wedge \Omega(u, v) + \int_{\mathcal{H}_0} \theta_{\mathcal{H}_0} \, i^*_{\mathcal{H}_0}\Omega(u_0, v_0) \geq 0. \tag{3.6}$$

Moreover, for any two spacelike hypersurfaces $\mathcal{H}_2$ in the future of $\mathcal{H}_1$

$$\int_{\mathcal{H}_2} i^*_{\mathcal{H}_2}\Omega(u, v) \leq \int_{\mathcal{H}_1} i^*_{\mathcal{H}_1}\Omega(u, v) < \infty. \tag{3.7}$$

REMARK 3.1. On the one-dimensional torus $M = T^1$, the case of general flux fields was covered earlier by LeFloch and Nédélec [17] under the assumption that f is strictly convex in $\overline{u}$. Then, entropy solutions satisfy the generalized Lax formula $u(t,x) = (\partial_u f)^{-1}\Big(\frac{x-y(t,x)}{t}\Big)$, in which the function $y = y(t,x)$ is determined by a minimization argument over the family of characteristics defined by

$$\partial_s X = (\partial_u f \circ f_\pm^{-1})\Big(\frac{c}{\overline{\omega}(X)}\Big), \qquad X(0) = y, \quad c \in \mathbb{R}.$$

Interestingly, it was established in [17] that any two entropy solutions $u, v : \mathbb{R}^+ \times T^1 \to \mathbb{R}$ satisfy

$$\|v(t_2) - u(t_2)\|_{L^1_\omega(T^1)} \leq \|v(t_1) - u(t_1)\|_{L^1_\omega(T^1)}, \qquad t_2 \geq t_1.$$

4. Intrinsic finite volume schemes.

4.1. General intrinsic approach. Using the framework based on n-differential forms, Amorim, LeFloch, and Okutmustur [2] have introduced finite volume schemes, which generate piecewise constant approximations on unstructured triangulations. One considers here the initial value problem associated with the conservation law (3.1) and discretize the weak form (3.2) of the conservation law. Indeed, the finite volume approach applies directly to the geometric weak form and, in particular, no local coordinates should be chosen at this stage. The finite volume discretization on a manifold M requires only the n-differential structure put forward in the previous section.

One introduces a family of triangulations made of curved polyhedra K, each of them having a past boundary e_K^-, a future boundary e_K^+, and "vertical boundary" elements $e \in \partial K^0$. For each boundary element e there is precisely one element K_e such that $K \cap K_e = e$. Then, one defines the notion of "total flux" along spacelike hypersurfaces $e_K^\pm$ by setting

$$\int_{e_K^\pm} i^*\omega(u) \approx \int_{e_K^\pm} i^*\omega\left(u_{e_K^\pm}\right) =: q_{e_K^\pm}\left(u_{e_K^\pm}\right),$$

where $u : M \to \mathbb{R}$ denotes the entropy solution of the initial value problem under consideration.

Then, applying Stokes theorem to the conservation law (3.1), we derive the **finite volume scheme**

$$q_{e_K^+}(u_{e_K^+}) = q_{e_K^-}(u_{e_K^-}) - \sum_{e \in \partial K^0} \mathbf{q}_{e,K}(u_{e_K^-}, u_{e_{K_e}^-}), \tag{4.1}$$

in which the **total flux** along vertical faces are *scalars* determined from Lipschitz continuous, numerical flux $\mathbf{q}_{e,K} : \mathbb{R}^2 \to \mathbb{R}$ satisfying

1. Consistency property:

$$\mathbf{q}_{K,e}(\overline{u}, \overline{u}) = q_e(\overline{u}),$$

2. Conservation property:

$$\mathbf{q}_{K,e}(\overline{u},\overline{v}) = -\mathbf{q}_{K_e,e}(\overline{v},\overline{u}),$$

3. Monotonicity property:

$$\partial_{\overline{u}}\mathbf{q}_{K,e} \geq 0, \qquad \partial_{\overline{v}}\mathbf{q}_{K,e} \leq 0$$

for all $\overline{u},\overline{v} \in \mathbb{R}$. For instance, Lax-Friedrich flux or Godunov flux may be used to determine the functions $\mathbf{q}_{e,K}$. The use of total flux leads to a geometrically natural setting, in which the volume of elements K or boundary elements e do not arise, so that the proposed formulation is both conceptually and technically simpler.

The convergence of finite volume schemes toward the entropy solution of the initial value problem is established in [2, 20]. Let us mention here the main difficulties: possible blow-up of the TV norm; derivation of uniform estimates tied to the geometry, only involving n-dimensional volumes. The main issues in our analysis are as follows:

1. Derivation of a discrete maximum principle.
2. Derivation of discrete entropy inequalities.
3. Derivation of entropy dissipation estimates.
4. The sequence of finite volume approximation generate an entropy measure-valued solution ν_p which, by our generalization of DiPerna's uniqueness theorem, coincides with $\nu_p = \delta_{u(p)}$, which implies that the scheme converges strongly.

DiPerna's measure-valued solutions were used to establish the convergence of schemes by Szepessy [23, 24], Coquel and LeFloch [8, 9, 10], and Cockburn, Coquel, and LeFloch [6, 7]. For many related results and a review about the convergence techniques for hyperbolic problems, we refer to Tadmor [25] and Tadmor, Rascle, and Bagnerini [26]. Further hyperbolic models, including also a coupling with elliptic equations, as well as many applications were successfully investigated by Kröners [14], and Eymard, Gallouet, and Herbin [12]. For higher-order schemes, see the paper by Kröner, Noelle, and Rokyta [15]. An alternative approach to the convergence of finite volume schemes was discovered by Westdickenberg and Noelle [27].

4.2. A second-order intrinsic scheme on the sphere. Ben-Artzi, Falcovitz, and LeFloch [5] adopted the fully intrinsic approach and extended the finite volume scheme to second-order accuracy (by using the generalized Riemann problem methodology), while implementing it for classes of hyperbolic conservation laws posed on the sphere $\mathbf{S}^2$. The sphere is the most important geometry for the applications to geophysical flows. Earlier works in this field were not fully geometric, and regarded the manifold as embedded in some Euclidian space and using projection techniques from the ambient space to the sphere. A fully intrinsic approach has definite advantages in terms of convergence and accuracy, and at the first-order at least, rigorous convergence analysis can be made.

We present here some features of the proposed scheme and refer to [5] for further discussions, especially on the implementation. We consider the conservation law

$$\partial_t u(t,x) + \nabla_{\mathbf{S}^2} \cdot (F(x,u(t,x))) = 0 \tag{4.2}$$

with flux vector tangent to the sphere, in the general form

$$F(x,u) = n(x) \wedge \Phi(x,u).$$

We are particularly interested in geometry-compatible flux vectors, and have proposed the following broad class of **gradient flux vector**

$$\Phi(x,u) = \nabla h(x,u).$$

In particular, this includes **homogeneous flux vectors** having $\Phi = \Phi(u)$. Interestingly enough, even in this case, the corresponding flux covers a broad and non-trivial class of interest.

Several classes of solutions of particular interest have been constructed, which exhibit very rich wave structure:

1. Spatially periodic solutions.
2. Non-trivial steady state solutions.
3. Confined solutions that remain compactly supported for all times.

Explicit formulas for such solutions together with various choices of corresponding flux vector fields can be found in [5]. These formulas are useful for numerical investigations.

An intrinsic triangulation was used, a *web-like mesh* made of segments of longitude and latitude lines. Artificial coordinate-singularities at the poles are coped with by suitably adapting the mesh in a non-conformal fashion as one approaches the North and South poles. To derive explicit formulas for the flux, for instance, in the case of an homogeneous flux vector $\Phi(u)$, one may introduce the following basis of the tangent space

$$i_\lambda = -\sin\lambda\, i_1 + \cos\lambda\, i_2, \qquad i_\phi = -\sin\phi\,\cos\lambda\, i_1 - \sin\phi\,\sin\lambda\, i_2 + \cos\phi\, i_3;$$

in terms of some Euclidian basis $i_1, i_2, i_3 in \mathbb{R}^3$. Then, by setting

$$\Phi(u) = f_1(u)\, i_1 + f_2(u)\, i_2 + f_3(u)\, i_3,$$

one finds

$$F(x,u) = F_\lambda(\lambda,\phi,u)\, i_\lambda + F_\phi(\lambda,\phi,u)\, i_\phi,$$

with

$$F_\lambda(\lambda,\phi,u) = f_1(u)\,\sin\phi\,\cos\lambda + f_2(u)\,\sin\phi\,\sin\lambda - f_3(u)\,\cos\phi, \tag{4.3}$$

and

$$F_\phi(\lambda, \phi, u) = -f_1(u) \sin\lambda + f_2(u) \cos\lambda. \tag{4.4}$$

The intrinsic finite volume discretization of the conservation law is then based on Stokes formula applied

$$\operatorname{div}_{\omega\, S^2} F = \frac{1}{\cos\phi}\Big(\frac{\partial}{\partial\lambda} F_\lambda + \frac{\partial}{\partial\phi}(F_\phi \cos\phi)\Big).$$

The flux are computed by explicit integration along longitude curves, or latitude curves. Length and areas are computed exactly, which allows one to derive a discrete version of the geometric compatibility condition.

5. Perspectives and open problems. In conclusion, based on our theoretical investigations (reported in this brief review), a shock-capturing numerical method was designed and implemented in [5], i.e. a fully intrinsic version of the second-order Godunov-type, finite volume scheme. This version, in its principle at first-order accuracy at least, extends to systems like the shallow water equations on the sphere with topography. Furthermore, the associated discrete properties (conservation, geometric compatibility, asymptotic-preserving) established in [5] extend to systems as well.

In the setup advocated here for the design numerical schemes, geometric terms in hyperbolic conservation laws are taken into account in a direct way by discretizing the geometric (weak) formulation of the conservation law, instead of using local coordinates and regarding the hyperbolic equation as an equation with variables coefficients and source-terms, or instead of viewing the manifold as embedded in a higher dimensional Euclidian space. This fully intrinsic approach was validated, both theoretically and numerically. It will be interesting to further investigate the properties of these finite volume schemes on general curved geometries, especially in the context of complex flows and geometries and to study the large-time asymptotics of solutions.

In the shallow water equations on the sphere, one unknown —the mass density— is a scalar field and can be treated as explained in the present work. However, the other unknown —the velocity— is a vector field, and it remains a challenging open problem to carry out the general Riemann problem methodology in this vector- or, more generally, tensor-valued context. This is an important issue, since understanding large-scale atmospheric and oceanic motions depends upon designing robust and accurate numerical techniques that, we advocate, should directly incorporate curved geometric effects.

The class of geometry-compatible conservation laws on spacetimes forms an analogue of the inviscid Burgers equation, which has served as an important simplified model for the development of shock-capturing methods. The treatment of boundary conditions on non-compact manifolds is an important issue, tackled in [13], while error estimates a la Kuznetsov were derived in [3, 18].

REFERENCES

[1] AMORIM P., BEN-ARTZI M., AND LEFLOCH P.G., *Hyperbolic conservation laws on manifolds: total variation estimates and the finite volume method,* Meth. Appl. Anal. **12** (2005), 291–324.

[2] AMORIM P., LEFLOCH P.G., AND OKUTMUSTUR B., *Finite volume schemes on Lorentzian manifolds,* Comm. Math. Sc. **6** (2008), 1059–1086.

[3] AMORIM P., LEFLOCH P.G., AND NEVES W., A geometric approach to error estimates for conservation laws posed on a spacetime, preprint ArXiv:1002.3137.

[4] BEN-ARTZI M. AND LEFLOCH P.G., *The well-posedness theory for geometry-compatible, hyperbolic conservation laws on manifolds,* Ann. Inst. H. Poincaré: Nonlin. Anal. **24** (2007), 989–1008.

[5] BEN-ARTZI M., FALCOVITZ J., AND LEFLOCH P.G., *Hyperbolic conservation laws on the sphere. A geometry-compatible finite volume scheme,* J. Comput. Phys. **228** (2009), 5650–5668.

[6] COCKBURN B., COQUEL F., AND LEFLOCH P.G., *Convergence of finite volume methods,* I.M.A. Series # **771**, Minneapolis, February 1991.

[7] COCKBURN B., COQUEL F., AND LEFLOCH P.G., *Convergence of finite volume methods for multidimensional conservation laws,* SIAM J. Numer. Anal. **32** (1995), 687–705.

[8] COQUEL F. AND LEFLOCH P.G., *Convergence of finite difference schemes for conservation laws in several space dimensions,* C.R. Acad. Sci. Paris Ser. I **310** (1990), 455–460.

[9] COQUEL F. AND LEFLOCH P.G., *Convergence of finite difference schemes for conservation laws in several space dimensions: a general theory,* SIAM J. Numer. Anal. **30** (1993), 675–700.

[10] COQUEL F. AND LEFLOCH P.G., *Convergence of finite difference schemes for conservation laws in several space dimensions: the corrected antidiffusive flux approach,* Math. Comp. **57** (1991), 169–210.

[11] DIPERNA R.J., *Measure-valued solutions to conservation laws,* Arch. Rational Mech. Anal. **88** (1985), 223–270.

[12] EYMARD R., GALLOUËT T., AND HERBIN R., *The finite volume method,* in "Handbook of Numerical Analysis", Vol. VII, North-Holland, Amsterdam, 2000, pp. 713–1020.

[13] GIESSELMAN J. AND LEFLOCH P.G., Hyperbolic conservation laws on spacetimes with boundary, in preparation.

[14] KRÖNER D., *Finite volume schemes in multidimensions,* in "Numerical analysis" 1997 (Dundee), Pitman Res. Notes Math. Ser. **380**, Longman, Harlow, 1998, pp. 179–192.

[15] KRÖNER D., NOELLE S., AND ROKYTA M., *Convergence of higher-order upwind finite volume schemes on unstructured grids for scalar conservation laws with several space dimensions,* Numer. Math. **71** (1995), 527–560.

[16] KRUZKOV S.N., *First-order quasilinear equations with several space variables,* Math. USSR Sb. **10** (1970), 217–243.

[17] LEFLOCH P.G. AND NÉDÉLEC J., *Explicit formula for weighted scalar nonlinear conservation laws,* Trans. Amer. Math. Soc. **308** (1988), 667–683.

[18] LEFLOCH P.G., NEVES W., AND OKUTMUSTUR B., *Hyperbolic conservation laws on manifolds. Error estimate for finite volume schemes,* Acta Math. Sinica **25** (2009), 1041–1066.

[19] LEFLOCH P.G. AND OKUTMUSTUR B., *Conservation laws on manifolds with limited regularity,* C.R. Acad. Sc. Paris, Ser. I **346** (2008), 539–543.

[20] LeFloch P.G. and Okutmustur B., *Hyperbolic conservation laws on spacetimes. A finite volume scheme based on differential forms,* Far East J. Math. Sci. **31** (2008), 49–83. (See also arXiv:0810.0255.)

[21] Panov E.Y., *On the Cauchy problem for a first-order quasilinear equation on a manifold,* Differential Equations **33** (1997), 257–266.

[22] Panov E.Y., *On the Dirichlet problem for first-order quasilinear equations on a manifold,* Trans. Amer. Math. Soc., 2010, to appear.

[23] Szepessy A., *Convergence of a shock-capturing streamline diffusion finite element method for a scalar conservation law in two space dimensions,* Math. Comp. **53** (1989), 527–545.

[24] Szepessy A., *Convergence of a streamline diffusion finite element method for scalar conservation laws with boundary conditions,* RAIRO Modél. Math. Anal. Numér. **25** (1991), 749–782.

[25] Tadmor E., *Approximate solutions of nonlinear conservation laws,* in "Advanced numerical approximation of nonlinear hyperbolic equations", Cetraro, 1997, Lect. Notes in Math., **1697**, Springer Berlin, 1998, pp. 1–149.

[26] Tadmor E., Rascle M., and Bagnerini P., *Compensated compactness for 2D conservation laws,* J. Hyperbolic Differ. Equ. **2** (2005), 697–712.

[27] Westdickenberg M. and Noelle S., *A new convergence proof for finite volume schemes using the kinetic formulation of conservation laws,* SIAM J. Numer. Anal. **37** (2000), 742–757.

REDUCED THEORIES IN NONLINEAR ELASTICITY

MARTA LEWICKA*

Abstract. The purpose of this note is to report on the recent development concerning the analysis and the rigorous derivation of thin film models for structures with nontrivial geometry. This includes: (i) shells with mid-surface of arbitrary curvature, and (ii) plates exhibiting residual stress at free equilibria. In the former setting, we derive a full range of models, some of them previously absent from the physics and engineering literature. The latter phenomenon has been observed in different contexts: growing leaves, torn plastic sheets and specifically engineered polymer gels. After reviewing available results, we list open problems with a promising angle of approach.

Key words. non-Euclidean plates, nonlinear elasticity, Gamma convergence, calculus of variations.

AMS(MOS) subject classifications. 74K20, 74B20.

Elastic materials exhibit qualitatively different responses to different kinematic boundary conditions or body forces. As a first step towards understanding the related evolutionary problem, one studies the minimizers of an appropriate nonlinear elastic energy functional. In sections 1-9 of this short review paper we focus on the recent development regarding the study of wavy patterns in experimentally attained configurations where the tissue endeavors to reach a non-attainable equilibrium through assuming a non-zero stress rest state. In sections 10 - 13 we present the available rigorous results for dimension reduction in theories of elastic shells of non-trivial mid-surface geometry. Section 14 contains a list of open problems.

1. Elastic energy of a growing tissue. Consider a sequence of thin 3d films $\Omega^h = \Omega \times (-h/2, h/2)$, with an open, bounded and simply connected mid-plate $\Omega \subset \mathbb{R}^2$. Each Ω^h undergoes a growth process, described instantaneously by a given smooth tensor $a^h = [a^h_{ij}] : \bar{\Omega} \to \mathbb{R}^{3\times 3}$ with the property: $\det a^h(x) > 0$. According to the formalism in [26], the following multiplicative decomposition is postulated:

$$\nabla u = Fa^h \tag{1.1}$$

for the gradient of a deformation $u : \Omega^h \to \mathbb{R}^3$. The tensor $F = \nabla u(a^h)^{-1}$ corresponds to the elastic part of ∇u, and accounts for the reorganization of Ω^h in response to the growth tensor a^h. Hence, the elastic energy of u depends only on F:

$$I^h_W(u) = \frac{1}{h}\int_{\Omega^h} W(\nabla u(a^h)^{-1})\,\mathrm{d}x, \quad \forall u \in W^{1,2}(\Omega^h, \mathbb{R}^3). \tag{1.2}$$

*University of Minnesota, Department of Mathematics, 206 Church St. S.E., Minneapolis, MN 55455 (lewicka@math.umn.edu); partially supported by the NSF grants DMS-0707275 and DMS-0846996.

The energy density $W : \mathbb{R}^{3\times 3} \longrightarrow \mathbb{R}_+$ is assumed to be $\mathcal{C}^2$ in a neighborhood of $SO(3)$, and to satisfy the following conditions of normalization, frame indifference and nondegeneracy:

$$\begin{aligned} \exists c>0 \quad \forall F \in \mathbb{R}^{3\times 3} \quad \forall R \in SO(3) \quad & W(R) = 0, \quad W(RF) = W(F) \\ & W(F) \geq c\, \mathrm{dist}^2(F, SO(3)). \end{aligned} \tag{1.3}$$

We remark that the validity of decomposition (1.1) into an elastic and inelastic part requires that it is possible to separate out a reference configuration. Thus this formalism is relevant for the description of processes such as plasticity, swelling and shrinkage in thin films, or plant morphogenesis. Although our results are valid for thin laminae that might be residually strained by a variety of means, we only consider the one-way coupling of growth to shape and ignore the feedback from shape back to growth. However, it would be very interesting to include this coupling in the model as well, once the basic coupling mechanisms are understood.

2. Non-Euclidean elasticity. We now compare the 'growth tensor' approach with the 'target metric' formalism [6, 21]. On each Ω^h we assume to be given a smooth Riemannian metric $g^h = [g^h_{ij}]$. A deformation u of Ω^h is said to be an orientation preserving realization of g^h, when $(\nabla u)^T \nabla u = g^h$ and $\det\nabla u > 0$, or equivalently when:

$$\nabla u(x) \in \mathcal{F}^h(x) = \Big\{R\sqrt{g^h}(x);\ R \in SO(3)\Big\} \qquad \text{a.e. in } \Omega^h. \tag{2.1}$$

It is hence instructive to study the energy functional:

$$\tilde{I}^h_{dist}(u) = \frac{1}{h}\int_{\Omega^h} \mathrm{dist}^2(\nabla u(x), \mathcal{F}^h(x))\,\mathrm{d}x \qquad \forall u \in W^{1,2}(\Omega^h, \mathbb{R}^3), \tag{2.2}$$

measuring the average pointwise deviation of u from the orientation preserving realizations of g^h. Note that $\tilde{I}^h_{dist}$ is comparable in magnitude with I^h_W, for $W = \mathrm{dist}^2(\cdot, SO(3))$. Indeed, the intrinsic metric of the material is transformed by a^h to the target metric $g^h = (a^h)^T a^h$ and, for isotropic W, only the symmetric positive definite part of a^h given by $\sqrt{g^h}$ plays the role in determining the deformed shape.

More generally, we also consider functionals $\tilde{I}^h_W = \int W(x, \nabla u(x))$ with the inhomogeneous densities W satisfying frame invariance, normalisation and the growth bound as in (1.3), with respect to energy well $\mathcal{F}^h$ in (2.1).

3. Residual stress. Note that one could define the energy as the difference between the pull-back metric of a deformation u and the given metric: $I^h_{str}(u) = \int |(\nabla u)^T\nabla u - g^h|^2\,\mathrm{d}x$. However, such 'stretching' functional is not appropriate from the variational point of view, because there always exists $u \in W^{1,\infty}$ such that $I^h_{str}(u) = 0$. Further, if the Riemann curvature tensor R^h associated to g^h does not vanish identically, say $R^h_{ijkl}(x) \neq 0$, then u has a 'folding structure' [11]; it cannot be orientation preserving (or reversing) in any open neighborhood of x.

As proved in [21], the functionals I_W^h, $\tilde{I}_W^h$ and $\tilde{I}_{dist}^h$ have strictly positive infima for non-flat g^h, which points to the existence of non-zero stress at free equilibria (in the absence of external forces or boundary conditions):

$$R^h \not\equiv 0 \quad \Leftrightarrow \quad \inf\left\{\tilde{I}_{dist}^h(u);\ u \in W^{1,2}(\Omega^h, \mathbb{R}^3)\right\} > 0.$$

It is conjectured that the same principle plays a role in the developmental processes of naturally growing tissues, where the process of growth provides a mechanism for the spontaneous formation of non-Euclidean metrics and consequently leads to complicated morphogenesis of the thin film exhibiting waves, ruffles and non-zero residual stress.

4. Relation to isometric immersions. Several questions arise in the study of the proposed energy functionals. A first one is to determine the scaling of the infimum energy in terms of the vanishing thickness $h \to 0$.

In [21] we considered the case where g^h is given by a tangential Riemannian metric $[g_{\alpha\beta}]$ on Ω, and is independent of h:

$$g^h = g(x', x_3) = \begin{bmatrix} \begin{bmatrix} g_{\alpha\beta}(x') \end{bmatrix} & \begin{matrix} 0 \\ 0 \end{matrix} \\ \begin{matrix} 0 & 0 \end{matrix} & 1 \end{bmatrix} \qquad \forall x' \in \Omega, \quad |x_3| < \frac{h}{2}. \tag{4.1}$$

Consequently, if $[g_{\alpha\beta}]$ has non-zero Gaussian curvature $\kappa_{[g_{\alpha\beta}]}$, then each $R^h \not\equiv 0$, and hence $\inf \tilde{I}_{dist}^h(u) > 0$ for every h. We proved that:

$$[g_{\alpha\beta}] \text{ has an isometric immersion } y \in W^{2,2}(\Omega, \mathbb{R}^3) \ \Leftrightarrow\ \inf \tilde{I}_{dist}^h \le Ch^2,$$

for a uniform constant C. It also follows that $\kappa_{[g_{\alpha\beta}]} \not\equiv 0$ if and only if $h^{-2} \inf \tilde{I}_{dist}^h \ge c > 0$. Existence of isometric immersions is a longstanding problem in differential geometry, depending heavily on the regularity [12]. Here, we deal with $W^{2,2}$ immersions not studied previously.

5. The prestrained Kirchhoff model. Let us remark that the particular choice of the metric (4.1) is motivated by the experiment reported in [14]. The experiment consisted in fabricating programmed flat disks of gels having a non-constant monomer concentration which induces a 'differential shrinking factor'. The disk was then activated in a temperature raised above a critical threshold, whereas the gel shrunk with a factor proportional to its concentration. This process defined a new target metric on the disk, of the described form and radially symmetric. The metric forced a 3d configuration in the initially planar plate, and one of the most remarkable features of this deformation is the onset of some transversal oscillations, which broke the radial symmetry.

Another interesting question in the setting, solved in [21], is to find the limiting zero-thickness theory under the obtained scaling law. Namely, we proved that any sequence of deformations u^h with $\tilde{I}_W^h(u^h) \leq Ch^2$, converges to a $W^{2,2}$ regular isometric immersion y of $[g_{\alpha\beta}]$. Conversely, every y with this property can be recovered as a limit of u^h whose energy scales like h^2. The Γ-limit of the rescaled energies is a curvature functional on the space of all $W^{2,2}$ realizations y of $[g_{\alpha\beta}]$ in $\mathbb{R}^3$:

$$\frac{1}{h^2}\tilde{I}_W^h \xrightarrow{\Gamma} \frac{1}{24}\int_\Omega \mathcal{Q}_2(x')\Big(\sqrt{[g_{\alpha\beta}]}^{-1}(\nabla y)^T\nabla\vec{n}\Big)\,\mathrm{d}x'.$$

Here $\vec{n}$ is the unit normal to the surface $y(\Omega)$, while $\mathcal{Q}_2(x')$ are the quadratic forms, nondegenerate and positive definite on the symmetric 2×2 tensors, which can be calculated explicitly (see [21]) from $\nabla^2 W(x', \sqrt{[g_{\alpha\beta}]})$.

A related reference in this context is [27], containing the derivation of Kirchhoff-like plate theory for heterogeneous multilayers from 3d nonlinear energies $\int W(x_3/h, \nabla u)$ with inhomogeneous density.

6. Rigidity estimate. As a ingredient of proofs, we give a generalization of the nonlinear rigidity estimate [8] to the non-Euclidean setting. Namely, for all $u \in W^{1,2}(\mathcal{U}, \mathbb{R}^n)$ there exists $Q \in \mathbb{R}^{n\times n}$ such that:

$$\int_\mathcal{U} |\nabla u(x) - Q|^2\,\mathrm{d}x \leq C\left(\int_\mathcal{U} \mathrm{dist}^2(\nabla u, \mathcal{F}(x))\,\mathrm{d}x + \|\nabla g\|^2_{L^\infty}(\mathrm{diam}\,\mathcal{U})^2|\mathcal{U}|\right)$$

where $\mathcal{F} = SO(n)\sqrt{g}$ is the energy well relative to a given metric g on an open, bounded, connected domain $\mathcal{U} \subset \mathbb{R}^n$. The constant C depends on $\|g\|_{L^\infty}$, $\|g^{-1}\|_{L^\infty}$, and on $\mathcal{U}$, uniformly for a family of domains which are uniformly bilipschitz equivalent.

Note that in case $g = \mathrm{Id}$, the infimum of I^h_{dist} in (2.2) is naturally 0 and is attained only by the rigid motions. In our setting, since there is no realization of $I^h_{dist}(u) = 0$ if the Riemann curvature of g is non-zero, we can only estimate the deviation of the deformation from a linear map at the expense of an extra term, proportional to the gradient of the metric.

7. A hierarchy of scalings. Given a sequence of growth tensors a^h defined on Ω^h, the general objective is now to analyse the behavior of the minimizers of the corresponding energies I_W^h as $h \to 0$.

As a first step in this direction we established in [17] a lower bound on $m_h = \inf I_W^h$ in terms of a power law: $m_h \geq ch^\beta$, for all values of β greater than a critical β_0 in (7.1). This critical exponent depends on the asymptotic behavior of the perturbation $a^h - \mathrm{Id}$ in terms of the thickness h. Under existence conditions for certain classes of isometries, it can be shown that actually $m_h \sim h^{\beta_0}$.

Namely, define: $Var(a^h) = \|\nabla_{tan}(a^h{}_{|\Omega})\|_{L^\infty(\Omega)} + \|\partial_3 a^h\|_{L^\infty(\Omega^h)}$, together with the scaling $\omega_1 = \sup\big\{\omega;\ \lim_{h\to 0} h^{-\omega} Var(a^h) = 0\big\}$ of the variations $Var(a^h)$ in h. Assume that: $\|a^h\|_{L^\infty(\Omega^h)} + \|(a^h)^{-1}\|_{L^\infty(\Omega^h)} \leq C$ and $\omega_1 > 0$. Further, assume that for some $\omega_0 \geq 0$, there exists the limit:

$$\epsilon_g(x') = \lim_{h\to 0} \frac{1}{h^{\omega_0}} \fint_{-h/2}^{h/2} a^h(x',t) - \mathrm{Id}\, dt \quad \text{in } L^2(\Omega, \mathbb{R}^{3\times 3}).$$

which satisfies: $\mathrm{curl}^T\mathrm{curl}\ (\epsilon_g)_{2\times 2} \not\equiv 0$, and that $\omega_0 < \min\{2\omega_1, \omega_1 + 1\}$. Then, for every β with:

$$\beta > \beta_0 = \max\{\omega_0 + 2, 2\omega_0\}, \tag{7.1}$$

there holds: $\limsup_{h\to 0} \frac{1}{h^\beta} \inf I^h_{dist} = +\infty$.

We expect it should be possible to rigorously derive a hierarchy of prestrained limiting theories, differentiated by the embeddability properties of the target metrics, encoded in the scalings of (the components of) their Riemann curvature tensors. This is in the same spirit as the different scalings of external forces lead to a hierarchy of nonlinear elastic plate theories displayed in [9]. For shells, that are thin films with mid-surface of arbitrary (non-flat) geometry, an infinite hierarchy of models was proposed, by means of asymptotic expansion in [22], and it remains in agreement with all the rigorously obtained results [7, 18, 19, 20].

8. The prestrained von Kármán model. Towards studying the dynamical growth problem, in [17] we considered the tensors of the form:

$$a^h(x', x_3) = \mathrm{Id} + h^2\epsilon_g(x') + hx_3\gamma_g(x'), \tag{8.1}$$

with given matrix fields $\epsilon_g, \gamma_g : \overline{\Omega} \to \mathbb{R}^{3\times 3}$. Note that the assumptions of the scaling result in the previous section do not hold, since in the present case: $\omega_0 = 2\omega_1 = \omega_1 + 1 = 2$.

We proved that, in this setting, $\inf I^h_W \leq Ch^4$, while the lower bound $h^{-4} \inf I^h_W \geq c > 0$ is equivalent to:

$$\mathrm{curl}((\gamma_g)_{2\times 2}) \not\equiv 0 \quad \text{or} \quad 2\mathrm{curl}^T\mathrm{curl}(\epsilon_g)_{2\times 2} + \det(\gamma_g)_{2\times 2} \not\equiv 0\ . \tag{8.2}$$

We use the following notational convention: for a matrix F, its $n \times m$ principle minor is denoted by $F_{n\times m}$ and the superscript ^{T} refers to the transpose of a matrix or an operator.

The two expressions in (8.2) are the (negated) linearized Gauss-Codazzi equations corresponding to the metric $I = \mathrm{Id} + h^2(\epsilon_g)_{2\times 2}$ and the second fundamental form $II = \frac{1}{2}h(\gamma_g)_{2\times 2}$ on Ω. Equivalently, the above conditions guarantee that the highest order terms in the expansion of the Riemann curvature tensor components R_{1213}, R_{2321} and R_{1212} of $g^h = (a^h)^T a^h$ do not vanish. Also, either of the conditions implies that $\inf \mathcal{I}_4 > 0$ (see definition below), which yields the lower bound on $\inf I^h_W$.

The Γ-limit of the rescaled energies is a function of the out-of-plane displacement $v \in W^{2,2}(\Omega, \mathbb{R})$ and in-plane displacement $w \in W^{1,2}(\Omega, \mathbb{R}^2)$:

$$\begin{aligned} &\frac{1}{h^4} I_W^h \xrightarrow{\Gamma} \mathcal{I}_4 \quad \text{where} \\ &\mathcal{I}_4(w,v) = \frac{1}{24}\int_\Omega \mathcal{Q}_2\big(\nabla^2 v + \frac{1}{2}(\gamma_g)_{2\times 2}\big) \\ &\qquad\qquad + \frac{1}{2}\int_\Omega \mathcal{Q}_2\big(\mathrm{sym}\nabla w + \frac{1}{2}\nabla v \otimes \nabla v - \frac{1}{2}(\epsilon_g)_{2\times 2}\big), \end{aligned} \tag{8.3}$$

with the quadratic nondegenerate form $\mathcal{Q}_2$, acting on matrices $F \in \mathbb{R}^{2\times 2}$:

$$\mathcal{Q}_2(F) = \min\{\mathcal{Q}_3(\tilde{F});\ \tilde{F} \in \mathbb{R}^{3\times 3}, \tilde{F}_{2\times 2} = F\}; \quad \mathcal{Q}_3(\tilde{F}) = D^2W(\mathrm{Id})(\tilde{F} \otimes \tilde{F}).$$

The two terms in (8.3) measure: the first order in h change of II, and the second order change in I, under the deformation $id + hve_3 + h^2 w$ of Ω. Moreover, any sequence of deformations u^h with $I_W^h(u^h) \leq Ch^4$ is, asymptotically, of this form.

9. The prestrained von Kármán equations. For the elastic energy W satisfying (1.3) which is additionally isotropic (that is $W(FR) = W(F)$ for all $F \in \mathbb{R}^{3\times 3}$ and $R \in SO(3)$) one can see [9] that the quadratic form $\mathcal{Q}_2$ in the expression of $\mathcal{I}_4$ is given explicitly as:

$$\mathcal{Q}_2(F_{2\times 2}) = 2\mu|\mathrm{sym}\ F_{2\times 2}|^2 + \frac{2\mu\lambda}{2\mu+\lambda}|\mathrm{tr}\ F_{2\times 2}|^2 .$$

Here, *tr* stands for the trace of a quadratic matrix, and μ and λ are the Lamé constants, satisfying: $\mu \geq 0$, $3\lambda + \mu \geq 0$.

Under these conditions, we rigorously derived the new von-Kármán-like system proposed in [23], written in terms of the out-of-plane displacement v and the Airy stress potential Φ. It describes the leading order displacements in a thin tissue which tries to adapt itself to an internally imposed metric $(a^h)^T a^h$, where a^h is of the form (8.1):

$$\Delta^2\Phi = -S(K_G + \lambda_g), \qquad B\Delta^2 v = [v, \Phi] - B\Omega_g.$$

Here $S = \mu(3\lambda + 2\mu)/(\lambda + \mu)$ is the Young's modulus, K_G the Gaussian curvature, $B = S/(12(1-\nu^2))$ the bending stiffness, and $\nu = \lambda/(2(\lambda+\mu))$ the Poisson ratio given in terms of the Lamé coefficients λ and μ. The corrections due to the prestrain are: $\lambda_g = \mathrm{curl}^T\mathrm{curl}\ (\epsilon_g)_{2\times 2}$, and $\Omega_g = \mathrm{div}^T\mathrm{div}\big((\gamma_g)_{2\times 2} + \nu\ \mathrm{cof}\ (\gamma_g)_{2\times 2}\big)$. More precisely, the Euler-Lagrange equations of $\mathcal{I}_4$ are equivalent, under a change of variables which replaces the in-plane displacement w by the Airy stress Φ, to :

$$\begin{aligned} \Delta^2\Phi &= -S\Big(\det\nabla^2 v + \mathrm{curl}^T\mathrm{curl}(\epsilon_g)_{2\times 2}\Big), \\ B\Delta^2 v &= \mathrm{cof}\nabla^2\Phi : \nabla^2 v - B\ \mathrm{div}^T\mathrm{div}\Big((\gamma_g)_{2\times 2} + \nu\ \mathrm{cof}(\gamma_g)_{2\times 2}\Big). \end{aligned}$$

The related (free) boundary conditions on $\partial\Omega$ are:

$$
\begin{aligned}
&\Phi = \partial_{\vec{n}}\Phi = 0,\\
&\tilde{\Psi} : (\vec{n}\otimes\vec{n}) + \nu\,\tilde{\Psi} : (\tau\otimes\tau) = 0,\\
&(1-\nu)\partial_\tau\Big(\tilde{\Psi} : (\vec{n}\otimes\tau)\Big) + \mathrm{div}\,\Big(\tilde{\Psi} + \nu\,\mathrm{cof}\tilde{\Psi}\Big)\,\vec{n} = 0,
\end{aligned}
$$

where $\vec{n}$ is the normal, τ the tangent to $\partial\Omega$, and: $\tilde{\Psi} = \nabla^2 v + \mathrm{sym}(\gamma_g)_{2\times 2}$. If additionally $\partial\Omega$ is a polygonal, then the above equations simplify to the equations (5) in [23].

10. Derivation of the elastic shell theories. We now turn to introducing some new results and a conjecture on the variational derivation of shell theories. They can be seen as generalizations of the results in [9], justifying a hierarchy of theories for nonlinearly elastic plates. This hierarchy corresponded to scaling of the applied forces (resulting in a scaling of the energy) in terms of thickness h, in the limit as $h \to 0$.

We consider a compact, connected, oriented 2d surface $S \subset \mathbb{R}^3$ with unit normal $\vec{n}$, and a family $\{S^h\}$ of thin shells around S:

$$
S^h = \{x + t\vec{n}(x);\ \ x\in S,\ \ -h/2 < t < h/2\}, \qquad 0 < h << 1.
$$

The elastic energy of a deformation $u^h \in W^{1,2}(S^h,\mathbb{R}^3)$ and its total energy in presence of applied force $f^h \in L^2(S^h,\mathbb{R}^3)$ are given, respectively, by:

$$
E^h(u^h) = \frac{1}{h}\int_{S^h} W(\nabla u^h), \qquad J^h(u^h) = E^h(u^h) - \frac{1}{h}\int_{S^h} f^h u^h \tag{10.1}
$$

where W is the nonnegative stored-energy density as in (1.3).

It can be shown [9] that if f^h scale like h^α, then $E^h(u^h)$ at (approximate) minimizers u^h of J^h scale like h^β, with $\beta = \alpha$ if $0 \le \alpha \le 2$ and $\beta = 2\alpha - 2$ if $\alpha > 2$. The main part of the analysis consists now of identifying the Γ-limit $\mathcal{I}_{\beta,S}$ of $h^{-\beta}E^h$ as $h \to 0$, for a given scaling $\beta \ge 0$, without making any a priori assumptions on the form of the deformations u^h.

In the case when $S \subset \mathbb{R}^2$ (i.e. a plate), such Γ-convergence was first established for $\beta = 0$ [15], and later [8, 9] for all $\beta \ge 2$. This last regime corresponds to a rigid behavior of the elastic material, since the limiting admissible deformations are isometric immersions (if $\beta = 2$) or infinitesimal isometries (if $\beta > 2$) of the mid-plate S. One particular case is $\beta = 4$, where the derived limiting theory turns out to be von Kármán theory [13]. The case $0 < \beta < 5/3$ was recently treated in [4], while the regime $5/3 \le \beta < 2$ remains open and is conjectured to be relevant for crumpling [29].

Much less was known when S is a surface of arbitrary geometry. The first result in [15] derived the membrane theory ($\beta = 0$) where the limit $\mathcal{I}_{0,S}$ depends only on the stretching and shearing produced by the deformation. Another study [7] analyzed the case $\beta = 2$, which reduces to the flexural shell model [3], i.e. a geometrically nonlinear purely bending theory, constrained to the isometric immersions of S. The energy $\mathcal{I}_{2,S}$ depends then on the change of curvature produced by such deformation.

11. Infinite hierarchy of shell theories. In [22] we derived (by means of asymptotic expansion) a precise conjecture on the form of $\mathcal{I}_{\beta,S}$ for $\beta > 2$ and arbitrary mid-surface S. Namely, $\mathcal{I}_{\beta,S}$ acts on the space of N-th order infinitesimal isometries $\mathcal{V}_N$, where N is such that:

$$\beta \in [\beta_{N+1}, \beta_N) \quad \text{and} \quad \beta_k = 2 + 2/k.$$

The space $\mathcal{V}_N$ consists of N-tuples $(V_1, \ldots, V_N)$ of displacements $V_i : S \to \mathbb{R}^3$ (with appropriate regularity), such that the deformations $u_\epsilon = \mathrm{id} + \sum_{i=1}^N \epsilon^i V_i$ preserve the metric on S up to order ϵ^N, i.e. $(\nabla u_\epsilon)^T \nabla u_\epsilon - \mathrm{Id} = \mathcal{O}(\epsilon^{N+1})$. Further, setting $\epsilon = h^{\beta/2-1}$, we have:

(i) When $\beta = \beta_{N+1}$ then $\mathcal{I}_{\beta,S} = \int_S \mathcal{Q}_2\,(x, \delta_{N+1} I_S) + \int_S \mathcal{Q}_2\,(x, \delta_1 II_S)$, where $\delta_{N+1} I_S$ is the change of metric on S of the order ϵ^{N+1}, generated by the family of deformations u_ϵ and $\delta_1 II_S$ is the first order (i.e. order ϵ) change in the second fundamental form.

(ii) When $\beta \in (\beta_{N+1}, \beta_N)$ then $\mathcal{I}_{\beta,S} = \int_S \mathcal{Q}_2\,(x, \delta_1 II_S)$.

(iii) The constraint of N-th order isometry $\mathcal{V}_N$ may be relaxed to that of $\mathcal{V}_M$, $M < N$, if S has the following $M \mapsto N$ matching property. For every $(V_1, \ldots V_M) \in \mathcal{V}_M$ there exist sequences of corrections $V_{M+1}^\epsilon, \ldots V_N^\epsilon$, uniformly bounded in ϵ, such that: $\tilde{u}_\epsilon = \mathrm{id} + \sum_{i=1}^M \epsilon^i V_i + \sum_{i=M+1}^N \epsilon^i V_i^\epsilon$ preserve the metric on S up to order ϵ^N.

This conjecture is supported by all the rigorously derived models. In particular, since plates enjoy the $2 \mapsto \infty$ matching property (i.e. [9] every $W^{1,\infty} \cap W^{2,2}$ member of $\mathcal{V}_2$ may be matched to an exact isometry, in the sense of (iii) above), all the plate theories for $\beta \in (2, 4)$ indeed collapse to a single theory (linearized Kirchhoff model of [9]).

12. The von Kármán theory for arbitrary shells. For $\beta = 4$, we proved [18, 19] that the Γ-limit $\mathcal{I}_{4,S}$ of $h^{-4}E^h$ acts on the 1-st order isometries $V \in \mathcal{V}_1 \cap W^{2,2}$ (hence the tensor field $A \in W^{1,2}(S, so(3))$, such that $\nabla V = A_{|T_xS}$, takes values in the space of skew-symmetric tensors $so(3)$) and finite strains $B \in \mathcal{B} = \mathrm{cl}_{L^2}\{\mathrm{sym}\nabla w;\ w \in W^{1,2}(S, \mathbb{R}^3)\}$:

$$\mathcal{I}_{4,S}(V, B) = \frac{1}{2}\int_S \mathcal{Q}_2(x)\Big(B - \frac{1}{2}(A^2)_{tan}\Big) + \frac{1}{24}\int_S \mathcal{Q}_2(x)\Big((\nabla(A\vec{n}) - AII)_{tan}\Big).$$

The two terms above correspond, in appearing order, to the stretching (2-nd order change in metric) and bending (1-st order change in the second fundamental form $II = \nabla\vec{n}$ on S) energies of a sequence of deformations $v^h = \mathrm{id} + hV + h^2 w^h$ of S, which is induced by a first order displacement $V \in \mathcal{V}_1$ and second order displacements w^h satisfying $\lim_{h\to 0} \mathrm{sym}\nabla w^h = B$.

We see that the out-of-plane displacements v (present in (8.3) or other plate theories) are replaced by the vector fields in $\mathcal{V}_1$; neither normal, nor tangential to S, but instead preserving the metric up to 1-st order.

13. The case of elliptic shells. It turns out that elliptic (i.e. strictly convex up to the boundary) surfaces enjoy a stronger than plates matching property of $1 \mapsto \infty$. Namely [20], for S elliptic, homeomorphic to a disk and $\mathcal{C}^{3,\alpha}$ ($\alpha > 0$) the following holds. Given $V \in \mathcal{V}_1 \cap \mathcal{C}^{2,\alpha}(\bar{S})$, there exists a sequence w_ϵ, equibounded in $\mathcal{C}^{2,\alpha}(\bar{S}, \mathbb{R}^3)$, and such that for all small $\epsilon > 0$ the field $u_\epsilon = \mathrm{id} + \epsilon V + \epsilon^2 w_\epsilon$ is an (exact) isometry.

Regarding the assumed above regularity of V (which is higher that the expected regularity $W^{2,2}$ of a limiting displacement) we note that the usual mollification techniques do not guarantee the density of smooth infinitesimal isometries in $\mathcal{V}_1 \cap W^{2,2}$, even for $S \in \mathcal{C}^\infty$. This is shown by an interesting example of Cohn-Vossen's surface [28], which is closed smooth and of non-negative curvature, and for which $\mathcal{C}^\infty \cap \mathcal{V}_1$ consists only of trivial fields with constant gradient, whereas $\mathcal{C}^2 \cap \mathcal{V}_1$ contains non-trivial elements.

However, we have the following result [20], valid for elliptic $S \in \mathcal{C}^{m+2,\alpha}$ ($\alpha \in (0,1)$ and integer $m > 0$). For every $V \in \mathcal{V}_1 \cap W^{2,2}$ there exists a sequence $V_n \in \mathcal{V}_1 \cap \mathcal{C}^{m,\alpha}(\bar{S}, \mathbb{R}^3)$ such that: $\lim_{n\to\infty} \|V_n - V\|_{W^{2,2}(S)} = 0$. In the proof of the quoted results we adapted techniques of Nirenberg used for Weyl's problem (immersability of all positive curvature metrics on S^2).

Ultimately, and as a consequence of the above statements, the main result of [20] asserts that for elliptic surfaces with sufficient regularity, the Γ-limit of the nonlinear elastic energies $h^{-\beta}E^h$ for any scaling regime $\beta > 2$ is given by the bending functional constrained to the 1-st order isometries:

$$\mathcal{I}_{lin}(V) = \frac{1}{24}\int_S \mathcal{Q}_2(x)\Big((\nabla(A\vec{n}) - AII)_{tan}\Big), \qquad A = \nabla V, \quad V \in \mathcal{V}_1.$$

14. Open problems. Here we list some open questions of diverse level of difficulty:

(a) identify scaling laws and Γ-limits of the 3d nonlinear energy functional (1.2) for other growth tensors than these considered in [21] and [17],

(b) derive the full range hierarchy of the limiting theories, whose order will depend on the scaling of curvature tensors of a general target metric,

(c) incorporate the feedback from the minimizing configuration (shape of the thin film) back to growth mechanism (growth tensor) and study the related time-dependent (quasistatic) problem,

(d) study existence, uniqueness and regularity of the critical points of the obtained models, including both minimizers and the non-minimizing equilibria (relating to the buckling phenomena),

(e) study convergence of non-minimizing equilibria to equilibria of the derived models (which does not follow directly from Γ-convergence), see [25, 16, 24] for some related preliminary results,

(f) discuss convergence of balance laws of the evolutionary non-Euclidean elasticity problem to the lower dimensional models, see [1] for the von Kármán plate case in the flat Euclidean geometry,

(g) justify the 3d continuous growth formalism rigorously from first principles, perhaps starting from the discrete model,

(h) study the implications of result in section 4 to existence of $W^{2,2}$ regular isometric immersions for a wide class of metrics, see [2, 10] for an interesting construction and a numerical study,

(i) use other variants of thin plate theory to highlight the self-similar structures that form near the edge of the domain [2, 5, 10].

(α) validate the conjecture by deriving limiting theories for $\beta \in (2,4)$ and other than convex types of surfaces; one difficulty here is to identify the elements $V \in \mathcal{V}_1$ which satisfy the 2nd order isometry condition $A^2 \in \mathcal{B}$,

(β) study matching, density and regularity of infinitesimal isometries (of different orders) on non-convex surfaces; e.g. hyperbolic or changing type; the analysis involves systems of PDEs of the same types,

(γ) give an example of S without flat regions and without the $1 \mapsto 2$ matching property; it will have the feature that $\mathcal{I}_{4,S}$ is again of the form as in section 12, rather than $\mathcal{I}_{lin}$; such surfaces are expected to be non-generic.

(δ) derive limiting theories for clamped shells; in the elliptic case this will imply that $\mathcal{V}_k = \{0\}$ for all k.

(ϵ) validate the obtained results for shells with varying thickness.

REFERENCES

[1] H. Abels, M.G. Mora, and S. Müller, *The time-dependent von Kármń plate equation as a limit of 3d nonlinear elasticity* , to appear in Calc. Var.

[2] P. Bella and R. Kohn, *Metric-induced wrinkling of an elastic thin film*, preprint (2010).

[3] P.G. Ciarlet, *Mathematical Elasticity*, North-Holland, Amsterdam (2000).

[4] S. Conti and F. Maggi, *Confining thin sheets and folding paper*, Arch. Ration. Mech. Anal. 187 (2008), no. 1, 1–48.

[5] J. Dervaux and M. Ben Amar, *Morphogenesis of growing soft tissues*, Phys. Rev. Lett. **101** (2008), no. 6, 068101.

[6] E. Efrati, E. Sharon, and R. Kupferman, *Elastic theory of unconstrained non-Euclidean plates*, J. Mechanics and Physics of Solids, **57** (2009), 762–775.

[7] G. Friesecke, R. James, M.G. Mora, and S. Müller, *Derivation of nonlinear bending theory for shells from three-dimensional nonlinear elasticity by Γ-convergence*, C.R. Math. Acad. Sci. Paris, **336** (2003), no. 8, 697–702.

[8] G. Friesecke, R. James, and S. Müller, *A theorem on geometric rigidity and the derivation of nonlinear plate theory from three dimensional elasticity*, Comm. Pure. Appl. Math., **55** (2002), 1461–1506.

[9] G. Friesecke, R. James, and S. Müller, *A hierarchy of plate models derived from nonlinear elasticity by gamma-convergence*, Arch. Ration. Mech. Anal., **180** (2006), no. 2, 183–236.

[10] J. Gemmer and S. Venkataramani, *Shape selection in non-Euclidean plates*, preprint (2010).

[11] M. Gromov, *Partial Differential Relations*, Springer-Verlag, Berlin-Heidelberg, (1986).

[12] Q. Han and J.-X. Hong, *Isometric embedding of Riemannian manifolds in Euclidean spaces*, Mathematical Surveys and Monographs, **130** AMS, Providence, RI (2006).

[13] T. von Kármán, *Festigkeitsprobleme im Maschinenbau*, in Encyclopädie der Mathematischen Wissenschaften. Vol. IV/4, pp. 311-385, Leipzig, 1910.

[14] Y. Klein, E. Efrati, and E. Sharon, *Shaping of elastic sheets by prescription of non-Euclidean metrics*, Science, **315** (2007), 1116–1120.

[15] H. LE DRET AND A. RAOULT, *The membrane shell model in nonlinear elasticity: a variational asymptotic derivation*, J. Nonlinear Sci., **6** (1996), 59–84.

[16] M. LEWICKA, *A note on the convergence of low energy critical points of nonlinear elasticity functionals, for thin shells of arbitrary geometry*, to appear in ESAIM: Control, Optimisation and Calculus of Variations (2009).

[17] M. LEWICKA, L. MAHADEVAN, AND M.R. PAKZAD, *Morphogenesis of growing elastic tissues: the rigorous derivation of von Kármán equations with residual stress*, to appear in Proceedings of the Royal Society A (2010).

[18] M. LEWICKA, M.G. MORA, AND M.R. PAKZAD, *Shell theories arising as low energy Γ-limit of 3d nonlinear elasticity*, Ann. Scuola Norm. Sup. Pisa Cl. Sci. (5) Vol. IX (2010), 1–43.

[19] M. LEWICKA, M.G. MORA, AND M.R. PAKZAD, *A nonlinear theory for shells with slowly varying thickness*, C.R. Acad. Sci. Paris, Ser I **347** (2009), 211–216

[20] M. LEWICKA, M.G. MORA, AND M.R. PAKZAD, *The matching property of infinitesimal isometries on elliptic surfaces and elasticity of thin shells*, submitted (2008).

[21] M. LEWICKA AND M.R. PAKZAD, *Scaling laws for non-Euclidean plates and the $W^{2,2}$ isometric immersions of Riemannian metrics*, submitted (2009).

[22] M. LEWICKA AND M.R. PAKZAD, *The infinite hierarchy of elastic shell models; some recent results and a conjecture*, to appear in Fields Institute Communications (2010).

[23] L. MAHADEVAN AND H. LIANG, *The shape of a long leaf*, Proc. Natl. Acad. Sci. USA, **106**, (2010) 22049–55.

[24] M.G. MORA AND L. SCARDIA, *Convergence of equilibria of thin elastic plates under physical growth conditions for the energy density*, preprint (2008).

[25] S. MÜLLER AND M.R. PAKZAD, *Convergence of equilibria of thin elastic plates – the von Kármán case*, Comm. Partial Differential Equations **33**, no. 4–6, 1018–1032, (2008).

[26] E.K. RODRIGUEZ, A. HOGER, AND A. MCCULLOCH, J. Biomechanics **27**, 455 (1994).

[27] B. SCHMIDT, *Plate theory for stressed heterogeneous multilayers of finite bending energy*, J. Math. Pures Appl. **88** (2007), 107–122.

[28] M. SPIVAK, *A Comprehensive Introduction to Differential Geometry, Vol V*, 2nd edition, Publish or Perish Inc. (1979).

[29] S. VENKATARAMANI, *Lower bounds for the energy in a crumpled elastic sheet – a minimal ridge*, Nonlinearity **17** (2004), no. 1, 301–312.

MATHEMATICAL, PHYSICAL AND NUMERICAL PRINCIPLES ESSENTIAL FOR MODELS OF TURBULENT MIXING

HYUNKYUNG LIM*, YAN YU†, JAMES GLIMM‡, AND DAVID H. SHARP§

Abstract. We propose mathematical, physical and numerical principles which are important for the modeling of turbulent mixing, especially the classical and well studied Rayleigh-Taylor and Richtmyer-Meshkov instabilities which involve acceleration driven mixing of a fluid discontinuity layer, by (respectively) a continuous acceleration or an impulsive delta function force.

The fundamental mathematical issue is the nonuniqueness and thus indeterminancy of solutions of the 3D compressible Euler equation. Verification (demonstration that the numerical solution of the equations is mathematically correct) is meaningless for such a model of turbulent mixing. Uniqueness requires physical fluid transport, i.e. the compressible (multifluid) Navier-Stokes equation.

The same fundamental issue, formulated in terms of physics, is that the properties of the mixing depend on dimensionless ratios of the transport coefficients, namely the Schmidt number (viscosity/mass diffusion) and Prandtl number (viscosity/thermal conductivity). Validation (meaning that the simulation equations correctly model the problem to be solved) is impossible without specification of fluid transport. It is in effect an effort to validate an answer for 0/0.

The fundamental issue, formulated in numerical terms is that the physical transport terms are so small that they cannot be resolved at feasible grid levels. Large eddy simulations (LES) are needed. For these, subgrid scale (SGS) terms must be added to the equations, to correctly reflect the influence of the unresolved transport on the grid scales that are resolved. In the absence of such an approach, numerical artifacts intrude, leading to apparently converged solutions, with answers that depend on the computer code.

Plainly, this issue, in its three guises, poses a challenge for verification and validation (V&V), and since V&V is a major scientific enterprise, it is of great importance.

Key words. Schmidt number, Prandtl number, mass diffusion, turbulence, multiphase flow.

Turbulent mixing is central to a number of problems, including turbulent combustion and inertial confinement fusion. Classic acceleration driven instabilities (Rayleigh-Taylor for continuous acceleration and Richtmyer-

*Department of Applied Mathematics and Statistics, Stony Brook University, Stony Brook, NY 11794-3600.

†Department of Applied Mathematics and Statistics, Stony Brook University, Stony Brook, NY 11794-3600.

‡Department of Applied Mathematics and Statistics, Stony Brook University, Stony Brook, NY 11794-3600. The work of the first three authors was supported in part by the U.S. Department of Energy, including grants DE-FC02-06ER25770, DE-FG07-07ID14889, DE-FC52-08NA28614, DE-AC07-05ID14517, DE-AC02-98CH1886 and DE-FG52-06NA26208, and the Army Research Office grant W911NF0910306.

§Los Alamos National Laboratory, Los Alamos, NM 87545. The simulations reported here were performed in part on the Galaxy linux cluster in the Department of Applied Mathematics and Statistics, Stony Brook University, and in part on New York Blue, the BG/L computer operated jointly by Stony Brook University and BNL. Los Alamos National Laboratory Preprint LA-UR 09-08281.

Meshkov for impulsive acceleration) have been a numerical challenge for 50 years. A satisfactory solution should

1. use a compressible code,
2. satisfy V&V standards,
3. achieve mesh converged joint probability distributions for temperature and mixture concentrations as is required for turbulent combustion,
4. not require full resolution (DNS) of all turbulent length scales,
5. apply over a range of physical parameters, including high Schmidt numbers,
6. provide sufficient intellectual understanding that others can follow and improve.

The progress summarized here is based on recent series of articles that achieve these six goals.

The compressible code is the LANL-Stony Brook code FronTier, whose main two innovative features are

- Front tracking to eliminate numerical mass diffusion across fluid interfaces and
- Dynamic subgrid scale (SGS) models to represent turbulent fluctuations at levels below the grid resolution.

The SGS models are free of adjustable parameters, and the entire simulation is thus free of adjustable parameters. Such SGS models are standard, but our use of them in flows with flow gradients that are strong at a mesh level appears to be original. Combustion, if based on these simulations, is also free of models and the adjustable parameters typically employed in model closure. In a sample simulation, we observe mesh convergence for the joint probability distributions and for a chemical reaction rate, which is parameter free, and (other than the SGS models) model free.

The main results are contained in a series of recent papers [8, 7, 9, 4, 5]. The summary here will emphasize point 6: the intellectual and conceptual innovations needed (including, but going beyond an advanced computer code) to obtain these results.

Commonly, Rayleigh-Taylor mixing simulations show a factor of two or more discrepancy between simulation and experiment. This discrepancy is conventionally attributed to long wave length noise in the experimental initial conditions, not measured; hence the simulations and (most) experiments, if this view is accepted, should not be compared. The Mueschke-Andrews experiments [13] (hot-cold water flowing over a splitter plate) are an exception: measured initial data shows very substantial long wave length initial perturbations. We duplicated the Mueschke-Schilling [14] fully resolved direct numerical simulation (DNS) results, with agreement between experiment, prior incompressible DNS simulation and our compressible LES simulation. We extended this agreement to the high Schmidt number case (salt-fresh water over a splitter plate), using a fully compressible code with SGS models, and a large eddy simulation (LES), i.e. not

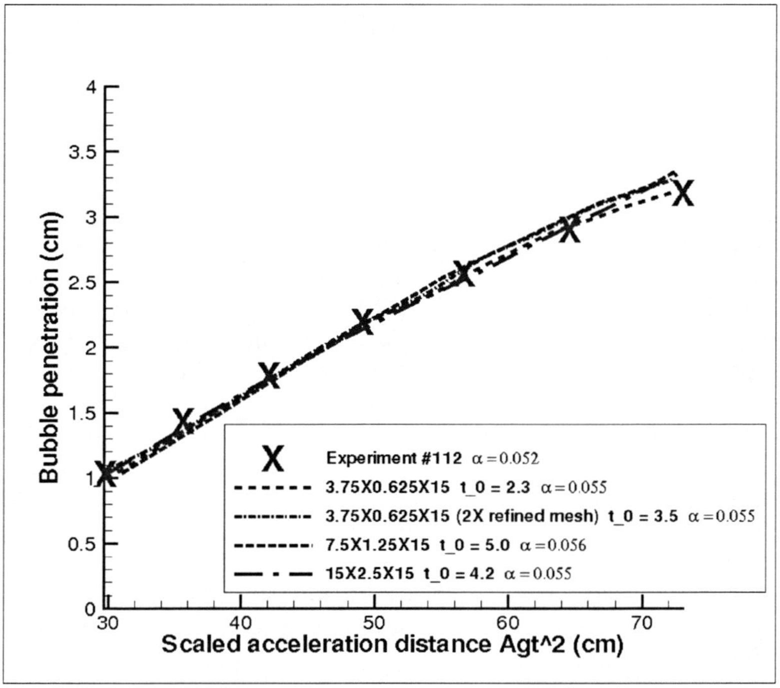

FIG. 1. *Plot of the light fluid (bubble) penetration distance vs. an acceleration length scale, Agt^2, where the Atwood number $A = (\rho_2 - \rho_1)/(\rho_2 + \rho_1)$ is a dimensionless buoyancy correction to the acceleration force g. The crosses are the experimental data points (experiment # 112 [18]), and the other curves show several simulations at different levels of grid and statistical resolution. Here statistical resolution refers to the number of initial unstable modes in the simulation. The entry t_0 in the figure legend is a time offset, as the initial times of the experiment and the simulation were not designed to coincide.*

requiring full DNS resolution. The high Schmidt number case is inaccessible to the DNS methods of [14]. See [4].

However, not all experiments have such a high level of noise. Our analysis [4] of the Smeeton-Youngs [18] rocket rig (fresh-salt water) experiments suggests that they were relatively free of long wave length noise. We simulated these, in agreement with experiment, and for the first time, with a simulation going through all experimental data points. See Fig.1.

Our results show a clear dependence of the mixing rates on the Schmidt number. A common simulation strategy is known as Implicit Large Eddy Simulation (ILES), whereby the numerical algorithm supplies what it will for the subgrid effects, while the physical transport parameters are either 0 or so small as to leave the physical transport under resolved for any feasible level of grid resolution, and negligible relative to numerical transport (numerical mass diffusion, heat conduction and viscosity). The Schmidt number is the ratio of viscosity to mass diffusivity. If both are set to zero, it is the indeterminant expression 0/0. If both are under resolved and thus set numerically by numerical algorithms and code artifacts, then it is a code dependent numerical Schmidt number which is determining the answer. The result (and this was observed in practice) is two well documented codes converging numerically to two distinct answers. Such non-uniqueness of solutions and the related phenomena of the dependence of mixing on the values of transport parameters in the absence of subgrid models is presented in [12, 6, 10]. Clearly this is a major challenge for V&V. The SGS models and a valid LES simulation cure this ambiguity, and allow mesh convergence, even for such sensitive and important observables as the joint probability distributions for temperature and species concentration, or for a chemical reaction rate.

The mathematical literature contains non-uniqueness theorems for the compressible Euler equations, i.e. fluid equations with fluid transport set to zero. See the 1993 paper of Scheffer [16], and refinements [17, 3]. Related numerical evidence for nonuniqueness is given in [11, 1]. Thus V&V for ILES simulations has hit a mathematical no-go theorem. Obviously, a change in strategy is needed, and is what we have provided.

In comparing the Smeeton-Youngs to the Mueschke-Andrews fresh-salt water Rayleigh-Taylor mixing experiments, we found six significant differences in the experimental conditions, and all of them had a significant effect on the mixing rate (α) [4]. Beyond the fluid transport parameters and long wave noise (in the Mueschke-Andrews experiment only), we found significant influence from: the change in Grashov number, initial mass diffusion layer width, and experimentally induced vs. ideal short wave length perturbations. The Grashov number is a dimensionless number characterizing the ratio of accelerating to viscous forces. For an ILES simulation, with viscosity zero or undefined, the Grashov number is meaningless and cannot be set to agree with specific experiments, nor can it be modified to reflect differences between experiments.

Even the short wave length perturbations were important. A small amplitude linearized theory leads to a set of equations called dispersion relations. Solutoin of the dispersion relation equations gives the instability growth rate as a function of the wave number or wave length. The dispersion relation growth rates depend on the range of physics included in the equations being linearized, and commonly incomplete physics is used (to give equations easier to solve), but unfortunately, this approximation gives an incorrect answer. The wave length which is most rapidly growing (the maximally instable wave length) can be used to set a range of length scales for the initial perturbations. When this is done for the Smeeton-Youngs experiments, and the solution evolves numerically to the second or later observation plates, where the wave lengths can be measured directly, the result is good agreement. Thus we assume that the Smeeton-Youngs experiment is initialized in accordance with results of dispersion theory.

However, the Mueschke-Andrews experiment, where the initial condition amplitudes (as a function of wave length) were measured, appears not to be initialized in accordance with dispersion theory. In other words, this experiment appears to be initialized at least in part by boundary layer effects related to the flow over the splitter plate. In this regard, the long wave length "noise" is large, equal to about 75% in amplitude of the dominant short wave length perturbation. But as observed above, even the short wave length signal is not driven purely by dispersion relations based on growth rates for Rayleigh-Taylor instabilities, and is evidently also "forced", i.e. also driven by or partially influenced by boundary layer effects.

Thus a central conclusion of our study is that there is nothing universal about Rayleigh-Taylor mixing. This view is a break with prevailing ideology and should force a re-examination of the basic mathematical, physical and numerical principles used for the solution of this problem.

Many details of the experiment are important. When modeled correctly, nearly perfect agreement between simulation and experiment can be obtained. Combined with mesh convergence, this is V&V.

We propose a scientific program for recovery in an ILES world. If front tracking to control or limit numerical mass diffusion is not part of the game plan, then it is necessary to increase the numerical viscosity so that the ratio of numerical viscosity to numerical mass diffusion agrees with the physical ratio, i.e. the numerical Schmidt number must equal the physical Schmidt number for a successful ILES simulation.

This program, while no doubt effective, will be painful to follow in practice. For high Schmidt number simulations, the poor control over numerical mass diffusion will lead to a need for substantial mesh refinement. For example, the fresh-salt water Schmidt number (about 700), will require additional mesh refinement by a factor of about $\sqrt{700} \approx 26$ for an increase of computational cost of $(\sqrt{700}^4 = 700^2 = 490,000$. Even for $Sc = 1$, common numerical methods have an excess of numerical mass diffusion over numerical viscosity, so that excess numerical viscosity (and additional mesh resolution) will be needed.

Even if the computational expense of this additional mesh refinement can be absorbed in the coming world of petaflop and beyond computing, there is a further practical problem with this advice. Simple scanning of the Handbook of Physics and Chemistry reveals many different values for Schmidt numbers. If these are used within a simulation, accurate calibration of the numerical transport (numerical viscosity and numerical mass diffusion) will be required. Very little has been reported on the effective Schmidt number of actual codes in use today, or for the algorithms on which they are based. It would seem that this will be a substantial enterprise. Even the calibration of an ILES code numerical viscosity to yield a quantified numerical Reynolds number (beyond asserting that it is "large") appears not to have been achieved.

Assuming that budgets will not allow the exercise of calibration of numerical algorithms to determine numerical Schmidt numbers from a calibration of numerical viscosity and mass species diffusion, the next best step is to allow the numerical Schmidt number to become an adjustable parameter, to be calibrated directly. Probably this will succeed. Experimentng with the required level of numerical viscosity will no doubt allow a simulation code to correctly predict Rayleigh-Taylor growth rates, especially if the other physics issues mentioned above are addressed, and if the mesh is refined sufficiently to compensate for added numerical viscosity. If the calibration of the numerical Schmidt number is problem independent, the calibration has to be done once only, and after that, the simulations remain predictive and free of adjusted (if not adjustable) parameters.

It was to avoid calibration complications and the necessary loss of resolution implied that we introduced our new algorithm. The proper physical and mathematical basis for its successful use, which apply in any case, has an independent value. Our recommended solution to the ILES indeterminancy problem is to merge such a code with FronTier. The full physics capabilities of major ILES codes and the sophisticated front and steep gradient resolving capabilities of FronTier are each extensive. It would be challenging to replicate either in the other, and more cost effective and efficient to combine the separate capabilities into a single code. To this end, an API to support merger of any external code with FronTier is under development.

It remains to discuss the mesh convergence of the joint probability density functions (PDFs) for temperature and species concentration [8]. Because this study called for extensive mess refinement and because we also wished to explore a range of physical parameters (high and low Schmidt and Prandtl) numbers, we conducted this study in 2D. The problem considered was a circular Richmyer-Meshkov problem, with reshock. The ingoing circular shock wave passes through a perturbed circular interface, reflects at the origin, and recrosses the interface. Hence the term reshock. After reshock, the flow becomes extremely chaotic, and in the absence of regularization, the interface has a “density” (within the mixing zone as defined by the inner and outer envelope of the mixing region) that is proportional to the mesh. In other words there is a constant density of mesh length per grid cell area. In the absence of regularization, there is no sign of convergence, except for statistical quantities.

This raises the question of what convergence should mean for a chaotic simulation. Plainly, some length scale must be introduced, and it should be based on physics, not numerics. The viscous length scale is too far from the grids used in practice to be useful. So we think of a measurement length scale and expect convergence only after averaging any observable over that length. For the present problem with its statistical circular symmetry, we average over angles. Additionally we average over the radius, within the mixing zone. There is some r dependence to the statistics, so it would be desirable to restrict the range of r averaging, but this was not done. Time averaging is possible but was not explored. Ensemble averaging can also be employed, and was investigated.

Once an averaging process is defined, the simulation generates a full PDF, as well as means, variances, and higher moments that describe approximately the PDF. We observe mesh convergence of the full PDF in the above sense. The PDF is defined by a spatial region over which data is collected. The choice of the spatial or space-time region is problem dependent. Mesh convergence of the PDF thus defined also depends on a norm for differences of the PDFs. We consider the difference of the probability distribution functions (the indefinite integrals of the PDFs), and for these we consider L_1 and L_∞ norms.

The emphasis on the joint PDF of concentration and temperature as micro scale observables is motivated by problems in turbulent combustion [15], where these variables affect the local flame speed and the overall flow. We have three main results concerning the joint temperature and species PDFs, and derived from these, the PDF for a chemical reaction rate. First, the PDFs are strongly bimodal, indicating that the mixing process is stirring, not diffusion dominated. Second, the PDFs depend strongly on the Schmidt and Prandtl numbers, and for LES, they depend on the SGS models to define turbulent Schmidt and Prandtl numbers. In the absence of SGS models (as with ILES), the PDFs depend on numerical aspects of the code design. Thirdly, the joint PDFs are convergent under mesh refinement.

We also develop a theoretical model for the concentration and temperature PDFs. The model derives the PDFs based on the 1D diffusion equation and a knowledge of the mixing geometry, namely the geometry of the 50% concentration isosurface, described statistically. The geometry is characterized in terms of the minimum exit distance from some arbitrary point to the interface location nearest to it. The statistics of these exit distances was studied previously [2], and satisfied an exponential distribution, as is also the case for the present simulations. Thus a single number, the length scale λ_C for decay of correlation in the geometry-phase information, characterizes the geometry statistically. The laminar and turbulent diffusion coefficients coming from the physics of the problem and from the SGS model coefficients (determined dynamically from the simulation) complete the parameterization of the model. With this model, we predict the PDFs with agreement about as good as the observed mesh errors.

The model has as its only input the location of the interface (50% concentration isosurface) at time t, the extremes of temperature and concentration within the mixing zone immediately after reshock and the time elapsed $t - t_0$ since reshock. It assumes no diffusion before reshock (and so it underpredicts diffusion). For each mesh block, we determine the minimum exit distance for the phase or component value at the mesh block center, as in the definition of λ_C. This distance and the time $t - t_0$ is inserted

TABLE 1

Mesh errors for the joint temperature-concentration PDFs to illustrate possibilities of mesh convergence. Mesh comparison is coarse (c) to fine (f) and medium (m) to fine with PDF's compared using the Kolmogorov-Smirnov metric. Model comparison is for the fime grid and uses an L_1 norm for the difference of distribution functions. Transport parameters typical of a liquid, gas and plasma are reported.

	liquid			**gas**			**plasma**		
Re	c to f	m to f error	model	c to f	m to f error	model	c to f	m to f error	model
≈ 300	0.24	0.13	0.06	0.08	0.06	0.06	0.57	0.31	0.05
≈ 6000	0.07	0.06	0.07	0.09	0.05	0.05	0.68	0.41	0.05
$\approx 600K$	0.26	0.06	0.07	0.22	0.09	0.08	0.14	0.10	0.06

into the solution of the one dimensional diffusion equation using combined laminar and turbulent transport coefficients. The result is a model for the joint temperature concentration PDF. We compare this model PDF to the actual computed one. We use the L_1 norm of the difference of their distribution functions. Since the model errors, such as the model assumption of no mixing prior to time t_0 are not mesh related, we do not study mesh dependence of the model errors, and we present the table of model errors for the finest grid only, see Table 1. Since the model depends only on the geometry of the time t interface and the values of the combined laminar and turbulent transport coefficients, we conclude that these two factors are the main determinants of the joint PDF. The model also serves as a kind of validation of the simulation, in that it uses a very reduced output from the simulation, namely the interface geometry and the turbulent transport coefficients, and it reproduces major conclusions of the simulation, namely the bimodal character of the pdf, especially for high Sc.

The mesh errors quoted are measured in the Kolmogorov-Smirnov metric, i.e., the L_∞ norm of the difference of the associated probability distribution functions. The model errors (determined for the fine grid only) are presented using an L_1 norm for the difference of the probability distribution functions. We compute with three distinct meshes (400×800, 800×1600 and 1600×3200) for the cases l, g, and p. Thus there are two levels of error for each case and one model to simulation comparison. The three cases refer to fluid parameters typical of a liquid (l), gas (g), and plasma (p).

From Table 1, we see that the mesh convergence error is about equal to the model error. Not shown is the ensemble fluctuation from randomness in the input perturbation. This is actually about an order of magnitude larger than either the mesh error or the model error for the PDFs. We conclude that the mesh convergence of the PDFs is probably satisfactory, as is the level of agreement between the simulation and the model.

Our success with points 1-6 introduced above open new directions for turbulent mixing. The progress has depended as much on a reconceptualization of the problem as on improvements in the numerical methods.

REFERENCES

[1] VOLKER ELLING. A possible counterexample to wellposedness of entopy solutions and to godunov convergence. *Math. Comp.*, **76**:1721–1733, 2006.

[2] J. GLIMM, J.W. GROVE, X.-L. LI, W. OH, AND D.H. SHARP. A critical analysis of Rayleigh-Taylor growth rates. *J. Comput. Phys.*, **169**:652–677, 2001.

[3] CAMILLO DE LELLIS AND LASZELO SZEKELYHIDI. On admissibility criteria for weak solutions of the Euler equations. *Archive for Rational Mechanics and Applications*, **195**:225–260, 2010.

[4] H. LIM, J. IWERKS, J. GLIMM, AND D.H. SHARP. Nonideal Rayleigh-Taylor mixing. *PNAS*, 2010. In press. Stony Brook Preprint SUNYSB-AMS-09-05 and Los Alamos National Laboratory preprint number LA-UR 09-06333.

[5] H. LIM, J. IWERKS, Y. YU, J. GLIMM, AND D.H. SHARP. Verification and validation for turbulent mixing. *Physica Scripta*, 2009. In press. Stony Brook Preprint SUNYSB-AMS-09-07 and Los Alamos National Laboratory preprint number LA-UR 09-07240.

[6] H. LIM, Y. YU, J. GLIMM, X.-L. LI, AND D.H. SHARP. Chaos, transport, and mesh convergence for fluid mixing. *Acta Mathematicae Applicatae Sinica*, **24**:355–368, 2008. Stony Brook University Preprint SUNYSB-AMS-07-09 Los Alamos National Laboratory preprint number LA-UR-08-0068.

[7] H. LIM, Y. YU, J. GLIMM, X.-L. LI, AND D.H. SHARP. Subgrid models in turbulent mixing. *Astronomical Society of the Pacific Conference Series*, **406**:42, 2008. Stony Brook Preprint SUNYSB-AMS-09-01 and Los Alamos National Laboratory Preprint LA-UR 08-05999.

[8] H. LIM, Y. YU, J. GLIMM, X.-L. LI, AND D.H. SHARP. Subgrid models for mass and thermal diffusion in turbulent mixing. *Physica Scripta*, 2009. Stony Brook Preprint SUNYSB-AMS-08-07 and Los Alamos National Laboratory Preprint LA-UR 08-07725; Submitted for Publication.

[9] H. LIM, Y. YU, J. GLIMM, AND D.H. SHARP. Nearly discontinuous chaotic mixing. *High Energy Density Physics*, **6**:223–226, 2010. Stony Brook University Preprint SUNYSB-AMS-09-02 and Los Alamos National Laboratory preprint number LA-UR-09-01364.

[10] H. LIM, Y. YU, H. JIN, D. KIM, H. LEE, J. GLIMM, X.-L. LI, AND D.H. SHARP. Multi scale models for fluid mixing. *Compu. Methods Appl. Mech. Engrg.*, **197**:3435–3444, 2008. Stony Brook University Preprint SUNYSB-AMS-07-05.

[11] M. LOPEZ, H. LOPEZ, J. LOWENGRUB, AND Y. ZHENG. Numerical evidence of nonuniqueness in the evolution of vortex sheets. *ESAIM:Mathematical Modeling and Numerical Analysis*, **40**:225–237, 2006.

[12] T.O. MASSER. *Breaking Temperature Equilibrium in Mixed Cell Hydrodynamics*. Ph.d. thesis, State University of New York at Stony Brook, 2007.

[13] NICHOLAS MUESCHKE. *Experimental and numerical study of molecular mixing dynamics in Rayleigh-Taylor unstable flows*. Ph.d. thesis, Texas A&M University, 2008.

[14] NICHOLAS MUESCHKE AND OLEG SCHILLING. Investigation of Rayleigh-Taylor turbulence and mixing using direct numerical simulation with experimentally measured initial conditions. i. Comparison to experimental data. *Physics of Fluids*, **21**:014106 1–19, 2009.

[15] HEINTZ PITSCH. Large-eddy simulation of turbulent combustion. *Annual Rev. Fluid Mech.*, **38**:453–482, 2006.

[16] V. SCHEFFER. An inviscid flow with compact support in space-time. *J. Geom. Anal.*, **3**:343–401, 1993.

[17] A. SHNIRELMAN. On the nonuniqueness of weak solutions of the Euler equations. *Comm. Pure Appl. Math.*, **50**:1261–1286, 1997.

[18] V.S. SMEETON AND D.L. YOUNGS. Experimental investigation of turbulent mixing by Rayleigh-Taylor instability (Part III). AWE Report Number 0 35/87, 1987.

ON THE EULER-POISSON EQUATIONS OF SELF-GRAVITATING COMPRESSIBLE FLUIDS

TAO LUO* AND JOEL SMOLLER†

Abstract. Compressible Euler-Poisson equations are used to model the evolution of stars of compressible fluids. In this paper, we survey some recent results on the stability of non-rotating and rotating star solutions and discuss some related open problems.

1. Introduction. In the framework of Newtonian Mechanics, the dynamics of stars which are made of compressible fluids is modeled by the compressible Euler-Poisson equations in three spatial dimensions :

$$\begin{cases} \rho_t + \nabla \cdot (\rho \mathbf{v}) = 0, \\ (\rho \mathbf{v})_t + \nabla \cdot (\rho \mathbf{v} \otimes \mathbf{v}) + \nabla p(\rho) = -\rho \nabla \Phi, \\ \Delta \Phi = 4\pi G \rho. \end{cases} \tag{1.1}$$

Here ρ is the density , $\mathbf{v} = (v_1, v_2, v_3)$ is the velocity p is the pressure and Φ is the gravitational potential given by

$$\Phi(x) = -G \int_{\mathrm{R}^3} \frac{\rho(y)}{|x-y|} dy, \tag{1.2}$$

and G is the Newtonian gravitational constant.

For the isentropic fluids the pressure function is given by $p = p(\rho)$ satisfying $p'(\rho) > 0$ for $\rho > 0$.

The pressure function for the polytropic stars are given by $p(\rho) = K\rho^\gamma$, $K > 0$ and $\gamma > 1$ are constants. When $\gamma = 4/3$, a star is called supermassive ([37]).

For a white dwarf star (a star in which gravity is balanced by electron degeneracy pressure), the pressure function $p(\rho)$ obeys the following asymptotics ([6], Chapter 10):

$$\begin{cases} p(\rho) = c_1 \rho^{4/3} - c_2 \rho^{2/3} + \cdots, & \rho \to \infty, \\ p(\rho) = d_1 \rho^{5/3} - d_2 \rho^{7/3} + O(\rho^3), & \rho \to 0, \end{cases} \tag{1.3}$$

where c_1, c_2, d_1 and d_2 are positive constants.

In the study of time-independent solutions of system (1.1), there are two cases, non-rotating stars and rotating stars. An important question concerns the stability of such solutions. Physicists call such star solutions stable provided that they are minima of an associated energy functional

*Department of Mathematics and Statistics, Georgetown University, Washington, DC 20057-1233.

†Department of Mathematics, University of Michigan, 525 East University Ave., Ann Arbor, MI 48109-1109 (smoller@umich.edu).

([37], p.305 & [33]). Mathematicians, on the other hand, consider dynamical nonlinear stability via solutions of the Cauchy problem. The existence theory for non-rotating white dwarf stars is classical provided the total mass M of the star is not greater than a critical mass M_c ($M \leq M_c$) ([6]). For rotating white dwarf stars with prescribed total mass and angular momentum distribution, Auchumuty and Beals ([1]) proved that if the angular momentum distribution is nonnegative, then existence holds if $M \leq M_c$. Friedman and Turkington ([11]) proved existence for any mass provided that the angular momentum distribution is everywhere positive; see Li ([21]), Chanillo & Li ([7]) and Luo & Smoller ([25]) for related results for rotating star solutions with prescribed constant angular velocity. For non-rotating stars, G. Rein ([30]) proved nonlinear dynamical stability for general perturbations. However, his result does not apply to white dwarf stars. For non-rotating white dwarf stars, the problem was formulated by Chandrasekhar [5] in 1931 (and also in [9] and [17]) and leads to an equation for the density which was called the " Chandrasekhar equation " by Lieb and Yau in [22]. This equation predicts the gravitational collapse at some critical mass ([5] and [6]). This gravitational collapse was also verified by Lieb and Yau ([22]) as the limit of Quantum Mechanics.

Other related results besides those mentioned above for compressible fluid rotating stars can be found in [2], [3], [10], and [25].

In the following, some recent results on the dynamical stability and (or) instability will be surveyed and some related open problems will be discussed.

2. Rotating star solutions. We now introduce some notation which will be used throughout this paper. We use $\int$ to denote $\int_{\mathbb{R}^3}$, and use $||\cdot||_q$ to denote $||\cdot||_{L^q(\mathbb{R}^3)}$. For any point $x=(x_1,x_2,x_3)\in\mathbb{R}^3$, let

$$r(x)=\sqrt{x_1^2+x_2^2},\ z(x)=x_3,\ B_R(x)=\{y\in\mathbb{R}^3,\ |y-x|<R\}. \tag{2.1}$$

For any function $f\in L^1(\mathbb{R}^3)$, we define the operator B by

$$Bf(x)=\int\frac{f(y)}{|x-y|}dy=f*\frac{1}{|x|}. \tag{2.2}$$

Also, we use ∇ to denote the spatial gradient, i.e., $\nabla=\nabla_x=(\partial_{x_1},\ \partial_{x_2},\ \partial_{x_3})$. C will denote a generic positive constant.

A rotating star solution $(\tilde\rho,\tilde{\mathbf{v}},\tilde\Phi)(r,z)$, where $r=\sqrt{x_1^2+x_2^2}$ and $z=x_3$, $x=(x_1,x_2,x_3)\in\mathbb{R}^3$, is an *axi-symmetric* time-independent solution of system (1.1), which models a star rotating about the x_3-axis. Suppose the angular momentum (per unit mass), $J(m_{\tilde\rho}(r))$ is prescribed, where

$$m_{\tilde\rho}(r)=\int_{\sqrt{x_1^2+x_2^2}<r}\tilde\rho(x)dx=\int_0^r 2\pi s\int_{-\infty}^{+\infty}\tilde\rho(s,z)dsdz, \tag{2.3}$$

is the mass in the cylinder $\{x = (x_1, x_2, x_3) : \sqrt{x_1^2 + x_2^2} < r\}$, and J is a given function. In this case, the velocity field $\tilde{\mathbf{v}}(x) = (v_1, v_2, v_3)$ takes the form

$$\tilde{\mathbf{v}}(x) = (-\frac{x_2 J(m_{\tilde{\rho}}(r))}{r^2}, \frac{x_1 J(m_{\tilde{\rho}}(r))}{r^2}, 0).$$

Substituting this in (1.1), we find that $\tilde{\rho}(r, z)$ satisfies the following two equations:

$$\begin{cases} \partial_r p(\tilde{\rho}) = \tilde{\rho}\partial_r(B\tilde{\rho}) + \tilde{\rho}L(m_{\tilde{\rho}}(r))r^{-3}, \\ \partial_z p(\tilde{\rho}) = \tilde{\rho}\partial_z(B\tilde{\rho}), \end{cases} \tag{2.4}$$

where the operator B is defined in (2.2), and

$$L(m_{\tilde{\rho}}) = J^2(m_{\tilde{\rho}})$$

is the square of the angular momentum. We define

$$A(\rho) = \rho \int_0^{\rho} \frac{p(s)}{s^2} ds. \tag{2.5}$$

It is easy to verify that (cf. [1]) (2.4) is equivalent to

$$A'(\tilde{\rho}(x)) + \int_{r(x)}^{\infty} L(m_{\tilde{\rho}}(s)s^{-3}ds - B\tilde{\rho}(x) = \lambda, \qquad \text{where } \tilde{\rho}(x) > 0, \tag{2.6}$$

for some constant λ. Here $r(x)$ and $z(x)$ are as in (2.1). Let M be a positive constant and let W_M be the set of functions ρ defined by,

$$\begin{aligned} W_M =& \{\rho : \mathbb{R}^3 \to \mathbb{R},\ \rho \text{ is axisymmetric},\ \rho \geq 0, a.e., \\ & \int \rho(x)dx = M, \\ & \int \left(A(\rho(x)) + \frac{\rho(x)L(m_\rho(r(x)))}{r(x)^2} + \rho(x)B\rho(x)\right) dx < +\infty\}. \end{aligned}$$

For $\rho \in W_M$, we define the **energy functional** F by

$$F(\rho) = \int [A(\rho(x)) + \frac{1}{2}\frac{\rho(x)L(m_\rho(r(x)))}{r(x)^2} - \frac{1}{2}\rho(x)B\rho(x)]dx. \tag{2.7}$$

In (2.7), the first term denotes the potential energy, the middle term denotes the rotational kinetic energy and the third term is the gravitational energy.

For a white dwarf star, the pressure function $p(\rho)$ satisfies the following conditions:

$$\lim_{\rho \to 0+} \frac{p(\rho)}{\rho^{4/3}} = 0, \ \lim_{\rho \to \infty} \frac{p(\rho)}{\rho^{4/3}} = \mathfrak{K}, \ p'(\rho) > 0 \text{ as } \rho > 0, \tag{2.8}$$

where $\mathfrak{K}$ is a finite positive constant. Assuming that the function $L \in C^1[0, M]$ and satisfies

$$L(0) = 0, \ L(m) \geq 0, \ for \ 0 \leq m \leq M, \tag{2.9}$$

Auchmuty and Beals (cf. [1]) proved the existence of a minimizer of the functional $F(\rho)$ in the class of functions $W_{M,S} = W_M \cap W_{sym}$, where

$$\begin{aligned} W_{sym} = \{\rho : \mathbb{R}^3 \to \mathbb{R}, \ \rho(x_1, x_2, -x_3) = \rho(x_1, x_2, x_3), \\ x_i \in \mathbb{R}, i = 1, \ 2, \ 3\}, \end{aligned} \tag{2.10}$$

for white dwarfs when the total mass is below a critical mass. The critical mass was first found by Chandrasekhar (cf.[6]) in the study of non-rotating white dwarf stars. Later on, it was proved by Friedman and Turkington ([11]) that, if the angular momentum satisfies the following condition

$$\begin{aligned} &J \in C^1([0, M]), \quad J'(m) \geq 0, \quad \text{for } 0 \leq m \leq M, \\ &J(0) = 0, \qquad\qquad J(m) > 0 \quad \text{for } 0 < m \leq M, \end{aligned} \tag{2.11}$$

where J is the angular momentum, then the restriction on the total mass can be removed.

We are interested in minimizers of functional F in the *larger* class W_M. By the same argument as in [1], it is easy to prove the following theorem on the regularity of a minimizer.

THEOREM 2.1. *Suppose that the pressure function p satisfies:*

$$\lim_{\rho \to 0+} \frac{p(\rho)}{\rho^{6/5}} = 0, \ \lim_{\rho \to \infty} \frac{p(\rho)}{\rho^{6/5}} = \infty, \ p'(\rho) > 0 \text{ as } \rho > 0, \tag{2.12}$$

and the angular momentum satisfies (2.9). Let $\tilde{\rho}$ be a minimizer of the energy functional F in W_M and let

$$\Gamma = \{x \in \mathbb{R}^3 : \ \tilde{\rho}(x) > 0\}, \tag{2.13}$$

then $\tilde{\rho} \in C(\mathbb{R}^3) \cap C^1(\Gamma)$. Moreover, there exists a constant λ such that

$$\begin{cases} A'(\tilde{\rho}(x)) + \int_{r(x)}^{\infty} L(m_{\tilde{\rho}}(s) s^{-3} ds - B\tilde{\rho}(x) = \lambda, & x \in \Gamma, \\ \int_{r(x)}^{\infty} L(m_{\tilde{\rho}}(s) s^{-3} ds - B\tilde{\rho}(x) \geq \lambda, & x \in \mathbb{R}^3 - \Gamma. \end{cases} \tag{2.14}$$

We call such a minimizer $\tilde{\rho}$ a *rotating star* solution with total mass M and angular momentum $\sqrt{L(m)}$.

3. General existence and stability theorems. For the angular momentum, besides the condition (2.9), we also assume that it satisfies the following conditions:

$$L(am) \geq a^{4/3} L(m), \ 0 < a \leq 1, \ 0 \leq m \leq M, \tag{3.1}$$

$$L'(m) \geq 0, \qquad 0 \leq m \leq M. \tag{3.2}$$

Condition (3.2) is called the Sölberg stability criterion ([35]).

3.1. Compactness of minimizing sequence. The following compactness result for the minimizing sequences of the functional F is crucial for the existence and stability analyses. The proof can be found in ([27]).

THEOREM 3.1. *Suppose that the square of the angular momentum L satisfies (2.9), (3.1) and (3.2), and the pressure function p satisfies the following conditions*

$$
\begin{aligned}
&p \in C^1[0, +\infty), \quad \int_0^1 \frac{p(\rho)}{\rho^2} d\rho < +\infty, \quad \lim_{\rho \to \infty} \frac{p(\rho)}{\rho^\gamma} = K, \\
&p(\rho) \geq 0, \quad p'(\rho) > 0 \quad \text{for } \rho > 0,
\end{aligned}
\tag{3.3}
$$

where $0 < K < +\infty$ and $\gamma \geq 4/3$. If
(1)

$$\inf_{\rho \in W_M} F(\rho) < 0, \tag{3.4}$$

and
(2) for $\rho \in W_M$,

$$\int [A(\rho)(x) + \frac{1}{2} \frac{\rho(x) L(m_\rho(r(x)))}{r(x)^2}] dx \leq C_1 F(\rho) + C_2, \tag{3.5}$$

for some positive constants C_1 and C_2, then the following hold:
(a) If $\{\rho^i\} \subset W_M$ is a minimizing sequence for the functional F, then there exist a sequence of vertical shifts $a_i \mathbf{e_3}$ ($a_i \in \mathbb{R}$, $\mathbf{e_3} = (0,0,1)$), a subsequence of $\{\rho^i\}$, (still labeled $\{\rho^i\}$), and a function $\tilde{\rho} \in W_M$, such that for any $\epsilon > 0$ there exists $R > 0$ with

$$\int_{|x| \geq R} T\rho^i(x) dx \leq \epsilon, \quad i \in \mathbb{N}, \tag{3.6}$$

and

$$T\rho^i(x) \rightharpoonup \tilde{\rho}, \text{ weakly in } L^\gamma(\mathbb{R}^3), \text{ as } i \to \infty, \tag{3.7}$$

where $T\rho^i(x) := \rho^i(x + a_i \mathbf{e_3})$.
Moreover
(b)

$$\nabla B(T\rho^i) \to \nabla B(\tilde{\rho}) \text{ strongly in } L^2(\mathbb{R}^3), \text{ as } i \to \infty. \tag{3.8}$$

(c) $\tilde{\rho}$ is a minimizer of F in W_M.

Thus $\tilde{\rho}$ is a rotating star solution with total mass M and angular momentum $\sqrt{L}$.

REMARK 1. i) The assumption (3.4) is crucial for our compactness and stability analysis. The physical meaning of this is that the gravitational energy, the negative part of the energy F, should be greater than the positive part, which means the gravitation should be strong enough to hold the star together. In section 4, we will verify this assumption. Roughly speaking, in addition to (3.3), if we require

$$\lim_{\rho\to 0+}\frac{p(\rho)}{\rho^{\gamma_1}}=\alpha, \tag{3.9}$$

for some constants $\gamma_1>4/3$ and $0<\alpha<+\infty$, then (3.4) holds for the following cases:
(a) when $\gamma=4/3$ (where γ is the constant in (3.3)), if the total mass M is less than a "critical mass" M_c, then (3.4) holds. This case includes white dwarf stars. For a white dwarf star, $\gamma_1=5/3$.
(b) When $\gamma>4/3$, (3.4) holds for arbitrary positive total mass M. This generalizes our previous result in [26] for the polytropic stars with $p(\rho)=\rho^{\beta}$, $\beta>4/3$.

3.2. Stability of rotating star solutions. In this section, we assume that the pressure function p satisfies

$$\begin{aligned} &p\in C^1[0,+\infty),\quad \lim_{\rho\to 0+}\tfrac{p(\rho)}{\rho^{6/5}}=0,\\ &\lim_{\rho\to\infty}\frac{p(\rho)}{\rho^{\gamma}}=K,\quad p'(\rho)>0 \text{ for } \rho>0, \end{aligned}\tag{3.10}$$

where $0<K<+\infty$ and $\gamma\ge 4/3$ are constants. It should be noticed that (3.10) implies both (2.15) and (3.3). We consider the Cauchy problem for (1.1) with the initial data

$$\rho(x,0)=\rho_0(x),\ \mathbf{v}(x,0)=\mathbf{v}_0(x). \tag{3.11}$$

We begin by giving the definition of a weak solution.

Definition 3.1. *Let $\rho\mathbf{v}=\mathbf{m}$. The triple $(\rho,\mathbf{m},\Phi)(x,t)$ $(x\in \mathrm{R}^3, t\in[0,T])$ $(T>0)$ and Φ given by (1.2), with $\rho\ge 0$, $p(\rho)$, $\mathbf{m}$, $\mathbf{m}\otimes\mathbf{m}/\rho$ and $\rho\nabla\Phi$ being in $L^1(\mathrm{R}^3\times[0,T])$, is called a weak solution of the Cauchy problem (1.1) and (3.11) on $\mathrm{R}^3\times[0,T]$ if for any Lipschitz continuous test function ψ with compact support in $\mathrm{R}^3\times[0,T]$,*

$$\int_0^T\int(\rho\psi_t+\mathbf{m}\cdot\nabla\psi+p(\rho)\nabla\psi)\,dxdt+\int\rho_0(x)\psi(x,0)dx=0, \tag{3.12}$$

and

$$\begin{aligned}\int_0^T\int\left(\mathbf{m}\psi_t+(p(\rho)\mathbb{I}+\frac{\mathbf{m}\otimes\mathbf{m}}{\rho})\nabla\psi\right)dxdt+\int\mathbf{m}_0(x)\psi(x,0)dx\\ =\int_0^T\int\rho\nabla\Phi\psi dxdt,\end{aligned}\tag{3.13}$$

where $\mathbb{I}$ *is the* 3×3 *unit matrix.*

The *total energy* of system (1.1) at time t is

$$\begin{aligned} E(t) &= E(\rho(t), \mathbf{v}(t)) \\ &= \int \left(A(\rho) + \frac{1}{2}\rho|\mathbf{v}|^2 \right)(x,t)dx - \frac{1}{8\pi}\int |\nabla\Phi|^2(x,t)dx, \end{aligned} \tag{3.14}$$

where as before,

$$A(\rho) = \rho \int_0^\rho \frac{p(s)}{s^2} ds. \tag{3.15}$$

For a solution of (1.1) without shock waves, the total energy is conserved, i.e., $E(t) = E(0)$ $(t \geq 0)$(cf. [35]). For solutions with shock waves, the energy should be non-increasing in time, so that for all $t \geq 0$,

$$E(t) \leq E(0), \tag{3.16}$$

due to the entropy conditions, which is described below.

DEFINITION 3.2. *A weak solution (defined above) on* $\mathbb{R}^3 \times [0,T]$ *is called an entropy weak solution of (1.1) if it satisfies the following "entropy inequality":*

$$\partial_t \eta + \sum_{j=1}^{3} \partial_{x_j} q_j + \rho \sum_{j=1}^{3} \eta_{m_j} \Phi_{x_j} \leq 0, \tag{3.17}$$

in the sense of distributions; i.e.,

$$\begin{aligned} &\int_0^T \int_{\mathbb{R}^3} \left(\eta\beta_t + \mathbf{q}\cdot\nabla\beta - \rho \sum_{j=1}^{3} \eta_{m_j} \Phi_{x_j} \beta \right) dxdt \\ &\qquad + \int_{\mathbb{R}^3} \beta(x,0)\eta(x,0)dx \geq 0, \end{aligned} \tag{3.18}$$

for any nonnegative Lipschitz continuous test function β *with compact support in* $[0,T) \times \mathbb{R}^3$. *Here the "entropy" function* η *and "entropy flux" functions* q_j *and* $\mathbf{q}$, *are defined by*

$$\begin{cases} \eta = \dfrac{|\mathbf{m}|^2}{2\rho} + \rho \displaystyle\int_0^\rho \frac{p(s)}{s^2} ds, \\ q_j = \dfrac{|\mathbf{m}|^2 m_j}{2\rho^2} + m_j \displaystyle\int_0^\rho \frac{p'(s)}{s} ds, \\ \mathbf{q} = (q_1,\ q_2,\ q_3). \end{cases} \tag{3.19}$$

Some properties of entropy weak solutions are given in the following theorem.

THEOREM 3.2. *If $(\rho, \mathbf{m}) \in L^\infty([0,T]; L^1(\mathbb{R}^3))$ satisfies the first equation in (1.1) in the sense of distributions, then*

$$\int_{\mathbb{R}^3} \rho(x,t)dx = \int_{\mathbb{R}^3} \rho(x,0)dx =: M, \qquad 0 < t < T. \tag{3.20}$$

Let $(\rho, \mathbf{m}, \Phi)$ be a weak solution defined in Definition 3.1. Suppose $(\rho, \mathbf{m}, \Phi)$ satisfies the entropy condition (3.17), $\rho \in L^\infty([0,T]; L^1(\mathbb{R}^3)) \cap L^\infty([0,T]; L^r(\mathbb{R}^3))$ for some r satisfying $r > 3/2$ and $r \geq \gamma$ ($\gamma \geq 4/3$ is the constant in 3.10), $\mathbf{m} \in L^\infty([0,T]; L^s(\mathbb{R}^3))$ ($s > 3$), $(\eta, \mathbf{q}) \in L^\infty([0,T]; L^1(\mathbb{R}^3))$, where η and $\mathbf{q}$ are given in (4.3). Moreover, we assume that $(\rho, \mathbf{m})$ has the following additional regularity:

$$\lim_{h\to 0} \int_0^t \int_{\mathbb{R}^3} |\rho(x,\tau+h) - \rho(x,\tau)| dx d\tau = 0, \quad t \in (0,T), a.e. \tag{3.21}$$

Then

$$E(t) \leq E(0), \qquad 0 < t < T, \tag{3.22}$$

where $E(t)$ is defined in (3.14). The proof of this theorem is the same as that for Theorem 5.1 in [26].

We consider axi-symmetric initial data, which takes the form

$$\begin{aligned} \rho_0(x) &= \rho(r,z), \\ \mathbf{v}_0(x) &= v_0^r(r,z)\mathbf{e}_r + v_0^\theta(r,z)\mathbf{e}_\theta + v_0^3(\rho,z)\mathbf{e}_3. \end{aligned} \tag{3.23}$$

Here $r = \sqrt{x_1^2 + x_2^2}$, $z = x_3$, $x = (x_1, x_2, x_3) \in \mathbb{R}^3$ (as before), and

$$\mathbf{e}_r = (x_1/r, x_2/r, 0)^{\mathrm{T}}, \ \mathbf{e}_\theta = (-x_2/r, x_1/r, \ 0)^{\mathrm{T}}, \ \mathbf{e}_3 = (0,0,1)^{\mathrm{T}}. \tag{3.24}$$

We seek axi-symmetric solutions of the form

$$\begin{aligned} \rho(x,t) &= \rho(r,z,t), \\ \mathbf{v}(x,t) &= v^r(r,z,t)\mathbf{e}_r + v^\theta(r,z,t)\mathbf{e}_\theta + v^3(r,z,t)\mathbf{e}_3, &(3.25) \\ \Phi(x,t) &= \Phi(r,z,t) = -B\rho(r,z,t). &(3.26) \end{aligned}$$

We call a vector field $\mathbf{u}(x,t) = (u_1, u_2, u_3)(x)$ ($x \in \mathbb{R}^3$) axi-symmetric if it can be written in the form

$$\mathbf{u}(x) = u^r(r,z)\mathbf{e}_r + u^\theta(r,z)\mathbf{e}_\theta + u^3(\rho,z)\mathbf{e}_3.$$

For the velocity field $\mathbf{v} = (v_1, v_2, v_3)(x,t)$, we define the angular momentum (per unit mass) $j(x,t)$ about the x_3-axis at (x,t) , $t \geq 0$, by

$$j(x,t) = x_1 v_2 - x_2 v_1. \tag{3.27}$$

For an axi-symmetric velocity field

$$\mathbf{v}(x,t) = v^r(r,z,t)\mathbf{e}_r + v^\theta(r,z,t)\mathbf{e}_\theta + v^3(\rho,z,t)\mathbf{e}_3, \tag{3.28}$$

$$v_1 = \frac{x_1}{r}v^r - \frac{x_2}{r}v^\theta,\ v_2 = \frac{x_2}{r}v^r + \frac{x_1}{r}v^\theta, v_3 = v^3, \tag{3.29}$$

so that

$$j(x,t) = rv^\theta(r,z,t). \tag{3.30}$$

In view of (3.28) and (3.30), we have

$$|\mathbf{v}|^2 = |v^r|^2 + \frac{j^2}{r^2} + |v^3|^2. \tag{3.31}$$

Therefore, the total energy at time t can be written as

$$\begin{aligned} E(\rho(t),\mathbf{v}(t)) = &\int A(\rho)(x,t)dx + \frac{1}{2}\int \frac{\rho j^2(x,t)}{r^2(x)}dx \\ &-\frac{1}{8\pi}\int |\nabla B\rho|^2(x,t)dx + \frac{1}{2}\int \rho(|v^r|^2 + |v^3|^2)(x,t)dx. \end{aligned} \tag{3.32}$$

We next make some assumptions on the initial data; namely, we assume that the initial data is such that the initial total mass and angular momentum are the same as those of the rotating star solution (those two quantities are conserved quantities). Therefore, we require

I$_1$)

$$\int \rho_0(x)dx = \int \tilde{\rho}(x)dx = M. \tag{3.33}$$

Moreover we assume

I$_2$) For the initial angular momentum $j(x,0) = rv_0^\theta(r,z) =: j_0(r,z)$ ($r = \sqrt{x_1^2 + x_2^2}$, $z = x_3$ for $x = (x_1,x_2,x_3)$, we assume $j(x,0)$ only depends on the total mass in the cylinder $\{y \in \mathbb{R}^3, r(y) \le r(x)\}$, i.e. ,

$$j(x,0) = j_0\left(m_{\rho_0}(r(x))\right). \tag{3.34}$$

(This implies that we require that $v_0^\theta(r,z)$ only depends on r.)
Finally, we assume that the initial profile of the angular momentum per unit mass is the same as that of the rotating star solution, i. e.,

I$_3$)

$$j_0^2(m) = L(m), \qquad 0 \le m \le M, \tag{3.35}$$

where $L(m)$ is the profile of the square of the angular momentum of the rotating star defined in Section 2.

In order to state our stability result, we need some notation. Let λ be the constant in Theorem 2.2, i.e.,

$$\begin{cases} A'(\tilde{\rho}(x)) + \int_{r(x)}^{\infty} L(m_{\tilde{\rho}}(s))s^{-3}ds - B\tilde{\rho}(x) = \lambda, \ x \in \Gamma, \\ \int_{r(x)}^{\infty} L(m_{\tilde{\rho}})(s))s^{-3}ds - B\tilde{\rho}(x) \geq \lambda, \qquad x \in \mathbb{R}^3 - \Gamma, \end{cases} \tag{3.36}$$

with A defined in (3.15) and Γ defined in (2.16).

For $\rho \in W_M$, we define,

$$d(\rho, \tilde{\rho}) = \int [A(\rho) - A(\tilde{\rho})] + (\rho - \tilde{\rho})\{\int_{r(x)}^{\infty} \frac{L(m_{\tilde{\rho}}(s))}{s^3} ds - \lambda - B\tilde{\rho}\}dx. \tag{3.37}$$

For $x \in \Gamma$, in view of the convexity of the function A (cf. (3.15)) and (3.36), we have,

$$\begin{aligned} &(A(\rho) - A(\tilde{\rho}))(x) + (\int_{r(x)}^{\infty} \frac{L(m_{\tilde{\rho}}(s))}{s^3} ds - \lambda - B\tilde{\rho}(x))(\rho - \tilde{\rho}) \\ &\qquad = (A(\rho) - A(\tilde{\rho}) - A'(\tilde{\rho})(\rho - \tilde{\rho}))(x) \geq 0. \end{aligned} \tag{3.38}$$

For $x \in \mathbb{R}^3 - \Gamma$, $\tilde{\rho}(x) = 0$, so we have $A(\tilde{\rho})(x)) = 0$. This is because since $A(0) = 0$ due to $p(0) = 0$ (cf. (3.3)) and (2.5). Therefore, by (3.36), we have, for $\rho \in W_M$ and $x \in \mathbb{R}^3 - \Gamma$,

$$\begin{aligned} &(A(\rho) - A(\tilde{\rho}))(x) + (\int_{r(x)}^{\infty} \frac{L(m_{\tilde{\rho}}(s))}{s^3} ds - \lambda - B\tilde{\rho}(x))(\rho - \tilde{\rho}) \\ &\qquad = A(\rho) \geq 0. \end{aligned} \tag{3.39}$$

Thus, for $\rho \in W_M$,

$$d(\rho, \tilde{\rho}) \geq 0. \tag{3.40}$$

We also define

$$\begin{aligned} d_1(\rho, \tilde{\rho}) = &\frac{1}{2} \int \frac{\rho(x)L(m_\rho(r(x))) - \tilde{\rho}(x)L(m_{\tilde{\rho}}(r(x))}{r^2(x)} dx \\ &- \int \int_{r(x)}^{\infty} s^{-3} L(m_{\tilde{\rho}}(s))ds(\rho(x) - \tilde{\rho}(x))dx, \end{aligned} \tag{3.41}$$

for $\rho \in W_M$. It is shown in [26] and [27] that $d_1 \geq 0$. The following stability result on the rotating star solutions is proved in [27].

THEOREM 3.3. *Suppose that the pressure function satisfies (3.10), and both (3.4), (3.5) hold. Let $\tilde{\rho}$ be a minimizer of the functional F in W_M, and assume that it is unique up to a vertical shift. Assume that I_1)- I_3),*

[(3.33)-(3.35)] hold. Moreover, assume that the angular momentum of the rotating star solution $\tilde{\rho}$ satisfies (2.9), (3.1) and (3.2). Let $(\rho, \mathbf{v}, \Phi)(x,t)$ be an entropy weak solution of the Cauchy problem (1.1) and (3.11) satisfying (3.20) and (3.22) with axi-symmetry. Under some physically reasonable assumption on the conservation of angular momentum (cf. [27]), then for every $\epsilon > 0$, there exists a number $\delta > 0$ such that if

$$\begin{aligned} & d(\rho_0, \tilde{\rho}) + \frac{1}{8\pi}||\nabla B\rho_0 - \nabla B\tilde{\rho}||_2^2 + |d_1(\rho_0, \tilde{\rho})| \\ & + \frac{1}{2}\int \rho_0(x)(|v_0^r|^2 + |v_0^3|^2)(x)dx < \delta, \end{aligned} \tag{3.42}$$

then for every $t > 0$, there is a vertical shift $a(t)\mathbf{e_3}$ ($a(t) \in \mathbb{R}$, $\mathbf{e_3} = (0,0,1)$) such that,

$$\begin{aligned} & d(\rho(t), T^{a(t)}\tilde{\rho}) + \frac{1}{8\pi}||\nabla B\rho(t) - \nabla BT^{a(t)}\tilde{\rho}||_2^2 + |d_1(\rho(t), T^{a(t)}\tilde{\rho})| \\ & + \frac{1}{2}\int \rho(x,t)(|v^r(x,t)|^2 + |v^3(x,t)|^2)dx < \epsilon, \end{aligned} \tag{3.43}$$

where $T^{a(t)}\tilde{\rho}(x) =: \tilde{\rho}(x + a(t)\mathbf{e_3})$.

REMARK 2. Without the uniqueness assumption for the minimizer of F in W_M, we can have the following type of stability result, as observed in [30] for the non-rotating star solutions. Suppose the assumptions in Theorem 3.3 hold. Let $\mathcal{S}_M$ be the set of all minimizers of F in W_M and $(\rho, \mathbf{v}, \Phi)(x,t)$ be an axi-symmetric weak entropy solution of the Cauchy problem (1.1) and (3.11)satisfying (3.20) and (3.22). Then for every $\epsilon > 0$, there exists a number $\delta > 0$ such that if

$$\begin{aligned} & \inf_{\tilde{\rho}\in\mathcal{S}_M}\left[d(\rho_0, \tilde{\rho}) + \frac{1}{8\pi}||\nabla B\rho_0 - \nabla B\tilde{\rho}||_2^2 + |d_1(\rho_0, \tilde{\rho})|\right] \\ & + \frac{1}{2}\int \rho_0(x)(|v_0^r|^2 + |v_0^3|^2)(x)dx < \delta, \end{aligned} \tag{3.44}$$

then for every $t > 0$, there is a vertical shift $a(t)\mathbf{e_3}$ ($a \in \mathbb{R}$, $\mathbf{e_3} = (0,0,1)$) such that

$$\begin{aligned} & \inf_{\tilde{\rho}\in\mathcal{S}_M}\left[d(\rho(t), T^{a(t)}\tilde{\rho}) + \frac{1}{8\pi}||\nabla B\rho(t) - \nabla BT^{a(t)}\tilde{\rho}||_2^2 + |d_1(\rho(t), T^{a(t)}\tilde{\rho})|\right] \\ & + \frac{1}{2}\int \rho(x,t)(|v^r(x,t)|^2 + |v^3(x,t)|^2)(x)dx < \epsilon, \end{aligned} \tag{3.45}$$

where $T^{a(t)}\tilde{\rho}(x) =: \tilde{\rho}(x + a(t)\mathbf{e_3})$. In the case of non-rotating stars, i.e. $L = 0$, the uniqueness of minimizers of the energy functional was proved by Lieb and Yau in [22]. There has been no uniqueness results for the case of rotating stars. It might be expected that this problem can be solved by using some ideas in [22].

4. Applications to white dwarf and supermassive stars. In this section, we verify the assumptions (3.4) and (3.5) in Theorem 3.1 for both white dwarfs and supermassive stars. Once we verify (3.4) and (3.5), we can apply Theorems 3.1 and 3.3. The following theorem verifies (3.4) and (3.5) for white dwarfs and polytropes with $\gamma > 4/3$, in both the rotating and non-rotating cases.

THEOREM 4.1. *Assume that the pressure function p satisfies (3.3). Then there exists a constant* $\mathfrak{M}_c$ *satisfying* $0 < \mathfrak{M}_c < \infty$ *if* $\gamma = 4/3$ *and* $\mathfrak{M}_c = \infty$ *if* $\gamma > 4/3$, *such that if* $M < \mathfrak{M}_c$, *then (3.4) and (3.5) hold for* $\rho \in W_M$. The proof of this theorem can be found in [27]).

5. Nonlinear dynamical stability of non-rotating white dwarf stars with general perturbations. The dynamical stability results in Section 3 apply for axi-symmetric perturbations. Moreover, the uniqueness of minimizers of the energy functional for rotating stars is not known.

For white dwarf stars, as mentioned before, the pressure function satisfies

$$\begin{aligned} &p \in C^1[0, +\infty), \ \lim_{\rho \to 0+} \frac{p(\rho)}{\rho^{\gamma_1}} = \beta, \\ &\lim_{\rho \to \infty} \frac{p(\rho)}{\rho^{\gamma}} = K, \ p'(\rho) > 0 \text{ for } \rho > 0, \end{aligned} \tag{5.1}$$

where $\gamma_1 > 4/3$, $0 < \beta < +\infty$ and $0 < K < +\infty$ are constants. In this section, we always assume that the pressure function satisfies (5.1). First, we define for $0 < M < +\infty$,

$$\begin{aligned} X_M = \{&\rho : \mathbb{R}^3 \to \mathbb{R}, \rho \geq 0, a.e., \int \rho(x)dx = M, \\ &\int [A(\rho(x)) + \frac{1}{2}\rho(x)B\rho(x)]dx < +\infty\}, \end{aligned} \tag{5.2}$$

where $A(\rho)$ is the function given in (2.5). For $\rho \in X_M$, we define the **energy functional** G for non-rotating stars by

$$G(\rho) = \int [A(\rho(x)) - \frac{1}{2}\rho(x)B\rho(x)]dx. \tag{5.3}$$

We begin with the following theorem.

THEOREM 5.1. *Suppose that the pressure function p satisfies (5.1). Let* $\tilde{\rho}_N$ *be a minimizer of the energy functional G in* X_M *and let*

$$\Gamma_N = \{x \in \mathbb{R}^3 : \ \tilde{\rho}_N(x) > 0\}, \tag{5.4}$$

then there exists a constant λ_N *such that*

$$\begin{cases} A'(\tilde{\rho}_N(x)) - B\tilde{\rho}_N(x) = \lambda_N, & x \in \Gamma_N, \\ -B\tilde{\rho}_N(x) \geq \lambda_N, & x \in \mathbb{R}^3 - \Gamma_N. \end{cases} \tag{5.5}$$

The proof of this theorem is well-known, cf. [30] or [1].

REMARK 3. 1) We call the minimizer $\tilde{\rho}_N$ of the functional G in X_M a non-rotating star solution.
2) It follows from [22] that the minimizer $\tilde{\rho}_N$ of the functional G in X_M is actually radial, and has a compact support.

Similar to Theorem 3.1, we have the following compactness theorem.

THEOREM 5.2. *Suppose that the pressure function p satisfies (5.1). There exists a constant M^c $(0 < M^c < \infty)$ such that if $M < M^c$, then the following hold:*
(1)

$$\inf_{\rho \in X_M} G(\rho) < 0, \tag{5.6}$$

(2) for $\rho \in X_M$,

$$\int A(\rho)(x)dx \le C_1 G(\rho) + C_2, \tag{5.7}$$

for some positive constants C_1 and C_2,
(3) if $\{\rho^i\} \subset X_M$ is a minimizing sequence for the functional G, then there exist a sequence of translations $\{x^i\} \subset \mathbb{R}^3$, a subsequence of $\{\rho^i\}$, (still labeled $\{\rho^i\}$), and a function $\tilde{\rho}_N \in X_M$, such that for any $\epsilon > 0$ there exists $R > 0$ with

$$\int_{|x| \ge R} T\rho^i(x)dx \le \epsilon, \quad i \in \mathbb{N}, \tag{5.8}$$

and

$$T\rho^i(x) \rightharpoonup \tilde{\rho}_N, \text{ weakly in } L^{4/3}(\mathbb{R}^3), \text{ as } i \to \infty, \tag{5.9}$$

where $T\rho^i(x) := \rho^i(x + x^i)$.
Moreover
(4)

$$\nabla B(T\rho^i) \to \nabla B(\tilde{\rho}_N) \text{ strongly in } L^2(\mathbb{R}^3), \text{ as } i \to \infty, \tag{5.10}$$

and
(5) $\tilde{\rho}_N$ is a minimizer of G in X_M.
(6) The minimizers of G in X_M are unique up to a translation $\rho_N(x) \to \rho_N(x + y)$.

The proof of this theorem can be found in ([27]). The uniqueness of minimizers is proved in [22]. For the stability, we consider the Cauchy problem (1.1) with the initial data (3.11). We *do not* assume that the initial data have any symmetry.

Let $\tilde\rho_N$ be the minimizer of G on X_M and λ_N be the constant in (5.5). For $\rho \in X_M$, we define

$$\begin{aligned} d(\rho, \tilde\rho_N) &= \int \{[A(\rho) - A(\tilde\rho_N)] - (\rho - \tilde\rho_N)(\lambda_N + B\tilde\rho_N\}dx, \\ &= \int \{[A(\rho) - A(\tilde\rho_N)] - B\tilde\rho_N(\rho - \tilde\rho_N)\}dx, \end{aligned} \tag{5.11}$$

where we have used the identity

$$\int \rho dx = \int \tilde\rho_N dx = M,$$

for $\rho \in X_M$.

$$d(\rho, \tilde\rho_N) \geq 0, \tag{5.12}$$

for any $\rho \in X_M$ (see [27]). The following nonlinear stability theorem of non-rotating white dwarf star solutions is proved in ([27]), which extends the results in [30] (The result in [30] does not apply to the white dwarf case).

THEOREM 5.3. *Suppose that the pressure function satisfies (5.1). Let $\tilde\rho_N$ the minimizer of the functional G in X_M. Let $(\rho, \mathbf{v}, \Phi)(x,t)$ be an entropy weak solution of the Cauchy problem (1.1) and (3.11) stated in Theorem 3.2 satisfying (3.20) and (3.22). If the initial data satisfies*

$$\int \rho_0(x) = \int \rho_N(x)dx = M,$$

then there exists a constant M^c $(0 < M^c < \infty)$ such that if $M < M^c$, then for every $\epsilon > 0$, there exists a number $\delta > 0$ such that if

$$d(\rho_0, \tilde\rho_N) + \frac{1}{8\pi}||\nabla B\rho_0 - \nabla B\tilde\rho_N||_2^2 + \frac{1}{2}\int \rho_0(x)(|v_0|^2)(x)dx < \delta, \tag{5.13}$$

then for every $t > 0$, there is a translation $y(t) \in \mathbb{R}^3$ such that,

$$\begin{aligned} &d(\rho(t), T^{y(t)}\tilde\rho_N) + \frac{1}{8\pi}||\nabla B\rho(t) - \nabla BT^{y(t)}\tilde\rho_N||_2^2 \\ &+\frac{1}{2}\int \rho(x,t)|v(x,t)|^2)dx < \epsilon, \end{aligned} \tag{5.14}$$

where $T^{y(t)}\tilde\rho_N(x) =: \tilde\rho_N(x + y(t))$.

6. Some related open problems. In this section, we list some open problems.

1) Uniqueness of rotating star solutions. This is to prove the uniqueness (up to a vertical shift) of the minimizers of the energy functional F in W_M,

assuming that the square of the angular momentum L satisfies (2.9), (3.1) and (3.2). The uniqueness of the non-rotating star solutions was proved by Lieb and Yau ([22]). For rotating star solutions, this problem is much more challenging.

2) Instability for the non-rotating star solutions for the polytropic star when $6/5 < \gamma < 4/3$. When $\gamma > 6/5$, the stationary radially symmetric solutions $\rho = \rho(r), \mathbf{v} = 0$ of the Euler-Poisson equations exist satisfying

$$(P(\rho))' = -\rho\Phi',\ \Phi'' + \frac{2}{r}\Phi' = 4\pi G\rho,$$

where $' = \frac{d}{dr}$, $r = |x|$. Moreover, $\rho(r)$ has a compact support when $\gamma > 6/5$. The stability when $\gamma > 4/3$ was proved by G. Rein in [30]. When $\gamma = 6/5$ (in this case ρ does not have a compact support but decay to zero as $r \to \infty$,) the instability was proved by Jang in [12]. For $6/5 < \gamma < 4/3$, the linearized instability was proved by Lin ([23]). The question here is to prove the nonlinear instability.

3) Free boundary problem to capture the physical vacuum boundary.

Consider system (1.1) in a moving open bounded domain $\Omega_t \subset \mathbb{R}^3$ which is occupied by a fluid at time $t \in [0,T]$. To make the free-boundary move with the fluid-velocity, we have

$$\partial_t + v \cdot \nabla \text{ is in the tangent space of } \cup_{t\in[0,T]} (\Omega_t \times \{t\}). \tag{6.1}$$

The boundary condition for pressure is given by

$$p(\rho) = 0 \ on\ \partial\Omega_t. \tag{6.2}$$

The initial conditions are

$$\Omega_0 = \Omega, (\rho, \mathbf{u}, S)(x, 0) = (\rho_0, \mathbf{u}_0, S_0)(x)\ x \in \Omega, \tag{6.3}$$

where $\Omega_0 \subset \mathbb{R}^3$ is an open bounded domain. $(\rho_0, \mathbf{u}_0, S_0)$ are given. We further require

$$\begin{aligned} &\rho_0(x) > 0 \quad for\ x \in \Omega_0, \\ &\rho_0(x) = 0 \quad for\ x \in \partial\Omega_0, \quad -\infty < \partial_{\mathbf{n}} c_0^2 < 0, \quad \text{on } \partial\Omega_0, \end{aligned} \tag{6.4}$$

where c_0 is the initial sound speed (the sound speed is given by $c = \sqrt{\frac{\partial p(\rho,S)}{\partial \rho}}$). The condition $-\infty < \partial_{\mathbf{n}} c_0^2 < 0$, on$\partial\Omega_0$ is to capture the physical vacuum condition. For the stationary solution of Euler-poisson equations, the sound speed is $C^{1/2}$- Holder continuous across the interface separating the gas and vacuum. This vacuum boundary condition poses a challenging problem for the local-in-time existence for the above free boundary problem, since the boundary is both characteristic and degenerate. The standard method of symmetric hyperbolic systems does not apply. For the 1-d isentropic compressible Euler equations, the local-in-time well-posedness of the free boundary problems was proved by D. Coutand & S. Shkoller ([4]) and Jang & Masmoudi ([13]).

Acknowledgments. Luo was supported in part by the National Science Foundation under Grants DMS-0839864. Smoller was supported in part by the National Science Foundation under Grant DMS-0603754.

REFERENCES

[1] G. Auchmuty and R. Beals, *Variational solutions of some nonlinear free boundary problems*, Arch. Rat. Mech. Anal. **43**, 255–271 (1971).

[2] G. Auchmuty, *The global branching of rotating stars*, Arch. Rat. Mech. Anal. **114**, 179–194 (1991).

[3] L. Caffarelli and A. Friedman, *The shape of axi-symmetric rotating fluid*, J. Funct. Anal. **694**, 109–142 (1980).

[4] D. Coutand and S. Shkoller, *Well-posedness in smooth function spaces for the moving-boundary 1-D compressible Euler equations in physical vacuum*, arXiv:0910.3136v1 [math.AP].

[5] S. Chandrasekhar, Phil. Mag. **11**, 592 (1931); Astrophys. J. **74**, p. 81 (1931); Monthly Notices Roy. Astron. Soc. **91**, p. 456 (1931); Rev. Mod. Phys. **56**, p. 137 (1984).

[6] S. Chandrasekhar, *Introduction to the Stellar Structure*, *University of Chicago Press* (1939).

[7] Sagun Chanillo and Yan Yan Li, *On diameters of uniformly rotating stars*, Comm. Math. Phys. **166**, no. 2, 417–430 (1994).

[8] Y. Deng, T.P. Liu, T. Yang, and Z.Yao, *Solutions of Euler-Poisson equations for gaseous stars*, Arch. Rat. Mech. Anal. **164**, no. 3, 261–285 (2002).

[9] R.H. Fowler, Monthly Notices Roy. Astron. Soc. **87**, p. 114 (1926).

[10] A. Friedman and B. Turkington, *Asymptotic estimates for an axi-symmetric rotating fluid*, J. Func. Anal. **37**, 136–163 (1980).

[11] A. Friedman and B. Turkington, *Existence and dimensions of a rotating white dwarf*, J. Diff. Eqns. **42**, 414–437 (1981).

[12] J. Jang, *Nonlinear instability in gravitational Euler-Poisson system for* $\gamma = \frac{6}{5}$, to appear in Arch. Rat. Mech. Anal.

[13] Juhi Jang and Nader Masmoudi, *Well-posedness for compressible Euler equations with physical vacuum singularity*, Comm. Pure Appl. Math. **62** no. 10, 1327–1385 (2009).

[14] Y. Guo and G. Rein, *Stable steady states in stellar dynamics*, Arch. Rat. Mech. Anal. **147**, 225–243 (1999).

[15] Y. Guo and G. Rein, *Stable models of elliptical galaxies*, Mon. Not. R. Astron. Soc. **344**, 1296–1306 (2002).

[16] D. Gilbarg and N. Trudinger, *Elliptic Partial Differentail Equations of Second Order* (2nd ed.), *Springer* (1983).

[17] L. Landau, Phys. Z. Sowjetunion **1**, p. 285 (1932).

[18] P. Lax, *Shock Waves and Entropy, Contribution to Nonlinear Functional Analysis*, E.A. Zarantonello, ed., *Academic Press,* New York, 603–634 (1971).

[19] N.R. Lebovitz, *The virial tensor and its application to self-gravitating fluids*, Astrophys. J. **134**, 500–536 (1961).

[20] N.R. Lebovitz and A. Lifschitz, *Short-wavelength instabilities of Riemann ellipsoids*, Philos. Trans. Roy. Soc. London Ser. A **354**, no. 1709, 927–950 (1996).

[21] Yan Yan Li, *On uniformly rotating stars*, Arch. Rat. Mech. Anal. **115**, no. 4, 367–393 (1991).

[22] E.H. Lieb and H.T. Yau, *The Chandrasekhar theory of stellar collapse as the limit of quantum mechanics*, Comm. Math. Phys. **112**, no. 1, 147–174 (1987).

[23] S.S. Lin, *Stability of gaseous stars in spherically symmetric motions*, SIAM J. Math. Anal. **28**, no. 3, 539–569 (1997).

[24] P.L. Lions, *The concentration-compactness principle in the calculus of variations. The locally compact case*, Part I, Ann. Inst. H. Poincaré Anal. Non Linéaire, **1**, 109–145 (1984).

[25] T. Luo and J. Smoller, *Rotating fluids with self-gravitation in bounded domains*, Arch. Rat. Mech. Anal. **173**, no. 3, 345–377 (2004).

[26] T. Luo and J. Smoller, *Existence and Nonlinear Stability of Rotating Star Solutions of the Compressible Euler-Poisson Equations*, Arch. Rational Mech. & Anal. **191**, no. 3, 447–496. (2009)

[27] T. Luo and J. Smoller, *Nonlinear Dynamical Stability of Newtonian Rotating and Non-rotating White Dwarfs and Rotating Supermassive Stars*, Comm. Math. Phys. **284**, 425-457 (2008).

[28] T. Makino, *Blowing up of the Euler-Poisson equation for the evolution of gaseous star*, Transport Theory and Statistical Physics **21**, 615–624 (1992).

[29] G. Rein, *Reduction and a concentration-compactness principle for energy-casimir functionals*, SIAM J. Math. Anal. Vol. **33**, no. 4, 896–912 (2001).

[30] G. Rein, *Non-linear stability of gaseous stars*, Arch. Rat. Mech. Anal. **168**, no. 2, 115–130 (2003).

[31] P. Secchi, *On the evolution equations of viscous gaseous stars*, Ann. Scuola Norm. Sup. Pisa Cl. Sci. (4) **18**, no. 2, 295–318 (1991).

[32] P. Secchi, *On the uniqueness of motion of viscous gaseous stars*, Math. Methods Appl. Sci. **13**, no. 5, 391–404 (1990).

[33] S.H. Shapiro and S.A. Teukolsky, *Black Holes, White Dwarfs, and Neutron Stars*, *WILEY-VCH* (2004).

[34] J. Smoller, *Shock Waves and Reaction-Diffusion Equations*, 2nd Ed.), Springer, Berlin, New York (1994).

[35] J.L. Tassoul, *Theory of Rotating Stars*, *Princeton University Press*, Princeton (1978).

[36] D. Wang, *Global Solutions and Stability for Self-Gravitating Isentropic Gases*, J. of Math. Anal. & Appl., **229**, 530–542 (1999).

[37] S. Weinberg, *Gravitation and Cosmology*, *John Wiley and Sons*, New York , 1972.

VISCOUS SYSTEM OF CONSERVATION LAWS: SINGULAR LIMITS

DENIS SERRE*

Abstract. We continue our analysis of the Cauchy problem for viscous system of conservation, under natural assumptions. We examine in which way does the existence time depend upon the viscous tensor $B(u)$. In particular, we consider singular limits, where the rank of the symbol $B(u;\xi)$ drops at the limit. This covers a lot of situations, for instance that of the limit of the Navier-Stokes-Fourier system towards the Euler-Fourier system, or that of the vanishing viscosity.

We emphasize the symmetry of the dissipation tensor, an hypothesis which is reminiscent to the Onsager's reciprocity relations. We find it useful in this asymptotic context, when establishing uniform estimates.

Key words. Systems of conservation laws; dissipative structure; entropy; singular limit; Onsager's reciprocity relations.

AMS(MOS) subject classifications. 35A02, 35K51, 35L65, 35M31, 35Q35, 35Q80, 76N17, 76W05.

1. Main results.

1.1. Dissipative viscous systems of conservation laws. We continue our work begun in [7]. We consider a system of PDEs in conservation form

$$\partial_t u + \mathrm{Div} f(u) = \mathrm{Div}(B(u)\nabla u) := \sum_{\alpha,\beta} \partial_\alpha (B^{\alpha\beta}(u)\partial_\beta u), \qquad (1.1)$$

in which $u : (0,T) \times \mathbb{R}^d \to \mathcal{U}$ is the unknown. The phase space $\mathcal{U}$ is an open convex subset of $\mathbb{R}^n$. The symbol ∂_α denotes the partial derivative with respect to the coordinate x_α. The nonlinearities are encoded in the smooth functions

$$f : \mathcal{U} \to \mathbf{M}_{n\times d}(\mathbb{R}), \qquad B^{\alpha\beta} : \mathcal{U} \to \mathbf{M}_n(\mathbb{R}).$$

They describe in mathematical terms the kind of physics that is modelled by the PDEs (1.1).

In the notation $B\nabla u$ in (1.1), the tensor B plays the role of a linear map from the space of $n \times d$ matrices into itself. The operator Div is the row-wise divergence which, to a field of $n \times d$ matrices, associates a vector field of dimension n. When needed, we use also the ordinary divergence operator div, which applies to d-vector fields and produces scalar functions.

---

*UMPA (UMR 5669 CNRS), ENS de Lyon, 46, allée d'Italie, F–69364 Lyon, cedex 07, FRANCE. The author thanks the staff of IMA for their warm and efficient hospitality, as well as the organizers of the summer program *Nonlinear Conservation Laws and Applications* for their kind invitation.

A paradigm for the class of such systems is the Navier-Stokes-Fourier system for a compressible fluid, in which the components of the unknown are the mass density ρ, the linear momentum ρv (v the mean velocity of the fluid) and the mechanical energy per unit volume:

$$u =: \begin{pmatrix} \rho \\ \rho v \\ \frac{1}{2}\rho|v|^2 + \rho e \end{pmatrix}.$$

The flux f and the tensor B are given by

$$f(u) := \begin{pmatrix} \rho v^T \\ \rho v \otimes v + p(\rho, e) I_d \\ \left(\frac{1}{2}\rho|v|^2 + \rho e + p(\rho, e)\right) v^T \end{pmatrix}$$

and

$$B(u)\nabla u :=$$
$$\begin{pmatrix} 0 \\ \mathcal{T} := \mu(\rho, e)(\nabla v + (\nabla v)^T) + (\zeta(\rho, e) - 2\mu(\rho, e))(\operatorname{div} v) I_d \\ (\mathcal{T} v + \kappa(\rho, e)\nabla\theta)^T \end{pmatrix}.$$

Hereabove, the exponent T denotes transposition, while θ, the temperature, is a prescribed function of (ρ, e). The dissipation is due to Newtonian viscosity (coefficients ζ and μ), and to heat diffusion (coefficient κ).

Other examples come from electromagnetism (in material media), magnetohydrodynamics (see below), viscoelasticity, ... Let us mention also the multi-component gas dynamics with chemistry; see [2] for an accurate modelling.

The common features of these examples are four-fold:

1. On the one hand, the left-hand side of (1.1), which is a first-order system of conservation laws, admits an entropy-flux pair (η, q) in which η, the entropy, is strongly convex ($\mathrm{D}^2\eta > 0_n$) over $\mathcal{U}$. In other words, the system

$$\partial_t u + \operatorname{Div} f(u) = 0$$

implies formally

$$\partial_t \eta(u) + \operatorname{div} q(u) = 0.$$

Let us recall that the differentials of η, f^α and q^α satisfy the identity

$$\mathrm{d}q^\alpha = \mathrm{d}\eta \, \mathrm{d}f^\alpha.$$

2. On the next hand, the entropy η is *dissipated* by the right-hand side of (1.1). In mathematical terms, this means the following. Multiplying (1.1) by the differential $d\eta$, we obtain

$$\partial_t \eta(u) + \operatorname{div} q(u) = d\eta(u) \sum_{\alpha,\beta} \partial_\alpha (B^{\alpha\beta}(u)\partial_\beta u).$$

The right-hand side can be recast as the difference between a divergence and a quadratic expression in ∇u :

$$\sum_{\alpha,\beta} \partial_\alpha (d\eta(u) B^{\alpha\beta}(u)\partial_\beta u) - \sum_{\alpha,\beta} D^2\eta(u)(\partial_\alpha u, B^{\alpha\beta}(u)\partial_\beta u),$$

where we view $D^2\eta(u)$ as a symmetric positive definite bilinear form. When u is a smooth solution, tending towards a constant $\bar{u}$ at infinity rapidly enough, we may assume that $\eta(\bar{u}) = 0$ and $d\eta(\bar{u}) = 0$ (just subtract an affine function to η). Then an integration over $\mathbb{R}^d$ suppresses the divergence terms and yields the identity

$$\frac{d}{dt}\int_{\mathbb{R}^d} \eta(u)\,dx + \int_{\mathbb{R}^d} \sum_{\alpha,\beta} D^2\eta(u)(\partial_\alpha u, B^{\alpha\beta}(u)\partial_\beta u)\,dx = 0. \quad (1.2)$$

We then say that the entropy is *strongly dissipated* if the quadratic term is non-negative and actually controls the dissipation flux. By this, we mean that

$$\sum_{\alpha,\beta} D^2\eta(u)(X_\alpha, B^{\alpha\beta}(u)X_\beta) \geq \omega \sum_\alpha \Big| \sum_\beta B^{\alpha\beta}(u)X_\beta \Big|^2 =: \omega |B(u)\mathbf{X}|^2,$$
$$\forall u \in \mathcal{U},\ \forall \mathbf{X} = X_1, \dots, X_d \in \mathbf{M}_{n\times d}(\mathbb{R}). \quad (1.3)$$

Hereabove, $\omega = \omega(u)$ is positive and may be chosen continuous. Applying (1.3) to (1.2), we have a differential inequality

$$\frac{d}{dt}\int_{\mathbb{R}^d} \eta(u)\,dx + \int_{\mathbb{R}^d} \omega(u)|B(u)\nabla_x u|^2 dx \leq 0,$$

from which we obtain an *a priori* estimate of $u(t,\cdot)$ in some Lebesgue–Orlicz space (associated to η) and another one of $\omega(u)^{1/2}B(u)\nabla_x u$ in $L^2_{t,x}$.

3. On a third leg, we observe that in physical systems, some components of the unknown field u obey first-order PDEs. This is the case for instance when one writes the conservation of mass

$$\partial_t \rho + \operatorname{div}(\rho v) = 0,$$

in which there cannot be any second-order derivative. This means that in the Navier-Stokes-Fourier equations, the first line of the matrices $B^{\alpha\beta}$ vanishes identically. For more general systems, we make the assumption

(**A**) The matrices $B^{\alpha\beta}$ have the block form

$$B^{\alpha\beta}(u) = \begin{pmatrix} 0_{p\times n} \\ b^{\alpha\beta} \end{pmatrix}.$$

In addition, we assume that this property is sharp, in the following sense. Defining the symbol

$$B(\xi;u) := \sum_{\alpha,\beta} \xi_\alpha \xi_\beta B^{\alpha\beta}(u), \qquad \forall \xi \in \mathbb{R}^d,$$

we ask that $B(\xi;u)$ have rank $n-p$ for every $\xi \neq 0$.

4. On the last leg, the dissipation tensor enjoys, in suitable coordinates, a symmetry property, which we discuss in Section 2. It is reminiscent to Onsager's reciprocity relations, and was considered by Kawashima [4] in his analysis of the Cauchy problem.

We recall the results obtained in [7]. For this, we split u into two blocks v and w of respective sizes p and $n-p$.

THEOREM 1.1. *Let the variables $z_{p+1}, \dots, z_n$ (dual to w) be defined by*

$$z_j = \frac{\partial \eta}{\partial u_j}.$$

Then the map

$$u = \begin{pmatrix} v \\ w \end{pmatrix} \quad \mapsto \quad U := \begin{pmatrix} v \\ z \end{pmatrix}$$

is a global diffeomorphism onto its image $\mathcal{V}$. The viscous flux $b(u)\nabla_x u$ can be written as $Z(u)\nabla_x z$. The tensor Z is uniquely defined and satisfies an inequality

$$\sum_{\alpha,\beta=1}^{d} \sum_{i,j\geq p+1} F_{i\alpha} F_{j\beta} Z_{ij}^{\alpha\beta}(u) \geq c_0(u) \|Z(u)F\|^2, \quad \forall F \in \mathbf{M}_{(n-p)\times d}(\mathbb{R}). \tag{1.4}$$

In addition, the operator $Z(u)\nabla_x$ is strongly elliptic:

$$\sum_{\alpha,\beta=1}^{d} \sum_{i,j\geq p+1} \xi_\alpha \lambda_i \xi_\beta \lambda_j Z_{ij}^{\alpha\beta}(u) \geq c_1(u) \|\xi\|^2 \|\lambda\|^2, \quad \forall \xi \in \mathbb{R}^d,\ \forall \lambda \in \mathbb{R}^{n-p}. \tag{1.5}$$

Hereabove, c_0 and $c_1(u)$ are positive continuous functions.

The kernel of $Z(u)$. The properties stated above do not imply that the linear map $F \mapsto Z(u)F$ be one-to-one over $\mathbf{M}_{(n-p)\times d}(\mathbb{R})$. At least, ellipticity tells us that its kernel does not intersect the cone of rank-one matrices. But for NSF, $\ker Z(u)$ consists of matrices $\begin{pmatrix} G \\ 0 \end{pmatrix}$ where G is skew-symmetric. We are thus led to Assumption

> (**B**) The kernel of $Z(u)$ does not depend upon $u \in \mathcal{U}$, and the derivatives of Z with respect to u (at suitable order) are bounded in terms of $Z(u)$ itself: for every derivative ∂ of order k, and every compact $K \subset \mathcal{U}$, there exists a finite $c_{k,K}$ such that

$$|\partial Z(u)F| \leq c_{k,K}|Z(u)F|. \tag{1.6}$$

This assumption is obviously met by NSF and by other systems of physical importance.

Another example: the MHD system. In MHD, the fluid is conducting but the electric field is stuck to the magnetic field by the constraint $q\vec{E}+v\wedge\vec{B}=0$, with q the (constant) density of charge. The fluid is thus described by $(\rho, v, e, \vec{B})$, where (ρ, v, e) are as in Navier-Stokes-Fourier. Actually, if $\vec{B} \equiv 0$ at initial time, this identity remains true forever and the system reduces to Navier-Stokes-Fourier.

Let ν be the electrical resistivity. Define the stress tensor Π and the total energy E by the formulæ

$$\Pi := \left(p + \frac{1}{2}|\vec{B}|^2\right) I_3 - \vec{B} \otimes \vec{B}, \qquad E := \rho\left(\frac{1}{2}|v|^2 + e\right) + \frac{1}{2}|\vec{B}|^2.$$

The MHD system is

$$\begin{aligned}
\rho_t + \operatorname{div}(\rho v) &= 0, \\
(\rho v)_t + \operatorname{Div}(\rho v \otimes v + \Pi) &= \operatorname{Div}\mathcal{T}, \\
\vec{B}_t + \operatorname{Div}(\vec{B} \otimes v - v \otimes \vec{B}) &= \operatorname{Div}(\nu(\nabla\vec{B} - \nabla\vec{B}^T)), \\
E_t + \operatorname{div}(Ev + \Pi v) &= \operatorname{div}(\mathcal{T}v + \kappa\nabla\theta), \\
\operatorname{div}\vec{B} &= 0,
\end{aligned}$$

The MHD system contains a differential constraint ($\operatorname{div}\vec{B} = 0$) and thus looks a bit different from the abstract form (1.1). However, this constraint is compatible with the evolution equations and thus can be ignored in a first instance. If $\vec{B}|_{t=0}$ satifies it, and $(\rho, v, e, \vec{B})$ is a solution of the evolutionary part, even in the weaker sense of distributions, then $\vec{B}$ is solenoidal forever.

Entropy. Linear combinations of the conservation laws, together with the chain rule imply

$$\rho(v_t + (v\cdot\nabla)v) + \operatorname{Div}\Pi = 0, \qquad \rho(e_t + (v\cdot\nabla)e) + p\operatorname{div}v = 0.$$

Notice that the last equation is the same as in the Euler equations. Then the chain rule yields $s_t + v\cdot\nabla s = 0$, where s is the usual entropy, satisfying $\theta \mathrm{d}s = \mathrm{d}e + p\mathrm{d}\frac{1}{\rho}$, where $\theta > 0$ is the temperature.

Dissipation. Instead, we have

$$\begin{aligned}\rho(v_t+(v\cdot\nabla)v)+\mathrm{Div}\Pi&=\mathrm{Div}\mathcal{T},\\ \rho(e_t+(v\cdot\nabla)e)+p\,\mathrm{div}v&=\mathrm{div}(\kappa\nabla\theta)+\frac{\mu}{2}\left|\nabla v+\nabla v^T\right|^2\\ &\quad+(\zeta-2\mu)(\mathrm{div}v)^2+\nu|\mathrm{curl}\vec{B}|^2.\end{aligned}$$

Whence

$$\theta\rho(s_t+(v\cdot\nabla)s)=\mathrm{div}(\kappa\nabla\theta)+\frac{\mu}{2}\left|\nabla v+\nabla v^T\right|^2+(\zeta-2\mu)(\mathrm{div}v)^2+\nu|\mathrm{curl}\vec{B}|^2.$$

Denoting again $\eta:=-\rho s$, we have

$$\begin{aligned}&\eta_t+\mathrm{div}(\eta v)+\mathrm{div}\frac{\kappa\nabla\theta}{\theta}+\kappa\left|\frac{\nabla\theta}{\theta}\right|^2+\frac{\mu}{2\theta}\left|\nabla v+\nabla v^T\right|^2\\ &+\frac{\zeta-2\mu}{\theta}(\mathrm{div}v)^2+\frac{\nu}{\theta}|\mathrm{curl}\vec{B}|^2=0.\end{aligned}$$

Strong dissipation requires $\kappa,\mu\geq 0$ and $\zeta\geq\frac{4}{3}\mu$.

1.2. Normal form and the Cauchy problem. In the new coordinates, the system (1.1) rewrites in the following quasilinear form

$$\partial_t U+\sum_\alpha\tilde{A}^\alpha(U)\partial_\alpha U=\begin{pmatrix}0\\ \mathrm{D}^2_{ww}\eta\sum_{\alpha,\beta}\partial_\alpha(Z^{\alpha\beta}\partial_\beta z)\end{pmatrix},\tag{1.7}$$

where $\tilde{A}^\alpha=(\mathrm{d}U)A^\alpha(\mathrm{dU})^{-1}$. Multiplying by the block diagonal, positive definite matrix

$$S_0(U):=\begin{pmatrix}\mathrm{D}^2_{vv}\eta-\mathrm{D}^2_{vw}\eta\left(\mathrm{D}^2_{ww}\eta\right)^{-1}\mathrm{D}^2_{wv}\eta & 0\\ 0 & \left(\mathrm{D}^2_{ww}\eta\right)^{-1}\end{pmatrix},$$

the system (1.7) is equivalent to

$$S_0(U)\partial_t U+\sum_\alpha S_\alpha(U)\partial_\alpha U=\begin{pmatrix}0\\ \sum_{\alpha,\beta}\partial_\alpha(Z^{\alpha\beta}\partial_\beta z)\end{pmatrix},\tag{1.8}$$

where S_α is symmetric. This symmetrization is discussed in details in [8].

We remark that the normal form (1.8) depends upon the range of the dissipation tensor B. This might be a source of difficulty in the analysis of singular limits, when this range drops.

Local existence. The main result of [8] is

THEOREM 1.2. *Consider a viscous system of conservation laws (1.1)*

$$\partial_t u+\sum_\alpha\partial_\alpha f^\alpha(u)=\sum_{\alpha,\beta}\partial_\alpha(B^{\alpha\beta}(u)\partial_\beta u).$$

Assume the following:

- *The maps* $u \mapsto f^\alpha(u)$ *and* $u \mapsto B^{\alpha\beta}(u)$ *are smooth over a convex open set* $\mathcal{U}$ *containing the origin,*
- *System (1.1) is strongly entropy-dissipative for some smooth strongly convex entropy* η*,*
- *(**A**) the range of the symbol matrix* $B(\xi; u)$ *does not depend neither on* $\xi \neq 0$ *in* $\mathbb{R}^d$*, nor on the state* u*.*
- *(**B**) the kernel of* $Z(u)$ *is independent of* u *and* $Z(u)$ *dominates its* u*-derivatives up to the order* $[s]+1$*.*

Then, given an initial data u_0 *in* $H^s(\mathbb{R}^d)$ *with* $s > 1 + d/2$*, there exists* $T > 0$ *and a unique solution in the class*

$$u \in C(0,T;H^s), \quad \partial_t u \in L^2(0,T;H^{s-1}).$$

In addition, the component v *belongs to* $C^1(0,T;H^{s-1})$ *and* $Z(u)\nabla z$ *is in* $L^2(0,T;H^s)$*.*

The local existence is actually proved for the more general class of system of the form (1.8).

2. Singular limits *vs* Onsager's relations.

2.1. Onsager's reciprocity relations for viscous conservation laws. In practice, the tensor Z in the normal form of the equations displays a symmetry property. For instance, in the case of the Navier-Stokes-Fourier system, we have

$$Z^{\alpha\beta}(u) =$$
$$\theta \begin{pmatrix} \mu(I_d + \mathbf{e}_\beta \mathbf{e}_\alpha^T) + (\zeta - 2\mu)\mathbf{e}_\alpha \mathbf{e}_\beta^T & \mu(v_\alpha \mathbf{e}_\beta + \delta_\alpha^\beta v) + (\zeta - 2\mu) v_\beta \mathbf{e}_\alpha \\ \mu(v_\beta \mathbf{e}_\alpha^T + \delta_\alpha^\beta v^T) + (\zeta - 2\mu) v_\alpha \mathbf{e}_\beta^T & \kappa\theta + \mu|v|^2 \delta_\alpha^\beta + (\zeta - \mu) v_\alpha v_\beta \end{pmatrix},$$

where the $\mathbf{e}_j$'s denote the vectors of the canonical basis of $\mathbb{R}^d$. We remark that these matrices satisfy

$$(Z^{\alpha\beta})^T = Z^{\beta\alpha}, \qquad \forall 1 \le \alpha, \beta \le d. \tag{2.1}$$

This symmetry is an example of Onsager's *reciprocity relations*, which originates in [5].

Onsager's relations (2.1) imply the symmetry of the symbol $Z(u;\xi)$ for every $\xi \in \mathbb{R}^d$. They are equivalent to saying that the linear map

$$F \mapsto G =: ZF, \qquad G_{\alpha i} := \sum_{\beta,j} Z_{ij}^{\alpha\beta} F_{\beta j}$$

is symmetric over $\mathbf{M}_{d\times(n-p)}(\mathbb{R})$, for the canonical scalar product $\langle F, G\rangle := \mathrm{Tr}(F^T G)$.

This property will be needed below. Let us remark that the optimal ellipticity constant ω_0 in

$$\langle F, ZF\rangle = \sum_{\alpha,\beta=1}^{d} \sum_{i,j\geq p+1} F_{\alpha i} F_{\beta j} Z_{ij}^{\alpha\beta} \geq \omega_0 \|ZF\|^2 \tag{2.2}$$

can be taken equal to the inverse of the norm of Z:

$$\omega_0(u) = \frac{1}{\|Z(u)\|} \tag{2.3}$$

in the symmetric case. This is easily seen by diagonalizing Z in an orthonormal basis of $\mathbf{M}_{d\times(n-p)}(\mathbb{R})$.

2.2. Stable families of dissipation tensors. We focus in this paper in the situation where the tensor B depends upon a small parameter, $B = B_\epsilon(u)$, and we let $\epsilon \to 0+$. When the rank of $B_\epsilon(\xi; u)$ drops at $\epsilon = 0$, we face a *singular limit*. Exemples of this situations are encountered frequently:

- In the vanishing viscosity limit, when $B_\epsilon = \epsilon B(u)$,
- In the Navier-Stokes-Fourier system, when the viscosity coefficients have the form $\epsilon\mu$ and $\epsilon\zeta$, while the heat conductivity κ remains independent of ϵ,
- In MHD, when either the viscosity coefficients, or the electric resistivity, or both have a factor ϵ,

An important question is whether the solution u^ϵ converges towards the solution u of the Cauchy problem for the limit problem, on some non-trivial time interval. This will occur if the *a priori* estimates given in Theorem 1.2 are uniform as $\epsilon \to 0$. To formulate our main result, we make the following definition.

DEFINITION 2.1. *The family $(B_\epsilon)_{\epsilon\in[0,1]}$ is* stable *if*

- *the range of $B_\epsilon(\xi; u)$ does not depend upon $\epsilon > 0$ (however it may, and does in general, be different when $\epsilon = 0$),*
- *$Z_\epsilon(u)$ is symmetric and the ellipticity parameter $\omega_0(u) > 0$ of Z_ϵ in (2.2) can be chosen independently of ϵ, and uniformly for u in compact subsets of $\mathcal{U}$,*
- *The kernel of $Z_\epsilon(u)$ does not depend upon $\epsilon > 0$ either,*
- *the partial derivatives of Z_ϵ are uniformly bounded in terms of Z_ϵ itself: For every multi-index ℓ of length less than s (s the regularity considered in Theorem 1.2)*

$$\|\partial_u^\ell Z_\epsilon(u)F\| \leq c_\ell(u)\|Z_\epsilon(u)F\|, \tag{2.4}$$

with c_ℓ independent of ϵ and bounded over compact sets of $\mathcal{U}$.

Discussion.

- In other words, we ask that Assumption (**B**) be satisfied uniformly in $\epsilon > 0$.

- Because of (2.3), the uniform ellipticity is equivalent to the uniform boundedness of $Z_\epsilon(u)$ for u in a compact and $\epsilon > 0$.
  This is satisfied in the case of the Navier-Stokes-Fourier system provided that the coefficients μ_ϵ, ζ_ϵ and κ_ϵ remain bounded. This is also true in the singular limit towards the Euler-Fourier system and in the vanishing viscosity limit, when $Z_\epsilon \equiv \epsilon Z$.
- Condition (2.4) amounts to saying that $\partial_u^\ell Z_\epsilon(u) Z_\epsilon(u)^\dagger$ remains bounded, where $Z^\dagger$ denotes the Moore-Penrose inverse. Since $Z(u)$ is symmetric and positive semi-definite, $Z(u)^\dagger$ coincides with the usual inverse on the range of $Z(u)$, and vanishes over $\ker Z(u)$. Because of the symmetry of $Z^\dagger$ and of $\partial_u^\ell Z_\epsilon(u)$, and the fact that the norm of operators is unchanged under transposition, this is equivalent to saying that

$$\|Z_\epsilon(u)^\dagger \partial_u^\ell Z_\epsilon(u)\| \leq c_\ell(u). \tag{2.5}$$

2.3. Uniformity of the existence time. With this in hand, we have

THEOREM 2.1. *Let us consider a system as in Theorem 1.2 with a viscous tensor parametrized by $\epsilon \in (0,1)$ and a flux f independent of ϵ. We make the block structure hypothesis* (**A**) *and we assume that the family $(B_\epsilon)_{\epsilon \in [0,1]}$ is stable in the sense defined above.*

Let u_0 in $H^s(\mathbb{R}^d)$ with $s > 1 + d/2$ be a given initial data, independent of ϵ. Let u^ϵ denote the solution obtained in Theorem 1.2. Then there exists $T > 0$ such that u^ϵ is defined over $(0,T)$ and the following sequences are bounded:

$$u^\epsilon \text{ in } C(0,T;H^s), \qquad \partial_t u^\epsilon \text{ in } L^2(0,T;H^{s-1})$$

and

$$v^\epsilon \text{ in } C^1(0,T;H^{s-1}), \qquad Z_\epsilon \nabla z^\epsilon \text{ in } L^2(0,T;H^s).$$

If in addition B_ϵ converges uniformly towards B as $\epsilon \to 0$, then u^ϵ converges towards the unique strong solution of the Cauchy problem associated to the viscous tensor B.

Comments.

- The stability assumption allows Z_ϵ to become singular at $\epsilon = 0$. For instance, we may have $Z_\epsilon = \epsilon Z$ (vanishing viscosity limit) and thus $Z_{\epsilon=0} = 0$. Therefore we do not expect that the components z^ϵ remain bounded in $L^2(0,T;H^{s+1})$.
- We do not discuss the behaviour of the *existence time*, defined as the maximal time for which the Cauchy problem has a solution in $C(0,T_{\max};H^s)$.

2.4. Proof of Theorem 2.1. The existence of a strong solution u^ϵ, respectively u in the limit problem, was proved in [8], where we obtained a positive lower bound T_ϵ (resp. T) of the existence time. This T_ϵ was that one for which we were able to get *a priori* estimates in the class described in Theorem 1.2. What we have to do here is to show that these estimates are uniform in ϵ. This will give us the first part of Theorem 2.1. The convergence follows from a classical compactness argument, plus the uniqueness of the strong solution of the limit problem.

We thus follow carefully the estimates in [8], Paragraph 3.1, and we examine where does they depend upon the dissipative tensor. We point out that since the existence has been proved yet, and because of the principle of continuation, we may work directly on the solutions u^ϵ, instead of dealing with approximate solutions. This means that we set $V = U$ in the estimates, simplifying a little bit the analysis.

The first estimate is obtained directly from the entropy inequality:

$$\int_{\mathbb{R}^d} \eta(u(T,x))\,dx + \int_0^T \int_{\mathbb{R}^d} \omega_0(u^\epsilon)|Z_\epsilon \nabla z^\epsilon|^2 dx\,dt \le \int_{\mathbb{R}^d} \eta(u_0(x))\,dx.$$

It gives a uniform bound of $Z_\epsilon \nabla z^\epsilon$ in $L^2((0,+\infty)\times\mathbb{R}^d)$, provided u^ϵ stays in some compact subset K of $\mathcal{U}$. This pointwise control is a part of the proof of Theorem 1.2, and we have to prove its uniformity in ϵ.

We now treat the higher order estimates. As in [8], we assume that $s = m > 1 + d/2$ is an integer. We point out that the change of variable $u \mapsto U$ does not depend upon ϵ, since the range of $B_\epsilon(\xi;u)$ is fixed for $\epsilon > 0$. In particular, the initial data U_0 is independent of ϵ. In the sequel, we employ the positive definite quadratic form

$$[[X]]^2 := X^T S_0(U^\epsilon) X,$$

which depends upon (x,t,ϵ) through the solution itself.

In order to estimate $\|\nabla_x^k U\|_{L^2}$, we denote by ∂ any spatial derivative of order $k \le m$ and apply ∂ to the equation (1.8). From now on, we adopt the convention of summation over repeated indices. Dropping the indices ϵ, we have

$$\begin{aligned} &S_0(U)\partial_t\partial U + S_\alpha(U)\partial_\alpha\partial U = \\ &\qquad \partial_\alpha(Y^{\alpha\beta}(U)\partial_\beta\partial U) + \partial_\alpha[\partial, Y^{\alpha\beta}(U)]\partial_\beta U + [S_\alpha(U),\partial]\partial_\alpha U \\ &\qquad +[S_0(U),\partial]\left(S_0(U)^{-1}\left\{\partial_\alpha(Y^{\alpha\beta}(U)\partial_\beta U) - S_\alpha(U)\partial_\alpha U\right\}\right). \end{aligned}$$

Multiplying scalarly by ∂U and using the symmetry of $S_0(U)$ and $S_\alpha(U)$, we derive

$$\begin{aligned} &\partial_t \frac{1}{2}[[\partial U]]^2 + \partial_\alpha\partial U \cdot Y^{\alpha\beta}\partial_\beta\partial U = \\ &\qquad \mathrm{div}_x(\cdots) + \frac{1}{2}\partial U(\partial_t S_0 + \partial_\alpha S_\alpha)\partial U - \partial_\alpha\partial U\cdot[\partial, Y^{\alpha\beta}]\partial_\beta U \qquad (2.6) \\ &\qquad +\partial U\cdot\left([S_\alpha,\partial]\partial_\alpha U + [S_0,\partial]\{S_0^{-1}(\partial_\alpha(Y^{\alpha\beta}\partial_\beta U) - S_\alpha\partial_\alpha U)\}\right), \end{aligned}$$

where we have pulled the dissipation rate in the left-hand side. We therefore deduce an inequality

$$\partial_t \frac{1}{2}[[\partial U]]^2 + \omega(K)|Z\nabla\partial z|^2 \leq \mathrm{div}_x(\cdots) + Q^{t\partial} + Q^{1\partial} + Q^{2\partial}, \tag{2.7}$$

with

$$Q^{t\partial} := \frac{1}{2}\partial U(\partial_t S_0)\partial U,$$
$$Q^{1\partial} := \frac{1}{2}\partial U(\partial_\alpha S_\alpha)\partial U + \partial U\cdot[S_\alpha,\partial]\partial_\alpha U - \partial U\cdot[S_0,\partial]\left(S_0^{-1}S_\alpha\partial_\alpha U\right)$$

and

$$Q^{2\partial} := -\partial_\alpha\partial U\cdot[\partial, Y^{\alpha\beta}]\partial_\beta U + \partial U\cdot[S_0,\partial]\left(S_0^{-1}\partial_\alpha(Y^{\alpha\beta}\partial_\beta U)\right).$$

Only $Q^{2\partial}$ involves the tensor $Y = Y_\epsilon$ and thus could be sensitive to ϵ. We estimate its integral in terms of the dissipation rate $\|Z\nabla^{k+1}z\|_{L^2}$ and of $\|u\|_{H^m}$. Because of the block form of S_0 and Y, it rewrites in terms of derivatives of z only. In compact form, we have

$$Q^{2\partial} := -\nabla\partial z\cdot[\partial, Z]\nabla z + \partial z\cdot[s_0,\partial]\left(s_0^{-1}\mathrm{Div}(Z\nabla z)\right),$$

where $s_0 := (\mathrm{D}^2_{ww}\eta)^{-1}$ is the second diagonal block in S_0.

To begin with, we isolate one factor $\nabla^{k+1}z$ when possible, since it can be absorbed by the left-hand side of (2.7), *via* the Young inequality. On the one hand, we have

$$|\nabla\partial z\cdot[\partial,Z]\nabla z| = |Z\nabla\partial z\cdot Z^\dagger[\partial,Z]\nabla z| \leq \frac{\omega}{4}|Z\nabla\partial z|^2 + \frac{1}{\omega}\left|Z^\dagger[\partial,Z]\nabla z\right|^2.$$

According to Faà di Bruno's Formula, the expression $Z^\dagger[\partial,Z]\nabla z$ is a polynomial in the differentials of ∇z up to order $k-1$, whose coefficients are of the form $Z^\dagger\partial_u^\ell Z$ with $1\leq|\ell|\leq k$. Because of (2.5), these coefficients are bounded as long as u^ϵ remains in a compact set K, *independently* of ϵ. Using Moser's inequalities (see [1], Chapter 10 and Appendix C for a description), and the fact that $1/\omega$ remains bounded over K, we therefore obtain

$$\int_{\mathbb{R}^d}\frac{1}{\omega}\left|Z^\dagger[\partial,Z]\nabla z\right|^2 \leq c(K)\|U(t)\|^4_{H^m},$$

where $c(K)$ is independent from ϵ.

The remaining term in the integral of $Q^{2\partial}$ is treating mainly as in [8]. This again involves Moser's type inequalities:

$$\begin{aligned}
&\int_{\mathbb{R}^d} \partial z \cdot [s_0, \partial] \left(s_0^{-1}\mathrm{Div}(Z\nabla z)\right) dx \\
&\quad \leq \|\partial z\|_{L^2} \|[s_0,\partial]\left(s_0^{-1}\mathrm{Div}(Z\nabla z)\right)\|_{L^2} \\
&\quad \leq c(K)\|\nabla s_0\|_{H^{m-1}}\|s_0^{-1}\|_{H^{m-1}}\|Z\nabla z\|_{H^m}[[\partial U]] \\
&\quad \leq \frac{\omega}{4m}\|Z\nabla z\|_{H^m}^2 + \frac{mc(K)^2}{\omega}\|\nabla s_0\|_{H^{m-1}}^2\|s_0^{-1}\|_{H^{m-1}}^2[[\partial U]]^2 \\
&\quad \leq \frac{\omega}{4m}\|Z\nabla z\|_{H^m}^2 + \frac{c'(K)}{\omega}\|U\|_{H^m}^6[[\partial U]]^2.
\end{aligned}$$

This is completed with

$$\begin{aligned}
\frac{1}{2}\|\nabla^m(Z\nabla z)\|_{L^2}^2 &\leq \|Z\nabla\nabla^m z\|_{L^2}^2 + \|[\nabla^m, Z]\nabla z\|_{L^2}^2 \\
&\leq \|Z\nabla\nabla^m z\|_{L^2}^2 + \|U\|_{H^m}^4,
\end{aligned}$$

where the first term will be absorbed by the dissipation after summing over k.

Finally, we obtained the same inequality as (15) in [8]:

$$\frac{dY}{dt} + \frac{\omega(K)}{2m}\|\nabla z\|_{H^m}^2 \leq c(K)\left(\|\partial_t U(t)\|_{H^{m-1}} + \|U(t)\|_{H^m}^2 + \|U(t)\|_{H^m}^6\right)Y,$$

where the coefficients *do not depend* upon ϵ, and

$$Y(t) := \int_{\mathbb{R}^d} \sum_{|\ell|\leq m} [[\partial^\ell U]]^2 dx.$$

We now close the loop as in Paragraph 3.2 of [8]. The only place where some care is needed is where we estimate $\partial_t U$ in $L^2(0,T;H^{m-1})$, since it involves explicitly $Z\nabla z$. However, the latter is controlled by (25), which writes here

$$\|Z\nabla z\|_{L^2(0,T;H^m)} \leq C(K)R_0 \exp\frac{c_\rho}{2}\left\{(R_m^2 + R_m^6)T + R_{m-1}\sqrt{T}\right\}. \quad (2.8)$$

This, together with

$$\partial_t U = S_0(V)^{-1}\left(\partial_\alpha(Y^{\alpha\beta}(V)\partial_\beta U) - S_\alpha(V)\partial_\alpha U\right)$$

yields (26). Once again, both estimates above are ϵ-free. The rest is as in [8] and yields the uniform estimates mentionned in our Theorem 2.1. This ends the proof of the theorem.

REFERENCES

[1] S. Benzoni-Gavage and D. Serre. *Multi-dimensional hyperbolic partial differential equations. First order systems and applications.* Oxford Mathematical Monographs, Oxford University Press (2007).

[2] G. Billet, V. Giovangigli, and G. de Gassowski. *Impact of volume viscosity on a shock-hydrogen-bubble interaction. Combustion Theory and Modelling,* **12** (2008), pp. 221–248.

[3] C. Dafermos and L. Hsiao. *Development of singularities in solutions of the equations of nonlinear thermoelasticity. Quarterly of applied mathematics,* **44** (1986), pp. 463–474.

[4] S. Kawashima. *Systems of a hyperbolic parabolic type with applications to the equations of magnetohydrodynamics.* PhD Thesis, Kyoto University (1983).

[5] L. Onsager. *Reciprocal relations in irreversible processes.* I. *Phys. Rev.,* **37** (1931), pp. 405–426.

[6] D. Serre. *Sur l'équation monodimensionnelle d'un fluide visqueux, compressible et conducteur de chaleur. C. R. Acad. Sci. Paris Sér. I Math.,* **303** (1986), pp. 703–706.

[7] D. Serre. *The structure of dissipative viscous system of conservation laws. Physica* **D 239** (2010), pp. 1381–1386.

[8] D. Serre. *Local existence for viscous system of conservation laws: H^s-data with $s > 1 + d/2$. Proceedings of C.A.S. program on PDEs, Oslo 2009.* H. Holden & K. Kalrsen eds. Contemporary Mathematics, Vol. **526**. American Mathematical Society, Providence, RI., pp. 339–358.

A TWO-DIMENSIONAL RIEMANN PROBLEM FOR SCALAR CONSERVATION LAWS

HAO YING* AND BARBARA LEE KEYFITZ†

Abstract. This paper provides an introduction to current research in self-similar problems for multidimensional conservation laws. This is presently an active area. To set the stage, we begin by solving a classic problem. We prove that the solution of the scalar equation $u_t + f(u)_x + g(u)_y = 0$ with Riemann initial data of the form $u(0, x, y) = u_0(\theta)$ $(0 \leq \theta \leq 2\pi)$ remains smooth outside a circle with center at the origin in the self-similar plane. Based on this approach to Riemann problems, and on recent research by a number of contributors, we give a list of open problems.

1. Introduction.

The theory of well-posedness for conservation laws in more than one space variable is still in its infancy. The Cauchy problem for a scalar equation is well-posed [9, 10], but scalar equations are not a good guide to the complexities of wave interactions for systems. And although one might expect well-posedness for systems modeling physics, on the grounds that the models have been thoroughly tested by numerical simulation and experimental verification, other points of view have been expressed in this workshop.

Lack of progress on general questions of existence has motivated the study of prototype cases, specifically self-similar data (two-dimensional Riemann problems) for systems inspired by physics, such as the compressible gas dynamics equations and related models.

This report has two parts. First, we present a new theorem on self-similar solutions. At the moment, it is proved only for the scalar case, but some generalization to systems is likely. The second half of the paper is a list – incomplete, but we hope useful – of open questions in this area.

2. What is a two-dimensional Riemann problem?

Our theorem provides a partial answer, and suggests a conjecture on the complete answer, to a question posed by Martin Kruskal [15] many years ago: What is the most general "two-dimensional Riemann problem", and what would its solution look like? Kruskal's point was that a two-dimensional system,

$$U_t + F(U)_x + G(U)_y = 0\,, \tag{2.1}$$

*Department of Mathematics, The Ohio State University, Columbus, OH 43210.
Research partially supported by the Department of Energy, Grant DE-SC0001285. Both authors thank the Institute for Mathematics and Its Applications for its hospitality during the 2009 Summer Program, 'Nonlinear Conservation Laws and Applications', where some of this research was completed.

†Department of Mathematics, The Ohio State University, Columbus, OH 43210.
Research partially supported by the Department of Energy, Grant DE-SC0001285, the National Science Foundation, Grant DMS-0968254.

with $U = U(t, x, y) : R^3 \to R^n$, and F, $G : R^n \to R^n$, ought to have a self-similar solution,

$$U(t, x, y) = V(\xi, \eta) = V\left(\frac{x}{t}, \frac{y}{t}\right),$$

for any data of the form

$$U(0, x, y) = U_0(\theta), \tag{2.2}$$

where θ is defined by $\cos\theta = x/(\sqrt{x^2 + y^2})$, $\sin\theta = y/(\sqrt{x^2 + y^2})$.

However, as is well-known, the standard conception of a two-dimensional Riemann problem posits data that is constant in quadrants; see, for example, [26]. Work of the second author of this paper, initiated with Sunčica Čanić in [3], following the lead of Brio and Hunter [2] and Tabak and Rosales [20], and since extended by Chen and Feldman [7] and many others, revised the notion to include sectorially constant data. To the best of our knowledge, no one has examined the 'Kruskal problem', with data (2.2).

To this end, we here consider scalar conservation laws in two dimensions:

$$u_t + f(u)_x + g(u)_y = 0 \tag{2.3}$$

with $f(u)$, $g(u) \in C^3(R^1)$ and the Riemann data

$$u(0, x, y) = u_0(\theta), \ (0 \le \theta \le 2\pi) \tag{2.4}$$

where θ is a polar angle in the (x, y)-plane and $u_0 \in C^2(R^1)$.

Since $u_0 \in C^2(R^1)$, $u(0, x, y)$ must be a bounded function having locally bounded variation in the sense of Tonelli-Cesari. Therefore, we can apply Conway and Smoller's results, [9, 10], to our case. So our problem (2.3) with initial data (2.4) has a unique solution. The solution of this problem must be self-similar; that is, $u = u(\xi, \eta)$, where $(\xi, \eta) = (x/t, y/t)$, since (2.3) and (2.4) are invariant under the dilation $(t, x, y) \to (ct, cx, cy)(c > 0)$.

Just as for problems in one space dimension, we expect shocks to form after a finite time, and the interactions of shocks and rarefaction waves may be quite complicated. Many people have tried to construct entropy solutions of this problem and to reveal the complicated structure of those interactions. Zhang and Zheng [25] constructed the exact solution for this problem with f and g satisfying some conditions and with four-constant-state Riemann data. And Wagner [24] constructed the entropy solution in the sense of Kružkov [16] when f and g are convex and $f \equiv g$.

In the present paper, we prove that the unique solution of (2.3) and (2.4) remains C^1 smooth outside a circle with finite radius in the self-similar plane.

2.1. Characteristics. As we know, the characteristic equation for the partial differential equation (2.3) is

$$\begin{cases} \dot{x}(t) = f'(u) \\ \dot{y}(t) = g'(u) . \end{cases} \tag{2.5}$$

So, along the characteristics,

$$\frac{d}{dt}u(t, x(t), y(t)) = u_t + f'(u)u_x + g'(u)u_y = 0,$$

when $u \in C^1$. Therefore, u is constant along the characteristics when $u \in C^1$.

When $u \in C^1$, so u is constant on each characteristic, we consider how u_x, u_y evolve along characteristics. We differentiate $u_x(t, x(t), y(t))$ and $u_y(t, x(t), y(t))$:

$$\begin{cases} \dot{u_x}(t) = u_{xt} + u_{xx}\dot{x} + u_{xy}\dot{y} \\ \dot{u_y}(t) = u_{yt} + u_{yx}\dot{x} + u_{yy}\dot{y} \end{cases} \tag{2.6}$$

We also differentiate (2.3) with respect to x and y:

$$\begin{cases} u_{tx} + f'(u)u_{xx} + g'(u)u_{yx} + (f''(u)u_x + g''(u)u_y)u_x = 0 \\ u_{ty} + f'(u)u_{xy} + g'(u)u_{yy} + (f''(u)u_x + g''(u)u_y)u_y = 0 . \end{cases} \tag{2.7}$$

Combining (2.5), (2.6) and (2.7), we get a system of ordinary differential equations for u_x and u_y:

$$\begin{cases} \dot{u_x}(t) = -(f''(u)u_x + g''(u)u_y)u_x \\ \dot{u_y}(t) = -(f''(u)u_x + g''(u)u_y)u_y . \end{cases} \tag{2.8}$$

When we solve this system, we get

$$\begin{cases} u_x(t) = \dfrac{\partial_x u_0}{[\partial_x u_0 f''(u_0) + \partial_y u_0 g''(u_0)]t + 1} \\ u_y(t) = \dfrac{\partial_y u_0}{[\partial_x u_0 f''(u_0) + \partial_y u_0 g''(u_0)]t + 1} , \end{cases} \tag{2.9}$$

where $u_0 = u(0, x(0), y(0))$ is the constant value of u along the characteristic $(t, x(t), y(t))$.

2.2. Region of smooth solutions. Now we can state and prove our main result:

THEOREM 2.1. *For problem* (2.3) *with initial data* (2.4), *there exists an* R *such that in the self-similar plane, the solution of* (2.3) *and* (2.4) *is* C^1 *smooth outside a circle of radius* R *with center at the origin.*

Proof. Since $u_0(\theta) \in C^1$, there exist M_0 such that $|u_0| \le M_0$ and C such that $|u_0'| \le C$.

We consider a point on a circle with radius r and center at the origin. Since u_0 is independent of r, we have (with $'$ denoting the derivative with respect to θ)

$$\partial_x u_0 = -\frac{1}{r} u_0' \sin\theta, \quad \partial_y u_0 = \frac{1}{r} u_0' \cos\theta.$$

Hence

$$|\partial_x u_0| \le \frac{C}{r}, \quad |\partial_y u_0| \le \frac{C}{r}.$$

In the introduction section we explained that our problem satisfies the conditions required in [9, 10], so we know the solution u is also bounded by M_0.

Hence, there exist constants M_1 and M_2 such that

$$M_1 = \max(|f'(u)|, |g'(u)|), \quad M_2 = \max(|f''(u), |g''(u)|),$$

for the solution u of (2.3) and (2.4).

We know that when a shock forms, u_x or u_y blows up. Now we try to estimate the minimum time for u_x and u_y to blow up.

From equations (2.9), we know that along a characteristic, if $\partial_x u_0 f''(u) + \partial_y u_0 g''(u)$ is negative, u_x and u_y will blow up together at time

$$t = -\frac{1}{\partial_x u_0 f''(u_0) + \partial_y u_0 g''(u_0)}.$$

Thus the smallest possible time will be no less than

$$\begin{aligned} \min\left|\frac{1}{\partial_x u_0 f''(u_0) + \partial_y u_0 g''(u_0)}\right| &= \frac{1}{\max|\partial_x u_0 f''(u_0) + \partial_y u_0 g''(u_0)|} \\ &\le \frac{1}{\frac{C}{r}M_2 + \frac{C}{r}M_2} = \frac{r}{2CM_2}. \end{aligned}$$

Thus, for any time t, new shocks can form only inside of the circle with radius $2CM_2 t$ and center at the origin.

According to the Rankine-Hugoniot relation, once a shock forms, it will travel at the speed

$$\sqrt{\left(\frac{f(u^+) - f(u^-)}{u^+ - u^-}\right)^2 + \left(\frac{g(u^+) - g(u^-)}{u^+ - u^-}\right)^2}$$

in the $x - y$ plane. So the speed is no more than $\sqrt{2}M_1$.

Hence, if we let $R = \max(2CM_2, \sqrt{2}M_1)$, then for any time t, all shocks will be inside the circle with radius Rt and center at the origin, which means that, in the self-similar plane, all shocks must be inside the circle with radius R and center at the origin. □

2.3. Further observations. Our result shows that in contrast to the solution of (2.3) with sectorially constant (but discontinuous) initial data, the solution to our problem has shocks only in a compact set. Of course, it is also possible to consider for data a function $u_0(\theta)$ that is piecewise smooth, with simple jump discontinuities. The solution will contain a set of shocks arising from the discontinuities in the data, interacting with another collection forming inside the circle of radius R as constructed in Theorem 2.1. As is well known, a scalar multi-dimensional conservation law cannot be genuinely nonlinear in all directions, and so the structure of the shocks may be quite complicated. Typically, shock/rarefaction fans will appear; see Zheng's book [27] for an example. We do not attempt to analyse the detailed shock structure in this paper.

For systems of two-dimensional conservation laws with similar initial data and conditions, we would expect a similar result to be true. In this case, there is no global existence result for weak solutions, so the strongest result that one could prove at the present time would be that classical solutions exist outside a circle of radius R when the data are sufficiently smooth and have bounded derivatives as in Theorem 2.1. Even here, an estimate of R would be more complicated, since for systems, unlike for a scalar equation, the data do not provide a priori pointwise bounds on the solution (even for classical solutions) and hence there are no estimates of the maximum speed of propagation of signals.

There is a sense in which the result, Theorem 2.1, is unsurprising. If one thinks of the one-dimensional Riemann problem as the problem of 'propagation of a point discontinuity', rather than as a 'resolution of piecewise constant data', then the Kruskal problem is the natural generalization to higher dimensions. But for any hyperbolic problem, the standard property of finite speed of propagation then suggests strongly that discontinuities in the solution will be confined to a compact set for any finite time.

We can relate this observation to the vision of a multi-dimensional Riemann problem. It was pointed out by Tai-Ping Liu in his Georgia Tech CBMS lectures [17] that the Riemann problem in one space dimension enjoys unique features: It describes wave propagation in the small (that is, it is a good template for solving problems locally), and also in the large (in that it provides a first asymptotic approximation to long-time global behavior of solutions). It would seem unlikely, according to Liu, that multi-dimensional Riemann problems would possess such universality. And now, as the example of the Riemann problem with Kruskal data shows, there does not seem to be a single type of problem in more than one space dimension. Rather, the different features that characterize Riemann problems in one space dimension may lead to different types of problems in two dimensions.

3. Challenges and open problems. The second part of the report provides a list of open problems. Some are related to the problem of Section 2; others, equally interesting, are independent of it. In particular, some are related to Keyfitz's talk at the IMA Conservation Laws Summer Program, [14].

1. It should be possible to generalize Theorem 2.1 to systems, using the H^s theory for short-time existence for quasilinear hyperbolic systems (see, for example, [13], [18, Ch 2], [21, page 360 ff]) to replace the Conley-Smoller theory that forms the basis of the scalar equation result. However, the lack of any existence theory for weak solutions of multidimensional systems, and the absence of a priori bounds for weak solutions constitute challenges. One approach would be to limit, by natural or artificial constraints, the speed of propagation of waves in the solution.
2. Granting that a system (2.1) with smooth or piecewise smooth Kruskal data (2.2) has a smooth or piecewise smooth solution outside some cone $x^2 + y^2 = R^2t^2$, one might adapt the classical (Glimm or Bressan) theory of hyperbolic systems in two independent variables to prove existence of a self-similar weak solution in a cone $r^2t^2 \leq x^2 + y^2 \leq R^2t^2$. One would need suitable structure conditions on F and G, to ensure strict hyperbolicity and genuine nonlinearity or linear degeneracy in each family, and a smallness assumption on the data U_0. The front-tracking construction would break down when hyperbolicy was lost, as one would expect to happen at some radius r for systems that resemble gas dynamics. Thus, currently available techniques would not provide a global solution, even for small data.
3. The study of how change of type occurs in the situation described just above, and what is the nature of the 'sonic line' that appears when it does, is of some interest in itself. (The paper [4] explains why it occurs in gas dynamics.) For a class of hyperbolic systems (2.1) with simple change of type properties (or for more general systems), it would be interesting to characterize the change of type locus, for smooth data $U_0(\theta)$ or for sectorially constant data.
4. One problem that is currently being studied by a number of researchers is to continue to solve wave interaction problems beyond the regular shock reflection problems currently being solved. For example, Yuxi Zheng has looked at rarefactions [28], and Čanić, Keyfitz and Kim at a model for a Mach stem [5]. The solution in [5] is incomplete, as the reflected shock, which would be another free boundary, is not modeled correctly. It seems likely that this reflection problem cannot be completed without a better understanding of weak shock reflection and Guderley Mach reflection discussed in topic 5 below. At this time, there are local results on

the unsteady small disturbance equation and the nonlinear wave system, for regular reflection (weak and strong), and global results for the steady and unsteady full potential equations, due to Chen and Feldman and co-authors [1, 8] and to Elling and Liu [11] – we do not attempt to give a complete survey here – but these prototype problems are only a sample of the many possibilities for equations and types of data. Recently, Jegdic has made progress on transonic regular reflection in the (non-potential) gas dynamics equations [12]. The methods needed to solve many of these problems appear to be beyond our grasp at the moment. Although it is not clear that the route to understanding multidimensional conservation laws lies through enlarging the catalogue of such examples, at the moment both analysis and numerical simulations are bringing new insights.

5. In the talk given at the IMA Summer Program, Keyfitz described a project with co-authors Tesdall, Payne and Popivanov [14], to shed light on the nature of the sonic line for a continuous solution of self-similar systems. Motivated by the phenomenon of Guderley Mach reflection, discovered first in the unsteady transonic small disturbance equation [22] and recently found in the gas dynamics equations [23] – so far only in numerical simulations – Keyfitz and co-authors have examined the transition between supersonic and subsonic flow in the steady transonic small disturbance equations, by constructing transitions that are perturbations of a straight-line sonic boundary. The ultimate goal of this research is an analysis of Guderley Mach reflection. Meanwhile, the formulation and solution of other classes of sonic line problems would be interesting.
6. Problems in steady transonic flow can be seen as a special case of two-dimensional self-similar problems. In fact, the free-boundary methods developed by Čanić, Keyfitz et al, and the related potential method of Chen, Feldman et al, were first achieved on steady problems, in [6] and [7] respectively. There remain many open problems in steady transonic flow. Here the program was begun, and further challenges laid out, by Cathleen Morawetz, [19].
7. When a number of groups began the analysis of two-dimensional Riemann problems, about two decades ago, a significant motivation was to find tractable problems in the face of our failure to make progress on multidimensional problems. At this point, it might be fruitful to understand the relationship between self-similar and general problems. Theorem 2.1 in this paper, and its generalization to systems (which seems likely to be true in some form), suggest a different way of looking at two-dimensional Riemann problems.

To repeat the first sentence of this paper, the field of multidimensional conservation laws is still in its infancy. The results that have been achieved, both in the references and the new result reported in this paper, are best thought of as suggestions for further research. An imaginative student is encouraged to read them in that spirit.

REFERENCES

[1] M. Bae, G.-Q. Chen, and M. Feldman. *Regularity of solutions to regular shock reflection for potential flow. Inventiones Mathematicae*, **175**:505–543, 2009.

[2] M. Brio and J.K. Hunter. *Mach reflection for the two dimensional Burgers equation. Physica D*, **60**:194–207, 1992.

[3] S. Čanić and B.L. Keyfitz. *An elliptic problem arising from the unsteady transonic small disturbance equation. Journal of Differential Equations*, **125**:548–574, 1996.

[4] S. Čanić and B.L. Keyfitz. *A useful class of two-dimensional conservation laws.* In K. Kirchgässner, O. Mahrenholtz, and R. Mennicken, editors, *Proceedings of ICIAM 95: Supplement 2: Applied Analysis; Mathematical Research*, Vol. **87**, pages 133–136. Akademie Verlag, Berlin, 1996.

[5] S. Čanić, B.L. Keyfitz, and E.H. Kim. *Free boundary problems for nonlinear wave systems: Mach stems for interacting shocks. SIAM Journal on Mathematical Analysis*, **37**:1947–1977, 2005.

[6] S. Čanić, B.L. Keyfitz, and G.M. Lieberman. *A proof of existence of perturbed steady transonic shocks via a free boundary problem. Communications on Pure and Applied Mathematics*, **LIII**:1–28, 2000.

[7] G.-Q. Chen and M. Feldman. *Multidimensional transonic shocks and free boundary problems for nonlinear equations of mixed type. Journal of the American Mathematical Society*, **16**:461–494, 2003.

[8] G.-Q. Chen and M. Feldman. *Existence and stability of multidimensional transonic flows through an infinite nozzle of arbitrary cross-sections. Archive for Rational Mechanics and Analysis*, **184**:185–242, 2007.

[9] E.D. Conway and J.A. Smoller. *Global solution of the Cauchy problem for quasi-linear first order equations in several space variables. Communications on Pure and Applied Mathematics*, **19**:95–105, 1966.

[10] E.D. Conway and J.A. Smoller. *Uniqueness and stability theorem for the generalized solution of the initial-value problem for a class of quasi-linear equations in several space variables. Archive for Rational Mechanics and Analysis*, **23**:399–408, 1966.

[11] V. Elling and T.-P. Liu. *Supersonic flow onto a solid wedge. Communications on Pure and Applied Mathematics*, **61**:1347–1448, 2008.

[12] K. Jegdić, B.L. Keyfitz, and S. Čanić. *A free boundary problem for the isentropic gas dynamics equations — transonic regular reflection. In preparation*, 2010.

[13] T. Kato. *The Cauchy problem for quasi-linear symmetric hyperbolic systems. Archive for Rational Mechanics and Analysis*, **58**:181–205, 1975.

[14] B.L. Keyfitz, A. Tesdall, K.R. Payne, and N.I. Popivanov. *The sonic line as a free boundary. In preparation*, 2010.

[15] M. Kruskal. *Private communication.*

[16] S.N. Kružkov. *First-order quasilinear equations in several independent variables. Mathematics of the USSR – Sbornik*, **10**:217–243, 1970.

[17] T.-P. Liu. *Hyperbolic and Viscous Conservation Laws.* CBMS-NSF Regional Conference Series in Applied Mathematics, **72**, SIAM, Philadelphia, 2000.

[18] A. Majda. *Compressible Fluid Flow and Systems of Conservation Laws in Several Space Variables.* Springer-Verlag, New York, 1984.

[19] C.S. Morawetz. *On a weak solution for a transonic flow problem. Communications on Pure and Applied Mathematics*, **XXXVIII**:797–817, 1985.

[20] E.G. Tabak and R.R. Rosales. *Focusing of weak shock waves and the von Neumann paradox of oblique shock reflection. Physics of Fluids A*, **6**:1874–1892, 1994.

[21] M.E. Taylor. *Partial Differential Equations III: Nonlinear Equations.* Springer, New York, 1996.

[22] A.M. Tesdall and J.K. Hunter. *Self-similar solutions for weak shock reflection. SIAM Journal on Applied Mathematics*, **63**:42–61, 2002.

[23] A.M. Tesdall, R. Sanders, and B.L. Keyfitz. *Self-similar solutions for the triple point paradox in gasdynamics. SIAM Journal on Applied Mathematics*, **68**:1360–1377, 2008.

[24] D.H. Wagner. *The Riemann problem in two space dimensions for a single conservation law. SIAM Journal on Mathematical Analysis*, **14**:534–559, 1983.

[25] T. Zhang and Y.-X. Zheng. *Two-dimensional Riemann problem for a scalar conservation law. Transactions of the American Mathematical Society*, **312**:589–619, 1989.

[26] T. Zhang and Y.-X. Zheng. *Conjecture on the structure of solutions of the Riemann problem for two-dimensional gas dynamics systems. SIAM Journal on Mathematical Analysis*, **21**:593–630, 1990.

[27] Y. Zheng. *Systems of Conservation Laws: Two-Dimensional Riemann Problems.* Birkhäuser, Boston, 2001.

[28] Y. Zheng. *A global solution to a two-dimensional Riemann problem involving shocks as free boundaries. Acta Mathematicae Applicatae Sinica, English Series*, **19**:559–572, 2003.

SEMI-HYPERBOLIC WAVES IN TWO-DIMENSIONAL COMPRESSIBLE EULER SYSTEMS

YUXI ZHENG*

Abstract. We describe a patch in the self-similar plane as a semi-hyperbolic wave which is locally hyperbolic, but not all characteristics in the patch leads to given boundary data, for the two-dimensional compressible Euler system of equations.

Key words. Bi-symmetric 4R interaction, characteristic decomposition, compressible, gas dynamics, Guderley reflection, patches, simple waves, transonic flow, two-dimensional Riemann problem.

AMS(MOS) subject classifications. Primary 35L65, 35J70, 35R35.

1. Introduction. We consider the **Euler system of equations**

$$\begin{cases} \rho_t + \nabla \cdot (\rho \mathbf{u}) = 0, \\ (\rho \mathbf{u})_t + \nabla \cdot (\rho \mathbf{u} \otimes \mathbf{u} + pI) = 0, \\ (\rho E)_t + \nabla \cdot (\rho E \mathbf{u} + p\mathbf{u}) = 0, \end{cases} \tag{1.1}$$

where ρ is the density, $\mathbf{u}$ is the velocity vector, p is the pressure, and $E = \frac{1}{2}|\mathbf{u}|^2 + \frac{p}{(\gamma-1)\rho}$ is the energy. The gas constant is $\gamma > 1$.

Twenty years ago, in 1989, at IMA, an earlier generation of us had a similar meeting devoted to multi-dimensional systems of conservation laws [4]. Progresses were clearly made over their earlier generation dated back to Richard Courant, von Neuman, et al., and new doable problems were proposed, which inspired and motivated many of us here, including me. As a student participant of the 1989 conference and a speaker of the current conference, I could not help but ponder these questions:

1. What have we achieved since then? What do we expect to achieve in the next 5–20 years?
2. What are the important issues? What are the basic tools needed?
3. Where are the good entry points to get into the field?

It is not easy to answer these questions to the satisfaction of many of us. I shall not attempt to do that here. Let me try, however, to present what I feel comfortable to do: To present my own or collaborators' recent work.

2. A two-dimensional Riemann problem. We consider two dimensional Riemann problems: That is, the initial data are constant in each of the four quadrants of the initial (x, y) plane. We seek solutions that depend on the self-similar variables: $\xi = \frac{x}{t}, \quad \eta = \frac{y}{t}$. We focus on a

*Department of Mathematics, The Pennsylvania State University, UP, PA 16802.
Research supported in part by NSF grant DMS-0908207.

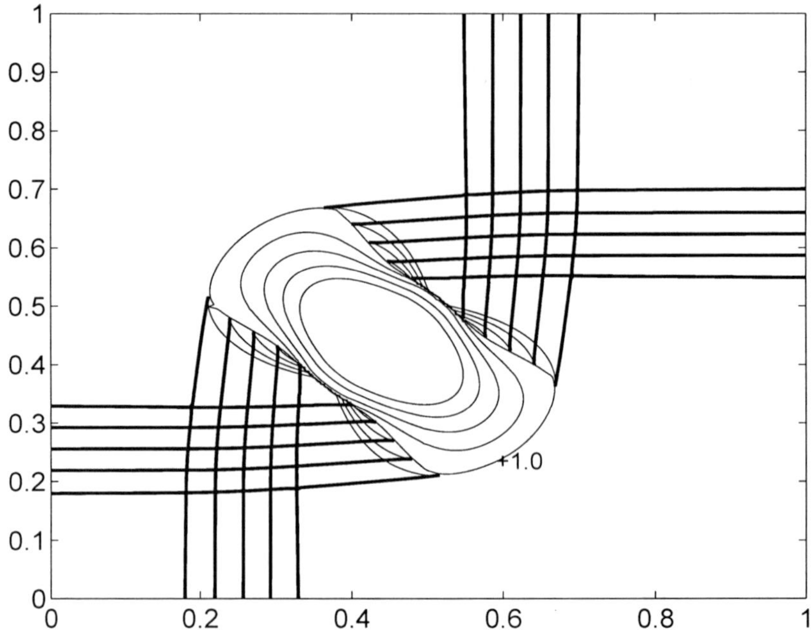

FIG. 1. *Interaction of two forward and two backward rarefaction waves with* $p_1 = 0.444, \rho_1 = 1.0, u_1 = v_1 = 0.00, \rho_2 = 0.5197$, *T (time)*$= 0.25$. *The five closed curves at the center covering the pseudo subsonic region are contour curves of the pseudo-Mach number. The twenty thick curves are characteristics, as are the sixteen short thin curves outside the closed sonic curve with pseudo Mach number 1.0. The data are centered at* $(0.5, 0.5)$ *rather than* $(0, 0)$ *in this figure. Each of the four regions with overlapping heavy and light curves is a semi-hyperbolic patch. Courtesy of SIAM: Glimm, et al. [3].*

basic case, called *bi-symmetric 4R interaction*, which consists of the interaction of two forward and two backward rarefaction waves. A numerical solution in the self-similar (ξ, η) plane from Glimm et. al. [3] is shown in Fig. 1, where characteristics and Mach number contour lines are shown.

Numerical simulations on the two-dimensional Riemann problems are performed by Chang et. al. [1], Schulz-Rinne et. al. [12], Lax and Liu [6], Kurganov and Tadmore [5] and Glimm et. al. [3]. In addition to the trivial regions of constant states and the complicated subsonic region, we identify three interesting regions:

1. Planar wave interactions;
2. Simple waves;
3. Semi-hyperbolic waves.

To handle them, we identify three main approaches:

1. Hodograph transform; (Omit it in this talk/writing, but see [9].)
2. Direct characteristic decomposition;
3. Selection of variables.

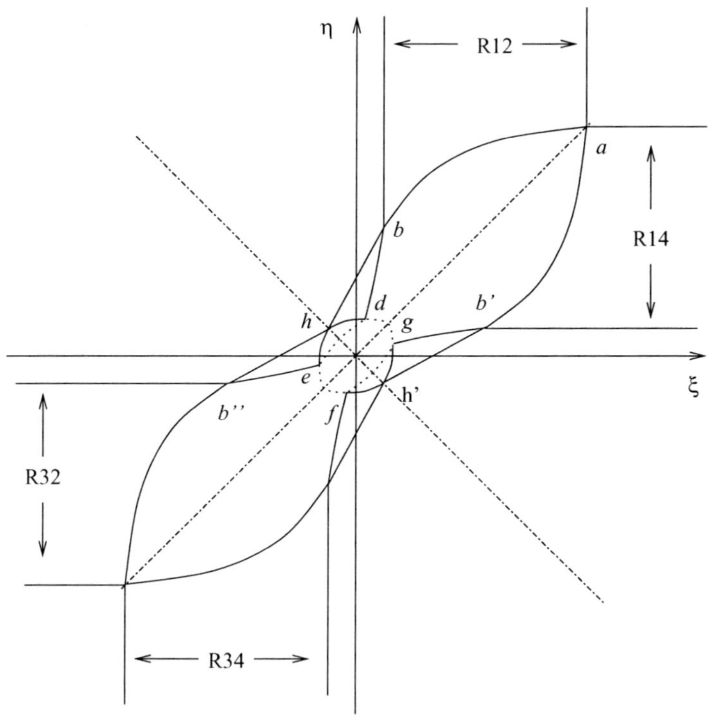

FIG. 2. *A global solution for bi-symmetric 4R case. The four rarefaction waves are R12, R32, R34, and R14. The two dash-dotted lines are lines of symmetry. The central region surrounded by the dotted curve defg is vacuum. The region surrounded by abdgb' is the interaction of R12 and R14. Region bhd is a simple wave while region dhe is the interaction of two simple waves.*

3. Binary interaction of planar waves. Let us report a global existence result. Let sound speed c be such that $c^2 = \gamma p/\rho$.

THEOREM 1 (Bi-symmetric 4 rarefaction Riemann problem, Li and Zheng [10]). *Consider the Riemann problem for the Euler system with initial data consisting of constant states (c_i, u_i, v_i) in the i-th quadrants $(i = 1, 2, 3, 4)$ so that states 1 and 2 form a forward rarefaction wave R_{12}^+, states 2 and 3 form a backward rarefaction wave R_{23}^-, states 3 and 4 form a forward rarefaction wave R_{34}^+, and states 4 and 1 form a backward rarefaction wave R_{41}^-. Then, this Riemann problem has global continuous solutions, cf. Fig. 2, provided $0 < c_2 < c_2^*(\gamma)c_1$, where $c_2^*(\gamma) \in (0,1)$ for $\gamma > 1 + \sqrt{2}$.*

Key steps of the proof of the theorem are the introduction of the *inclination angles of characteristics* (α, β): $\tan\alpha = \Lambda_+;\quad \tan\beta = \Lambda_-$ where

$$\Lambda_{\pm} = \frac{(u-\xi)(v-\eta) \pm c\sqrt{(u-\xi)^2+(v-\eta)^2-c^2}}{(u-\xi)^2-c^2} \tag{3.1}$$

are the eigenvalues, and the pseudo Mach angle $\omega = (\alpha - \beta)/2$. In the work of Xiao Chen and Zheng [2] and Li, Yang and Zheng [7] is the key diagonalized form of the isentropic Euler system

$$\begin{cases} \bar{\partial}^{+}(-\beta+\psi(\omega)) = \dfrac{\sin^2\omega[\cos(2\omega)-\kappa]}{c(\kappa+\sin^2\omega)}, \\ \bar{\partial}^{-}(\alpha+\psi(\omega)) = \dfrac{\sin^2\omega[\cos(2\omega)-\kappa]}{c(\kappa+\sin^2\omega)}, \\ \bar{\partial}^{0}[c^2(1+\kappa M^2)] = 2c\kappa M, \end{cases} \tag{3.2}$$

where

$$\begin{aligned} &\psi(\omega) := \sqrt{\frac{\gamma+1}{\gamma-1}}\arctan\left(\sqrt{\frac{\gamma-1}{\gamma+1}}\cot\omega\right), \\ &\kappa = \frac{\gamma-1}{2}, \qquad \nu := \frac{\gamma+1}{2(\gamma-1)}, \end{aligned} \tag{3.3}$$

and the pseudo Mach number M is related to ω as

$$\frac{1}{\sin\omega} = M := \sqrt{(u-\xi)^2+(v-\eta)^2}/c\,. \tag{3.4}$$

The **Riemann variables** $\psi - \beta$ and $\psi + \alpha$ correspond to the classical Riemann invariants for homogeneous systems. The symbols $\bar{\partial}^{+}$ etc. denote normalized directional derivatives. Then Theorem 1 is a consequence of Theorem 2, stated below.

THEOREM 2 (X. Chen and Zheng [2], Li, Yang and Zheng [7]). *There exists a classical solution for the interaction of two planar rarefaction waves. Characteristics are either convex or concave, cf. Fig. 3.*

Proof. See the papers [2, 7]. □

4. Simple waves. Next in line is the simple wave region: We have

THEOREM 3 (J. Li, T. Zhang, YZ [8]). *Adjacent to a constant state is a simple wave in which one family of characteristics are straight lines along which the physical variables (u, v, p, ρ) are constant.*

The proof is based on a nice characteristic decomposition:

$$\begin{cases} \partial^{+}(\partial^{-}u) + \frac{\partial^{+}\Lambda_{-}-\partial^{-}\Lambda_{+}}{\Lambda_{+}-\Lambda_{-}}\partial^{-}u = \frac{\Lambda_{+}\Lambda_{-}}{\Lambda_{+}-\Lambda_{-}}\left[\frac{\partial^{-}\Lambda_{-}}{\Lambda_{-}^2}\partial^{+}u - \frac{\partial^{+}\Lambda_{+}}{\Lambda_{+}^2}\partial^{-}u\right], \\ \partial^{-}(\partial^{+}u) + \frac{\partial^{+}\Lambda_{-}-\partial^{-}\Lambda_{+}}{\Lambda_{+}-\Lambda_{-}}\partial^{+}u = \frac{\Lambda_{+}\Lambda_{-}}{\Lambda_{+}-\Lambda_{-}}\left[\frac{\partial^{-}\Lambda_{-}}{\Lambda_{-}^2}\partial^{+}u - \frac{\partial^{+}\Lambda_{+}}{\Lambda_{+}^2}\partial^{-}u\right]. \end{cases}$$

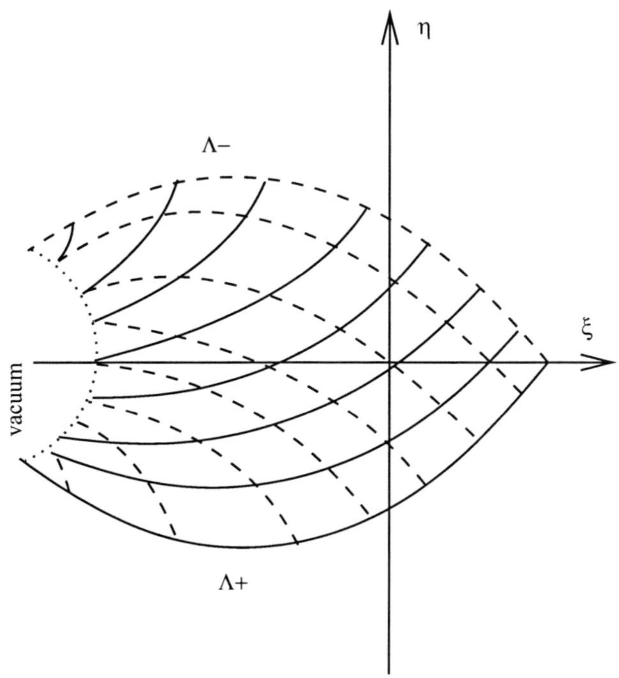

FIG. 3. *A global solution for the interaction of two planar waves. The solid curve boundary and the dashed curve boundary are given characteristic boundaries, the solution exists in the enclosed region with the dotted curve boundary which is vacuum.*

$$\partial^{\pm}\Lambda_{\pm} = \left[\partial_U \Lambda_{\pm} - \Lambda_{\mp}^{-1}\partial_V \Lambda_{\pm} - (\gamma - 1)\partial_{c^2}\Lambda_{\pm}(U - \Lambda_{\mp}^{-1}V)\right]\partial^{\pm}u.$$

The notation here is $\partial^{\pm} = \partial_\xi + \Lambda_{\pm}\partial_\eta$. Once $\partial^{+}u$ is zero at a point, then it is equal to zero along an entire minus characteristic curve. From a plus characteristic curve bordering a constant region, one must have $\partial^{+}u = 0$ in the neighboring region which implies that the plus characteristics are all straight lines. And that is a simple wave.

5. Semi-hyperbolic patches. We now deal with the *semi-hyperbolic patches*. These patches refer to the four small regions around the subsonic domain in Fig. 1. In terms of characteristic orientation, similar patches appear in ducts of moving air if the ducts have narrow necks, see Fig. 4.

In work of J. Hunter, B. L. Keyfitz, R. Sanders and A. Tesdall, see e.g. article by Keyfitz in this volume, these semi-hyperbolic waves appear in the so-called Guderley reflection.

In joint work with K. Song [13] (for pressure gradient system) and Mingjie Li [11] we find the important decomposition for the Euler system:

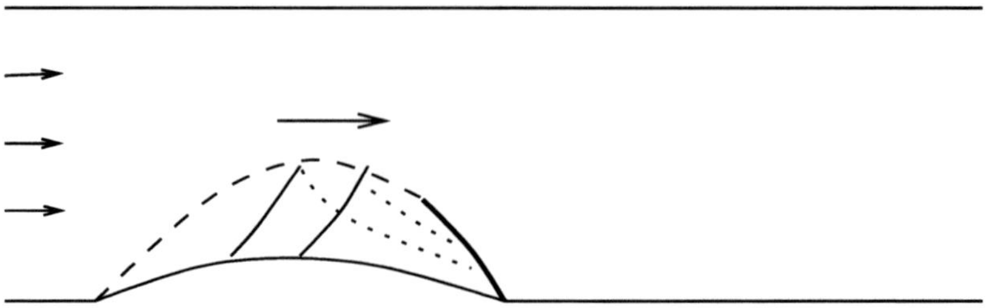

FIG. 4. *A small supersonic (hyperbolic) region in a mostly subsonic duct. The supersonic region is semi-hyperbolic in the terminology of this paper, since one set of characteristics (i.e., the dotted ones) originate from, and terminate as well, at the sonic boundary (the dashed curve) or the shock curve boundary (the heavy curve) of the subsonic region.*

$$\begin{cases} c\bar\partial^+\Big[\dfrac{\bar\partial^- c}{\sin^2\omega}\Big]+\Big[\dfrac{8\sin^2\omega}{\gamma-1}+2\Big]\bar\partial^+ c\Big[\dfrac{\bar\partial^- c}{\sin^2\omega}\Big]=\dfrac{\nu}{\cos^2\omega}\Big[\dfrac{\bar\partial^- c}{\sin^2\omega}\Big]\Big[\bar\partial^+ c+\bar\partial^- c\Big], \\ c\bar\partial^-\Big[\dfrac{\bar\partial^+ c}{\sin^2\omega}\Big]+\Big[\dfrac{8\sin^2\omega}{\gamma-1}+2\Big]\bar\partial^- c\Big[\dfrac{\bar\partial^+ c}{\sin^2\omega}\Big]=\dfrac{\nu}{\cos^2\omega}\Big[\dfrac{\bar\partial^+ c}{\sin^2\omega}\Big]\Big[\bar\partial^- c+\bar\partial^+ c\Big]. \end{cases}$$

THEOREM 4 (Song, Li, Zheng[13, 11]). *Suppose the pressure is smooth and increasing along both the plus characteristic BA and the minus characteristic BC, with both points A and C being sonic. Then there exists a classical solution in a region ABC with sonic boundary AC, but a shock wave forms in the other direction, cf. Fig. 5.*

The proof depends on the estimates on derivatives, provided by a lemma.

LEMMA 1 ((**Uniform bounds of** $\pm\bar\partial^\pm c$). *For a smooth solution in the hyperbolic domain ABC, the maximum of $\pm\bar\partial^\pm c$ in ABC are subject to:*

$$\max_{ABC}\{\bar\partial^+ c, -\bar\partial^- c\}\le 2\max_{\widehat{AB},\widehat{BC}}\{\bar\partial^+ c, -\bar\partial^- c\}.$$

Proof. Note the notation $\bar\partial^+=\cos\alpha\partial_\xi+\sin\alpha\partial_\eta$, etc. Let

$$M=\max_{\widehat{AB},\widehat{BC}}\Big\{\frac{\bar\partial^+ c}{\sin^2\omega}, \frac{-\bar\partial^- c}{\sin^2\omega}\Big\}.$$

First, consider the case that $-\bar\partial^- c/\sin^2\omega\ge\bar\partial^+ c/\sin^2\omega$ in the whole D_ϵ, which denotes the region ABC but an $\epsilon>0$ distance away from the sonic curve. Let T be a point on $\widehat{BC}$. We note that $-\bar\partial^- c/\sin^2\omega\,|_T>0$ implies $-\bar\partial^- c/\sin^2\omega>0$ along the positive characteristic curve Y_+ starting from T. Then we have

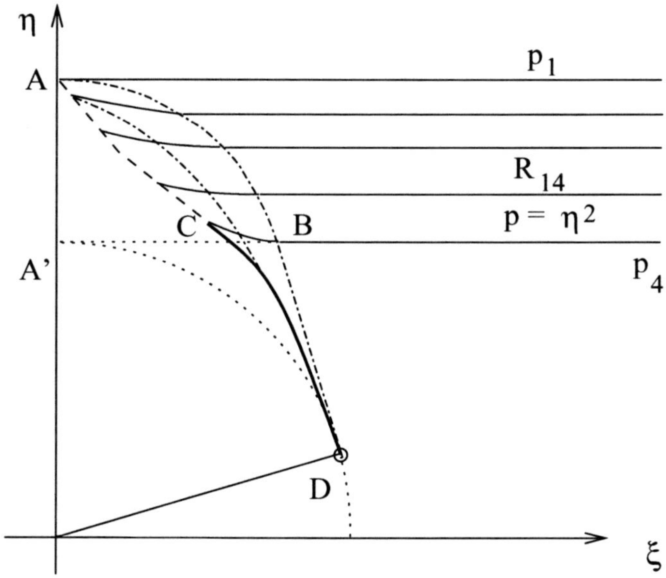

FIG. 5. *Notation and setting for a semi-hyperbolic wave region ABC. The parallel lines and their natural extensions are characteristics of the minus family, while the dash-dotted lines are characteristics of the plus family. The characteristic curves BA and BC are given, where points A and C are sonic. The pressure is increasing from B to A and from B to C. The goal is to show the existence of a classical solution in the region ABC, where the dashed curve AC is sonic, and a shock forms at the heavy curve CD along a boundary of the region BCD.*

$$\begin{aligned} c\bar\partial^+\Big(\frac{-\bar\partial^- c}{\sin^2\omega}\Big) &= -\Big[\frac{8\sin^2\omega}{\gamma-1}+2\Big]\bar\partial^+ c\Big(\frac{-\bar\partial^- c}{\sin^2\omega}\Big)\\ &\quad +\nu\frac{\sin^2\omega}{\cos^2\omega}\Big(\frac{-\bar\partial^- c}{\sin^2\omega}\Big)\Big[\frac{\bar\partial^+ c}{\sin^2\omega}-\Big(\frac{-\bar\partial^- c}{\sin^2\omega}\Big)\Big]\le 0, \end{aligned}$$

which implies that $-\bar\partial^- c/\sin^2\omega$ is decreasing along the positive characteristic curve Y_+. Then for any interior point S in D_ϵ

$$0<\frac{-\bar\partial^- c}{\sin^2\omega}\,|_S\le\frac{-\bar\partial^- c}{\sin^2\omega}\,|_T\le M.$$

Therefore, in the region D_ϵ

$$0<\frac{\bar\partial^+ c}{\sin^2\omega}\le\frac{-\bar\partial^- c}{\sin^2\omega}\le M.$$

Second, if $-\bar\partial^- c/\sin^2\omega\ge\bar\partial^+ c/\sin^2\omega$ in D_ϵ is not true, then there must be some point $Z\in D_\epsilon$ satisfying $\bar\partial^+ c/\sin^2\omega\,|_Z> -\bar\partial^- c/\sin^2\omega\,|_Z$. Then by the continuity property of $-\bar\partial^- c/\sin^2\omega$ and $\bar\partial^+ c/\sin^2\omega$ in D_ϵ, there exists a neighborhood N_0 of Z such that $-\bar\partial^- c/\sin^2\omega\le\bar\partial^+ c/\sin^2\omega$

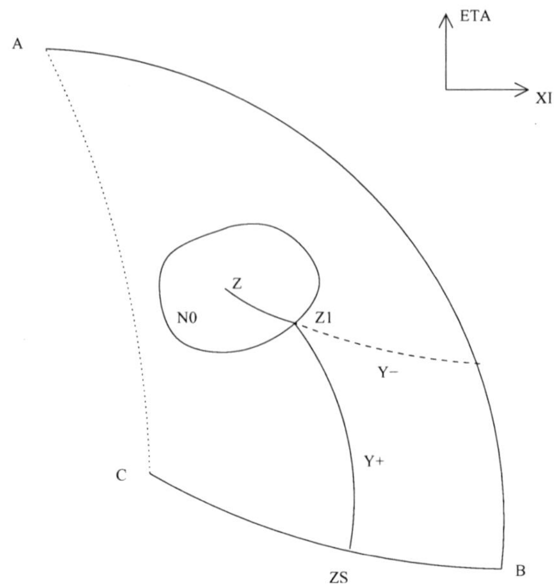

FIG. 6. *An imagined situation, case 1. Here ZS denotes* Z_*.

in N_0. We see that $\bar{\partial}^+ c/\sin^2\omega$ is decreasing on N_0 along the negative characteristic curve Y_- passing through Z. Then

$$\frac{-\bar{\partial}^- c}{\sin^2\omega}\,|_Z < \frac{\bar{\partial}^+ c}{\sin^2\omega}\,|_Z \le \frac{\bar{\partial}^+ c}{\sin^2\omega}\,|_{Z_1},$$

where Z_1 lies on both the boundary of N_0 and the negative characteristic curve Y_-. After the intersection point Z_1, two derivatives $-\bar{\partial}^- c/\sin^2\omega$ and $\bar{\partial}^+ c/\sin^2\omega$ may cross each other many times along the negative characteristic curve until the boundary point on $\widehat{AB}$. So we investigate the behavior of $-\bar{\partial}^- c/\sin^2\omega$ and $\bar{\partial}^+ c/\sin^2\omega$ by considering the following two subcases at Z_1.

Case 1: Assume at Z_1, we can draw a positive characteristic up to the boundary $\widehat{BC}$ without crossing of $-\bar{\partial}^- c/\sin^2\omega$ and $\bar{\partial}^+ c/\sin^2\omega$ (see Fig. 6). Then by the decreasing property of $-\bar{\partial}^- c/\sin^2\omega$ along the positive characteristic curve, we obtain

$$\frac{\bar{\partial}^+ c}{\sin^2\omega}\,|_Z \le \frac{\bar{\partial}^+ c}{\sin^2\omega}\,|_{Z_1} = \frac{-\bar{\partial}^- c}{\sin^2\omega}\,|_{Z_1} \le \frac{-\bar{\partial}^- c}{\sin^2\omega}\,|_{Z_*},$$

where $Z_\star$ is an intersection point of the positive characteristic curve and $\widehat{BC}$.

Case 2: Assume at Z_1, we cannot draw a positive characteristic up to the boundary $\widehat{BC}$ without crossing of $-\bar{\partial}^- c/\sin^2\omega$ and $\bar{\partial}^+ c/\sin^2\omega$ or we cannot draw a positive characteristic outside of N_0 (see Fig. 7). Then we

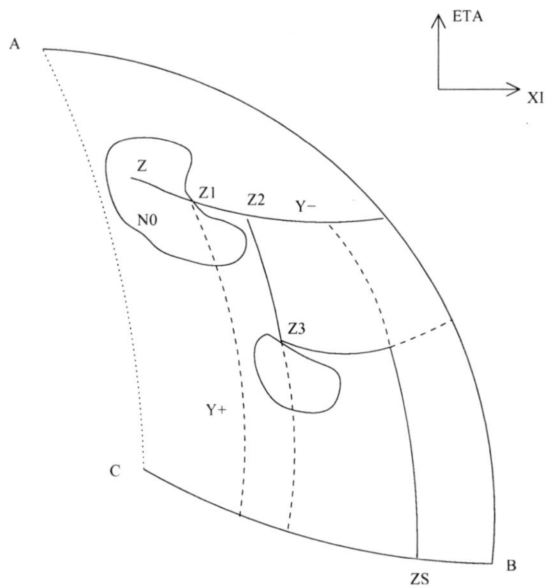

FIG. 7. *Case 2. Here ZS denotes Z_*.*

use a negative characteristic, called Y_-, on the region of $-\bar{\partial}^- c/\sin^2\omega \geq \bar{\partial}^+ c/\sin^2\omega$ for a while or up to and including the boundary $\widehat{AB}$. Since the two derivatives $-\bar{\partial}^- c/\sin^2\omega$ and $\bar{\partial}^+ c/\sin^2\omega$ meet at Z_1 and

$$\begin{aligned}-c\bar{\partial}^-\Big(\frac{\bar{\partial}^+ c}{\sin^2\omega}\Big) &= -\Big[\frac{8\sin^2\omega}{\gamma-1}+2\Big](-\bar{\partial}^- c)\Big(\frac{\bar{\partial}^+ c}{\sin^2\omega}\Big)\\ &\quad +\nu\frac{\sin^2\omega}{\cos^2\omega}\Big(\frac{\bar{\partial}^+ c}{\sin^2\omega}\Big)\Big[\Big(\frac{-\bar{\partial}^- c}{\sin^2\omega}\Big)-\frac{\bar{\partial}^+ c}{\sin^2\omega}\Big],\end{aligned}$$

so we obtain

$$-c\bar{\partial}^-\Big(\frac{\bar{\partial}^+ c}{\sin^2\omega}\Big) < 0.$$

Hence we can select a point Z_2 on the negative characteristic Y_- so that

$$\frac{-\bar{\partial}^- c}{\sin^2\omega}\,|_{Z_1} = \frac{\bar{\partial}^+ c}{\sin^2\omega}\,|_{Z_1} < \frac{\bar{\partial}^+ c}{\sin^2\omega}\,|_{Z_2} \leq \frac{-\bar{\partial}^- c}{\sin^2\omega}\,|_{Z_2}\,.$$

That is, along the negative characteristic starting from a point in $\widehat{BA}$, $-\bar{\partial}^- c/\sin^2\omega$ should decrease after Z_2 to meet $\bar{\partial}^+ c/\sin^2\omega$ at Z_1. Thus

$$\frac{\bar{\partial}^+ c}{\sin^2\omega}\,|_{Z} \leq \frac{\bar{\partial}^+ c}{\sin^2\omega}\,|_{Z_1} = \frac{-\bar{\partial}^- c}{\sin^2\omega}\,|_{Z_1} < \frac{-\bar{\partial}^- c}{\sin^2\omega}\,|_{Z_2}\,.$$

If Z_2 is on $\widehat{BA}$, we have finished. If not, we change moving direction at Z_2 to take a positive characteristic. If we can move along a positive characteristic at Z_2 up to the boundary $\widehat{BC}$ without crossing of $-\bar{\partial}^- c/\sin^2\omega$ and $\bar{\partial}^+ c/\sin^2\omega$, then we have

$$\frac{\bar{\partial}^+ c}{\sin^2\omega}\,|_Z \le \frac{\bar{\partial}^+ c}{\sin^2\omega}\,|_{Z_1} = \frac{-\bar{\partial}^- c}{\sin^2\omega}\,|_{Z_1} < \frac{-\bar{\partial}^- c}{\sin^2\omega}\,|_{Z_2} \le \frac{-\bar{\partial}^- c}{\sin^2\omega}\,|_{Z_*}\ .$$

Otherwise, there may be crossing of $-\bar{\partial}^- c/\sin^2\omega$ and $\bar{\partial}^+ c/\sin^2\omega$ while we are moving along the positive characteristic from Z_2. Then we take a negative characteristic at the cross point Z_3 where we cannot preserve the inequality of $-\bar{\partial}^- c/\sin^2\omega$ and $\bar{\partial}^+ c/\sin^2\omega$ which has been valid from the point Z_2 (see Fig. 7). Then we again apply the first step in Case 2. Properly applying these steps in Case 2 continuously, eventually we have

$$\frac{-\bar{\partial}^- c}{\sin^2\omega}\,|_Z \le \frac{\bar{\partial}^+ c}{\sin^2\omega}\,|_Z \le \frac{-\bar{\partial}^- c}{\sin^2\omega}\,|_{Z_1} < \frac{-\bar{\partial}^- c}{\sin^2\omega}\,|_{Z_2} \le \cdots \le \frac{-\bar{\partial}^- c}{\sin^2\omega}\,|_{Z_*},$$

where $Z_* \in \widehat{AB} \cup \widehat{BC}$. So the boundary values of $-\bar{\partial}^- c/\sin^2\omega$ and $\bar{\partial}^+ c/\sin^2\omega$ dominant the values of $-\bar{\partial}^- c/\sin^2\omega$ and $\bar{\partial}^+ c/\sin^2\omega$ in the interior of the domain. But we know that

$$\frac{1}{\sin^2\omega} \le \frac{1}{\sin^2(\frac{\pi}{4})} = 2, \quad on\ \ \widehat{AB} \cup \widehat{BC}.$$

Therefore

$$0 < -\bar{\partial}^- c,\ \bar{\partial}^+ c \le M \le 2 \max_{\widehat{AB},\widehat{BC}} \{\bar{\partial}^+ c, -\bar{\partial}^- c\}.$$

The proof of Lemma 1 is complete. □

LEMMA 2 (Envelope formation). *For a given monotone and strictly convex curve $\widehat{BC}$, we draw the positive characteristics starting on the curve $\widehat{BC}$, downward, as shown in Fig. 5. Then the positive characteristics form an envelope before their sonic points.*

Proof. Recall that

$$\bar{\partial}^+\bar{\partial}^- c = \frac{\bar{\partial}^- c}{c}\Big[-\sin(2\omega) + \frac{\nu}{\cos^2\omega}\bar{\partial}^- c + \Big(\nu\Omega\cos(2\omega) + 1\Big)\bar{\partial}^+ c\Big].$$

Since $\bar{\partial}^+ c = 0$ in the simple wave area with the positive characteristics, we obtain

$$\bar{\partial}^+\bar{\partial}^- c = \frac{-\sin(2\omega)}{c}\bar{\partial}^- c + \frac{\nu}{c\cos^2\omega}(\bar{\partial}^- c)^2.$$

So we have

$$-\bar{\partial}^+\Big(\frac{1}{-\bar{\partial}^- c}\Big) = -\frac{\sin(2\omega)}{c}\Big(\frac{1}{-\bar{\partial}^- c}\Big) - \frac{\nu}{c\cos^2\omega}.$$

Notice that the right-hand side of the above is always negative, while on the boundary

$$\frac{1}{-\bar{\partial}^- c}\,|_{\widehat{BC}} > 0.$$

Then $1/(-\bar{\partial}^- c)$ is decreasing along the positive characteristic in the direction from B to D. Furthermore, we observe the inhomogeneous term

$$-\frac{\nu}{c\cos^2\omega}$$

in the above blows up inverse quadratically at the least. So $1/(-\bar{\partial}^- c)$ reaches zero before the characteristic gets to its sonic point. The blow-up of $\bar{\partial}^- c$ yields that the positive characteristics form an envelope before their sonic points. □

6. Summary. In summary, we can fit the above waves together to construct the solutions outside the sonic curve. We have obtained solutions for any binary interaction of planar waves —- one extreme case is a global solution without a subsonic region – a central vacuum instead. The potentially very useful quantities are the **Riemann variables**: $-\beta + \psi(\omega)$ and $\alpha + \psi(\omega)$ where

$$\psi(\omega) := \sqrt{\frac{\gamma+1}{\gamma-1}}\arctan\left(\sqrt{\frac{\gamma-1}{\gamma+1}}\cot\omega\right).$$

Riemann problems are idealized situations, but their studies are helpful to applications in Mach (Guderley) reflection, channel flow, flow around airfoil, de Laval nozzle, etc. With regards to numerics, we see a mutual movement – they challenge and promote each other in details.

7. Discussion. This workshop allows for a separate hour for discussions on the topic of the speaker and related subjects. Dehua Wang, Volker Elling, Kris Jenssen, and Allen Tesdall were invited to present their results and open problems. One of several aspects of the multi-dimensional conservation laws in Dehua's presentation is the existence of a weak solution in a bounded domain for the two-dimensional steady potential flow, of which the assumption of no stagnation points is used. One open question is to remove the assumption of no stagnation points. For details, see his article in this volume.

Volker Elling presented his numerical simulations indicating two physical solutions for the Euler system in two space dimensions. James Glimm offered opinion that points to multiple entropy solutions in his work associated with numerical solutions to physical processes. For details, see Elling's and Glimm's articles in this volume.

Kris Jenssen presented issues for symmetrical Euler systems in multi-dimensions. Numerical domains of determinacy and ranges of influence are considered in light of theoretical counterparts. For details, see Jenssen's article in this volume.

Allen Tesdall presented his numerical simulations done with collaborators on Guderley reflection. The numerical simulations reveal with more detail than ever the complex structure at the triple point of shock interaction. For more details, see Keyfitz and Hunter's articles in this volume.

Related work not covered here are for potential equations (see the talks from Guiqiang Chen, M. Feldman, T.P. Liu, and V. Elling in this workshop); Unsteady transonic small disturbance equation (UTSD) and a nonlinear wave system (see papers by Canic, Keyfitz, Kim, Tesdall, and Hunter).

REFERENCES

[1] T. Chang, G. Chen, and S.Yang, *On the 2-D Riemann problem for the compressible Euler equations, I. Interaction of shocks and rarefaction waves*, Disc. Conti. Dyna. Syst., **1** (1995), 555–584; *II. Interaction of contact discontinuities*, Disc. Conti. Dyna. Syst., **6** (2000), 419–430.

[2] Xiao Chen and Yuxi Zheng, *The Interaction of Rarefaction Waves of the Two-Dimensional Euler Equations*, Indiana University Math. Jour., **59** (2010), 231–256.

[3] J. Glimm, Xiaomei Ji, Jiequan Li, Xiaolin Li, Peng Zhang, Tong Zhang, and Y. Zheng, *Transonic shock formation in a rarefaction Riemann problem for the 2-D compressible Euler equations*, SIAM: J. Appl. Math., **69** (2008), 720–742.

[4] J. Glimm and A. Majda, *Multidimensional Hyperbolic Problems and Computations*, IMA Vol. **29**, Springer-Verlag, 1991.

[5] A. Kurganov and E. Tadmor, *Solution of two-dimensional Riemann problems for gas dynamics without Riemann problem solvers,* Numerical Methods for Partial Differential Equations, **18** (2002), 548–608.

[6] P. Lax and X. Liu, *Solution of two-dimensional Riemann problems of gas dynamics by positive schemes*, SIAM J. Sci. Comp., **19** (1998), 319–340.

[7] Jiequan Li, Zhicheng Yang, and Yuxi Zheng, *Characteristic Decompositions and Interactions of Rarefaction Waves of 2-D Euler Equations*, J. Differential Equations (2010), doi:10.106/j.jde.2010.07.009.

[8] J. Li, T. Zhang, and Y. Zheng, *Simple waves and a characteristic decomposition of the two dimensional compressible Euler equations*, Comm. Math. Phys., **267** (2006), pp. 1–12.

[9] Jiequan Li and Y. Zheng, *Interaction of Rarefaction Waves of the Two-Dimensional Self-Similar Euler Equations*, Arch. Rat. Mech. Anal., **193** (2009), 623–657.

[10] Jiequan Li and Y. Zheng, *Interaction of Four Rarefaction Waves in the Bi-Symmetric Class of the Two-Dimensional Euler Equations*, Comm. Math. Phys., **296** (2010), 303–321.

[11] Mingjie Li and Yuxi Zheng, *Semi-Hyperbolic Patches of Solutions of the Two-Dimensional Euler equations*, preprint 2009.

[12] C.W. Schulz-Rinne, J.P. Collins, and H.M. Glaz, *Numerical solution of the Riemann problem for two-dimensional gas dynamics*, SIAM J. Sci. Comp., **4** (1993), 1394–1414.

[13] K. Song and Y. Zheng, *Semi-hyperbolic patches of solutions of the pressure gradient system*, Disc. Cont. Dyna. Syst., series A, **24** (2009), pp. 1365–1380.

IMA SUMMER PROGRAM PARTICIPANTS

Nonlinear Conservation Laws and Applications

- Luigi Ambrosio, Scuola Normale Superiore
- Fabio Ancona, Dipartimento di Matematica Pura ed Applicata, Università di Padova
- Myoungjean Bae, Department of Mathematics, University of Wisconsin
- Santiago Ignacio Betelu, Department of Mathematics, University of North Texas
- Stefano Bianchini, International School for Advanced Studies (SISSA/ISAS)
- Michael Blaser, Eidgenössische TH Hönggerberg
- Alberto Bressan, Department of Mathematics, Pennsylvania State University
- Sunčica Čanić, Department of Mathematics, University of Houston
- Gui-Qiang G. Chen, Mathematical Institute, University of Oxford
- Bin Cheng, Department of Mathematics, University of Michigan
- Antoine Choffrut, Department of Mathematics, University of Minnesota
- Rinaldo Mario Colombo, Dipartimento di Matematica, Università di Brescia
- Gianluca Crippa, Dipartimento di Matematica, Università di Parma
- Constantine Dafermos, Division of Applied Mathematics, Brown University
- Xue Mei Deng, School of Mathematical Sciences, Northwestern University
- Qian Ding, Mathematics Department, Northwestern University
- Liviu Florin Dinu, Institute of Mathematics, Romanian Academy of Sciences
- Marina Ileana Dinu, Department of Mathematics III, Polytechnic University of Bucharest
- Carlotta Donadello, Department of Mathematics, Northwestern University
- Wolf-Patrick Duell, Courant, New York University
- Volker Wilhelm Elling, Department of Mathematics , University of Michigan
- Nadine Even, Institut für Mathematik, Bayerische-Julius-Maximilians-Universität Würzburg

- Beixiang Fang, Department of Mathematics, Shanghai Jiaotong University
- Eduard Feireisl, Mathematical Institute of the Academy of Sciences of the Czech Republic
- Mikhail Feldman, Department of Mathematics, University of Wisconsin
- Jin Feng, Department of Mathematics, University of Kansas
- Hermano Frid, Instituto de Matemática Pura e Aplicada - IMPA Estrada Dona Castorina
- Frederico C. Furtado, Department of Mathematics, University of Wyoming
- Shu Gao, Department of Mathematics, Northwestern University
- Mauro Garavello, Dipartimento di Scienze e Tecnologie Avanzate, Università del Piemonte Orientale "Amedeo Avogadro"
- James G. Glimm, Department of Applied Mathamatics and Statistics, Stony Brook University
- G.D. Veerappa Gowda, Centre for Applicable Mathematics, Tata Institute of Fundamental Research
- Graziano Guerra, Dipartimento di Matematica e Applicazioni, Universit di Milano - Bicocca
- Arvind Kumar Gupta, Birla Institute of Technology and Science
- Michael Herty, Department of Mathematics, RWTH Aachen
- Luan Thach Hoang, Department of Mathematics and Statistics, Texas Tech University
- David Hoff, Department of Mathematics, Indiana University
- Jingwei Hu, Department of Mathematics, University of Wisconsin
- Xianpeng Hu, Department of Mathematics, University of Pittsburgh
- John K. Hunter, Department of Mathematics, University of California at Davis
- Juhi Jang, Department of Mathematics, University of California, Riverside
- Katarina Jegdic, Department of Computer and Mathematical Sciences, University of Houston-Downtown
- Helge Kristian Jenssen, Department of Mathematics, Pennsylvania State University
- Joseph W. Jerome, Department of Mathematics, Northwestern University
- Peng Jiang, Department of Mathematics, University of Pittsburgh Mathew A. Johnson, Department of Mathematics, Indiana University
- Kayyunnapara Thomas Joseph, School of Mathematics, Tata Institute of Fundamental Research
- Kenneth Hvistendahl Karlsen, University of Oslo

- Markus Keel, Institute for Mathematics and its Applications, University of Minnesota
- Barbara Lee Keyfitz, Department of Mathematics, The Ohio State University
- Christian F. Klingenberg, Institut für Mathematik, Bayerische-Julius-Maximilians-Universitt Wrzburg
- Alexander Kurganov, Department of Mathematics, Tulane University
- Bongsuk Kwon, Department of Mathematics, Indiana University
- Marc Laforest, Mathématiques et Génie Industriel École Polytechnique de Montréal
- Philippe G. LeFloch, Laboratoire Jacques-Louis Lions, and Centre National de la Recherche Scientifique, Université Pierre et Marie Curie (Paris 6)
- Nicholas Matthew Leger, Department of Mathematics, University of Texas at Austin
- Marta Lewicka, School of Mathematics, University of Minnesota
- Lu Li, School of Mathematics, University of Minnesota
- Tian-Hong Li, Academy of Mathematics and Systems Science, Chinese Academy of Sciences
- Tong Li, Department of Mathematics, University of Iowa
- Yachun Li, Department of Mathematics, Shanghai Jiaotong University
- Chun Liu, Institute for Mathematics and its Applications, University of Minnesota
- Hailiang Liu, Department of Mathematics, Iowa State University
- Tai-Ping Liu, Department of Mathematics, Stanford University
- Maria Lukacova, Institute of Numerical Simulation, Technische Universitt Hamburg-Harburg
- Tao Luo, Department of Mathematics, Georgetown University
- Pierangelo Marcati, Department of Pure and Applied Mathematics, Universitá di L'Aquila
- Francesca Marcellini, Department of Mathematics and Applications, Universit di Milano - Bicocca
- Dan Marchesin, Instituto de Matematica Pura e Aplicada
- Andrea Marson, Department of Pure and Applied Mathematics, Università di Padova
- Nader Masmoudi, Courant Institute of Mathematical Sciences, New York University
- Francesca Monti, Department of Mathematics and Application, Università di Milano - Bicocca
- Sarka Necasova, Czech Academy of Sciences (AVČR)
- Toan Nguyen, Department of Mathematics, Indiana University
- Truyen V. Nguyen, Department of Theoretical and Applied Mathematics, University of Akron

- Shinya Nishibata, Department of Mathematical and Computer Sciences, Tokyo Institute of Technology
- Reza Pakzad, Department of Mathematics, University of Pittsburgh
- Ronghua Pan, School of Mathematics, Georgia Institute of Technology
- Evgeniy Panov, Department of Mathematical Analysis, Novgorod State University
- Arshad Ahmud Iqbal Peer, Department of Mathematics, University of Mauritius
- Mikhail Perepelitsa, Department of Mathematics, Vanderbilt University
- Benoit Perthame, Laboratoire Jacques-Louis Lions, Université de Paris VI (Pierre et Marie Curie)
- Tomasz Piotr Piasecki, Polish Academy of Sciences
- Bojan Popov, Department of Mathematics, Texas A&M University
- Roger Robyr, Institute of Mathematics, Universität Zürich
- Olga Rozanova, Department of Differential Equations, Moscow State University
- Fadil Santosa, Institute for Mathematics and its Applications, University of Minnesota
- Denis Serre, École Normale Supérieure de Lyon
- Michael Shearer, Department of Mathematics, North Carolina State University
- Wen Shen, Department of Mathematics, Pennsylvania State University
- Chi-Wang Shu, Division of Applied Mathematics, Brown University
- Marshall Slemrod, Faculty of Mathematics and Computer Science, The Weizmann Institute of Science
- Kyungwoo Song, Department of Mathematics, Kyung Hee University
- Laura Valentina Spinolo, Centro De Giorgi, Scuola Normale Superiore
- Shaowei Su, Mathematics Department, Northwestern University
- Eitan Tadmor, Department of Mathematics, Institute for Physical Science and Technology, Center of Scientific Computation And Mathematical Modeling (CSCAMM), University of Maryland
- Allen Tesdall, Department of Mathematics, College of Staten Island, CUNY
- Monica Torres, Department of Mathematics, Purdue University
- Konstantina Trivisa, Department of Mathematics, University of Maryland
- Lev Truskinovsky, Laboratoire de Mcanique des Solides, École Polytechnique

- Athanasios E. Tzavaras, Department of Applied Mathematics, University of Crete
- Alexis F. Vasseur, Department of Mathematics, University of Texas at Austin
- David H. Wagner, Department of Mathematics, University of Houston
- Dehua Wang, Department of Mathematics, University of Pittsburgh
- Ya-Guang Wang, Department of Mathematics, Shanghai
- Jiaotong University
- Ying Wang, Department of Mathematics, Ohio State University
- Michael Westdickenberg, School of Mathematics, Georgia Institute of Technology
- Mark Williams, Department of Mathematics, University of North Carolina
- Meng Xu, Department of Mathematics, Northwestern University
- Jue Yan, Department of Mathematics, Iowa State University
- Hao Ying, Department of Mathematics, The Ohio State University
- Robin Young, Department of Mathematics and Statistics, University of Massachusetts
- Matthias Youngs, Department of Mathematics, Indiana University
- Cheng Yu, Department of Mathematics, Indiana University
- Haijun Yu, Department of Mathematics, Purdue University
- Shih-Hsien Yu, Department of Mathematics, National University of Singapore
- Hairong Yuan, Department of Mathematics, East China Normal University
- Yanni Zeng, Department of Mathematics, University of Alabama at Birmingham
- Tianyou Zhang, Department of Mathematics, Pennsylvania State University
- Yongqian Zhang, School of Mathematical Sciences, Fudan University
- Kun Zhao, Department of Mathematics, Georgia Institute of Technology
- Yuxi Zheng, Department of Mathematics, Pennsylvania State University
- Dianwen Zhu, Department of Mathematics, University of California, Davis
- Kevin Zumbrun, Department of Mathematics, Indiana University

1998–1999	Mathematics in Biology
1999–2000	Reactive Flows and Transport Phenomena
2000–2001	Mathematics in Multimedia
2001–2002	Mathematics in the Geosciences
2002–2003	Optimization
2003–2004	Probability and Statistics in Complex Systems: Genomics, Networks, and Financial Engineering
2004–2005	Mathematics of Materials and Macromolecules: Multiple Scales, Disorder, and Singularities
2005–2006	Imaging
2006–2007	Applications of Algebraic Geometry
2007–2008	Mathematics of Molecular and Cellular Biology
2008–2009	Mathematics and Chemistry
2009–2010	Complex Fluids and Complex Flows
2010–2011	Simulating Our Complex World: Modeling, Computation and Analysis
2011–2012	Mathematics of Information

IMA SUMMER PROGRAMS

1987	Robotics
1988	Signal Processing
1989	Robust Statistics and Diagnostics
1990	Radar and Sonar (June 18–29) New Directions in Time Series Analysis (July 2–27)
1991	Semiconductors
1992	Environmental Studies: Mathematical, Computational, and Statistical Analysis
1993	Modeling, Mesh Generation, and Adaptive Numerical Methods for Partial Differential Equations
1994	Molecular Biology
1995	Large Scale Optimizations with Applications to Inverse Problems, Optimal Control and Design, and Molecular and Structural Optimization
1996	Emerging Applications of Number Theory (July 15–26) Theory of Random Sets (August 22–24)
1997	Statistics in the Health Sciences
1998	Coding and Cryptography (July 6–18) Mathematical Modeling in Industry (July 22–31)
1999	Codes, Systems, and Graphical Models (August 2–13, 1999)
2000	Mathematical Modeling in Industry: A Workshop for Graduate Students (July 19–28)
2001	Geometric Methods in Inverse Problems and PDE Control (July 16–27)
2002	Special Functions in the Digital Age (July 22–August 2)

2003 Probability and Partial Differential Equations in Modern Applied Mathematics (July 21–August 1)
2004 n-Categories: Foundations and Applications (June 7–18)
2005 Wireless Communications (June 22–July 1)
2006 Symmetries and Overdetermined Systems of Partial Differential Equations (July 17–August 4)
2007 Classical and Quantum Approaches in Molecular Modeling (July 23-August 3)
2008 Geometrical Singularities and Singular Geometries (July 14-25)
2009 Nonlinear Conservation Laws and Applications (July 13-31)

IMA "HOT TOPICS/SPECIAL" WORKSHOPS

- Challenges and Opportunities in Genomics: Production, Storage, Mining and Use, April 24–27, 1999
- Decision Making Under Uncertainty: Energy and Environmental Models, July 20–24, 1999
- Analysis and Modeling of Optical Devices, September 9–10, 1999
- Decision Making under Uncertainty: Assessment of the Reliability of Mathematical Models, September 16–17, 1999
- Scaling Phenomena in Communication Networks, October 22–24, 1999
- Text Mining, April 17–18, 2000
- Mathematical Challenges in Global Positioning Systems (GPS), August 16–18, 2000
- Modeling and Analysis of Noise in Integrated Circuits and Systems, August 29–30, 2000
- Mathematics of the Internet: E-Auction and Markets, December 3–5, 2000
- Analysis and Modeling of Industrial Jetting Processes, January 10–13, 2001
- Special Workshop: Mathematical Opportunities in Large-Scale Network Dynamics, August 6–7, 2001
- Wireless Networks, August 8–10 2001
- Numerical Relativity, June 24–29, 2002
- Operational Modeling and Biodefense: Problems, Techniques, and Opportunities, September 28, 2002
- Data-driven Control and Optimization, December 4–6, 2002
- Agent Based Modeling and Simulation, November 3–6, 2003
- Enhancing the Search of Mathematics, April 26–27, 2004
- Compatible Spatial Discretizations for Partial Differential Equations, May 11–15, 2004
- Adaptive Sensing and Multimode Data Inversion, June 27–30, 2004
- Mixed Integer Programming, July 25–29, 2005
- New Directions in Probability Theory, August 5–6, 2005
- Negative Index Materials, October 2–4, 2006

- The Evolution of Mathematical Communication in the Age of Digital Libraries, December 8–9, 2006
- Math is Cool! and Who Wants to Be a Mathematician?, November 3, 2006
- Special Workshop: Blackwell-Tapia Conference, November 3–4, 2006
- Stochastic Models for Intracellular Reaction Networks, May 11–13, 2008
- Multi-Manifold Data Modeling and Applications, October 27–30, 2008
- Mixed-Integer Nonlinear Optimization: Algorithmic Advances and Applications, November 17–21, 2008
- Higher Order Geometric Evolution Equations: Theory and Applications from Microfluidics to Image Understanding, March 23–26, 2009
- Career Options for Women in Mathematical Sciences, April 2-4, 2009
- MOLCAS, May 4-8, 2009
- IMA Interdisciplinary Research Experience for Undergraduates, June 29-July 31, 2009
- Research in Imaging Sciences, October 5-7, 2009
- Career Options for Underrepresented Groups in Mathematical Sciences, March 25-27, 2010
- Physical Knotting and Linking and its Applications, April 9, 2010
- IMA Interdisciplinary Research Experience for Undergraduates, June 14 - July 16, 2010
- Kickoff Workshop for Project MOSAIC, June 30-July 2, 2010
- Finite Element Circus Featuring a Scientific Celebration of Falk, Pasciak, and Wahlbin, November 5-6, 2010
- Integral Equation Methods, Fast Algorithms and Applications, August 2-5, 2010
- Medical Device-Biological Interactions at the Material-Tissue Interface, September 13-15, 2010
- First Abel Conference A Mathematical Celebration of John Tate, January 3-5, 2011
- Strain Induced Shape Formation: Analysis, Geometry and Materials Science, May 16-20, 2011
- Macaulay2, July 25-29, 2011
- Instantaneous Frequencies and Trends for Nonstationary Nonlinear Data, September 7-9, 2011

SPRINGER LECTURE NOTES FROM THE IMA:

The Mathematics and Physics of Disordered Media
Editors: Barry Hughes and Barry Ninham
(Lecture Notes in Math., Volume 1035, 1983)

Orienting Polymers
Editor: J.L. Ericksen
(Lecture Notes in Math., Volume 1063, 1984)

New Perspectives in Thermodynamics
Editor: James Serrin
(Springer-Verlag, 1986)

Models of Economic Dynamics
Editor: Hugo Sonnenschein
(Lecture Notes in Econ., Volume 264, 1986)

The IMA Volumes
in Mathematics and its Applications

Current Volumes:

Volume 1: Homogenization and Effective Moduli of Materials and Media
Editors: Jerry Ericksen, David Kinderlehrer, Robert Kohn, and J.-L. Lions

Volume 2: Oscillation Theory, Computation, and Methods of Compensated Compactness
Editors: Constantine Dafermos, Jerry Ericksen, David Kinderlehrer, and Marshall Slemrod

Volume 3: Metastability and Incompletely Posed Problems
Editors: Stuart Antman, Jerry Ericksen, David Kinderlehrer, and Ingo Muller

Volume 4: Dynamical Problems in Continuum Physics
Editors: Jerry Bona, Constantine Dafermos, Jerry Ericksen, and David Kinderlehrer

Volume 5: Theory and Applications of Liquid Crystals
Editors: Jerry Ericksen and David Kinderlehrer

Volume 6: Amorphous Polymers and Non-Newtonian Fluids
Editors: Constantine Dafermos, Jerry Ericksen, and David Kinderlehrer

Volume 7: Random Media
Editor: George Papanicolaou

Volume 8: Percolation Theory and Ergodic Theory of Infinite Particle Systems
Editor: Harry Kesten

Volume 9: Hydrodynamic Behavior and Interacting Particle Systems
Editor: George Papanicolaou

Volume 10: Stochastic Differential Systems, Stochastic Control Theory, and Applications
Editors: Wendell Fleming and Pierre-Louis Lions

Volume 11: Numerical Simulation in Oil Recovery
Editor: Mary Fanett Wheeler

Volume 12: Computational Fluid Dynamics and Reacting Gas Flows
Editors: Bjorn Engquist, M. Luskin, and Andrew Majda

Volume 13: Numerical Algorithms for Parallel Computer Architectures
Editor: Martin H. Schultz

Volume 14: Mathematical Aspects of Scientific Software
Editor: J.R. Rice

Volume 15: Mathematical Frontiers in Computational Chemical Physics
Editor: D. Truhlar

Volume 16: Mathematics in Industrial Problems
by Avner Friedman

Volume 17: Applications of Combinatorics and Graph Theory to the Biological and Social Sciences
Editor: Fred Roberts

Volume 18: q-Series and Partitions
Editor: Dennis Stanton

Volume 19: Invariant Theory and Tableaux
Editor: Dennis Stanton

Volume 20: Coding Theory and Design Theory Part I: Coding Theory
Editor: Dijen Ray-Chaudhuri

Volume 21: Coding Theory and Design Theory Part II: Design Theory
Editor: Dijen Ray-Chaudhuri

Volume 22: Signal Processing: Part I - Signal Processing Theory
Editors: L. Auslander, F.A. Grünbaum, J.W. Helton, T. Kailath, P. Khargonekar, and S. Mitter

Volume 23: Signal Processing: Part II - Control Theory and Applications of Signal Processing
Editors: L. Auslander, F.A. Grünbaum, J.W. Helton, T. Kailath, P. Khargonekar, and S. Mitter

Volume 24: Mathematics in Industrial Problems, Part 2
by Avner Friedman

Volume 25: Solitons in Physics, Mathematics, and Nonlinear Optics
Editors: Peter J. Olver and David H. Sattinger

Volume 26: Two Phase Flows and Waves
Editors: Daniel D. Joseph and David G. Schaeffer

Volume 27: Nonlinear Evolution Equations that Change Type
Editors: Barbara Lee Keyfitz and Michael Shearer

Volume 28: Computer Aided Proofs in Analysis
Editors: Kenneth Meyer and Dieter Schmidt

Volume 29: Multidimensional Hyperbolic Problems and Computations
Editors: Andrew Majda and Jim Glimm

Volume 30: Microlocal Analysis and Nonlinear Waves
Editors: Michael Beals, R. Melrose, and J. Rauch

Volume 31: Mathematics in Industrial Problems, Part 3
by Avner Friedman

Volume 32: Radar and Sonar, Part I
by Richard Blahut, Willard Miller, Jr., and Calvin Wilcox

Volume 33: Directions in Robust Statistics and Diagnostics: Part I
Editors: Werner A. Stahel and Sanford Weisberg

Volume 34: Directions in Robust Statistics and Diagnostics: Part II
Editors: Werner A. Stahel and Sanford Weisberg

Volume 35: Dynamical Issues in Combustion Theory
Editors: P. Fife, A. Liñán, and F.A. Williams

Volume 36: Computing and Graphics in Statistics
Editors: Andreas Buja and Paul Tukey

Volume 37: Patterns and Dynamics in Reactive Media
Editors: Harry Swinney, Gus Aris, and Don Aronson

Volume 38: Mathematics in Industrial Problems, Part 4
by Avner Friedman

Volume 39: Radar and Sonar, Part II
Editors: F. Alberto Grünbaum, Marvin Bernfeld, and Richard E. Blahut

Volume 40: Nonlinear Phenomena in Atmospheric and Oceanic Sciences
Editors: George F. Carnevale and Raymond T. Pierrehumbert

Volume 41: Chaotic Processes in the Geological Sciences
Editor: David A. Yuen

Volume 42: Partial Differential Equations with Minimal Smoothness and Applications
Editors: B. Dahlberg, E. Fabes, R. Fefferman, D. Jerison, C. Kenig, and J. Pipher

Volume 43: On the Evolution of Phase Boundaries
Editors: Morton E. Gurtin and Geoffrey B. McFadden

Volume 44: Twist Mappings and Their Applications
Editors: Richard McGehee and Kenneth R. Meyer

Volume 45: New Directions in Time Series Analysis, Part I
Editors: David Brillinger, Peter Caines, John Geweke, Emanuel Parzen, Murray Rosenblatt, and Murad S. Taqqu

Volume 46: New Directions in Time Series Analysis, Part II
Editors: David Brillinger, Peter Caines, John Geweke, Emanuel Parzen, Murray Rosenblatt, and Murad S. Taqqu

Volume 47: Degenerate Diffusions
Editors: Wei-Ming Ni, L.A. Peletier, and J.-L. Vazquez

Volume 48: Linear Algebra, Markov Chains, and Queueing Models
Editors: Carl D. Meyer and Robert J. Plemmons

Volume 49: Mathematics in Industrial Problems, Part 5
by Avner Friedman

Volume 50: Combinatorial and Graph-Theoretic Problems in Linear Algebra
Editors: Richard A. Brualdi, Shmuel Friedland, and Victor Klee

Volume 51: Statistical Thermodynamics and Differential Geometry of Microstructured Materials
Editors: H. Ted Davis and Johannes C.C. Nitsche

Volume 52: Shock Induced Transitions and Phase Structures in General Media
Editors: J.E. Dunn, Roger Fosdick, and Marshall Slemrod

Volume 53: Variational and Free Boundary Problems
Editors: Avner Friedman and Joel Spruck

Volume 54: Microstructure and Phase Transitions
Editors: David Kinderlehrer, Richard James, Mitchell Luskin, and Jerry L. Ericksen

Volume 55: Turbulence in Fluid Flows: A Dynamical Systems Approach
Editors: George R. Sell, Ciprian Foias, and Roger Temam

Volume 56: Graph Theory and Sparse Matrix Computation
Editors: Alan George, John R. Gilbert, and Joseph W.H. Liu

Volume 57: Mathematics in Industrial Problems, Part 6
by Avner Friedman

Volume 58: Semiconductors, Part I
Editors: W.M. Coughran, Jr., Julian Cole, Peter Lloyd, and Jacob White

Volume 59: Semiconductors, Part II
Editors: W.M. Coughran, Jr., Julian Cole, Peter Lloyd, and Jacob White

Volume 60: Recent Advances in Iterative Methods
Editors: Gene Golub, Anne Greenbaum, and Mitchell Luskin

Volume 61: Free Boundaries in Viscous Flows
Editors: Robert A. Brown and Stephen H. Davis

Volume 62: Linear Algebra for Control Theory
Editors: Paul Van Dooren and Bostwick Wyman

Volume 63: Hamiltonian Dynamical Systems: History, Theory, and Applications
Editors: H.S. Dumas, K.R. Meyer, and D.S. Schmidt

Volume 64: Systems and Control Theory for Power Systems
Editors: Joe H. Chow, Petar V. Kokotovic, and Robert J. Thomas

Volume 65: Mathematical Finance
Editors: Mark H.A. Davis, Darrell Duffie, Wendell H. Fleming, and Steven E. Shreve

Volume 66: Robust Control Theory
Editors: Bruce A. Francis and Pramod P. Khargonekar

Volume 67: Mathematics in Industrial Problems, Part 7
by Avner Friedman

Volume 68: Flow Control
Editor: Max D. Gunzburger

Volume 69: Linear Algebra for Signal Processing
Editors: Adam Bojanczyk and George Cybenko

Volume 70: Control and Optimal Design of Distributed Parameter Systems
Editors: John E. Lagnese, David L. Russell, and Luther W. White

Volume 71: Stochastic Networks
Editors: Frank P. Kelly and Ruth J. Williams

Volume 72: Discrete Probability and Algorithms
Editors: David Aldous, Persi Diaconis, Joel Spencer, and J. Michael Steele

Volume 73: Discrete Event Systems, Manufacturing Systems, and Communication Networks
Editors: P.R. Kumar and P.P. Varaiya

Volume 74: Adaptive Control, Filtering, and Signal Processing
Editors: K.J. Åström, G.C. Goodwin, and P.R. Kumar

Volume 75: Modeling, Mesh Generation, and Adaptive Numerical Methods for Partial Differential Equations
Editors: Ivo Babuska, Joseph E. Flaherty, William D. Henshaw, John E. Hopcroft, Joseph E. Oliger, and Tayfun Tezduyar

Volume 76: Random Discrete Structures
Editors: David Aldous and Robin Pemantle

Volume 77: Nonlinear Stochastic PDE's: Hydrodynamic Limit and Burgers' Turbulence
Editors: Tadahisa Funaki and Wojbor A. Woyczynski

Volume 78: Nonsmooth Analysis and Geometric Methods in Deterministic Optimal Control
Editors: Boris S. Mordukhovich and Hector J. Sussmann

Volume 79: Environmental Studies: Mathematical, Computational, and Statistical Analysis
Editor: Mary Fanett Wheeler

Volume 80: Image Models (and their Speech Model Cousins)
Editors: Stephen E. Levinson and Larry Shepp

Volume 81: Genetic Mapping and DNA Sequencing
Editors: Terry Speed and Michael S. Waterman

Volume 82: Mathematical Approaches to Biomolecular Structure and Dynamics
Editors: Jill P. Mesirov, Klaus Schulten, and De Witt Sumners

Volume 83: Mathematics in Industrial Problems, Part 8
by Avner Friedman

Volume 84: Classical and Modern Branching Processes
Editors: Krishna B. Athreya and Peter Jagers

Volume 85: Stochastics Models in Geosystems
Editors: Stanislav A. Molchanov and Wojbor A. Woyczynski

Volume 86: Computational Wave Propagation
Editors: Bjorn Engquist and Gregory A. Kriegsmann

Volume 87: Progress in Population Genetics and Human Evolution
Editors: Peter Donnelly and Simon Tavaré

Volume 88: Mathematics in Industrial Problems, Part 9
by Avner Friedman

Volume 89: Multiparticle Quantum Scattering with Applications to Nuclear, Atomic and Molecular Physics
Editors: Donald G. Truhlar and Barry Simon

Volume 90: Inverse Problems in Wave Propagation
Editors: Guy Chavent, George Papanicolaou, Paul Sacks, and William Symes

Volume 91: Singularities and Oscillations
Editors: Jeffrey Rauch and Michael Taylor

Volume 92: Large-Scale Optimization with Applications, Part I: Optimization in Inverse Problems and Design
Editors: Lorenz T. Biegler, Thomas F. Coleman, Andrew R. Conn, and Fadil Santosa

Volume 93: Large-Scale Optimization with Applications, Part II: Optimal Design and Control
Editors: Lorenz T. Biegler, Thomas F. Coleman, Andrew R. Conn, and Fadil Santosa

Volume 94: Large-Scale Optimization with Applications, Part III: Molecular Structure and Optimization
Editors: Lorenz T. Biegler, Thomas F. Coleman, Andrew R. Conn, and Fadil Santosa

Volume 95: Quasiclassical Methods
Editors: Jeffrey Rauch and Barry Simon

Volume 96: Wave Propagation in Complex Media
Editor: George Papanicolaou

Volume 97: Random Sets: Theory and Applications
Editors: John Goutsias, Ronald P.S. Mahler, and Hung T. Nguyen

Volume 98: Particulate Flows: Processing and Rheology
Editors: Donald A. Drew, Daniel D. Joseph, and Stephen L. Passman

Volume 99: Mathematics of Multiscale Materials
Editors: Kenneth M. Golden, Geoffrey R. Grimmett, Richard D. James, Graeme W. Milton, and Pabitra N. Sen

Volume 100: Mathematics in Industrial Problems, Part 10
by Avner Friedman

Volume 101: Nonlinear Optical Materials
Editor: Jerome V. Moloney

Volume 102: Numerical Methods for Polymeric Systems
Editor: Stuart G. Whittington

Volume 103: Topology and Geometry in Polymer Science
Editors: Stuart G. Whittington, De Witt Sumners, and Timothy Lodge

Volume 104: Essays on Mathematical Robotics
Editors: John Baillieul, Shankar S. Sastry, and Hector J. Sussmann

Volume 105: Algorithms for Parallel Processing
Editors: Robert S. Schreiber, Michael T. Heath, and Abhiram Ranade

Volume 118: Dynamics of Algorithms
Editors: Rafael de la Llave, Linda R. Petzold, and Jens Lorenz

Volume 119: Numerical Methods for Bifurcation Problems and Large-Scale Dynamical Systems
Editors: Eusebius Doedel and Laurette S. Tuckerman

Volume 120: Parallel Solution of Partial Differential Equations
Editors: Petter Bjørstad and Mitchell Luskin

Volume 121: Mathematical Models for Biological Pattern Formation
Editors: Philip K. Maini and Hans G. Othmer

Volume 122: Multiple-Time-Scale Dynamical Systems
Editors: Christopher K.R.T. Jones and Alexander Khibnik

Volume 123: Codes, Systems, and Graphical Models
Editors: Brian Marcus and Joachim Rosenthal

Volume 124: Computational Modeling in Biological Fluid Dynamics
Editors: Lisa J. Fauci and Shay Gueron

Volume 125: Mathematical Approaches for Emerging and Reemerging Infectious Diseases: An Introduction
Editors: Carlos Castillo-Chavez with Sally Blower, Pauline van den Driessche, Denise Kirschner, and Abdul-Aziz Yakubu

Volume 126: Mathematical Approaches for Emerging and Reemerging Infectious Diseases: Models, Methods and Theory
Editors: Carlos Castillo-Chavez with Sally Blower, Pauline van den Driessche, Denise Kirschner, and Abdul-Aziz Yakubu

Volume 127: Mathematics of the Internet: E-Auction and Markets
Editors: Brenda Dietrich and Rakesh V. Vohra

Volume 128: Decision Making Under Uncertainty: Energy and Power
Editors: Claude Greengard and Andrzej Ruszczynski

Volume 129: Membrane Transport and Renal Physiology
Editors: Harold E. Layton and Alan M. Weinstein

Volume 130: Atmospheric Modeling
Editors: David P. Chock and Gregory R. Carmichael

Volume 131: Resource Recovery, Confinement, and Remediation of Environmental Hazards
Editors: John Chadam, Al Cunningham, Richard E. Ewing, Peter Ortoleva, and Mary Fanett Wheeler

Volume 132: Fractals in Multimedia
Editors: Michael F. Barnsley, Dietmar Saupe, and Edward Vrscay

Volume 133: Mathematical Methods in Computer Vision
Editors: Peter J. Olver and Allen Tannenbaum

Volume 134: Mathematical Systems Theory in Biology, Communications, Computation, and Finance
Editors: Joachim Rosenthal and David S. Gilliam

Volume 135: Transport in Transition Regimes
Editors: Naoufel Ben Abdallah, Anton Arnold, Pierre Degond, Irene Gamba, Robert Glassey, C. David Levermore, and Christian Ringhofer

Volume 136: Dispersive Transport Equations and Multiscale Models
Editors: Naoufel Ben Abdallah, Anton Arnold, Pierre Degond, Irene Gamba, Robert Glassey, C. David Levermore, and Christian Ringhofer

Volume 137: Geometric Methods in Inverse Problems and PDE Control
Editors: Christopher B. Croke, Irena Lasiecka, Gunther Uhlmann, and Michael S. Vogelius

Volume 138: Mathematical Foundations of Speech and Language Processing
Editors: Mark Johnson, Sanjeev Khudanpur, Mari Ostendorf, and Roni Rosenfeld

Volume 139: Time Series Analysis and Applications to Geophysical Systems
Editors: David R. Brillinger, Enders Anthony Robinson, and Frederic Paik Schoenberg

Volume 140: Probability and Partial Differential Equations in Modern Applied Mathematics
Editors: Jinqiao Duan and Edward C. Waymire

Volume 141: Modeling of Soft Matter
Editors: Maria-Carme T. Calderer and Eugene M. Terentjev

Volume 142: Compatible Spatial Discretizations
Editors: Douglas N. Arnold, Pavel B. Bochev, Richard B. Lehoucq, Roy A. Nicolaides, and Mikhail Shashkov

Volume 143: Wireless Communications
Editors: Prathima Agrawal, Daniel Matthew Andrews, Philip J. Fleming, George Yin, and Lisa Zhang

Volume 144: Symmetries and Overdetermined Systems of Partial Differential Equations
Editors: Michael Eastwood and Willard Miller, Jr.

Volume 145: Topics in Stochastic Analysis and Nonparametric Estimation
Editors: Pao-Liu Chow, Boris Mordukhovich, and George Yin

Volume 146: Algorithms in Algebraic Geometry
Editors: Alicia Dickenstein, Frank-Olaf Schreyer, and Andrew J. Sommese

Volume 147: Symmetric Functionals on Random Matrices and Random Matchings Problems by Grzegorz A. Rempała and Jacek Wesołowski

Volume 148: Software for Algebraic Geometry
Editors: Michael E. Stillman, Nobuki Takayama, and Jan Verschelde

Volume 149: Emerging Applications of Algebraic Geometry
Editors: Mihai Putinar and Seth Sullivant

Volume 150: Mathematics of DNA Structure, Function, and Interactions
Editors: Craig John Benham, Stephen Harvey, Wilma K. Olson, De Witt L. Sumners, and David Swigon

Volume 151: Nonlinear Computational Geometry
Editors: Ioannis Z. Emiris, Frank Sottile, and Thorsten Theobald

Volume 152: Towards Higher Categories
Editors: John C. Baez and J. Peter May

Volume 153: Nonlinear Conservation Laws and Applications
Editors: Alberto Bressan, Gui-Qiang Chen, Marta Lewicka, and Dehua Wang

CPSIA information can be obtained at www.ICGtesting.com
225712LV00007B/1/P

9 781441 995537